Molecular Encapsulation

Molecular Encapsulation

Organic Reactions in Constrained Systems

Editors

Udo H. Brinker

Institute of Organic Chemistry, University of Vienna, Vienna, Austria

Jean-Luc Mieusset

Institute of Organic Chemistry, University of Vienna, Vienna, Austria

WILEY

A John Wiley & Sons, Ltd., Publication

This edition first published 2010

Registered office
John Wiley & Sons Ltd, The Atrium, Southern Gate, Chichester, West Sussex, PO19 8SQ, United Kingdom

For details of our global editorial offices, for customer services and for information about how to apply for permission to reuse the copyright material in this book please see our website at www.wiley.com.

Library of Congress Cataloging-in-Publication Data

Molecular encapsulation : organic reactions in constrained systems / editors, Udo H. Brinker, Jean-Luc Mieusset.
p. cm.
Includes bibliographical references and index.
ISBN 978-0-470-99807-6 (cloth : alk. paper)
1. Nanochemistry. 2. Microencapsulation. I. Brinker, Udo H. II. Mieusset, Jean-Luc.
QC176.8.N35M653 2010
547′.2–dc22

2010004249

A catalogue record for this book is available from the British Library.

ISBN 9780470998076

Set in 10 on 12 pt Times by Toppan Best-set Premedia Limited.
Printed in The United Kingdom by Antony Rowe Ltd, Chippenham, Wiltshire.

Contents

Preface

The inclusion of small guest molecules within suitable host compounds results in constrained systems that imbue novel properties upon the incarcerated organic substrates. For example, the chromophoric nature of dye molecules is considerably altered upon complexation. And the targeted delivery of active pharmaceutical ingredients can be facilitated *in vivo*. So, understandably, supramolecular tactics are becoming widely employed and this treatise spotlights them. Indeed, the knowledge gleaned from constructing such assemblies can be applied to develop innovative molecular devices. Insofar as molecular constraint influences chemical reactivity, including chemo-, regio-, and stereoselectivity, new synthetic routes become plausible. Often, the impact of encapsulation on product formation is substantial. Therefore, special attention must be paid to the stabilization and chemistry of short-lived, high-energy intermediates, since they are key compounds that govern the outcome of reactions. Common among these transient species is their ability to subsequently undergo an array of reaction mechanisms that include the generation of unusual or strained products. However, the conventional drawback has been that high product yields are sacrificed due to a lack of selectivity across low-energy barriers. The use of a constrained system, however, offers the means to solve this problem by steering the reaction trajectory along a desired pathway. In general, we provide a broad overview of many different approaches that aim to manipulate chemical reactions. We anticipate that this endeavor will be thought provoking and ultimately facilitate the exchange of ideas between various scientific disciplines.

This volume comprises 17 chapters. It is spearheaded by an account of molecular recognition being used to precisely control photochemical reactions and foster enantiomeric excess. The supramolecular reaction proceeds under mild conditions, wherein a highly reactive excited-state intermediate readily generates products that are difficult to obtain thermally. Such inductions of enantioselectivity by chiral auxiliary are especially appealing. The following three chapters expound upon the cyclodextrin macrocyclic hosts, whose wide availability and ability to form inclusion complexes makes them miniature reaction vessels *par excellence*. Chapter 2 deals with the use of natural and derivatized Schardinger dextrins to accelerate reactions, such as transesterifications and Diels–Alder reactions, and to mimic enzymes, *e.g.*, ribonucleases and transaminases. Chapter 3 focuses on the development of molecular reactors. The primary role of these containers is to increase the selectivity of a chemical transformation, which also may occur by decreasing the overall rate of the process. In chapter 4, special cyclodextrin-mediated reactions are discussed. Since guest complexation offers the possibility of performing organic reactions in aqueous media, microsolvent effects are feasible, the protection of water-sensitive intermediates is achieved, and conformational control during the reaction is realized. Indeed, this approach has great appeal for environmental chemistry. Chapter 5 demonstrates the potential of nanoreactor technology, using zeolites as an example. These inorganic inclusion compounds are widely used on an industrial scale during petroleum refining, for example, because they catalyze the isomerization and oligomerization of alkenes. The utility of these materials is due to the presence of active sites in well-defined cavities that enable customized control of reaction selectivity. And though these natural catalysts are readily available, easily reusable, and therefore cost-effective, the facile preparation of a container of sufficient size suitable for the regulation of a chemical reaction is an ongoing challenge. To fashion cages as appropriate nanoreactors, a novel approach avoiding tedious multistep synthesis is becoming widespread. It entails the self-assembly of noncovalent capsules. The results obtained using these organized media are highlighted in chapter 6.

Another promising method to augment selectivity is described in chapter 7 and is based on the concept of concave reagents whose reactive moiety is situated inside a cleft or macrocycle and thereby sterically shielded. Another popular group of hosts used to control reactions are the calixarenes (chapter 8). Their appeal derives from the ease of the macrocyclic ring-closure during their synthesis and the capability of performing selective derivatization to both increase the size of the cavity and introduce functional groups. Using this method, a cage can be built wherein the reacting guest is completely imprisoned and may only escape at elevated temperatures. These hosts, known as hemicarcerands, have allowed isolation and spectroscopic observation of traditionally ephemeral intermediates. Remarkable results are presented in chapter 9. In chapter 10, an overview is given regarding the generation and associated change in reactivity of highly reactive intermediates, like radicals, carbenes, and nitrenes, within cavitands. This exposition is further elaborated by a presentation of the actual developments in dye encapsulation where improvements in the photoproperties and chemical stability are especially prized (chapter 11) and by a description of research on organic cations (chapter 12).

The subsequent chapters concern methodologies that are not only inspired by biology and enzyme mimicry, but actually employ modified biopolymers to achieve catalysis. In the first approach, the goal is to improve the enantioselectivity of a classic organometallic catalyst through its incorporation into a host protein that generates a secondary coordination sphere environment responsible for a rise in selectivity. These artificial metalloenzymes are showcased in chapter 13. The second technique is summarized in chapter 14 and details the generation of artificial RNA enzymes that are able to catalyze various chemical reactions, including the Diels–Alder reaction.

Finally, the assembly of more complex systems requires that the exact supramolecular reaction site be known. For this purpose, micelles and liposomes are very attractive due to the sequestering of reagents, the possibility of conducting various reactions in a confined space, and the similarity with living cells. These particular organized media are examined in detail in chapter 15, which describes the catalytic effect of micelles and supramolecular systems based on cyclophanes. In addition, the encapsulation process of vesicles and their catalytic effect on organic reactions is outlined in chapter 16. And, finally, chapter 17 previews efforts to assemble a semi-synthetic cell.

Many authors contributed chapters for this book, whose scope covers an expansive interdisciplinary area. We are profoundly grateful for their input and expertise. Indeed, we expect that this monograph will be useful to students and researchers alike who want to be informed about the latest developments in the burgeoning field of supramolecular chemistry. Not only will it help them to master established reactions, but new synthetic opportunities and exciting applications of these methodologies will surely arise.

Vienna, January 2010
Udo H. Brinker
Jean-Luc Mieusset

List of Contributors

Werner Abraham, Institut für Chemie, Humboldt-Universität zu Berlin, D-12489 Berlin, Germany

Easwaran Arunkumar, Molecular Targeting Technologies Inc, PA 19380, USA

Ronald Breslow, Department of Chemistry, Columbia University, New York, NY 10027, USA

Udo H. Brinker, Institut für Organische Chemie, Universität Wien, A–1090 Wien, Austria

Roberta Cacciapaglia, IMC CNR Sezione Meccanisma di Reazione, c/o Dipartimento di Chimica, Università La Sapienza, I-00185 Roma, Italy

Stefano Di Stefano, IMC CNR Sezione Meccanisma di Reazione, c/o Dipartimento di Chimica, Università La Sapienza, I-00185 Roma, Italy

Christopher Easton, Research School of Chemistry, Australian National University, Canberra ACT 0200, Australia

Jan B.F.N. Engberts, Stratingh Institute of Chemistry, University of Groningen, 9747 AG Groningen, The Netherlands

Jeremiah J. Gassensmith, Department of Chemistry and Biochemistry, 251 Nieuwland Science Hall, University of Notre Dame, Notre Dame, IN 46556, USA

Lutz Grubert, Institut für Chemie, Humboldt-Universität zu Berlin, D-12489 Berlin, Germany

Yoshihisa Inoue, Department of Applied Chemistry, Osaka University, Suita 565-0871, Japan

Andres Jäschke, Institut für Pharmazie und Molekulare Biotechnologie (IPMB), Universität Heidelberg, D-69120 Heidelberg, Germany

Chenfeng Ke, Department of Chemistry, Nankai University, Tianjing, China

Tehila S. Koblenz, Supramolecular and Homogeneous Catalysis, Van 't Hoff Institute for Molecular Sciences, University of Amsterdam, 1018 WV Amsterdam, The Netherlands

Alexander Konovalov, A.E. Arbuzov Institute of Organic and Physical Chemistry of the Russian Academy of Science, 420088 Kazan, Russia

Ludmila Kudryavtseva, A.E. Arbuzov Institute of Organic and Physical Chemistry of the Russian Academy of Science, 420088 Kazan, Russia

Yu Liu, Department of Chemistry, Nankai University, Tianjing, China

Pier Luigi Luisi, Biology Department, University of Roma Tre, 00146 Rome, Italy

Ulrich Lüning, Otto-Diels-Institut für Organische Chemie, D-24098 Kiel, Germany

Luigi Mandolini, IMC CNR Sezione Meccanisma di Reazione, c/o Dipartimento di Chimica, Università La Sapienza, I-00185 Roma, Italy

Jincheng Mao, Department of Chemistry, University of Basel, CH-4056, Basel Switzerland

Jean-Luc Mieusset, Institut für Organische Chemie, Universität Wien, A–1090 Wien, Austria

Alla Mirgorodskaya, A.E. Arbuzov Institute of Organic and Physical Chemistry of the Russian Academy of Science, 420088 Kazan, Russia

Hideki Onagi, Research School of Chemistry, Australian National University, Canberra ACT 0200, Australia

Joost N. H. Reek, Supramolecular and Homogeneous Catalysis, van 't Hoff Institute for Molecular Sciences, University of Amsterdam, 1018 WV Amsterdam, The Netherlands

Bradley D. Smith, Department of Chemistry and Biochemistry, 251 Nieuwland Science Hall, University of Notre Dame, Notre Dame, IN 46556, USA

Jean Sommer, Faculté de Chimie, Université Louis Pasteur, Strasbourg, F-67008 Strasbourg, France

Pasquale Stano, Biology Department, University of Roma Tre, 00146 Rome, Italy

Marc C.A. Stuart, Groningen Biomolecular Sciences and Biotechnology Institute and Stratingh Institute of Chemistry, University of Groningen, 9747 AG Groningen, The Netherlands

Keiko Takahashi, Department of Nanochemistry, Faculty of Engineering, Tokyo Polytechnic University, Atsugi, Kanagawa 243-0297, Japan

Jarl Ivar van der Vlugt, Supramolecular and Homogeneous Catalysis, Van 't Hoff Institute for Molecular Sciences, University of Amsterdam, 1018 WV Amsterdam, The Netherlands

Thomas R. Ward, Department of Chemistry, University of Basel, CH-4056 Basel, Switzerland

Ralf Warmuth, Rutgers The State University of New Jersey, Department of Chemistry and Chemical Biology, Piscataway, NJ 08854, USA

Jeroen Wassenaar, Supramolecular and Homogeneous Catalysis, Van 't Hoff Institute for Molecular Sciences, University of Amsterdam, 1018 WV Amsterdam, The Netherlands

Stéphane Walspurger, Energy Research Centre of The Netherlands, Petter, The Netherlands

Cheng Yang, PRESTO (JST) and Department of Applied Chemistry, Osaka University, Suita 565-0871, Japan

Lucia Zakharova, A.E. Arbuzov Institute of Organic and Physical Chemistry of the Russian Academy of Science, 420088 Kazan, Russia

Elena Zhiltsova, A.E. Arbuzov Institute of Organic and Physical Chemistry of the Russian Academy of Science, 420088 Kazan, Russia

1

Reaction Control by Molecular Recognition – A Survey from the Photochemical Perspective

Cheng Yang[1], *Chenfeng Ke*[2], *Yu Liu*[2], *and Yoshihisa Inoue*[1]
[1] PRESTO (JST) and Department of Applied Chemistry, Osaka University, Suita 565-0871, Japan
[2] Department of Chemistry, Nankai University, Tianjing, China

1.1 Introduction

Molecular recognition through non-covalent interactions between two or more molecules has attracted much attention from a broad spectrum of chemists for a long period of time and has already found many applications in various areas of science and technology. The concept of molecular recognition was first developed for biomolecular systems such as enzyme, antibody and DNA, which can selectively bind the specific target molecules through non-covalent weak interactions, including hydrogen bonding, van der Waals, dipole–dipole, charge–dipole and hydrophobic interactions.[1–3] Recent studies on artificial host–guest systems have revealed that molecular recognition is the essential conceptual basis for supramolecular chemistry and nanotechnology.[4,5]

Reaction control through complexation of substrate by supramolecular host is a relatively new idea compared to the conventional approaches that involve simple collisional attack or coordination of substrate to metal. Multiple non-covalent interactions in supramolecular assembly bind and locate a site-specific substrate in the right position, orientation and conformation near the catalyst or active site, stabilize the high-energy transition state, and eventually make the reaction faster and more selective. Typical examples are found in enzymatic reactions, which proceed with high specificity and efficiency in aqueous solutions under mild conditions. These observations in natural systems have inspired researchers to develop novel research areas such as supramolecular chemistry, biomimetic chemistry and bio-inspired materials science and technology.[6,7]

Contrary to the thermal counterpart, photochemical reactions in supramolecular system have been less investigated and therefore of current interest. Photochemistry is a powerful tool in synthetic chemistry as a complementary method for achieving compounds that are difficult to obtain through thermal reactions due to high strain, low stability, and orbital symmetry reasons. Unlike thermal reactions, photoreactions deal with excited-state molecules that are usually short-lived but experience much lower energy barriers and exhibit high reactivities even at low temperatures. As a consequence of these features, the precise control of a photoreaction is more difficult to achieve than that of thermal one. This is one of the reasons why most asymmetric photoreactions result in only relatively low enantioselectivities. In this context, supramolecular approaches to the photochemical asymmetric synthesis enable the more precise control of the orientation and conformation of substrates and, as a result, the enantioselectivity of photoproducts, by utilizing the non-covalent interactions in both ground and excited states.

Supramolecular photochemistry is a relatively new interdisciplinary area of science and may be tracked back to the work in early 1980s, where the spectroscopic properties of ions were manipulated by crown ethers.[8–10] The rapid development of supramolecular chemistry in the last two decades accelerated the application of supramolecular systems to organic photochemistry[11,12] and more recently to asymmetric synthesis,[13,14] leading to a great number of publications on reaction control by molecular recognition. Consequently, not all of these areas will be covered, but the concentration will be rather on the representative supramolecular photoreactions conducted primarily in solution. This will help idenitify the crucial concepts, strategies and conclusions as well as the major factors and mechanisms that govern the supramolecular photochemistry in different systems, and also provide the possible applications and future perspectives of this interesting area of supramolecular chemistry.

1.2 Photochemical Reactions Mediated by Macrocyclic Compounds

1.2.1 Supramolecular Photoreactions with Crown Ethers

Crown ethers, a family of cyclic oligomers of ethylene oxide, are artificial macrocyclic hosts which have been synthesized and utilized since the early days of supramolecular chemistry.[15,16] Besides various metal ions that are complexed by crown ethers mainly through ion–dipole interaction, primary and secondary organic ammonium ions also form stable complexes with larger sized crown ethers through ion–dipole and hydrogen-bonding interactions. Stoddart and co-workers used crown ethers that can simultaneously bind two organic ammonium guests to facilitate photodimerization.[17] As illustrated in Scheme 1.1, *trans*-stilbene derivative **2** forms a doubly encircled, doubly threaded 2:2 complex with bis-*p*-phenylene-34-crown-10 **1** to give a centrosymmetric [4]pseudorotaxane in the solid state. In addition to the hydrogen-bonding interactions between **1** and **2**, the complex is also stabilized by π–π stacking interactions between the two *trans*-stilbene units with mean interplanar and centroid-centroid separations of 3.57 and 4.33 Å, respectively. The close arrangement of stilbenes accelerates the photodimerization upon irradiation to

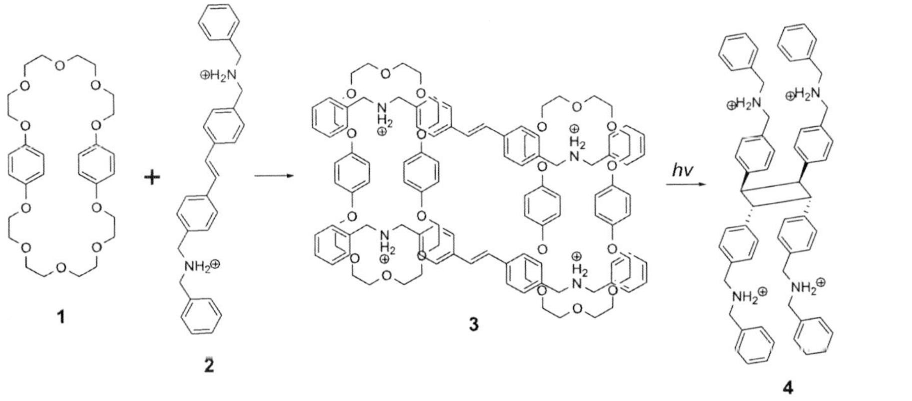

Scheme 1.1

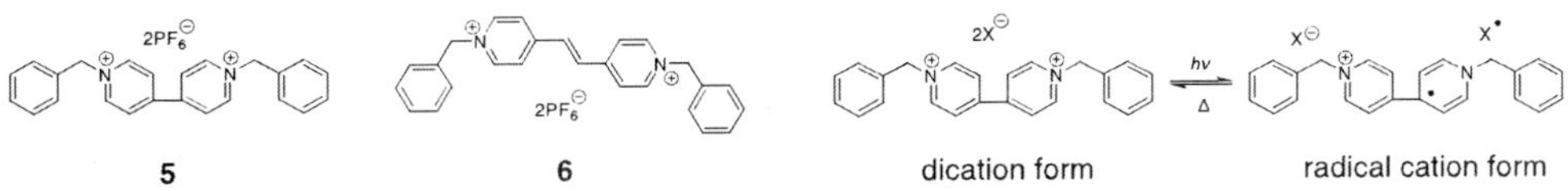

Scheme 1.2

exclusively give a single cyclobutane isomer **4** with a *syn–anti–syn* conformation, as confirmed by X-ray crystallographic analysis. A control experiment showed that no photodimerization but only *trans*-to-*cis* isomerization took place in the absence of crown ether **1**.

The photochromic behavior of viologens is greatly affected upon complexion with crown ether. Viologens show photoinduced colour change in the absence of any additional reagents when dispersed in isotropic thin polymer films.[18] Irradiation of benzylviologen **5** (Scheme 1.2) incorporated in a polymer matrix caused a colour change from colourless to blue as a result of the reduction of **5** from dication to radical cation, which was reverted to dication within 2 h in the dark. Crown ether **1** forms a 1 : 1 complex with **5** and also with **6** in acetone with association constants of *ca.* $200\,M^{-1}$. The charge-transfer (CT) interaction between viologen **5** or **6** and **1** led to the formation of yellow-coloured CT complex with a CT absorption band at 453 nm and 421 nm for **5** and **6**, respectively. The crown ether complexes of photoreduced viologens showed much accelerated bleaching rates than the corresponding free viologen radical cations. A similar phenomenon was observed also for $\mathbf{2} \subset \mathbf{1}$ complex in a polymer matrix.

1.2.2 Supramolecular Photoreactions with Calixarenes

Calix[*n*]arenes are a class of macrocycles that are normally made up of phenol units linked with methylene bridges and possess cavities of various sizes that can accommodate small organic molecules primarily driven by hydrophobic interactions.

Water soluble *p*-sulfonatocalix[6]arene (SCA[6]) and [8]arene (SCA[8]) (Scheme 1.3) have dimensions of 15.9 × 11.8 Å and 20.4 × 16.7 Å at the upper rim and 6.5 × 3.3 Å and

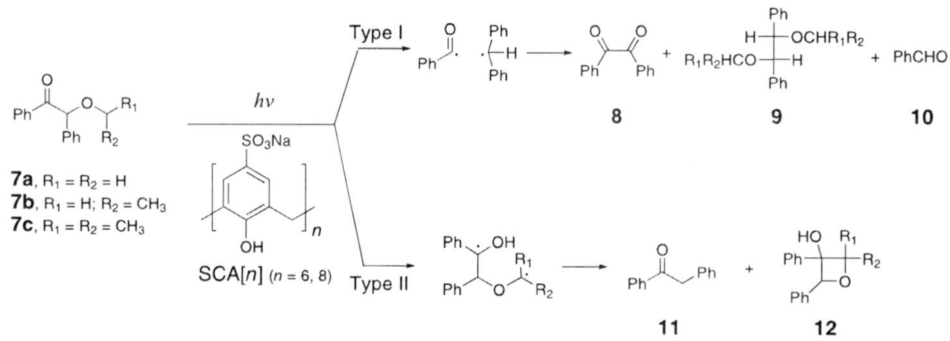

Scheme 1.3

8.6 × 5.9 Å at the lower rim, respectively,[19,20] and provide suitable cavities for accommodating quaternary ammonium ions,[21,22] native amino acids,[23] and small neutral organic molecules.[24] Using SCA[6] and SCA[8][25] as supramolecular hosts, Ramamurthy *et al.* investigated the photochemistry of benzoin alkyl ethers **7a–c**.[26–28] These SCAs formed the corresponding 1 : 1 complexes with **7a–c** in aqueous solution, and **7a** showed modest association constants of 137 M^{-1} and 386 M^{-1} with SCA[6] and SCA[8], respectively. Upon photolysis, these benzoin alkyl ethers underwent both the Norrish type I (α-cleavage) and type II (γ-hydrogen abstraction) reactions to give photoproducts **8–12** (Scheme 1.3). In the absence of SCAs, pinacol ether **9** was obtained as in 92% relative yield in aqueous solution at pH 7, along with deoxybenzoin **11** (8% relative yield), for which the α-cleavage ($k_\alpha \approx 10^{10}\,s^{-1}$) faster than the γ-hydrogen abstraction ($k_\gamma \approx 10^{9}\,s^{-1}$) is responsible. However, when the photoreaction was performed in the presence of SCA, product **11** became the major product. The relative yield of the type II products **11** and **12** depends critically on the cavity size of hosts and association constants. Photoirradiation of **7a** in the presence of an 8-fold excess amount of SCA[8] or SCA[6] gave **11** and **12** in 96% or 70% combined yield, respectively. The dominant formation of type II products suggests that **7** adopts a conformation favourable for γ-hydrogen absorption in the cavity. Choosing a suitable substrate with higher binding affinity for SCA is likely to be crucial in improving the stereoselectivity of photoreaction.

Photoreaction of stilbazoles **13a–f** (Scheme 1.4) has also been investigated in the presence of SCA[*n*]. Upon irradiation in dilute aqueous HCl solution without a host, *trans*-stilbazoles **13a–f** gave predominantly *cis*-isomers **16a–f**, along with only small amounts of *anti-head-to-tail* (HT) dimers **14a–f** and *anti-head-to-head* (HH) dimers **15a–f**.[29] In aqueous solution, stilbazoles were found to form 2 : 1 complexes with SCA[8] and SCA[6], irradiation of which afforded the corresponding *anti*-HT dimer in 66–86% yield in the presence of SCA[8] and in 60–76% yield in the presence of SCA[6]. The *anti*-HH dimer was given in only 16% yield, which was attributed to the electrostatic repulsion between two pyridinium components. The hydrophobic nature of SCA cavity should be responsible for the inclusion of stilbazoles. It is thus concluded that SCAs control the orientation of stilbazoles through complex formation to eventually enhance the distribution of HT dimer.

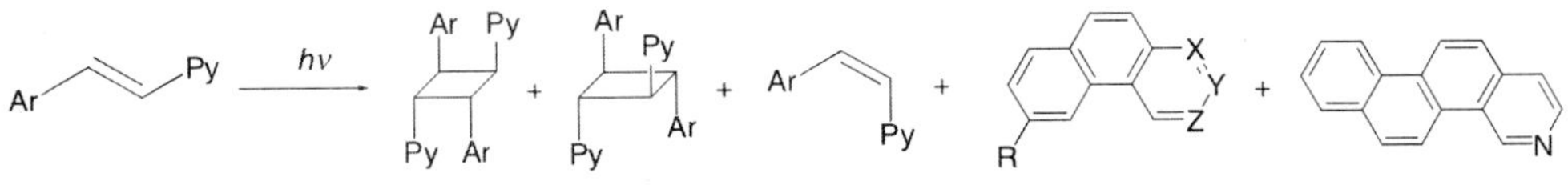

a: Ar = phenyl; Py = 4-pyridyl
b: Ar = 4'Cl-phenyl; Py = 4-pyridyl
c: Ar = 4'-OCH_3-phenyl; Py = 4-pyridyl
d: Ar = 1-naphthyl; Py = 4-pyridyl
e: Ar = phenyl; Py = 2-pyridyl
f: Ar = phenyl; Py = 3-pyridyl

a: R = H, X = CH, Y = CH, Z = N
b: R = Cl, X = CH, Y = CH, Z = N
c: R = OCH_3, X = CH, Y = CH, Z = N
e: R = H, X = N, Y = CH, Z = CH
f: R = H, X = CH, Y = N, Z = CH

Scheme 1.4

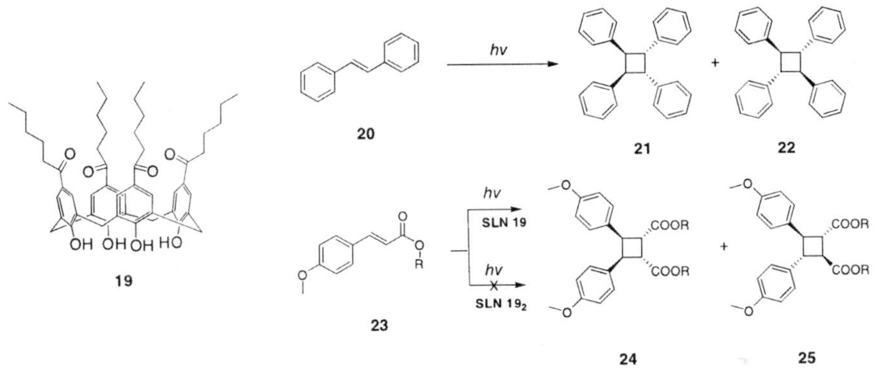

Scheme 1.5

Unlike water-soluble calixarenes that encapsulate guests by hydrophobic, electrostatic and van der Waals interactions, amphiphilic calixarenes in solid state form container- or capsule-like structures and include guest molecules predominantly through van der Waals interactions. Photochemical behavior of stilbenes included in amphiphilic molecular capsules of *p*-hexanoylcalix[4]arene **19** (Scheme 1.5) in the solid state was studied by Ananchenko *et al.*[30] Possessing diverse uni- and bimolecular reaction routes, including geometrical isomerization, cyclization and cyclodimerization, stilbenes are good model compounds for examining the effects of calixarene capsules on photoreactivity. In the solid state, two molecules of **19** self-assemble in a head-to-head fashion to form a hydrophobic capsule, which can accommodate two *cis*-stilbene molecules to give a 2 : 2 complex. This 2:2 complex is not very stable and can release one *cis*-stilbene molecule. On the other hand, *trans*-stilbene **20** is co-included with an ethanol molecule in the capsule to give a 1 : 1 : 2 complex. Irradiation of the complex of *cis*-stilbene first gave rise to *trans*-stilbene, and then to [2+2] dimers **21** and **22** in 5 : 1 ratio upon prolonged irradiation. Irradiation of *trans*-stilbene complex gave *cis*- and *trans*-stilbene in 1 : 1 ratio, along with a trace amount of dimers **21** and **22**, confirming that the initial capsule is occupied primarily by only one *trans*-stilbene molecule.

Notably, calixarene **19** readily forms solid lipid nanoparticles (SLNs), which can serve as a potential carrier system.[31] 2-Ethylhexyl *trans*-4-methoxycinnamate **23** can be loaded by a solvent displacement method into the SLNs based on **19**. Comparison of the powder X-ray diffraction (XRD) patterns of unloaded SLN and SLN loaded with **23** ([**23**]/[**19**] = 0.6) with that of crystalline complex **23** $\subset$ **19**$_2$ revealed that the complexation pattern of the SLN loaded with **23** is similar to that of the crystalline complex. No dimerization product was obtained upon irradiation of **23** loaded in SLN of **19**, and only photoisomerization occurred to give a 4:6 mixture of *cis*- and *trans*-isomer. However, when the stoichiometry was changed to [**23**]/[**19**] = 1.2, photolysis of **23** loaded in SLN yielded HH dimers **24** and **25** in a combined yield of *ca.* 10%. This regioselectivity of cyclodimerization of **23** is attributed to the fact that two **23** molecules preferably align in the **19's** cavity in HH orientation. These results suggest that one can efficiently manipulate the photochemical behavior of **23** by simply adjusting the molar ratio of **23** loaded to the SLNs of amphiphilic **19**.

1.2.3 Supramolecular Photoreactions with Cyclodextrins

Cyclodextrins (CDs) are most frequently employed hosts for supramolecular photoreactions by virtue of the fact that they are readily available, UV-transparent, water-soluble and capable of accommodating a wide range of organic guests in their hydrophobic cavities. Organic guests can form inclusion complexes with α-, β- and γ-CDs in various stoichiometries both in solution and in the solid state.

Unimolecular Reactions The cavity of β-CD is smaller than that of γ-CD, and normally includes one organic molecule of suitable size to form 1:1, 2:2 or 2:1 host–guest complexes. Asymmetric intramolecular *meta*-photocycloaddition of phenoxyalkenes (Scheme 1.6) has been studied in the presence of β-CD by Eycken and co-workers[32]. The inclusion complexes of **26**, **28**, **30** and **32** with β-CD were collected from the deposit formed in a warm aqueous solution containing β-CD and substrates by centrifugation and decantation. For the complexes of **26a**, **28b**, **30a** and **30b** with β-CD, the host-guest stoichiometries were determined as 2:1 by UV spectroscopy, while the rest of the substrates formed 1:1 complexes. Irradiation of the solid complexes for 24 h led to *ca.* 55–82% conversion, which are appreciably lower than that obtained in the solution-phase photoreactions. In all cases, *meta*-photocycloaddition led exclusively to only one regioisomer of 1,5-bridged dihydrosemibullvalenes **27**, **31** and **33** respectively from **26**, **30** and **32**. On the contrary, irradiation of complex **28** $\subset$ CD afforded two *meta* adducts **29a,b** and **29c,d**. The regioselectivity of adducts **29a** and **29c** is about 1:3 in the presence of β-CD against 1:1 in the absence of β-CD. Photocyclization of 'linear' photosubstrates **26a,b** gave insignificant enantiomeric excess (ee) of less than 3%, which is probably due to the less restricted orientation of *p*-substituted substrate in the cavity. Photocyclization of **28a** afforded minor product **29a** in much better 17% ee but the major product **29c** in only 2% ee. These results may indicate the trade-off relationship between chemical and optical yield: thus, larger steric hindrance leads to low chemical but high optical yield, and vice versa.

It is well documented that direct or sensitized photolysis of (*Z*)-cyclooctene **34Z** leads to geometrical isomerization to planar chiral (*E*)-isomer **34E** (Scheme 1.7).[33] Due to the comprehensive work by Inoue *et al.*, the enantiodifferentiating photoisomerization of **34Z** sensitized by optically active benzenecarboxylates has become a benchmark reaction for evaluating chiral photosensitizing system.[34–36] The first supramolecular enantiodifferentiating photoisomerization of **34Z** was examined in direct photolysis of its β-CD complex

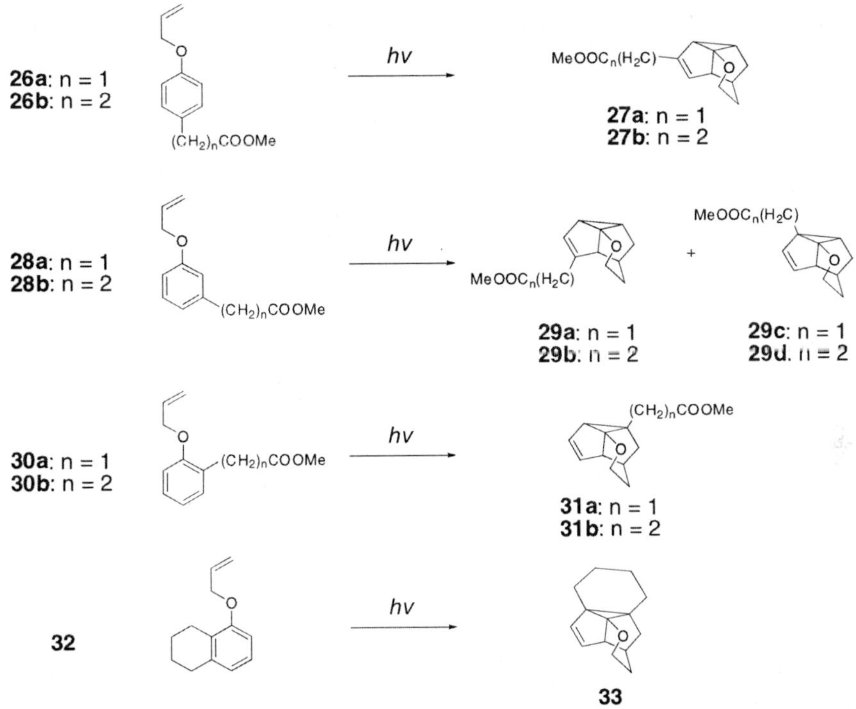

Scheme 1.6

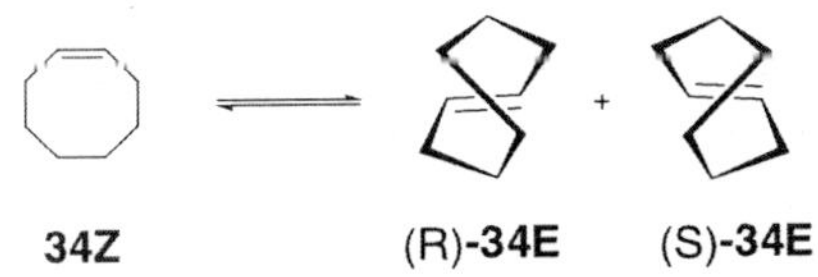

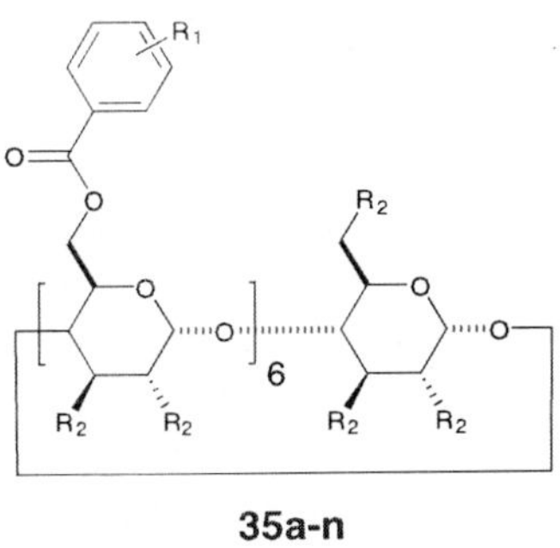

a, R_1 = H ; R_2 = OH
c, R_1 = *m*-CO_2Me; R_2 = OH
d, R_1 = *p*-CO_2Me; R_2 = OH
e, R_1 = H ; R_2 = OMe
f, R_1 = *o*-CO_2Me; R_2 = OMe
g, R_1 = *m*-CO_2Me; R_2 = OMe
h, R_1 = *p*-CO_2Me; R_2 = OMe
i, R_1 = *o*-Me; R_2 = OH
j, R_1 = *m*-Me; R_2 = OH
k, R_1 = *p*-Me; R_2 = OH
l, R_1 = *o*-OMe; R_2 = OH
m, R_1 = *m*-OMe; R_2 = OH
n, R_1 = *p*-OMe; R_2 = OH

Scheme 1.7

in the solid state. Photoproduct obtained from this complex was almost racemic with only 0.24% ee, demonstrating that the cavity interior of native β-CD does not possess sufficient enantiodifferentiating ability for the photoisomerization of **34Z**.

A more sophisticated strategy for achieving higher levels of enantiodifferentiation has been established by performing the photolysis of **34Z** with sensitizing chiral hosts **35** (Scheme 1.7), in which a chromophore is introduced to the primary rim of β-CD. A clear merit of this supramolecular photosensitization system is that the energy of excited sensitizer can be efficiently transferred to **34Z** included in CD cavity, while the undesirable, less-enantiodifferentiating sensitization outside the CD cavity can be prevented. Furthermore, this sensitizing host strategy allows the asymmetric photoreaction to occur with only a catalytic amount of chiral host, achieving photochemical chirality amplification.

In contrast to the small *E*/*Z* ratios around 0.25 reported for non-supramolecular photosensitizations by conventional alkyl benzoates,[37] an *E*/*Z* ratio of up to 0.8 were obtained at the photostationary state in 1:1 methanol–water mixture by using **35**. By adjusting the solvent composition, good enantiomeric excesses of up to 24% were obtained upon photosensitization with **35a–d**. The enantioselectivity of **34E** obtained in the photoisomerization of **34Z** with non-methylated CD is in general insensitive to the temperature variations as a result of the insignificant role of entropy in the β-CD-based supramolecular photochirogenesis system. However, the photoisomerization of **34Z** in permethylated CD derivatives **35e–h** turned out to be highly temperature dependent, frequently leading to a switching of the product chirality by changing the reaction temperature.[38] The more flexible skeleton of permethylated CD due to the lack of hydrogen-bonding net at the secondary rim is responsible for the temperature-dependent enantioselectivity.

Recently, further efforts were done to elucidate the effects of chromophore substitution on the ee of **34E**,[39] and much better enantiomeric excesses of up to 46% were reported upon photosensitized isomerization mediated by **35m**. Analogous hosts **35a**, **35l** and **35n** afforded only modest ee's in a range of 4%–10% under optimized conditions.[40] The ee value falls drastically to 1.7% when the *meta*-substituent is altered from methoxy (**35m**) to methyl in **35j**. This result reveals the critical dependence of the efficiency of chirality transfer on the sensitizer structure in the supramolecular photosensitized isomerization system.

Bimolecular Reactions β-CD is known to hinder the [2+2] photodimerization of stilbenes through complexation within its cavity.[41] In contrast, γ-CD can simultaneously include two stilbenes in its larger cavity in both solution and solid phase to significantly facilitate the [2+2] photodimerization reaction.

Wenz and co-workers examined the complexation and photochemical behavior of water-soluble stilbene **36g** with γ-CD in aqueous solution.[41] γ-CD forms 1:2 host–guest complex with **36g** in buffer solution (pH 5.7) at 25 °C with stepwise association constants of $K_1 = 385\,M^{-1}$ and $K_2 = 2730\,M^{-1}$. Photoirradiation of this ternary complex gave cyclobutanes *cis*-**37g** in 19% and *trans*-**39g** in 79% relative yield. The association constants of adducts **37g** and **39g** with γ-CD were $K_{37g \subset \gamma\text{-}CD} = 520\,M^{-1}$ and $K_{39g \subset \gamma\text{-}CD} = 18000\,M^{-1}$, respectively, revealing a great difference in stability between the isomeric complexes of $[\mathbf{36g}]_2 \subset \gamma$-CD.

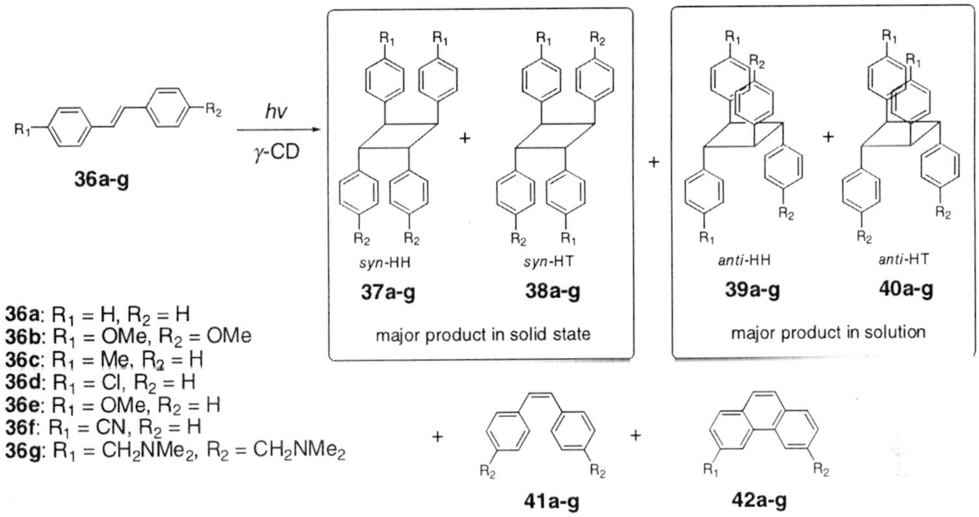

Scheme 1.8

The inclusion complexes of **36a–f** with γ-CD were prepared by the co-crystallization method.[42] The host/guest ratio varies from 1 : 1.60 to 1 : 1.95, indicating that most of the guests formed 1 : 2 host–guest complexes. However, the *cis*-isomer **41a** gave only 1 : 1 complex with γ-CD under the same condition. Photolysis of **36** ⊂ γ-CD for 24 h gave *syn*-dimers **37a–f** and **38a–f** in 59-79% relative yield. This observation is in sharp contrast to the results obtained in the solution-phase photolysis of **37g** mentioned above, and is attributed to the rate-limiting diffusion process prior to the reaction. The large distance and nonparallel orientation of the olefinic double bonds of the stilbenes in crystal lattices prohibit the dimerization of *anti*-isomers.

Tamaki *et al.* reported that water-soluble anthracene derivatives, such as 2-anthracenesulfonate and 2-anthracenecarboxylate, form 2 : 1 complexes with γ-CD in aqueous solution.[43,44] The [4+4] photodimerization of 2-anthracenecarboxylate (AC) accommodated in γ-CD and its derivatives was investigated in detail.[45] Inoue and co-workers have recently reinvestigated the photodimerization of AC from the chiral point of view by using a variety of chiral hosts. γ-CD forms very stable 1 : 2 host-guest complex with AC in aqueous buffer solution with association constants of $K_1 = 160\,M^{-1}$ and $K_2 = 38500\,M^{-1}$.[46]

Photolysis of AC gives four isomeric cyclodimers: *anti*- and *syn*-HT dimers **43** and **44** and *anti*- and *syn*-HH dimers **45** and **46**, of which **44** and **45** are chiral (Scheme 1.9). Irradiation of γ-CD complex of AC at 0 °C gives the HT dimers **43** and **44** as major products in 88% combined yield and HH dimers **45** and **46** as minor products in 12% yield. HT dimer **44** is given in a moderate enantioselectivity of 37% ee, while poor enantioselectivity of less than 5% ee is given for HH dimer **45**.

Further efforts to improve the HT/HH and enantioselectivity of the photoreaction were carried out by using modified γ-CDs. Ikeda *et al.* reported that, by using dipyridinio-appended γ-CDs **49a–d** (Scheme 1.10) as chiral hosts, **45** is given in enhanced ee of 13%, which is significantly higher than that obtained with native γ-CD.[47]

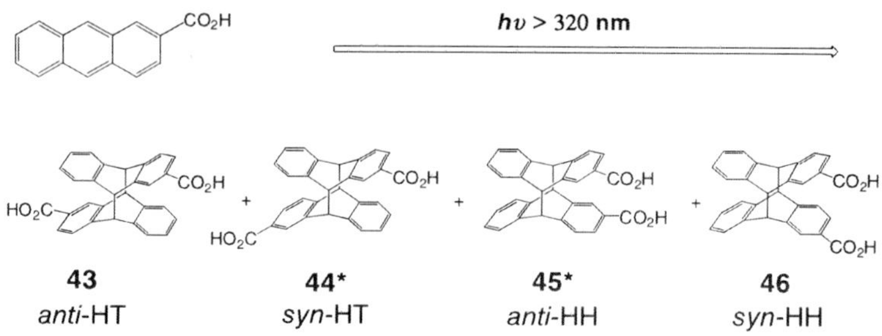

Scheme 1.9

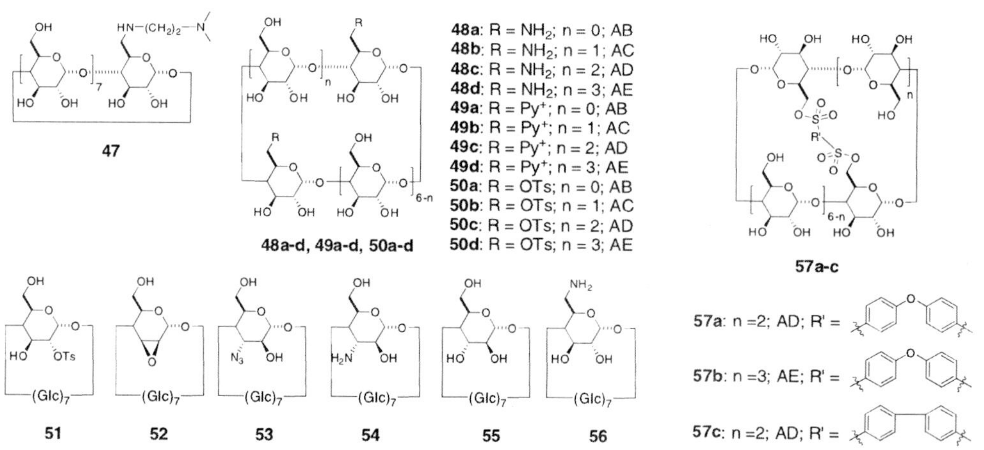

Scheme 1.10

γ-CD derivative **47** that tethered with a flexible dicationic sidearm on the primary rim was also employed for the purpose to improve the chemical and optical yield of HH dimers.[48] In aqueous solution, photoirradiation of AC with **47** gave the cyclodimers in relative yield and ee similar to those obtained with native γ-CD. However, when the photoreaction was performed in a 1 : 1 mixture of water/methanol at −45 °C, the chemical yield of HH dimers was dramatically enhanced to 72% with ee of **45** being significantly improved to 40%. The *anti*/*syn* (**45**/**46**) ratio of HH dimers obtained in the photocyclodimerization of AC mediated by regioisomeric 6A,6X-diamino-γ-CDs **48a–d** gradually increased with increasing distance between the two ammonium groups, showing a nice structure–function relationship in supramolecular chirogenesis system.[49] Furthermore, a series of secondary-rim-substituted and skeleton-modified γ-CDs **51–56** were synthesized in order to examine the effect of modification of CD skeleton on the photocyclodimerization of AC.[50] The ee of HT dimer **44** was enhanced up to 71% and the combined yield of HT dimers reached 93% by the use of **54** at 210 MPa and −21.5 °C.

Recently, the effects of flexible and rigid caps introduced to the primary rim of γ-CD (**50a–d**, **57a–c**) on the photodimerization of AC were investigated, as capping is known to significantly modify the binding and reaction behavior of native CDs.[51,52] The chemical

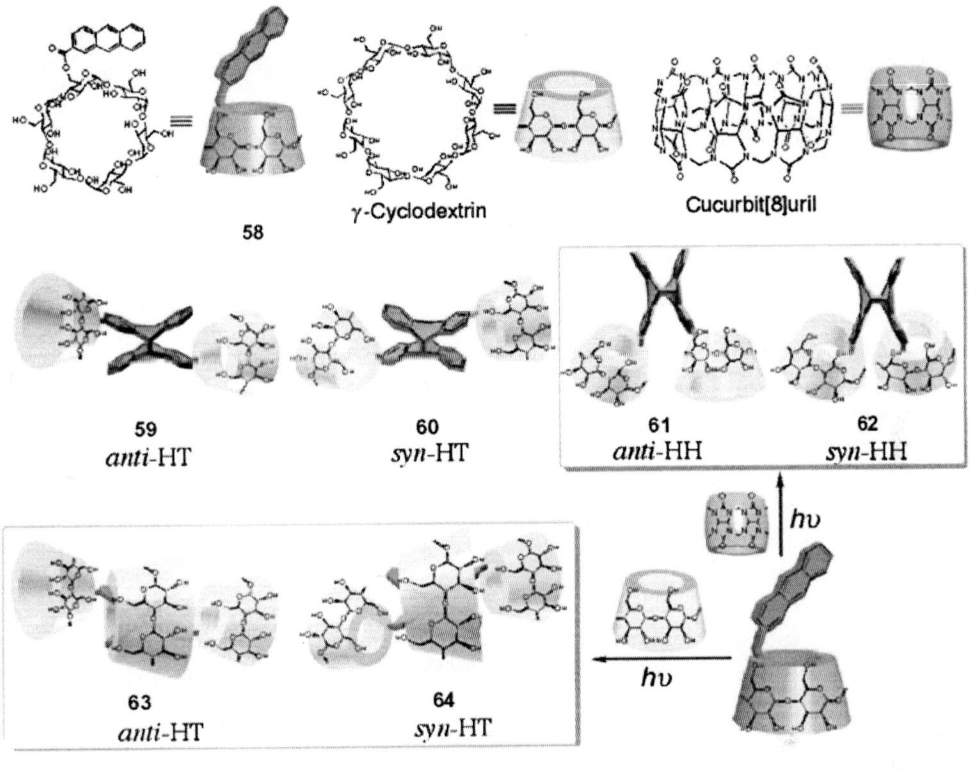

Scheme 1.11

yield and ee of HH dimer **45** were considerably improved by using capped γ-CDs **57a–c**. The enantioselectivity of HT dimer **44** was found to critically rely on the rigidity of the capping moiety. Thus, flexibly capped γ-CD **50a** afforded **44** in moderate ee around 40%, whereas γ-CD **57c** that has a rigid biphenyl cap gave the antipodal **44** in 58% ee.

The main motif in the photocyclodimerization of AC mentioned above is to enhance the HT/HH and enantioselectivity through modification of γ-CD to provide an optimized microenvironment. An interesting approach has recently been explored to influence the photoreaction inside a host by manipulating the moiety existing outside the cavity.[53] Instead of directly using AC as substrate, 6-*O*-(2-anthracenecarbonyl)-α-CD **58** was employed as a photosubstrate that possesses a bulky α-CD to be located outside the host cavity upon complexation. Two anthracene moieties of **58** can be included in γ-CD cavity to give the stepwise 1 : 1 and 1 : 2 association constants of 270 and 21700 M^{-1}, respectively. Photoirradiation of **58** yielded dimerization products **59–62** in the absence of host, which were hydrolyzed in alkaline solution to give **43–46** for evaluating the yield and enantioselectivity. Direct irradiation of **58** gave HT dimers in 76% combined yield, with low ee's for both HT dimer **44** (5% ee) and HH dimer **45** (16% ee). Photoirradiation of **58** in the presence of γ-CD afforded α-CD-stopped rotaxanes **63** and **64** in a high combined yield of 98%. Furthermore, the ee of HT dimer **44** was greatly enhanced to 91% under a pressure of 210 MPa at −20 °C, suggesting that the α-CD that locates outside the γ-CD cavity can critically affect the stereo- and enantioselectivity of the [4 + 4] photocyclodimerization of AC occurring inside the cavity.

1.2.4 Supramolecular Photoreactions with Cucurbiturils

Cucurbit[*n*]urils (CB[*n*]s) are a family of macrocyclic compounds comprising 5–10 glycoluril units. CB[6], CB[7] and CB[8] are similar in cavity volume to α-CD, β-CD and γ-CD, respectively. Differing from CDs, CBs are achiral host molecules of D_{nh} symmetry and therefore cannot provide chiral microenvironment.

Although the solubility of CB[7] in water is the highest in the CB family, the relative small cavity, compared to that of CB[8], appears to have limited its application as a template in photochemistry. Up to now, only one work on [4 + 4] photodimerization of a small cationic guest, i.e. protonated 2-aminopyridine **65** (Scheme 1.12), by CB[7] in aqueous solution has been reported by Macartney and co-workers.[54] The [4 + 4] photodimerization of **65** in acidic aqueous solution in the absence of CB[7] gave *anti-* and *syn-trans*-4,8-diamino-3,7-diazatricyclo[4.2.2.2]-dodeca-3,7,9,11-tetraenes **66** and **67** in 4 : 1 ratio. CB[7] can form complex with **65** in 1 : 1 and 2 : 1 stoichiometries, depending on the amount of **65** added. Irradiation of 2 : 1 complex showed a greatly improved reaction rate and gave exclusively *anti-trans*-**66** in 90% yield without accompanying any other products. Such a high stereoselectivity suggests that two **65** molecules in an *anti-trans* alignment are stabilized in CB[7] cavity. Interestingly, the inclusion of photoproduct **66** in the CB[7] cavity efficiently prohibits its thermal re-aromatization.

Kim *et al.* reported that CB[8] could accommodate two aromatic guest molecules to form 1 : 2 host-guest complexes or 1 : 1 : 1 ternary complexes. The solubility of CB[8] is quite poor in water and organic solvents. Fortunately, by adding some metal cations, the solubility of CB[8] could be considerably increased in water. This observation enabled scientists to investigate intermolecular photoreactions in CB[8] cavity.

Stilbene and its derivatives, which had been widely investigated in CDs and CAs, can be included in CB[8] cavity to form 2 : 1 complexes.[55] Two (*E*)-diaminostilbene dihydrochloride **68** (Scheme 1.13) can insert into CB[8] cavity to give a 2 : 1 host-guest complex which has a good solubility in water. Photoirradiation of the complex gave [2+2] photodimerization product 1α,2α,3β,4β-tetrakis(4-aminophenyl)cyclobutane **69** as major product, along with a trace amount of 1α,2β,3α,4β-tetrakis(4-aminophenyl)cyclobutane **70**. No formation of the isomerization product (*Z*)-**68** was observed, which is significantly different from the photoreaction of (*E*)-**68** in the absence of CB[8]. The stereoselectivity of photodimerization in the presence of CB[8] (*syn*/*anti* = 95/5) is much better than that obtained with γ-CD (*syn*/*anti* = 80/20).

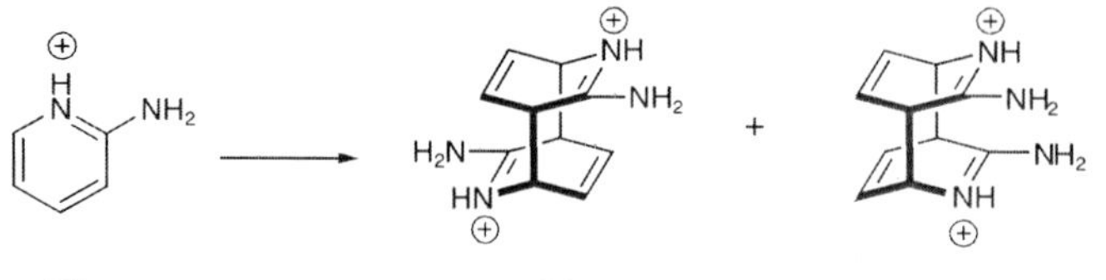

Scheme 1.12

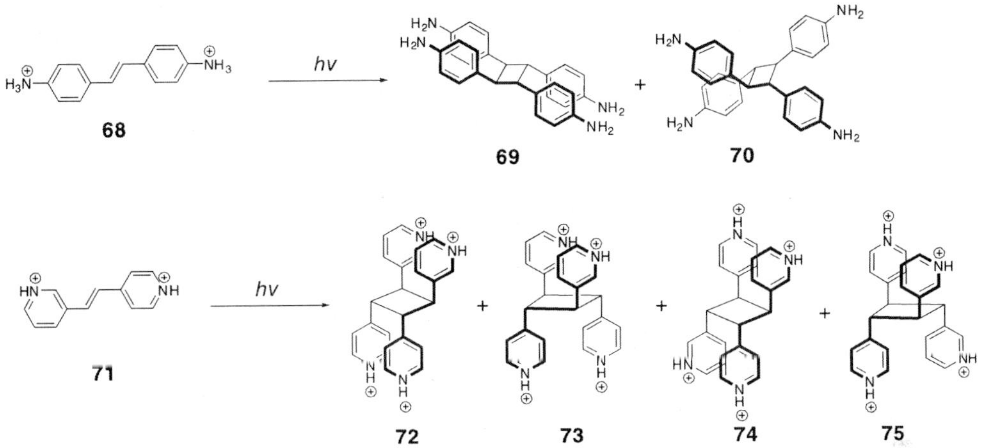

Scheme 1.13

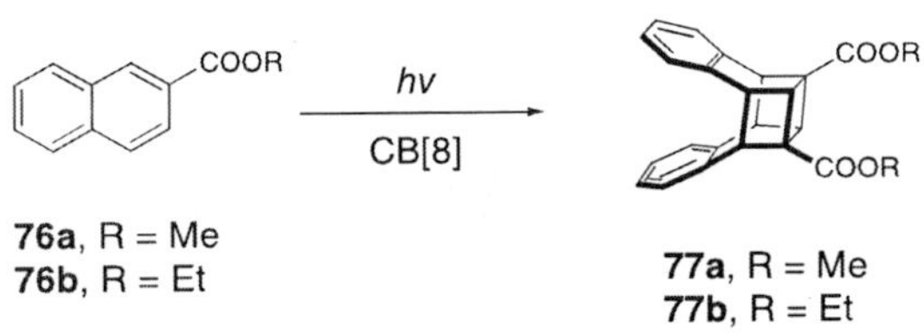

Scheme 1.14

Irradiation of *trans*-1,2-bis(4-pyridyl)ethylene dihydrochloride in the presence of CB[8] also yields the *syn*-dimer as major product in 90% relative yield and *anti*-dimer in only 4% relative yield.[56] However, in the presence of CB[7], no [2+2] dimerization product was obtained and only *cis*-1,2-bis(4-pyridyl)ethylene was formed. This result is attributed to the different complexation stoichiometry of 1,2-bis(4-pyridyl)ethylene with CB[7] (1 : 1) and CB[8] (2 : 1).

Recently, photodimerization of unsymmetrical bispyridylethylenes **71** was investigated in the cavity of CB[8].[57] Irradiation of **71** in the absence of CB[8] gave no [2+2] photodimer, and only *trans*-to-*cis* isomerization is observed for the photoreaction in the presence of CB[7]. Interestingly, irradiation of **71** with CB[8] gave exclusively *anti*-HT **74** and *syn*-HH **72** dimers in 80% and 15% relative yield, respectively, among four possible [2+2] dimers of *syn*-HH **72**, *syn*-HT **73**, *anti*-HT **74** and *anti*-HH **75**.

Wu *et al.* investigated the photodimerization of alkyl 2-naphthoates **76a,b** (Scheme 1.14) in the presence of CB[8].[58] The solubility of **76a,b** in water was enhanced upon addition of CB[8]. As a result of 1 : 2 host–guest complexation between CB[8] and **76**, photodimerization reaction rate was greatly accelerated in aqueous solution to give cubane-like photodimers **77a,b**.

Inoue and co-workers also examined the [4+4] photocyclodimerization of α-CD appended anthracene derivative **58** mediated by CB[8] (Scheme 1.11).[58] Photodimerization of AC using CB[8] as template gave photodimers in relative yields similar to those obtained in the absence of CB[8]. However, the use of CB[8] as a host for the photodimerization of **58** totally inverted the selectivity observed with γ-CD to give HH dimers **45** and **46** in 99% combined yield. The steric repulsion between two α-CD does not appear to play an important role in the determination of the HH/HT ratio but significantly affects the *anti/syn* ratio of HH dimers, revealing a dual role of the bulky α-CD in the stereodifferentiation step of the supramolecular photodimerization. On the other hand, α-CD located outside the cavity could even transfer its chiral information to the photodimerization occurring within the achiral cavity of CB[8] to yield **44** and **45** in modest 17% and 10% ee, respectively.

Large-sized host molecules derived from calixarene and cyclotriveratrylenes have been reported recently. Gibb and co-workers reported some nanoscale host molecules based on resorcinarenes and studied their shape complementarities and the complexation behavior with guest molecules.[59–61] Host **78** (Scheme 1.15), which provides a well-defined hydrophobic environment in water, forms a capsular complex $[\mathbf{78}]_2$ with a cavity dimension of 1 nm wide and 2 nm long.[59] Ramamurthy *et al.* examined the cage effect of host $[\mathbf{78}]_2$ on the photoreaction of 1-phenyl-3-*p*-tolyl-2-propanone,[62] which exhibit significant cage effect comparable to that reported in the solid state.[63,64]

This host molecule can also form 2:2 complex with anthracene (Scheme 1.15)[65]. However, no anthracene dimer was observed even after 10 h irradiation of the complex.

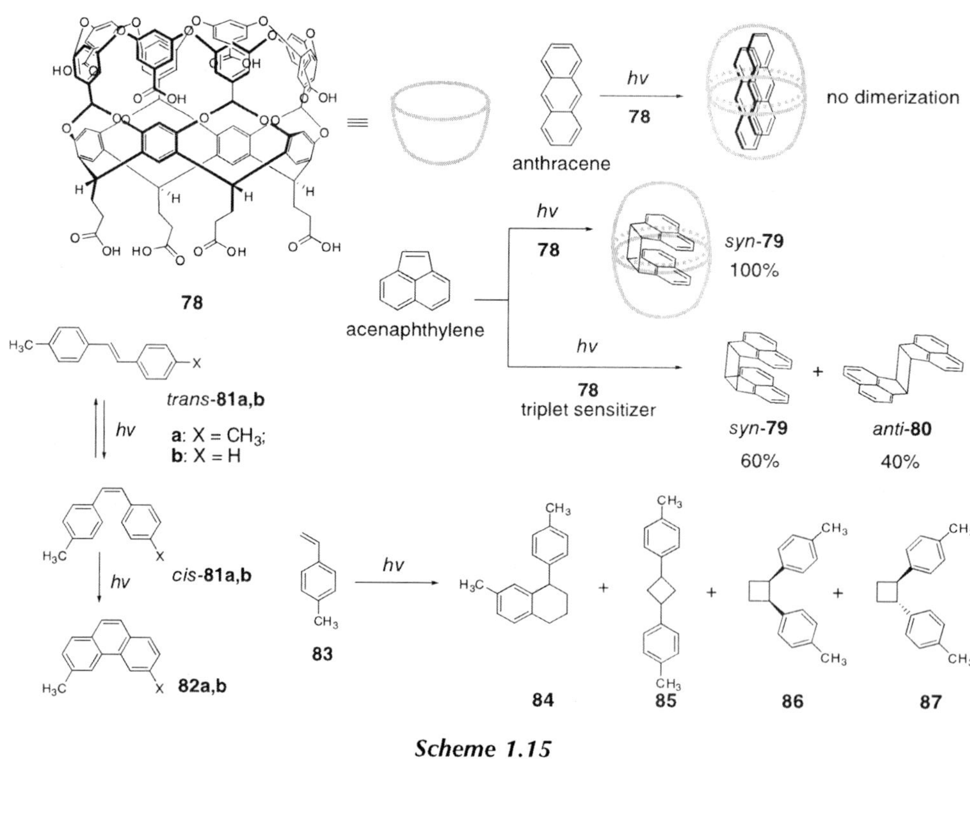

Scheme 1.15

Steady-state fluorescence spectrum of anthracene without **78** was completely different from that in the presence of **78**. Thus, anthracene in aqueous solution showed only monomer fluorescence at 425 nm, but the complex with $[\mathbf{78}]_2$ exhibited excimer emission of anthracene with a maximum at 510 nm and a long lifetime of 263 ns. This suggests that the capsule $[\mathbf{78}]_2$ confines two anthracene molecules in its cavity in staggered orientation.

Similar to anthracene, acenaphthylene can also be encapsulated by $[\mathbf{78}]_2$ to form a 2:2 complex.[66] NMR studies suggested that two acenaphthylene molecules are located in the middle of the capsule cavity and oriented in a parallel fashion. Irradiation of acenaphthylene in the absence of any host yielded *syn*- and *anti*-dimers, with the latter being the major product. Within the $[\mathbf{78}]_2$ cavity, only *syn*-dimer **79** was obtained upon irradiation. In the presence of triplet sensitizer eosin-Y, irradiation of the complex gave *syn*- and *anti*-dimers in a ratio 60:40.

Olefins were also examined as guest substrates for this nanocapsule.[67] In borate buffer solution, *trans*- and *cis*-4,4′-dimethylstilbene **81a** formed 1:2 complex with $[\mathbf{78}]_2$. Irradiation of *trans*-**81a** $\subset [\mathbf{78}]_2$ led to very slow isomerization, while *cis*-**81a** $\subset [\mathbf{78}]_2$ converted to *trans*-isomer very fast. For stilbene **81b** with a smaller size, the *trans*-to-*cis* isomerization became much easier and was almost unaffected by the capsule. Furthermore, 4-methylstyrene **83**, which has a half size of **81a**, was irradiated in the capsule to explore its photochemical behavior. 4-Methylstyrene **83** forms 2:2 complex with **78**. Irradiation of this complex afforded **84** and **85** in 45% and 55% relative yield, respectively.

1.3 Photochemical Reactions with Biomolecules

Biomolecules, such as protein, antibody and DNA, are macromolecules that possess binding sites for organic guests with well-defined three-dimensional geometries and precisely arranged functional groups. As a result of their structural properties, biomolecules are characterized by the significant specificity for complexing and catalyzing their substrates. For example, it is well known that enzyme can accelerate the reaction of its substrate by a factor of one million or even more compared to the control reaction in the absence of enzyme. Their binding sites are highly specific to substrates and can usually catalyze a set of closely related substrates or reactions, but are silent to those mismatched. Since most biomolecules are water-soluble and heat-sensitive, performing a reaction with biomolecules definitely means the use of mild and environmentally benign conditions. On the other hand, light energy is known to be environmentally benign and can be efficiently converted into chemical energy through biophotochemical process, such as photosynthesis. These properties are superior in many aspects to those of artificial hosts and have attracted chemists to pursue efficient and selective chemical reactions in binding sites of biomolecules.

1.3.1 Photochemical Reactions Templated by Deoxyribonucleic Acid (DNA)

DNAs are important molecules that carry the genetic information and are capable of self-replication in living organism. The basic unit of DNA is nucleotide that comprises of a nucleobase (adenine, thymine, cytosine and guanine), deoxyribose, and a phosphate tail (Scheme 1.16). The precise complementarity between adenine and thymine and between cytosine and guanine renders two long chains of nucleotides coiled to give rise to a twisted

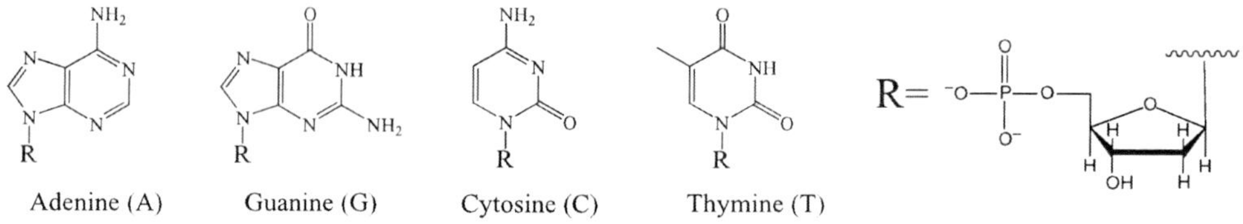

Scheme 1.16

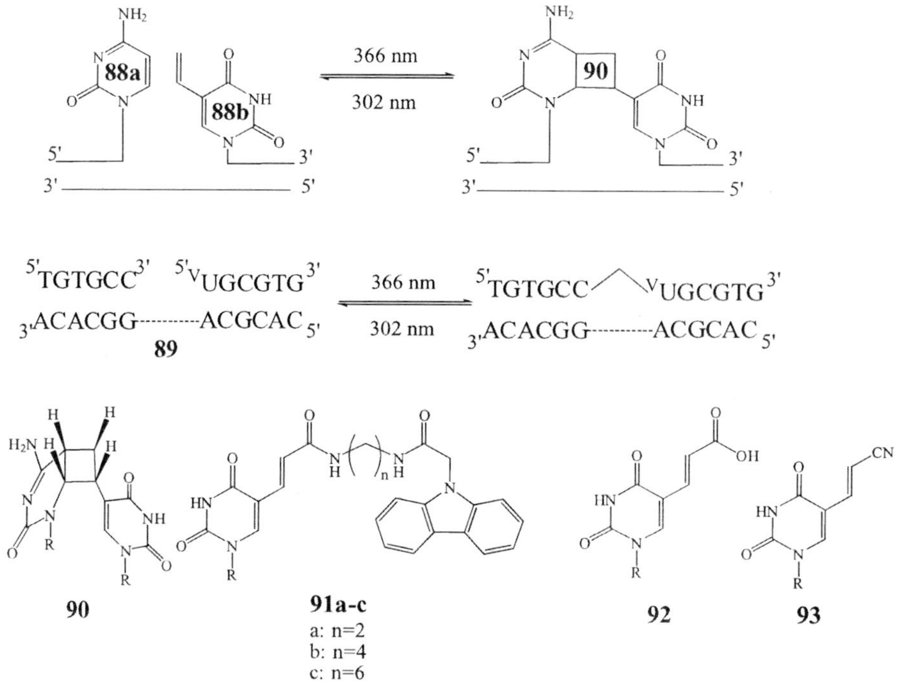

Scheme 1.17

double helix, which is driven predominantly by hydrogen bonding and π π stacking interactions. It is therefore reasonable to anticipate that the DNA chain itself can serve as a good template for precisely positioning the substrates linked to it.

Photoaddition of substrates covalently-linked to DNA at its terminus has been studied for the purpose of exploring photoinduced ligation of oligonucleotides. Photoligation has evident advantages that the ligation can be conducted in water without adding any external reagent in a non-enzymatic fashion and the original DNA can be recovered by wavelength-selective irradiation.

Saito and co-workers investigated a series of photoligation reactions of DNA using native and terminus-functionalized oligonucleotides.[68–76] As illustrated in Scheme 1.17, synthetic oligodeoxynucleotides **88a** and **88b**, which bear a cytosine at the 3′-end and a 5-vinyldeoxyuridine at the 5″-end, are complementary to an oligodeoxynucleotide template **89**.[68] In the absence of **89**, no photoadduct was given upon photoirradiation at 366 nm, at which **88b** has only weak absorption of $\varepsilon_{366} = 3.3$. However, photoligated oligodeoxynucleotide **90** was obtained in up to 96% yield when photolysis was carried out for 12 h in the presence of **89**. Only one [2+2] photoadduct that has a *cis-syn* stereochemistry was afforded. The significantly enhanced reactivity and selectivity are attributed to the well-arranged orientation of two oligodeoxynucleotides when complexed with the template **89**. Vinyldeoxyuridine introduced at the 5′-end underwent the same [2+2] photoaddition with thymine at the 3″-end but not with adenine and guanine. The ligated oligodeoxynucleotide can be photodecomposed to give the original 6-mers upon irradiation at 302 nm but better recovered by further irradiation at 366 nm, suggesting photochemical reversibility of the ligation process.

However, the split of ligated DNA with light at 302 nm may also result in damage to DNA due to the dimerization of thymine. To avoid this disadvantage, Fujimoto and coworkers synthesized 3″-terminus-modified oligodeoxynucleotide **91** (Scheme 1.17), in which carbazole is tethered to the 5-carboxyvinyluracil moiety as a sensitizer for photodecomposition.[69] Photolysis of oligodeoxynucleotides with a template gave ligated oligodeoxynucleotide with an efficiency order **91c** > **91b** > **91a**. A *cis-syn* [2+2] adduct was given in 89% yield after the photolysis of **91c** at 366 nm for 6 h at room temperature in the presence of the template. The same reaction performed at 70 °C, where **91c** was decomplexed from the template, afforded no photoligated product, indicative of the importance of preorientation of photosubstrate by supramolecular host. On the other hand, photoirradiation of the isolated adduct in the absence of the template at 366 nm for 6 h afforded **91c** in 90% yield as a result of the photosensitized split.

In order to improve the photoreactivity of template-directed photochemical synthesis of branched oligodeoxynucleotides, the 5-vinyldeoxyuridine component was replaced by 5-carboxyvinyldeoxyuridine **92**, which has a much higher extinction coefficient ($\varepsilon_{366} = 76$).[70] Photoirradiation of **92** for 1 h in the presence of a template gave a *cis-syn* [2+2] adduct in good yield of 93%. Similar [2+2] photocycloaddition between an α-5-cyanovinyldeoxyuridine **93** at the 3′-end and a thymine at the 5′-end also occurred smoothly in the presence of a template.[71] A single mismatch in oligodeoxynucleotide chain led to a great decrease of reaction rate, indicating a critical dependence of photoligation on sequence specificity.

Using **94** as a template DNA, Ihara *et al.* investigated the photodimerization of anthracene tethered to complementary oligodeoxynucleotides **95** and **96** (Scheme 1.18).[77] The reactivity of the photodimerization mediated by **94** was found to be dependent on the distance between oligodeoxynucleotides and anthracene, and decrease in an order of **95a/96a** > **95a/95b** > **95b/96b** > **96a/96b**. The reaction efficiency is also significantly affected by the introduction of an additional or mismatched nucleotide, indicating that this DNA-templated photoligation is highly position- and distance-dependent.

Photodimerization of 1-substituted anthracene derivatives **97a** and **97b** (Scheme 1.19) templated by a full-match DNA **98** selectively afforded only two isomeric dimers among the eight dimers that are possibly formed from the dimerization of **97a** and **97b**. Photoreaction with the full-match template was found to be much more efficient than those with mismatch templates.[78,79]

There are two kinds of grooves in DNA helixes, i.e. major and minor grooves, which can interact with metal ions or organic guests. Since these grooves are well-structured

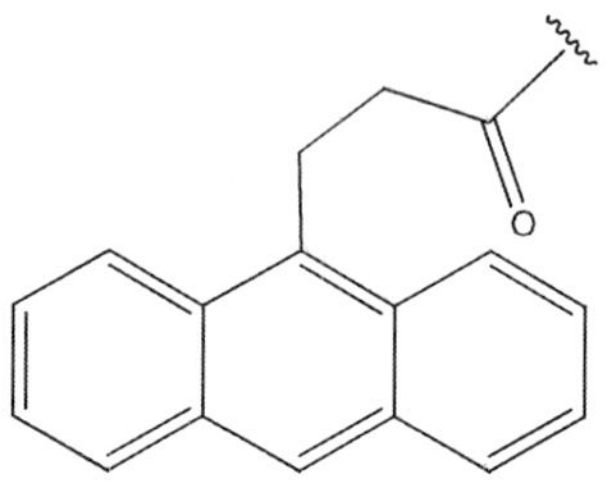

TTAGGGTTAGCG

94

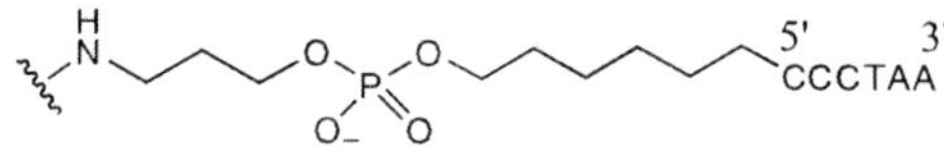

95a

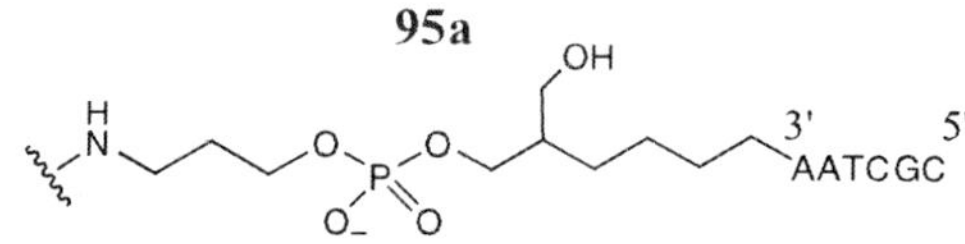

95b

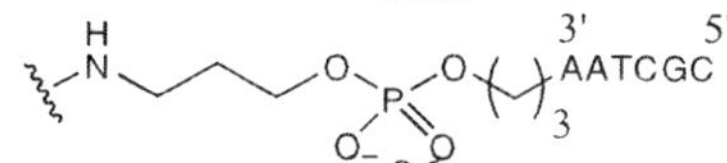

96a

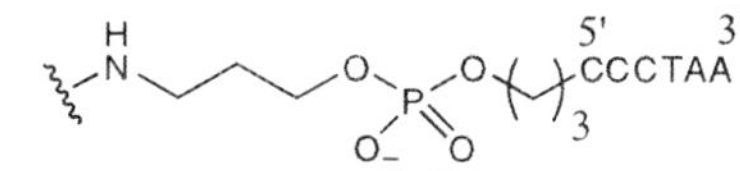

96b

Scheme 1.18

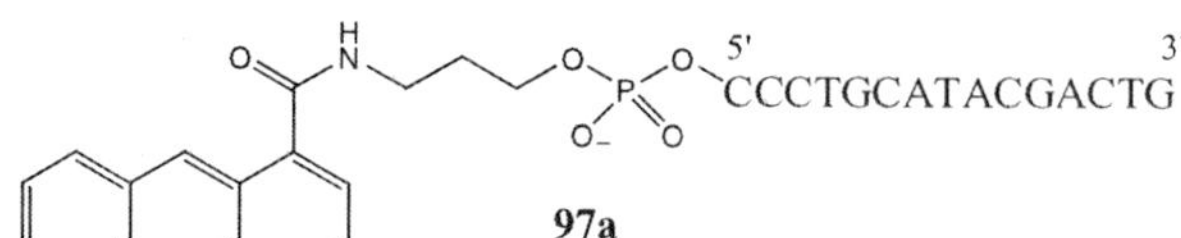

97a

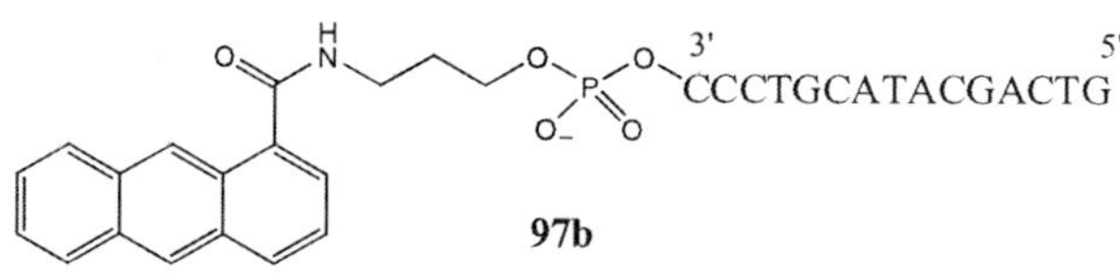

97b

5' CAGTCGTATGCAGGGGGGACGTATGCTGAC 3'

98

Scheme 1.19

and chiral in nature, they can be used as ideal stereodifferentiating binding sites. Inoue *et al.* performed the enantiodifferentiating photoisomerization of **34Z** (Scheme 1.7) complexed and sensitized by calf thymus DNA (ctDNA).[80] Photosensitized isomerization of **34Z** with pyrimidine nucleosides as chiral sensitizers afforded **34E** in 5.2% ee at the photostationary state. However, much better ee of up to 15.2% was obtained when the photolysis was carried out in the presence of ctDNA in aqueous solution. Supramolecular interaction of **34Z** with hydrophobic minor groove of ctDNA evidently plays an important role in the photochirogenesis, as only negligible ee was obtained for **34E** when the photosensitization was performed in 50% aqueous methanol, in which the supramolecular interaction of **34E** with ctDNA was significantly reduced.

1.3.2 Photochemical Reactions Mediated by Proteins

Proteins, the most abundant and versatile macromolecules in living organisms, are polymers built from about twenty essential L-α-amino acids. Driven by a variety of non-

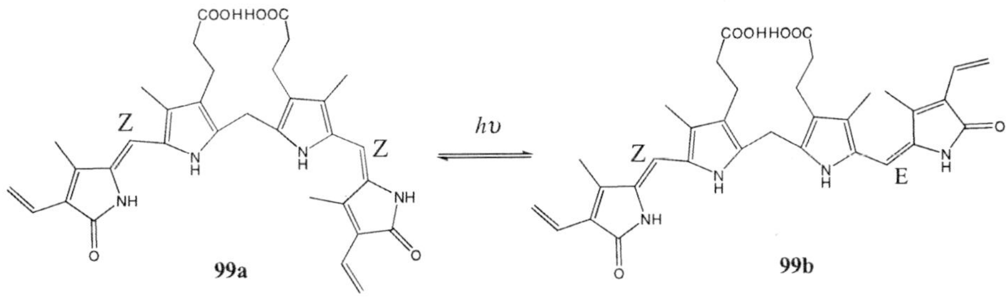

99a 99b

Scheme 1.20

covalent interactions, such as ionic, hydrogen binding, van der Waals and hydrophobic interaction, these biopolymers fold into specific conformations to achieve various biological functions. The binding sites of defined shape and size, as well as the chiral nature, render proteins excellent supramolecular hosts for specific binding and recognition. A series of photochemical reactions have been investigated with various proteins, mainly targeting the photochemical chiral discrimination and generation based on the inherently chiral character of binding sites in protein. Serum albumin is the most abundant protein in plasma which binds and transports not only endogenous but also exogenous compounds, and has therefore been most frequently employed as biosupramolecular host for photochemical reactions.

McDonagh and co-workers investigated the photoisomerization of bilirubin-IIIa **99a,b** (Scheme 1.20) in the presence of human serum albumin (HSA). The *E-Z* photoisomerization between (*Z,Z*)-**99a** and (*E,Z*)-**99b** is not inhibited even in the binding cavity of HSA. Bilirubin-IIIa shows intense bisignate circular dichroism spectra in the chiral pocket of HSA.[81,82] The negative exciton coupling signal, which was observed upon irradiation of (Z,Z)-bilirubin-IIIa/HSA complex with a broad-band blue light of λ_{max} = 430 nm, was dramatically switched to a positive exciton coupling signal after irradiation with green light at 544 nm.[83] This signal switching, which is controllable by changing the irradiation wavelength, renders the system a potential chiroptical switch.

Zandomeneghi *et al.* reported the enantioselective photoinduced decomposition of 1,1′-binaphthol **100** (Scheme 1.21) included in bovine serum albumin (BSA). Complexation of (*R*)-(+)- and (*S*)-(–)-**100** with BSA revealed a remarkable difference in stability between the two diastereomeric complexes, and the BSA complex of (*S*)-(–)-**100** exhibited a significantly bathochromic shift in the absorption spectrum. By selectively irradiating the (*S*)-(–)-**100** complex at longer wavelengths, (*R*)-(+)-**100** of up to 99.5% ee was obtained after 77% of the starting material (racemic **100**) were decomposed.[84–86]

Miranda *et al.* studied the excited-state behavior of chiral drugs bound to BSA and HSA.[87–89] Inclusion of carprofen **101** (Scheme 1.22) in HSA significantly lengthened the triplet-state lifetime of **101** as a result of more rigid surroundings and the suppression of self-quenching of **101**. The complexation of enantiomeric **101** with HSA caused a significant difference in triplet lifetime, and the lifetime of (*S*)-**101**-HSA complex is 4 times shorter than that of (*R*)-**101**-HSA complex. Irradiation of **101** alone in aqueous buffer solution led to a polymerization of **101**. However, dechlorinated carprofen **102** was

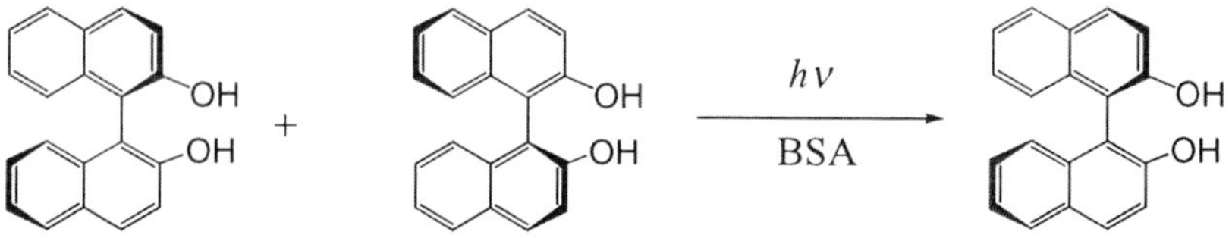

Scheme 1.21

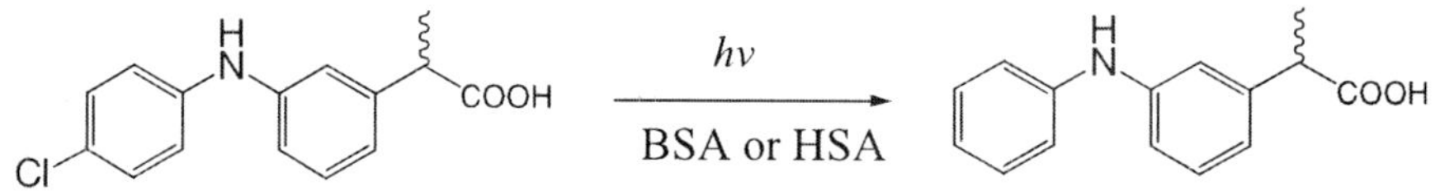

Scheme 1.22

formed in the presence of HSA, and (*S*)-**101** was 1.5 times more efficient than (*R*)-**101** in producing the corresponding (*S*)- and (*R*)-**102**.

The complexation behavior of organic substrates with proteins is usually very complicated as a consequence of the diversified binding sites with distinctive size, shape, hydrophobicity and binding affinity. Since these binding sites differ in reactivity and selectivity, elucidation of binding behaviour is essential for understanding and manipulating the photoreaction performed in the binding sites of proteins. Inoue and co-workers have comprehensively studied the binding behaviour of 2-anthracenecarboxylate (AC) (Scheme 1.9) to the hydrophobic pockets of BSA and HSA by a combined use of spectroscopic techniques, i.e. UV/Vis, fluorescence and circular dichroism spectroscopy.[90–92] Although HSA and BSA share more than 95% of amino acid residues in common (i.e., 26 out of *ca.* 600 amino acids), the two albumins display entirely different binding, photophysical and photochemical behavior. BSA possesses four binding sites, which accommodate 1, 3, 2 and 3 AC molecules with individual binding constants of 5.3×10^7, 1.3×10^5, 1.4×10^4 and $3 \times 10^3\,M^{-1}$, respectively. In contrast, HSA provides five binding sites that accommodate 1, 1, 3, 5 and >10 AC molecules in the order of reducing binding ability.[93–97] Although multiple AC molecules are bound to sites 2–4 of BSA and sites 3–5 of HSA, they do not appear to be very closely located to each other, since no exciton coupling was seen in the circular dichroism spectrum and no static quenching of fluorescence was detected. AC bound to site 1 of BSA shows highly structured fluorescence with dual lifetimes of 4.8 and 2.1 ns, which are assignable to two kinds of AC that differ in position or orientation in highly hydrophobic environment. However, ACs in site 2 of BSA gave a much longer lifetime of 13.3 ns and the fluorescence spectra became broader; the lifetimes of AC in sites 3 and 4 are practically indistinguishable from that in bulk water (15.9 ns). Almost the same is true for HSA, except for the fact that not only the first but also the second site bind only one AC and are highly hydrophobic.

Each binding site in BSA and HSA exhibits very different photochemical behaviour. Thus, the HT/HH ratio and enantioselectivity obtained in albumin-mediated AC photocyclodimerization are dynamic functions of the AC/BSA or AC/HSA ratio. For example, the HT dimers are produced as the major products in the photodimerization of AC in host-free bulk water and also in the presence of HSA, whereas the HH dimers become the dominant products in the BSA-mediated photodimerization. In the presence of BSA (AC/BSA = 1.3), chiral HT and HH dimers, **44** and **45**, were obtained in 29% and 41% ee, respectively. Interestingly, the ee's for **44** and **45** were further improved up to 38% and 58% by adding nitromethane as a site-selective quencher of AC located in such binding sites that are more accessible but less enantioselective.

In the case of HSA, much higher 79% and 88% ee were reported for **44** and **45**, respectively, at 25 °C and AC/HSA = 3. Temperature was found to have a modest effect on the enantioselectivity to give better ee's of 82% and 90% for **44** and **45**, respectively, by decreasing the temperature to 5 °C.

1.4 Photochemical Reactions with Confined Cages Based on Inorganic and Organic–Inorganic Hybrid Materials

1.4.1 Photochemical Reactions with Zeolites

Besides the natural and artificial molecular hosts that are primarily based on organic compounds, photochemical reactions have also been investigated in the confined spaces of inorganic nano- or mesoarchitecture. Zeolites, which are porous crystalline aluminosilicates composed of tetrahedral AlO_4 and SiO_4 building blocks and usually possess cages and channels of 4–14 Å, have been extensively used as hosts for mediating photochemical reactions. The ratio of SiO_4 to AlO_4 can be changed from 1 to infinite, and metal cations are commonly introduced to balance the negative charge of AlO_4 units. The faujasite and pentasil family of zeolites, which form roughly spherical supercages of *ca.* 13 Å and tubular channels with lengths of *ca.* 50 Å, are most commonly used zeolites for catalyzing chemical reactions. Zeolites have a wide UV/Vis transparent region as a result of Al–O–Si framework that is similar to glass and quartz, and therefore are compatible with most organic compounds. The size and free volume zeolite supercages are adjustable by the number and sort of countercations. By virtue of these properties, zeolites have attracted much attention of photochemist as confining media in the last two decades.

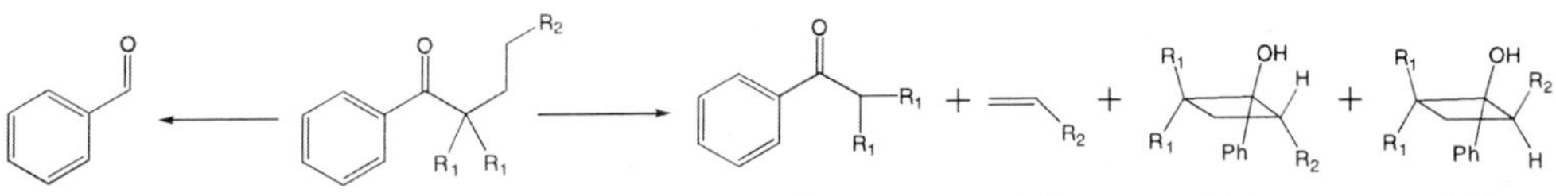

104 **103** **105** **106** **107** **108**

a: R_1=H; R_2= CH_3
b: R_1=H; R_2= $(CH_2)_3CH_3$
c: R_1= CH_3; R_2= CH_3

Scheme 1.23

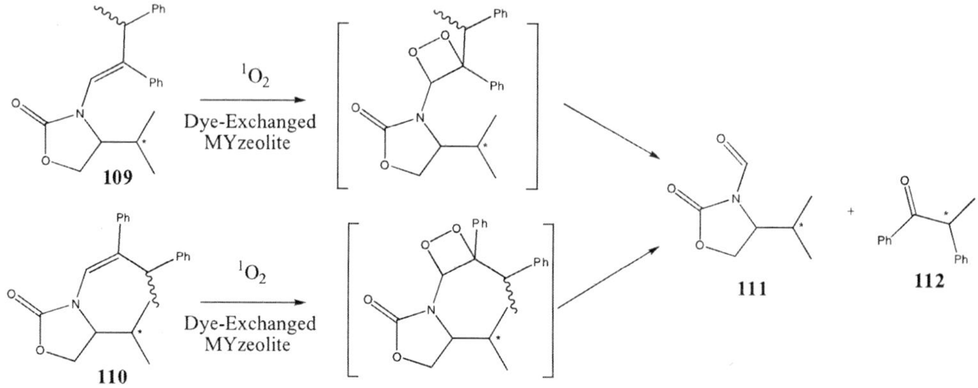

Scheme 1.24

Studies on photochemical reactions within zeolites have already been summarized in several reviews.[98–102] The earliest work in this field appeared in 1980s.[103–106] Turro and co-workers investigated Norrish type I and type II photoreactions of phenyl alkyl ketones **103** in the supercage of faujasite-type zeolites. Photolysis of **103** in solution or solid state was known to give a variety of photoproducts **104–108**, depending on the molecular mobility, conformation and orbital overlap in the excited state. Inclusion of **103** in zeolites can significantly modify the above factors and lead to a switching of reaction selectivity. Photolysis of **103** in supercages of NaY zeolite gave a mixture of products with modest selectivities; **105**/(**107** + **108**) = 0.72–0.83. The ratio was improved to 3.2–3.8 by using NaX zeolite, and was dramatically enhanced up to >50 by using silicalite (S-115) zeolite,[105] suggesting that this photoreaction is highly dependent on the cavity size of zeolite.

Photooxygenation of (*Z*)-enecarbamate **109** (Scheme 1.24) in solution gave methyl deoxybenzoin **112** in only a modest enantioselectivity even at a very low temperature, while a much better optical yield was obtained by starting from the (*E*)-isomer **110**.[107–111] The confinement in zeolite supercage greatly restricts the orientation and conformation of photosubstrate and leads to a significant switching of reaction selectivity from that observed in isotropic media. Thus, singlet oxygenation of (*Z*)-**109** in NaY zeolite loaded with cationic dye sensitizer methylene blue (which is readily introduced into zeolite supercage through cation exchange and generates $^1O_2$ from $^3O_2$ upon irradiation) gave **112** in up to 80% ee. This is possibly due to the restricted space of zeolite supercage that aligns the substrate in an orientation suitable for π-face-selective attack of the C=C bond by $^1O_2$. Compared to the high ee obtained in methylene-blue-exchanged NaY, the same reaction performed in LiY zeolite gave **112** in only 35% ee, for which the larger available volume in the supercage of LiY zeolite is possibly responsible.

Besides the well-confined and rigid framework of zeolites, cation species also plays an important role in manipulating photochemical reaction occurring in the cavity.[112–121] The free volume in zeolite cavity relies on the number and sort of cations located in the cavity. Differing from isotropic media, in which the direction and magnitude of electric field fluctuate around a solvated molecule, cations in zeolite cavity generate a stronger, anisotropic and more stable electric field. Such an electric field can polarize the included

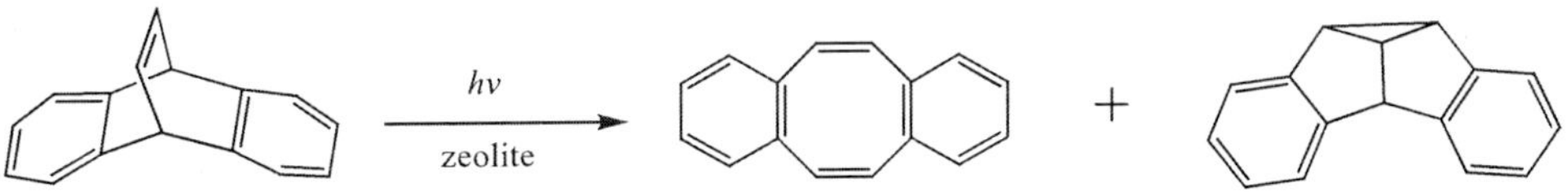

113 114 115

Scheme 1.25

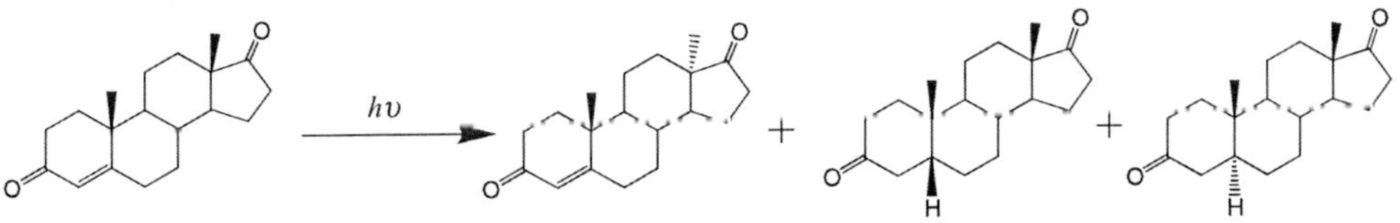

116 117 118 119

Scheme 1.26

molecules, distort their electron density, and even facilitate heterolytic bond cleavage. Indeed, it has been observed that the intensity and wavelength of absorption and emission of included aromatic compounds were significantly changed in the cavity of zeolite.[122–125] Ramamurthy *et al.* reported that photoreaction of dibenzobarrelene **113** (Scheme 1.25) in Li^+ and Na^+ exchanged zeolites resulted in triplet photoproduct **115** in 67% and 62% yield, respectively. This observation indicates that the intersystem crossing is significantly accelerated even by light atoms. However, this effect disappeared when the zeolites were equilibrated with water before use. The electric field of the cations in the cavity, which perturbs the symmetric nature of photosubstrate and facilitates the intersystem crossing, may be responsible for this 'light atom' effect. On the other hand, the heavy atom effect is clearly seen even for hydrated cations, and >99% triplet photoproduct was observed with Tl^+-incorporated zeolite.[126–129]

Photoreactivity of androstenedione **116** (Scheme 1.26) was demonstrated to be dramatically affected by the electric field of zeolite.[130] The epimerization of **116** to yield **117** is the major reaction in isotropic solution such as hexane, methylene chloride, methanol and cyclohexane. The reduction of the cyclohexenone A-ring can be observed in only propanol, with **118** and **119** being given in 14% combined yield. However, irradiation of **116** in NaY afforded exclusively the reduction products in >85% yield, and no product due to the reaction at the D-ring was observed. This reactivity change observed for the androstenedione included in zeolite cavity was ascribed to the lowering of the π,π^* excited state of the A-ring due to the electric field in zeolite cavity.

Interactions between cation and photosubstrate, such as cation–π, cation–dipole and cation–lone pair interactions, may restrict the mobility and conformation of photosubstrate, and therefore endow the cation with versatile roles in manipulating the reactivity and selectivity of photoreaction in zeolite cavity.[131–142] Ramamurthy and co-workers have studied stereoselective photoisomerization of diphenylcyclopropane derivatives in zeolite supercage. It was found that the photoisomerization of *trans*-diphenylcyclopropane **121a** occurred in the cavity of alkali cation-exchanged Y zeolite to give *cis*-**120a** in high

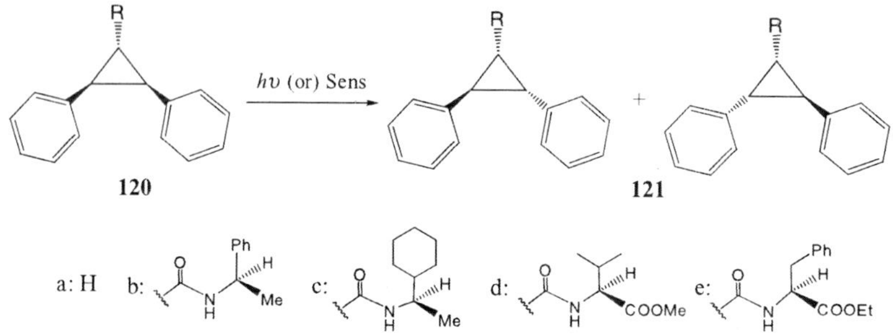

Scheme 1.27

efficiency, while the photolysis of *cis*-**120a** failed to give any *trans*-isomer.[134] This is ascribed to the formation of cation–π complex in which two phenyl groups of *cis*-**121a** sandwich an alkali metal ion. The *cis*-isomer is thus stabilized and the isomerization reaction is inhibited.

However, *trans*-isomers can be given upon photoirradiation of zeolite-mediated *cis*-diphenylcyclopropane derivatives **120b–e** due to the stronger cation–dipole interaction between cation and carboxamide group than cation–π interaction. Photolysis of amide **120b–e** in hexane-methylene chloride solution afforded the corresponding *trans*-isomers in diastereomeric excess (de) less than 2%. In sharp contrast to the result obtained in isotropic solution, photoreaction of **120b–e** adsorbed in zeolite MY gave *trans*-isomers in 20–83% de. The dramatic improvement of de is believed to occur as a result of cation–carbonyl and cation–nitrogen dipolar interactions that fix the conformation of reactants, with asymmetric centres of the chiral amide moiety being located closer to the chirogenic center.[114,132,134,138,143]

Chiral photochemistry in zeolite is an intriguing topic that has attracted intensive attention. Although theoretically zeolites can be chiral, until now no chiral zeolite has been separated in an enantiomerically pure form. Chiral photochemistry in zeolite is conducted primarily through two strategies: the first one is to introduce a chiral auxiliary to photosubstrate and the diastereodifferentiating photoreaction is conducted in intact zeolite, as exemplified in the diastereoselective photoisomerization of diphenylcyclopropane derivatives mentioned above, and the second one is to immobilize optically active compound to chirally modify the zeolite supercage and the enantiodifferentiating photoreaction of prochiral substrate is conducted in this chiral supercage.

Zeolites modified with norephedrine, ephedrine and pseudoephedrine as chiral inductors have been employed for the enantioselective photocyclization of a series of tropolone derivatives **122**.[113,144,145] Photocyclization of **122b** conducted in NaY zeolite modified with (–)-ephedrine gave **123b** in up to 78% ee. The enantioselectivity of the photoreaction was found to be critically controlled by the cationic species and water content in zeolites.

The optical yield of photoreaction in chirally modified zeolite relies on several factors such as cation, chiral inductor and its loading level, which is usually kept at a high level to maximize the number of chiral inductors located around the photosubstrate. An exceptional example may be found in the enantiodifferentiating photoisomerization of **34Z** (Scheme 1.7) sensitized by optically active (poly)alkyl benzoate derivatives immobilized

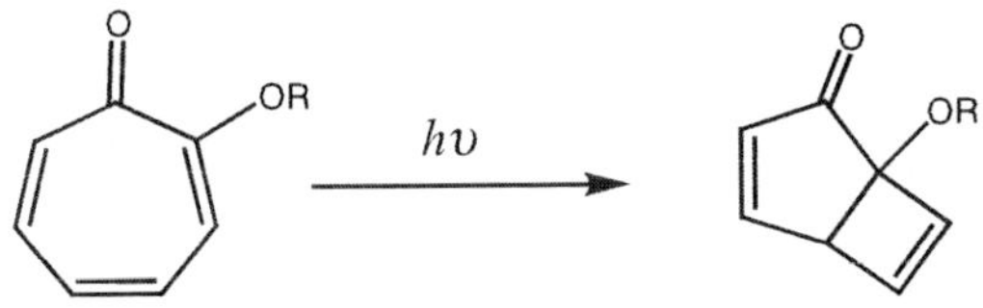

122 **123**

a. R = CH_3
b. R = CH_2CH_2Ph
c. R = CH_2COOCH_3
d. R = CH_2CH_3
e. R = $CH_2CH_2CH_3$
f. R = CH_2Ph
g. R = $CH_2CH_2CH_2Ph$

Scheme 1.28

in NaY zeolite reported by Inoue and co-workers.[146] The photoreaction was allowed to occur in the presence of a catalytic amount of sensitizer, and **34E** was obtained in a modest ee of 4.5% upon photosensitization with (–)-methylheptyl benzoate-immobilized NaY zeolite. In contrast, the same sensitizer gave racemic **34E** in homogenous solution.

1.4.2 Photochemical Reactions in Mesoporous Materials

Mesoporous materials have recently attracted considerable attention for the use in photochemical reactions. As defined by the IUPAC nomenclature rules, mesoporous materials have a pore diameter in the range of 2.0–50.0 nm, which is much larger in general than that of zeolite and is certainly beyond the size of normal organic substrates. Therefore, photoreactions can freely proceed without suffering any restriction. In fact, studies on photochemical reaction in mesoporous materials focus more on construction of functional materials rather than the chemical selectivity. For example, azobenzene, which is structurally rigid and has a van der Waals volume of 144 Å^3,[147] requires a large extra volume of 127 Å^3 for the photoisomerization to occur. Taking advantage of this property, azobenzene was loaded into nanocomposite films to be used for and showed a potential application in adjusting the cavity volume of mesoporous materials.[148] Mal *et al.* constructed a photochemically controlled system by anchoring coumarin to the pore openings of MCM-41 for taking up or releasing organic guests by controlling the reversible photodimerization of coumarin.[149]

The size of mesoporous silica MCM-41 can be adjusted by changing the number of carbon atoms in surfactant micelles used in the hydrothermal synthesis. Iwamoto and co-workers have reported photocyclization of diarylethenes **124** in different-sized MCM-41. Only the anti-parallel conformation of open form **124** can undergo cyclization upon irradiation to give closed form **125**. The reaction rate of **124** was found to be remarkably dependent on the amount of **124** loaded as well as the pore diameter of MCM-41s, but be independent of the organic groups on the surface of MCM-41s.[150]

Recently, Inoue and co-workers studied the enantiodifferentiating photocyclodimerization of AC in the channel of mesoporous silica that is modified with γ-CD.[151] Photocyclodimerization of AC with mesoporous without γ-CD gave HH photodimers

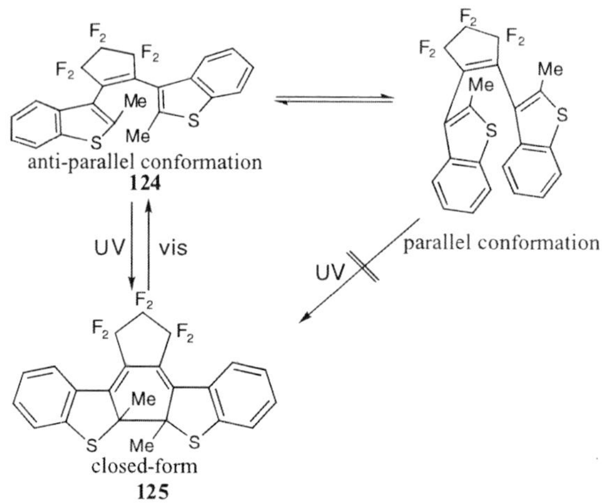

Scheme 1.29

with a low conversion and enantioselectivity. In remarkable contrast to the γ-CD-mediated photoreaction performed in aqueous solution, the photocyclodimerization of AC in γ-CD-modified mesoporous silica led to the major formation of HH photodimers, in particular *anti*-HH dimer **45** in 45% yield and 24.1% ee.

1.4.3 Photochemical Reaction with Self-Assembled Molecular Cages

Organic ligands bearing two or more coordinating sites can form well-structured aggregates upon coordination to metals. Some of these ligand–metal self-assemblies form structurally well-confined cavity that can accommodate organic guests of suitable size. These cages generally have polyhedron structure with rigid skeleton, and the motion of guest(s) included in the cavity is greatly inhibited. Thermal and photochemical reactions conducted in these molecular cages often show remarkably accelerated rates and significantly high selectivities.[152,153]

Compounds **126** and **127** shown in Scheme 1.30 represent two typical coordination cages that are self-assembled from six palladium ions and four tridentate ligands. These coordination cages are well soluble in water and can selectively encapsulate organic substrates in the nano-sized cavity in a well-arranged fashion, mainly through hydrophobic and aromatic interactions. So far, a number of different types of photoreactions have been examined within the cavities of these molecular cages.[154–160] As shown in Scheme 1.31, four stereoisomers **129a-d** are theoretically possible to be formed upon photocyclodimerization of **128**. Strikingly, photocyclodimerization of 0.5 mM **128** in the cavity of **126a** gave the HT *syn*-isomer **129a** as a sole product in >98% yield after 3 h irradiation. On the contrary, photoirradiation of **128** in benzene in the absence of host **126a** gave no adducts even at a high concentration of 150 mM. In the presence of **127a** as a host, photocyclodimerization of **130** in aqueous solution exclusively afforded *syn*-dimer **131a**. This is in sharp contrast to the same photoreaction conducted in benzene without any host, where *anti*-dimer **131b** (21% yield) was favored over *syn*-dimer **131a** (2% yield).

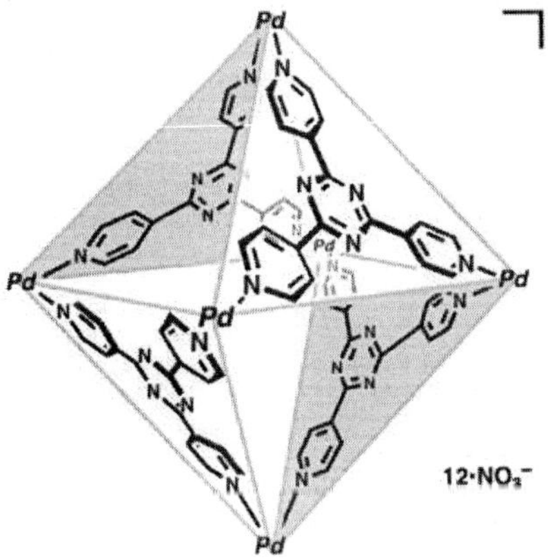

126

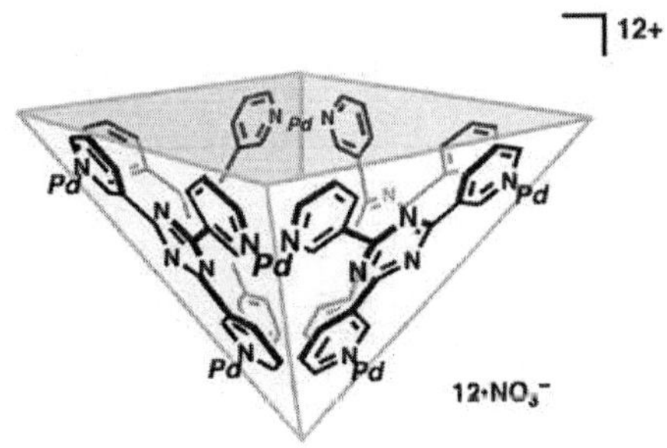

127

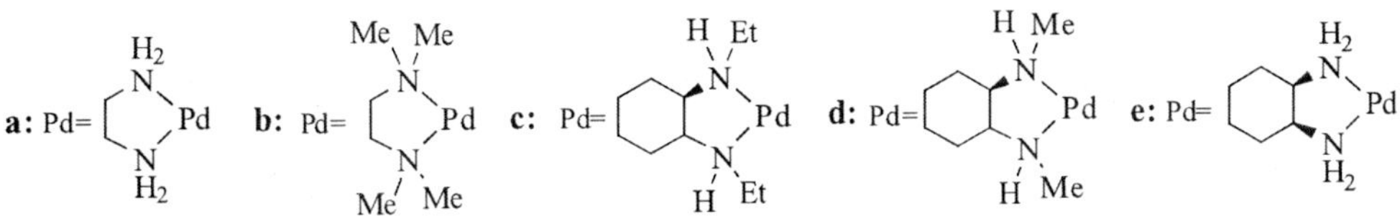

Scheme 1.30

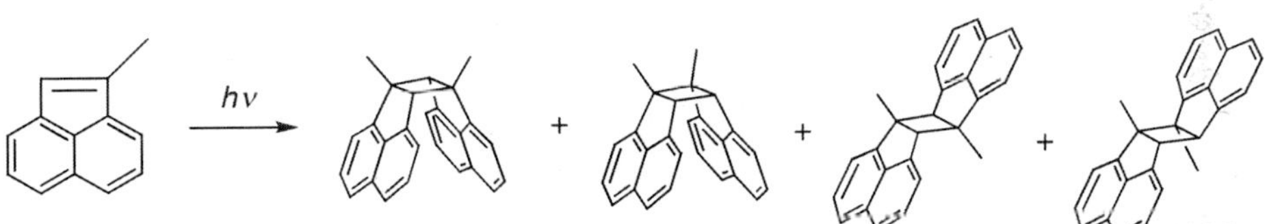

128 **129a** **129b** **129c** **129d**

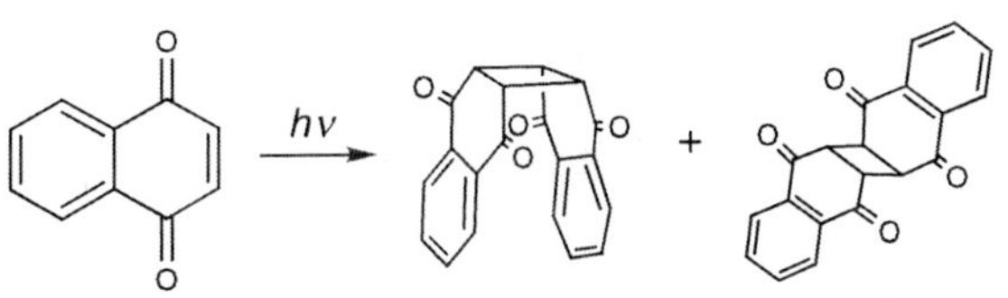

130 **131a** **131b**

Scheme 1.31

The remarkable reactivity and selectivity observed with these molecular vessels are attributed to the greatly increased concentration and the strictly regulated orientation of substrates in the well-confined cavity.

The cavity of molecular cage **126a** is achiral in origin. Interestingly, a chiral cavity can be constructed simply by replacing the peripheral ethylenediamine with optically active diamines, with the original electronic nature, size and T symmetry element being

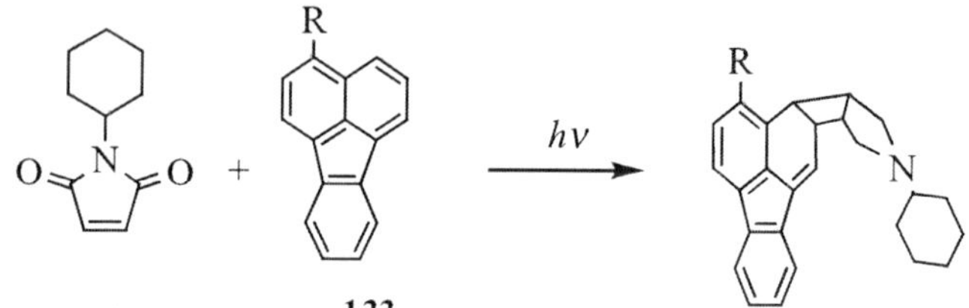

132

133

134

a: R= H
b: R=CH_3

Scheme 1.32

maintained on the whole. Cross addition of **132** and **133** (Scheme 1.32) in the cavity of **126b–e** (Scheme 1.30) gave a single product **134**, which is ascribed to steric control arising from the confined cavity rather than orbital control.[158] The optical yields of this photoaddition reaction increased upon increasing the steric bulk of the *N*-substituent on the chiral diamine and finally gave **134b** in 50% ee with **126c**. This observation is indicative that peripheral chiral auxiliaries, which locate far away from the reaction center and induce only a slight deformation of the triazine panel, are possible to produce considerable asymmetric induction for the photoreaction occurring in the coordinated cages.

1.5 Photochemical Reactions with other Artificial Hosts

1.5.1 Photochemical Reactions with Dendrimers

Dendrimers are highly branched globular molecules which are generally well-defined, effectively monodispersed and highly symmetric.[161–163] A dendrimer is composed of three architectural region: a core moiety, branched repeating units derived from the core and end groups on the periphery of the molecule. Properly designed high-generation dendrimers possess a distinct interior space that is sterically protected from solvent molecules, and therefore dendrimers can serve as container molecules. An elegant 'dendritic box' that can accommodate small organic molecules and control their release by modifying the steric crowding of the dendritic periphery has been described by Meijer *et al.*[164]

Poly(alky aryl ether) dendrimers **135** (Scheme 1.33), possessing carboxylic acid or hydroxyl group at their periphery, were synthesized and used as hosts for several photoreactions.[165] Both **135a** and **135b** are water soluble under the base condition and capable of including organic molecules mainly through hydrophobic interaction. Photolysis of 1-phenyl-3-*p*-tolyl-2-propanone **136** led to α-cleavage to form radical pairs, which after decarbonylation recombine to give diarylethanes **137a**, **137b** and **137c**. The photolysis of **136** in hexane solution gave the product **137a**, **137b** and **137c** in a ratio 2.4 : 1.0 : 1.4, indicating a zero 'cage effect,' where the 'cage effect' was defined as ([**137a**]–[**137b**]–[**137c**])/([**137a**] + [**137b**] + [**137c**]). Interestingly, photoreaction of **136** mediated by G1, G2 and G3 acid dendrimers **135b** showed the gradually increasing cage effects of 0.09, 0.18 and 0.50, respectively, suggesting stronger confinement with increasing dendrimer

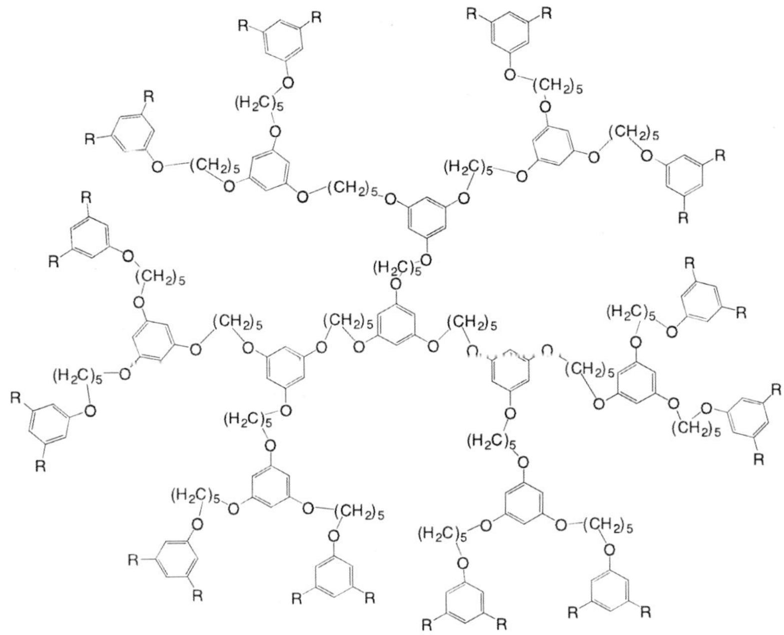

135a: R=OH, b: R=COOH

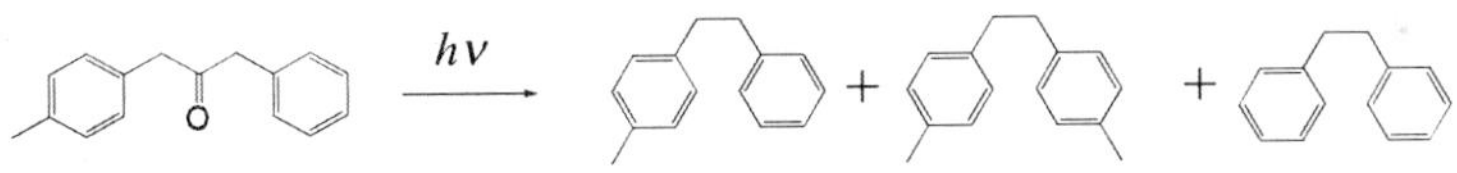

136 **137a** **137b** **137c**

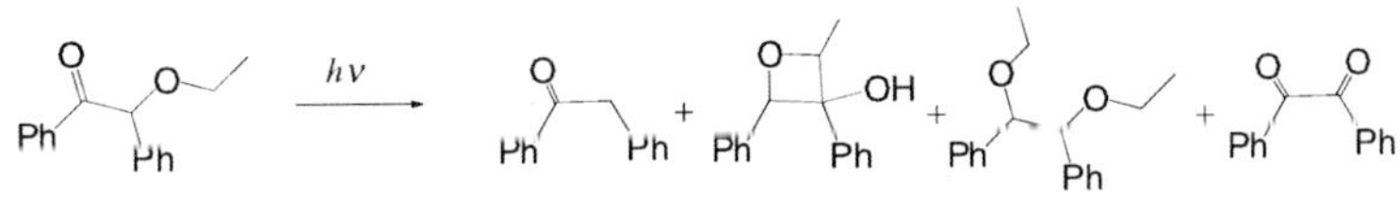

138 **139a** **139b** **139c** **139d**

Scheme 1.33

generation. Photolyses of **136** encapsulated by G3 of **135a** gave a larger cage effect of 0.77, revealing that the phenolic dendrimer **135a** is more strongly confining than the acid dendrimer **135b**.

Photolysis of benzoin ethyl ether **138** was demonstrated to be sensitive to the reaction media employed. It affords Norrish type I products **139c** and **139d** in more than 94% combined yield in hexane but type II products **139a** and **139b** in 73% yield in aqueous alkaline solution. Photoreaction of **138** with dendrimer **135b** predominantly gave the type II products, particularly **139a** in 95% yield. However, when the photoreaction was mediated by **135a**, **139a** and **139b** were produced in 21% and 57% yield, respectively, suggesting that the end group plays an important role in dictating the interior character of the dendrimers.

1.5.2 Photochemical Reactions with Hydrogen-Bonding Templates

Hydrogen bonding interaction plays vital roles in regulating biochemical processes of proteins and deoxyribonucleic acids, and has been widely utilized for constructing supramolecular architectures. This is due to the largest interaction energy (5–30 $kJ\,mol^{-1}$) among the non-covalent weak interactions as well as the ability to assemble molecules at right distance and angle. As a result of these advantages, a series of photochemical reactions have been investigated using hydrogen-bonding templates.

Bassani and co-workers investigated the complexation behavior of melamine and barbituric acid derivatives, and examined the photodimerization of styrene, stilbene and cinnamate mediated by hydrogen bonding templates.[166–170] Melamine derivatives **140** (Scheme 1.34), which have a hydrogen-bonding motif complementary to barbituric acid derivatives such as **141**, are possible to sequentially form 1 : 1 and 1 : 2 hydrogen-bonded complexes with **141**. As illustrated in Scheme 1.35, **140** and **141** can form two kinds of 1 : 1 complexes and three kinds of 1 : 2 complexes, among which only complex C is expected to accelerate the photodimerization of **140**. Each of the binding constants K_1 and K_2 should be a sum of these binding constants weighted by their population. A positive cooperativity ($K_2 > K_1$) was observed upon complexation of **141b** with **140a,b**, presumably as a combined result of allosteric and electronic effects. Photoirradiation of **140** without any template led to a fast *E-to-Z* isomerization to give the *Z*/*E* ratio of 1.1–5.5 at the photostationary state, although elongated irradiations resulted in a formation of dimers. The existence of barbiturate **141** considerably improved the efficiency and selectivity of the photoreaction. The quantum yields of **142a** and **142b** were increased respectively from 0.7×10^{-3} and 0.1×10^{-3} in the absence of a template to 2.3×10^{-3} and 0.8×10^{-3} in the presence of **141a**.[170] The same strategy was applied to a barbituric acid derivative of fullerene to give a photodimer of fullerene, which could not been obtained without the hydrogen-bonding template.[169]

Bach and co-workers investigated a series of chiral photoreactions that were mediated by Kamp's triacid derivatives **143** working as chiral hydrogen-bonding templates.[171–177] As illustrated in Scheme 1.37, these chiral templates possess an amide (lactam) group to bind an amide substrate through dual hydrogen bonds and also a bulky substituent to shield one of the enantiofaces of substrate when bound to the template. Although these

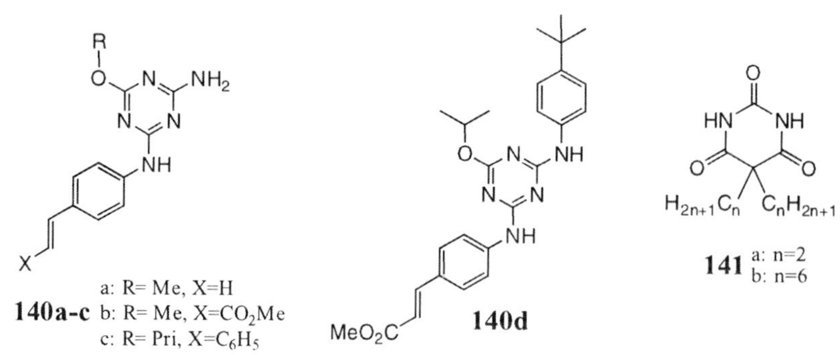

Scheme 1.34

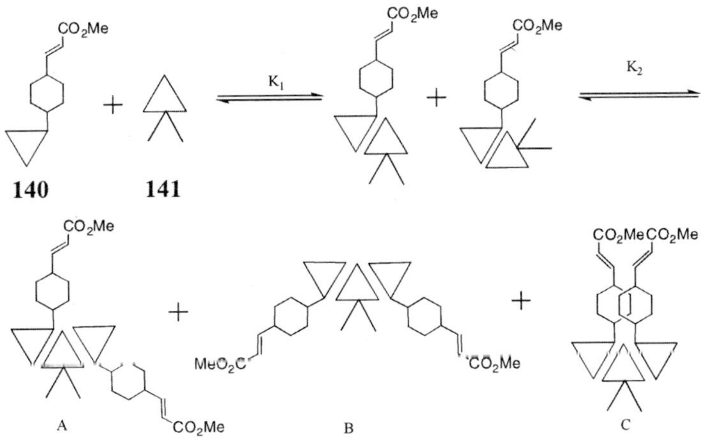

Scheme 1.35

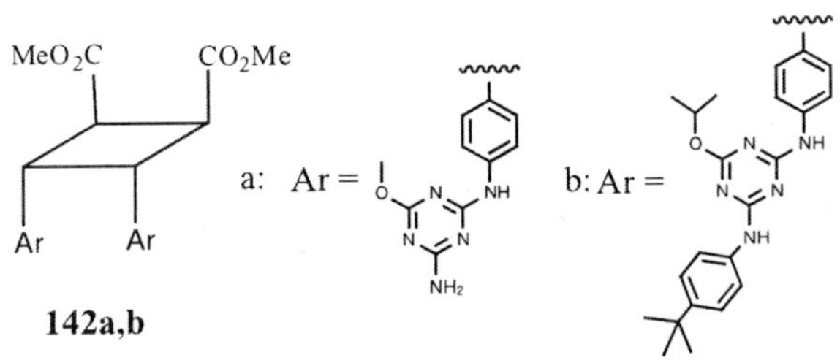

Scheme 1.36

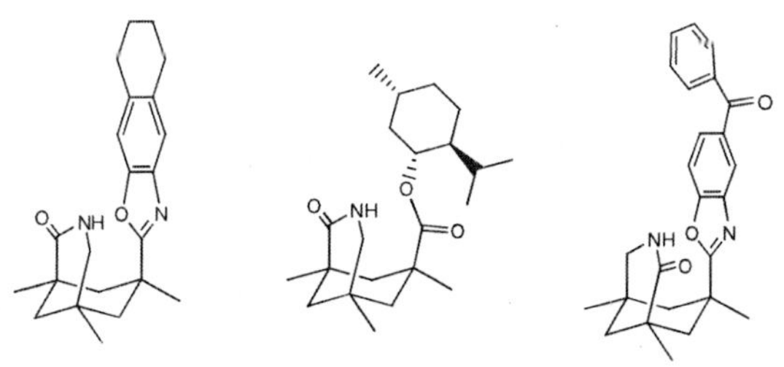

143a **143b** **143c**

Scheme 1.37

templates can form hydrogen-bonded homodimers, the association constants are low in general due to the steric repulsion between the bulky moieties.

The intramolecular photocycloaddition of prochiral 2-quinolone **144** (Scheme 1.38) was examined in the presence of chiral templates **143a,b**.[178] The amide group of chiral templates forms dual hydrogen bonds with the quinolone moiety of **144**, with the bulky tetrahydronaphthalene moiety preventing the approach of olefinic double bond from the

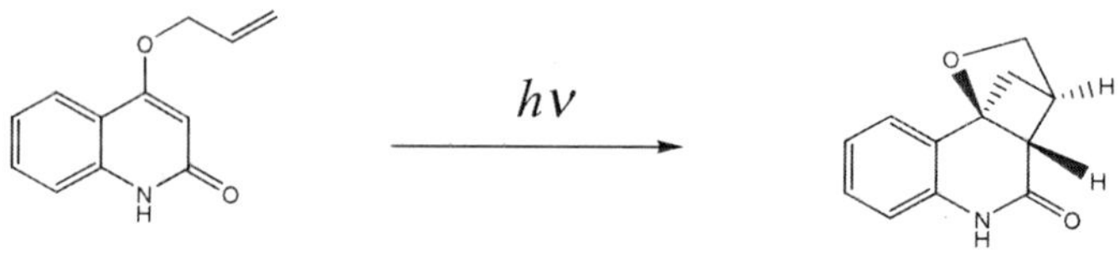

144 145

Scheme 1.38

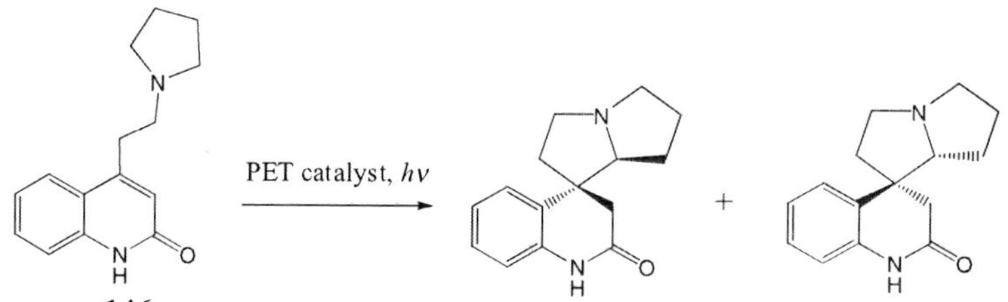

146 147

Scheme 1.39

facing side. Increasing the amount of chiral template or lowering the reaction temperature, by which the ratio of bound photosubstrate is greatly increased, significantly improved the enantioselectivity in the photocycloaddition of **144**. By performing the photoreaction in the presence of **143b** at −60 °C, **145** was given in up to 93% ee. In contrast, the same reaction sensitized by a chiral sensitizer lacking the hydrogen-bonding site gave only racemic photoproducts.[179]

Since photoexcitation of free substrate existing in the bulk solution will give racemic photoproducts, at least an equimolar amount of chiral template is necessary in order to obtain photoproduct with good enantioselectivity. A recent report by Bach *et al.* presented a good solution for this shortcoming by covalently grafting a sensitizer to a chiral template. Thus, sensitizing template **143c** (Scheme 1.37) bearing a photoinduced electron transfer sensitizer was synthesized for the use in photosensitized cycloaddition of **146**. Photolysis of **146** in the presence of 0.1 equivalent amount of **143c** afforded **147** in 52-64% yield with up to 70% ee (Scheme 1.39).[180]

1.5.3 Photochemical Reactions Templated by Cationic Ion

Metal ion can form complex through coordination with unsaturated compounds such as olefin. Through coordination with metal ion, two olefins can be positioned in close proximity, and olefin photochemistry may be caused upon irradiation at longer wavelengths by exciting the charge-transfer band of the resultant complex. Evers and Mackor investigated the intramolecular [2+2] photocycloaddition of diene **148**, which is photochemically catalyzed by copper(I) trifluoromethanesulphonate.[181] In the presence of the copper

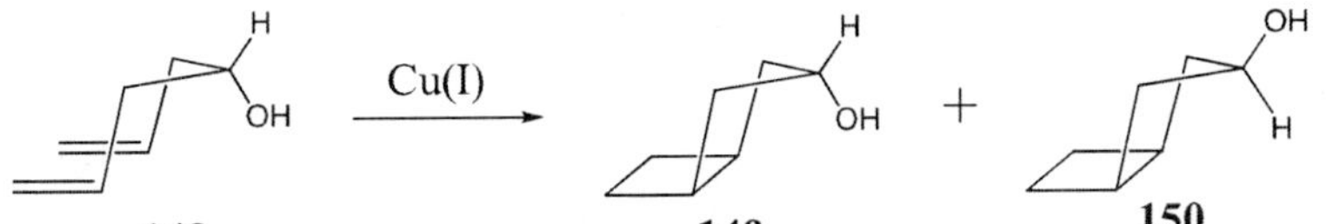

148 149 150

Scheme 1.40

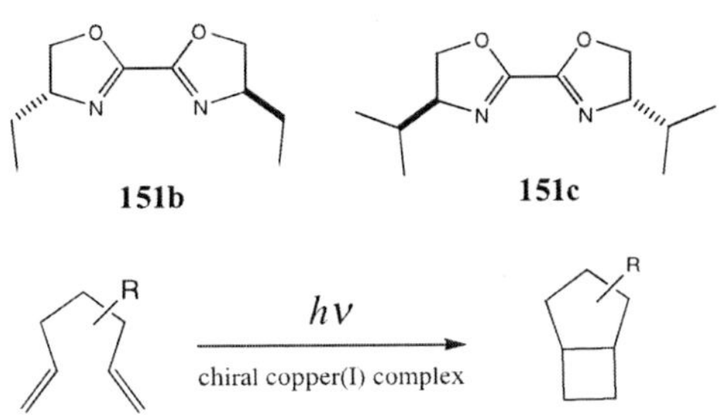

151b 151c

152 153

Scheme 1.41

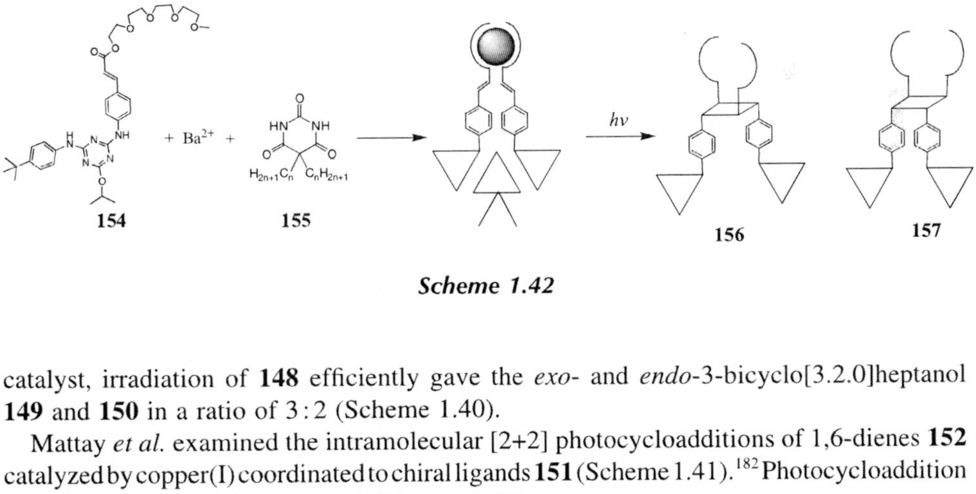

154 155 156 157

Scheme 1.42

catalyst, irradiation of **148** efficiently gave the *exo*- and *endo*-3-bicyclo[3.2.0]heptanol **149** and **150** in a ratio of 3 : 2 (Scheme 1.40).

Mattay *et al.* examined the intramolecular [2+2] photocycloadditions of 1,6-dienes **152** catalyzed by copper(I) coordinated to chiral ligands **151** (Scheme 1.41).[182] Photocycloaddition of **152** catalyzed by **151c** gave **153** in up to 5% ee.

Bassani and co-workers studied the photodimerization of **154** using a synergistic effect of hydrogen bonding and metal ion complexation.[183] Photodimerization of **154** was accelerated by a factor of 2 in the presence of one equivalent of KPF_6, and a factor of 5 in the coexistence of KPF_6 and **155**. Among the metal ions examined, Ba^{2+} showed the most significant cooperative effect. While **155** enhanced the photodimerization by a factor of 1.4–1.8, a simultaneous use of **155** and Ba^{2+} gave the quantum yield for the formation of **156** and **157** in more than three orders of magnitude greater than that obtained in the absence of template, and the ratio of **156/157** was greatly improved from 1 to 3.7.

1.6 Photoreaction Control by External Variants

While most studies on supramolecular photochemistry have focused on the design and synthesis of suitable hosts that provide desirable interaction or environment for aimed photosubstrates, recent studies by Inoue *et al.* suggest that the external factors, such as temperature, solvent and pressure, may also play a crucial role in controlling the reactivity and selectivity of photoreactions in supramolecular systems. Changing environmental variants, which is much easier to practice than the optimization of host structure involving tedious syntheses, can provide not only a convenient, yet useful, tool for controlling photochemical outcomes, but also important information regarding the mechanism of supramolecular photochemical reaction.

The effect of temperature has been well established in the photosensitized isomerization of cycloalkenes with conventional sensitizers.[184–187] The change of optical yield upon temperature variation is ascribed to the non-zero differential entropy change for the formation of enantiomeric photoproducts. The entropy effect on supramolecular photochirogenesis systems was found to be closely related to the structural properties of supramolecular hosts. The enantioselectivity shows only small changes with the variation of temperature in photoisomerization of **34Z** sensitized by β-CD-based chromophores.[37] This insignificant effect of temperature is attributed to the low entropy environment in the cavity of β-CD due to its rigid skeleton arising from the hydrogen-bonding network at the secondary rim.

On the contrary, photoisomerization of **158ZZ** with sensitizer-appended γ-CD, which has a more flexible framework than γ-CD, is more sensitive to the temperature alternation. An inversion of product chirality was observed by lowering the temperature in the photoisomerization of **158ZZ** mediated by **159**.[188] Similar phenomenon was observed in photocyclodimerization of AC (Scheme 1.9) included in the cavity of γ-CD derivatives.[189] The enantioselectivity for HH dimer in the photocyclodimerization of AC with diamino- γ -CDs **48** in aqueous methanol solution decreased greatly by lowering the reaction temperature and even showed a dramatic inversion to give the antipodal product. When γ-CD derivatives **50a–d** with flexible caps were used, the ee values obtained for **44** were highly dependent on the temperature. However, in the presence of γ-CD derivatives **57a–c**, which possess a rigid aryldisulfonate cap, the enantiodifferentiation become much less temperature-dependent.[52] On the other hand, when photoisomerization of **34Z** was carried out

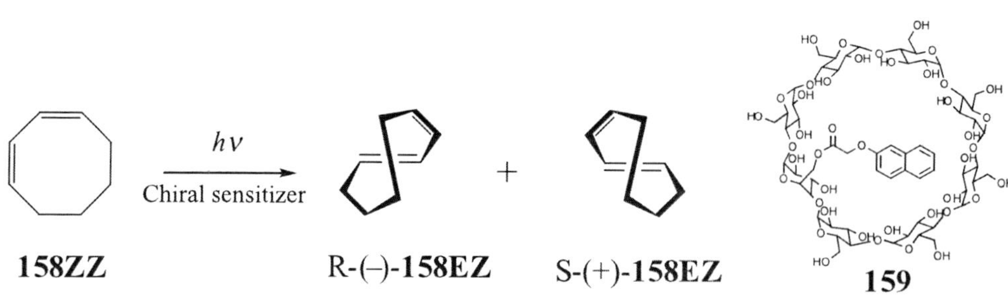

Scheme 1.43

with permethylated β-CD sensitizers **35e–h**, in which the hydrogen bonding network is broken by methylation to give a flexible skeleton, the temperature effect became pronounced to afford a chirality inversion by changing the temperature.

These observations suggest that the entropy effect is a function of the flexibility/rigidity of the skeleton of supramolecular hosts, and is significant for hosts with flexible framework but is insignificant with rigid hosts. Indeed, works of the other research groups have also demonstrated that hosts bearing rigid skeleton such as zeolite show only slight temperature-dependence.

Solvation and desolvation are of paramount importance for molecular recognition in solution. Solvent influences the binding strength of host–guest equilibrium in a diverse way for various host–guest systems. For example, the complexation strength between 18-crown-6 and potassium ion decreases by a factor of $>10^4$ on going from propylene carbonate to water.[190] This significant solvent dependence mainly comes from the difference in desolvation energy of the metal cation in different solvent. In some cases, solvent may play a role of filler upon co-inclusion with the guest substrate to significantly modify the orientation of substrate in the host cavity. It is known that CD bind organic guests mainly through hydrophobic interactions, and addition of methanol to the aqueous solution greatly reduces the binding constant as a result of increased solvation to both of substrate and host. The plot of log Ks for the complexation of **35** and **34Z** against the methanol content showed good straight lines, suggesting that no selective solvation by water or methanol to CD derivatives occurs and water–methanol mixture behaves as a bulk solvent of continuously changing hydrophilicity. The enantioselectivity in photoisomerization of **34Z** with **35** decreases with increasing content of methanol, and appears to be a nice function of host occupancy. On the other hand, addition of methanol decreases the solvent polarity and thus increases the electrostatic interaction of ion pairs. In aqueous solution, the photodimerization of AC mediated by **48a–d** shows only subtle effect of the ammonium cations introduced to the primary rim of γ-CD, but the electrostatic interaction dominates the photodimerization to give the chiral HH dimer as the major product with a greatly enhanced enantioselectivity

Hydrostatic pressure is also known to play crucial roles in conventional chiral photosensitization of **34Z**, as demonstrated by the fact that product chirality can be manipulated or even inverted by simply changing the pressure when the differential activation volume for the formation of the enantiomeric or diastereomeric pair is not equal to zero. Under high pressure, less volume-demanding structures and processes are absolutely favoured and hence the host-guest association process is advantageous in principle. If there are two or more possible complex structures in a supramolecular system, dramatic equilibrium and selectivity shifts may occur from a shallow, loose and/or bulky complex to a deeper, tighter and compact ones under high pressure. In other words, we acquire an additional tool, that is pressure as an environmental variant, for manipulating the recognition ability and selectivity of a given guest. Another somewhat unexpected great advantage of high pressure applied to supramolecular systems performed in aqueous solutions is the significant lowering of the melting temperature of water, which goes down to –25 °C at 210 MPa, which enables us to perform photochemical reactions below zero Celsius even in water. Thus, the chemical yield and ee of HT dimer **44** were greatly enhanced in the photocyclodimerization of AC mediated by γ-CD derivatives to reach 53% yield and 71% ee by using host **54** in water at –22 °C at 210 MPa.

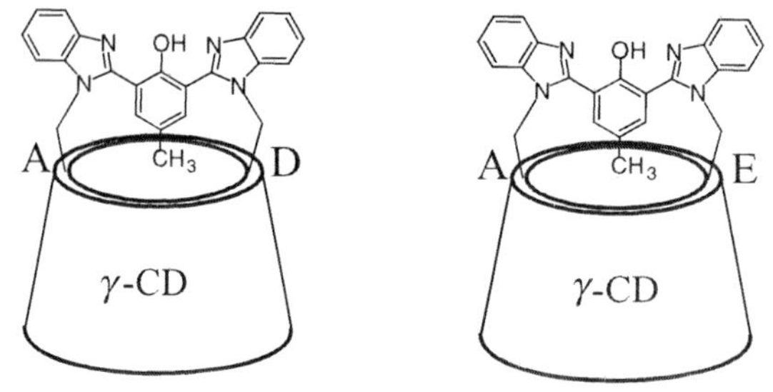

160a **160b**

Scheme 1.44

More recently, the effect of pH was investigated in the supramolecular photochirogenic dimerization of AC with *p*-cresolbisbenzimidazole-capped γ-CDs **160a,b**.[191] Thus, photodimerization of AC with AE-capped γ-CD **160b** gave *anti*-HH dimer in 5% ee at pH 11 but the antipodal product in 28% ee at pH 6. The conformational change of the ionizable capping moiety upon the pH variation is responsible for the pH-dependent enantioselectivity.

1.7 Conclusions

Controlling photochemistry through molecular recognition has drawn significant interest from a broad spectrum of scientists and met great progress in recent decades. A number of supramolecular hosts have been successfully exploited for conducting various types of photochemical reactions to reveal the distinct advantages and disadvantages peculiar to each supramolecular system. The multiple non-covalent interactions and the confinement effect operating in host cavity are two major factors that control the supramolecular photochemistry. Successful supramolecular photoreaction requires reasonable design and synthesis of host molecules and a careful selection of suitable photoreaction and substrate. The complementary fitting of size, shape and interaction motif between host and guest substrate are crucially important in choosing a host–guest pair for specific supramolecular photochemistry. In general, supramolecular hosts that have more confined binding sites usually show specific complexation ability and significant reaction selectivity, while those with flexible binding sites are susceptive to the external control by environmental factors.

The supramolecular approach to photochemistry is undoubted a smart efficient way to manipulate and enhance its reactivity and selectivity. In addition to the electronic and orbital properties governing the conventional photoreactions in isotropic media, the steric and conformational effects play heavier roles in supramolecular photochemistry. As a result, photoreactions in supramolecular systems often show unique behavior distinctively different from that of conventional one. On the other hand, the confinement effect of supramolecular hosts can hold two or more photosubstrates in close proximity and consequentially promote bi- and multimolecular photoreactions that are silent in conventional photochemistry under normal conditions.

Despite that supramolecular complexation can crucially affect the steric and/or electronic properties of photosubstrates in both ground and excited state, the stereochemical control of photochemical reaction in supramolecular systems is often considered to be dominated by the prearrangement of photosubstrate in the ground state. A photosubstrate is captured by a host for a period of time that is much longer than the excited-state lifetime. It has been demonstrated that the rate constant for a photosubstrate to get into the cavity of host with a sufficiently large opening is normally in the order 10^{-7}–10^{-8} M^{-1} S^{-1}.[192] An organic guest may stay in the cavity for much longer than 1 µs if it has a binding constant of 1000 M^{-1} with the host, which is significantly longer than the lifetime of excited molecules commonly encountered. Therefore, the complexation behaviour of photosubstrate in the ground state is of significant importance in supramolecular photochemistry.

Optimizing the structure and interactions of a supramolecular system still requires sophisticated design, synthesis, and experimental feedback even in the ground state, and should be much more difficult for supramolecular photochemical reactions. In this context, the combined use of external factors, such as temperature, solvent and pressure, provides a versatile and convenient tool for manipulating photochemical reactions in supramolecular system. The role of these external factors is closely correlated with the properties, in particular rigidity, of supramolecular host employed, and the outcomes are often significantly different from those observed in conventional photoreactions, which would be another reason for performing photochemical reaction in supramolecular system.

The number and quality of publications on supramolecular photochirogenesis are rapidly growing in the last decade. Laying in the interdisciplinary area of supramolecular chemistry, photochemistry and asymmetric synthesis, this new area of science is still challenging but appears promising, proposing a novel concept and methodology for overcoming the relatively weak and short-lived interactions in the excited state through multiple supramolecular interactions in both ground and excited states. Indeed, the highest enantioselectivity in solution-phase photoreaction has been achieved by supramolecular photochirogenesis.[193]

In summary, the control of photochemical reactions by molecular recognition becomes one of the most active topics in supramolecular chemistry in recent years. However, supramolecular photochemistry is still conceptually less established, mechanistically less understood, and experimentally less explored compared to the conventional supramolecular chemistry in the ground state. We believe that further comprehensive studies in this area will reveal more intriguing features of supramolecular photochemistry and strength our capability to control photochemical reactions.

Acknowledgements

This work was supported by PRESTO, Japan Science and Technology Agency.

References

1. M. Mammen, S.-K. Chio, G. M. Whitesides, *Angew. Chem., Int. Ed.*, 1998, **37**, 2755–2794.
2. E. R. Jamieson, S. J. Lippard, *Chem. Rev.*, 1999, **99**, 2467–2498.

3. E. A. Meyer, R. K. Castellano, F. Diederich, *Angew. Chem., Int. Ed.*, 2003, **42**, 1210–1250.
4. E. Di Cera, *Chem. Rev.*, 1998, **98**, 1563–1591.
5. M. Lahav, L. Leiserowitz, *Angew. Chem., Int. Ed.*, 1999, **38**, 2533–2536.
6. R. Breslow, *Acc. Chem. Res.*, 1995, **28**, 146–153.
7. R. Breslow, Ed, *Artificial Enzymes*, Wiley-VCH, Weinheim, Germany, 2005.
8. M. F. Manfrin, L. Moggi, V. Castelvetro, V. Balzani, M. W. Hosseini, J. M. Lehn, *J. Am. Chem. Soc.*, 1985, **107**, 6888–6892.
9. V. Balzani, L. Moggi, F. Scandola, *NATO ASI Ser., Ser. C*, 1987, **214**, 1–28.
10. J. M. Lehn, *NATO ASI Ser., Ser. C*, 1987, **214**, 29–43.
11. J. Lehn, *Supramolecular chemistry: concepts and perspectives*, VCH New York, 1995.
12. N. Turro, *Proc. Natl. Acad. Sci. USA*, 2002, **99**, 4805–4809.
13. C. Yang, Y. Inoue, *Supramolecular Photochirogenesis with Cyclodextrin*, in *Cyclodextrin materials photochemistry, photophysics and photobiology*; Douhal, A., Ed.; 2006.
14. Y. Inoue, V. Ramamurthy, *Chiral Photochemistry*, CRC Press, 2004.
15. C. J. Pedersen, *J. Am. Chem. Soc.*, 1967, **89**, 2495–2496.
16. C. J. Pedersen, *Science*, 1988, **241**, 536–540.
17. D. G. Amirsakis, M. A. Garcia-Garibay, S. J. Rowan, J. F. Stoddart, A. J. P. White, D. J. Williams, *Angew. Chem. Int. Ed.*, 2001, **40**, 4256–4261.
18. T. Kuwabara, M. Sugiyama, M. Nanasawa, *Photochem. Photobiol.*, 2001, **73**, 469–472.
19. S. J. Dalgarno, M. J. Hardie, J. L. Atwood, J. E. Warren, C. L. Raston, *New. J. Chem.*, 2005, **29**, 649–652.
20. J. L. Atwood, S. J. Dalgarno, M. J. Hardie, C. L. Raston, *Chem. Commun.*, 2005, 337–339.
21. G. Arena, A. Casnati, A. Contino, G. G. Lombardo, D. Sciotto, R. Ungaro, *Chem. Eur. J.*, 1999, **5**, 738–744.
22. G. Arena, A. Casnati, L. Mirone, D. Sciotto, R. Ungaro, *Tetrahedron Lett.*, 1997, **38**, 1999–2002.
23. G. Arena, A. Contino, F. G. Gulino, A. Magri, F. Sansone, D. Sciotto, R. Ungaro, *Tetrahedron Lett.*, 1999, **40**, 1597–1600.
24. G. Arena, A. Contino, F. G. Gulino, A. Magri, D. Sciotto, R. Ungaro, *Tetrahedron Lett.*, 2000, **41**, 9327–9330.
25. R. Kaliappan, L. Kaanumalle, V. Ramamurthy, *Chem. Commun.*, 2005, 4056.
26. L. Kaanumalle, J. Nithyanandhan, M. Pattabiraman, J. Narayanaswamy, V. Ramamurthy, *J. Am. Chem. Soc.*, 2004, **126**, 8999–9006.
27. R. T. Lauterbach, H. G. Heine, W. Hartmann, H. Rudolph, *J. Am. Chem. Soc.*, 1975, **97**, 1519–1525.
28. S. P. Pappas, A. Chattopadhyay, *J. Am. Chem. Soc.*, 1973, **95**, 6484–6485.
29. R. Kaliappan, L. Kaanumalle, A. Natarajan, V. Ramamurthy, *Photochem. Photobiol. Sci.*, 2006, **5**, 925–930.
30. G. Ananchenko, K. Udachin, J. Ripmeester, T. Perrier, A. Coleman, *Chem. Eur. J.*, 2006, **12**, 2441–2447.
31. M. Pojarova, G. Ananchenko, K. Udachin, M. Daroszewska, F. Perret, A. Coleman, J. Ripmeester, *Chem. Mater.*, 2006, **18**, 5817–5819.
32. K. Vízvárdi, K. Desmet, I. Luyten, P. Sandra, G. Hoornaert, E. Van der Eycken, *Org. Lett.*, 2001, **3**, 1173–1175.
33. A. C., Cope, C. R. Ganellin, H. W. J. Johnson, H.J.S. Winkler, *J. Am. Chem. Soc.*, 1963, **85**, 3276–3279.
34. Y. Inoue, Y. Kunitomi, S. Takamuku, H. Sakurai, *Chem. Commun.*, 1978, 1024–1025.
35. Y. Inoue, N. Sugahara, T. Wada, *Pure. Appl. Chem.*, 2001, **73**, 475–480.
36. Y. Inoue, T. Wada, S. Asaoka, H. Sato, J. P. Pete, *Chem. Commun.*, 2000, 251–259.
37. Y. Inoue, T. Wada, N. Sugahara, K. Yamamoto, K. Kimura, L. H. Tong, X. M. Gao, Z. J. Hou, Y. Liu, J. *Org. Chem.*, 2000, **65**, 8041–8050.
38. G. Fukuhara, T. Mori, T. Wada, Y. Inoue, *J. Org. Chem.*, 2006, **71**, 8233–8243.
39. R. Lu, C. Yang, Y. Cao, Z. Wang, T. Wada, W. Jiao, *Chem. Commun.*, 2008, 374–376.
40. R. Lu, C. Yang, Y. Cao, L. Tong, W. Jiao, T. Wada, Z. Wang, T. Mori, Y. Inoue, *J. Org. Chem.*, 2008, **73**, 7695–7701.

41. W. Herrmann, S. Wehrle, G. Wenz, *Chem. Commun.*, 1997, 1709–1710.
42. K. S. Rao, S. M. Hubig, J. N. Moorthy, J. K. Kochi, *J. Org. Chem.*, 1999, **64**, 8098–8104.
43. T. Tamaki, T. Kokubu, K. Ichimura, *Tetrahedron Lett.*, 1987, **43**, 1485–1494.
44. T. Tamaki, T. Kokubu, *J. Incl. Phenom. Macro.*, 1984, **2**, 815–822.
45. Y. Ishida, Y. Kai, S. Y. Kato, A. Misawa, S. Amano, Y. Matsuoka, K. Saigo, *Angew. Chem. Int. Ed.*, 2008, 8421–8425.
46. A. Nakamura, Y. Inoue, *J. Am. Chem. Soc.*, 2003, **125**, 966–972.
47. H. Ikeda, T. Nihei, A. Ueno, *J. Org. Chem.*, 2005, **70**, 1237–1242.
48. A. Nakamura, Y. Inoue, *J. Am. Chem. Soc.*, 2005, **127**, 5338–5339.
49. C. Yang, G. Fukuhara, A. Nakamura, Y. Origane, K. Fujita, D. Yuan, T. Mori, T. Wada, Y. Inoue, *J. Photoch. Photobio. A*, 2005, **173**, 375–383.
50. C. Yang, A. Nakamura, G. Fukuhara, Y. Origane, T. Mori, T. Wada, Y. Inoue, *J. Org. Chem.*, 2006, **71**, 3126–3136.
51. C. Yang, A. Nakamura, T. Wada, Y. Inoue, *Org. Lett.*, 2006, **8**, 3005–3008.
52. C. Yang, T. Mori, Y. Inoue, *J. Org. Chem.*, 2008, **73**, 5786–5794.
53. C. Yang, T. Mori, Y. Origane, Y. H. Ko, N. Selvapalam, K. Kim, Y. Inoue, *J. Am. Chem. Soc.*, 2008, **130**, 8574–8575.
54. R. Wang, L. Yuan, D. H. Macartney, *J. Org. Chem.*, 2006, **71**, 1237–1239.
55. S. Jon, Y. Ko, S. Park, H. Kim, K. Kim, *Chem. Commun.*, 2001, 1938–1939.
56. M. Pattabiraman, A. Natarajan, R. Kaliappan, J. T. Mague, V. Ramamurthy, *Chem. Commun.*, 2005, 4542–4544.
57. M. V. Maddipatla, L. S. Kaanumalle, A. Natarajan, M. Pattabiraman, V. Ramamurthy, *Langmuir*, 2007, **23**, 7545–7554.
58. L. Lei, L. Luo, X. Wu, G. Liao, L. Wu, C. Tung, *Tetrahedron Lett.*, 2008, **49**, 1502–1505.
59. C. L. Gibb, B. C. Gibb, *J. Am. Chem. Soc.*, 2004, **126**, 11408–11409.
60. Z. R. Laughrey, C. L. Gibb, T. Senechal, B. C. Gibb, *Chem. Eur. J.*, 2003, **9**, 130–139.
61. C. L. Gibb, E. D. Stevens, B. C. Gibb, *J. Am. Chem. Soc.*, 2001, **123**, 5849–5850.
62. L. S. Kaanumalle, C. L. Gibb, B. Gibb, V. Ramamurthy, *J. Am. Chem. Soc.*, 2004, **126**, 14366–14367.
63. C. L. Gibb, A. K. Sundaresan, V. Ramamurthy, B. C. Gibb, *J. Am. Chem. Soc.*, 2008, **130**, 4069–4080.
64. A. K. Sundaresan, V. Ramamurthy, *Org. Lett.*, 2007, **9**, 3575–3578.
65. L. S. Kaanumalle, C. L. Gibb, B. Gibb, V. Ramamurthy, *J. Am. Chem. Soc.*, 2005, **127**, 3674–3675.
66. L. S. Kaanumalle, V. Ramamurthy, *Chem. Commun.*, 2007, 1062–1064.
67. A. Parthasarathy, L. S. Kaanumalle, V. Ramamurthy, *Org. Lett.*, 2007, **9**, 5059–5062.
68. K. Fujimoto, S. Matsuda, N. Takahashi, I. Saito, *J. Am. Chem. Soc.*, 2000, **122**, 5646–5647.
69. K. Fujimoto, H. Yoshino, T. Ami, Y. Yoshimura, I. Saito, *Org. Lett.*, 2008, **10**, 397–400.
70. K. Fujimoto, N. Ogawa, M. Hayashi, S. Matsuda, I. Saito, *Tetrahedron Lett.*, 2000, **41**, 9437–9440.
71. M. Ogino, Y. Yoshimura, A. Nakazawa, I. Saito, K. Fujimoto, *Org. Lett.*, 2005, **7**, 2853–2856.
72. K. Fujimoto, S. Matsuda, M. Hayashi, I. Saito, *Tetrahedron Lett.*, 2000, **41**, 7897–7900.
73. K. Fujimoto, Y. Ikeda, S. Ishihara, I. Saito, *Tetrahedron Lett.*, 2002, **43**, 2243–2245.
74. I. Saito, Y. Miyauchi, Y. Saito, K. Fujimoto, *Tetrahedron Lett.*, 2004, **46**, 97–99.
75. K. Fujimoto, S. Matsuda, Y. Yoshimura, T. Ami, I. Saito, *Chem. Commun.*, 2007, 2968–2970.
76. Y. Saito, K. Matsumoto, S. S. Bag, S. Ogasawara, K. Fujimoto, K. Hanawa, I. Saito, *Tetrahedron*, 2008, **64**, 3578–3588.
77. T. Ihara, T. Fujii, M. Mukae, Y. Kitamura, A. Jyo, *J. Am. Chem. Soc.*, 2004, **126**, 8880–8881.
78. P. Arslan, T. Ihara, M. Mukae, A. Jyo, *Anal Sci*, 2008, **24**, 173–176.
79. P. Arslan, T. Ihara, M. Mukae, A. Jyo, *Nucleic Acids Symp. Ser.*, 2007, 237–238.
80. T. Wada, N. Sugahara, M. Kawano, Y. Inoue, *Chem. Lett.*, 2000, 1174–1175.
81. R. V. Person, B. R. Peterson, D. A. Lightner, *J. Am. Chem. Soc.*, 1994, **116**, 42–59.

82. D. A. Lightner, J. K. Gawronski, W. M. D. Wijekoon, *J. Am. Chem. Soc.*, 1987, **109**, 6354–6362.
83. G. Agati, A. F. McDonagh, *J. Am. Chem. Soc.*, 1995, **117**, 4425–4426.
84. C. Festa, N. Levi-Minzi, M. Zandomeneghi, *Gazz. Chim. Ital.*, 1996, **126**, 599–603.
85. N. Levi-Minzi, M. Zandomeneghi, *J. Am. Chem. Soc.*, 1992, **114**, 9300–9304.
86. A. Ouchi, G. Zandomeneghi, M. Zandomeneghi, *Chirality*, 2002, **14**, 1–11.
87. V. Lhiaubet-Vallet, S. Encinas, M. A. Miranda, *J. Am. Chem. Soc.*, 2005, **127**, 12774–12775.
88. V. Lhiaubet-Vallet, Z. Sarabia, F. Bosca, M. A. Miranda, *J. Am. Chem. Soc.*, 2004, **126**, 9538–9539.
89. V. Lhiaubet-Vallet, F. Bosca, M. A. Miranda, *J. Phys. Chem. B*, 2007, **111**, 423–431.
90. T. Wada, M. Nishijima, T. Fujisawa, N. Sugahara, T. Mori, A. Nakamura, Y. Inoue, *J. Am. Chem. Soc.*, 2003, **125**, 7492–7493.
91. M. Nishijima, T. C. S. Pace, A. Nakamura, T. Mori, T. Wada, C. Bohne, Y. Inoue, *J. Org. Chem.*, 2007, **72**, 2707–2715.
92. M. Nishijima, T. Wada, T. Mori, T. C. S. Pace, C. Bohne, Y. Inoue, *J. Am. Chem. Soc.*, 2007, **129**, 3478–3479.
93. U. Kragh-Hansen, *Pharmacol. Rev.*, 1981, **33**, 17–53.
94. T. Peters, *All About Albumin: Biochemistry, Genetics, and Medical Applications*, Academic Press, 1996.
95. S. Asaoka, T. Wada, Y. Inoue, *J. Am. Chem. Soc.*, 2003, **125**, 3008–3027.
96. T. C. S. Pace, M. Nishijima, T. Wada, Y. Inoue, C. Bohne, J. Phys, *Chem. B.*, 2009, **113**, 10445–10453.
97. M. Nishijima, T. Wada, K. Nagamori, Y. Inoue, *Chem. Lett.*, 2009, **38**, 726–727.
98. V. Ramamurthy, J. Shailaja, L. S. Kaanumalle, R. B. Sunoj, J. Chandrasekhar, *Chem. Commun.*, 2003, 1987–1999.
99. S. Hashimoto, *J. Photochem. Photobiol., C*, 2003, **4**, 19–49.
100. H. Garcia, H. D. Roth, *Chem. Rev.*, 2002, **102**, 3947–4007.
101. V. Ramamurthy, *J. Photochem. Photobiol., C*, 2000, **1**, 145–166.
102. A. Joy, V. Ramamurthy, *Chem. Eur. J.*, 2000, **6**, 1287–1293.
103. N. J. Turro, *Pure Appl. Chem.*, 1986, **58**, 1219–1228.
104. N. J. Turro, P. Wan, *J. Am. Chem. Soc.*, 1985, **107**, 678–682.
105. N. J. Turro, P. Wan, *Tetrahedron Lett.*, 1984, **25**, 3655–3658.
106. S. L. Suib, A. Kostapapas, D. Psaras, *J. Am. Chem. Soc.*, 1984, **106**, 1614–1620.
107. T. Poon, J. Sivaguru, R. Franz, S. Jockusch, C. Martinez, I. Washington, W. Adam, Y. Inoue, N. J. Turro, *J. Am. Chem. Soc.*, 2004, **126**, 10498–10499.
108. W. Adam, S. G. Bosio, N. J. Turro, B. T. Wolff, *J. Org. Chem.*, 2004, **69**, 1704–1715.
109. T. Poon, N. J. Turro, J. Chapman, P. Lakshminarasimhan, X. Lei, W. Adam, S. G. Bosio, *Org. Lett.*, 2003, **5**, 2025–2028.
110. W. Adam, S. G. Bosio, N. J. Turro, *J. Am. Chem. Soc.*, 2002, **124**, 8814–8815.
111. W. Adam, S. G. Bosio, N. J. Turro, *J. Am. Chem. Soc.*, 2002, **124**, 14004–14005.
112. J. Sivaguru, H. Saito, M. R. Solomon, L. S. Kaanumalle, T. Poon, S. Jockusch, W. Adam, V. Ramamurthy, Y. Inoue, N. J. Turro, *Photochem. Photobiol.*, 2006, **82**, 123–131.
113. A. Joy, L. S. Kaanumalle, V. Ramamurthy, *Org. Biomol. Chem.*, 2005, **3**, 3045–3053.
114. J. Sivaguru, R. B. Sunoj, T. Wada, Y. Origane, Y. Inoue, V. Ramamurthy, *J. Org. Chem.*, 2004, **69**, 6533–6547.
115. J. Sivaguru, T. Poon, R. Franz, S. Jockusch, W. Adam, N. J. Turro, *J. Am. Chem. Soc.*, 2004, **126**, 10816–10817.
116. M. Stratakis, D. Kalaitzakis, D. Stavroulakis, G. Kosmas, C. Tsangarakis, *Org. Lett.*, 2003, **5**, 3471–3474.
117. J. Sivaguru, J. Shailaja, V. Ramamurthy, *Handb. Zeolite Sci. Technol.*, 2003, 515–589.
118. J. Sivaguru, A. Natarajan, L. S. Kaanumalle, J. Shailaja, S. Uppili, A. Joy, V. Ramamurthy, *Acc. Chem. Res.*, 2003, **36**, 509–521.
119. M. Stratakis, G. Kosmas, *Tetrahedron Lett.*, 2001, **42**, 6007–6009.
120. J. Shailaja, J. Sivaguru, S. Uppili, A. Joy, V. Ramamurthy, *Microporous Mesoporous Mater.*, 2001, **48**, 319–328.

121. A. Joy, S. Uppili, M. R. Netherton, J. R. Scheffer, V. Ramamurthy, *J. Am. Chem. Soc.*, 2000, **122**, 728–729.
122. B. Coughlan, W. Carroll, P. O'Malley, J. Nunan, *J. Chem. Soc. Faraday Trans.* **1**, 1981, **77**, 3037–3047.
123. I. V. Matyash, M. A. Piontkovskaya, G. I. Denisenko, A. M. Kalinichenko, *Zh. Strukt. Khim.*, 1971, **12**, 13–18.
124. G. I. Denisenko, M. A. Piontkovskaya, I. E. Neimark, *Ukr. Khim. Zh.*, 1970, **36**, 260–264.
125. G. I. Denisenko, *Zh. Prikl. Spektrosk.*, 1968, **9**, 307–313.
126. K. Pitchumani, M. Warrier, L. S. Kaanumalle, V. Ramamurthy, *Tetrahedron*, 2003, **59**, 5763–5772.
127. K. Pitchumani, M. Warrier, V. Ramamurthy, J. R. Scheffer, *Chem. Commun.*, 1998, 1197–1198.
128. A. Joy, R. J. Robbins, K. Pitchumani, V. Ramamurthy, *Tetrahedron Lett.*, 1997, **38**, 8825–8828.
129. K. Pitchumani, V. Ramamurthy, *Tetrahedron Lett.*, 1996, **37**, 5297–5300.
130. V. J. Rao, S. R. Uppili, D. R. Corbin, S. Schwarz, S. R. Lustig, V. Ramamurthy, *J. Am. Chem. Soc.*, 1998, **120**, 2480–2481.
131. M. Leibovitch, G. Olovsson, G. Sundarababu, V. Ramamurthy, J. R. Scheffer, J. Trotter, *J. Am. Chem. Soc.*, 1996, **118**, 1219–1220.
132. E. Cheung, K. C. W. Chong, S. Jayaraman, V. Ramamurthy, J. R. Scheffer, J. Trotter, *Org. Lett.*, 2000, **2**, 2801–2804.
133. A. Joy, S. Uppili, M. R. Netherton, J. R. Scheffer, V. Ramamurthy, *J. Am. Chem. Soc.*, 2000, **122**, 728–729.
134. P. Lakshminarasimhan, R. B. Sunoj, J. Chandrasekhar, V. Ramamurthy, *J. Am. Chem. Soc.*, 2000, **122**, 4815–4816.
135. K. C. W. Chong, J. Sivaguru, T. Shichi, Y. Yoshimi, V. Ramamurthy, J. R. Scheffer, *J. Am. Chem. Soc.*, 2002, **124**, 2858–2859.
136. L. S. Kaanumalle, J. Sivaguru, R. B. Sunoj, P. H. Lakshminarasimhan, J. Chandrasekhar, V. Ramamurthy, *J. Org. Chem.*, 2002, **67**, 8711–8720.
137. A. Natarajan, A. Joy, L. S. Kaanumalle, J. R. Scheffer, V. Ramamurthy, *J. Org. Chem.*, 2002, **67**, 8339–8350.
138. J. Sivaguru, T. Shichi, V. Ramamurthy, *Org. Lett.*, 2002, **4**, 4221–4224.
139. S. Uppili, V. Ramamurthy, *Org. Lett.*, 2002, **4**, 87–90.
140. J. Sivaguru, R. B. Sunoj, T. Wada, Y. Origane, Y. Inoue, V. Ramamurthy, *J. Org. Chem.*, 2004, **69**, 6533–6547.
141. J. Shailaja, L. S. Kaanumalle, K. Sivasubramanian, A. Natarajan, K. J. Ponchot, A. Pradhan, V. Ramamurthy, *Org. Biomol. Chem.*, 2006, **4**, 1561–1571.
142. J. Sivaguru, H. Saito, M. R. Solomon, L. S. Kaanumalle, T. Poon, S. Jockusch, W. Adam, V. Ramamurthy, Y. Inoue, N. J. Turro, *Photochem. Photobiol. Sci.*, 2006, **82**, 123–131.
143. J. Sivaguru, R. B. Sunoj, T. Wada, Y. Origane, Y. Inoue, V. Ramamurthy, *J. Org. Chem.*, 2004, **69**, 5528–5536.
144. A. Joy, L. S. Kaanumalle, V. Ramamurthy, *Org. Biomol. Chem.*, 2005, **3**, 3045–3053.
145. A. Joy, J. R. Scheffer, V. Ramamurthy, *Org. Lett.*, 2000, **2**, 119–121.
146. T. Wada, M. Shikimi, Y. Inoue, G. Lem, N. J. Turro, *Chem. Commun.*, 2001, 1864–1865.
147. J. G. Victor, J. M. Torkelson, *Macromolecules*, 1987, **20**, 2241–2250.
148. N. Liu, Z. Chen, D. Dunphy, Y. Jiang, R. Assink, C. Brinker, *Angew. Chem.*, 2003, **115**, 1773–1776.
149. N. Mal, M. Fujiwara, Y. Tanaka, *Nature*, 2003, **421**, 350–353.
150. H. Okada, N. Nakajima, T. Tanaka, M. Iwamoto, *Angew. Chem. Int. Ed.*, 2005, **44**, 7233–7236.
151. H. Qiu, C. Yang, Y. Inoue, S. Che, *Org. Lett.*, 2009, **11**, 1793–1796.
152. D. Fiedler, D. H. Leung, R. G. Bergman, K. N. Raymond, *Acc. Chem. Res.*, 2005, **38**, 349–358.
153. M. Fujita, M. Tominaga, A. Hori, B. Therrien, *Acc. Chem. Res.*, 2005, **38**, 369–378.

154. T. Furusawa, M. Kawano, M. Fujita, *Angew. Chem. Int. Ed.*, 2007, **46**, 5717–5719.
155. T. Yamaguchi, M. Fujita, *Angew. Chem. Int. Ed.*, 2008, **47**, 2067–2069.
156. M. Yoshizawa, Y. Takeyama, T. Kusukawa, M. Fujita, *Angew. Chem. Int. Ed.*, 2002, **41**, 1347–1349.
157. K. Takaoka, M. Kawano, T. Ozeki, M. Fujita, *Chem. Commun.*, 2006, 1625–1627.
158. Y. Nishioka, T. Yamaguchi, M. Kawano, M. Fujita, *J. Am. Chem. Soc.*, 2008, **130**, 8160–8161.
159. M. Yoshizawa, S. Miyagi, M. Kawano, K. Ishiguro, M. Fujita, *J. Am. Chem. Soc.*, 2004, **126**, 9172–9173.
160. M. Yoshizawa, Y. Takeyama, T. Okano, M. Fujita, *J. Am. Chem. Soc.*, 2003, **125**, 3243–3247.
161. A. W. Bosman, H. M. Janssen, E. W. Meijer, *Chem. Rev.*, 1999, **99**, 1665–1688.
162. S. M. Grayson, J. M. J. Frechet, *Chem. Rev.*, 2001, **101**, 3819–3868.
163. F. Zeng, S. C. Zimmerman, *Chem. Rev.*, 1997, **97**, 1681–1712.
164. J. Jansen, de E. Brabander-van den Berg, E. Meijer, *Science*, 1994, **266**, 1226–1229.
165. L. S. Kaanumalle, R. Ramesh, V. S. N. Murthy Maddipatla, J. Nithyanandhan, N. Jayaraman, V. Ramamurthy, *J. Org. Chem.*, 2005, **70**, 5062–5069.
166. D. M. Bassani, X. Sallenave, V. Darcos, J.-P. Desvergne, *Chem. Commun.*, 2001, 1446–1447.
167. F.-C. Fang, C.-C. Chu, C.-H. Huang, G. Raffy, A. D. Guerzo, K.-T. Wong, D. M. Bassani, *Chem. Commun.*, 2008, 6369–6371.
168. D. M. Bassani, V. Darcos, S. Mahony, J.-P. Desvergne, *J. Am. Chem. Soc.*, 2000, **122**, 8795–8796.
169. N. D. McClenaghan, C. Absalon, D. M. Bassani, *J. Am. Chem. Soc.*, 2003, **125**, 13004–13005.
170. V. Darcos, K. Griffith, X. Sallenave, J.-P. Desvergne, C. Guyard-Duhayon, B. Hasenknopf, D. M. Bassani, *Photochem. Photobiol. Sci.*, 2003, **2**, 1152–1161.
171. T. Bach, H. Bergmann, K. Harms, *Angew. Chem. Int. Ed.*, 2000, **39**, 2302–2304.
172. T. Bach, T. Aechtner, B. Neumuller, *Chem. Eur. J.*, 2002, **8**, 2464–2475.
173. B. Grosch, C. N. Orlebar, E. Herdtweck, M. Kaneda, T. Wada, Y. Inoue, T. Bach, *Chem. Eur. J.*, 2004, **10**, 2179–2189.
174. T. Bach, T. Aechtner, B. Neumuller, *Chem. Commun.*, 2001, 607–608.
175. T. Bach, H. Bergmann, B. Grosch, K. Harms, *J. Am. Chem. Soc.*, 2002, **124**, 7982–7990.
176. T. Bach, H. Bergmann, K. Harms, *J. Am. Chem. Soc.*, 1999, **121**, 10650–10651.
177. C. Muller, A. Bauer, T. Bach, *Angew. Chem. Int. Ed.*, 2009, **48**, 6640–6642.
178. S. Breitenlechner, T. Bach, *Angew. Chem. Int. Ed.*, 2008, **47**, 7957–7959.
179. D. F. Cauble, V. Lynch, M. J. Krische, *J. Org. Chem.*, 2003, **68**, 15–21.
180. A. Bauer, F. Westkaemper, S. Grimme, T. Bach, *Nature*, 2005, **436**, 1139–1140.
181. J. T. M. Evers, A. Mackor, *Tetrahedron Lett.*, 1978, 821–824.
182. K. Langer, J. Mattay, *J. Org. Chem.*, 1995, **60**, 7256–7266.
183. Y. Pol, R. Suau, E. Perez-Inestrosa, D. Bassani, *Chem. Commun.*, 2004, 1270–1271.
184. Y. Inoue, E. Matsushima, T. Wada, *J. Am. Chem. Soc.*, 1998, **120**, 10687–10696.
185. Y. Inoue, *Chem. Rev.*, 1992, **92**, 741–770.
186. Y. Inoue, H. Tsuneishi, T. Hakushi, A. Tai, *J. Am. Chem. Soc.*, 1997, **119**, 472–478.
187. T. Inoue, K. Matsuyama, Y. Inoue, *J. Am. Chem. Soc.*, 1999, **121**, 9877–9878.
188. C. Yang, T. Mori, T. Wada, Y. Inoue, *New J. Chem.*, 2007, **31**, 697–702.
189. C. Ke, C. Yang, T. Mori, T. Wada, Y. Liu, Y. Inoue, *Angew. Chem. Int. Ed.*, 2009, **48**, 6675–6677.
190. V. P. Solov'ev, N. N. Strakhova, O. A. Raevsky, V. Rudiger, H.-J. Schneider, *J. Org. Chem.*, 1996, **61**, 5221–5226.
191. C. Yang, C. Ke, K. Fujita, D.-Q. Yuan, T. Mori, Y. Inoue, *Austr. J. Chem.*, 2008, **61**, 565–568.
192. X. Zhang, G. Gramlich, X. Wang, W. M. Nau, *J. Am. Chem. Soc.*, 2002, **124**, 254–263.
193. C. Muller, T. Bach, *Austr. J. Chem.*, 2008, **61**, 557–564.

2

Cyclodextrins

Ronald Breslow
Department of Chemistry, Columbia University, New York, NY 10027

2.1 Introduction

Cyclodextrins are cyclic oligomers of glucose (Figure 2.1). The most common are α-cyclodextrin with six glucoses in a ring (**1**), β-cyclodextrin with seven glucoses (**2**), and γ-cyclodextrin (**3**) with eight glucoses in the ring. Because of all the sugar hydroxyls these compounds are reasonably soluble in water but the interior cavities are relatively non-polar, with about the polarity of dioxane. Thus in water the cyclodextrins can reversibly bind hydrophobic substrates into their cavities, and within the complexes chemical reactions can occur.

Friedrich Cramer did the first work in this area,[1] and early work was done by Myron Bender.[2,3] In Bender's studies, a bound ester reacted with a hydroxyl group on the rim of the cyclodextrin to undergo a transesterification, with reasonable geometric selectivity and some rate acceleration. This was followed up with substrates better designed to be accelerated by such a process, and this will be the first part of the review. Then there are some reactions in which cyclodextrin promotes a process but is not itself transformed, and the first example of this was work we reported on selective aromatic substitution, the second section of this review. I also describe the use of cyclodextrins to catalyse Diels–Alder reactions, in which both the diene and the dienophile can bind into the cyclodextrin cavity in water and in 'water-like' solvents.

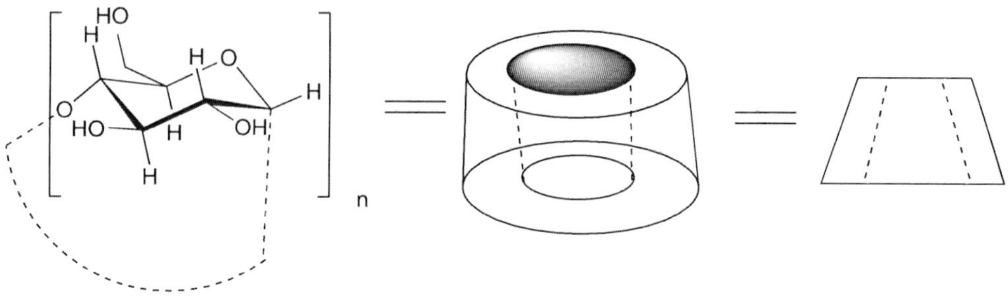

n = 1 alpha-cyclodextrin 1
n = 2 beta-cyclodextrin 2
n = 3 gamma-cyclodextrin 3

Figure 2.1 *α, β, and γ cyclodextrins in three representations*

Then systems are described in which the cyclodextrin binding group has catalytic groups attached to it, to mimic enzymes whose binding groups and catalytic groups are not the same. The first section describes catalysed hydrolysis reactions with metal ions bound to ligands that are attached to one or two cyclodextrins. Then mimics of ribonuclease enzymes in which two catalytic groups are attached in well-defined positions to a cyclodextrin are described. There follows brief descriptions of studies on simple binding by cyclodextrin dimers and trimers.

Extensive studies on mimics of transaminase enzymes follow, and then mimics of enzymes that use thiamine pyrophosphate as their coenzyme. Studies on mimics of enzymes that perform aldol condensations are then described. After a brief section on mimics of enzymes that use coenzyme B_{12} as a coenzyme, there is a description of mimics of cytochrome P-450 oxidizing enzymes, in which as many as four cyclodextrins are involved in binding.

We have written a number of reviews of our work in this field.[4–16] Our chapter in Chemical Reviews[14] was particularly comprehensive, and covered the work in many other laboratories besides our own. There are actually two other chapters on reactions in cyclodextrins in this book.[17,18] It has been agreed that each of our laboratories will concentrate chiefly on the work that we ourselves have done.

2.2 Acylations of the Cyclodextrins by Bound Substrates

Myron Bender had reported that a meta-*t*-butylphenyl acetate (**4**) acetylated *β*-cyclodextrin in water with a rate 250 times as fast as that for hydrolysis of that same substrate at the same pH.[2] We had shown that the same reaction was even faster in a mixed DMSO/water solvent, but still the acceleration was not what one would have hoped for.[19] Model-building suggested that in the acylation reaction the tetrahedral intermediate is partly pulled out of the cavity, so cyclodextrin binding is to some extent fighting against the reaction rate. Thus we made a new substrate, the *p*-nitrophenyl ester of ferroceneacrylic acid (**5**), and saw that it acylated *β*-cyclodextrin with a rate acceleration of 51,000 compared with the hydrolysis rate in free solution.[20] With this substrate there was no need to pull the binding ferrocene group up relative to its preferred geometry in the cavity.

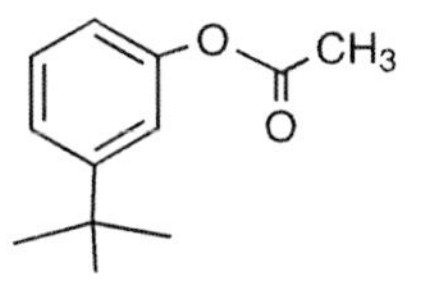

4

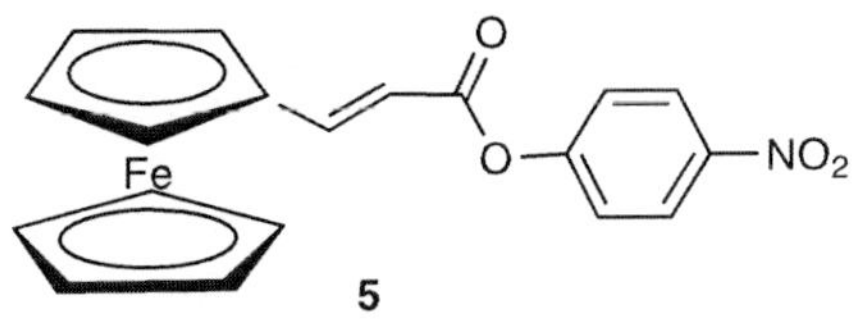

5

The cyclodextrin ring does not have a well-defined depth, so we adopted a simple procedure to turn it from an open tube into a cup. We attached some flexible groups on the primary carbons of the cyclodextrin that could partially invade the cavity and produce a flexible floor.[21] Later we studied acylation reactions with cyclodextrins that had been modified to include a floor on the bottom of the cavity – and also with improved substrates and a somewhat better solvent – that led to a relative rate of 750,000 compared with the uncatalysed hydrolysis rate at the same pH.[22] When we looked at related compounds in which we had frozen out one of the degrees of freedom of the substrate the improvement came up to a 3,200,000-fold acceleration compared with the hydrolysis rate in water, and with some selectivity among the geometries of the complex.[23]

In a later study on the optimization of metallocene substrates for such cyclodextrin reactions we saw that the two different enantiomers (**6**) of a fused ring system – in which there was no substrate flexibility – was 62-fold faster for one enantiomer than the other and showed a rate 150 million times as fast for acylation as was the rate of hydrolysis in water at the same pH.[24] We then addressed this with molecular modeling calculations, and saw that quantitative theory accounted well for these findings.[25]

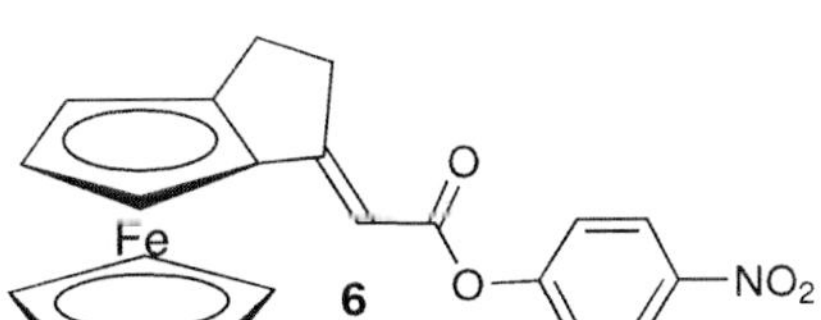

6

When the leaving group on these esters was less reactive than *p*-nitrophenoxide, we and Menger had seen that the rate acceleration decreased strongly.[26,27] We reasoned that a completely rigid substrate has a problem if the rate-determining step involves forward decomposition of the tetrahedral intermediate, since it is necessary for the ester carbonyl to rotate in the final product. This explained why reactions whose rate-determining step involved the formation of the tetrahedral intermediate were very fast, while those in which forward decomposition of the intermediate to the product was rate determining were not nearly so well accelerated. We solved this problem by putting in one degree of rotational freedom in the substrate, which permitted the needed rotation, and saw that now there was no abnormal behaviour of the *p*-nitrophenoxide group.[26]

In a different experimental approach to the question of geometric changes during the acylation reaction of cyclodextrin bound substrates, we collaborated with Le Noble in a study of pressure effects on the reaction rates.[28] The volume changes that this technique indicated – as starting materials proceeded to the transition state – were consistent with the conclusions we had reached from earlier studies of rate dependences on substrate geometries.

2.3 Catalytic Reactions in Cyclodextrin Cavities: Aromatic Substitution

In the previous reactions the cyclodextrin acted as a reactant, not a catalyst. However, there are some excellent examples in which true catalysis occurs with simple binding into a cyclodextrin cavity. Here we will describe the cases where the cyclodextrin has not been modified, while in later sections we will discuss cases in which additional catalytic groups have been added to the cyclodextrin, and mimics of metalloenzymes and of enzymes with co-enzymes have been achieved.

In our earlier study we saw that we could achieve geometric selectivity in the chlorination of anisole by HOCl in water solution.[29] In simple water solution we saw a 1.5 ratio of *para* chlorination to *ortho* chlorination of the anisole, while in a 9 mM concentration of α-cyclodextrin the *para-* to *ortho*-ratio was 29.6. Under these conditions the anisole was only 72% bound into the cavity, so these results indicated that the cyclodextrin not only covered the *ortho* positions, it also actively catalysed the *para* chlorination. The most reasonable idea about this was that one of the hydroxyl groups of the cyclodextrin reacted with HOCl to become a hypochlorite group, and that this then delivered the chlorine to the *para* position of the bound anisole (Figure 2.2).

In a subsequent detailed study we saw that the chlorination of unbound anisole involved two molecules of HOCl, surely as their very reactive product Cl_2O, while the chlorination in the complex involved only a single molecule of HOCl. The obvious conclusion from this is that the HOCl reacts with a hydroxyl group of the cyclodextrin to make a hypochlorite that is able to donate a chlorine atom to the *para* position of bound anisole, which is the one accessible when the anisole is fully bound. Proximity makes up for the weak chlorinating ability of a hypochlorite.

In a full paper we examined this system in more detail.[30] We found that a substituting reagent, a diazonium ion that could not be delivered by a cyclodextrin hydroxy, was not catalysed by the cyclodextrin, since the diazonium group could not be covalently linked to a hydroxyl of the cyclodextrin and still be an electrophile. As another piece of evidence, when we looked at *para*-methoxytoluene we saw that binding to the cyclodextrin now inhibited chlorination, which of course occurred on the *ortho*-chloro position. However, with *para*-cresol there was again chlorination ortho to the oxygen, but this was now catalysed by the cyclodextrin binding. In this case, in contrast to the methoxytoluene, the polar hydroxyl group would be expected to protrude from the cavity, bringing the *ortho* position within reach of chlorines covalently linked to hydroxyl groups.

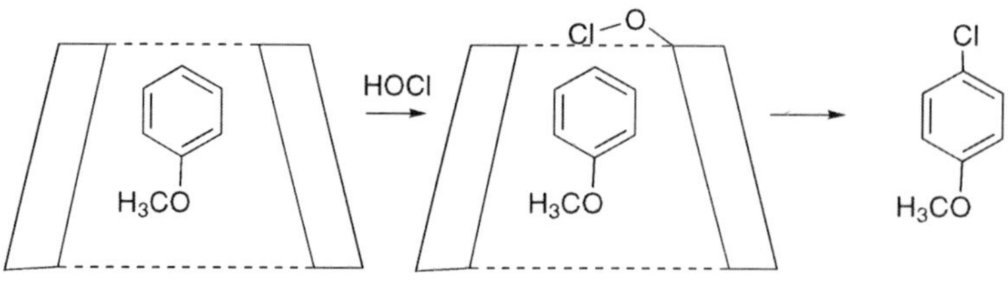

Figure 2.2 *α-Cyclodextrin catalyses the para chlorination of anisole by hypochlorous acid*

In another paper related to this chemistry we examined α-cyclodextrin in which all of the C-6 and C-2 hydroxyl groups were methylated.[31] We found that this was also an excellent selective catalyst for the chlorinations, and in particular because anisole binds more strongly into this somewhat more hydrophobic cavity we saw that the selectivity was even higher, reflecting a higher proportion of bound substrate. This demonstrates that the C-3 hydroxyl can be the one that delivers the chlorine to the substrate, although it does not rule out the possibility of other hydroxyls being involved in the case where we do not have them selectively methylated. In the same paper we also prepared a polymer of α-cyclodextrin by cross-linking it with epichlorhydrin. We saw that this polymer was an excellent catalyst for the selective chlorination of anisole, and when a water solution of anisole and HOCl was simply run down a column containing the polymer the product was essentially completely chlorinated in the *para* position.

One extra point might be mentioned about these catalysts. There is an enzyme, chlorinase, that will also chlorinate anisole but with only the random 60/40 distribution between *para* and *ortho* chlorination. The enzyme obviously has not been optimized for this process, but it is fair to say that our very simple artificial enzyme, α-cyclodextrin, is actually better at selective chlorination of this substrate than the enzyme is.

2.3.1 Catalytic Reactions in Cyclodextrin Cavities: Diels–Alder Reactions

We also examined Diels–Alder reactions in a cyclodextrin complex.[32] In the reaction of cyclopentadiene with 2-butenone (Figure 2.3) or with acrylonitrile we saw that there was catalysis in water with β-cyclodextrin, but not with α-cyclodextrin, which was an inhibitor. The cyclopentadiene bound into both of the cyclodextrins, but only in the β-cyclodextrin is there also room for the dienophile to fit. In this same publication[32] we saw that the Diels–Alder reaction is also strongly accelerated simply in water solution, which favours the transition state for the reaction in which a smaller total amount of hydrophobic surface is exposed to the water solvent. This latter finding stimulated us, and others, to study many other examples of catalytic effects in water alone with hydrophobic reagents. However, the cyclodextrin effect was added on top of this in the case where both reagents could fit into the pocket.

2.4 Other Solvents than Water

In a later study we examined Diels–Alder reactions in non-aqueous polar solvents, both in the solvents alone and in the solvents with β-cyclodextrin.[33] We found that DMSO did not promote β-cyclodextrin binding, but that it was promoted by both ethylene glycol and

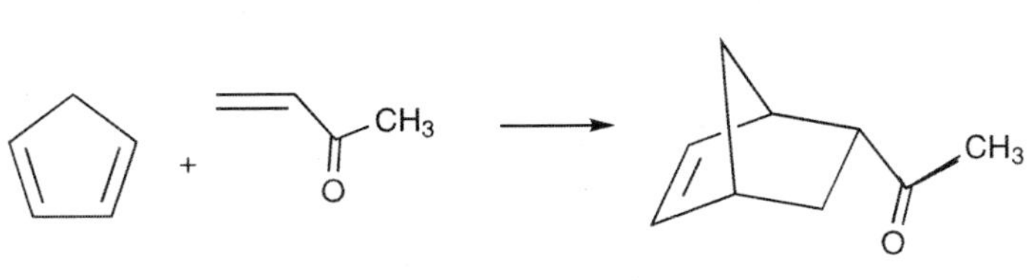

Figure 2.3 The Diels–Alder reaction of 1,3-cyclopentadiene with 2-butenone, which is catalysed in the cavity of β-cyclodextrin

formamide. Thus the 'hydrophobic' binding induced by water can be duplicated to a lesser extent with these other water-like solvents.

In a previous study on simple cyclodextrin binding we examined the ability of cyclodextrin to bind various substrates in dimethyl sulfoxide solution rather than in water.[19] We saw that indeed hydrophobic substances were bound into the cyclodextrin, and also that DMSO itself was a solvent in which the acylation of a cyclodextrin hydroxyl group by a bound ester could be observed. Cyclodextrin binding is not exclusively limited to water solutions, as had been suggested by others previously, but water is so far the best solvent to see such binding and catalytic processes.

2.5 Catalytic Reactions Produced by Cyclodextrins With Covalently Attached Catalytic Groups

2.5.1 Catalysed Hydrolysis Reactions

Combining a cyclodextrin for substrate binding with catalysis of a reaction by some attached groups seems to be the best approach so far for the production of artificial enzymes that imitate the situation in natural enzymes, where binding and catalysis normally involve different enzyme groups. Catalytic groups can be attached to either the primary or the secondary face of a cyclodextrin, and on the secondary face there are both carbon-2 and carbon-3 positions to which attachment can be made. We synthesized some β-cyclodextrin compounds carrying phosphate groups at the carbon-2, carbon-3, or carbon-6 positions and compared their effectiveness as general acids or as general base catalysts with bound substrates.[34] All were to some extent catalytic.

The first compound described as an 'artificial enzyme' in the literature was the one we reported in which we attached a metal ion binding group to α-cyclodextrin.[35] We found that this would bind *p*-nitrophenyl acetate into the cavity and a bound nickel ion then catalysed the hydrolysis of the substrate. This was a direct hydrolysis, not an acylation of a cyclodextrin hydroxyl (which is not in reach with the *para* esters). This type of catalyst then extends metal-catalysed reactions to substrates that do not intrinsically bind to metal ions, which was formerly required for such catalysis.

Even better rates of hydrolysis were seen when a cyclodextrin dimer was used in which the linking group had a metal ion bound in a position to be able to catalyse hydrolysis of an appropriate substrate. In our first example (Figure 2.4) we showed that a dipeptide could bind across such a metal binding linker, and that its affinity for binding was increased when the linker carried a zinc ion that could coordinate to the peptide group.[36] However, this was not powerful enough to catalyse the cleavage of the peptide. With

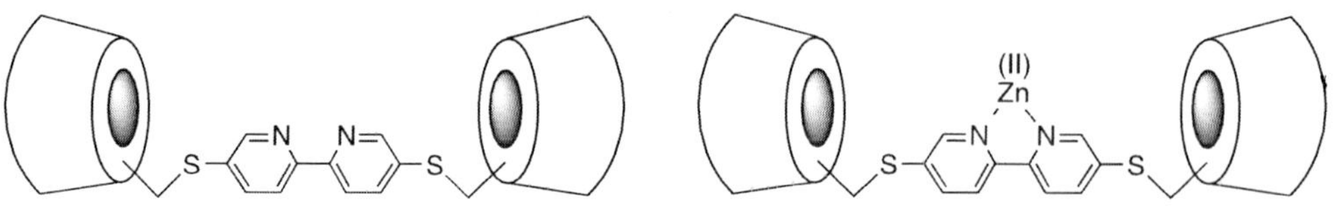

Figure 2.4 *A cyclodextrin dimer that can bind a diester with two hydrophobic ends, and catalyse its hydrolysis as the zinc complex of the dimer*

esters involving two hydrophobic groups we saw outstanding acceleration of catalysis, with acceleration 220,000-fold faster than the simple rate of uncatalysed hydrolysis under the same conditions. This was comparable to the rate acceleration seen at the time by the best catalytic antibodies.

Interestingly, in a study of the binding of ditopic substrates to such cyclodextrin dimers we saw that the binding was dominated by an improved enthalpy, rather than entropy.[37] The simplest ideas about chelate binding would have suggested an entropy advantage, but in solution enthalpy–entropy compensation can be seen if the binding or release of water molecules is also considered.

In a later study we examined the ability of cyclodextrin dimers linked by a bipyridyl unit to catalyse the hydrolysis of a phosphate ester, *bis-p*-nitrophenyl phosphate.[38] With a bound lanthanum ion the hydrolysis was accelerated by 300 million-fold, a huge acceleration that could well be of practical interest in the hydrolysis of phosphate esters generally. A full paper described this work in some detail.[39]

In earlier work we had demonstrated that a zinc complex of pyridyl-2-carboxaldoxime (**7**) could be effective in cleaving esters. The interesting point is that the oxime anion is available as a nucleophile and the zinc as an electrophile, but they are not coordinated to each other – which would of course destroy the catalytic effect. To amplify catalysis we attached such oxime-zinc complexes to β-cyclodextrin on both the secondary and primary faces of the cyclodextrin and examined their reaction with *p*-nitrophenyl acetate.[40] We observed burst kinetics, in which there was an extremely rapid release of one mole of nitrophenoxide ion, followed by a slower release in a second phase. This indicated that we first rapidly produced the acetate of the oxime, and this then slowly hydrolysed to regenerate the oxime anion for further catalytic reaction. Such burst kinetics is very commonly seen in enzymatic reactions of *para*-nitrophenyl acetate, reflecting the same kind of two-step overall mechanism.

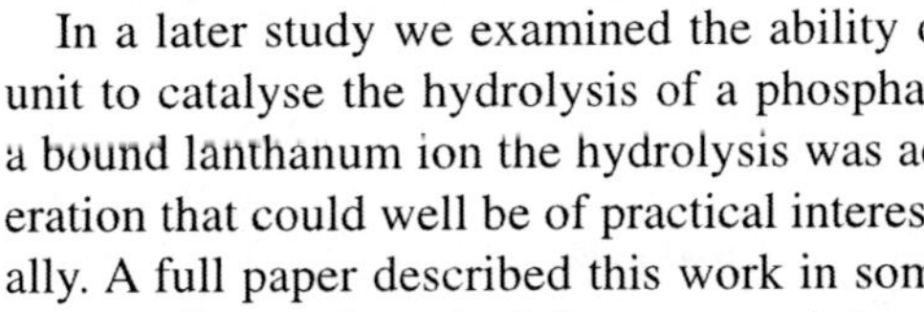

7

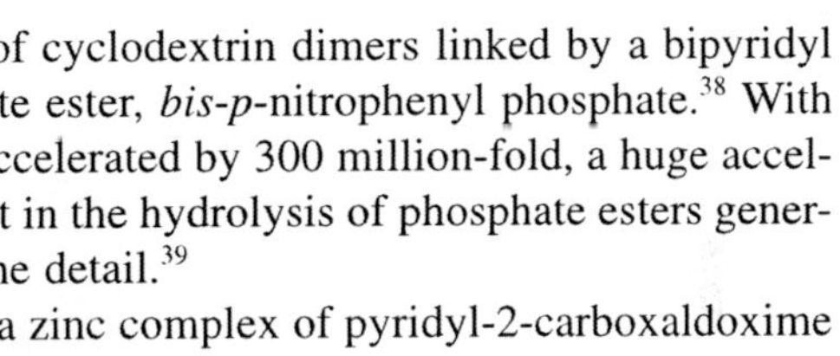

8

One problem with such studies is that *p*-nitrophenyl acetate is a highly reactive ester, and it is more of a challenge to catalyse the hydrolysis of ordinary esters or of amides. In a move in this direction we showed that an appropriate cyclodextrin dimer with a bound copper ion (**8**) could indeed catalyse the hydrolysis of an ordinary ester group, not a phenyl ester.[41] The acceleration was 18,000-fold, certainly a respectable catalytic result.

2.5.2 Ribonuclease Mimics

The class of enzymes known as ribonucleases catalyses two separate reactions (Figure 2.5). In the first one a phosphate diester between two RNA units is cleaved by attack of the C-2 hydroxyl to convert the species into a cyclic phosphate with release of the other RNA unit, so the chain is cleaved. In a second reaction ribonuclease catalyses the conversion of the cyclic phosphate back to an open-chain phosphate monoester. In both reactions

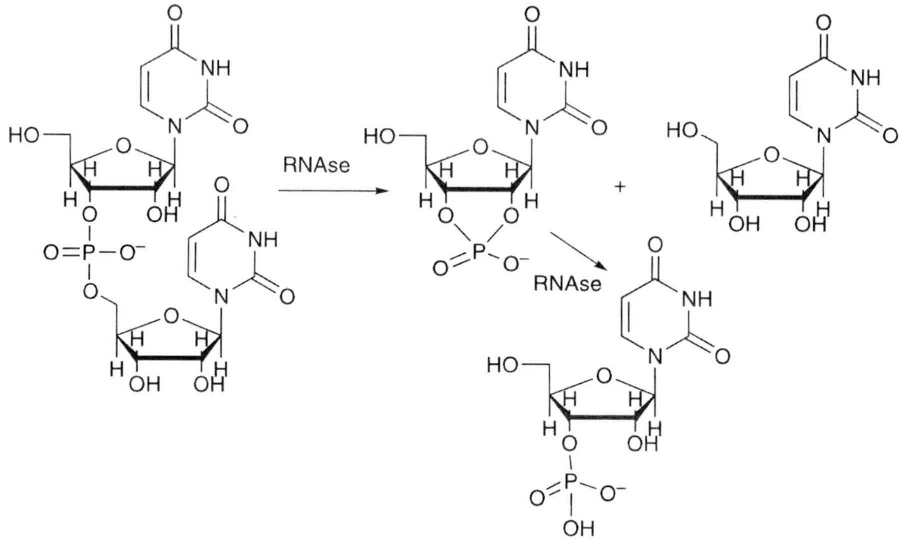

Figure 2.5 The enzyme ribonuclease catalyses a two-step cyclization and cleavage of RNA, shown with the simple dinucleotide substrate uridyluridine

ribonuclease A uses two catalytic groups – the imidazoles of histidine 12 and histidine 119 – using one as a base catalyst and the other, in its protonated form, as an acid catalyst.

In the cyclization-cleavage, histidine-12 is the base that removes the proton from the C-2 hydroxyl group of ribose and histidine-119 is the acid group that eventually protonates the leaving oxygen of the second nucleoside. In the hydrolysis of the cyclic phosphate these roles are reversed, with the now deprotonated histidine-119 acting as a base to deliver a water molecule while the protonated histidine-12 now acts to protonate the leaving C-2 oxygen that had originally attacked in the formation of the cyclic phosphate. This reversal of roles explains why the cyclic phosphate is not cleaved as rapidly by the enzyme as would normally be required for a reactive intermediate. The cyclic phosphate presented to the enzyme in its starting protonation state would not have the acid and base groups in the right position for the second cleavage step. In a sense this is almost a see-saw mechanism, in which first a base and acid combination produces cyclic phosphate and then the resulting new base and acid combination catalyses the second step.

The cleavage of the cyclic phosphate is particularly interesting because it is relatively well defined in geometry, compared to the somewhat flexible phosphate diester group. For this reason we undertook a study of catalysed hydrolysis of a rigid cyclic phosphate, a mimic of that formed with RNA itself. In our initial study we used the cyclic phosphate of *t*-butyl-catechol (**9**) as the substrate, and cyclodextrins carrying two attached imidazole rings as catalysts.[42] Tabushi reported the preparation of compound in which *β*-cyclodextrin was converted to a disulfonate derivative on the primary side.[43] We have used this compound to make a cyclodextrin *bis*-imidazole (**10**) by simple displacement with imidazole, and used it as a catalyst for the hydrolysis of the catechol cyclic phosphate **9**. All of our evidence indicated that this disulfonation was not specific, so **10** was a mixture of what we call the AB and AC *bis*-imidazoles, in which the glucose rings of *β*-cyclodextrin are lettered sequentially from A through G.

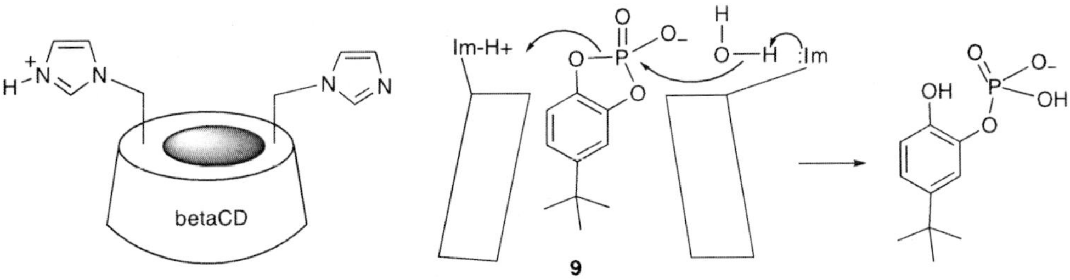

9

11

10, a mixture of the AC and AD positional isomers, as the mono-protonated species.

a possible catalytic bifunctional mechanism for cleavage of **9** by catalyst **10**

Figure 2.6 *A cyclodextrin carrying two imidazole groups, one of them protonated, catalyses the hydrolysis of a catechol cyclic phosphate with selectivity*

We saw that this *bis*-imidazole mixture was a powerful catalyst for the hydrolysis of the cyclic phosphate, showing two exciting new aspects. First of all it was in fact a bifunctional catalyst whose plot of pH *vs.* rate was a bell-shaped curve, just as with the natural enzyme ribonuclease itself. Furthermore, it was a much better catalyst than cyclodextrin mono-imidazole, obtained from simple cyclodextrin tosylate. In addition, the reaction was highly specific with respect to the product, cleaving the cyclic phosphate so as to leave the phosphate on oxygen 3 and the new 4-hydroxy group *para* to the *t*-butyl group in **11**. This is the selectivity expected if the imidazole is able to deliver a water molecule to the cyclic phosphate in a line more or less perpendicular to the rotational axis of the cyclodextrin, as models suggested. In this hydrolysis the two imidazoles are playing the same role as they do in the enzyme, with the basic imidazole delivering the water molecule to the phosphate by acting as a general base to remove the proton while the imidazolium ion acts to furnish a proton for the cleavage (Figure 2.6).

At the time we assumed that this used the mechanism commonly proposed for such processes, in which the proton was simply delivered to the leaving group. However, other work, to be described shortly, indicated that the mechanism of this cleavage is more interesting than this simple idea suggests. In a subsequent study we examined a cyclodextrin *bis*-imidazole in which the imidazole rings were held further from the cyclodextrin through a sulfur-carbon bridge.[44] In this case the water was delivered in a somewhat different direction, and now the cleavage of the cyclic phosphate involved a preferential formation of the monophosphate with the phosphorus on the *para* oxygen while the meta oxygen was present as the free hydroxyl group. In this paper we also indicated that the imidazolium ion could be coordinated to one of the phosphate oxygens rather than simply delivering a proton to the leaving group, a bit of foresight that was eventually proven to be correct.

A striking result was observed when we synthesized a new isomer of the cyclodextrin *bis*-imidazole, one in which the imidazole groups were attached to the primary carbons of two neighboring glucose units, the so-called AB isomer.[45] By the ordinary mechanism – in which a water is delivered by a general base imidazole to the phosphorus as the imidazolium ion of the second catalyst protonates the leaving group – this isomer should have been essentially inactive, since the 180 degree requirement for the previous mechanism is impossible with these two neighboring imidazoles. However, this turned out to

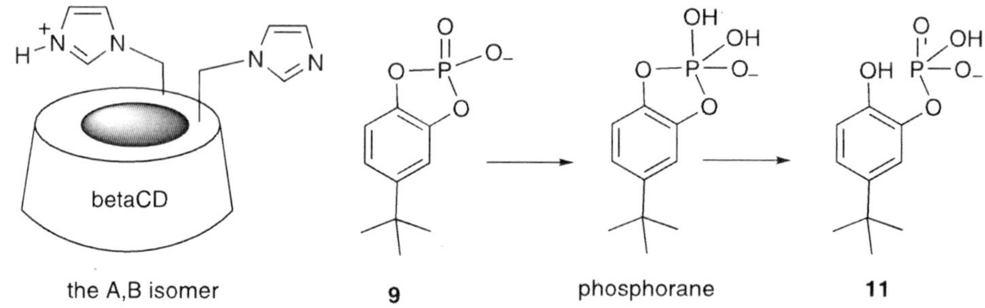

Figure 2.7 The AB isomer of the cyclodextrin bis-imidazole is the best catalyst for the hydrolysis of the catechol phosphate **9**, indicating that the reaction proceeds through a phosphorane intermediate

be the most potent of all three *bis*-imidazoles (we had in the meantime learned to prepare both the AC and AD isomers as pure materials, not as a mixture).

The observation that the AB isomer of cyclodextrin bisimidazole is the most potent catalyst requires a new mechanism (Figure 2.7). The imidazolium cation is putting the proton on a phosphate oxygen anion instead of on the leaving group, so the two protons that are moving need to be much closer, more or less 90 degrees apart rather than 180 degrees. Such a mechanism requires that the cleavage be a two-step process in which the first step involves the addition of water to the phosphate to make a five-coordinate phosphorus, a phosphorane anion. In a second step this anion has to cleave to liberate the free ribose hydroxyl group with assistance by an imidazolium cation.

This mechanism is really unambiguous for the model system, so we were interested to see whether we could find any evidence for it as well in the cleavage of RNA by imidazole buffers. In work that we will not describe further in this review, since it does not involve cyclodextrin chemistry, we were indeed able to get convincing evidence that imidazole buffers cleave the simple dinucleotide uridyluridine through such a phosphorane intermediate rather than by direct cleavage.[46] We also proposed that this might well be preferred by the enzyme itself, and offered some arguments for this, although it is not universally accepted as the mechanism for the enzyme.

We had invoked a simultaneous two proton transfer mechanism rather than a sequential mechanism – in which one catalytic group followed the other in the overall process – and were able to test this with a technique called proton inventory.[47] We examined the original cyclodextrin *bis*-imidazole in mixtures of water and D_2O and saw that the rate constant as a function of deuterium concentration followed a curved line, indicating that the isotope effect involved two different protons rather than a single one. To validate this, we also examined the same kind of plot with the cyclodextrin mono-imidazole, in which only one proton would be expected to be moving in the transition state, and this indeed followed a linear plot supporting a single proton motion in the isotope effect.

Some of this work is described in reviews that include the cyclodextrin work.[48–51] We also examined a comparison of catalysts with the AB placement of imidazoles but based on either α-cyclodextrin, β-cyclodextrin or γ-cyclodextrin. And these were also compared with substrates carrying various groups other than *t*-butyl. The conclusion from this work was that by far the best rates and selectivities were seen in cases where there was a very tight fit between substrate and catalyst cyclodextrin ring, giving the best definition of the geometry of the system and resulting in the highest rate constant.[52]

Some years ago we had studied the cleavage of the simple nucleotide uridyluridine by a combination of imidazole buffer and zinc ion. We found that they were very effective cooperative catalysts for the process, much more effective that simple imidazole buffer alone.[53] Thus we prepared a bifunctional catalyst based on β-cyclodextrin in which we attached one imidazole ring and one metal ligand group, and saw that they were effective catalysts for the cleavage of the *t*-butylcatechol cyclic phosphate.[54] This is a model for the many metalloenzymes containing not only a metal ion but also a basic group as part of the catalytic centre. In this paper we reported that the combination of a zinc ion chelate and an imidazole were more effective than simple imidazole buffer in cyclizing a model for RNA and then cleaving it. This is an attractive lead for the synthesis of a related cyclodextrin compound, which has not yet been studied.

2.6 Binding by Cyclodextrins and their Dimers and Trimers

We saw very strong binding of lithocholic acid to β-cyclodextrin.[55] The AB *cis* geometry in this substrate made it fit particularly well into the cyclodextrin cavity. This study was related to the studies on steroid oxidations that we will describe later. We took up a study of the binding by cyclodextrin dimers.[56] With various linkers between the cyclodextrins, which were attached on both the primary and the secondary faces, we saw cooperative very strong binding of a number of substrates with two groups that could bind into a cyclodextrin cavity. Interestingly, cholesterol as a substrate is large enough to be able to bridge into the two rings of some cyclodextrin dimers.[57] We saw evidence for such binding in both well-defined cyclodextrin dimers and also in cyclodextrin polymers in which there were of course dimeric sections that could encompass both ends of the cholesterol in binding.

Even more strikingly, when we linked two cyclodextrins with a pair of linkers attached to the A and B glucose residues we saw two dimers that had very different binding properties (Figure 2.8).[58] When the linkage was from the A residue of one to the B residue of the other the result was a structure we described as the 'clamshell', in which it was possible for a substrate to bind simultaneously into the two rings. When the linkage was from the A residue of one cyclodextrin to the A residue of the other, and B to B, the result was that we had a structure we described as a 'loveseat' in which no substrate could occupy both cyclodextrins at the same time because the linkers interfered with such binding. We

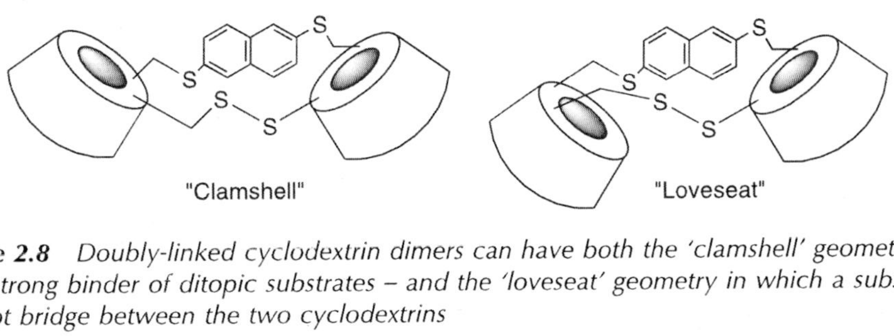

Figure 2.8 Doubly-linked cyclodextrin dimers can have both the 'clamshell' geometry – a very strong binder of ditopic substrates – and the 'loveseat' geometry in which a substrate cannot bridge between the two cyclodextrins

also described the clamshell version as 'occlusive' while the loveseat was described as 'aversive'. The aversive isomer showed normal binding of substrates with two binding ends, with no cooperative binding of both ends into the dimer. On the other hand the occlusive isomer had a binding constant as high as $10^{10}\,M^{-1}$. This corresponded to a free energy larger than additive, as expected for chelate binding.

We have found that what we called 'antihydrophobic' agents could increase the solubilities of hydrocarbon groups in water and would also decrease their ability to bind into cyclodextrin. In a quantitative study we saw that the free energies corresponding to these two processes were the same, indicating that the decreased binding principally resulted from effects on the substrate, not on the cyclodextrin.[59] We also examined these effects on the binding of cyclodextrin dimers. As described earlier,[60] we have looked at cyclodextrin dimers with catalytic groups in the linker as catalysts for hydrolysis processes. A full paper summarized some of the results of these binding studies.[61]

We also examined the synthesis and binding properties of some cyclodextrin trimers.[62] With substrates that might be able to occupy all three of the binding cavities we did see improved binding compared with that for cyclodextrin dimers, but not by as large a factor as one might have expected. It seems likely that we have not yet achieved the correct non-flexible geometry in the case of the trimers. I will describe another example of binding to a cyclodextrin trimer in the section on mimics of cytochrome P-450 enzymes.

We have examined some of possible applications of the strong binding by cyclodextrin dimers. In one study we examined the use of cyclodextrin dimers with a light-cleavable link as a possible carrier for the photosensitizer in photodynamic tumor therapy.[63,64] The idea, which was demonstrated, is that irradiating the complex would cause singlet oxygen to cleave the linker, whereupon the substrate would be released from the dimer since it did not have the advantage of chelate binding. If this occurred at the tumor site, because irradiation was directed there, this could lead to concentration of the photodynamic material right next to the tumor. This should avoid the problem that photodynamic processes can otherwise occur throughout the body, usually with unpleasant or toxic effects.

We also examined the binding of various peptides to β-cyclodextrin. Simple β-cyclodextrin itself showed selective binding of particular dipeptides, reflecting the sequence of hydrophobic side chains.[65] Protein aggregation in solution normally involves the binding to each other of hydrophobic surfaces in the protein monomers, so cyclodextrin binding might well be expected to disrupt this. In one study we saw that indeed cyclodextrin dimers could disrupt protein aggregation, a finding that could lead to selectively useful processes.[66] In a cooperative study we saw that some of our dimers and even trimers acted as inhibitors of amyloid peptide aggregation, the process that is damaging in Alzheimer's disease.[67]

Helical structures are important in proteins, and the extent of helix formation in a given polypeptide can be influenced by external factors. In a study related to this question we saw that with appropriate cyclodextrin dimers we could induce helix formation in some oligopeptides if the helix structure presented hydrophobic side chains in a geometry such that our dimer could bind to them, stabilizing the helix.[68]

2.6.1 Transaminase Mimics

Transaminase enzymes react pyridoxamine phosphate with ketoacids to form pyridoxal phosphate and amino acids (Figure 2.9). This is the first step in an overall transamination

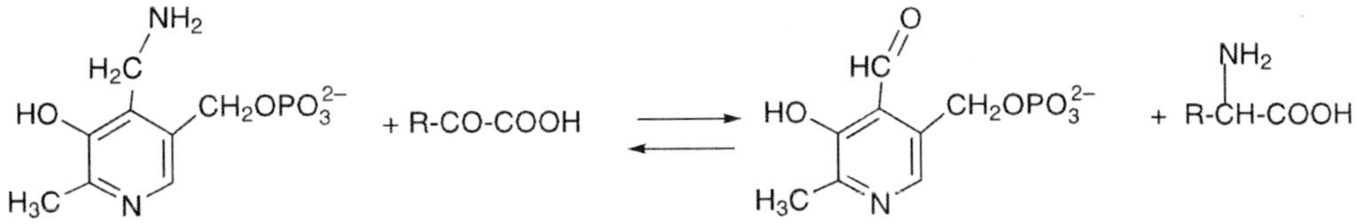

Figure 2.9 *In transaminase enzymes, pyridoxamine phosphate converts a ketoacid to an amino acid while being itself converted to pyridoxal phosphate*

process in which the pyridoxal phosphate is then converted back to pyridoxamine phosphate by the sacrifice of a different amino acid, which is converted to its ketoacid. We have imitated the first part of this process, the conversion of pyridoxamine to pyridoxal while making a ketoacid into an amino acid, and studied how we could promote this process with the help of binding to cyclodextrins. In later work we have shown how some polyamine substances can also catalyse such a process, and can as well catalyse a reverse process in which the pyridoxamine is regenerated, so the overall process is in fact catalytic. We will describe this regeneration of pyridoxamine at the end of this section.

In our first study we attached a pyridoxamine unit to a primary carbon of *β*-cyclodextrin (structure **12**).[69] We saw that pyridoxamine alone is able to transaminate pyruvic acid to form alanine, phenylpyruvic acid to form phenylalanine, and indolepyruvic acid to form tryptophan, all with equal reactivity by competition experiments. However, when the cyclodextrin was attached to the pyridoxamine there was a 200-fold preference for the indolepyruvate over pyruvate in one-to-one competition, forming greater than 98% of tryptophan, and in the competition with phenylpyruvate and pyruvate the phenylalanine was formed in greater than 98% as well. Thus the ability of the substrates to bind into the cyclodextrin cavity led to striking selectivities. In addition there was some chiral induction in these processes, since *β*-cyclodextrin is itself chiral, but the magnitudes of the induction were quite modest.

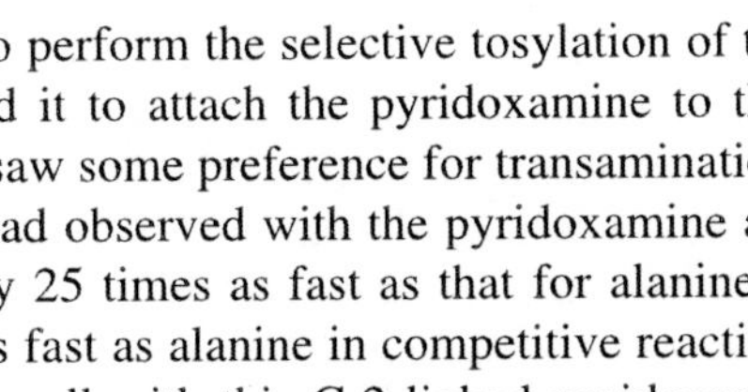

12

We devised a method to perform the selective tosylation of the C-2 hydroxyl group in *β*-cyclodextrin,[70] and used it to attach the pyridoxamine to this secondary side of the cyclodextrin.[71] Again we saw some preference for transamination of the aromatic ketoacids, but by less than we had observed with the pyridoxamine attached to C-6. The tryptophan synthesis was only 25 times as fast as that for alanine, while the phenylalanine formation was 18 times as fast as alanine in competitive reactions. Modest enantioselectivities were observed as well with this C-2 linked pyridoxamine, and they differed in detail from those produced with the C-6 isomer that we had reported earlier.

We also examined an interesting modified version in which the β-cyclodextrin had a pyridoxamine attached to the primary C-6 carbon but the other six hydroxymethyl groups on the primary side were deoxygenated to become methyl groups. This leads to quite a different hydrophobic cavity,[72] but it showed properties similar to those of the original compound that still had its primary hydroxyls unmodified.

In a full paper we described this work in detail, including some work on pyridoxamine derivatives that did not involve cyclodextrins and other work in which a synthetic binding group was used instead of the cyclodextrin.[73]

We also set about to see whether we could get good chiral induction in the product amino acids, not by the simple accident of the chirality of the cyclodextrin but by basic groups that could direct the proton transfer involved in transamination so as to give a preference for one enantiomer of the product amino acid. We described some of this work,[74] and in it also referred to work reported by Tabushi in which the same general principle was applied. In our work only relatively modest selectivities were seen; the largest optical ratio (L/D) in the product was only 6.8.

In the meantime the Tabushi laboratory had prepared a somewhat different version of this idea and had reported outstanding optical ratios.[75] In his paper my name was included simply because I had proposed the original idea, but I did not have any interaction with the experimental evidence. In our later work we found that attempts to repeat the work reported in reference 75 did not lead to the reported results.

The pyridoxamine units linked to the cyclodextrins in these previous cases are rather flexibly attached, and do not perhaps have the optimum defined geometry. Thus we prepared a new series of compounds in which a pyridoxamine unit carried two sulfurs in a well-defined geometry so that they could link to the A and B primary carbons of β-cyclodextrin, using the same mechanism by which we had originally made the 6A, 6B bisimidazole cyclodextrin. With the new pyridoxamines we observed very high selectivities for hydrophobic substrates, and excellent control of their detailed geometries.[76]

In a full paper we described detailed studies of our various approaches to chiral induction in the amino acid products of transaminations.[77] In that paper we also pointed out the odd fact that the Tabushi laboratory had reported the use of large concentrations of buffer in their reaction, which would have been expected to interfere with the selective proton transfer by an internal catalytic group if the buffer itself could start playing the role of protonating catalyst.

In our subsequent work we studied compounds in which a pyridoxamine was attached to C-6 of β-cyclodextrin and an imidazole unit was attached to C-6 of the neighbouring glucose residue.[78] The two isomers are diastereomers, reflecting the chirality of their attachment and the chirality of the glucose units. We found that we could in fact get a reversal of the optical induction reflecting the different geometry of these two enantiomers. Interestingly, we found that the imidazole as placed did not catalyse selective proton transfer, but instead performed a selective blocking of one face of the reaction intermediate so that protonation by the medium occurred on the side away from the imidazole ring. It was in this paper that we also reported the re-examination of the Tabushi case and our finding that his isomers showed little enantioselectivity, unrelated to the position of the so-called protonating group.

Pyridoxal phosphate is actually a catalyst for a number of transformations of amino acids that do not involve transamination. It is the catalyst in amino acid racemases, for instance, and in some interesting rearrangements of amino acids. Thus we examined a pyridoxal unit attached to β-cyclodextrin to see whether we could imitate the substrate binding and reactions involved in such pyridoxal catalysis. Pyridoxal catalyses the condensation of serine with indole to produce tryptophan, by a process in which the imine formed between pyridoxal and serine dehydrates to form a double bond to which the indole adds. We thought it likely that we could bind the indole into the cyclodextrin cavity and have it couple with the species that pyridoxal forms. We found that indeed this catalyst could couple d,l-β-chloroalanine to indole, in which the imine formed between pyridoxal and this alanine derivative eliminated HCl to form the unsaturated intermediate that reacts with indole to form tryptophan.[79] The results indicated some cooperativity in the process, but it was not very effective.

Pyridoxal phosphate is also the catalyst for the process by which glycine and formaldehyde are reversibly condensed in an aldol reaction to form serine. In a system that did not involve substrate binding with cyclodextrins we were able to show that we could duplicate that process, and that with sufficient geometric control we could obtain reasonable optical selectivity.[80] We also examined transamination in which the group that bound the substrate was a polyamine with a hydrophobic core, rather than a cyclodextrin.[81] With this system we saw that we could reverse the transamination process – in the sense of converting the pyridoxal derivate back to pyridoxamine – by using a sacrificial α-methyl amino acid that would undergo irreversible decarboxylation and transamination to afford the pyridoxamine and a ketone derived from the original sacrificial amino acid.[82]

2.7 Mimics of Enzymes that Use Thiamine Pyrophosphate as a Co-Enzyme

Thiamine pyrophosphate (**13**) is the co-factor for a number of enzymes that can be described as stabilizing hypothetical acyl anion intermediates. For instance, it is the co-enzyme for the enzyme carboxylase that catalyses the conversion of pyruvic acid to acetaldehyde. We had early shown that this mechanism involves a thiazolium anion (**14**) whose second resonance form (**15**) is a carbene.[83] Ionization of the C-2 proton of the thiazolium ring generates this species that can add nucleophilically to carbonyl groups such as that in pyruvic acid, forming an intermediate whose decarboxylation generates a stabilized anion.

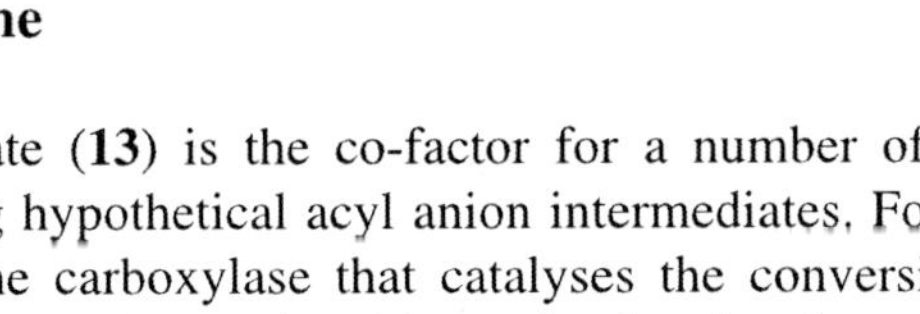

13

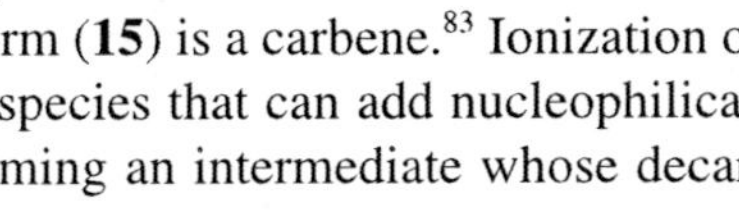

14 **15**

In early studies of this chemistry we had examined the ability of such thiazolium salts to catalyse the benzoin condensation, a process which also formally involves an acyl anion but which is really of course the anion in which the thiazolium salt has been added to the carbonyl group (Figure 2.10). In this sense the thiazolium anion is very much like cyanide anion, the normal catalyst for simple benzoin condensations. Benzaldehyde would be expected to bind into a β-cyclodextrin cavity, so we attached a thiazolium salt to a primary carbon of β-cyclodextrin and examined it as a catalyst.[84] We found that this was not a better catalyst for the benzoin condensation, apparently because there was no room in the β-cyclodextrin cavity for the binding of two benzaldehyde molecules. However it was clear that at least the reaction intermediate was being formed; we got very rapid tritium exchange from the aldehyde by formation of the thiazolium adduct, and as well a very rapid oxidation of *para-t*-butyl benzaldehyde by ferricyanide ion since it was able to oxidize the reaction intermediate formed when the bound *t*-butyl benzaldehyde underwent addition of the thiazolium ring.

We then prepared essentially the same compound but based it on γ-cyclodextrin, the molecule with eight glucose residues that has a cavity large enough to hold two benzaldehyde groups.[85] This gave a 7-fold to 9-fold acceleration of the formation of benzoin relative to a simple thiazolium salt without the attached γ-cyclodextrin binding group. In subsequent work we have shown that in this system the use of the polyamine with a hydrophobic core as a catalyst has led to even larger rate accelerations, up to 2000-fold.

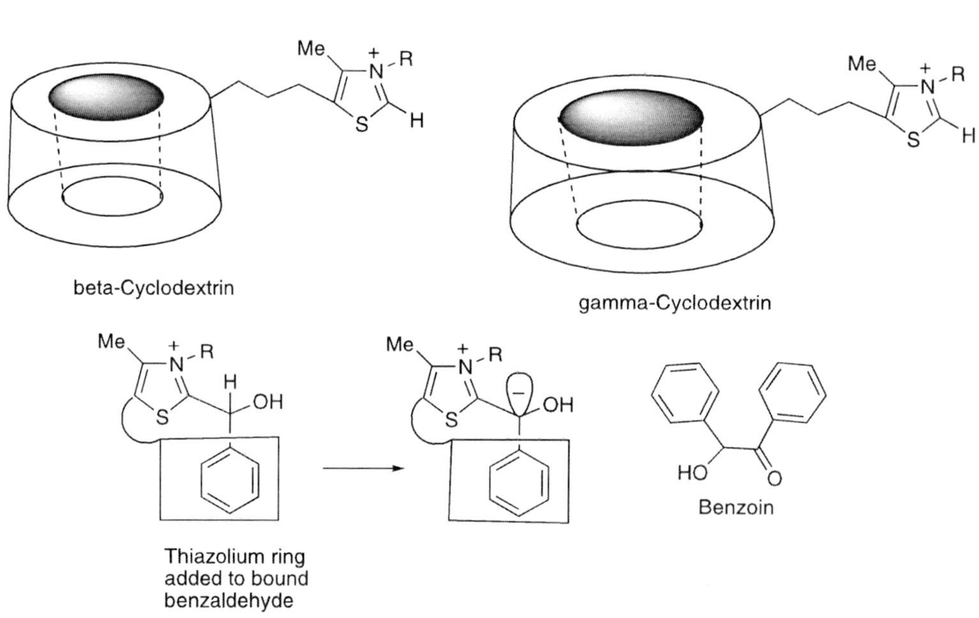

Figure 2.10 *Thiazolium salts catalyse the benzoin condensation, and with an extra rate when they are linked to γ-cyclodextrin, which can bind two benzaldehydes into its large cavity*

2.8 Aldol Condensations Catalysed by Cyclodextrin Derivatives

We prepared a series of catalysts with relatively simple groups attached to a cyclodextrin that would be able to catalyse an aldol condensation between acetone and some bound aromatic aldehydes, when the acetone formed an enamine with a catalytic group such as a simple amine.[86] We observed catalyses as high as 270-fold relative to the uncatalysed aldol condensation with some of these cyclodextrin derivatives.

We also took advantage of our synthesis of three isomers of β-cyclodextrin *bis*-imidazoles – those with the imidazoles on neighboring glucose units (AB), with one intervening glucose unit (AC) and with two intervening glucoses (AD) – to examine the detailed geometry involved in the enolization of ketones, in which a proton is removed from an α carbon by a general base while another proton is added to the carbonyl oxygen by a general acid. We examined the rate of exchange of the methyl group of *p*-*t*-butylacetophenone with deuterium in D_2O catalysed by these three *bis*-imidazole isomers and by β-cyclodextrin 6-monotosylate.[87]

The best catalyst was the AD isomer, the reverse of what we had seen in our ribonuclease model studies. The pH/rate profile indicated that there was bifunctional catalysis by imidazole and imidazolium groups. This indicated that proton removal has a preferential direction of attack by base that is not directly along the C-H bond line, but instead pushes the electrons toward the carbonyl group (Figure 2.11). Such subtle information is available only because of the defined geometry in molecular complexes of cyclodextrin *bis*-imidazoles.

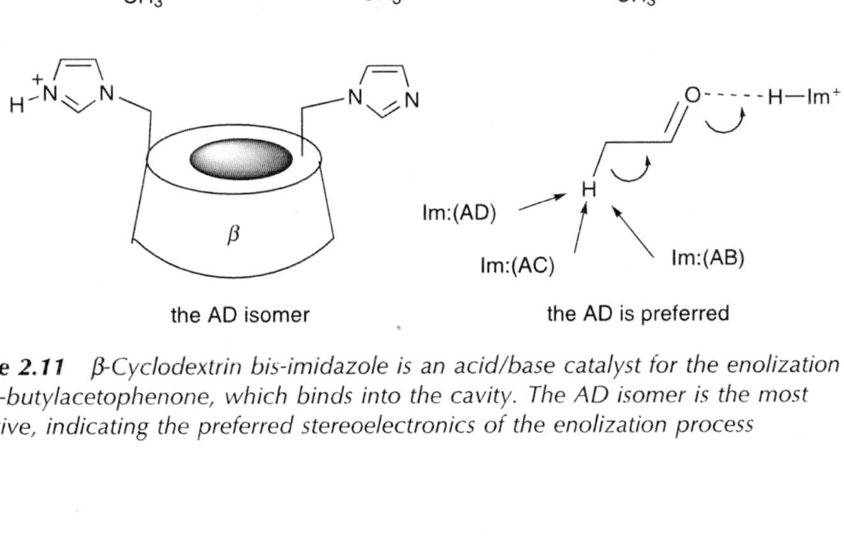

Figure 2.11 β-Cyclodextrin bis-imidazole is an acid/base catalyst for the enolization of p-tert-butylacetophenone, which binds into the cavity. The AD isomer is the most effective, indicating the preferred stereoelectronics of the enolization process

Subsequently we used the β-cyclodextrin *bis*-imidazoles to perform aldol condensations. In our first such study we examined the aldol condensation of ketoaldehyde **16**.[88] We saw that the best catalyst for this process was again the AD isomer of the *bis*-imidazole, but there were some special features. First of all, the catalyst promoted deuterium exchange into the CH_2 next to the ketone group, so enolization was reversible and the addition of this enol to the aldehyde group was rate determining. Thus the bifunctional catalyst not only speeded the enolization to the point at which it was reversible and not rate-determining, it also catalysed the addition of the enol to the aldehyde group. Furthermore, the catalyst reversed the selectivity of enolization, promoting enolization next to the accessible keto group even more rapidly than next to the normally more reactive aldehyde group.

We also examined aldol condensations of the dialdehyde **17**.[89] Without the special catalysis afforded by the cyclodextrin *bis*-imidazoles there was an almost random reaction to form compounds **18** and **19**, as either aldehyde acted as the enolizing group. However, the cyclodextrin imidazole catalysts directed the selective formation of products **19**, with no selectivity among its stereoisomers. Interestingly, the least selective catalyst for **17** was the cyclodextrin mono-imidazole, the AD isomer of the *bis*-imidazole was more selective, and the most selective was the AB isomer. Obviously these results indicate that the cyclodextrin imidazole catalysts promote enolization of the aldehyde group closest to the cyclodextrin, as expected, but the subtlety of preferences among the *bis*-imidazole isomers is not yet understood in this case.

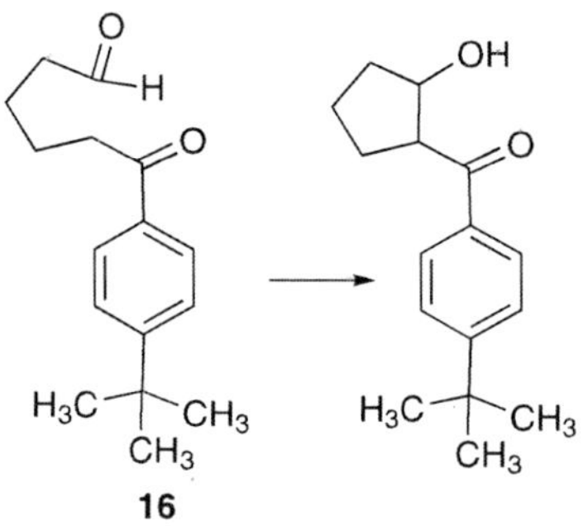

16

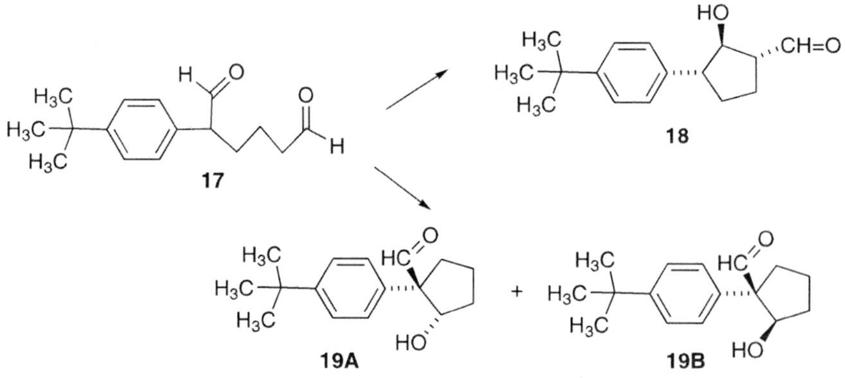

17 **18** **19A** **19B**

2.9 Mimics of Enzymes Using Coenzyme B_{12} as a Cofactor

Coenzyme B_{12} is involved in some remarkable catalytic processes, including in particular the reversible conversion of a succinic acid derivative to a methylmalonyl derivative and the reversible conversion of *β*-methylaspartic acid to glutamic acid. In these reactions vitamin B_{12} is covalently linked at its cobalt atom with an adenosine group, and the catalytic chemistry involves dissociation of the carbon-cobalt bond and subsequent reactions promoted by the resulting adenosyl radical. In our first study we synthesized a compound (**20**) in which the cobalt of B_{12} is covalently linked to one of the primary methylene groups of cyclodextrin.[90] We saw that the carbon-cobalt bond easily dissociated to form a cyclodextrinyl radical. This could then form a substrate radical by group transfer from a substrate bound in the cyclodextrin cavity, although not yet by hydrogen transfer.

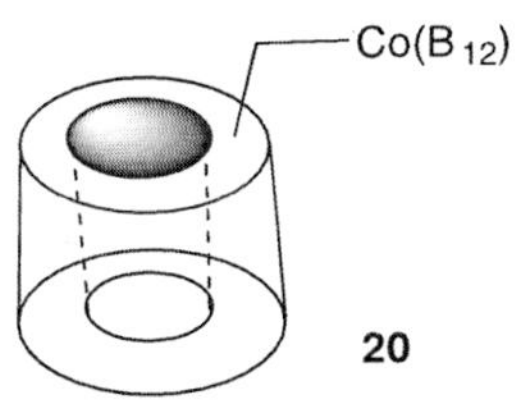

20

In a subsequent paper we linked B_{12} to the cyclodextrin not with a carbon–cobalt bond but with a bond to one of the B_{12} side chains.[91] We examined the rearrangement and reaction of a substrate that could bind into the cyclodextrin cavity, and for which the B_{12} component could initiate a free radical process with a rearrangement related to those one sees in the biochemical processes. Some such preference was seen for the binding substrate compared to the non-binding substrate, indicating that this was a promising beginning to an interesting approach to B_{12} mimics.

2.10 Mimics of Cytochrome P-450

Cytochrome P-450 describes a group of enzymes that use a heme group to perform oxidations, including in particular oxidations of carbon-hydrogen bonds. Many model systems had been examined for such processes, but they had not included any defined binding of a substrate in such a way as to achieve pre-selected functionalization of particular carbon atoms. Thus we initiated a series of studies of such systems in which we used cyclodextrins as binding groups for the substrates to hold them, in water, in such a position as to achieve selective functionalizations.

In general we used steroids as the substrates. They are rigid and make it easier to achieve selectivity without concern about the substrate conformation, and the selective oxidation of steroids in particular positions is actually of practical interest. Currently there are steroid hormones that are produced by biochemical oxidation using the enzymes P-450, but we wished to produce mimics that would not have the disadvantage of enzymatic reactions – with their problems in dealing with proteins and fermentation.

Cytochrome P-450 models are generally used to insert an oxygen atom into either a double bond or a carbon-hydrogen bond, although we also demonstrated that such a model could insert nitrogen atoms instead, producing interesting and selective ways of synthesizing quite useful compounds.[92] We also showed that such nitrogen insertion can be done by the actual biological enzyme itself under some conditions.[93]

We synthesized metalloporphyrins (**21**) carrying four β-cyclodextrin rings attached to phenyl groups on the meso carbons of the porphyrin, and examined their use for selective oxidations. In our first work we used this system, as its manganese complex, to insert oxygens into double bonds of substrate that could doubly bind to cyclodextrins across the center of the porphyrin ring.[94] We saw quite good selectivity. As oxidant we use iodosobenzene as an oxygen transfer species, putting the oxygen on the manganese which then transferred it to the substrate. We also added a coordinating bulky carboxylic acid to the system so that it would bind to one face of the manganese, and direct both the substrate and the oxygen to the other face.

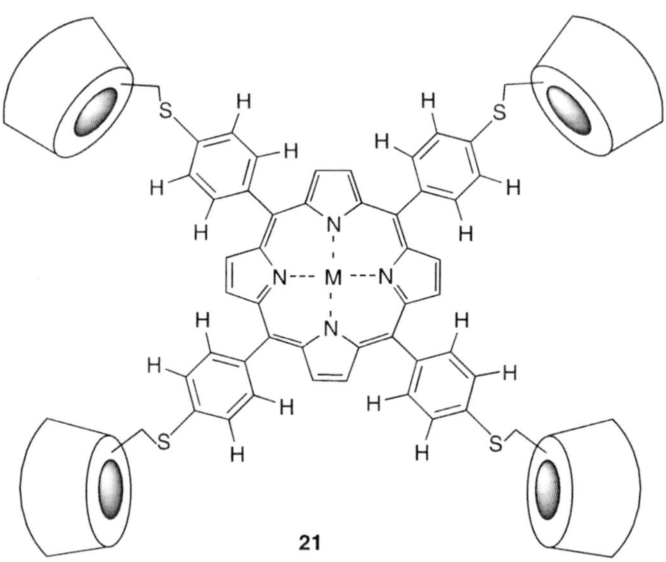

21

We then used this same manganese porphyrin system carrying four β-cyclodextrins to perform the hydroxylation of a saturated carbon in a steroid substrate.[95,96] The steroid (**22**) had binding groups attached to positions 3 and 17 in rings A and D, so it could dissolve in water and use these binding groups to insert into cyclodextrins on opposite sides of the porphyrin system. The catalyst was remarkably selective, inserting oxygen only into C-6 of the steroid to make a steroidal alcohol (**23**). Furthermore, this alcohol was not further oxidized to a carbonyl group, which would normally be an easier oxidation than the insertion into a C–H bond. However, molecular models indicated that we could insert into the α-face of the steroid generating the α C-6 alcohol, but that we could not reach the axial β-hydrogen on that carbon. Apparently the oxidation of such alcohols to ketones by this oxidation class occurs by hydrogen abstraction from the carbinol carbon, and that is geometrically impossible in this case.

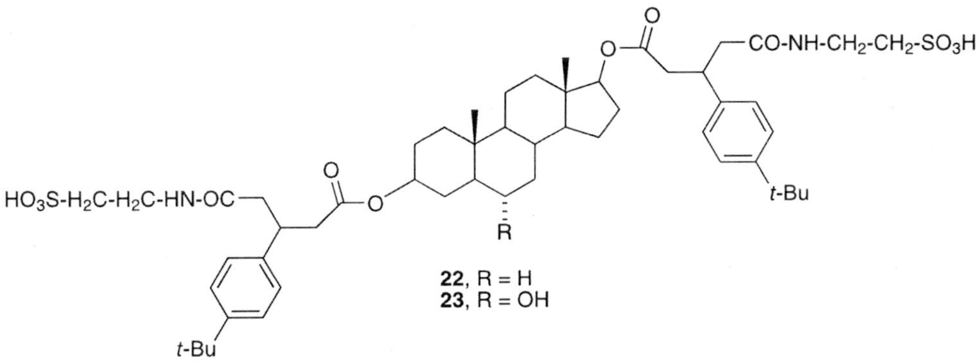

22, R = H
23, R = OH

One problem with our system was that the catalyst itself was oxidized fairly easily, so we did not get many catalytic turnovers. Others working on cytochrome models had shown that the catalyst was much more oxidatively stable when the phenyl rings were perfluorinated. Thus we synthesized a perfluorinated version (**24**) of our previous compound, easily prepared from the porphyrin with four pentafluorophenyl groups by simple reaction with β-cyclodextrin 6-thiol, which selectively replaced the *para* fluorines in the phenyl rings.[97] While with the previous catalyst we only achieved three to five turnovers before the catalyst was destroyed, the new fluorinated catalyst performed the same selective hydroxylation of C-6 in the steroid with 187 turnovers.

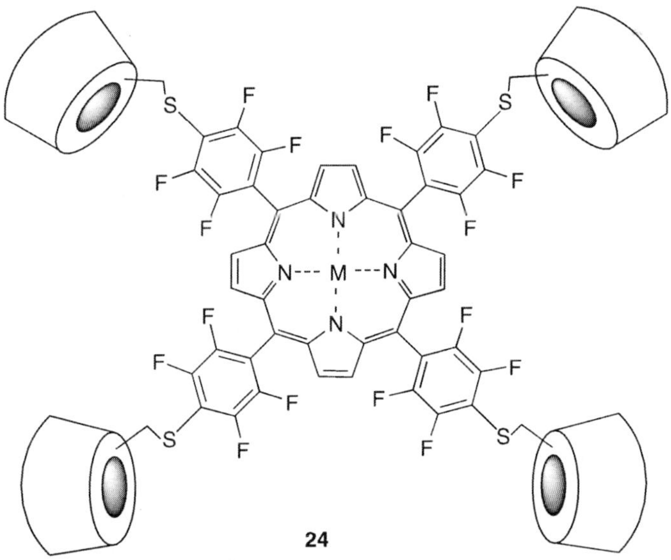

24

The selective hydroxylation at C-6 of an androstane ring was striking, but not necessarily the most desired target. We really wanted to functionalize either C-9 or C-11 in order to be able to convert a simple steroid to a corticosteroid, a process that is currently performed by biological fermentation. Models indicated that if we could get triple binding of a steroid derivative with binding groups at carbons 3, 6 and 17, the hydrogen on carbon 9 would be in reach of the oxygen atom on the manganese, so we examined this system.

We synthesized the substrate (**25**) using the C-6 hydroxyl group that we had previously inserted, and performed its oxidation with our fluorinated catalyst. Indeed we saw that the result was the product **26** with the selective hydroxylation of C-9 – in line with molecular models indicating that the preferred geometry of the catalyst/substrate complex put the hydrogen at C-9 in van der Waals contact with the oxygen on the manganese atom.[98] We described these studies in more detail and indicated that in some cases as many as 3,000 catalytic turnovers were achieved.[99]

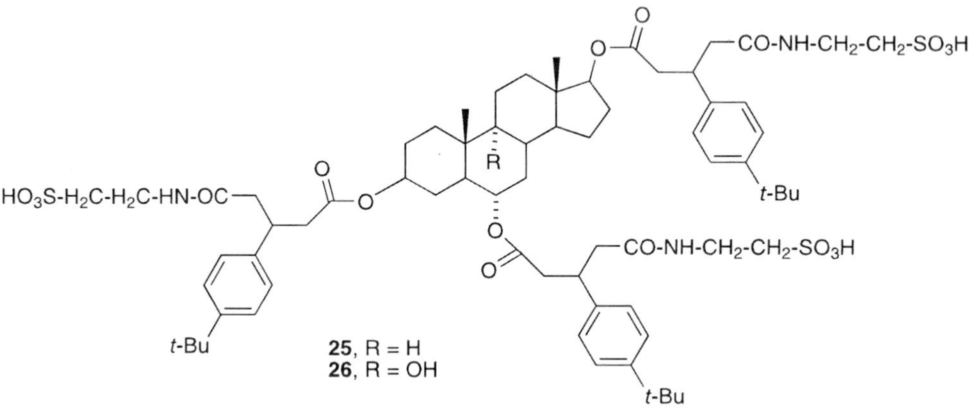

25, R = H
26, R = OH

Models suggested that we could achieve the desired hydroxylation at C-9 using only two binding groups at carbons 3 and 6 of the steroid, but when this was tried with our original catalyst we saw that we obtained not only hydroxylation at C-9 but also some hydroxylation at C-15. This indicated that we had to adjust the geometry of the complex a bit by using a related molecule in which the cyclodextrins were attached to the meta positions of the phenyls in the porphyrin, not the *para* positions. To achieve the same geometry with an easier synthesis we prepared the porphyrin carrying tetrafluoropyridine rings instead of phenyl rings. The cyclodextrins were then attached by substitution next to the nitrogen of the pyridine ring, achieving the desired geometry. We found that with new catalyst **27** we were able to achieve selective hydroxylation of C-9 starting with oxygens only at C-3 and C-6 of the steroid in a doubly bound – not triply bound – complex.[100] The entire system was described further in a full paper summarizing the work with iodosobenzene as an oxidant and porphyrins carrying four β-cyclodextrin units as the catalyst.[101]

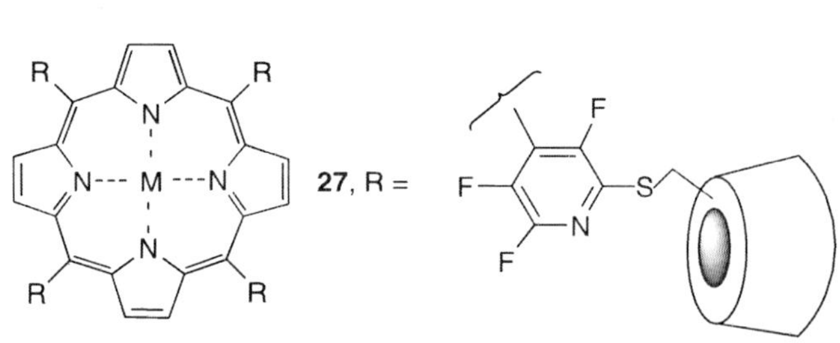

The use of iodosobenzene is not attractive, but our experiments with other oxidants and our original catalyst system indicated that we were not successful in using simple oxidants such as hydrogen peroxide or potassium persulfate. Such oxidants had been used previously in oxidations by metalloporphyrins, but not in an aqueous system. However we needed the water system both because we were interested in mimicking such aqueous chemistry and also because we needed the hydrophobic binding of the substrates into the catalyst cyclodextrin units. In the actual enzyme there is a thiol ligand coordinated to the iron in the porphyrin ring, and other work suggested that a thiol ligand as the thiolate anion made the porphyrin system a stronger reducing agent able to accept oxygen from simpler oxygen donor species. Thus we synthesized a system in which we coordinated a phenylthiolate anion to the manganese of the porphyrin and saw that we were able to use hydrogen peroxide as the effective oxidant in our water system.[102] We did this in two ways.

In one approach we synthesized a compound with a covalently linked bridge across the porphyrin carrying the phenyl thiol group, and with two cyclodextrins on the other two positions. In the other approach we synthesized a phenylthiol derivative with two linkers to binding groups, and used it with our original fluorinated porphyrin. It bound to two cyclodextrins on one face of the porphyrin, while the substrate bound to the other two cyclodextrins. This is much like the enzyme in which the thiol comes from a cysteine of the protein, not from a covalent link to the porphyrin. In both cases we saw that we could now use hydrogen peroxide as the oxidant for our selective C-6 hydroxylation. Thus a major problem in the previous work – the use of an expensive oxidant – was solved.

Acknowledgements

I gratefully acknowledge the experimental and intellectual contributions of my many coworkers, who are named in the references. Financial support of our work was provided over the years by the NIH and NSF.

References

1. F. Cramer, W. Saenger, H. C. Spatz, Inclusion compounds. XIX. The formation of inclusion compounds of α-cyclodextrin in aqueous solutions. Thermodynamics and kinetics, *J. Am. Chem. Soc.*, 1967, **89**, 14–20.
2. R. L. VanEtten, G. A. Clowes, J. F. Sebastian, M. L. Bender, The mechanism of the cycloamylose-accelerated cleavage of phenyl esters, *J. Am. Chem. Soc.*, 1967, **89**, 3253–3262.
3. R. L. VanEtten, J. F. Sebastian, G. A. Clowes, M. L. Bender, Acceleration of phenyl ester cleavage by cycloamyloses. A model for enzymic specificity, *J. Am. Chem. Soc.*, 1967, **89**, 3242–3253.
4. R. Breslow, Adjusting the lock and adjusting the key, in *Biomimetic Chemistry*, D. Dolphin, C. McKenna, Y. Murakami, I. Tabushi (Eds) Am. Chem. Soc., Washington D.C., 1–15, 1980.
5. R. Breslow, Enzyme models related to inclusion compounds, in *Inclusion Compounds*, J. L. Atwood, J. E. Davies (Eds) Academic Press, Inc., Orlando, FL, Vol. 3, 473–508, 1984.
6. R. Breslow, Attempts to mimic enzymes – report from the battle front, in *Chemical Reactions in Organic and Inorganic Constrained Systems*, R. Setton, (Ed.) D. Reidel Publishing Company, 17–28, 1986.
7. R. Breslow, Artificial enzymes and enzyme models, in *Advances in Enzymology and Related Areas of Molecular Biology*, A. Meister (Ed.) John Wiley & Sons, Inc., **58**, 1–60, 1986.

8. R. Breslow, Enzyme mimics, *Ciba Foundation Symposium* 1991, **158**, 115–127.
9. R. Breslow, Bifunctional binding and catalysis in host-guest chemistry, *Israel Jour. Chem.*, 1992, **32**, 23–30.
10. R. Breslow, Selective binding and catalysis by cyclodextrins, *Minutes of The Sixth International Symposium on Cyclodextrins*, Editions de Santé, Paris, 625–630 1992.
11. R. Breslow, Molecular recognition, *Proc. Natl. Acad. Sci. USA*, 1993, **90**, 1139.
12. R. Breslow, Bifunctional binding and catalysis, *Supramol. Chem.*, 1993, **1**, 111–118.
13. R. Breslow, Biomimetic chemistry and artificial enzymes: catalysis by design, *Acc. Chem. Res.*, 1995, **28**, 146–153.
14. R. Breslow, S. D. Dong, Biomimetic reactions catalysed by cyclodextrins and their derivatives, *Chem. Rev.*, 1998, **98**, 1997–2011.
15. R. Breslow, S. Belvedere, L. Gershell, D. Leung, The chelate effect in binding, catalysis, and chemotherapy, *Pure Appl. Chem.*, 2000, **72**, 333–342.
16. R. Breslow, Cyclodextrin derivatives as mimics of enzymes and antibodies, *Biolog. J. Armenia*, 2001, **53**, 117–120.
17. C. J. Easton, Cyclodextrins as molecular reactors, next chapter in this book.
18. K. Takahashi, Reactions mediated by cyclodextrins, the chapter after next in this book.
19. B. Siegel, R. Breslow, Lyophobic binding of substrates by cyclodextrins in nonaqueous solvents, *J. Am. Chem. Soc.*, 1975, **97**, 6869–6870.
20. M. F. Czarniecki, R. Breslow, Very fast acylation of *β*-cyclodextrin by bound *p*-nitrophenyl ferrocinnamate, *J. Am. Chem. Soc.*, 1978, **100**, 7771–7772.
21. J. Emert, R. Breslow, Modification of the cavity of *β*-cyclodextrin by flexible capping, *J. Am. Chem. Soc.*, 1975, **97**, 670–672.
22. R. Breslow, M. F. Czarniecki, J. Emert, H. Hamaguchi, Improved acylation rates within cyclodextrin complexes from flexible capping of the cyclodextrin and from adjustment of the substrate geometry, *J. Am. Chem. Soc.*, 1980, **102**, 762–770.
23. G. Trainor, R. Breslow, High acylation rates and enantioselectivity with cyclodextrin complexes of rigid substrates, *J. Am. Chem. Soc.*, 1981, **103**, 154–158.
24. W. J. le Noble, S. Srivastava, R. Breslow, G. Trainor, Effect of pressure on two cyclodextrin-promoted ester hydrolyses, *J. Am. Chem. Soc.*, 1983, **105**, 2745–2748.
25. H.-J. Thiem, M. Brandl, R. Breslow, Molecular modeling calculations on the acylation of *β*-cyclodextrin by ferroceneacrylate esters, *J. Am. Chem. Soc.*, 1988, **110**, 8612–8616.
26. R. Breslow, S. Chung, Additional flexibility solves the leaving group problem in cyclodextrin acylation, *Tetrahedron Lett.*, 1990, **31**, 631–634.
27. F. M. Menger, M. Ladika, Origin of rate accelerations in an enzyme model: the *p*-nitrophenyl ester syndrome. *J. Am. Chem. Soc.*, 1987, **109**, 3145–3146.
28. W. J. le Noble, S. Srivastava, R. Breslow, G. Trainor, Effect of pressure on two cyclodextrin-promoted ester hydrolyses, *J. Am. Chem. Soc.*, 1983, **105**, 2745–2748.
29. R. Breslow, P. Campbell, Selective aromatic substitution within a cyclodextrin mixed complex, *J. Am. Chem. Soc.*, 1969, **91**, 3085.
30. R. Breslow, P. Campbell, Selective aromatic substitution by hydrophobic binding of a substrate to a simple cyclodextrin catalyst, *Bioorg. Chem.*, 1971, **1**, 140–156.
31. R. Breslow, H. Kohn, B. Siegel, Methylated cyclodextrin and a cyclodextrin polymer as catalysts in selective anisole chlorination, *Tetrahedron Lett.*, 1976, 1645–1646.
32. D. Rideout, R. Breslow, Hydrophobic acceleration of Diels–Alder reactions, *J. Am. Chem. Soc.*, 1980, **102**, 7816–7817.
33. R. Breslow, T. Guo, Diels–Alder reactions in nonaqueous polar solvents. kinetic effects of chaotropic and antichaotropic agents and of *β*-cyclodextrin, *J. Am. Chem. Soc.*, 1988, **110**, 5613–5617.
34. B. Siegel, A. Pinter, R. Breslow, Synthesis of cycloheptaamylose 2–, 3–, and 6–phosphoric acids, and a comparative study of their effectiveness as general acid or general base catalysts with bound substrates, *J. Am. Chem. Soc.*, 1977, **99**, 2309–2312.
35. R. Breslow, L. E. Overman, An 'artificial enzyme' combining a metal catalytic group and a hydrophobic binding cavity, *J. Am. Chem. Soc.*, 1970, **92**, 1075–1077.

36. R. Breslow, B. Zhang, Very fast ester hydrolysis by a cyclodextrin dimer with a catalytic linking group, *J. Am. Chem. Soc.*, 1992, **114**, 5882–5883.
37. B. Zhang, R. Breslow, Enthalpic domination of the chelate effect in cyclodextrin dimers, *J. Am. Chem. Soc.*, 1993, **115**, 9353–9354.
38. R. Breslow, B. Zhang, Cleavage of phosphate esters by a cyclodextrin dimer catalyst that binds the substrates together with La^{3+} and hydrogen peroxide, *J. Am. Chem. Soc.*, 1994, **116**, 7893–7894.
39. B. Zhang, R. Breslow, Ester hydrolysis by a catalytic cyclodextrin dimer enzyme mimic with a metallobipyridyl linking group, *J. Am. Chem. Soc.*, 1997, **119**, 1676–1681.
40. R. Breslow, N. Nesnas, Burst kinetics and turnover in an esterase mimic, *Tetrahedron Lett.*, 1999, **40**, 3335–3338.
41. J. Yan, R. Breslow, An enzyme mimic that hydrolyzes an unactivated ester with catalytic turnover, *Tetrahedron Lett.*, 2000, **41**, 2059–2062.
42. R. Breslow, J. Doherty, G. Guillot, C. Lipsey, β-cyclodextrinyl-bisimidazole, a model for ribonuclease, *J. Am. Chem. Soc.*, 1978, **100**, 3227–3229.
43. I. Tabushi, K. Shimokawa, N. Shimizu, H. Shirakata, K. Fujita, Capped cyclodextrin, *J. Am. Chem. Soc.*, 1976, **98**, 7855–7856.
44. R. Breslow, P. Bovy, C. Lipsey Hersh, Reversing the selectivity of cyclodextrin bisimidazole ribonuclease mimics by changing the catalyst geometry, *J. Am. Chem. Soc.*, 1980, **102**, 2115–2117.
45. E. Anslyn, R. Breslow, Geometric evidence on the ribonuclease model mechanism, *J. Am. Chem. Soc.*, 1989, **111**, 5972–5973.
46. R. Breslow, S. D. Dong, Y. Webb, R. Xu, Further studies on the buffer-catalysed cleavage and isomerization of uridyluridine. Medium and ionic strength effects on catalysis by morpholine, imidazole, and acetate buffers help clarify the mechanisms involved and their relationship to the mechanism used by the enzyme ribonuclease and by a ribonuclease mimic, *J. Am. Chem. Soc.*, 1996, **118**, 6588–6600.
47. E. Anslyn, R. Breslow, Proton inventory of a bifunctional ribonuclease model, *J. Am. Chem. Soc.*, 1989, **111**, 8931–8932.
48. R. Breslow, E. Anslyn, D.-L. Huang, Ribonuclease mimics, *Tetrahedron*, 1991, **47**, 2365–2376.
49. R. Breslow, How do imidazole groups catalyse the cleavage of RNA in enzyme models and in enzymes? Evidence from 'negative catalysis', *Acc. Chem. Res.*, 1991, **24**, 317–324.
50. R. Breslow, Bifunctional acid-base catalysis by imidazole groups in enzyme mimics, *J. Mol. Cat.*, 1994, **91**, 161–174.
51. R. Breslow, S. D. Dong, Y. Webb, R. Xu, Further studies on the buffer-catalysed cleavage and isomerization of uridyluridine. Medium and ionic strength effects on catalysis by morpholine, imidazole, and acetate buffers help clarify the mechanisms involved and their relationship to the mechanism used by the enzyme ribonuclease and by a ribonuclease mimic, *J. Am. Chem. Soc.*, 1996, **118**, 6588–6600.
52. R. Breslow, C. Schmuck, Goodness of fit in complexes between substrates and ribonuclease mimics: effects on binding, catalytic rate constants, and regiochemistry, *J. Am. Chem. Soc.*, 1996, **118**, 6601–6605.
53. R. Breslow, D.-L. Huang, E. Anslyn, On the mechanism of action of ribonucleases: dinucleotide cleavage catalysed by imidazole and Zn^{2+}, *Proc. Natl. Acad. Sci. USA*, 1989, **86**, 1746–1750.
54. S. D. Dong, R. Breslow, Bifunctional cyclodextrin metalloenzyme mimics, *Tetrahedron Lett.*, 1998, **39**, 9343–9346.
55. Z. Yang, R. Breslow, Very strong binding of lithocholic acid to β-cyclodextrin, *Tetrahedron Lett.*, 1997, **38**, 6171–6172.
56. R. Breslow, N. Greenspoon, T. Guo, R. Zarzycki, Very strong binding of appropriate substrates by cyclodextrin dimers, *J. Am. Chem. Soc.*, 1989, **111**, 8296–8297.
57. R. Breslow, B. Zhang, Cholesterol recognition and binding by cyclodextrin dimers, *J. Am. Chem. Soc.*, 1996, **118**, 8495–8496.

58. R. Breslow, S. Chung, Strong binding of ditopic substrates by a doubly linked occlusive C1 'clamshell' as distinguished from an aversive C2 'loveseat' cyclodextrin dimer, *J. Am. Chem. Soc.*, 1990, **112**, 9659–9660.
59. R. Breslow, S. Halfon, Quantitative effects of antihydrophobic agents on binding constants and solubilities in water, *Proc. Natl. Acad. Sci. USA*, 1992, **89**, 6916–6918.
60. B. Zhang, R. Breslow, Enthalpic domination of the chelate effect in cyclodextrin dimers, *J. Am. Chem. Soc.*, 1993, **115**, 9353–9354.
61. R. Breslow, S. Halfon, B. Zhang, Molecular recognition by cyclodextrin dimers, *Tetrahedron*, 1995, **51**, 377–388.
62. D. K. Leung, J. H. Atkins, R. Breslow, Synthesis and binding properties of cyclodextrin trimers, *Tetrahedron Lett.*, 2001, **42**, 6255–6258.
63. A. Ruebner, Z. Yang, D. Leung, R. Breslow, A cyclodextrin dimer with a photocleavable linker as a possible carrier for the photosensitizer in photodynamic tumor therapy, *Proc. Natl. Acad. Sci. USA*, 1999, **96**, 14692–14693.
64. S. P. D. Baugh, Z. Yang, D. K. Leung, D. M. Wilson, R. Breslow, Cyclodextrin dimers as cleavable carriers of photodynamic sensitizers, *J. Am. Chem. Soc.*, 2001, **123**, 12488 –12494.
65. M. Maletic, H. Wennemers, D. Q. McDonald, R. Breslow, W. C. Still, Selective binding of the dipeptides L-Phe-D-Pro and D-Phe-L-Pro to β-cyclodextrin, *Angew. Chem. Int. Ed. Engl.*, 1996, **35**, 1490–1492.
66. D. K. Leung, Z. Yang, R. Breslow, Selective disruption of protein aggregation by cyclodextrin dimers, *Proc. Natl. Acad. Sci. USA*, 2000, **97**, 5050–5053.
67. S. Moore, J. A. Askew, G. L. Gibson, A. Aborgrein, O. El-Agnaf, D. Allsop, D. K. Leung, R. Breslow, Novel cyclodextrin dimers and trimers as inhibitors of amyloid peptide aggregation, *Neurobiology of Aging*, 2002, **23**, S105–397.
68. D. Wilson, L. Perlson, R. Breslow, Helical templating of oligopeptides by cyclodextrin dimers, *Bioorg. Med. Chem.*, 2003, **11**, 2649–2653.
69. R. Breslow, M. Hammond, M. Lauer, Selective transamination and optical induction by a β-cyclodextrin-pyridoxamine artificial enzyme, *J. Am. Chem. Soc.*, 1980, **102**, 421–422.
70. A. Ueno, R. Breslow, Selective sulfonation of a secondary hydroxyl group of β-cyclodextrin, *Tetrahedron Lett.*, 1982, **23**, 3451–3454.
71. R. Breslow, A. W. Czarnik, Transaminations by pyridoxamine selectively attached at C-3 in β-cyclodextrin, *J. Am. Chem. Soc.*, 1983, **105**, 1390–1391.
72. A. W. Czarnik, R. Breslow, Hexadeoxycycloheptaamylose-pyridoxamine, an artificial transaminase with a 'deeper' binding pocket, *Carbohydr. Res.*, 1984, **128**, 133–139.
73. R. Breslow, A. W. Czarnik, M. Lauer, R. Leppkes, J. Winkler, S. Zimmerman, Mimics of transaminase enzymes, *J. Am. Chem. Soc.*, 1986, **108**, 1969–1979.
74. R. Breslow, J. Chmielewski, D. Foley, B. Johnson, N. Kumabe, M. Varney, R. Mehra, Optically active amino acid synthesis by artificial transaminase enzymes, *Tetrahedron*, 1988, **44**, 5515–5524.
75. I. Tabushi, Y. Kuroda, M. Yamada, H. Higashimura, R. Breslow, 'A-(Modified B-6)-B[omega-amino(ethylamino)]-β-cyclodextrin as an artificial B-6 enzyme for chiral aminotransfer reaction', *J. Am. Chem. Soc.*, 1985, **107**, 5545–5546.
76. R. Breslow, J. W. Canary, M. Varney, S. T. Waddell, D. Yang, Artificial transaminases linking pyridoxamine to binding cavities: controlling the geometry, *J. Am. Chem. Soc.*, 1990, **112**, 5212–5219.
77. R. Breslow, J. Chmielewski, D. Foley, B. Johnson, N. Kumabe, M. Varney, R. Mehra, Optically active amino acid synthesis by artificial transaminase enzymes, *Tetrahedron*, 1988, **44**, 5515–5524.
78. E. Fasella, S. D. Dong, R. Breslow, Reversal of optical induction in transamination by regioisomeric bifunctionalized cyclodextrins, *Bioorg. Med. Chem.*, 1999, **7**, 709–714.
79. W. Weiner, J. Winkler, S. C. Zimmerman, A. W. Czarnik, R. Breslow, Mimics of tryptophan synthetase and of biochemical dehydroalanine formation, *J. Am. Chem. Soc.*, 1985, **107**, 4093–4094.
80. J. T. Koh, L. Delaude, R. Breslow, Geometric control of a pyridoxal-catalysed aldol condensation, *J. Am. Chem. Soc.*, 1994, **116**, 11234–11240.

81. W. Zhou, L. Liu, R. Breslow, Transamination by polymeric enzyme mimics, *Helv. Chim. Acta*, 2003, **86**, 3560–3567.
82. L. Liu, W. Zhou, J. J. Chruma, R. Breslow Transamination reactions with multiple turnovers catalysed by hydrophobic pyridoxamine cofactors in the presence of polyethylenimine polymers, *J. Am. Chem. Soc.*, 2004, **126**, 8136–8137.
83. R. Breslow, On the mechanism of thiamine action. IV. Evidence from studies on model systems, *J. Am. Chem. Soc.*, 1958, **80**, 3719–3726.
84. D. Hilvert, R. Breslow, Functionalized cyclodextrins as holoenzyme mimics of thiamine-dependent enzymes, *Bioorg. Chem.*, 1984, **12**, 206–220.
85. R. Breslow, E. Kool, A γ-cyclodextrin thiazolium salt holoenzyme mimic for the benzoin condensation, *Tetrahedron Lett.*, 1988, **29**, 1635–1638.
86. D.-Q. Yuan, S. D. Dong, R. Breslow, Cyclodextrin-based class I aldolase enzyme mimics to catalyse crossed aldol condensations, *Tetrahedron Lett.*, 1998, **39**, 7673–7676.
87. R. Breslow, A. Graff, Geometry of enolization using a bifunctional cyclodextrin-based catalyst, *J. Am. Chem. Soc.*, 1993, **115**, 10988–10989.
88. J. M. Desper, R. Breslow, Catalysis of an intramolecular aldol condensation by imidazole-bearing cyclodextrins, *J. Am. Chem. Soc.*, 1994, **116**, 12081–12082.
89. R. Breslow, J. Desper, Y. Huang, A selective intramolecular aldol condensation directed by a bifunctional enzyme mimic, *Tetrahedron Lett.*, 1996, **37**, 2541–2544.
90. R. Breslow, P. J. Duggan, J. P. Light, Cyclodextrin-B_{12}, a potential enzyme-coenzyme mimic, *J. Am. Chem. Soc.*, 1992, **114**, 3982–3983.
91. M. Rezac, R. Breslow, A mutase mimic with cobalamin linked to cyclodextrin, *Tetrahedron Lett.*, 1997, **38**, 5763–5766.
92. R. Breslow, S. H. Gellman, Tosylamidation of cyclohexane by a cytochrome P-450 model', *J. Chem. Soc. Chem. Comm.*, 1982, 1400–1401.
93. E. W. Svastits, J. H. Dawson, R. Breslow, S. Gellman, 'Functionalized nitrogen atom transfer catalysed by cytochrome P-450', *J. Am. Chem. Soc.*, 1985, **107**, 6427–6428.
94. R. Breslow, X. Zhang, R. Xu, M. Maletic, R. Merger, Selective catalytic oxidation of substrates that bind to metalloporphyrin enzyme mimics carrying two or four cyclodextrin groups and related metallosalens, *J. Am. Chem. Soc.*, 1996, **118**, 11678–11679.
95. R. Breslow, X. Zhang, Y. Huang, Selective catalytic hydroxylation of a steroid by an artificial cytochrome P-450 enzyme', *J. Am. Chem. Soc.*, 1997, **119**, 4535–4536.
96. R. Breslow, Y. Huang, X. Zhang, J. Yang, An artificial cytochrome P450 that hydroxylates unactivated carbons with regio- and stereoselectivity and useful catalytic turnovers, *Proc. Natl. Acad. Sci. USA*, 1997, **94**, 11156–11158.
97. R. Breslow, B. Gabriele, J. Yang, Geometrically directed selective steroid hydroxylation with high turnover by a fluorinated artificial cytochrome P-450, *Tetrahedron Lett.*, 1998, **39**, 2887–2890.
98. J. Yang, R. Breslow, Selective hydroxylation of a steroid at C-9 by an artificial cytochrome P-450, *Angew. Chem. Int. Ed.*, 2000, **39**, 2692–2694.
99. R. Breslow, J. Yang, J. Yan, Biomimetic hydroxylation of saturated carbons with artificial cytochrome P-450 enzymes – liberating chemistry from the tyranny of functional groups, *Tetrahedron*, 2002, **58**, 653–659.
100. R. Breslow, J. Yan, S. Belvedere, Catalytic hydroxylation of steroids by cytochrome P-450 mimics. Hydroxylation at C-9 with novel catalysts and steroid substrates, *Tetrahedron Lett.*, 2002, **43**, 363–365.
101. J. Yang, B. Gabriele, S. Belvedere, Y. Huang, R. Breslow, Catalytic oxidations of steroid substrates by artificial cytochrome P-450 enzymes, *J. Org. Chem.*, 2002, **67**, 5057–5067.
102. Z. Fang, R. Breslow, A thiolate ligand on a cytochrome P-450 mimic permits the use of simple environmentally benign oxidants for biomimetic steroid hydroxylation in water, *Bioorg. Med. Chem. Lett.*, 2005, **15**, 5463–5466.

3

Cyclodextrins as Molecular Reactors

Christopher J. Easton and Hideki Onagi
Research School of Chemistry, Australian National University, Canberra ACT 0200, Australia

3.1 Introduction

Cyclodextrins have been known for more than a century[1] and throughout that time they have been the subjects of continuous study, due mainly to their ability to act as molecular hosts and form inclusion complexes with hydrophobic guests in aqueous solution (Figure 3.1).[2] In recent decades these investigations have culminated in numerous applications of cyclodextrins in the pharmaceutical industry, as food additives, and in cosmetics and other personal care and household products.[3] One theme of our interest in these compounds, which is the subject of this chapter, has been their use as molecular reactors.[4] Like the other applications, this depends on the formation of inclusion complexes. In this particular case, the cyclodextrins are used as miniature reaction vessels and their tendency to form inclusion complexes is exploited to manipulate the assembly of reagents in the vessels in order to affect the outcomes of chemical transformations at the molecular level. In this regard the cyclodextrins are quite similar to the large scale vats and chemical reactors used for industrial chemical processing, beakers and flasks used in laboratory syntheses, and pots and saucepans used to prepare food – they are simply vessels for chemical reactions. However, the important distinction is that only the molecular containers provide a means by which the individual chemical processes may be regulated. Where a variety of

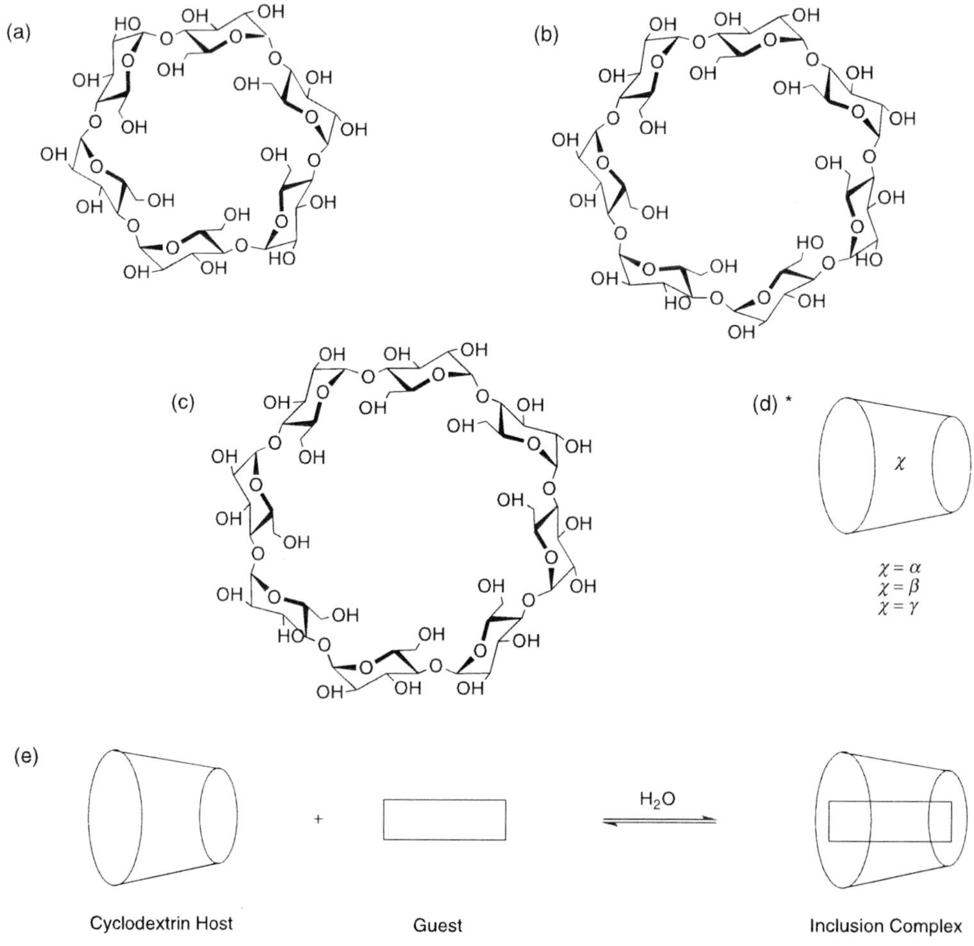

*A substituent drawn at the narrow end of the cone indicates that it replaces one of the cyclodextrin primary hydroxyl groups

Figure 3.1 *Structures of a) α-cyclodextrin, b) β-cyclodextrin, c) γ-cyclodextrin, d) their representation as truncated cones, and e) their formation of inclusion complexes with hydrophobic guests in aqueous solution*

different chemical reactions would take place in bulk media, molecular reactors provide a means to favour one or a particular combination.

This behaviour is in many ways like that displayed by enzymes and other authors have therefore used the term artificial enzymes instead of molecular reactors.[5] However, the need for a different term has become more important since recent rapid advances in molecular biology have provided tools for protein engineering to produce entirely new catalysts that might be more obviously classed as artificial enzymes, while molecular reactors encompass a much broader range of species. Another important distinction is that while enzymes achieve their effects because they are catalysts and enhance reaction rates, molecular reactors frequently bias the balance between various chemical reactions to affect the overall outcome by slowing down reactions, but selectively to a greater or lesser extent. That is, a molecular reactor may accomplish a particular result either by enhancing the rate of a desired process or by decreasing the rate of an undesired one.

The molecular reactor concept is common to many parts of this volume, which describe a wide variety of such reaction vessels. Many groups other than ours have studied cyclodextrins and other chapters, especially those by Breslow and Takahashi, also deal with their application in this context. This chapter is therefore focussed on our particular perspective and illustrated mainly with examples from our own laboratories. Much of our work has involved modified cyclodextrins[2] because these provide a greater range of functional groups and corresponding chemistry, and more opportunities to template reaction geometries, but it would be inappropriate not to acknowledge the related pioneering work with the natural cyclodextrins that was carried out by the research groups of Cramer,[6] Bender,[7] and others, to provide the essential foundations of this field.

3.2 Regiocontrolled Electrophilic Aromatic Substitutions

The most straightforward examples of cyclodextrin molecular reactors are probably those that involve a change in the regioselectivity of a reaction, as a result of a substrate being included in a cyclodextrin in such a way as to restrict or control the geometry of access of a reagent. Breslow *et al.*,[8,9] in particular have developed this approach to substituting hydrogen for functional groups in both aliphatic and aromatic compounds. In an illustrative example we developed,[10] but which is derivative of related aromatic chlorinations,[8,11] bromination of acetanilide with pyridinium dichlorobromate in water at room temperature gave a 56:44 ratio of the corresponding *ortho*- and *para*-substituted monobromides in the absence of a cyclodextrin, but that ratio changed to 79:21 when β-cyclodextrin was added and, within the limits of detection, only the *para*-isomer was produced in the presence of α-cyclodextrin (Figure 3.2a). It appears that acetanilide includes in the cyclodextrins with an orientation that shields the *ortho*-positions, but leaves the *para*-position exposed to attack by the brominating agent. The benefit provided by the cyclodextrins in the synthesis of *para*-bromoacetanilide is not just to increase the yield, even though with α-cyclodextrin this is almost doubled. Particularly for reactions carried out on a large scale, it is also relevant that less *ortho*-substituted by-product is formed, reducing waste and simplifying product isolation by obviating fractional crystallisation or chromatography. There are several other advantageous features of this system common to many involving cyclodextrins: the processes occur in water, which is attractive as an environmentally benign solvent; the cyclodextrins substantially increase the solubility of the substrate in water, thereby reducing reaction volumes; and an aqueous solution of the cyclodextrin can be reused repeatedly in a batch-type process whereby the substrate and reagent are added and, after reaction, the products are separated by extraction, to leave the cyclodextrin solution which can be recycled.

The effect of the cyclodextrins on the regioselectivity of bromination of anisole is similar to that seen with acetanilide (Figure 3.2b). With systems more activated to electrophilic substitution, such as 3-methylacetanilide and 3-methylanisole, the cyclodextrins

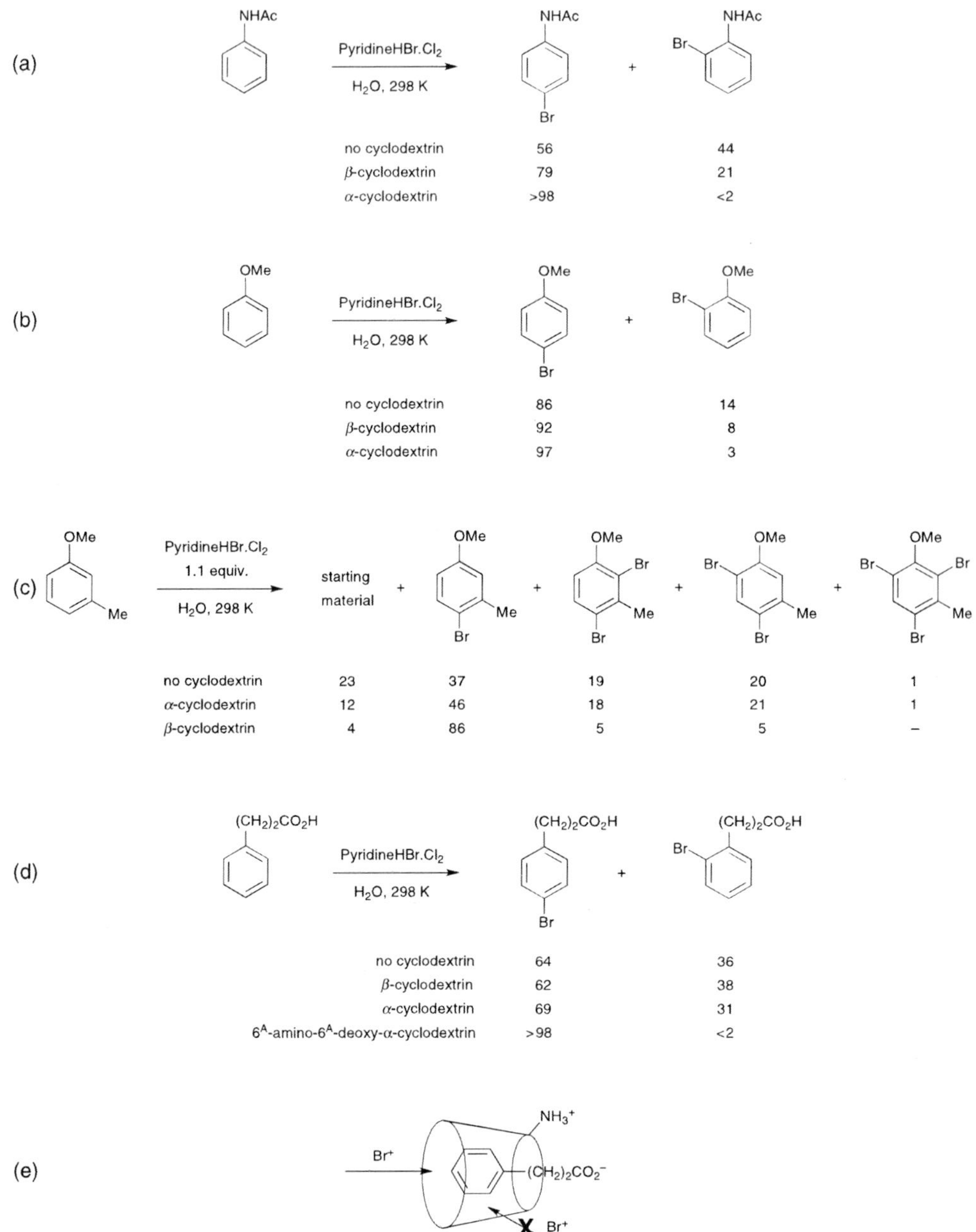

Figure 3.2 *Effect of cyclodextrins on the pyridinium dichlorobromate bromination of a) acetanilide,*[10] *b) anisole,*[10] *c) 3-methylanisole*[10] *and d) 3-phenylpropanoic acid,*[12] *and e) the orientation of 3-phenylpropanoic acid in an amino cyclodextrin to shield the* ortho-*positions from reaction and leave the* para-*position exposed*

also limit the extent of further bromination of the 4-bromosubstituted initial product (Figure 3.2c). For example, under conditions where treatment of 3-methylanisole with 1.1 equivalents of the brominating reagent afforded starting material and the 4-, 2,4-, 4,6- and 2,4,6-substituted products in the ratio 23 : 37 : 19 : 20 : 1, addition of *α*-cyclodextrin changed this ratio to 12 : 46 : 18 : 21 : 1, while in the presence of *β*-cyclodextrin only the starting material and the 4-, 2,4- and 4,6-substituted products were detected, in the ratio 4 : 86 : 5 : 5. Thus *β*-cyclodextrin increases the yield of 4-bromo-3-methylanisole from 37 to 86% and proportionately decreases the amount of by-products from 63 to 14%.

The choice of which cyclodextrin is best to use to achieve a particular outcome is difficult to predict. It depends on a number of factors such as the extent of inclusion of substrates, reagents and products, that can be calculated from the association constants of the various inclusion complexes, but also on the orientation of inclusion and the effect of the cyclodextrin on reaction intermediates and the geometry of interactions between substrates and reagents, which are not easily determined. Interestingly, *α*-cyclodextrin was more effective than *β*-cyclodextrin in controlling the bromination of acetanilide and anisole, while the opposite was seen with the corresponding reactions of the 3-methyl-substituted substrates. In some cases a more significant effect is achieved with a modified cyclodextrin. For example, the bromination of 3-phenylpropanoic acid gave the *para*- and *ortho*-monobromides in the ratio 64 : 36 in the absence of a cyclodextrin, and in the ratios 62 : 38 and 69 : 31 in the presence of *β*- and *α*-cyclodextrin, respectively, but using the same concentration of the amino cyclodextrin derivative having an amino substituent in place of a primary hydroxyl group, only 3-(4-bromophenyl)propanoic acid was produced (Figure 3.2d).[12] The substantially greater effect of the modified cyclodextrin in this system can be attributed to an ionic interaction between the amino group of the host and the carboxyl group of the guest, that are both ionised under the neutral experimental conditions, to stabilise the inclusion complex and control the alignment of the host within the guest (Figure 3.2e).

3.3 Catalysis of Hydrolytic Reactions

Some of the earliest examples of cyclodextrin molecular reactors involved their use as catalysts for hydrolysis of carboxylic acid derivatives, as demonstrated by Bender and co-workers[13] with phenyl esters and the natural cyclodextrins, in the original papers in this area. The reactions involve transient acylation of the cyclodextrin similar to that seen with serine and cysteine proteases. They exhibit kinetic characteristics, such as substrate saturation, non-productive binding and competitive inhibition, which are reminiscent of enzyme-catalysed processes. In addition, the discrimination displayed by cyclodextrins in binding guests and promoting their reactions is analogous to the substrate selectivity displayed by enzymes. One form of this type of substrate selectivity displayed by the cyclodextrins is chiral discrimination[14] and, for example, *β*-cyclodextrin reacts diastereoselectively with the (*R*)-enantiomer of the acid chloride of the non-steroidal anti-inflammatory drug Ibuprofen (Figure 3.3).[15] The cyclodextrin ester formed through this acylation undergoes subsequent hydrolysis to release the drug, and the stereoselectivity of the hydrolysis is complementary to that of the acylation, also favouring reaction of the diastereomer derived from (*R*)-Ibuprofen. For the combined synthesis and hydrolysis of

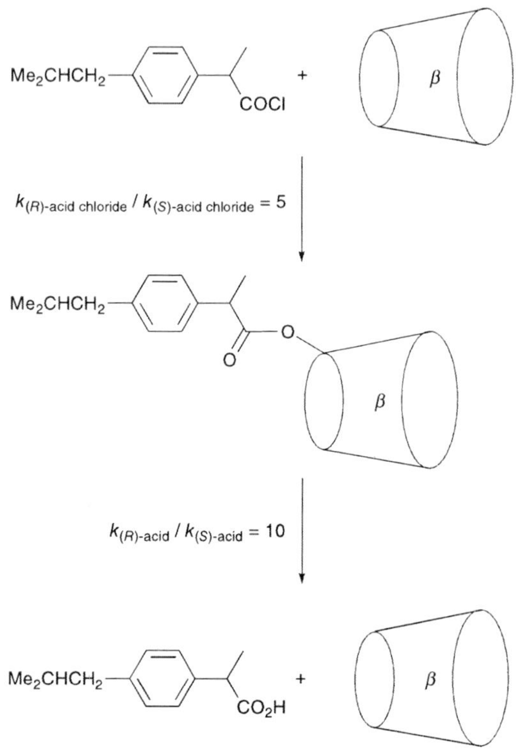

Figure 3.3 *Complementary diastereoselectivity in the acylation of β-cyclodextrin with the acid chloride of Ibuprofen and in the hydrolysis of the corresponding cyclodextrin ester*[15]

the cyclodextrin ester, the overall chiral discrimination in favour of the (*R*)-enantiomer is a factor of approximately fifty.

With the natural cyclodextrins, hydroxyl groups are the only functionality available to promote reactions of included guests. However, the introduction of a diverse range of new functional groups through modification of the natural cyclodextrins has resulted in catalysts that mimic the entire range of enzyme behaviour. In one particular example, Breslow *et al.*,[16] constructed a series of bis-imidazole-substituted β-cyclodextrins that catalyse the hydrolysis of 4-*tert*-butylcatechol cyclic phosphate. With the $6^A,6^B$-disubstituted cyclodextrin, which has imidazole groups in place of the primary hydroxyl groups of two adjacent glucopyranose residues, there is good evidence that the reaction involves simultaneous bifunctional acid-base catalysis, where one imidazole ring acts as a base and the other, which is protonated and therefore present as the imidazolium ion, acts as an acid. In this case the controlled alignment of both functional groups and the cyclodextrin annulus optimises the geometry for binding and reaction of the included substrate.

The hydrolysis of organophosphate triesters is a particular challenge that is of interest in the remediation of soils and waters contaminated by insecticides of this type, and in the destruction of related chemical warfare agents. The natural and most modified cyclo-

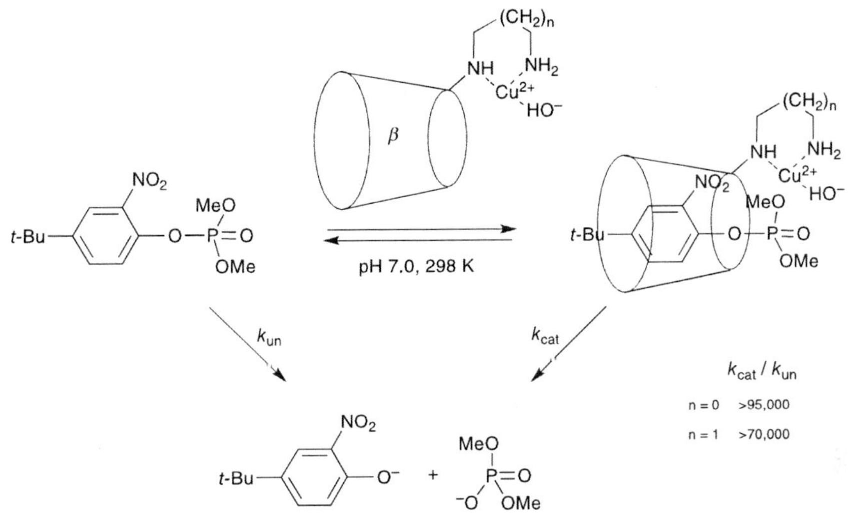

Figure 3.4 *Metallocyclodextrin-catalysed hydrolysis of a phosphate triester*[17]

dextrins either inhibit or only modestly accelerate breakdown of these phosphates in aqueous solution. However, the copper complexes of the *β*-cyclodextrin derivatives with 2-aminoethylamino and 3-aminopropylamino groups in place of one of the primary hydroxyl substituents efficiently catalyse the hydrolysis of 4-*tert*-butyl-2-nitrophenyl dimethyl phosphate at pH 7.0, increasing the rate of reaction of the cyclodextrin-bound species by more than 95,000 and 70,000 times, respectively (Figure 3.4).[17] The metallocyclodextrins are substrate selective and behave as true catalysts, in that there is multiple turnover of the triester. The reaction is most probably brought about by metal-bound hydroxide. Water bound to copper in the complex of the aminopropylamino-substituted cyclodextrin has a pK_a of 7.84.

3.4 A Molecular Reactor for the Synthesis of Indigoid Dyes

In addition to molecular reactors that affect both the regioselectivity of introduction and rates of interconversion of functional groups, others have been developed to control the outcomes of carbon-carbon bond forming reactions. In this area, one of the systems that we have studied in some detail involves the use of a cyclodextrin dimer to template the synthesis of indirubin.[18,19] While this dye is of considerable interest in its own right,[18] it is most commonly known as a by-product of the formation of indigo, the compound used to colour denim. Indirubin has a ruby red colour and the presence of this material in indigo or indigo-treated denim results in a pinkish-purple tinge often associated with low quality.

Indoxyl, which is readily obtained by hydrolysis of indoxyl acetate, gives indigo through oxidative dimerisation, but competitive oxidation of indoxyl affords isatin, that also reacts with indoxyl to give indirubin (Figure 3.5). To bias these competing processes in favour of indirubin production, a dimer comprised of two *β*-cyclodextrins joined by substituting a primary hydroxyl group of each one with a urea linkage was used as a

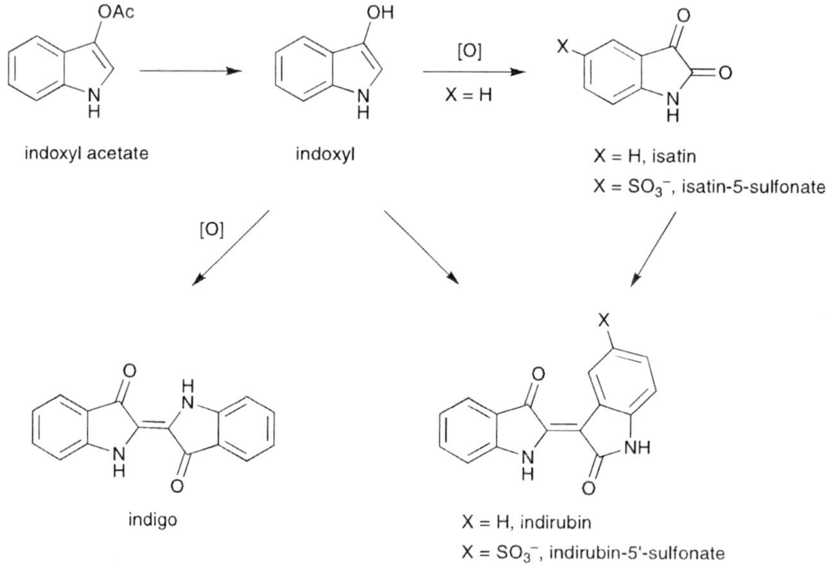

Figure 3.5 *Competing reactions of indoxyl and isatin/isatin-5-sulfonate to give indigo and indirubin/indirubin-5′-sulfonate*

template. In principle such a template could have been chosen based on its ability to selectively bind the desired product but that was not possible to determine in this case, due to the very low solubility of both indirubin and indigo in water. Instead, the cyclodextrin dimer was selected because it showed particularly strong binding of Tropaeolin 000 No. 2, a water-soluble dye similar in shape to indirubin. The expectation was then that since the binding of the Tropaeolin showed the dimer readily adopted the shape of indirubin, it would complex both indoxyl and isatin in the orientation to form indirubin (Figure 3.6), but less readily include two molecules of indoxyl aligned to form indigo. In the event, under conditions in which isatin and indoxyl derived from indoxyl acetate reacted in the absence of a cyclodextrin to give indigo and indirubin in yields of 16 and 13%, respectively, the corresponding yields of the dyes formed in the presence of the urea-linked cyclodextrin dimer were 0.03 and 1.0% (Table 3.1). The cyclodextrin dimer changed the ratio of production of the dyes by a factor of more than forty, in favour of indirubin. *β*-Cyclodextrin and *β*-cyclodextrin dimers joined by oxalamide and succinamide groups also reduced the yields of both dyes but did not show the same bias towards indirubin formation.

The problem with this system is that the molecular reactor offers very little in the way of a real practical advantage! The ratio of formation of the dyes has been changed in favour of indirubin but only by reducing the overall yield of the two dyes by a factor of almost thirty and the yield of indirubin by thirteen times. This is almost certainly the result of complexation of indoxyl and isatin by the cyclodextrin, increasing their effective steric bulk and reducing the frequency of their productive collisions. At the same time the rates of hydrolytic decomposition of indoxyl and isatin are largely unaffected, so these processes become more dominant. Similar effects are likely to be common with molecular reactors unless complexation increases the desired reactivity in some other way.

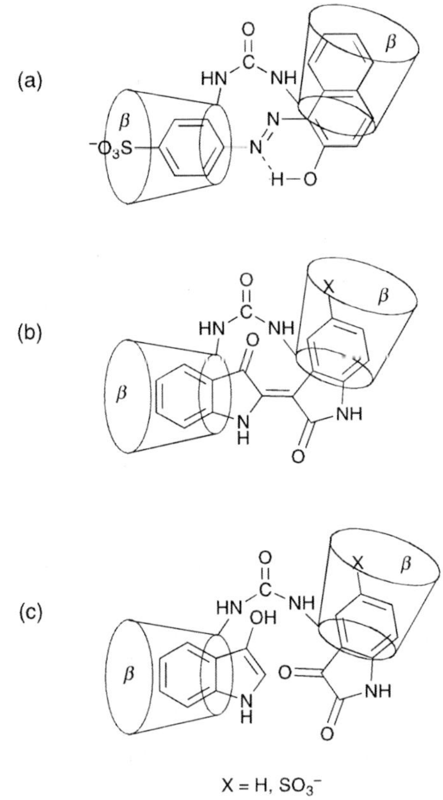

Figure 3.6 *Illustrations of a urea-linked β-cyclodextrin dimer a) complexing Tropaeolin 000 No. 2, and matching b) the orientation required for binding indirubin/indirubin-5′-sulfonate, and therefore c) the orientation required to template the reaction of indoxyl with isatin/isatin-5-sulfonate*

Table 3.1 *Affects of cyclodextrins on the percentage yields of indigo and a) indirubin or b) indirubin-5′-sulfonate, formed under otherwise identical conditions*[19]

	a)	Yield (%)	b)	Yield (%)
	indigo	indirubin	indigo	indirubin-5′-sulfonate
No cyclodextrin	16	13	25	1.4
β-CD	2.5	2.5	1.6	11
β-CDNHCO(CH_2)$_2$CONH-β-CD	0.5	0.7	6.0	16
β-CDNHCOCONH-β-CD	0.2	0.6	1.8	36
β-CDNHCONH-β-CD	0.03	1.0	<0.1	22

In the present case, this was accomplished with isatin-5-sulfonate instead of isatin, using the urea-linked cyclodextrin dimer to alter the ratio of formation of indigo, through oxidative dimerisation of indoxyl, to indirubin-5′-sulfonate, from indoxyl and the isatin-sulfonate (Figure 3.5). The isatinsulfonate exists in solution in equilibrium with its hydrate, which deprotonates with a pK_a of 9.55 to give the corresponding anion

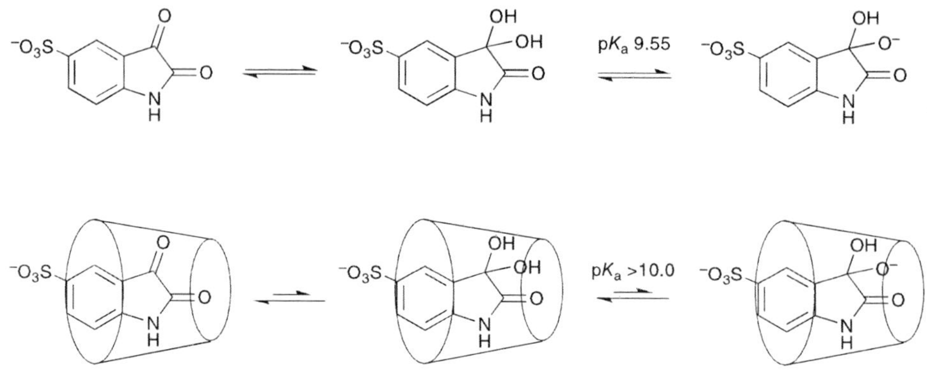

Figure 3.7 *Affect of a cyclodextrin on the equilibrium between isatin-5-sulfonate, its hydrate and its deprotonated hydrate*[19]

(Figure 3.7). Neither the hydrate nor the anion reacts with indoxyl. Complexation by a cyclodextrin decreases the extent of hydration of the isatinsulfonate and increases the pK_a of the hydrate to above 10.0. For the condensation reaction carried out at pH 10.0, the cyclodextrin therefore substantially increases the percentage of the isatinsulfonate present at equilibrium in the neutral, non-hydrated, reactive form. Accordingly, under conditions in which indigo and indirubin-5′-sulfonate are formed in yields of 25 and 1.4%, respectively, in the absence of a cyclodextrin, addition of the urea-linked cyclodextrin dimer results in a 22% yield of the indirubin and no indigo was formed within the limits of detection (<0.1%) (Table 3.1). Consequently the molecular reactor changes the ratio of formation of the dyes by a factor of more than three and a half thousand, with very little loss of yield. In this case β-cyclodextrin and the dimers joined by oxalamide and succinamide groups also favour formation of the indirubinsulfonate by factors of approximately one hundred, fifty and three hundred, respectively, because their complexation of the isatinsulfonate also increases its reactivity, but their effects are less than those seen with the urea dimer because they do not simultaneously template the interaction between the sulfonate and indoxyl.

3.5 Manipulation of Cycloadditions

Cyclodextrins have also been studied as molecular reactors for cycloadditions. To some extent reactions of this class are obvious candidates for such investigations since they are generally solvent-independent, non-catalysed processes, and factors such as reaction rates and regio- and stereo-selectivity are only affected by the concentrations of the reacting species and how those align to react. Through inclusion of the reactants in a cyclodextrin, their effective local concentrations can be increased and their relative alignment can be altered, to increase the reaction rates and change the product ratios. β-Cyclodextrin has been found to alter the diastereoselectivity of Diels–Alder reactions of cyclopentadiene with maleic and fumaric acid derivatives,[20] and accelerate the rates of such reactions by factors ranging from fifteen to more than seven thousand.[21] Different natural cyclodextrins

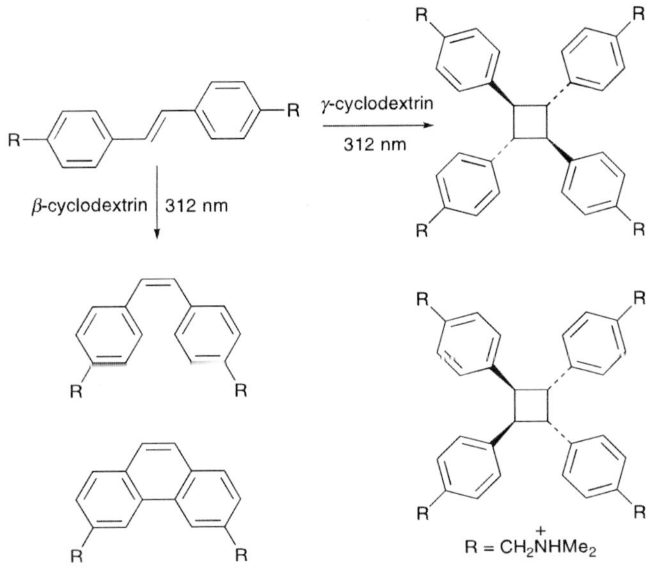

Figure 3.8 *Changes to the outcome of photolysis of a stilbene derivative through complexation with either β- or γ-cyclodextrin*[22]

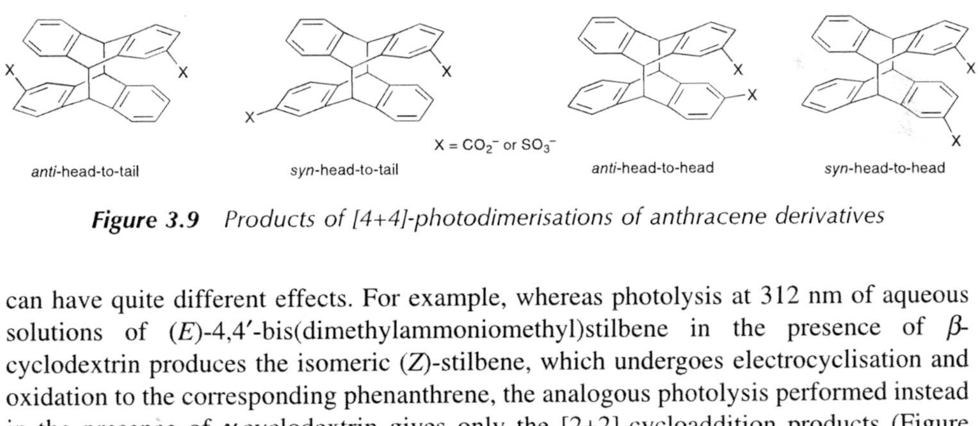

Figure 3.9 *Products of [4+4]-photodimerisations of anthracene derivatives*

can have quite different effects. For example, whereas photolysis at 312 nm of aqueous solutions of (*E*)-4,4′-bis(dimethylammoniomethyl)stilbene in the presence of β-cyclodextrin produces the isomeric (*Z*)-stilbene, which undergoes electrocyclisation and oxidation to the corresponding phenanthrene, the analogous photolysis performed instead in the presence of γ-cyclodextrin gives only the [2+2]-cycloaddition products (Figure 3.8).[22] The difference reflects the formation of a 1 : 1 host-guest inclusion complex of the stilbene with β-cyclodextrin compared with a 1 : 2 complex with γ-cyclodextrin.

The effects of cyclodextrins on [4+4]-photodimerizations of anthracene derivatives have been studied quite extensively.[23,24] For example, it has been found that such reactions of anthracene-2-sulfonate and -2-carboxylate in water are accelerated by both β- and γ-cyclodextrin. γ-Cyclodextrin has very little effect on the ratio of the four possible cycloadducts but, with each of the anthracene derivatives, β-cyclodextrin increases the proportion of the *anti*-head-to-tail cycloadduct from less than 45% to more than 85% (Figure 3.9).[23] The effects observed with the natural cyclodextrins generally arise from the fortuitous way the reagents self-assemble in the cyclodextrin annulus. By comparison, modified cyclodextrins offer the additional benefit of providing a means to control the

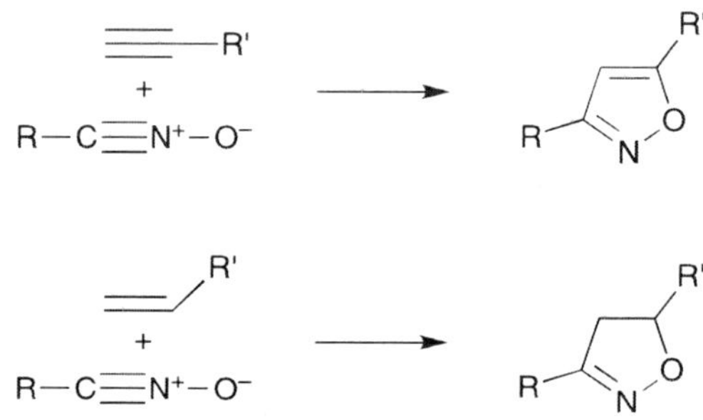

Figure 3.10 *Regioselectivity of reactions of nitrile oxides with monosubstituted alkynes and alkenes*

assembly of the reacting species, so as to manipulate the outcomes of cycloadditions by design. Accordingly, bispyridinio-appended and capped γ-cyclodextrin derivatives have been identified that induce a two- to five-fold increase in the proportion of the *anti*-head-to-head cycloadduct formed from anthracene-2-carboxylate.[24] They also afford that dimer with enantiomeric excesses as high as 35% and there seems to be a general correlation between the yield and enantiomeric excess obtained using a particular cyclodextrin derivative.

We have exploited modified cyclodextrins to reverse the regioselectivity of [3+2]-dipolar cycloadditions. Nitrile oxides (dipoles) undergo dipolar cycloadditions with alkynes and alkenes (dipolarophiles) to give isoxazoles and isoxazolines (cycloadducts), respectively. With monosubstituted dipolarophiles, in principle both 3,4- and 3,5-disubstituted cycloadducts could form but, in practice, the reactions generally produce mainly the 3,5-disubstituted-ring system (Figure 3.10).[25] Steric effects determine the regioselectivity, with the more hindered end of the dipolarophile bonding preferentially to the oxygen of the nitrile oxide. For example, the reaction of 4-*tert*-butylbenzonitrile oxide with propynamide in water at room temperature produces the corresponding 3,5- and 3,4-disubstituted isoxazoles in the ratio 4 : 1. However, β-cyclodextrin can act as a molecular scaffold to change the outcome. By attaching propynamide to the cyclodextrin through substitution of one of its primary hydroxy that is should be hydroxyl groups, reaction with the nitrile oxide then affords the 3,5- and 3,4-cycloadducts in the ratio 1 : 15 (Figure 3.11).[26] Whereas 80% of the 3,5-disubstituted species is produced in the absence of the cyclodextrin, more than 90% of the opposite regioisomer is formed when it is present. This is consistent with the formation of an inclusion complex between the nitrile oxide and the modified cyclodextrin, in which the relative orientations of the dipole and dipolarophile are controlled (Figure 3.12). The importance of this inclusion to the reaction outcome is highlighted by the observation that the effect of the cyclodextrin is much less both when the reactions are carried out in *N,N*-dimethylformamide instead of water, where the extent of host–guest complexation is much less, and in water with nitrile oxides other than the *tert*-butylbenzo derivative, that are less hydrophobic and less inclined to include.

Similar effects have been observed with azide dipolar cycloadditions.[27] In the system directly analogous to that described in the preceding paragraph, reaction of 4-*tert*-butylphenylazide with propynamide produces a 2 : 1 ratio of the 1,4- and 1,5-disubstituted-1,2,3-triazoles, whereas tethering the propynamide to β-cyclodextrin changes the ratio

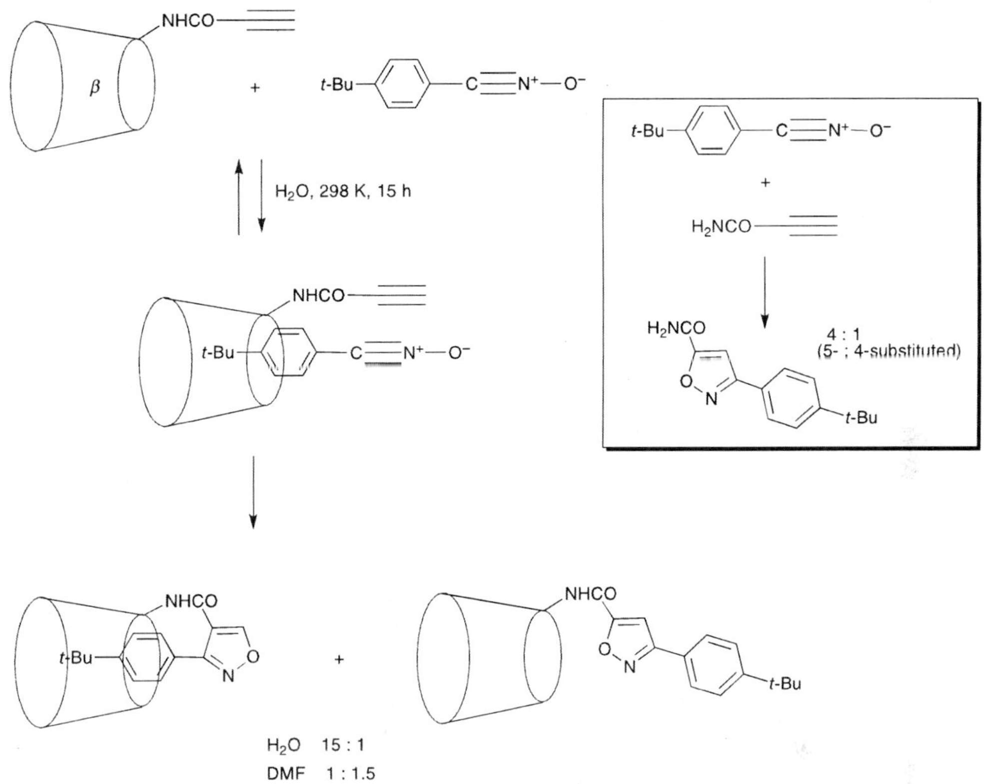

Figure 3.11 *Effect of a cyclodextrin to reverse the regioselectivity of a nitrile oxide cycloaddition*[26]

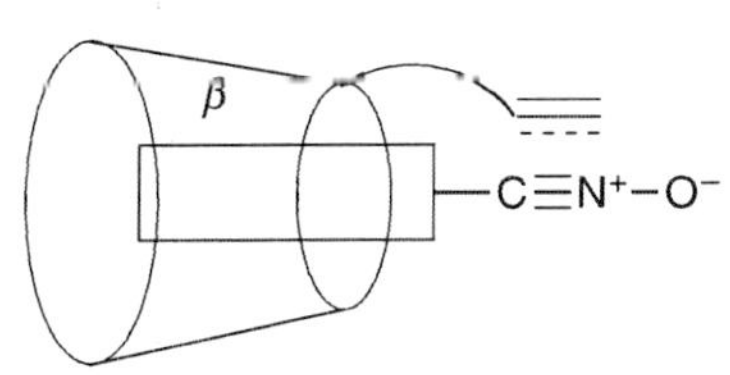

Figure 3.12 *Use of a cyclodextrin to template the relative orientation of a dipole and a dipolarophile for cycloaddition*[26]

of the corresponding cycloadducts to 1:25 (Figure 3.13). That is, while 67% of the 1,4-disubstituted species is produced in the absence of the cyclodextrin, more than 95% of the opposite regioisomer is formed when the cyclodextrin is present. Either the dipolarophile or the dipole may be attached to the cyclodextrin to affect the regioselectivity. With the azido-β-cyclodextrin where azide is substituted for one of the primary hydroxyl groups, the reaction with *N*-(4-*tert*-butylphenyl)propynamide gave an 8:1 mixture of the 1,4- and 1,5-disubstituted triazoles, presumably through formation of a host-guest inclusion complex of the reactants, but with *N*-(3,5-dimethylphenyl)propynamide, where the bulk of the aromatic group restricts complexation, the ratio was swapped to 1:6.

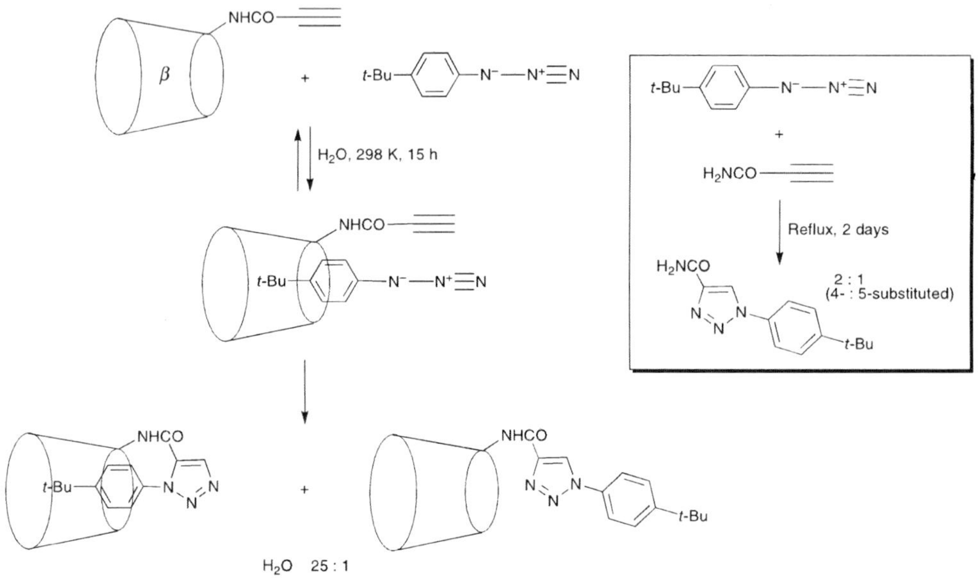

Figure 3.13 *Effect of a cyclodextrin to reverse the regioselectivity of an azide cycloaddition*[27]

Qualitatively, the cyclodextrin increases the rate of these cycloadditions as well as affecting the regioselectivity. The extent of reaction of 4-*tert*-butylphenylazide with propynamide in refluxing toluene after two days is less than that of the corresponding reaction of the cyclodextrin-substituted propynamide at room temperature in water after fifteen hours. A semi-quantitative analysis of this type of effect was performed in water using various nitrile oxides each at 1 mM concentration, in competing reactions with propynamide (20 mM) and the cyclodextrin-substituted analogue (2 mM). Despite the use of a ten-fold excess of propynamide, benzo-, 4-phenylbenzo- and 4-*tert*-butylbenzo-nitrile oxide each reacts selectively with the cyclodextrin derivative, by factors of more than two, fifty and three hundred, respectively. This selective molecular recognition between alkynes by the nitrile oxides is complemented by a similar selection between nitrile oxides by the cyclodextrin-substituted alkyne. The cyclodextrin (25 μM) shows a seventy five-fold selectivity for reaction with 4-*tert*-butylbenzonitrile oxide (50 μM) in preference to benzonitrile oxide (500 μM), despite the use of a ten-fold excess of the latter.

These trends correspond to pseudo-first order rate constants of 0.5×10^{-3}, 2.0×10^{-3}, $>650 \times 10^{-3}$ and $>30 \times 10^{-3}\,min^{-1}$ for production of the 3,4- and 3,5-disubstituted isoxazoles from reaction of 4-*tert*-butylbenzonitrile oxide (1 mM) with propynamide (5 mM) and the cyclodextrin derivative (5 mM), respectively (Figure 3.14). Thus, the cyclodextrin moiety increases the rate of formation of the 3,4-disubstituted isoxazole by more than three orders of magnitude. Even the rate of formation of the 3,5-disubstituted isoxazole is increased more than fifteen times by the cyclodextrin, despite templating of the other isomer being preferred. Under analogous conditions, the pseudo-first order rate constants for production of the 1,5- and 1,4-disubstituted triazoles from reaction of 4-*tert*-butylphenylazide with propynamide and the cyclodextrin derivative are $<0.1 \times 10^{-5}$,

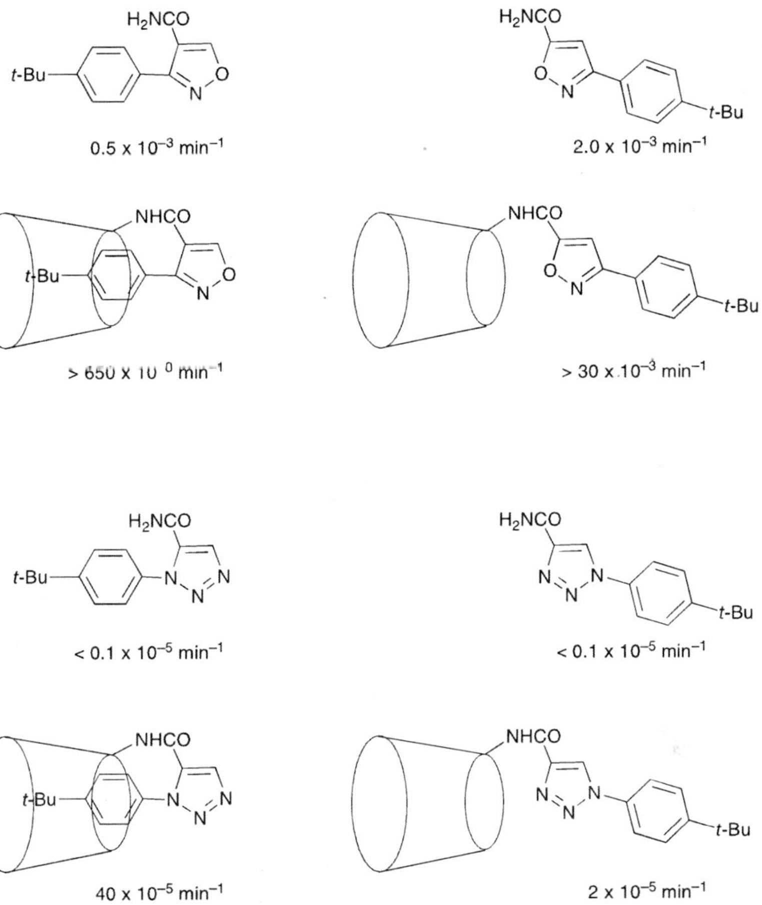

Figure 3.14 *Pseudo-first order rate constants for reactions of azides and nitrile oxides with alkynes to give the cycloadducts illustrated*[26,27]

$<0.1 \times 10^{-5}$, 40×10^{-5} and $2 \times 10^{-5}\,min^{-1}$, respectively (Figure 3.14). Therefore, in this case, the cyclodextrin moiety increases the rate of formation of the 1,5-disubstituted triazole by a factor of more than four hundred, and even the rate of formation of the 1,4-isomer is accelerated by a factor of at least twenty.

The effects of β-cyclodextrin on the regioselectivity of cycloaddition reactions with alkynes are also seen with alkenes, as well as when the dipolarophiles are attached to the cyclodextrin through ester instead of amide linkages. Under conditions where 4-*tert*-butylbenzonitrile oxide reacts with methyl acrylate to give the corresponding 3,5- and 3,4-disubstituted isoxazolines in the ratio 20:1, the acrylate ester formed with β-cyclodextrin through one of its primary hydroxyl groups produces the corresponding cycloadducts in the ratio 1:20. In this case the cyclodextrin changes isoxazoline formation from more than 95% of the 3,5-disubstituted species to more than 95% of the alternative cycloadduct. Similar changes to the ratios of the cycloadducts are observed with the esters of α-methylacrylic and crotonic acid.

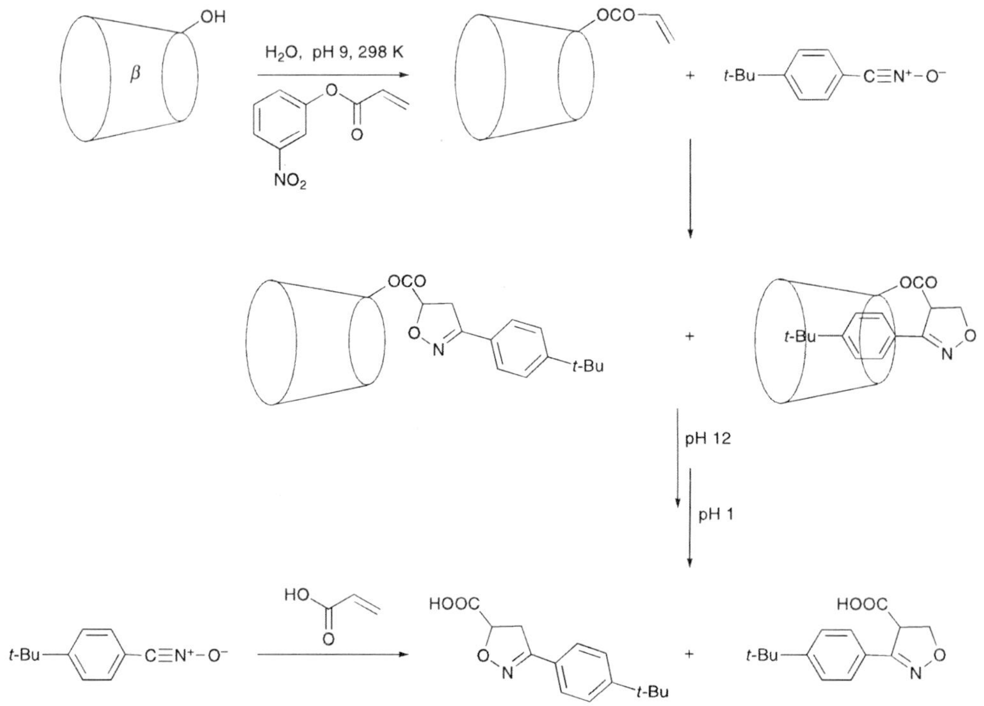

Figure 3.15 *Effect on a cycloaddition of the transient attachment of the dipolarophile to a cyclodextrin*[28]

These reactions can be incorporated into a cycle where the cyclodextrin is used and then regenerated, although the effect of the cyclodextrin on the regioselectivity is somewhat reduced under these circumstances (Figure 3.15).[28] Accordingly, acylation of β-cyclodextrin with 3-nitrophenyl acrylate, in aqueous solution at pH 9.0 and room temperature for eight hours, gives the cyclodextrin acrylate, which undergoes cycloaddition with 4-*tert*-butylbenzonitrile oxide. After a further hour at pH 12.0, to allow for ester hydrolysis, adjustment to pH 1.0 gives the 3,5- and 3,4-disubstituted cycloadducts, which extract into organic solvent leaving the aqueous solution of the reformed cyclodextrin. The ratio of the 3,5- and 3,4-disubstituted isoxazolines obtained in this way is 1 : 2, compared with 20 : 1 from acrylic acid and 1 : 20 from the cyclodextrin acrylate. The intermediate value indicates that not all of the reaction of the 3-nitrophenyl ester proceeds through the cyclodextrin acrylate.

3.6 Conclusion

The examples described above are just a selection of those reported, chosen to illustrate what has already been achieved using cyclodextrin-based molecular reactors. The concepts have been clearly demonstrated and it is reasonable to expect that practical applica-

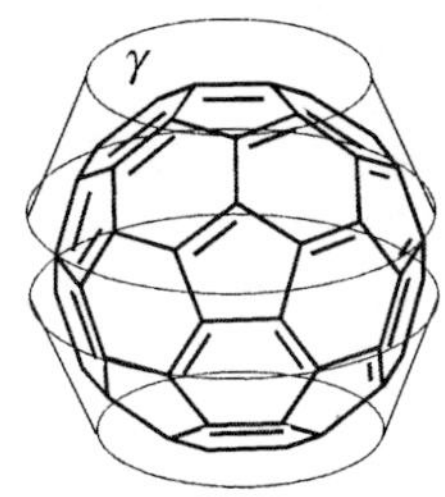

Figure 3.16 *The 1 : 2 complex of C_{60} with γ-cyclodextrin that catalyses nitrogen fixation in an aqueous bisulfite solution irradiated under nitrogen at atmospheric pressure*[30]

tions will follow, once the methodology becomes accepted as a way to address particular types of problems faced in synthetic chemistry. Meanwhile, there remains a lot of scope to develop further methodology in the field. In the last few years entirely new applications of cyclodextrin molecular reactors have been developed, for example, in the ring-opening polymerisation of lactones,[29] while perhaps the most interesting and potentially important report in this area is of the fixation of nitrogen, through the use of a 1 : 2 complex of buckminsterfullerene (C_{60}) with γ-cyclodextrin (Figure 3.16) in aqueous sodium bisulfite, irradiated under nitrogen at atmospheric pressure.[30] These reports indicate that molecular reactors, and artificial enzymes as they were known earlier, still have considerable potential for further development and their study continues to be a fruitful area for research.

Acknowledgements

The authors acknowledge the receipt of financial support from the Australian Research Council for research in this area.

References

1. A. Villiers, Fermentation of starch by the butyric ferment, *Compt. Rend.*, 1891, **112**, 536–538; F. Schardinger, Formation of crystalline non-reducing polysaccharides from starch by microbic action, *Cent. Bakt. Parasitenk. [II]*, 1909, **22**, 98–103; F. Schardinger, The formation of crystallin polysaccharides (dextrins) from starch paste by microorganisims, *Cent. Bakt. Parasitenk. [II]*, 1911, **29**, 188–197.
2. C. J. Easton, S. F. Lincoln, *Modified Cyclodextrins, Scaffolds and Templates for Supramolecular Chemistry*, Imperial College Press, London, 1999.
3. D. Duchêne, *New Trends in Cyclodextrins and Derivatives*, Editions Santé, Paris, 1991; *Cyclodextrins and Their Complexes*, H. Dodziuk (Ed.), Wiley-VCH Verlag, Weinheim, 2008.
4. C. J. Easton, S. F. Lincoln, L. Barr, H. Onagi, Molecular reactors and machines: applications, potential and limitations, *Chem. Eur. J.*, 2004, **10**, 3120–3128.
5. V. T. D'Souza, Modification of cyclodextrins for use as artificial enzymes, *Supramol. Chem.*, 2003, **15**, 221–229; W. B. Motherwell, M. J. Bingham, Y. Six, Recent progress in the design and synthesis of artificial enzymes, *Tetrahedron*, 2001, **57**, 4663–4686.
6. F. Cramer, H. Hettler, Inclusion compounds of cyclodextrins, *Die Naturwissenschaften*, 1967, **54**, 625–632; F. Cramer, Cyclodextrins – old friends from a new viewpoint, *Starch/Staerke*, 1983, **35**, 203–206.

7. M. L. Bender, M. Komiyama, *Cyclodextrin Chemistry*, Springer-Verlag, New York, 1978.
8. R. Breslow, P. Campbell, Selective aromatic substitution within a cyclodextrin mixed complex, *J. Am. Chem. Soc.*, 1969, **91**, 3085; R. Breslow, P. Campbell, Selective aromatic substitution by hydrophobic binding of a substrate to a simple cyclodextrin catalyst, *Bioorg. Chem.*, 1971, **1**, 140–156; R. Breslow, H. Kohn, B. Siegel, Methylated cyclodextrin and a cyclodextrin polymer as catalysts in selective anisole chlorinations, *Tetrahedron Lett.*, 1976, 1645–1646.
9. R. Breslow, X. Zhang, R. Xu, M. Maletic, R. Merger, Selective catalytic oxidation of substrates that bind to metalloporphyrin enzyme mimics carrying two to four cyclodextrin groups and related metallosalens, *J. Am. Chem. Soc.*, 1996, **118**, 11678–11679; R. Breslow, Y. Huang, X. Zhang, J. Yang, An artificial cytochrome P450 that hydroxylates unactivated carbons with regio- and stereoselectivity and useful catalytic turnovers, *Proc. Nat. Acad. Sci. USA*, 1997, **94**, 11156–11158; R. Breslow, X. Zhang, Y. Huang, Selective catalytic hydroxylation of a steroid by an artificial cytochrome P450 enzyme, *J. Am. Chem. Soc.*, 1997, **119**, 4535–4536.
10. P. G. Dumanski, C. J. Easton, S. F. Lincoln, J. S. Simpson, Effect of cyclodextrins on electrophilic aromatic bromination in aqueous solution, *Aust. J. Chem.*, 2003, **56**, 1107–1111.
11. R. Chênevert, G. Ampleman, Chlorination of acetanilide and benzanilide in cyclodextrins or amylase, *Can. J. Chem.*, 1987, **65**, 307–310.
12. S. Bowen, H. Onagi, C. J. Easton, Unpublished observations.
13. R. L. VanEtten, J. F. Sebastian, G. A. Clowes, M. L. Bender, Acceleration of phenyl ester cleavage by cycloamyloses. A model for enzymic specificity, *J. Am. Chem. Soc.*, 1967, **89**, 3242–3253; R. L. VanEtten, G. A. Clowes, J. F. Sebastian, M. L. Bender, The mechanism of the cycloamylose-accelerated cleavage of phenyl esters, *J. Am. Chem. Soc.*, 1967, **89**, 3253–3262.
14. C. J. Easton, S. F. Lincoln, Chiral discrimination by modified cyclodextrins, *Chem. Soc. Rev.*, 1996, 163–170.
15. J. H. Coates, C. J. Easton, S. J. van Eyk, B. L. May, P. Singh, S. F. Lincoln, Chiral differentiation in the deacylation of 6^A-*O*-(2-(4-(2-methylpropyl)phenyl)propanoyl)-β-cyclodextrin, *J. Chem. Soc., Chem. Commun.*, 1991, 759–760; J. H. Coates, C. J. Easton, N. L. Fryer, S. F. Lincoln, Complementary diastereoselectivity in the synthesis and hydrolysis of acylated cyclodextrins, *Chem. Lett.*, 1994, 1153–1156.
16. E. Anslyn, R. Breslow, Geometric evidence on the ribonuclease model mechanism, *J. Am. Chem. Soc.*, 1989, **111**, 5972–5973; E. Anslyn, R. Breslow, Proton inventory of a bifunctional ribonuclease model, *J. Am. Chem. Soc.*, 1989, **111**, 8931–8932.
17. L. Barr, C. J. Easton, K. Lee, S. F. Lincoln, J. S. Simpson, Metallocyclodextrin catalysts for hydrolysis of phosphate triesters, *Tetrahedron Lett.*, 2002, **43**, 7797–7800.
18. L. Meijer, N. Guyard, L. A. Skaltsounis, G. Eisenbrand (Eds.), *Indirubin, the Red Shade of Indigo*, Life in Progress Editions, Roscoff, 2006.
19. C. J. Easton, J. B. Harper, S. F. Lincoln, *N,N'*-Bis(6^A-deoxy-α-cyclodextrin-6^A-yl)urea as a molecular template in the formation of indigoid dyes, *New J. Chem.*, 1998, **22**, 1163–1165; J. B. Harper, C. J. Easton, S. F. Lincoln, A cyclodextrin-based molecular reactor to template the formation of indigoid dyes, *Tetrahedron Lett.*, 2003, **44**, 5815–5818.
20. H.-J. Schneider, N. K. Sangwan, Diels–Alder reactions in hydrophobic cavities: a quantitative correlation with solvophobicity and rate enhancements by macrocycles, *J. Chem. Soc., Chem. Commun.*, 1986, 1787–1789; H.-J. Schneider, N. K. Sangwan, Changes of stereoselectivity in Diels–Alder reactions by hydrophobic solvent effects and by β-cyclodextrin, *Angew. Chem. Int. Ed. Engl.*, 1987, **26**, 896–897.
21. N. K. Sangwan, H.-J. Schneider, The kinetic effects of water and of cyclodextrins on Diels–Alder reactions. Host-guest chemistry. Part 18, *J. Chem. Soc., Perkin Trans. II*, 1989, 1223–1227.
22. W. Herrmann, S. Wehrle, G. Wenz, Supramolecular control of the photochemistry of stilbenes by cyclodextrins, *J. Chem. Soc., Chem. Commun.*, 1997, 1709–1710.
23. T. Tamaki, T. Kokubu, K. Ichimura, Regio- and stereoselective photodimerization of anthracene derivatives included by cyclodextrins, *Tetrahedron*, 1987, **43**, 1485–1494.
24. H. Ikeda, T. Nihei, A. Ueno, Template-assisted stereoselective photocyclodimerization of 2-anthracenecarboxylic acid by bispyridinio-appended γ-cyclodextrin, *J. Org. Chem.*, 2005, **70**,

1237–1242; C. Yang, C. Ke, F. Kahee, D.-Q. Yuan, T. Mori, Y. Inoue, pH Controlled supramolecular enantiodifferentiating photocyclodimerization of 2-anthracenecarboxylate with capped α-cyclodextrins, *Aust. J. Chem.*, 2008, **41**, 565–568; C. Yang, T. Mori, Y. Inoue, Supramolecular enantiodifferentiating photocyclodimerization of 2-anthracenecarboxylate mediated by capped α-cyclodextrins: critical control of enantioselectivity by cap rigidity, *J. Org. Chem.*, 2008, **73**, 5786–5794.

25. C. J. Easton, C. M. M. Hughes, G. P. Savage, G. W. Simpson, Cycloaddition reactions of nitrile oxides with alkenes, *Adv. Heterocyclic Chem.*, 1994, **60**, 261–327.
26. A. G. Meyer, C. J. Easton, S. F. Lincoln, G. W. Simpson, A cyclodextrin to reverse the regiochemistry in nitrile oxide cycloaddition to a terminal alkene, *J. Chem. Soc., Chem. Commun.*, 1997, 1517–1518; A. G. Meyer, C. J. Easton, S. F. Lincoln, G. W. Simpson, β-Cyclodextrin as a scaffold for supramolecular chemistry, to reverse the regioselectivity of nitrile oxide cycloadditions, *J. Org. Chem.*, 1998, **63**, 9069–9075.
27. L. Barr, S. F. Lincoln, C. J. Easton, A cyclodextrin molecular reactor for the regioselective synthesis of 1,5-disubstituted-1,2,3-triazoles, *Supramol. Chem.*, 2005, **17**, 547–555.
28. L. Barr, S. F. Lincoln, C. J. Easton, Reversal of regioselectivity and enhancement of rates of nitrile oxide cycloadditions through transient attachment of dipolarophiles to cyclodextrins, *Chem. Eur. J.*, 2006, **12**, 8571–8580.
29. M. Osaki, Y. Takashima, H. Yamaguchi, A. Harada, Polymerization of lactones initiated by cyclodextrins: effects of cyclodextrins on the initiation and propagation reactions, *Macromolecules*, 2007, **40**, 3154–3158.
30. Y. Nishibayashi, M. Saito, S. Uemura, S. Takekuma, H. Takekuma, Z. Yoshida, A non-metal system for nitrogen fixation, *Nature*, 2004, **428**, 279–280.

4

Reactions Mediated by Cyclodextrins

Keiko Takahashi
Department of Nanochemistry, Faculty of Engineering, Tokyo Polytechnic University, Atsugi, Kanagawa 243-0297, Japan

4.1 Introduction

Cyclodextrins (CDs) are naturally occurring torus-shaped molecules composed of 6, 7, or 8 D-glucose units. A fascinating property of CDs is their ability to incorporate other organic compounds into their cavities in aqueous solution.[1] Since their discovery, CDs have served as prototypes for novel host compounds and catalysts. The use of CDs as *microvessels* to perform chemical reactions has attracted the interest of chemists since the 1960s.[2] Effects of CDs on reactions are mainly divided into two categories. The first effect is the catalytic effect, designated as the 'enzyme model.' CD and the reactant initially form a CD-reactant reaction intermediate involving a covalent bond, which then proceeds to product. The second effect does not involve a covalent bond. The cavity of CD presents a new reaction environment to the reactant, an '*extra reaction field*', in which the reactivity, such as rate or selectivity, changes. In these cases, the role of the CD does not always catalyze, but *mediates* the reaction. The CD cavity is less polar than the bulk aqueous medium. The permittivity in the CD cavity is known to be nearly the same as that of dioxane. The microenvironment around the reactant in the CD cavity is different from that in the reaction media. Three distinct microenvironmental effects are expected: a) microsolvent effects, b) the protection of unstable intermediates or products, and c) the solubilization of the reactant. In addition, conformational effects can also be expected:

d) control of the reactant conformation, e) control of the orientation between reactants, and f) control of the size of molecule. A combination of several of these effects is evident in organic reactions.

Many organic, photochemical, radiochemical, electrochemical, and biochemical reactions have been carried out with CD cavities under various conditions, such as in the solid phase, in a heterogeneous suspension of water, or in organic solvents. These reactions were reported in a previous review.[3] Although the main driving force for formation of an inclusion complex between a CD cavity and a guest molecule is the hydrophobic interaction, appreciable selectivity has been achieved in organic solvents such as pyridine, DMF, THF, toluene, and CCl_4. The results in the solid state and in organic solvents suggest that other interactions must also be considered.

On the other hand, water is of particular interest, because it is a safe, inexpensive, and environmentally friendly solvent, which makes its use favorable both in academic laboratories and in industry. The pioneering studies of Diels–Alder reactions in water by Breslow[4] in the early 1980s triggered a more widespread interest in the use of water as a medium for organic synthesis, not only because these reactions eliminate the necessity of vigorously drying solvents and substrates, but also because of the unique reactivity and selectivity found in water. Thus, considerable effort has been made in developing water-based synthetic organic reactions.[5] Studying organic reactions mediated by CD is important not only in supramolecular chemistry but also in green chemistry. The number of papers reporting reactions mediated by CDs has been increasing exponentially and it would be impossible to cover them in the present limited space. In this chapter, the author wishes to indicate the origin of the concept and some representative reactions mediated by CDs in water. The 'now' and 'then' in CD science will be indicated.

4.2 The Inclusion Phenomena of Cyclodextrins

Before describing organic reactions mediated by CDs, some explanation of the inclusion phenomena with CD is warranted, focusing on the complexation mode and interactions. Most of our understanding of CD inclusion phenomena has been derived from studying the aqueous phase in an equilibrium state. In this state, hydrophobic forces are assumed to be responsible for driving a guest into the CD's hydrophobic interior, where a 1 : 1 host–guest complex usually forms and where there is no interaction with other CD molecules.

The complexation mode: *stoichiometries, conformations, and dynamics*: The possibility of simultaneous formation of complexes with different stoichiometries must also be considered. Cooperative binding, a bi-molecular process, has been observed, as well as the uni-molecular process, producing a 1 : 1 complex in either a 'head-to-tail (HT),' 'tail-to-tail (TT)' or 'head-to-head (HH)' state. Moreover, a study of the fluorescence decay of a number of anilinonaphthalene sulfonates in the presence of CD supported the view that a 1 : 1 complex can be present in several slightly different conformations. There are, in fact, various conformations and orientations of reactants as well as varying mobilities in reactant-CD complexes.

Interactions: The complexation is based on a combination of several intermolecular interactions depending on the solvent and the nature of the host and guest: steric fit, van der Waals interactions, dispersive forces, dipole–dipole interactions, charge-transfer interactions, electrostatic interactions, and hydrogen bonding. The choice of reaction media used in CD systems also depends on the above combination of interactions. The study of organic reactions mediated by CDs means the study of interactions in the CD molecule. It should be noted that CD can also include guest molecules in the solid state. Such host–guest complexes must clearly be discriminated from other host compounds, i.e., clathrates only in the solid state and crown ethers only in solution. Inclusion complex formation in the gas phase is expected to be different from that in liquid phases. Short-range London forces or polar interactions are responsible for host–guest complexation in nonaqueous environments. Clearly, there is the possibility of many kinds of interactions to comprise the inclusion phenomena both in solid and solution states. This is the advantage of using CDs for supramolecular functions.

4.3 Origin of Microvessels as Molecular Flasks

4.3.1 Ternary Complex Formation with γ-CD

γ-CD has a large cavity, which can accommodate two molecules of naphthalene derivatives. This two-guest inclusion phenomenon in CD chemistry was first reported in γ-CD-induced excimer fluorescence.[6] Solutions containing only sodium α-naphthylacetate (NA) at a high concentration exhibit a structureless band appearing at a longer wavelength than the normal band in the fluorescence spectrum. This band can be attributed to excimer emission since the ratio of intensities of excimer to monomer fluorescence is proportional to the concentration of naphthalene derivatives. Upon addition of α-, β-, and γ-CD, both normal and excimer fluorescence intensities were markedly enhanced only upon increasing γ-CD concentration. The accompanying enhancement of the normal fluorescence might be attributed to the formation of a 1 : 1 complex, but this rationalization was inconsistent with the observation that a dilute solution of NA (10^{-4} mol L^{-1}) showed no enhancement despite the presence of excess γ-CD (10^{-2} mol L^{-1}). Thus, the enhancement of normal fluorescence was attributed to a 1 : 2 complex formation. This is the first example in which two identical molecules were included into a CD cavity. The excimer fluorescence depended on the structure of the naphthyl derivative. It is impossible to predict what kind of monosubstituted naphthalene derivatives would be included in the γ-CD cavity with 1 : 2 stoichiometry. (Figure 4.1a). The fluorescence study also suggested that two different kinds of guest molecules could be included in the γ-CD cavity.[7] The fluorescence intensity of α-naphthyloxyacetic acid in aqueous solution was slightly enhanced by the addition of γ-CD, but markedly enhanced in the presence of both γ-CD and cyclohexanol, showing a ternary complex formation among γ-CD, α-naphthyloxyacetic acid and cyclohexanol (Figure 4.1b). These properties of γ-CD enabled us to make use of γ-CD as a reaction medium in which two different kinds of molecules, such as a catalyst and a substrate, could be brought together effectively. Ueno commented that this phenomenon showed that CD *mediates* the reactions, in other words, 'CD acts as molecular flask' in water.[8]

4.3.2 Organic Reactions Mediated by γ-CD

A modified γ-CD bearing an *O*-2-naphthylacetyl substituent produced larger overall ester hydrolysis rates in water (pH 8.7)[9] than native γ-CD. A 12-fold increase was observed for *p*-nitrophenyl acetate, arising from both an increased rate of intracomplex hydrolysis

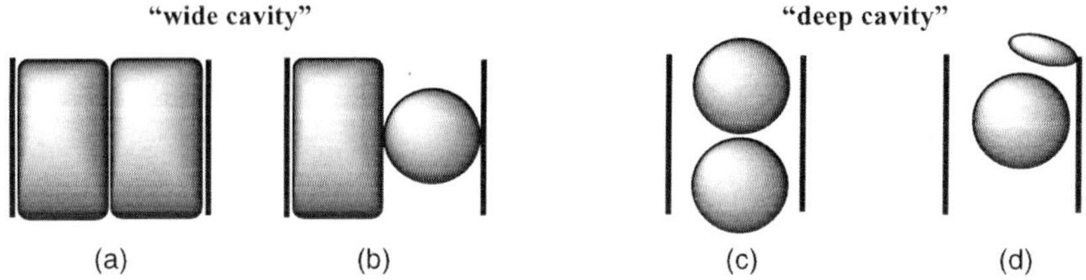

Figure 4.1 *Ternary complex formation with 'wide cavity (γ-CD)' and 'deep cavity (β-CD)'*

and stronger binding, and an 11-fold increase for *m*-nitrophenyl acetate, arising from stronger binding. The results were presumed to be due to spacer effects of the appended naphthalene moiety, which narrows the large γ-CD cavity.

Redox reactions can be controlled by complexes between one host and two guest molecules. For example, the redox reaction between *N*-benzylnicotinamide and ninhydrin was affected by the addition of CD. γ-CD facilitated the formation of the ternary complex of catalyst-substrate-γ-CD in a 1 : 1 : 1 ratio, in which the redox reaction proceeded. However, β-CD inhibited the formation of the catalyst-substrate complex, decreasing the reaction rate. This is an example in which the CD cavity controlled the catalyst-substrate complex formation.[10]

When γ-CD has one naphthyl pendant, it shows normal fluorescence whose intensity is enhanced by the addition of (–)-borneol, reflecting that the naphthyl pendant is included with the guest molecule in the γ-CD cavity. When naphthalene is added to a solution of the host, excimer emission appears as the result of the face-to-face interaction between the naphthalene pendant and the naphthalene guest in the cavity. When two molecules of the anthracene derivative were attached to the primary hydroxyl side of γ-CD, photoirradiation yielded the capped γ-CD with a rigid anthracene dimer on the primary hydroxyl side of γ-CD. The bis(anthracene-9-carbonyl)-γ-CD system gives only the *trans*-photodimer as a stable photodimer, but it may be possible to produce several stable photodimers if an anthracene-1-carbonyl unit was used in place of an anthracene-9-carbonyl unit. From this viewpoint, Ueno *et al.* prepared a series of bisanthracene modified-γ-CDs and used photodimerization in a γ-CD as a template. The anthracene groups of γ-CDs AB, AC, AD, and AE isomers underwent reversible dimerization upon irradiation, and afforded capping compounds. Although the *trans*-head-tail (HT) or *trans*-head-head (HH) dimer is usually the favored configuration in C-9-substituted, photoinduced anthracene dimers, the product from AB dimers displays a *cis*-head-head (HH) or *cis*-head-tail (HT) geometry only. AB-capped CD was the least stable of the AB, AC, AD and AE substituted CDs, owing to the strain within the short bridge. As a result, it was demonstrated that the stereochemistry of these bimolecular reactions can be controlled by this γ-CD method. Inoue has also reported that native γ-CD afforded HT photodimers in good yield and enantioexcess (ee).[11] The stereoselectivity for HH photoproducts improved in the anthracenecarbonyl-γ-CD system and depended on external effectors, such as temperature, solvent and pressure (Figure 4.2).[12]

Higher selectivity has been reported using 3,6-anhydro-γ-CD instead of native γ-CD.[13] It has been known that anhydro-CD has a distorted cavity. Interestingly, the isomer- and

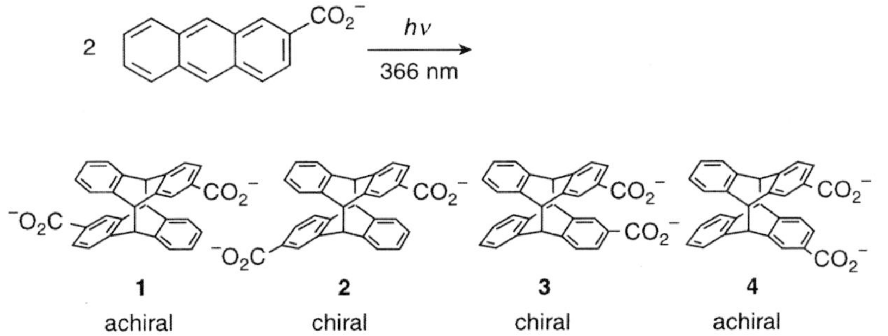

Figure 4.2 Enantioselective production of syn HT 2 anthracenecarboxylic acid dimer with AD capped γ-CD. Reprinted with permission from [12]. Copyright 2005 American Chemical Society

enantioselectivity of the photocyclodimerization are switched in the solid state. Solid-state photolyses gave inherently disfavored HH photodimers in much higher chemical and optical yields than in aqueous solutions. Circular dichroism (CD) spectra indicated that the switching of the stereoselectivity is primarily attributable to the varied arrangement of anthracene derivative pairs in the ground state.

Recently, the efficiency of CD for the enzymatic synthesis of neoglycolipids, which are barely water soluble, was described.[14] Increasing the concentration of γ-CD led to an increase in the limitation of the concentration of the sugar peptide substrate, which suggested that CDs did not work as surfactants or phase transfer systems but as molecular recognition systems. CD did not affect the HPLC purification and did not cause any lather. These merits are superior to the surfactants commonly used in such cases. In this study, a bridge between oligosaccharide synthesis using cell function *in vivo* and enzymatic synthesis *in vitro* was constructed which established the possibility of cellular-chemoenzymatic synthesis (Figure 4.3).

4.3.3 Ternary Complex Formation with β-CD

Since the cavity size of γ-CD is wider than that of β-CD, there should be more reactions mediated by γ-CD. However, there have been fewer reports of 'mediation' with γ-CD than with β-CD. Why? To mediate organic reactions with CD, at least two molecules must aggregate closely in the CD cavity. We have searched for reports showing a narrower cavity sized CD including two or more guest molecules. Azo-capped CD (AzCD), which was prepared from β-CD and 4,4′-bis(chlorocarbonyl)azobenzene, can form a ternary complex with some organic guest molecules (Figure 4.4).[15] These remarkable complexation properties were studied with the help of CD spectra. The *trans-cis* mixture in a photostationary state and the *trans*-isomer have distinctly different CD traces, as can be seen in Figure 4.5. The CD bands lose some of their intensity in the presence of guest molecules. From the concentration dependence of the ellipticity, conclusions can be drawn on both the stoichiometry and the association constant of the inclusion compound. The guest molecule, e.g., toluene, cyclohexanol, or methoxybenzene, and the configurationally isomeric terpenes, e.g., nerol and geraniol, were included in *cis*-AzCD with a 1 : 2 host–guest stoichiometry, but only in a 1 : 1 ratio in the *trans*-AzCD. The complexes of

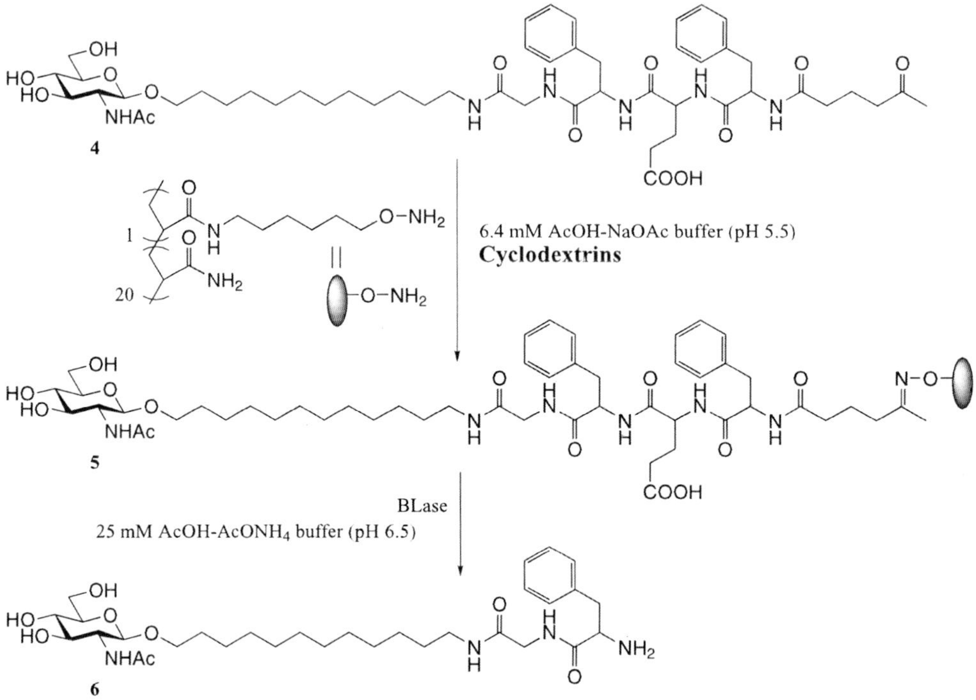

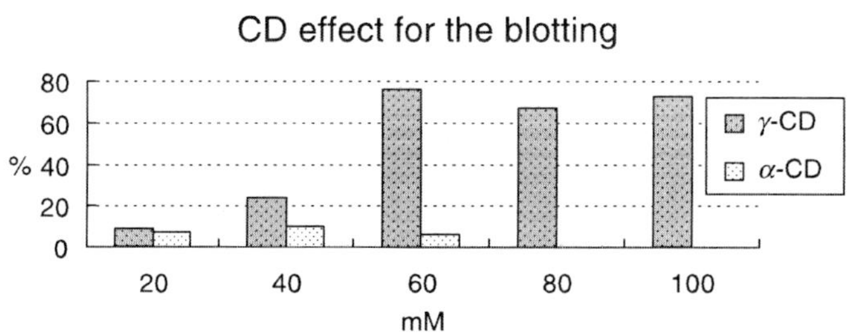

Figure 4.3 *Preliminary experiment for a model primer blotting to and releasing from the water-soluble polymer CDs, which were added in the blotting stage only. Reprinted with permission from [14]. Copyright 2008 Elsevier*

cis-AzCD are more stable than those of the corresponding native β-CD, whereas the *trans*-isomer shows less binding properties. It is interesting that the 1 : 1 *cis*-AzCD-geraniol complex is more stable than the corresponding nerol compound, but that in the 1 : 2 complexes the situation is reversed.

The different complexation properties of the *trans*- and *cis*-AzCD are attributed to the difference in cavity size induced by the configuration of the azo double bond. It is assumed that the cavity of *cis*-AzCD is deeper than both that of *trans*-AzCD and native CD. The cavity of *trans*-AzCD might be shallow and broader. This is the reason why the first guest molecule binds more strongly than the second. *p*-Nitrophenylacetate was hydrolyzed

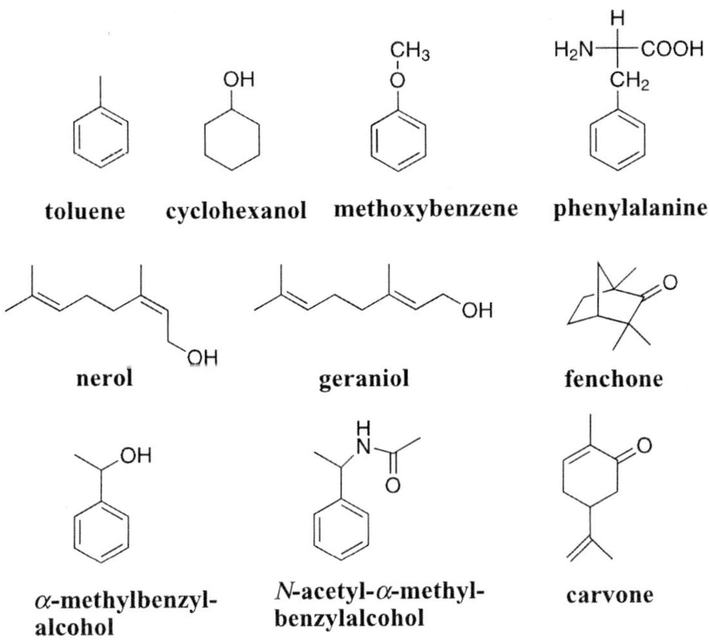

Figure 4.4 *Various guests included in AzCD forming 1 : 2 host–guest complex. Reprinted with permission from [15]. Copyright 1979 American Chemical Society*

preferentially by *cis*-AzCD over the *trans*-isomer, as deduced from the 5-fold increase in the apparent overall hydrolysis rate.[16] In addition, AzCD displays catalytic activity in the hydrolysis of *m*-nitrophenylacetate, whereas native β-CD has *m*-selectivity. This inversion is triggered by the deeper cavity space provided by *cis*-AzCD, which allows binding of the *p*-isomer in *cis*-AzCD.

With chiral guests, both *trans*- and *cis*-AzCD were found to form 1 : 1 and 1 : 2 host–guest complexes,[17] with the *cis*-complex being much more stable. The inclusion behavior towards chiral guests was of particular interest. The formation constants for 1 : 2 complexes are summarized in Table 4.1. Those for the 1 : 1 complex could not be determined since the values are too large to be measured. Differences in the formation constants of the enantiomeric guest molecules were observed, where the *trans*- and *cis*-AzCD show exactly opposite enantioselectivities. These results could be attributed to the changed depth of the CD cavity, from a shallow cavity to a deep one upon *trans* to *cis* photoisomerization of the cap azobenzene. The stereoselectivity observed might arise from the interactions of the guest molecule with the CD rim or wall corresponding to the shallow or deep depth of the cavity, although the chiral guest itself might also participate in the chiral discrimination. It is assumed that interactions of the guest with the rim or the wall of the CD are at the origin of the selectivity. Even though the enantioselectivity of chiral guests with AzCD complexes is relatively poor, this can nonetheless be called photocontrolled chiral recognition by complexation.

These results hold particular interest in the reversed photoregulated system of molecular functions. Thus, the AzCD results suggested that the cavity of β-CD can include more

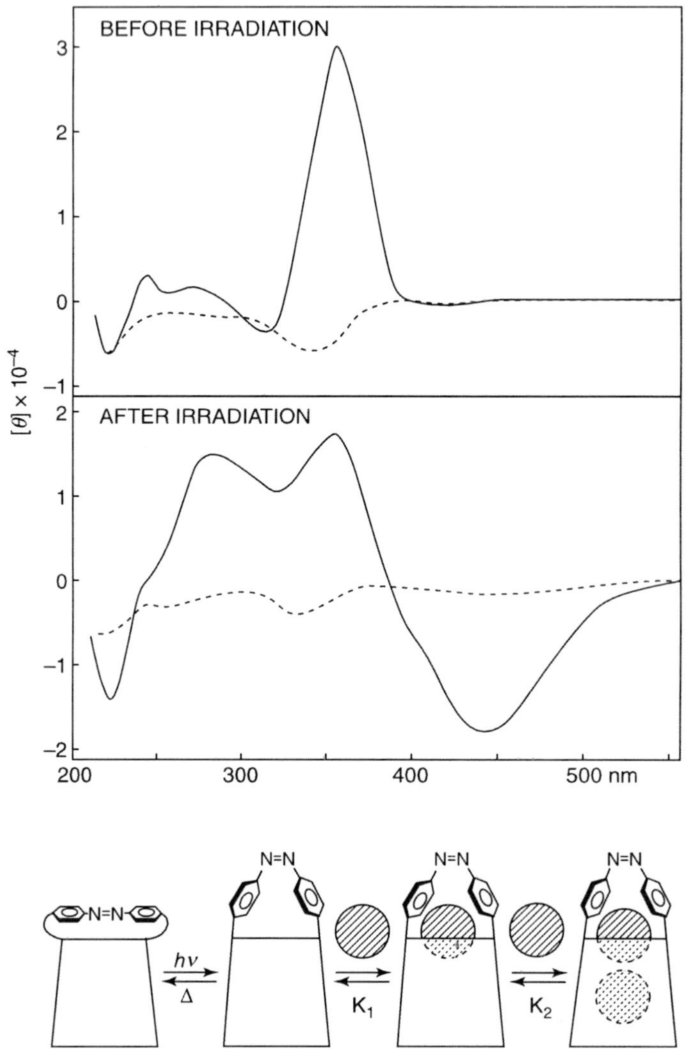

Figure 4.5 *Circular dichroism spectra of AzCD. (a) (—), trans-AzCD; (b) (----), trans-AzCD with a 2000-fold excess of cyclohexanol; (c) (—), cis-AzCD; (d) (----), cis-AzCD with a 2000-fold excess of cyclohexanol*

than two molecules in different conformations as in the wide cavity sized *γ*-CD (Figure 4.1). In addition, modification with hydrophobic groups, such as phenyl rings, supports the inclusion with interactions with the wall or cap of the cavity. One of the developing fields in CD chemistry is 'self-guest modification',[18] in which a *β*-CD sensoring system has been reported,[19] as well as molecular recognition systems or formation of suprapolymers.

Table 4.1 The formation constants of various guests with trans- and cis-AzCD

Guest	Enantiomeric form	Host	Formation constant[a] $K_2(M^{-1})$	K(L)/K(D)
Carvone	L(–)	*trans*-AzCD	204	1.29
	D(+)		158	
	L(–)	*cis*-AzCD	1550	0.92
	D(+)		1680	
Fenchone	L(–)	*trans*-AzCD	604	1.13
	D(+)		536	
	L(–)	*cis*-AzCD	1230	0.86
	D(–)		1480	
Phenylalanine	L	*trans*-AzCD	1.20	0.82
	D		1.47	
	L	*cis*-AzCD	45	1.22
	D		37	
α-methylbenzyl-alcohol	L	*trans*-AzCD	16.1	1.02
	D		15.8	
	L	*cis*-AzCD	145	1.29
	D		112	
N-acetyl-α-methyl-benzylamine	L	*trans*-AzCD	27.5	0.73
	D		37.5	
	L	*cis*-AzCD	2090	2.23
	D		939	

[a] pH 7.2, Tris-HCl buffer; $K_2 = [CD\text{-}G_2]/[CD\text{-}G][G]$.

4.4 Organic Reactions Mediated by CD in Water

The main characteristic effect of CDs is improvement of reaction selectivity. Herein, several examples of reactions mediated by CD are introduced. The reactions proceed in natural and mild conditions, in the presence of CD with water as the solvent. The reactions were achieved in good yields (over 80%) and CD could be reused in the system. The reaction conditions are described in the Schemes.

4.4.1 Catalytic Systems Based on Metal Complexes

Recently, there have been many reports on the effects of CD on metal catalytic systems. A potentially interesting approach for the modification of catalytic systems based on metal complexes, such as Zn, Mo, Fe, In, Sn, Pd or Ru, involves the use of CD as a protective agent. There are many reports on the interaction between CDs and metallic nanoparticles.[20] Immobilized CD on Pd nanoparticles was employed as an efficient mediator in aqueous biphasic hydrogenation reactions (Figure 4.6a,b).[21]

Randomly hydroxylated and methylated CD (RM-β-CD) with different cavity sizes have been used as biphasic aqueous catalysts in a palladium catalyzed Tsuji–Trost reaction with water-insoluble alkylallylcarbonates and alkylallylurethanes as substrates (Figure 4.6c).[22] The reaction rate and substrate selectivity strongly depended on the cavity size of the CD. The cavity size is a crucial factor in controlling the substrate selectivity. Moreover, phosphanes that did not interact with the methylated CD were the most

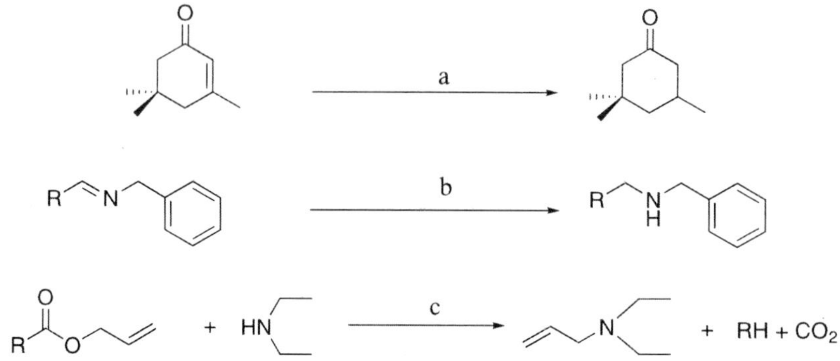

Figure 4.6 *The reactions mediated by β-CD in the presence of metal in aqueous solution.* ***a****: H_2, Pd, 25 °C, 1–5 h;* ***b****: H_2, Pd, 25 °C, 1–3 h, R = Ar;* ***c****: $Pd(OAc)_2$, TPPTS, R = R′O or R′R″N*

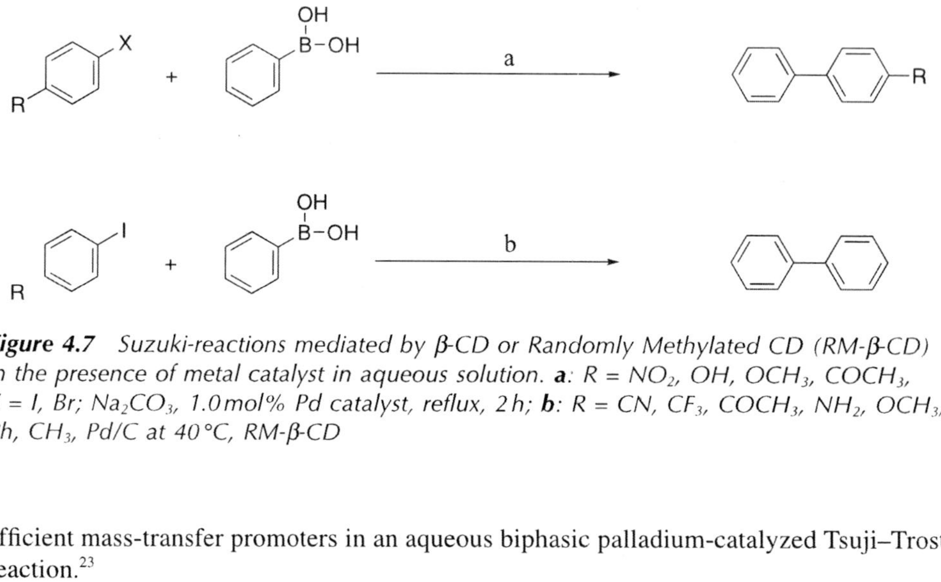

Figure 4.7 *Suzuki-reactions mediated by β-CD or Randomly Methylated CD (RM-β-CD) in the presence of metal catalyst in aqueous solution.* ***a****: R = NO_2, OH, OCH_3, $COCH_3$, X = I, Br; Na_2CO_3, 1.0 mol% Pd catalyst, reflux, 2 h;* ***b****: R = CN, CF_3, $COCH_3$, NH_2, OCH_3, Ph, CH_3, Pd/C at 40 °C, RM-β-CD*

efficient mass-transfer promoters in an aqueous biphasic palladium-catalyzed Tsuji–Trost reaction.[23]

The Suzuki reaction is the reaction of an aryl or vinyl boronic acid with an aryl or vinyl halide catalyzed by a palladium(0) complex. It is widely used to synthesize polyolefins, styrenes, and substituted biphenyls, and has been extended to incorporate alkyl bromides. Experimentally, the coupling reactions between aryl halides and arylboronic derivatives were mainly performed using a phosphine-palladium catalyst in homogeneous, heterogeneous, or biphasic systems.[24] RM-β-CDs have shown polyvalent properties in a biphasic Pd/C-catalyzed Suzuki–Miyaura reaction (Figure 4.7).[25] In addition to their mass transfer ability, CDs favored the dispersion of the catalyst in water. With the RM-β-CDs, the gains in initial activities were multiplied by factors between 3.8 and 343 depending on the nature of the substrates. This could be explained by assuming adsorption of the CD on the surface of the Pd/C catalyst. Indeed, adsorption of CD increases the hydrophilic character of the support, making its dispersion easier in the aqueous phase. The performance of this catalytic system is likely a consequence of a combination of two effects: the well-known mass transfer promoter properties of the CDs and their dispersing role on the Pd/C catalyst. This was the first example of CD-stabilized Pd/C particles for a catalytic application. PPV-based polyrotaxanes were prepared by coupling vinyl boronic acids to aryl iodides in the presence of α- or β-CDs. The crystal structure of the [2]rotaxane of this type was also determined.[26]

Chiral ruthenium complexes with a monoaminoalcohol modified β-CD were shown to catalyze the reduction of aliphatic ketones with up to 97% ee and in excellent yields in the presence of HCO_2Na.[27] The ratio of catalyst, substrate and sodium formate was 1:10:100. Although a number of highly enantioselective ruthenium-based hydrogen-transfer catalysts are known,[28] including one example that functions in water, none of these systems have been shown to reduce unconjugated ketones.[29] CD made the complex soluble in water and played an important role in the enantioselectivity through preorganization of the substrates in the hydrophobic cavity.

The use of CDs in molybdenum-catalyzed olefin epoxidation may be particularly advantageous in cases where the 'free' catalyst is initially highly active but loses activity during the reaction as a result of decomposition. CD may help to stabilize the metal catalyst and also facilitate subsequent recycling. η-$C_5H_5Fe(CO)_2Cl$, η-C_5H_5-$Mo(CO)_3Cl$, and η-C_5H_5-$Mo(CO)_3CH_2CONH_2$ are included in β-CD with a 1:1 stoichiometry. The latter could be used directly as precursors of catalysts for the epoxidation of cyclooctane with *t*-BuOOH[30] or H_2O_2. β-CD adducts have the potential to be used both in heterogeneous solid-liquid systems and biphasic liquid-liquid systems. The structures of single crystals of the η-$C_5H_5Fe(CO)_2Cl$ and permethylated-β-CD, heptakis-(2,3,6-*O*-methyl)-β-CD, were described. The metal-complex–CD exhibited significant catalytic activity in the olefin epoxidation system with H_2O_2, with cyclooctane oxide being the only observed reaction product.[31]

Homoallylic alcohols were directly prepared from aromatic acetals and dioxolanes ($\geq$ 80%) using a Barbier-type allylation, Zn/allyl bromide/NH4Cl, in the presence of β CD in water at 50 °C (Figure 4.8).[32] The synthesis of enantiomerically enriched homoallylic alcohols is an important goal in organic synthesis. An enantioselective metal-mediated allylation of substituted benzaldehydes with β-CD was achieved.[33] The reaction proceeds in short reaction times (1 h) at room temperature. The homoallylic alcohols are obtained in moderate yields with up to 93% enantioexcess.

4.4.2 Ring Opening Reactions

Reductive ring opening of epoxides with $NaBH_4$ has also been influenced by the addition of CDs (Figures 4.9 and 4.10). Styrene oxide smoothly underwent a ring-opening reaction with $NaBH_4$ and 2 mol equiv. of a β-CD to give 1-phenylethanol with up to 90% selectivity.[34] Chiral recognition in the complex formation was excluded as the reason for the enrichment of the enantioselectivity. The reaction rate of the $NaBH_4$ reduction for the β-CD-*R* form is faster than that for the β-CD-*S* form, this rate difference giving preferentially the *S* form. Similar phenomena have been reported in oxidations in crystalline complexes.[35] The reactions of other aryloxyepoxides were affected by the presence of β-CD to proceed regioselectively.[36] Highly regioselective ring opening of epoxides to

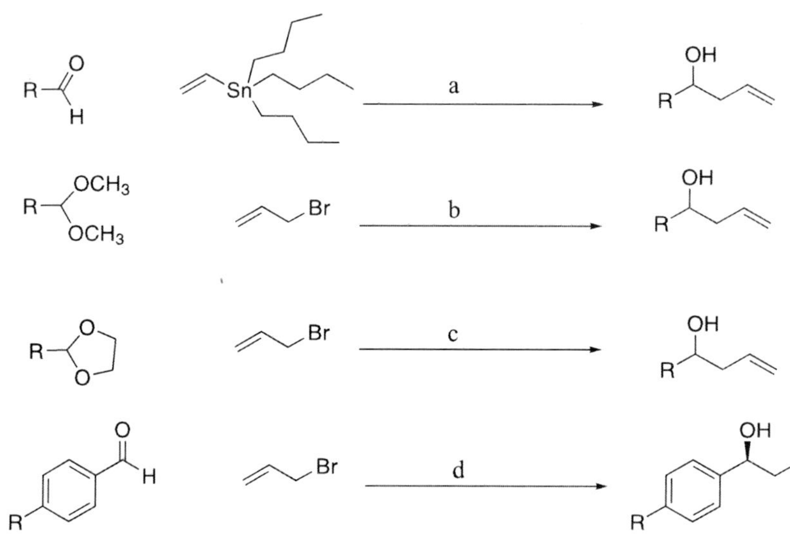

Figure 4.8 *Allylation mediated by β-CD in the presence of metal in aqueous solution.* **a**: *R = aryl, alkyl, naphthyl, HCl, 60 °C, 3 h;* **b,c**: *R = aryl, naphthyl, Zn · NH_4Cl, 50 °C;* **d**: *R = H, OCH_3, CH_3Cl, Metal (Zn, In or Sn), 1 h.*[32]

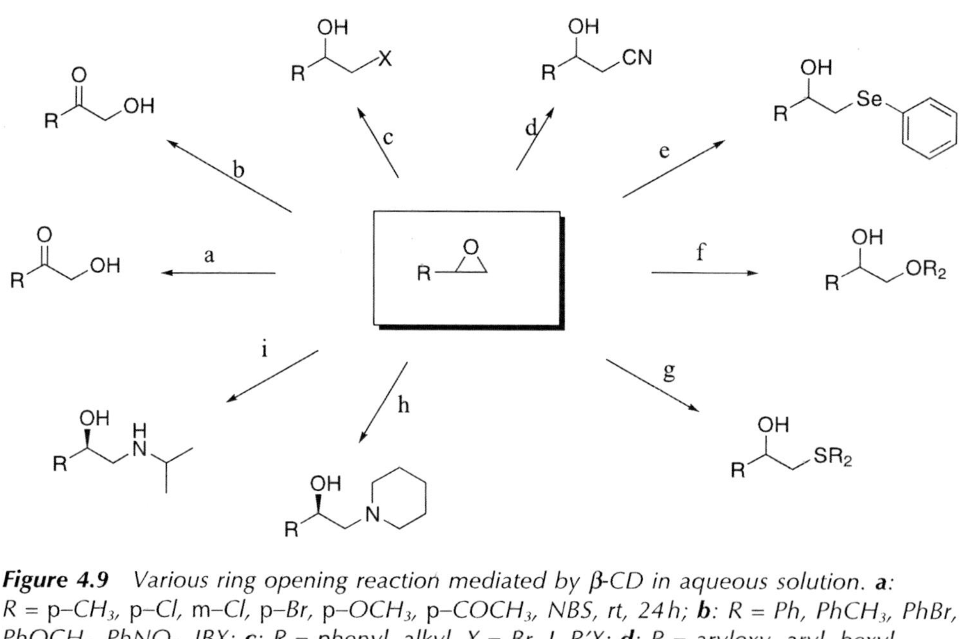

Figure 4.9 *Various ring opening reaction mediated by β-CD in aqueous solution.* **a**: *R = p-CH_3, p-Cl, m-Cl, p-Br, p-OCH_3, p-$COCH_3$, NBS, rt, 24 h;* **b**: *R = Ph, $PhCH_3$, PhBr, $PhOCH_3$, $PhNO_2$, IBX;* **c**: *R = phenyl, alkyl, X = Br, I, R′X;* **d**: *R = aryloxy, aryl, hexyl, NaCN, rt, 12 h;* **e**: *R = aryloxy, aryl, hexyl, PhSeH, rt, 25–40 min;* **f**: *R_1 = aryloxy, aryl, hexyl, R_2 = Ph, PhCl, $PhOCH_3$, 60 °C, 8 h;* **g**: *R_1 = aryloxy, aryl, hexyl, R_2 = Ph, PhCl, $PhOCH_3$, ArSNa, rt;* **h**: *R = H, Cl, CH_3, $CH_2CH_2OCH_3$,* **i**: *R = H, Cl, CH_3, $CH_2CH_2OCH_3$, $H_2NCH(CH_3)_2$, 3–12 h*

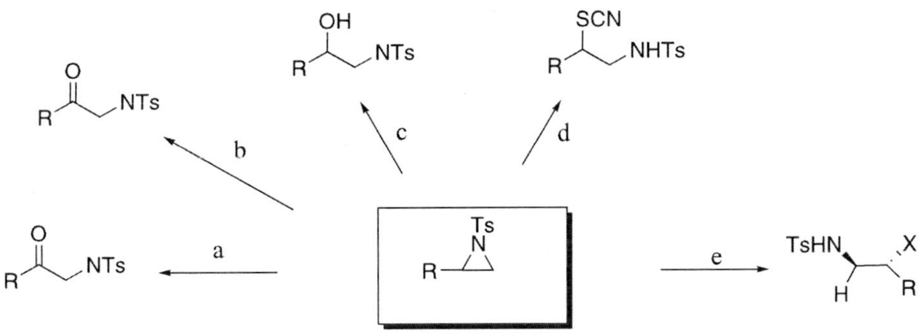

Figure 4.10 Various ring opening reaction mediated by β-CD in aqueous solution. **a**: R = p-CH_3, p-Cl, m-Cl, p-Br, p-OCH_3, p-$COCH_3$, NBS, rt, 24 h, **b**: R = Ph, $PhCH_3$, PhIBr, $PhOCH_3$, $PhNO_2$, IBX; **c**: R = aryloxy, aryl, hexyl, 50–60 °C; **d**: R = p–CH_3, p–Cl, p–Br, p–OCH_3, p–$COCH_3$, KSCN, rt; **e**: R = phenyl, alkyl, X = Br, I, TBAX, rt, 3–6 h

halohydrins has been carried out with hydrogen and lithium halides in the presence of β-CD.[37]

The β-CD inclusion complexes of aryloxyepoxides with amines afforded aminoalcohols enantioselectively in the solid state.[38] When carried out in water as the reaction medium, nearly racemic aminoalcohols were produced.

N-Activated aziridines are versatile intermediates for the synthesis of many biologically active compounds. Aziridines can be easily prepared and the inherent ring strain leads to high reactivity with various nucleophiles, such as organometallic reagents, Wittig reagents, amines, hydroxyl compounds, and metal halides (InX_3, NaX, $MgBr_2$, $CeCl_3/7H_2O/NaI$, TMSI). In the presence of β-CD, a variety of *N*-tosylaziridines underwent regioselective ring opening with tetrabutylammonium halides (TBAX) at pH 4 and room temperature to afford the corresponding haloamines (>84%).[39]

α-Tosylaminoketones could be generated directly from easily accessible aziridines with β-CD and *N*-bromosuccinimide (NBS) (≥85%) in a single step.[40] With the combination of β-CD and NBS, α-hydroxymethylarylketones were prepared from oxiranes at room temperature within 1–2 days.[41] α-Aminoarylketones provide starting materials for the synthesis of biologically active β-aminoalcohols. Only a few direct methods have been reported for their synthesis. However, these methods have various limitations such as the use of transition metal catalysts, organic solvents, controlled temperatures, long reaction times, and hazardous reagents. Thus, there is still a need to develop cleaner synthetic methodologies for the generation of α-aminoketones.

NBS is a commonly used oxidizing reagent. Selective oxidation of sulfides and addition to olefins was also mediated by CD. A simple and highly selective oxidation of sulfides to sulfoxides with NBS in the presence of β-CD has been reported (Figure 4.11a).[42] A series of sulfides were oxidized selectively without overoxidation to sulfones (≥90%). The reaction proceeded under neutral and mild conditions and could be carried out easily with recycling of CD and regeneration of NBS (Figure 4.11b).

vic-Halohydrins (bromo- and iodohydrins) were synthesized regioselectively in a one-step procedure by treating the corresponding olefins at room temperature with NBS and

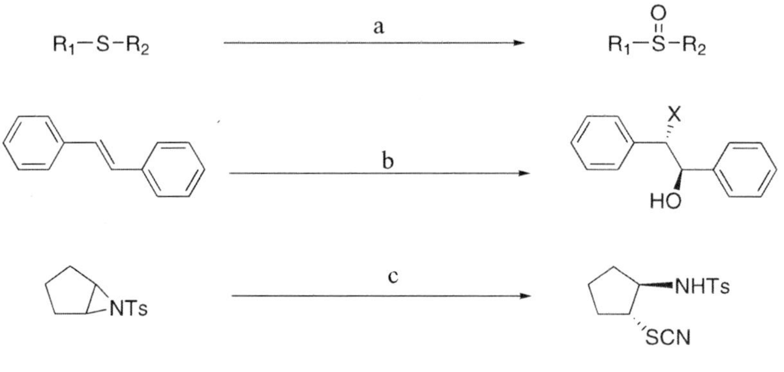

Figure 4.11 Reaction mediated by CD with NXS and KSCN. **a**: R_1 = aryl, naphthyl, R_2 = alkyl, NBS, rt, 6–8 h; **b**: X = Br, I, NXS, rt, 40 min; **c**: KSCN, rt

Figure 4.12 Preparation of carbonyl compounds mediated by CD. R = H, CH_3, Br, Cl, OCH_3, NO_2, IBX, rt, 40–120 min

NIS (*N*-halosuccinimides), respectively.[43] With KSCN instead of NBS, aziridines were cleaved regioselectivity (Figure 4.11c).[44]

The direct one-pot synthesis of *β*-hydroxy selenides was reported. In the presence of *β*-CD, highly regioselective ring-opening of oxiranes with benzenesenol was achieved (≥75%).[45] *β*-CD could also be recovered and reused.

In the presence of *β*-CD, ring opening of epoxides with various thiophenols proceeds with high regioselectivity. *β*-CD can also be recovered and reused in various runs without affecting the efficiency of the process.[46] Highly regioselective ring opening of oxiranes to *β*-hydroxy ethers with phenoxides has been achieved (≥94%).[47] *β*-hydroxynitriles and *β*-aminoalcohols have been synthesized regioselectively from epoxides and *N*-tosyl azidines in the presence of *β*-CD.[48]

2-Iodoxybenzoic acid (IBX) is attractive due to its low toxicity, ease of handling, and moisture stability. Various epoxides and aziridines were cleaved with IBX to afford *α*-hydroxyketones and *α*-aminoketones, respectively.[49] Carbonyl compounds were also prepared from the corresponding tetrahydropyranyl (THP) ethers directly with IBX and *β*-CD. The yields obtained were up to 96% (Figure 4.12).[50] No overoxidized products were detected in the case of aldehydes. In these reactions, the role of CD appears to be to activate the THP ethers by hydrogen bonding and thereby facilitating the hydrolysis. Since the *β*-CD cavity is hydrophobic, it may also be forming reversible complexes with the THP ethers. When the reaction was performed without IBX, only deprotection occurred and no oxidized product was isolated. Without CD, no deprotection took place and hence no oxidation occurred.

4.4.3 Addition

Allylation of carbonyl compounds is an important carbon–carbon bond forming reaction in organic chemistry for the preparation of homoallylic alcohols. Homoallylic alcohols are useful tools for the construction of complex molecules and are important building blocks for the synthesis of natural products. A highly efficient allylation of aldehydes with allyltributyltin has been developed mediated by *β*-CD in the presence of HCl without any metal catalysts to afford the corresponding homoallylic alcohols (≥88%: Figure 4.8a).[51] CD can be recovered and reused for a number of runs without significant loss of activity. The key process is inclusion in the CD cavity. This process is more important than hydrogen bonding between the aldehyde and hydroxyl group on the rim of the CD cavity. As described earlier, homoallylic alcohols could be prepared by adding CD to the metal catalyzed reaction systems (Figure 4.8).

The aza-Michael reaction, which is an important reaction for the synthesis of heterocycles containing a *β*-amino carbonyl unit, could be mediated by CD in water to produce the corresponding *β*-amino compounds (>80%) under mild conditions, at room temperature and for 6–8 h reaction time (Figure 4.13a).[52] Interestingly, Michael additions of nitroalkanes to methyl vinyl ketone are also known to occur in the presence of sugars.[53] No reaction was observed without CD. Hydrogen bonding of amines with the CD hydroxyl makes the N–H bond weaker, enhancing the nucleophilicity of nitrogen for addition to electron-deficient alkenes.

Michael addition of thiols from the secondary side of *β*-CD to *α*,*β*-unsaturated compounds at the primary side was described (up to 97%: Figure 4.13b).[54] Products of undesirable side reactions resulting from polymerization were not observed. In addition, the use of CD precludes the use of either acid or base and the catalyst can be recovered and reused.

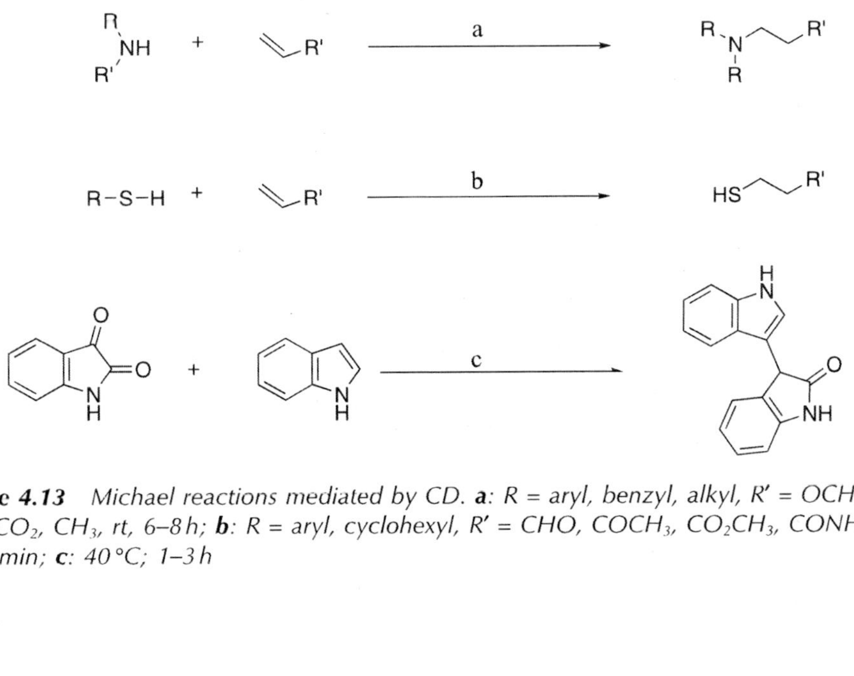

Figure 4.13 *Michael reactions mediated by CD.* ***a****: R = aryl, benzyl, alkyl, R′ = OCH_3, CN, CO_2, CH_3, rt, 6–8 h;* ***b****: R = aryl, cyclohexyl, R′ = CHO, $COCH_3$, CO_2CH_3, $CONH_2$, rt, 5–45 min;* ***c****: 40 °C; 1–3 h*

3-Indolyl-3-hydroxy oxindoles were prepared by β-CD complexes with indoles under neutral conditions in water to give up to 94% yields (Figure 4.13c).[55]

Phosphorus–carbon bond formation has attracted growing attention because of its application in organic synthesis and bioorganic chemistry. α-Functionalized phosphonic acids are valuable intermediates for the preparation of medical compounds and synthetic intermediates. 1-Aminophosphonic acids are important substrates for the corresponding α-amino acids in biological systems. The Kabachnik–Fields synthesis, which is the most convenient method to synthesize 1-aminophosphonic acids, is catalysed by base or acids such as $SnCl_2$, $BF_3.Et_2O$ and $MgBr_2$ and cannot be carried out in a one-step operation. However, α-aminophosphonic esters were obtained in a one-pot synthesis with β-CD (45–82%) with good purity under mild conditions by the reaction of diethyl phosphate with a mixture of aldehyde and amine at reflux for 12–24 h (Figure 4.14).[56]

The Strecker reaction that is the nucleophilic addition of trimethylsilyl cyanide (TMSCN) to imines in water has been developed in the presence of β-CD to afford α-aminonitriles (Figure 4.15a).[57] The use of CD precludes the use of either acid or base, and the catalyst can be recycled a number of times without loss in activity. No reaction

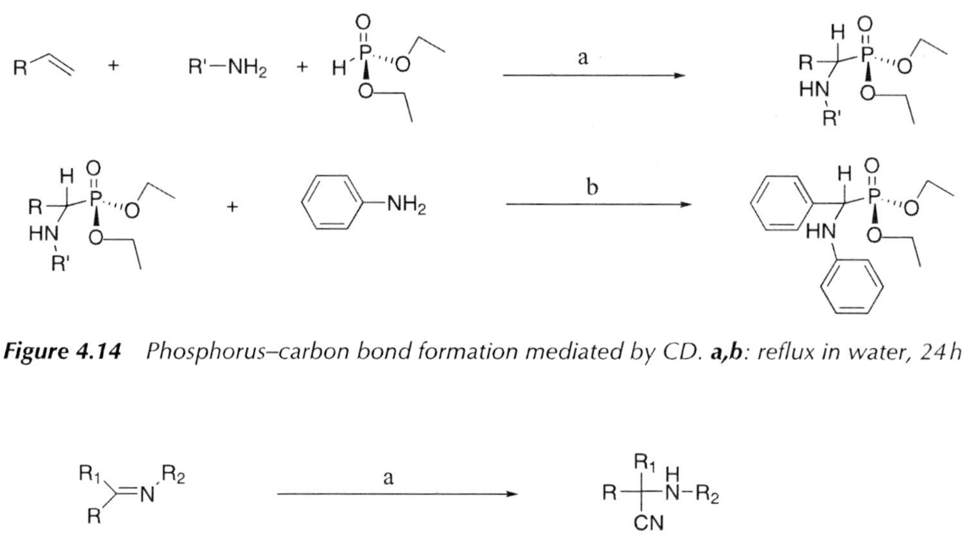

Figure 4.14 *Phosphorus–carbon bond formation mediated by CD.* ***a,b****: reflux in water, 24 h*

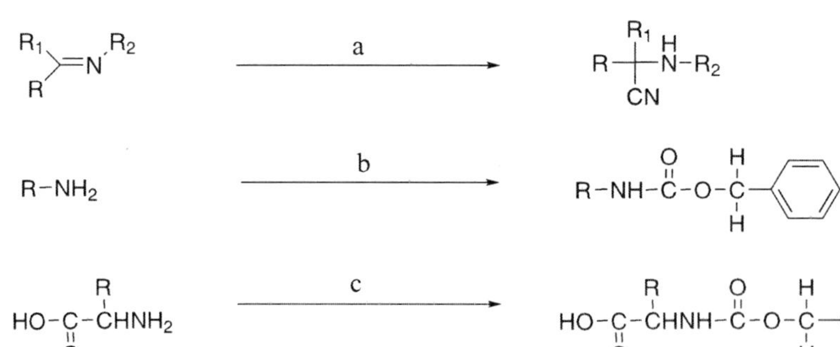

Figure 4.15 *Nitrogen–carbon bond formation mediated by CD.* ***a****: R = aryl, naphthyl, R_1 = H, CH_3, R_2 = aryl, benzyl, tosyl, TMSCN, 1 h;* ***b****: Cbz, rt, 1–4 min;* ***c****: Cbz, rt, 1–15 min*

was observed without CD. The mechanism was suggested as follows: hydrogen bonding of the CD hydroxyl with the nitrogen of the imine increases the electrophilicity of the imine carbon, thus activating it for attack by the cyanide ion.

Monoprotection of amines/amino acids with Cbz–Cl proceeded in only 1 h in high yield (≥90%) with 0.1 mol% of β-CD.[58] This reaction proceeds without the formation of by-products and has advantages over existing methods (Figure 4.15b and 4.15c).

The thiazole ring system is a useful structural motif found in numerous biologically active molecules. Addition of thioamide or thiourea to the aqueous solution of phenacyl bromide-β-CD complex gave the corresponding thiazole or aminothiazole, respectively (≥82%), without the formation of by-products or rearranged products (Figure 4.16).[59] The role of CD appears to be to activate and solubilize the phenacyl bromide, and drive the reaction to completion in decreased reaction times. Without CD, the reaction took place but the yields were poor (20%). Selenazoles were also prepared from α-bromoketones and selenourea in the presence of β-CD at 50 °C under atmospheric pressure.[60]

The aldol reaction is an important carbon–carbon bond formation reaction in organic chemistry. To get high enantioselectivity, the use of a chiral secondary amine, such as proline, has been reported. Interestingly, proline-catalyzed aldol reactions do not show any non-linear effects. This strategy allows the otherwise challenging cross-aldol reaction between two aldehydes. Proline modified β-CD catalyzes the asymmetric aldol reaction.[61] The aldol reaction of various aromatic aldehydes with cyclohexanone is mediated by the inclusion complex of a proline derivative and β-CD, yielding hydroxy-ketones with an *anti/syn* ratio of up to 99 : 1 and ee values well above 90%.[62] The asymmetric catalytic system mediated by sulfated β-CD, which can bind an organo-catalyst of *tert*-butylphenoxyproline and associated hydrophobic reactant in a water–organic solvent phase system, has been developed.[63] With this system, only 10 mol% of CD induces the highly enantiopure aldol products in near quantitative yields.

The use of CD provides a unique way to study the reactivity of cation radicals of small oligothiophenes and oligopyrroles in water,[64] especially in the case of bithiophene and terthiophene for which the lack of water solubility does not allow such a study. After photochemical oxidation, the cation radical leaves the CD host to undergo coupling with another cation radical, similar to reactions occurring in organic media, with no reaction between cation radicals.

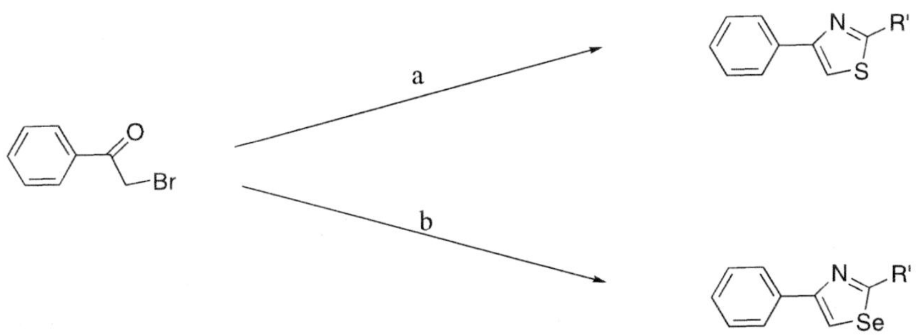

Figure 4.16 *Thiazole (a) and selenazole (b) ring formation mediated by CD. 40 °C, 1–2 h*

Figure 4.17 Polymerization mediated by RM-CD at room temperature

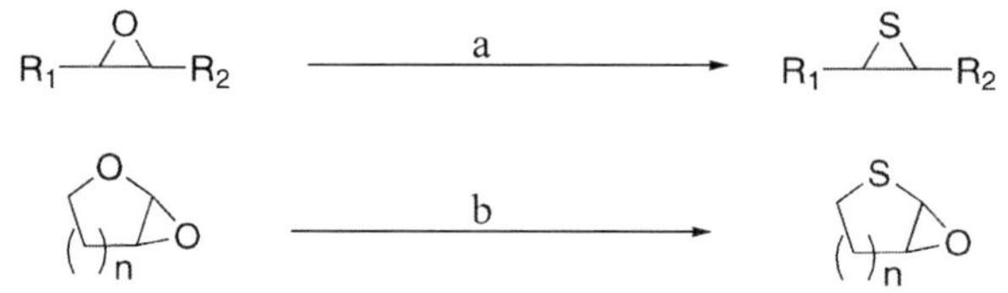

Figure 4.18 Conversion of oxiranes to thiiranes mediated by CD at room temperature within 3–6 h

Fluorinated monomers, 4-(*N*-adamantylamino)-2,3,5,6-tetrafluorostyrene and 2,3,4,5,6-pentafluorostyrene, were polymerized in water after complexation with RM-*β*-CD without the use of surfactants or cosolvents (Figure 4.17).[65] CD induced high reactivity and formation of stable poly(2,3,4,5,6-pentafluorostyrene) latex particles. The CD ring has a strong influence on copolymerization parameters compared to the uncomplexed monomers.

Thiiranes, which are the simplest sulfur heterocycles, are useful from both theoretical and synthetic points of view. The most important method for their synthesis is the conversion of oxiranes to thiiranes by an oxygen–sulfur exchange reaction. Oxiranes react smoothly with thiourea in the presence of *β*-CD to afford the corresponding thiiranes and the *β*-CD can be recycled (Figure 4.18).[66]

4.4.4 Oxidation and Reduction

After precomplexation with *β*-CD, a variety of alcohols, including aromatic alcohols, were oxidized to their corresponding carbonyl compounds in good yields with NaOCl–KBr in aqueous solution.[67] A substrate-selective and transition metal-free oxidation of benzoic and allylic alcohols with NaOCl oxidant mediated by *β*-CD in water was developed.[68] In the presence of one molar equivalent of *β*-CD, benzyl alcohol, 4-methoxybenzyl alcohol and some primary aromatic alcohols were oxidated to form benzaldehyde, 4-methoxybenzaldehyde and aromatic aldehydes, respectively, at 50 °C for 1–4 h. When 20% of acetone was added to the reaction system, the yield of aldehyde was dramatically decreased.

$NaBH_4$ is the most generally used reduction reagent in aqueous solution, albeit with poor selectivity. The highly asymmetric reduction of ketones with $NaBH_4$ has been performed with crystalline CD complexes since the 1970s.[69,70] It was suggested that hydrogen bonding between the carbonyl group of the substrate and the *β*-CD secondary hydroxyl groups may result in the enantiomeric face-selective attack of the hydride anion.

Doussot *et al.* have reported that regio- and stereoselectivity in the reduction of substituted epoxides and aromatic ketones with $NaBH_4$ in the presence of CDs depended mainly on interactions with the wider rim, the secondary hydroxyl side, of the CD. This indicates that an alkoxyborohydride intermediate is formed in the first step of the reaction.[71]

Various α-azido alkyl ketones were reported to be reduced with good enantiospecificity by $NaBH_4$ in the presence of β-CD. High enantioselectivity was induced by the interaction between the substrate and hydroxyl groups on the wider rim of CD. The high selectivity may be due to the stability of the CD–substrate complexes and the orientation of the carbonyl group within the complex. Sodium dithionate, which is also easily handled and is a water-soluble reducing reagent, has been used with β-CD for the stereoselective reduction of (R)-(−)-carvone,[72] menthone,[73] and compounds having carbon–carbon double bonds next to the α,β-carbonyl group.[74] Interestingly, a remarkable effect by CD on the menthone product was found in mixed-solvent systems. Aqueous DMF (1 : 1) increased the yield from 14% without CD to 76% with β-CD in water, the menthol/neomenthol ratio being 3.6/1.0 under phase-transfer conditions. Lipophilic CD, heptakis(2,6-*O*-methyl)-β-CD (DMCD), in a water–benzene mixture gave an 82.0% yield with good stereoselectivity (the menthol/neomenthol ratio being 2.0/1.0). DMCD also affected the reaction of compounds that have carbon–carbon double bonds next to the α,β-carbonyl group in a water–toluene mixture (1 : 1). Both complexation and reduction seem to occur mainly in the organic phase, leading to a very high yield.

In aqueous or organic solvents, it is possible that the reagent or substrate can still form a host–guest complex through specific interactions with the CD. Adequate modification of the CD to form new interaction points or adequate change of the cavity can be expected to result in appreciable ee in an asymmetric reaction.

Higher enantioselectivity in the reduction of benzoylformic acid (BFA) using $NaBH_4$ has been reported using mono-6-deoxyamino-β-CD (ACD) as the host, instead of β-CD, in neutral aqueous solution.[75] The phenyl group is used to control cavity size and the sp^3 carbons between the parent CD cavity and the phenyl group act as a flexible arm. As shown in Figure 4.19, the enantioselectivity was related to the orientation of the substrate in the presence of a 'self-guest group'. In the case of BFA, when the substrate is held between the cavity and the bulky hydrophobic substituent of the modified CD, a higher ee is obtained. When BFA cannot participate in self-inclusion, then lower ee's are observed. In the case of an oval-shaped substrate, such as indol-3-pyruvic acid (IPA), which cannot participate in the self-inclusion, but can form an outside complex near the rim of the CD derivative, high enantioselectivities were observed with almost all of the amino-CD derivatives, irrespective of the position of the substituent.[76] These results suggest that the rim of the CD cavity also may lead to substantial improvements in organic reactions.

Reduction of mononitroarenes mediated by β-CD has been reported[77] using hydroxide ion as a reductant. Ordinarily the reducing ability of OH^- in water is very low as a result of its stabilization by hydration. Reductions by OH^- have only been observed in aprotic organic solvents. CD includes the reactant (nitrobenzene) and the reaction is carried out near the rim of the cavity.

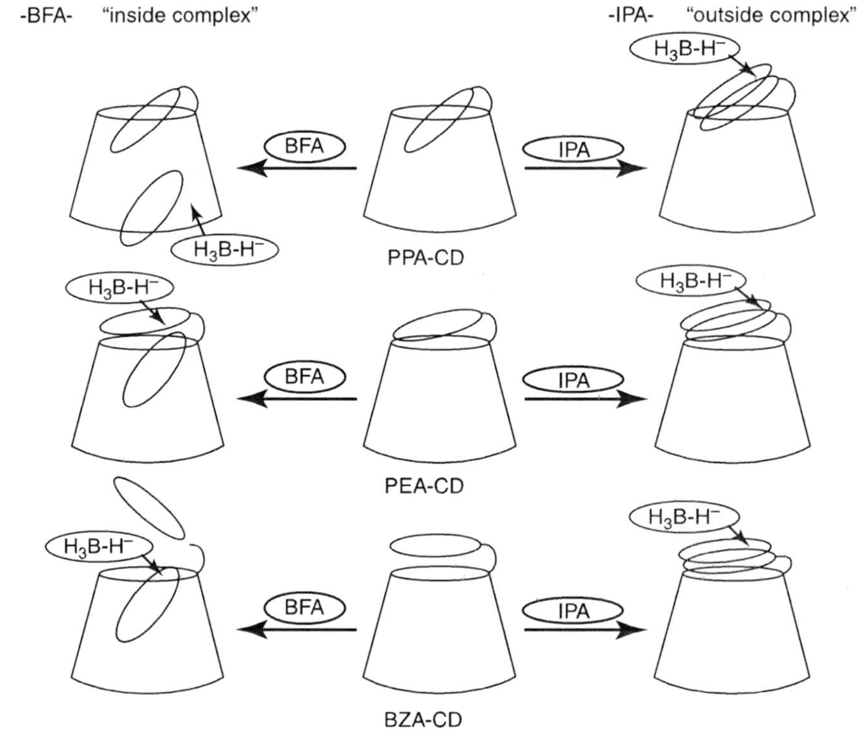

Figure 4.19 Suggested orientation of substituent-modified CD complex

4.5 Conclusion

Many organic reactions have been carried out with CD cavities in water. In the early stages of CD chemistry, the subject was limited to organic chemistry. Many reactions related to the inorganic field have since been reported. The most attractive advantage of CD is its 'selectivity', however, the solvation effect cannot be ignored. To develop CD-mediated reactions, it is necessary to clarify the reaction mechanism in detail. Complexation certainly induces selectivity. Although it is difficult to elucidate the role of the CD, complex orientation in the reaction transition state must be made clear at the molecular level.

The early stages in CD chemistry presented many chemists with a very impressive concept, that of a host–guest complex which has been described as a ball in a bottomless pail. We have shown that the ball or guest is not always in the pail cavity. Other interactions play sufficiently significant roles in that they recognize the ball and thus mediate the reaction. Indeed, hydrogen bonding between the guest and secondary hydroxyl groups at the wider rim of the cavity are suggested to explain the reaction mechanism. The so-called 'rims' of the pail also have to be considered. The rims have great potential to interact with the guest molecules through hydrogen bonds polar interaction supported by the hydrophobic cavity. Without rims, CD cannot mediate that many reactions. This is

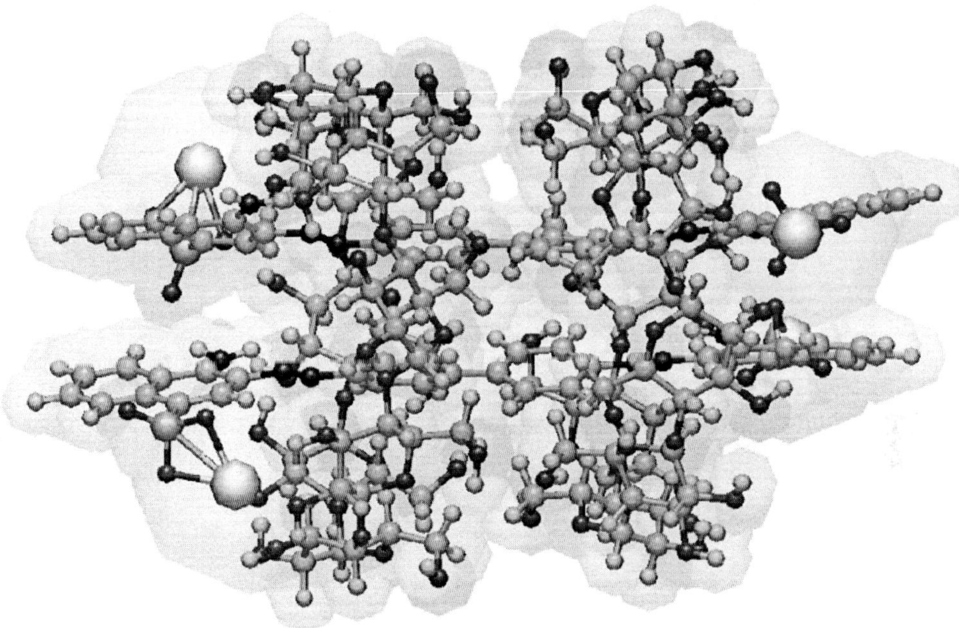

Figure 4.20 Molecular structure of Congo red-γ-CD 2:2 complex optimized by MOPAC

the reason why β-CD derivatives can mediate many reactions. More recently, computational chemistry has played a role in understanding CD chemistry. However, the relationship between CD chemistry and molecular modelling has not yet been established. A simple and easy method to express the molecular structure based on the practical results is needed. CD is not a bottomless pail. Where are the 'hydrophobic fields'? Which CD has a symmetrical round cavity? CD has carbon, hydrogen and oxygen atoms forming glucose rings. The molecular structure of CD which is drawn in molecular level can provide new applications in supramolecular chemistry. For example, molecular drawing of the inclusion complex between two γ-CDs and two Congo red molecules with MOPAC are indicated in Figure 4.20.[78] The methods with CD are very simple, generating high yields in water. Water is a safe, economical and environmentally friendly solvent. Since green chemistry has become such an important issue in the 21st century, reactions mediated by CD warrant increased attention.

References

1. M. L. Bender, M. Komiyama, *Cyclodextrin Chemistry*, Springer, Berlin, 1978.
2. R. Van Etten, J. F. Sebastian, G. A. Clowes, M. L. Bender, Acceleration of phenyl ester cleavage by cycloamyloses, a model for enzymatic specificity, *J. Am. Chem. Soc.*, 1967, **89**, 3242–3253; R. Van Etten, J. F. Sebastian, G. A. Clowes, M. L. Bender, The mechanism of the cycloamylose-accelerated cleavage of phenyl esters, *J. Am. Chem. Soc.*, 1967, **89**, 3253–3261.

3. K. Takahashi, Organic reactions mediated by cyclodextrins, *Chem. Rev.*, 1998, **98**, 2013–2034.
4. D. C. Rideout, R. Breslow, Hydrophobic acceleration of Diels-Alder reactions, *J. Am. Chem. Soc.*, 1980, **102**, 7816–7817.
5. C. J. Li, T. H. Chan, *Organic Reactions in Aqueous Media*, Wiley, New York, 1997; U. M. Lindstrom, *Organic Reactions in Water: Principles, Strategies and Applications*, Blackwell Publishing, Oxford, 2007.
6. A. Ueno, K. Takahashi, T. Osa, One host-two guests complexation between γ-cyclodextrin and sodium-α-naphthylacetate as shown by excimer fluorescence, *J. Chem. Soc., Chem. Commun.*, 1980, 921–922.
7. A. Ueno, K. Takahahsi, Y. Hino, T. Osa, Fluorescence enhancement of α-naphthyloxyacetic acid in the cavity of γ-cyclodextrin, assisted by a space-regulating molecule, *J. Chem. Soc., Chem. Commun.*, 1981, 194–195.
8. A. Ueno, *Supramolecular Science*, Sangyo Tosho, Tokyo 1993.
9. A. Ueno, F. Moriwaki, Y. Hino, T. Osa, Ester hydrolysis by a 2–naphthylacetyl-substituted γ-cyclodextrin, *J. Chem. Soc., Perkin Trans. 2*, 1985, 921–923.
10. K. Takahashi, *Studies on the Control of Various Aqueous Reactions by Host–Guest Complexation of Cyclodextrins*, Pharmaceutical Institute Tohoku University, 1986.
11. A. Nakamura, Y. Inoue, Supramolecular catalysis of the enentiodifferentiating [4+4]photocyclodimerization of 2–anthracenecarboxylate by γ-cyclodextrin, *J. Am .Chem. Soc.*, 2003, **125**, 966–972.
12. A. Nakamura, Y. Inoue, Electrostatic manipulation of enentiodifferentiating photocyclodimerization of 2–anthracenecarboxylate within γ-cyclodextrin cavity through chemical modification. Inverted product distribution and enhanced enantioselectivity, *J. Am. Chem. Soc.*, 2005, **127**, 5338–5339; C. Yang, A. Nakamura, T. Wada, Y. Inoue, Enantiodifferentiating photocyclodimerization of 2–anthracenecarboxylic acid mediated by γ-cyclodextrins with a flexible or rigid cap, *Org. Lett.*, 2006, **8**, 3005–3008; C. Yang, A. Nakamura, G. Fukuhara, Y. Origane, T. Mori, T. Wada, Y. Inoue, Pressure and temperature-controlled enantiodifferentiating [4+4] photocyclodimerization of 2–anthracenecarboxylate mediated by secondary face- and skeleton-modified γ-cyclodextrins, *J. Org. Chem.*, 2006, **71**, 3126–3236.
13. C. Yang, M. Nishijima, A. Nakamura, T. Mori, T. Wada, Y. Inoue, A remarkable stereoselectivity switching upon solid-state versus solution-phase enantiodifferentiating photocyclodimerization of 2–anthracenecarboxylic acid mediated by native and 3,6–anhydro-γ-cyclodextrins, *Tetrahedron Lett.*, 2007, **48**, 4357–4360.
14. I. Nagashima, H. Shimizu, T. Matsushita, S. Nishimura, Chemical and enzymatic synthesis of neoglycolipids in the presence of cyclodextrin, *Tetrahedron Lett.*, 2008, **49**, 3413–3418.
15. A. Ueno, H. Yoshimura, R. Saka, T. Osa, Photocontrol of binding ability of capped cyclodextrin, *J. Am. Chem. Soc.*, 1979, **101**, 2779–2780.
16. A. Ueno, K. Takahashi, T. Osa, Photocontrol of catalytic activity of capped cyclodextrin, *J. Chem. Soc., Chem. Commun.*, 1981, 94–95.
17. A. Ueno, R. Saka, K. Takahashi, T. Osa, Photocontrol of chiral recognition by capped cyclodextrin, *Heterocycles*, 1981, **15**, 671–674.
18. A. Ueno, T. Kuwabara, A. Nakamura, A modified cyclodextrin as a guest responsive colour-change indicator, *Nature*, 1992, **356**, 136–137.
19. A. Ueno, Fluorescent sensors and color-change indicators for molecules, *Adv. Materials*, 1993, **5**, 132–134; A. Ueno, Fluorescent cyclodextrins for molecular sensing, *Supramol. Sci.*, **3**, 31–36; T. Aoyagi, A. Nakamura, H. Ikeda, T. Ikeda, H. Mihara, A. Ueno, Alizarin yellow-modified β-cyclodextrin as a guest-responsive absorption change sensor, *Anal. Chem.*, 1997, **69**, 659–663.
20. A. Nowicki, Y. Zhang, B. Leger, J.-P. Rolland, H. Bricout, *Chem. Commun.*, 2006, 296–298, S. Guiffrida, G. Ventimiglia, S. Petralia, S. Conoci, S. Sortino, *Inorg. Chem.*, 2006, **45**, 508–510.
21. S. C. Mhadgut, K. Palaniappan, M. Thimmaiah, S. A. Hackney, B. Torok, J. Liu, A metal nanoparticle-based supramolecular approach for aqueous biphasic reactions, *Chem. Commun.*, 2005, 3207–3209.

22. C. Torque, H. Bricout, F. Hapiot, E. Monflier, Substrate-selective aqueous organometallic catalysis. How size and chemical modification of cyclodextrin influence the substrate selectivity, *Tetrahedron*, 2004, **60**, 6487–6493.
23. M. Ferreira, H. Bricout, A. Sayede, A. Ponchel, S. Foumentin, S. Tilloy, E. Monflier, Biphasic aqueous organometallic catalysis promoted by cyclodextrins: how to design the water-soluble phenylphosphane to avoid interaction with cyclodextrin, *Adv. Synth. Catal.*, 2008, **350**, 609–618.
24. N. Miyaura, A. Suzuki, *Chem. Rev.*, 1995, **95**, 2457–2483.
25. A. Cassez, A. Ponchel, F. Hapiot, E. Monflier, Unexpected multifunctional effect of methylated cyclodextrins in a palladium charcoal-catalyzed Suzuki-Miyaura reaction, *Org. Lett.*, 2006, **8**, 4823–4826; L. Strimbu, J. Liu, A. E. Kaifer, Cyclodextrin-capped palladium nanoparticles as catalysts for the Suzuki reaction, *Langmuir*, 2003, **19**, 483–485.
26. J. Terao, A. Tang, J. Michels, A. Krivokapic, L. Anderson, Synthesis of poly(*para*-phenylenevinylene) rotaxanes by aqueous Suzuki coupling, *Chem. Commun.*, 2004, 56–57.
27. A. Schlatter, M. K. Kundu, W.-D. Woggon, Enantioselective reduction of aromatic and aliphatic ketones catalyzed by ruthenium complexes attached to *β*-cyclodextrin, *Angew. Chem. Int. Ed.*, 2004, **43**, 6731–6734.
28. R. Noyori, Asymmetric catalysis: science and opportunities (Nobel lecture), *Angew. Chem. Int. Ed.*, 2002, **41**, 2008–2022.
29. H. Y. Rhyoo, H.-J. Park, W. H. Suh, Y. K. Chung, *Tetrahedron Lett.*, 2002, **43**, 269.
30. S. S. Braga, S. Gago, J. D. Seixas, A. A. Valente, M. Pillinger, T. M. Santos, I. S. Gonçalves, C. C. Romao, *Inorg. Chim. Acta*, 2006, **359**, 4757–4764.
31. S. S. Balula, A. C. Coelho, S. S. Braga, A. Hazell, A. A. Valente, M. Pillinger, J. D. Seixas, C. C. Romao, I. S. Gonçalves, Influence of cyclodextrins on catalytic olefin epoxidation with metal-carbonyl compounds, crystal structure of the TRIMEB complex with $CpFe(CO)_2Cl$, *Organometallics*, 2007, **26**, 6857–6863.
32. K. Surendra, N. S. Krishnaveni, K. R. Rao, Direct Barbier-type allylation of aromatic acetals and dioxolanes in the presence of *β*-cyclodextrin in water, *Tetrahedron Lett.*, 2006, **47**, 2133–2136.
33. H. R. Appelt, J. B. Limberger, M. Weber, O. E. D. Rodrigues, J. S. Oliveira, D. S. Ludtke, A. L. Braga, Carbohydrates in asymmetric synthesis: enantioselective allylation of aldehydes, *Tetrahedron Lett.*, 2008, **49**, 4956–4957.
34. H. Hu, A. Uno, A. Harada, S. Takahashi, *Chem. Lett.*, 1990, 797; H. Hu, A. Uno, A. Harada, S. Takahashi, *Bull. Chem. Soc. Jpn.*, 1991, **64**, 1884.
35. H. Sakuraba, S. Ushiki, *Tetrahedron Lett.*, 1990, **31**, 5349.
36. L. R. Reddy, N. Bhanumathi, K. R. Rao, Dynamic kinetic asymmetric synthesis of *β*-aminoalcohols from racemic epoxides in cyclodextrin complexes under solid conditions, *Chem. Commun.*, 2000, 2321–2322.
37. M. A. Reddy, K. Suredra, N. Bhanumathi, K. R. Rao, Highly facile biomimetic regioselective ring opening of epoxides to halohydrins in the presence of *β*-cyclodextrin, *Tetrahedron*, 2002, **58**, 6003–6008.
38. L. R. Reddy, N. Bhanumathi, K. R. Rao, Dynamic kinetic asymmetric synthesis of *β*-aminoalcohols from racemic epoxides in cyclodextrin complexes under solid state conditions, *Chem. Commun.*, 2000, 2321–2322, M. A. Reddy, N. Bhanumathi, K. R. Rao, Asymmetric synthesis of 2–azido-1–arylethanols from azido aryl ketone-*β*-cyclodextrin complexes and sodium borohydride in water, *Chem. Commun.*, 2001, 1974–1975.
39. M. Narender, K. Surenda, N. S. Krishnaveni, M. S. Reddy, K. R. Rao, A facile regioselective ring opening of aziridines to haloamines using tetrabutylammonium halides in the presence of *β*-cyclodextrin in water, *Tetrahedron Lett.*, 2004, **45**, 7995–7997.
40. M. S. Reddy, M. Narender, K. Rama Rao, A mild efficient synthesis of *α*-tosylamino ketones from aryl aziridines in the presence of *β*-cyclodextrin and NBS in water, *Tetrahedron Lett.*, 2005, **46**, 1299–1301.
41. M. A. Reddy, N. Bhanumathi, K. Rama Rao, A mild and efficient biomimetic synthesis of *α*-hydroxymethylarylketones from oxiranes in the presence of *β*-cyclodextrin in water, *Tetrahedron Lett.*, 2002, **43**, 3237–3238.

42. K. Surendra, N. S. Krishnaveni, V. P. Kumar, R. Sridhar, K. R. Rao, Selective and efficient oxidation of sulfides to sulfoxides with *N*-bromosuccinimide in the presence of *β*-cyclodextrin in water, *Tetrahedron Lett.*, 2005, **46**, 4581–4583.
43. M. Narender, M. S. Reddy, Y. B. D. Nageswar, K. R. Rao, Aqueous phase synthesis of *vic*-halohydrins from olefins and *N*-halosuccinimides in the presence of *β*-cyclodextrin, *J. Mol. Catal. A: Chem.*, 2006, **258**, 10–14.
44. M. S. Reddy, M. Narender, Y. V. D. Nageswar, K. R. Rao, Regioselective ring-opening of aziridines with potassium thiocyanate in the presence of *β*-cyclodextrin in water, *Tetrahedron Lett.*, 2005, **46**, 6437–6439.
45. R. Sridhar, B. Srinivas, K. Surendra, N. S. Krishnaveni, K. R. Rao, Synthesis of *β*-hydroxy selenides using benzeneselenol and oxiranes under supramolecular catalysis in the presence of *β*-cyclodextrin in water, *Tetrahedron Lett.*, 2005, **46**, 8837–8839.
46. M. S. Raddey, B. Srinivas, R. Sridhar, M. Narender, K. R. Rao, Highly regioselective thiolysis of oxiranes under supramolecular catalysis involving *β*-cyclodextrin in water, *J. Mol. Catal. A: Chem.*, 2006, **255**, 180–183.
47. K. Surendra, N. S. Krishnaveni, Y. V. D. Nageswar, K. R. Rao, Highly regioselective ring opening of oxiranes with phenoxides in the presence of *β*-cyclodextrin in water, *J. Org. Chem.*, 2003, **68**, 4994–4995.
48. B. Srinivas, V. P. Kumar, R. Sridhar, K. Surendra, Y. V. D. Nageswar, K. R. Rao, Regioselective nucleophilic opening of epoxides and aziridines under neutral conditions in the presence of *β*-cyclodextrin in water, *J. Mol. Catal. A: Chem.*, 2007, **261**, 1–5.
49. K. Surendra, N. S. Krishnaveni, M. A. Reddy, Y. V. D. Nageswar, K. R. Rao, Highly selective cleavage of *β*-cyclodextrin-epoxide/aziridine complexes with IBX in water, *J. Org. Chem.*, 2003, **68**, 9119–9121.
50. M. Narender, M. S. Reddy, V. P. Kumar, Y. V. D. Nageswar, K. R. Rao, Direct synthesis of carbonyl compounds from THP ethers with IBX in the presence of *β*-cyclodextrin in water, *Tetrahedron Lett.*, 2005, **46**, 1971–1973.
51. N. S. Krishnaveni, K. Suredra, V. P. Kumar, B. Srinivas, C. S. Reddy, K. R. Rao, *β*-Cyclodextrin promoted allylation of aldehydes with allylbutyltin under supramolecular catalysis in water, *Tetrahedron Lett.*, 2005, **46**, 4299–4301.
52. K. Surendra, N. Srilakshni, R. Sridhar, K. R. Rao, *β*-Cyclodextrin promoted aza-Michael addition of amines to conjugated alkenes in water, *Tetrahedron Lett.*, 2006, **47**, 2125–2127.
53. A. Lubineau, J. Auge, *Tetrahedron Lett.*, 1992, **33**, 8073–8074.
54. N. S. Krishnaveni, K. Surendra, K. R. Rao, Study of the Michael addition of *β*-cyclodextrin-thiol complexes to conjugated alkenes in water, *Chem. Commun.*, 2005, 669–671.
55. V. P. Kumar, V. P. Reddy, S. Sridhar, B. Srinivas, M. Narender, K. R. Rao, Supramolecular synthesis of 3–indolyl-3–hydroxy oxindoles under neutral conditions in water, *J. Org. Chem.*, 2008, **73**, 1646–1648.
56. B. Kaboudin, M. Sorbium, *β*-Cyclodextrin as an efficient catalyst for the one-pot synthesis of 1–aminophosphonic esters in water, *Tetrahedron Lett.*, 2007, **48**, 9015–9017.
57. K. Surendra, N. S. Krishnaveni, A. Mahesh, K. R. Rao, Supramolecular catalysis of Strecker reduction in water under neutral conditions in the presence of *β*-cyclodextrin, *J. Org. Chem.*, 2006, **71**, 2532–2534.
58. V. P. Kumar, M. S. Reddy, M. Narender, K. Surendra, Y. V. D. Nageswar, K. R. Rao, Aqueous phase mono-protection of amines and amino acids as *N*-benzyloxycarbonyl derivatives in the presence of *β*-cyclodextrin, *Tetrahedron Lett.*, 2006, **47**, 6393–6396.
59. M. Narender, M. S. Reddy, R. Sridhar, Y. V. Nageswar, K. R. Rao, Aqueous phase synthesis of thiazoles and aminothiazoles in the presence of *β*-cyclodextrin, *Tetrahedron Lett.*, 2005, **46**, 5953–5955.
60. M. Narender, M. S. Reddy, V. P. Kumar, Y. V. D. Nageswar, K. R. Rao, Supramolecular synthesis of selenourea in water in the presence of *β*-cyclodextrin under atmospheric pressure, *J. Org, Chem.*, 2007, **72**, 1849–1851.
61. Z. Shen, J. Ma, Y. Liu, C. Jiao, M. Li, Y. Zhang, *β*-cyclodextrin-immobilized (4S)-phenoxy-(S)-proline as a catalyst for direct asymmetric aldol reactions, *Chirality*, 2005, **17**, 556–558.

62. K. Liu, D. Haussinger, W.-D. Woggon, Aldol reactions in water using a β-cyclodextrin-binding proline derivative, *Synlett.*, 2007, **3**, 2298–2300.
63. J. Huang, X. Zhang, D. W. Armstrong, Highly efficient asymmetric direct stoichiometric Aldol-reactions on/in water, *Angew. Chem. Int. Ed.*, 2007, **46**, 9073–9077.
64. P. Hapiot, C. Lagrost, S. Aeiyach, M. Jouini, J.-C. Lacroix, Oxidative coupling of small oligothiophenes and oligopyrroles in water in the presence of cyclodextrin: flash photolyses investigations, *J. Phys. Chem. B*, 2002, **106**, 3622–3628.
65. H. Cinar, O. Kretschmann, H. Ritter, Synthesis of novel fluorinated polymers via cyclodextrin complexes in aqueous solution, *Macromolecules*, 2005, **38**, 5078–5082.
66. K. Surendra, N. S. Krishnaveni, K. R. Rao, A new and efficient method for the synthesis of thiiranes from oxirane-β-cyclodextrin complexes and thiourea in water, *Tetrahedron Lett.*, 2004, **45**, 6523–6526.
67. K. Surendra, N. S. Krishnaveni, K. R. Rao, A simple biomimetic protocol for the oxidation of alcohols with sodium hypochlorite in the presence of β-cyclodextrin in water, *Can. J. Chem.*, 2004, **82**, 1230–1233.
68. H-B. Ji, D.-P. Shi, M. Shao, Z. Li, L.-F. Wang, Transition metal-free and substrate-selective oxidation of alcohols using water as an only solvent in the presence of β-cyclodextrin, *Tetrahedron Lett.*, 2005, **46**, 2517–2520.
69. H. Sakuraba, N. Inomata, Y. Tanaka, Asymmetric reduction of ketones with crystalline cyclodextrin complexes of amine-boranes, *J. Org. Chem.*, 1989, **54**, 3482.
70. Y. Kawajiri, N. Motohashi, Strong asymmetric induction without covalent bond connection, *J. Chem. Soc., Chem. Commun.*, 1989, 1336.
71. J. Doussot, A. Guy, R. Garreau, A. Falguieres, J. Cossy, C. Amsterdamsky, Selective synthesis of substituted benzylic alcohols from aromatic substrates included in β-cyclodextrin, *Bull. Soc. Chim. Fr.*, 1996, **133**, 161–166.
72. R. Fornasier, F. Marcuzzi, M. Parmagnani, U. Tonellato, Stereoselective reduction of (R)-(−)-carvone with sodium dithionite in the presence of cyclomaltoheptaose (β-cyclodextrin) and its heptakis(2,6–di-*O*-methyl) derivative, *Carbohydr. Res.*, 1991, **217**, 245.
73. R. Ravichandran, S. Divakar, Stereoselective reduction of menthone with sodium dithionite in the presence of β-cyclodextrin, *J. Mol. Catal.*, 1994, **93**, 247–251.
74. R. Fornasier, F. Marcuzzi, M. Parmagnani, U. Tonellato, Stereoselective reduction with sodium dithionite of conjugated enones in the presence of β-cyclodextrin and its heptakis(2,6–di-*O*-methyl) derivative as host compounds or phase transfer agents, *J. Inclusion Phenom. Mol. Recognit. Chem.*, 1994, **18**, 81–92.
75. K. Hattori, K. Takahashi, M. Uematsu, N. Sakai, Multiple interactions between host cyclodextrin and guest compound assisting asymmetrically selective reduction with $NaBH_4$ in aqueous media, *Chem. Lett.*, 1990, **19**, 1463–1466; K. Hattori, K. Takahashi, N. Sakai, Enantioface differentiating reduction of keto-acid in the presence of 6–deoxy-6–amino-β-cyclodextrin with $NaBH_4$ in aqueous media. *Bull. Chem. Soc. Jpn.*, 1992, **65**, 2690–2696; K. Hattori, K. Takahashi, Asymmetric reduction of prochiral inclusion complex in aqueous media, *Supramol. Chem.*, 1993, **2**, 209–213.
76. K. Takahashi, H. Yokomizo, K. Ishiyama, M. Kitsuta, M. Ohashi, New aspects of cyclodextrin chemistry induced by outside type complex formation; asymmetric reduction of indol-3–pyruvic acid with $NaBH_4$, *J. Incl. Phenom. Macrocyc. Chem.*, 2006, **56**, 95–99.
77. Y. Lu, J. Liu, G. Diffee, D. Liu, B. Liu, Reduction of mononitroarenes by hydroxide ion in water catalyzed by β-cyclodextrin: enhanced reactivity of hydroxide ion, *Tetrahedron Lett.*, 2006, **47**, 4597–4599.
78. K. Takahashi, T. Narusawa, *unpublished data*.

5

Reactions in Zeolites

Stéphane Walspurger[1] *and Jean Sommer*[2]
[1]Energy Research Centre of The Netherlands, The Netherlands
[2]Institut de Chimie, Université de Strasbourg, France

Zeolites are naturally occurring minerals found in volcanic rocks where they have been formed by hydrothermal processes. The term 'zeolite' has been introduced by the Swedish mineralogist Cronstedt who gave the name according to the observation of the large amount of water they release upon heating. 'Zeolite' is indeed derived from the Greek 'zein' (to boil) and 'lithos' (stone).

Zeolites are crystalline aluminosilicates materials that possess ordered and interconnected microporous channels with diameters ranging from 0.2–20 Å. Their unique properties (microporosity, high surface area, acid–base character, shape) have made them a material of choice in a great number of applications. Zeolites are intensively used in gas separation due to their ability to adsorb selectively a large variety of molecules and are also known as molecular sieves. Furthermore, these materials are also used as ion exchangers (water softeners) and catalysts in petrochemistry. Currently, the world's annual production of natural zeolite is about 4 million tons. Of this quantity, 2.6 million tons are shipped to Chinese markets to be used in the concrete industry. The amount of synthetic zeolites produced is about 1.5 million tons (Figure 5.1).

The number of known zeolitic structures is currently 174 but, according to theoretical work based on these existing stable structures, there might be several million structures available which would be more or less stable. Hence designing a zeolitic material for a

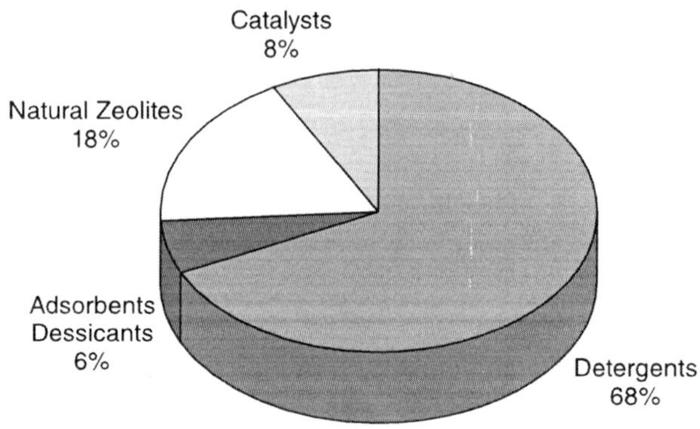

Figure 5.1 *Main uses of zeolites in North America, Western Europe and Japan*

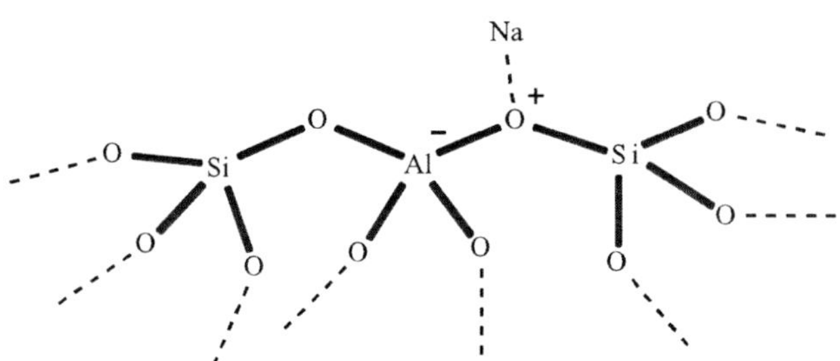

Scheme 5.1 $M_{x/n}[(Al_2O_3)_x, (SiO_2)_y]_z H_2O$; M = Na, Ca, Mg

chemical reaction or chemical storage should be feasible and may open new opportunities for the use of these materials in the future.[1] Existing computational methods may allow to select a zeolite for a targeted reaction product.[2] For instance in biomass conversion, novel large pores zeolites may be very adapted to convert oligosaccharide to fuels.

Zeolites tridimensional porous structure arises from a framework of $[SiO_4]^{4-}$ and $[AlO_4]^{5-}$ tetrahedra TO_4 linked together by their corners. During the synthesis, these primary building blocks assemble by sharing O atoms, thus creating an infinite lattice with identical unit cells. Each Si ion has its 4+ charge balanced by the four surrounding O, so that Si tetrahedron is electrically neutral. Hence pure siliceous zeolites are electrically neutral. By replacing part of Si ions with Al (with a formal charge 3+), a negatively charged tetrahedron is created. Therefore each Al ion requires a 1+ charge from a cation which can be Na^+ in most of the case and also NH_4^+ or Ca^{2+}, K^+, and so on.

These ions can be reversibly exchanged giving zeolites their most important application in ion exchange such as water softener in detergent (largest use) and water purification. Among others, H^+ counterions may compensate the negative charge as well. The protons are formally assigned to be bonded to the bridging O of a Si—O—Al bond, forming hydroxyl groups that act as a Brønsted acid. The so-called Brønsted acid sites are characterized by their acid strength, which depends on their environment, i.e. on the chemical composition (density of protonic sites, heteroatoms such as P) and the structure of the zeolite (geometry, cages and configuration).

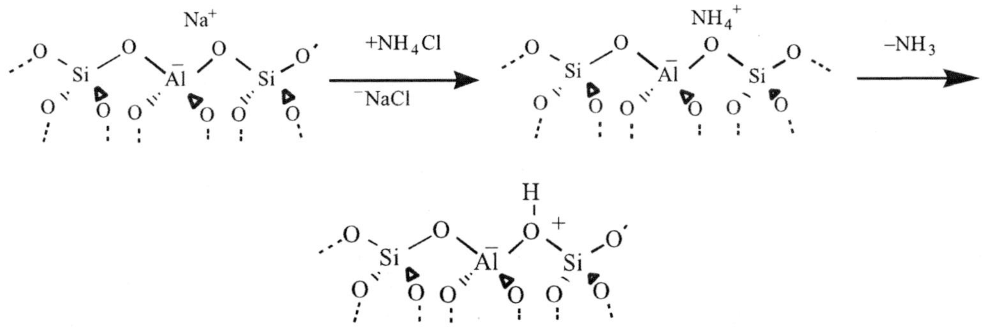

Scheme 5.2 *From basic to acidic zeolites*

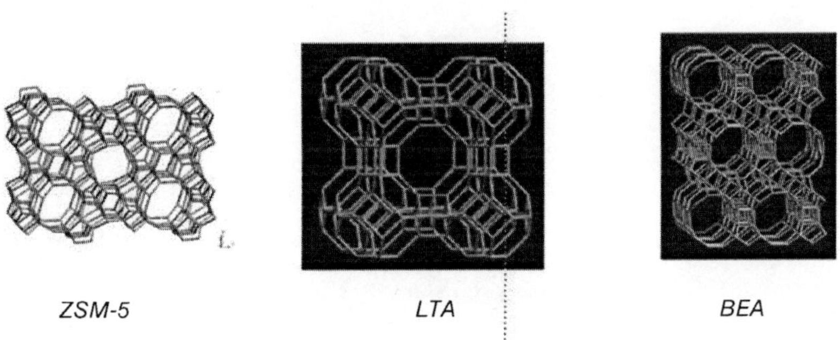

Figure 5.2 *Tridimensional schematic structure of zeolite ZSM-5 (MFI), L (LTA) and β (BEA)*

The local environment of the acid site is determined by the zeolite topology that imposes a defined coordination of the TO_4 tetrahedra in the framework. This leads to different amounts of topologically non-equivalent TO_4 sites which induce heterogeneity in the acid strength distribution as well as a difference in accessibility. Together, the acid character of these materials and the pore size of the channels that falls in the range of the kinetic radii of many organic molecules give a unique synergy for the use of zeolites as acid catalysts in numerous industrial key reactions.

As catalysts, zeolites have found their most important application in petroleum refining processes. Their acid function is used in Fluid Catalytic Cracking (FCC), in hydroisomerisation of light alkane fraction as well as in oligomerisation and isomerisation steps to upgrade the liquid fuels into gasoline and diesel. The combination of two different zeolites in the same industrial process is illustrated in the Shell–UOP TIP process: an acidic zeolite, MOR, is used for isomerisation and the neutral LTA is used as molecular sieve for separation as shown in the scheme below (Figure 5.3).

In 1999, Tanabe and Hoelderich[3] have reviewed 74 industrial processes ranging from cracking to fine chemical synthesis in which zeolites are used as a catalyst.

Numerous books and reviews have already been published concerning organic reactions catalysed by zeolites.[4–6] Due to their pore sizes, their internal shapes and their compositions,[7] zeolites are well known for their selectivity properties, discriminating reactants, intermediates, transition states or products.[8]

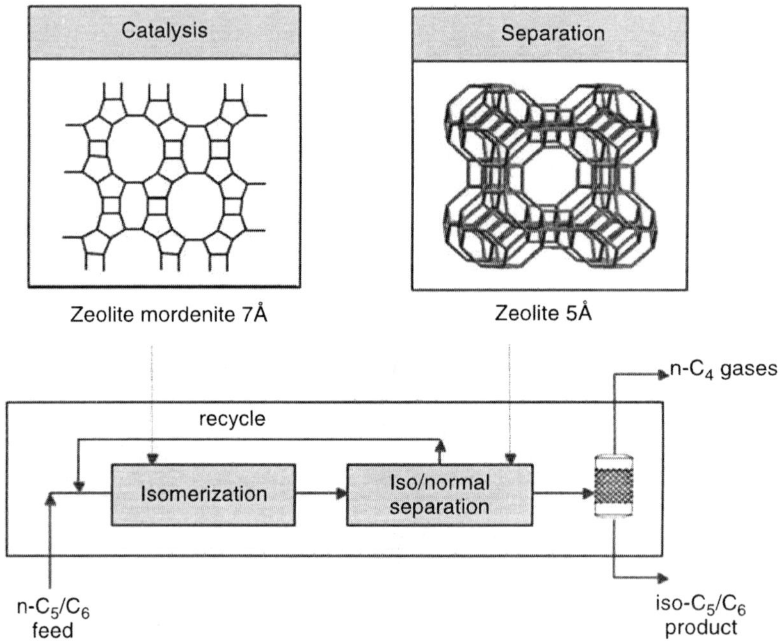

Figure 5.3 *The TIP process*

Zeolites can be considered as microreactors and one may find some analogy with enzymatic-like systems:[9]

- the differential diffusion in the zeolite channels between products and reactants of different size allows a control on the selectivity on chemical reactions,
- the preferential adsorption of molecules in the channels as a function of their polarity offers a selective activation of chemical compounds. The electrostatic field inside the cavities and channels, depending strongly on the Si/Al ratio, can be tuned by isomorphous substitution of Si and Al in the framework. Hence one can drive the adsorption selectivity towards hydrophilic/hydrophobic and polar/non-polar molecules by the synthesis of a zeolite with an appropriate Si/Al ratio,
- molecules that are absorbed in a channel or cavity tend to optimize their configuration towards the most favourable Van der Waals interaction with the cavity or channels walls under the confinement effect. When the molecular size of an adsorbate is close to the dimension of the cavity or channels, its frontier molecular orbitals HOMO–LUMO are strongly affected. It has been evidenced that adsorbed molecules inside a zeolite channel or cavity show increased basic character and greater reactivity.

Last but not least, as easily recoverable and reusable catalysts, zeolites also allow to perform acid-mediated transformations with a clear environmental benefit, since it avoids the formation of huge amounts of salts which is typically encountered in acid catalyzed processes. Noteworthy, zeolites can also host active catalytic species such as metals or organometallic complexes for fine chemical synthesis.

5.1 The Confinement Effect

The so-called *confinement effect in microporous and mesoporous materials* is known to strongly affect diffusion and catalytic properties.[10] Various approaches have been used to characterise this effect. Denayer and Baron, in a chromatographic study of the adsorption properties of linear alkanes in various mesoporous materials, proposed to define a confinement factor Z as the ratio of adsorption entropy over adsorption enthalpy.[11] For non-polar molecules, the adsorption enthalpy reflects the energetic interaction between the molecule and the zeolite force field and increases with decreasing pore size. The adsorption entropy relates to the amount of freedom lost by going from bulk state to adsorbed state. Using calorimetry, gravimetry and in-situ infrared spectroscopy, Lercher and Eder showed that the apolar alkanes are very locally adsorbed via hydrogen bonding. By monitoring the disappearance of infrared absorption bands of O—H groups from the zeolite, they were able to distinguish a precise population of protons that are involved in the adsorption of alkanes. For instance, with *n*-hexane and zeolite Mordenite (MOR), they convincingly showed that only protons in 12 and 10 membered ring pores were involved in hydrogen bonding with the alkane whereas the protons located in 8 membered ring pores the sorption was hindered. The hydrogen bonding between the hydroxyl group of the zeolite and the hydrocarbons induces a dipole in the hydrocarbon that is at the origin of their reactivity.[12,13]

The confinement effect has been evidenced both by quantum chemical calculations as well as spectroscopic investigations on naphthalene molecule entrapped within a zeolite by the group of A. Corma.[9,14–17] As an example, the aromaticity of anthracene was strongly perturbed due to limitation of the π-orbital spatial extension induced by the proximity of the pore walls depending on the size of the pores. This 'electronic confinement effect' is particularly large when a tight host–guest configuration occurs. It has been suggested that this effect implies important consequences in terms of reactivity, particularly for the reactions involving electric charge donation from the guest molecule to the zeolite framework atoms. Reactions, like proton transfer for example, will be favoured in confined systems where the rise in the frontier occupied molecular orbitals will make the electronic transfer easier. An other example is provided in a computer modelling study of acetonitrile adsorption in Mordenite:[19] the nitrogen atom of the molecule in the side pockets favours interaction with the proton of an OH group in these confined sites. The role of side pockets has also been demonstrated in many other cases.[18,19]

Remarkably, the confinement in the small channels of Mordenite zeolite (side pocket) polarizes significantly the acetonitrile molecule toward protonation. It leads to a long residence time of the adsorbate in the neighbourhood of the acidic hydroxyl group, and modifies greatly the charge distribution (even in a silicic zeolite), contributing to the explanation of the preferred protonation of acetonitrile in side pockets of acidic Mordenites.

In a more intuitive way, Derouane[20] proposed that zeolites should be considered as solid solvents, a concept very familiar to organic chemists.

By the characterisation of the acidic OH groups in zeolites of different types, Hunger[18] interpreted NH_3–TPD results in the light of confinement effects: in the Faujasite (FAU) structure, a bimodal distribution of adsorption energies in both the supercages and the sodalite cages was found, with the stronger acid sites being more common in the supercages. In contrast, essentially the same acid strength for different regions in MOR is seen,

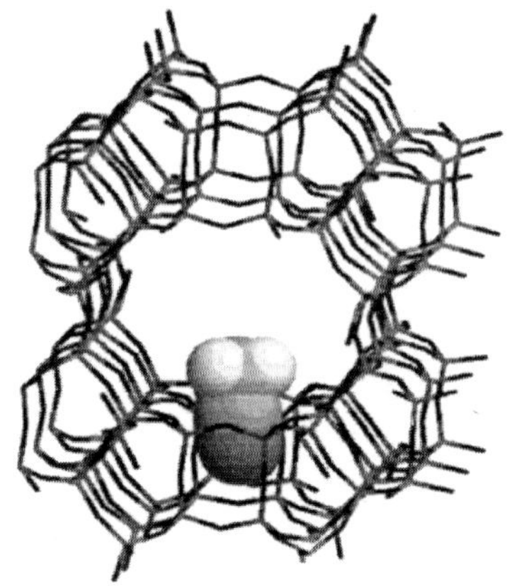

Figure 5.4 *Acetonitrile in side pockets of acidic Mordenites*[19]

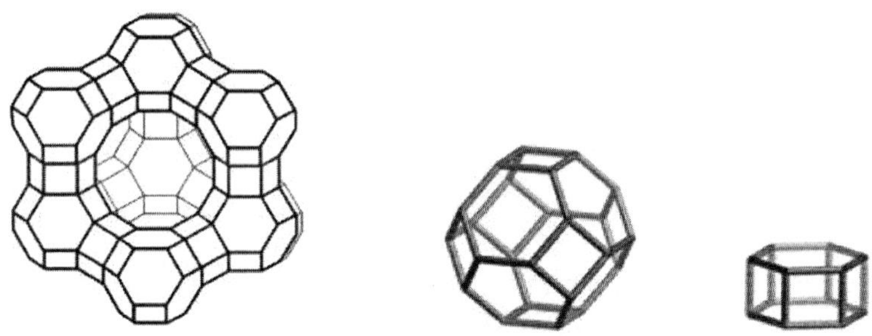

Figure 5.5 *Faujasite structure (left) and sodalite cage with double 6 connectors (right)*

though the acid strength for the sites located in the side pockets may be slightly higher. For FER and MFI, the TPD gives unimodal distributions and therefore prohibits distinction between different regions.

Using a mesostructured aluminosilicate of the MCM-41 type with the same composition but varying the pore size from 2.3 to 9.3 nm, Tanchoux *et al.*[21] measured the adsorption energetics and the catalytic isomerisation of 1-hexene. The authors could demonstrate that the geometry-dependent contributions dominated catalytic behaviour over all other factors.

However, for the understanding of catalytic hydrocarbon conversion over solid acid catalysts, there is still a sharp contrast between the general agreement on the similarity with carbocation chemistry and the lack of experimental information on the mode of formation or the nature of the carbocationic intermediates.[22,23] Many years ago, on the basis of the initial product distribution in the cracking of light alkanes at high temperature, the Haag–Dessau mechanism[24,25] (sigma-bond protolysis as in superacids[26–28]) was suggested, but hydride abstraction, oxidation and alkene impurities have also been suggested as possible first step in the formation of the reaction intermediates.[29–31]

It has been shown that simple carbenium ions can only be considered as transient species (unless with a large charge distribution) within the zeolite channels and that the most stable intermediates are the framework-bound alkoxy species[32–34]. However, the existence of carbocations as short-lived reaction intermediates[35] in equilibrium with the alkoxy species is supported by the Koch reaction,[36,37] halogen switch reaction[38] or H/D exchange reactions[29,30] observed in reactions catalysed by zeolites.

Based on *ab initio* calculations using a FAU supercage as model, Van Santen and Pitko[39] demonstrated that the role of the zeolite framework in the photooxidation of alkenes with molecular oxygen is the complexation of the hydrocarbon and O_2 to the exchanged cations, resulting in confinement of these molecules with a specific relative orientation. This leads to the formation of a π–π intermolecular complex. The interaction between alkene and oxygen in this complex occurs with a finite overlap of the involved orbitals. The formation of such a complex results in a significant transition moment of the intermolecular charge transfer. The role of the electrostatic field of the zeolite is only indirect. It stabilizes the reagents in a suitable 'pre-transition state' configuration. For the model DFT calculations a $Mg_2Al_{14}Si_{12}O_{20}H_{24}$ cluster was chosen.

Partial oxidation of aromatics is one of the most challenging reactions in the field of organic synthesis. Oxidation of benzene to phenol using N_2O as oxidant on ZSM-5-type catalysts between 300 and 400 °C could be achieved with a selectivity close to 100%.[40,41] A substitution mechanistic scheme involving the formation of a hydroxyl group bonded to the zeolite surface (HO–ZSM-5) and its further attack on benzene was proposed.[42]

The density functional theory study considering the possible reaction intermediates also confirmed the possible formation of protonated nitrous oxide, leading to a Wheland-type intermediate, thus supporting an electrophilic aromatic substitution assisted by the confined environment provided by the active zeolite framework.

Acid form of zeolite H–ZSM-5 (MFI type) has particularly attracted attention since researchers at Mobil have developed the methanol to gasoline process. In this process, small pores zeolites with 8- to 12-membered rings have shown outstanding catalytic activities in the transformation of methanol (C_1 alcohol) into alkanes in the gasoline range

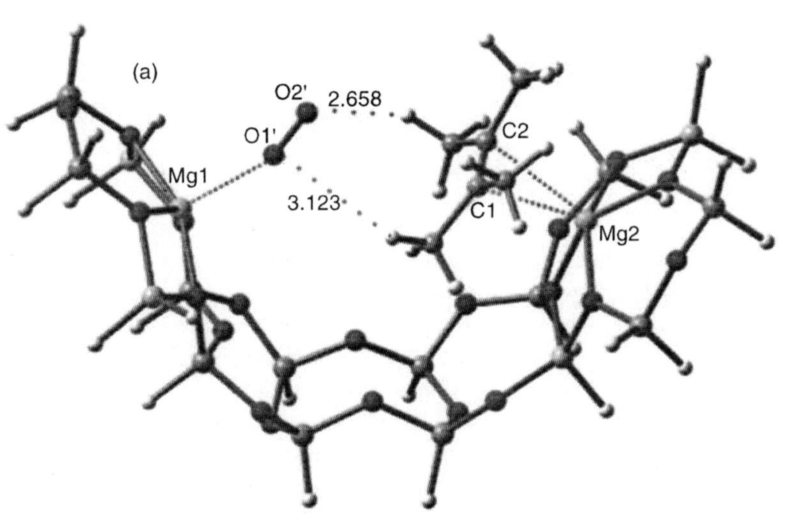

Figure 5.6 *Coadsorption of 2,3-dimethylbutene-2 and O_2 on Mg-Zeolite*[39]

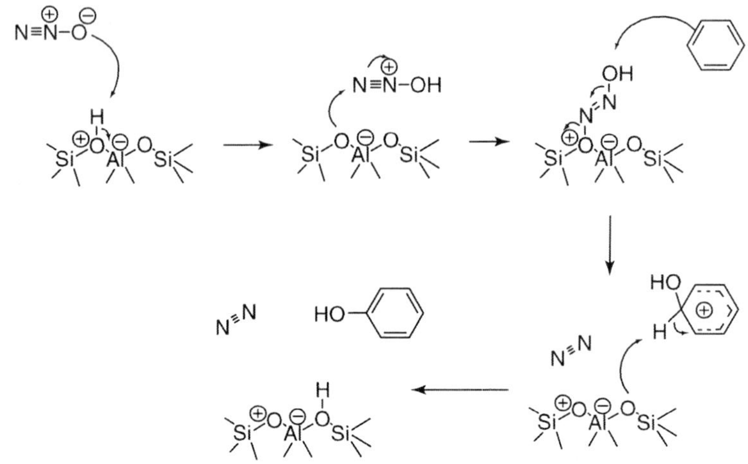

Scheme 5.3 Mechanism of oxidation of benzene to phenol using N_2O as oxidant[42]

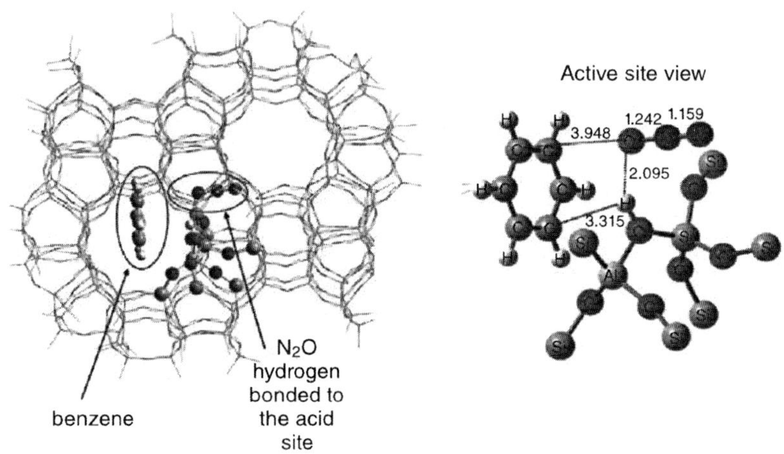

Figure 5.7 Detailed view of the simultaneous adsorption of benzene in the N_2O, H-ZSM-5 π-complex[42]

but also into olefins in the methanol to olefin reaction (MTO).[43,44] Interestingly, the product distribution can be controlled by the zeolite properties among other parameters. For instance, the reaction could be directed towards the production of light olefins such as the key industrial ethylene and propylene using a small pore zeolite with a modified chemical composition (silica–alumina–phosphate known as SAPO, AlPO…). As a typical example, the SAPO-34 being an 8-membered ring zeolite (chabazite-type, CHA) is less acidic compared to its silica–alumina (H–CHA) equivalent and allows a very high selectivity to ethylene and propylene when used in a properly designed reactor. Hydro/UOP process, which is developed in Norway, allows the selective production of ethylene or propylene using a silico–alumino–phosphate zeolite.[45] Methanol conversion to valuable

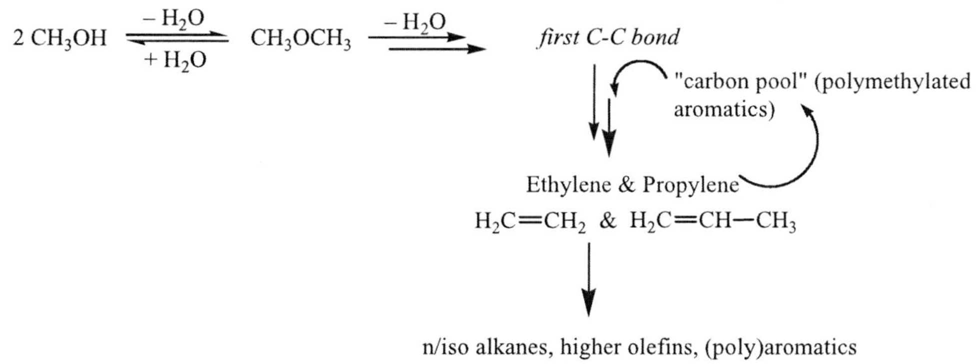

Scheme 5.4 *Methanol to olefins general simplified reaction scheme*

products is obviously of primary importance and will probably gain even more importance as oil resources are progressively depleted.[46]

From a mechanistic point of view, the C—C coupling is of utmost interest because it may eventually allow the selective synthesis of valuable alkanes and olefins (Scheme 5.4). Despite experimental observation of formation of methyl groups on the surface of zeolites at the early stage of the reaction[47–50] and an ongoing research on the possible impurities triggering the first C—C formation, there is still a debate over the molecular mechanisms involved at the beginning of the reaction. Although computational studies usually exclude the formation of oxonium ylide[51] (Scheme 5.5) due to the lack of basicity of the zeolites framework, the role and the nature of impurities at the origin of the so called carbon pool are still quite obscure.

For the latter steps, there is a general agreement that the production of light alkanes and olefins in zeolite is auto catalyzed by large aromatic rings[52] that are formed within the zeolites channels and have been evidenced by spectroscopy.[53–56] Although a similar chemistry is also observed on non-zeolitic materials,[57–60] the selective production of ethylene and propylene may probably be enhanced by the zeolites cage effects.

Strong effects of the pore size and the shape of zeolites have been clearly evidenced in the case of alkanes having comparable molecular dimensions. Remarkable contributions have shown that the adsorption enthalpy of alkanes is strongly dependent on both the topology of the zeolite and the branching and geometry of the alkanes[12]. The adsorption enthalpy becomes higher as the number of carbons in the alkane increases. Some critical carbon numbers have been recently found by Denayer *et al.* depending on the cage size of the zeolite. These authors have rationalised their results by the spatial demand of the alkane within the zeolite.[61] As long as the alkane can fit inside a single zeolite cage, they have observed the well known linear increase of enthalpy of adsorption with increased number of carbons. At some point, they showed that the enthalpy does not increase linearly anymore and they have argued that the spatial constraint within the zeolitic cage impose a geometrical rearrangement of the alkane itself. Further, for higher carbon numbers, the linear increase of adsorption enthalpy has again been observed with increasing carbon numbers. That phenomenon was attributed to the adsorption of large alkane in two neighbouring cages.

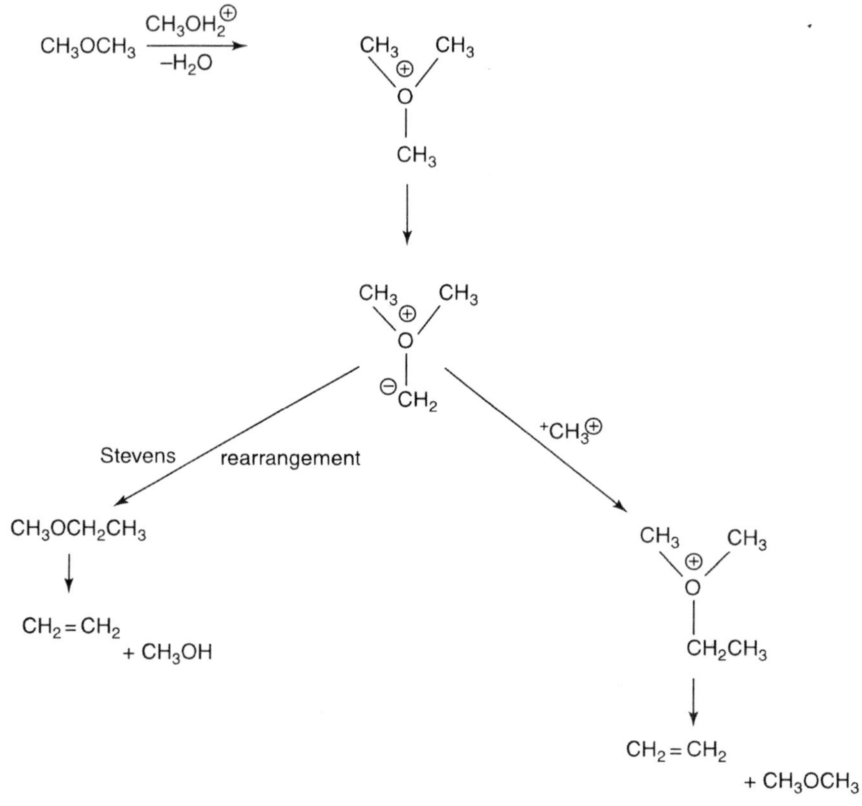

Scheme 5.5 *First C—C bond formation by Stevens rearrangement (left) or oxonium ylide route*[42]

Another remarkable demonstration of the confinement effect was described in the carbonylation of dimethyl ether by Iglesia and co-workers.[62] Surface methoxy groups are the simplest and most stable alkoxide species because they cannot deprotonate to give more stable products under mild conditions. The reactivity of surface methoxy with CO is very different from one zeolite to another. In their experimental work Iglesia *et al.* have observed that the rate of carbonylation of dimethyl ether is much higher in the case of H–MOR than for other zeolites. H–MOR consists of 12-membered ring main channels intersecting 8-membered ring channels (also known as 'side pockets'). They have been able to show experimentally that methyl groups located in these side pockets are much more prone to the carbonylation than those located in the 12-membered ring main channels. The molecular architecture of the side pockets may be particularly favourable to stabilize the acylium like transition state or may impose the reaction between approaching CO and surface methyl groups because of the constrained space. These studies clearly demonstrated the role of the local structure of the zeolites on the reactivity of methoxy groups with CO which acts as a nucleophile in this case. Furthermore, Iglesia *et al.* have published a note about the thermodynamics in small alkanes (C3–C6) C—C bond activation. In monomolecular cracking of these alkanes the turnover rate increases with the number of carbons of the alkane chain while the activation energy remains size independent. The increase of turnover rates of the cracking reaction was explained by the concomitant increase of activation entropies.[63]

Last but not least, the versatility of zeolites is demonstrated by exchanging the acidic proton with deuteron which enables investigation of interesting mechanisms related to catalysis and by exchanging the proton with transition metal cations, such as Cu(I), and opens new areas of environmentally friendly organic chemistry. For these reasons, we are including in this chapter acidic-zeolite catalysed reactions from our own work which can be best understood as examples of confinement effects: superelectrophilic, Cu(I) catalysed Click chemistry, and specific H/D exchange reactions.

5.2 Superelectrophilic Activation in Zeolites

The term 'Superelectrophile' was suggested by Olah in the early seventies when it was noticed that acylation of chlorobenzene by acetyl tetrafluoroborate was substantially accelerated in the presence of the corresponding Brønsted acid.[64]

Superelectrophilic intermediates[64] are typically generated when a cationic electrophile is further protonated or coordinated by a Lewis acid to produce a dicationic species.

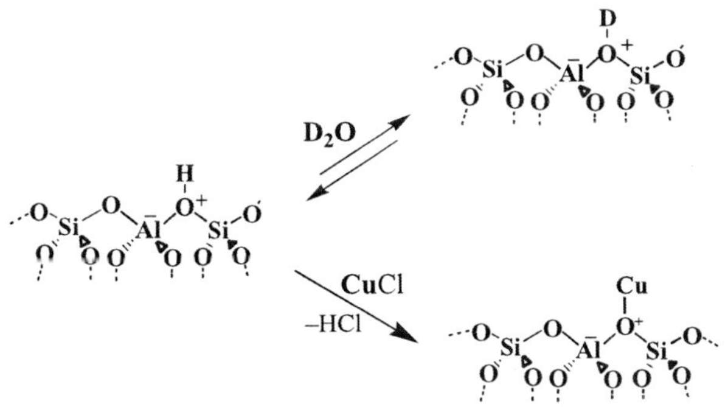

Scheme 5.6 Exchange of acidic protons by deuterons or copper(I) cations

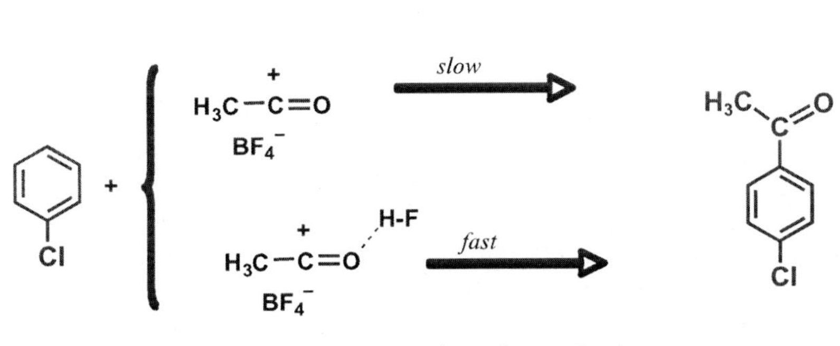

Scheme 5.7 Increase of rate by acid solvation

$$E^{+} + HA \rightleftharpoons A^{-}, HE^{2+}$$

Electrophile Superelectrophile

Scheme 5.8 *Superelectrophilicity*

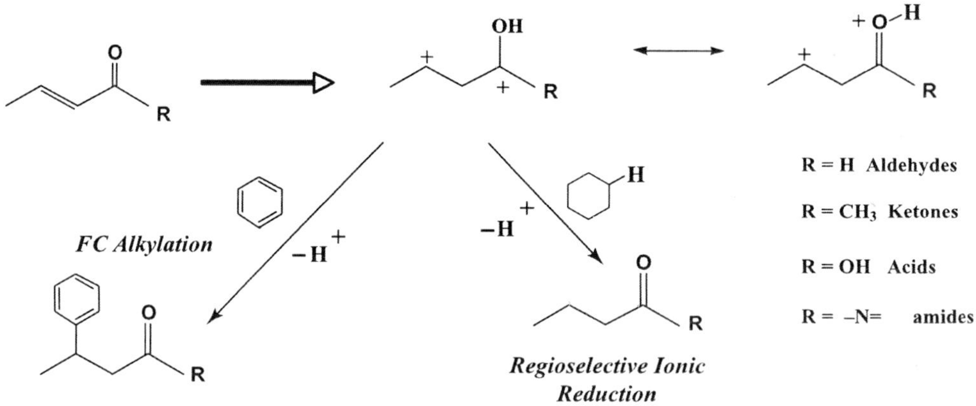

Scheme 5.9 *Superelectrophilic activation of α,β-unsaturated carbonyl compounds*

Similarly, superelectrophilic solvation[65] may also involve weaker interactions without forming limiting dications resulting in enhanced electrophilic reactivity. In order to be involved in superelectrophilic activation the starting molecule is best bifunctional.

In the last two decades many useful reactions and synthetic approaches involve superelectrophilic activation of reagents mediated by liquid superacids or by excess of aluminium halides. As an example, α,β-unsaturated aldehydes, ketones, acids or amides, readily condense with aromatics such as benzene and dichlorobenzene which are poor nucleophiles, but only in the presence of a large excess of aluminium chloride or triflic acid. This reactivity has been interpreted by the intermediacy of dicationic species.

Reaction kinetics,[66] theoretical calculations[67] and finally the observation of such species by NMR in superacids[68] as stable dications point out the reliability of this mechanistical hypothesis under superacidic conditions.

Zeolites have been successfully applied in similar reactions. As the acidity of H–zeolites[69–70], is generally considered to be well below that of superacids[26] ($-12 < Ho < -24$), the hypothesis of identical dicationic intermediates may be easily discarded, but we assume that confined space may allow a strong activation by imposing a strong interaction between protons of the surface and the electrophile. In other words, the framework of the solid acid may provide close proximity of acidic sites enabling the formation of diprotonated (or coordinated) species. Moreover, an effective compensation of comparatively low proton acidity of zeolites may be obtained due to nucleophilic assistance of the lattice oxygens in the transition state of a concerted mechanism resulting in an increase of the protonation rate.[71] Similarly, the enzyme mediated protonation of a heterocyclic carbocation[72] provides an example showing that structural factors of acids can have a profound influence in electrophilic activation.

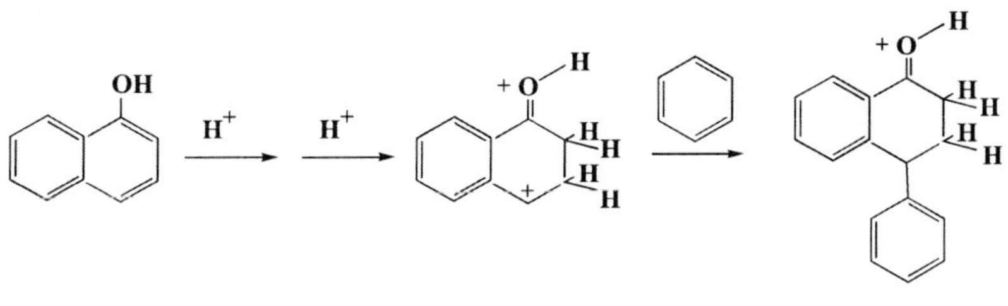

Scheme 5.10 *Arylation of β-naphthol*

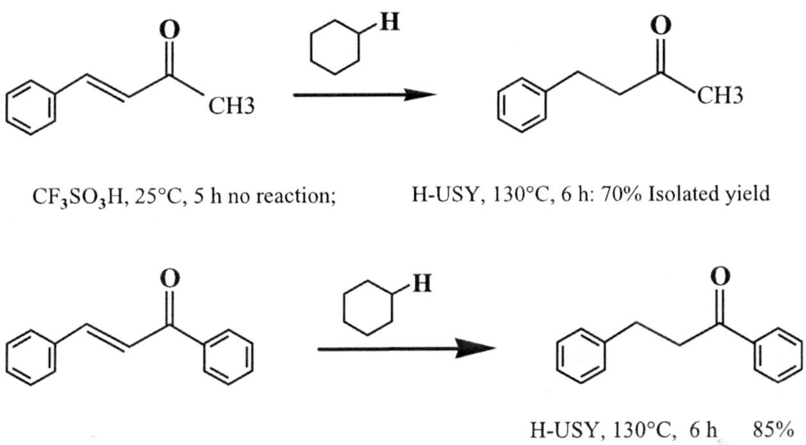

Scheme 5.11 *Regioselective reduction of α,β-unsaturated carbonyl compounds*

The arylation of naphthol, which does not occur in the presence of triflic acid was successfully achieved by using benzene as solvent in a pressure tube at 130 °C with H-USY with 72% yield.[73]

As a further example, α,β-unsaturated carbonyl compounds possess two sites which can undergo reduction: the carbonyl group and the adjacent C—C-double bond. While an abundant literature is available on the selective reduction of the carbonyl group, often achieved with relative ease, the selective hydrogenation of the double bond suffers from the lack of simple methods using practical chemicals. Selective ionic hydrogenation of α,β-unsaturated carbonyl compounds with cyclohexane was previously known to proceed only in superacidic conditions due to the necessity of dicationic, superelectrophilic activation of the enones. H-form zeolites with acidity well below superacidity are however also able to induce the reduction of α,β-unsaturated carbonyl compounds with cyclohexane, in analogy, or sometimes even better than the 'parent', superacid mediated reactions.[52,74]

Furthermore, arylvinylketones were efficiently converted to indanones using zeolite to promote cyclization.[75] Indanones are indeed one of the most useful family of compounds, which can be obtained from arylvinylketones, since they are the basis of many biologically active compounds including indacrinones, indanoyl isoleucine conjugates, indanocines and other medicinally important products.

The zeolite nature and the substitution pattern in the starting materials proved critical: 2-substituted arylvinylketones reacted more readily in H–ZSM5, giving high yield (>90%)

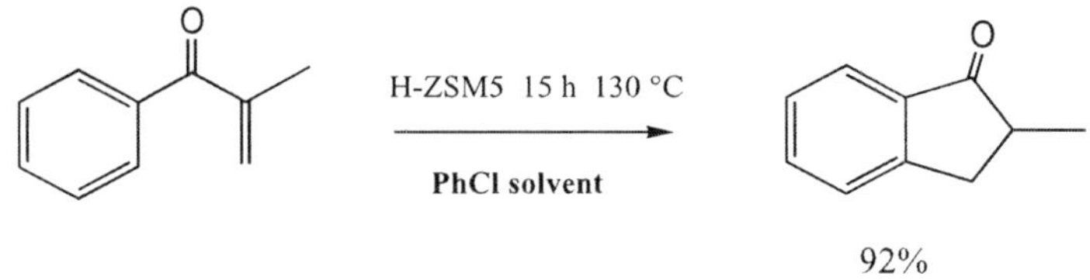

Scheme 5.12

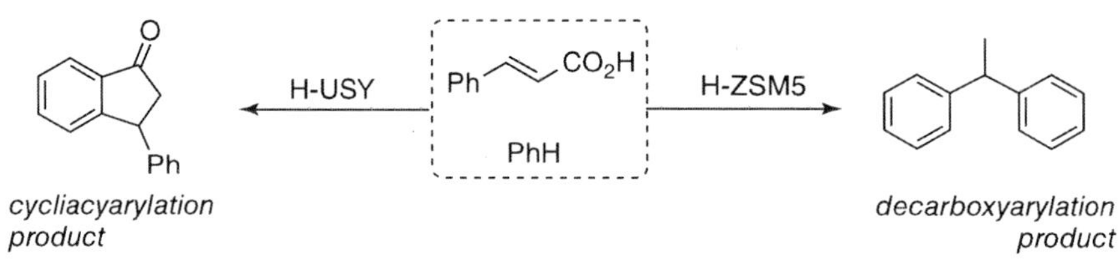

Scheme 5.13

of the corresponding indanones, while the 3-substituted are more efficiently converted to the corresponding indanones by H–USY. However in the presence of a more nucleophilic aromatic solvent such as toluene competitive alkylation is observed.

The interaction of α,β-unsaturated carboxylic acids with benzene derivatives was investigated in H–zeolites and led to two distinct but competing processes, cycliacyarylation and decarboxyarylation. Interestingly, H–USY selectively induced the cycliacyarylation cascade reaction, whereas H–ZSM5 selectively promoted decarboxyarylation.[76]

Recently, for the first time, copper(I)-exchanged zeolites were developed as catalysts in organic synthesis.[77] These solid materials proved to be versatile and efficient heterogeneous ligand-free catalytic systems. Cheap and easy to prepare, these catalysts exhibited a wide scope and a good compatibility with functional groups. They are very simple to use, easy to remove (by simple filtration) and they are recyclable.

5.3 Huisgen [3+2]-Cycloadditions

The first application we found was the Huisgen [3+2]-cycloaddition, between a terminal alkyne and an azide, to date the most practical and useful 'click' reaction, regioselectively affording 1,4-disubstituted 1,2,3-triazoles (Scheme 5.14).

Investigations with deuterated alkynes and deuterated zeolites proved that this Cu(I)–zeolite-catalysed 'click' reaction exhibited a mechanism different from the one reported for the Meldal–Sharpless version. which relies on the intermediate formation of copper acetylides (Scheme 5.15). Therefore, if such species were also formed within zeolites, deuteroalkynes would not give deuterated triazoles as observed. (Scheme 5.16)

We have also shown that copper(I)-modified zeolites can be used as catalysts for the [3+2] cycloaddition of pyrazol-3-one ylides with alkynes.[78] This heterogeneous method offers a mild and efficient access to tetrahydropyrazolo[1,2-a]pyrazol-3-ones with a reasonably wide scope, tolerating various functional groups, and with high regioselectivity.

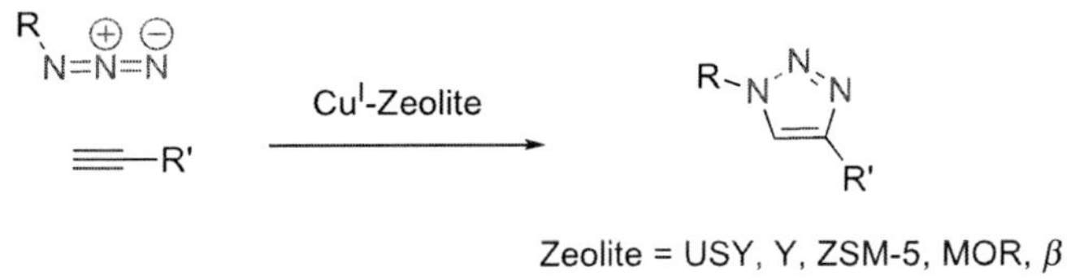

Scheme 5.14

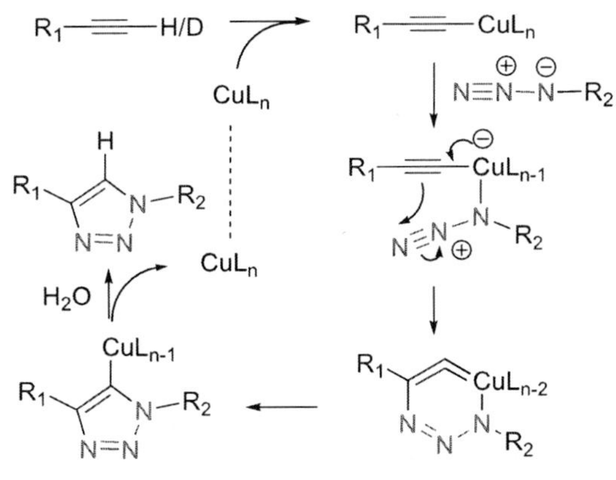

Scheme 5.15 *Meldal–Sharpless version*

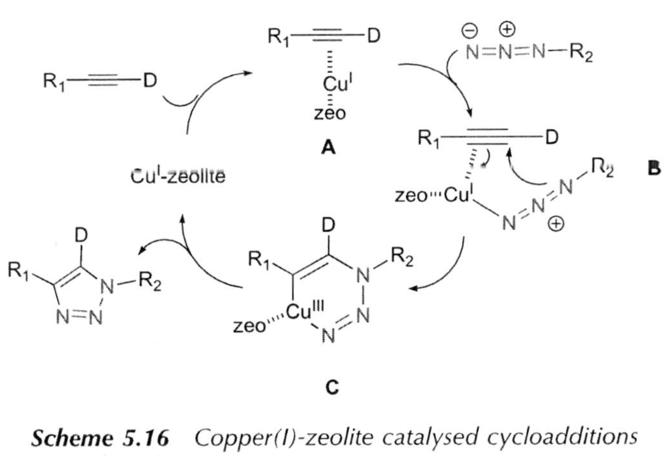

Scheme 5.16 *Copper(I)-zeolite catalysed cycloadditions*

Pyrazoline heterocycles and specially dihydropyrazolones are widely used as dyes for various applications in food, textile, photography and cosmetics industries. Some of these heterocycles exhibit useful biological properties, and for example, phenazone was one of the first synthetic drugs. The corresponding saturated pyrazolidin-3-ones exhibit similar properties, as exemplified with the anti-inflammatory drug phenylbutazone. Moreover, this heterogeneous copper(I)-modified zeolite catalyst can be reused six times without significant loss of activity.

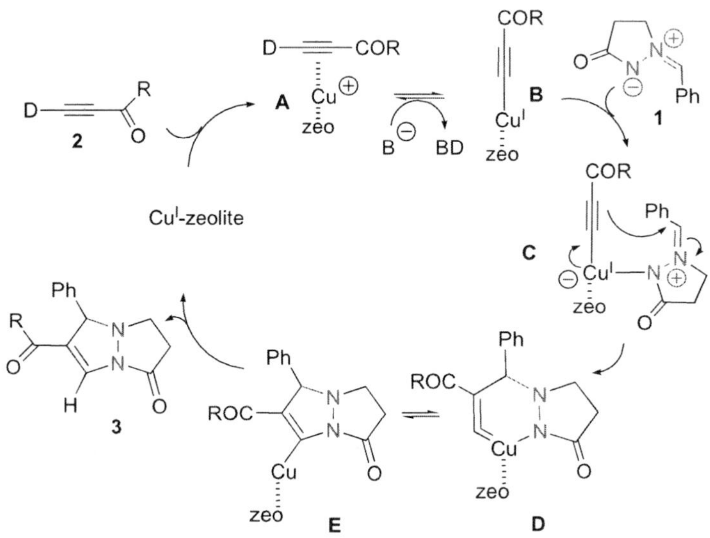

Scheme 5.17 *Mechanism for copper(I)-zeolite catalysed cycloaddition of azomethine ylide with activated alkyne*[79]

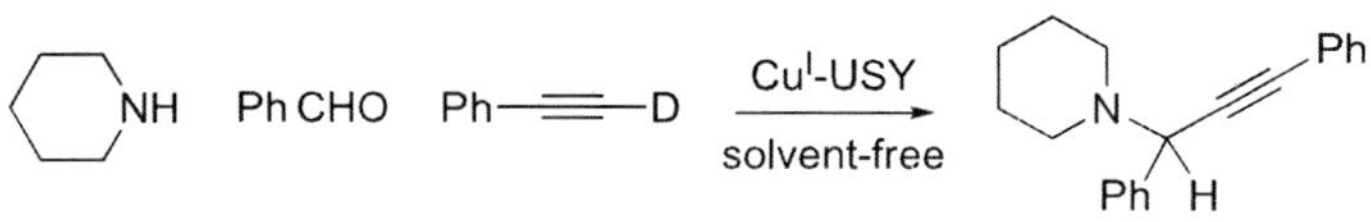

Scheme 5.18 *Cu^{I}-USY catalyzed reaction between piperidine, benzaldehyde and 1-deutero-2-phenylacetylene*

Experiments with labelled materials suggested in this case involvement of copper acetylide as intermediate (Scheme 5.17).

5.4 Multicomponent Reactions

One of the fundamental aspects in 'Green Chemistry' is linked to the number of steps in organic synthesis as well as atom economy. Multicomponent reactions (MCR) are thus becoming a more and more important class of reactions since they allow combining several starting materials in usually a single compound and in a one-flask operation.

Cu(I)-modified zeolites, especially Cu(I)–USY, proved to be very efficient catalyst in multicomponent reaction.[80] Such catalysts allowed for an efficient, *solvent-free* synthesis of propargylamines from aldehydes, amines and terminal alkynes with high yields.

Propargylic amines are high value building blocks in organic synthesis, and the corresponding structural motif has been found in various natural products, and in compounds of pharmaceutical or phytoprotective relevance. They can be obtained by addition of alkynes to imines, but since imines are easily formed from aldehydes and amines,

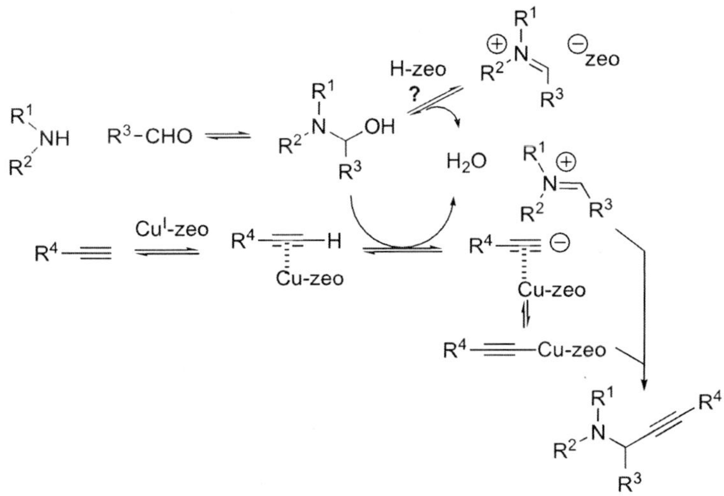

Scheme 5.19 *Proposed mechanism for MCR catalysed by Cu^I-USY*

3-component versions are known, either as such or promoted by various transition metals. A few supported versions or versions based on heterogeneous catalysts have recently been described, but zeolite-catalyzed version has only been reported recently.

Based on the use of deuteroalkynes the mechanism was suggested to involve copper–acetylides (Scheme 5.19).

In our experimental work on small alkanes activation on solid acids and zeolites we have also observed a very close relationship between zeolitic structure and alkane reactivity. The first observable interaction of hydrocarbons with the acidic protons on the surface of the catalyst consists of a fast hydron exchange that can be monitored using isotopic labelling.[81] This technique has been applied with various small alkanes as probes in order to understand the activation mechanism. For instance in the case of isobutane[29,82,83] a regiospecific exchange on the methyl groups was observed in the H/D exchange reaction with solid acids being in agreement with a general mechanism based on intermediate carbenium ions (classical trivalent carbocation). Once the carbocation is generated it is in equilibrium with its corresponding olefin on the surface by a fast deprotonation/reprotonation step. Similarly, we[84,85] and others[86,87] have studied more recently both the H/D exchange and the skeletal rearrangement in the reaction between propane and sulfated zirconia as well as zeolite using deuterium and carbon-13 as labels. With the help of *in situ* solid state MAS NMR the intra- and intermolecular rearrangements in propane were fully characterized providing strong evidences for a mechanism based on carbenium ions. Again the fast deprotonation/reprotonation process involving the equilibrium between the *sec*-propyl carbenium and propene, explains the Markovnikov-type incorporation of deuterium in the methyl group. In the aim to correlate better the acidity to the proton transfer it is preferable to study alkanes that cannot form an olefin by deprotonation such as methane and neopentane (2,2-dimethylpropane). As methyl cations are not generated under mild condition, due to their high energy, methane does undergo H/D exchange on solid acids without secondary reaction.[88,89] This exchange can only be possible via a concerted mechanism in which the transition state resembles that of a carbonium ion presenting a 2electron-3centre bond as in protonated methane. This species can be considered as the conjugated acid of methane. Non-classical cations were earlier suggested as transition state or strongly solvated intermediates in the H/D exchange reaction between methane and superacids DF/SbF_5[90] or DSO_3F/SbF_5.[91] While the reactivity of hydrocarbons in liquid superacids is clearly due to the σ-basicity of their C—H bonds,[66,92] the mechanistic pathway involves the surface of solid acids and the participation of nucleophilic oxygen lone pairs compensating the lack of superacidity and stabilizing charges.[93–95]

2,2-Dimethylpropane (neopentane) is a 'limiting' case on the σ-basicity scale for alkanes[96] because protolysis can only take place either on a primary C—H bond or on a C—C bond. Its behaviour in superacids has been studied in the late 1960s by Olah *et al.* in HSO_3F/SbF_5[97,98] and Hogeveen *et al.* in HF/SbF_5.[99] Depending on the acidity and steric factors, protolytic cleavage takes place either on the C—H bond or on a C—C bond, demonstrating the close basic character of both σ-bonds. Neopentane cracking has been previously studied over solid acids such as silica–aluminas, protonic zeolites[100–102] metal exchanged zeolites,[103,104] promoted and unpromoted sulfated zirconia.[105] The analysis of products distribution has revealed a protolytic cleavage of the C—C bond yielding methane and the adsorbed *t*-butyl cation as primary products.

The C—H bond activation characterized by H/D exchange with solid acids allows ranking solid acids by their activity and acidity.[106] Moreover neopentane being a bulky probe[107] is sensible to steric hindrance and confinement effect.[108]

In full agreement with Haag and Dessau hypothesis,[24] the first step consists in the protolysis of a C—C bond leading to the carbonium species (I) as a discrete intermediate or a transition state. In the next step methane and the adsorbed *t*-butyl cation are simultaneously formed and further deprotonation of the latter leads to various oligomerisation/cracking reactions (Scheme 5.20). Intermediate II was generally discarded in the literature, mainly because it would generate the primary highly energetic neopentyl cation. While this second pathway was highlighted in superacidic media by the formation of *t*-amyl cation resulting from the 1,2 methyl shift, several authors described the C—C bond protolysis as the major pathway on solid acids rather than the C—H bond protolysis.[106–111] We have shown that the species II should not be neglected as a direct exchange between primary protons and the solid acid was observed.[106]

5.4.1 H/D Exchange Between Neopentane and Zeolites

Neopentane was recirculated on deuterated zeolites (D–MOR, D–FAU, D–ZSM-5, D–BEA) at temperatures between 473 and 573 K. H/D exchange took place without any side reaction at the beginning of the reaction and the first order rate of deuteration of neopentane itself was determined at short contact times. It should be pointed out that the isotopomers of neopentane are formed in a consecutive way: D1→D2→D3→D4 … and so on. This corresponds to an H/D exchange reaction where one hydron is exchanged at a time. At higher temperatures, the competitive C—C bond cleavage takes place in a non negligible rate. For this reason we have limited our study to the H/D exchange of neopentane as follows: on D–USY between 523 and 573 K, D–BEA 513–548 K as well as D–ZSM5 (513–543 K) following the same experimental procedure.

5.4.2 H/D Exchange Between Neopentane and Zirconia

Sulfated zirconia is known to activate hydrocarbons at low temperature.[109] In the case of the well studied *n*-butane isomerization, but also for propane rearrangement the first step is most probably an oxidative dehydrogenation.[110,111] The synthesis of the material constitutes also a crucial point for its activity towards alkanes.[112,113] The catalyst we used in this study readily exchanged its deuterons with the hydrons of neopentane in the temperature range of 483–513 K. At higher temperature C—C protolysis competes significantly with H/D exchange. The experimental results have highlighted the ability of acidic sites of sulfated zirconia to generate a carbonium ion (species II, scheme 5.20) at such moderate temperature. Sulfated zirconia apart from its well known oxidative properties exhibits sufficiently strong acid sites allowing the generation of a pentacoordinated carbon similarly to carbonium-type ions in liquid superacids.

5.4.3 H/D Exchange and Acidity

The analysis of kinetic data of the H/D exchange allowed us to determinate the activation energies for each solid acid. The Eyring equation separates the free enthalpy from the entropic contribution for a better comparison and comprehension of the transition state involved in the proton transfer process.

Figure 5.8 shows that the free enthalpy of activation $\Delta H\#$ of the H/D exchange between neopentane and zeolites is very close for all zeolites tested. The similarity of these values is consistent with a common transition state in the H/D exchange step on zeolites. Further analysis of the entropic term (–222, –226, –228 and –236 $J\,mol^{-1}\,K^{-1}$ for MOR, BEA,

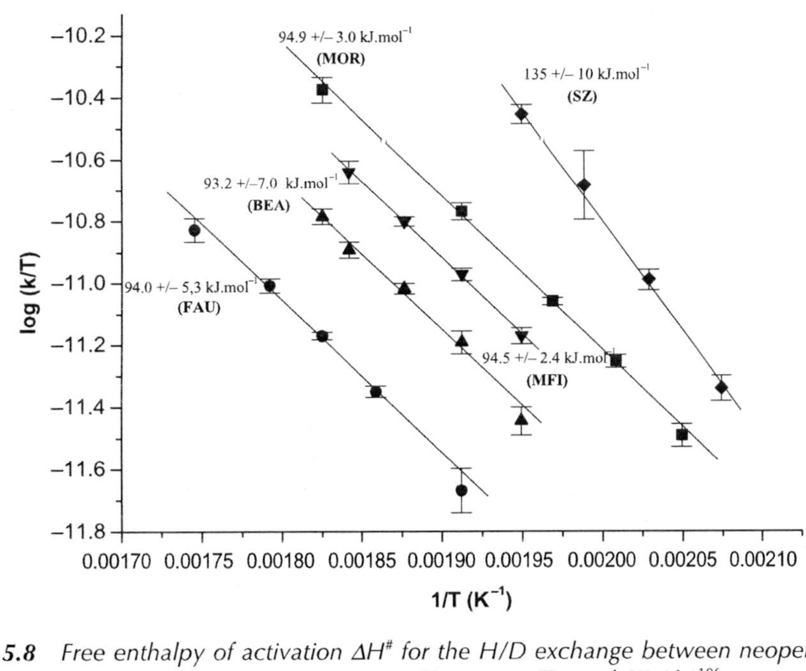

Figure 5.8 *Free enthalpy of activation $\Delta H^{\#}$ for the H/D exchange between neopentane and D_2O exchanged FAU (●), BEA (▲), MFI (▼), MOR (■) and SZ (◆)*[106]

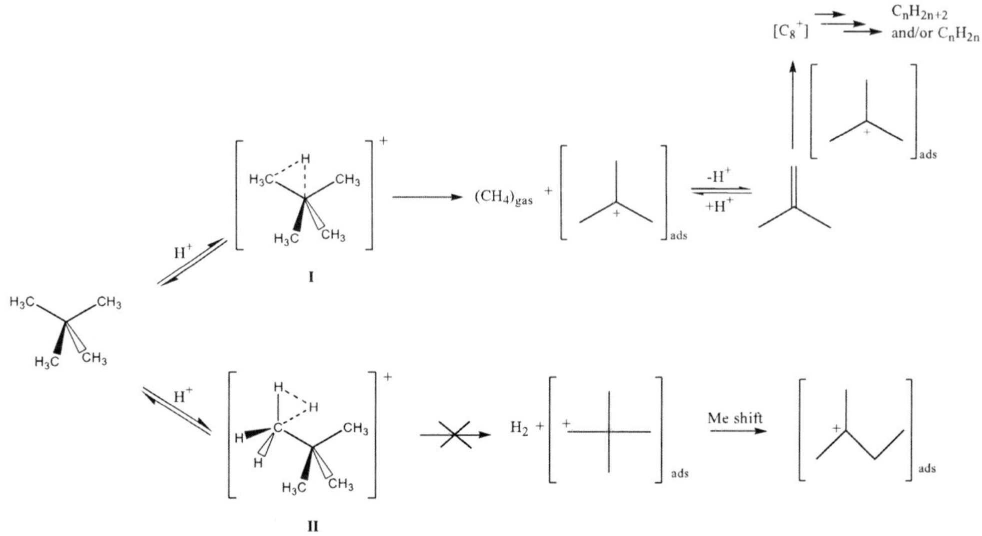

Scheme 5.20 *Protolysis and H/D exchange in the activation of neopentane on solid acids*[106]

MFI, USY, respectively) revealed an important participation of the surface in the transition state, constraining the transition in a defined geometry. In other words, the significant loss of entropy evidenced a strongly concerted mechanism with the surface.

Species II (Scheme 5.20) should rather be represented as an activated complex in which the surface of the solid acid delocalizes extensively the charge as proposed earlier.[97–99] Accordingly, the value $\Delta H\#$ calculated for sulfated zirconia differed markedly from those of zeolites, most probably due to the difference in chemical composition of the surface. This result constitutes a further proof of the strong involvement of the surface in the transition state. This situation is nevertheless very similar to what happens in liquid superacids for the activation of methane. In a previous study, the H/D exchange between methane and DSO_3F/SbF_5 – varying the SbF_5 concentration – showed not only a unique transition state for all systems but also a strong participation of the anionic part in the transition state. Hence, an analogy between these superacidic media and zeolites was proposed on the basis of kinetic analysis, and the rate of exchange at a given temperature should correlate directly with the acidity of the system.

Remarkably, at 523 K the rates of exchange were found to be in the same order of magnitude for all zeolites tested. In agreement with the previous studies on neopentane cracking, mordenite showed the highest activity among zeolites. Again, in full analogy with previous results in superacids,[98] since the transition state is unique for all zeolites as well as in the case of the exchange between CH_4 and DSO_3F/SbF_5, we suggested that the activity in H/D exchange can be related to the acidity of zeolites. The order of activity appears to be the following: MOR>MFI>BEA>FAU. Generally the Si/Al ratio is considered as a major parameter to characterize the acidity of the zeolites. Nevertheless Si/Al ratio difference in this study was not sufficient to explain the difference observed for the activity in H/D exchange. In fact the activity of zeolites seemed to fit more with the size of the pores. Considering the most confined space or the smallest pore of zeolite allows finding the order of activity of zeolites in H/D exchange with neopentane. The closer the molecular size of the probe fits with the pores, the more basic the probe becomes, thus facilitating the acid catalysis. Busca *et al.* and Thibault-Starzyck *et al.* have very recently published their results concerning the adsorption of nitriles and pyridines in zeolite.[114–117] It is very interesting to note that acetonitrile can enter such small pores as the 'side pockets' of Mordenite (3.4 × 4.8 Å) at low temperature. In contrast these authors demonstrated by IR spectroscopy that pivalonitrile could not enter these small pores and remains in the main channel of Mordenite. Neopentane is an isolobal analog of pivalonitrile having a diameter of 6.2 Å which prevents the entry in the 'side pockets'. However, computational studies showed clearly that a methyl group having a critical diameter of 4 Å enters the 'side pockets' of mordenite when temperature is increased.[117] Indeed the increase of temperature implies a greater flexibility of the framework and allows the distortion of the molecular probe.[118,119] On the same basis Nascimento has also shown that neopentane can enter the MFI channel (5.3 × 5.6 Å).[120] Taking into account these results, it seems that the order of activity of the zeolites we present here emphasizes the confinement effect. Mordenite is the most active catalyst for the H/D exchange of neopentane as well as for its cracking and H/D exchange takes place at lower temperature than with the other zeolites. Considering the 'side pockets' as ideal sites where the methyl group of neopentane can be optimally protonated can explain these experimental data. Therefore the order of activity found for H/D exchange between zeolites and neopentane seems to be strongly related to the size of the pores.

5.5 Conclusion

Never has the need for Green Chemistry and clean technology using processes that generate little or no waste, been greater than at present.

Zeolites are important molecular sieves or heterogeneous catalysts which may operate under mild and environmentally benign conditions preferably without use of non-aqueous or any other solvents. They present the advantages, in comparison with previously used alumina-based catalysts, of having a better long-term stability, and yielding higher product selectivity. The higher selectivity of this class of catalysts originates from the well-defined micropore structures of the zeolite crystals, which lead to size- and shape-dependent adsorption, reaction, or diffusion of the reactants or products. It is therefore of interest to study the behaviour of well-documented probes with novel zeolitic materials in order to assess whether new topological effects, not previously found in conventional zeolites, can be observed. The adsorption strength is strongly dependent on the molecule to pore size ratio via the so-called '*confinement effect*.' Because in different types of zeolites the OH groups, as well as the metals, are located in pores, channels, and cages of different size and shape, the *confinement effect* should be considered when comparing the properties of zeolites on the basis of adsorption of probe molecules. Zeolite micropores have also a stabilizing effect on the reaction transition states. This stabilization has been demonstrated to be of a short-range electrostatic nature and to involve polarization of zeolite oxygen atoms, which is of course related to the molecule to pore size ratio. Nevertheless the problem of selecting the optimal pore size for a given reaction has no unique solution.

In this chapter we provided the reader with some examples, mostly from our own work, which underline the ability of zeolites to operate selectively under mild, environmentally benign conditions, in a field which is still under rapid development.

References

1. J. M. Thomas, Heterogeneous catalysis: enigmas, illusions, challenges, realities, and emergent strategies of design, *J. Chem. Phys.*, 2008, **128**, 182502 (19 pages).
2. B. Smit, T. L. M. Maesen, Towards a molecular understanding of shape selectivity, *Nature*, 2008, **451**, 671–678.
3. K. Tanabe, W. Hoelderich, Industrial application of solid acid–base catalysts, *Appl. Catal. A*, 1999, **181**, 399–434.
4. R. A. Sheldon, J. A. Elings, S. K. Lee, H. E. B. Lempers, R. S. Downing, Zeolite-catalyzed rearrangements in organic synthesis, *J. Mol. Catal. A: Chemical*, 1998, **134**, 129–135.
5. W. F. Hoelderich, H. van Bekkum, Zeolites in organic syntheses, in 'Introduction into Zeolite Science and Practice' 631–726, Elsevier, 1991.
6. M. Guisnet, J.-P. Gilson, *Zeolites for cleaner technologies*, Imperial College Press, 2002.
7. C. Baerlocher, W. M. Meier, D. H. Olson, *Atlas of Zeolite Framework Types*, 5th ed., Elsevier, Amsterdam, 2001.
8. N. Y. Chen, W. E. Garwood, F. G. Dwyer, *Shape Selective Catalysis in Industrial Applications*, Marcel Dekker, New York, 1989.
9. M. Boronat, C. Martinez-Sanchez, D. Law, A. Corma, Enzyme-like specificity in zeolites: a unique site position in mordenite for selective carbonylation of methanol and dimethyl ether with CO, *J. Am. Chem. Soc.*, 2008, **130**, 16316–16323.
10. F. Goettmann, C. Sanchez, How does confinement affect the catalytic activity of mesoporous materials?, *J. Mater. Chem.*, 2007, **17**, 24–30.
11. J. F. M. Denayer, G. V. Baron, The confinement factor: a thermodynamic parameter to characterize microporous adsorbents, *Adsorption*, 2005, **11**, 85–90.
12. F. Eder, J. A. Lercher, On the role of the pore size and tortuosity for sorption of alkanes in molecular sieves, *J. Phys. Chem. B*, 1997, **101**, 1273–1278.
13. F. Eder, M. Stockenhuber, J. A. Lercher, Brønsted acid site and pore controlled siting of alkane sorption in acidic molecular sieves, *J. Phys. Chem. B*, 1997, **101**, 5414–5419.
14. C. M. Zicowich-Wilson, A. Corma, P. M. Viruela, Electronic confinement of molecules in microscopic pores. A new concept which contributes to the explanation of the catalytic activity of zeolites, *J. Phys. Chem.*, 1994, **98**, 10863–10870.
15. A. Corma, H. Garcia, G. Sastre, P. M. Viruela, Activation of molecules in confined spaces: an approach to zeolite–guest supramolecular systems, *J. Phys. Chem. B*, 1997, **101**, 4575–4582.
16. F. Marquez, C. M. Zicowich-Wilson, A. Corma, E. Palomares, H. Garcia, Naphthalene included within all-silica zeolites: influence of the host on the naphthalene photophysics, *J. Phys. Chem. B*, 2001, **105**, 9973–9979.
17. M. S. Galletero, A. Corma, B. Ferrer, V. Fornes, H. Garcia, Confinement effects at the external surface of delaminated zeolites (ITQ-2): an inorganic mimic of cyclodextrins, *J. Phys. Chem. B*, 2003, **107**, 1135–1141.
18. B. Hunger, M. Heuchel, L. A. Clark, R. Q. Snurr, Characterization of acidic OH groups in zeolites of different types: an interpretation of NH_3–TPD results in the light of confinement effects, *J. Phys. Chem. B*, 2002, **106**, 3882–3889.
19. K. S. Smirnov, F. Thibault-Starzyk, Confinement of acetonitrile molecules in Mordenite. A computer modeling study, *J. Phys. Chem. B*, 1999, **103**, 8595–8601.
20. E. G. Derouane, Zeolites as solid solvents, *J. Mol. Catal.*, 1998, **134**, 29–45.
21. S. Pariente, P. Trens, F. Fajula, F. Di Renzo, N. Tanchoux, Heterogeneous catalysis and confinement effects: the isomerization of 1–hexene on MCM-41 materials, *Appl. Catal. A*, 2006, **307**, 51–57.

22. A. Corma, Inorganic solid acids and their use in acid-catalysed hydrocarbon reactions, *Chem. Rev.*, 1995, **95**, 559–614.
23. M. Boronat, A. Corma, Are carbenium and carbonium ions reaction intermediates in zeolite-catalyzed reactions? *Appl. Catal. A*, 2008, **336**, 2–10.
24. W. O. Haag, R. M. Dessau, Duality of mechanism of acid-catalyzed paraffin cracking, in *Proceedings of the 6th International Congress on Catalysis*, Berlin, VZ, 305–308, 1984.
25. S. Kotrel, H. Knözinger, B. C. Gates, The Haag–Dessau mechanism of protolytic cracking of alkanes, *Micropor. Mesopor. Mater.*, 2000, **35–36**, 11–20.
26. G. A. Olah, A. Molnar, G. K. S. Prakash, J. Sommer, *Superacid chemistry*, J. Wiley, Oborok, N.J., 2009.
27. J. Sommer, J. Bukala, Selective electrophilic activation of alkanes, *Acc. Chem. Res.*, 1993, **26**, 370–376.
28. J. Sommer, J. Bukala, M. Hachoumy, R. Jost, Reversible protonation of isobutane in liquid superacids in competition with protolytic ionization, *J. Am. Chem. Soc.*, 1997, **119**, 3274–3279.
29. B. Schoofs, J. Schuermans, R. A. Schoonheydt, Hydrogen–deuterium exchange reactions with isobutane over acid zeolites, *Micropor. Mesopor. Mater.*, 2000, **35–36**, 99–111.
30. J. Sommer, R. Jost, M. Hachoumy, Activation of small alkanes on strong solid acids: mechanistic approaches, *Catal. Today*, 1997, **38**, 309–319.
31. J. Huang, Y. Jiang, V. R. R. Marthala, M. Hunger, Insight into the mechanisms of the ethylbenzene disproportionation: transition state shape selectivity on zeolites, *J. Am. Chem. Soc.*, 2008, **130**, 12642–12644.
32. J. F. Haw, J. B. Nicholas, T. Xu, L. W. Beck, D. B. Ferguson, Physical organic chemistry of solid acids: lessons from in Situ NMR and theoretical chemistry, *Acc. Chem. Res.*, 1996, **29**, 259–267.
33. J. F. Haw, Zeolite acid strength and reaction mechanisms in catalysis, *Phys. Chem. Chem. Phys.*, 2002, **4**, 5431–5441.
34. J. Sauer, C. Tuma, Protonated isobutene in zeolites: tert-butyl cation or alkoxide? *Angew. Chem. Int. Ed.*, 2005, **44**, 4769–4771.
35. M. Boronat, A. Corma, Reaction intermediates in acid catalysis by zeolites: prediction of the relative tendency to form alkoxides or carbocations as a function of hydrocarbon nature and active site structure, *J. Am. Chem. Soc.*, 2004, **126**, 3300–3309.
36. A. G. Stepanov, M. V. Luzgin, V. N. Rommanikov, K. I. Zamaraev, NMR observation of the Koch reaction in zeolite H-ZSM-5 under mild conditions, *J. Am. Chem. Soc.*, 1995, **117**, 3615–3616.
37. M. V. Luzgin, V. N. Rommanikov, A. G. Stepanov, K. I. Zamaraev, Interaction of olefins with carbon monoxide on zeolite H-ZSM-5. NMR observation of the Friedel–Crafts acylation of alkenes at ambient temperature, *J. Am. Chem. Soc.*, 1996, **118**, 10890–10891.
38. M. Franco, N. Rosenbach Jr., G. B. Ferreira, A. C. O. Guerra, W. B. Kover, C. C. Turci, C. J. A. Mota, Rearrangement, nucleophilic substitution, and halogen switch reactions of alkyl halides over NaY zeolite: formation of the bicyclobutonium cation inside the zeolite cavity, *J. Am. Chem. Soc.*, 2008, **130**, 1592–1600.
39. E. A. Pidko, R. A. van Santen, Confined space-controlled olefin–oxygen charge transfer in zeolites, *J. Phys. Chem. B*, 2006, **110**, 2963–2967.
40. M. Iwamoto, J. Hirata, K. Matsukami, S. Kagawa, Catalytic oxidation by oxide radical ions. 1. One-step hydroxylation of benzene to phenol over group 5 and 6 oxides supported on silica gel, *J. Phys. Chem.*, 1983, **87**, 903–905.
41. E. Suzuki, K. Nakashiro, Y. Ono, Hydroxylation of benzene with dinitrogen monoxyde over H-ZSM-5 zeolite, *Chem. Lett.*, 1988, 953–956.
42. P. M. Esteves, B. Louis, Experimental and DFT study of the partial oxidation of benzene by N_2O over H-ZSM-5: acid catalyzed mechanism, *J. Phys. Chem. B*, 2006, **110**, 16793–16800.
43. M. Stöcker, Methanol-to-hydrocarbons: catalytic materials and their behavior, *Micropor. Mesopor. Mater.*, 1999, **29**, 3–48.

44. J. Q. Chen, A. Bozzano, B. Glover, T. Fuglerud, S. Kvisle, Recent advancements in ethylene and propylene production using the UOP/Hydro MTO process, *Catal. Today*, 2005, **106**, 103–107.
45. http://www.uop.com/objects/26%20MTO%20process.pdf
46. G. A. Olah, A. Goeppert, G. K. S. Prakash, *Beyond Oil and Gas: The Methanol Economy*, Wiley-VCH, Weinheim, 2006.
47. Y. Ono, T. Mori, Kinetic model for methanol conversion over Mg-ZSM-5 zeolite, *J. Chem. Soc., Faraday Trans. 1*, 1981, **77**, 2209–2221.
48. C. M. Zicovich-Wilson, P. Viruela, A. Corma, Formation of surface methoxy groups on H-zeolites from methanol. A quantum chemical study, *J. Phys. Chem.*, 1995, **99**, 13224–13231.
49. Y. Jiang, M. Hunger, W. Wang, On the reactivity of surface methoxy species in acidic zeolites, *J. Am. Chem. Soc.*, 2006, **128**, 11679–11692.
50. Y. Jiang, W. Wang, V. R. R. Marthala, J. Huang, B. Sulikowski, M. Hunger, Effect of organic impurities on the hydrocarbon formation via the decomposition of surface methoxy groups on acidic zeolite catalysts, *J. Catal.*, 2006, **238**, 21–27.
51. D. Lesthaeghe, V. van Speybroeck, G. B. Marin, M. Waroquier, The rise and fall of direct mechanisms in methanol-to-olefin catalysis: an overview of theoretical contributions, *Ind. Eng. Chem. Res.*, 2007, **46**, 8832–8838.
52. D. M. McCann, D. Lesthaeghe, P. W. Kletnieks, D. R. Guenther, M. J. Hayman, V. van Speybroeck, M. Waroquier, J. F. Haw, A complete catalytic cycle for supramolecular methanol-to-olefins conversion by linking theory with experiment, *Angew. Chem. Int. Ed.*, 2008, **47**, 5179–5182.
53. I. M. Dahl, S. Kolboe, On the reaction mechanism for propene formation in the MTO reaction over SAPO-34, *Catal. Lett.*, 1993, **20**, 329–336.
54. M. Hunger, In situ NMR spectroscopy in heterogeneous catalysis, *Catal. Today*, 2004, **97**, 3–12.
55. J. F. Haw, W. Song, D. M. Marcus, J. B. Nicholas, The mechanism of methanol to hydrocarbon catalysis, *Acc. Chem. Res.*, 2003, **36**, 317–326.
56. Z. M. Cui, Q. Liu, W. Song, L. J. Wan, Insights into the mechanism of methanol-to-olefin conversion at zeolites with systematically selected framework structures, *Angew. Chem. Int. Ed.*, 2006, **45**, 6512–6515.
57. Y. Ono, T. Baba, J. Sakai, T. Keii, J., Conversion of methanol into hydrocarbons catalysed by metal salts of heteropolyacids, *Chem. Soc., Chem. Commun.*, 1981, 400–401.
58. G. A. Olah, H. Doggweiler, J. D. Felberg, S. Frohlich, M. J. Grdina, R. Karpeles, T. Keumi, S. I. Inaba, W. M. Ip, K. Lammertsma, G. Salem, D. C. Tabor, Onium ylide chemistry. 1. Bifunctional acid-base-catalyzed conversion of heterosubstituted methanes into ethylene and derived hydrocarbons. The onium ylide mechanism of the C1 $\rightarrow$ C2 conversion, *J. Am. Chem Soc.*, 1984, **106**, 2143–2149.
59. J. E. Bercaw, P. L. Diaconescu, R. H. Grubbs, R. D. Kay, S. Kitching, J. A. Labinger, X. Li, P. Mehrkhodavandi, G. E. Morris, G. J. Sunley, P. Vagner, On the mechanism of the conversion of methanol to 2,2,3–trimethylbutane (triptane) over zinc iodide, *J. Org. Chem.*, 2006, **71**, 8907–8917.
60. S. Walspurger, G. K. S. Prakash, G. A. Olah, Zinc catalyzed conversion of methanol–methyl iodide to hydrocarbons with increased formation of triptane, *Appl. Catal. A*, 2008, **336**, 48–53.
61. J. F. M. Denayer, L. I. Devriese, S. Couck, J. Martens, R. Singh, P. A. Webley, G. V. Baron, Cage and window effects in the adsorption of n-alkanes on chabazite and SAPO-34, *J. Phys. Chem. C*, 2008, **112**, 16593–16599.
62. A. Bhan, E. Iglesia, A link between reactivity and local structure in acid catalysis on zeolites, *Acc. Chem. Res.*, 2008, **41**, 559–567.
63. A. Bhan, R. Gounder, J. Macht, E. Iglesia, Entropy considerations in monomolecular cracking of alkanes on acidic zeolites, *J. Catal.*, 2008, **253**, 221–224.
64. G. A. Olah, Superelectrophiles, *Angew. Chem. Int. Ed. Engl.*, 1993, **32**, 767–788.
65. G. A. Olah, D. A. Klumpp, Superelectrophilic solvation, *Acc. Chem. Res.*, 2004, **37**, 211–220.

66. T. Ohwada, N. Yamagata, K. Shudo, Friedel-Crafts-type reactions involving di- and tricationic species. Onium-allyl dications and O,O-diprotonated aci-nitro species bearing a protonated carbonyl group, *J. Am. Chem. Soc.*, 1991, **113**, 1364–1373.
67. R. Rendy, Y. Zhang, A. McElrea, A. Gomez, D. A. Klumpp, Superacid-catalyzed reactions of cinnamic acids and the role of superelectrophiles, *J. Org. Chem.*, 2004, **69**, 2340–2347.
68. K. Yu, Koltunov, I. B. Repinskaya, Reaction of phenols and their derivatives with aromatic compounds in the presence of acidic reagents. XIV. α,β-Enone protonation in superacidic $HF\text{-}SbF_5\text{–}SO_2ClF$ system, *Russ. J. Org. Chem.*, 1994, **30**, 97–100.
69. R. S. Drago, N. Kob, Acidity and reactivity of sulfated zirconia and metal-doped sulfated zirconia, *J. Phys. Chem. B*, 1997, **101**, 3360–3364.
70. J. F. Haw, Zeolite acid strength and reaction mechanisms in catalysis, *Phys. Chem. Chem. Phys.*, 2002, **4**, 5431–5441.
71. G. J. Kramer, R. A. van Santen, An ab initio study of D/H exchange between CD_4 and the H-forms of zeolites FAU and MFI, *J. Am. Chem. Soc.*, 1995, **117**, 1766–1776.
72. A. Berkessel, R. K. Thauer, On the mechanism of catalysis by a metal-free hydrogenase from methanogenic archaea: enzymatic transformation of H_2 without a metal and its analogy to the chemistry of alkanes in superacidic solution, *Angew. Chem. Int. Ed.*, 1995, **34**, 2247–2250.
73. K. Yu. Koltunov, S. Walspurger, J. Sommer, Superelectrophilic activation of polyfunctional organic compounds using zeolites and other solid acids, *Chem. Commun.*, 2004, 1754–1755.
74. K. Yu. Koltunov, S. Walspurger, J. Sommer, Selective, C,C-double bond reduction of α,β-unsaturated carbonyl compounds with cyclohexane using zeolites, *J. Mol. Catal. A*, 2006, **245**, 231–234.
75. A. Sani-Souna-Sido, S. Chassaing, P. Pale, J. Sommer, Behavior of arylvinylketones in zeolites: a systematic study, *Appl. Catal. A*, 2008, **336**, 101–108.
76. S. Chassaing, M. Kumarraja, P. Pale, J. Sommer, Zeolite-directed cascade reactions of unsaturated carboxylic acids with aromatics: cycliacyarylation *vs* decarboxyarylation, *Org. Lett.*, 2007, **9**, 3889–3892.
77. S. Chassaing, A. Sani Souna Sido, M. Kumarraja, P. Pale, J. Sommer, 'Click chemistry' in Cu(I)-zeolites: the Huisgen cycloaddition, *Org. Lett.*, 2007, **9**, 883–886.
78. S. Chassaing, A. Sani Souna-Sido, A. Alix, M. Kumarraja, P. Pale, J. Sommer, Click chemistry in zeolites: copper(I) zeolites as new heterogeneous and ligand-free catalysts for the cycloaddition, *Chem. Eur. J.*, 2008, **14**, 6713–6721.
79. M. Keller, A. Sani Souna Sido, P. Pale, J. Sommer, Copper(I)-zeolites as new heterogeneous and ligand-free catalysts: [3+2]-cycloaddition of azomethine imines, *Eur. J. Org. Chem.*, 2009, **15**, in press.
80. M. K. Patil, M. Keller, B. M. Reddy, P. Pale, J. Sommer, Copper-zeolites as green catalysts for multicomponent reactions: an efficient and green synthesis of propargyl amines, *Eur. J. Org. Chem.*, 2008, **26**, 4440–4445.
81. T. F. Narbeshuber, M. Stockenhuber, A. Brait, K. Seshan, J. A. Lercher, Hydrogen/deuterium exchange during-butane conversion on H-ZSM-5, *J. Catal.*, 1996, **160**, 183–189.
82. J. Sommer, D. Habermacher, R. Jost, A. Sassi, A. G. Stepanov, M. V. Luzgin, D. Freude, H. Ernst, J. Martens, Activation of small alkanes on solid acids. An H/D exchange study by liquid and solid-state NMR: the activation energy and the inhibiting effect of carbon monoxide, *J. Catal.*, 1999, **181**, 265–270.
83. W. Hua, A. Sassi, A. Goeppert, F. Taulelle, C. Lorentz, J. Sommer, H/D exchange reaction between isobutane and acidic USY Zeolite: a mechanistic study by mass spectrometry and in situ NMR, *J. Catal.*, 2001, **204**, 460–465.
84. M. Haouas, S. Walspurger, J. Sommer, Regioselective H/D isotope exchange and skeletal rearrangement reactions of propane over strong solid acids, *J. Catal.*, 2003, **215**, 122–128.
85. M. Haouas, S. Walspurger, F. Taulelle, J. Sommer, The initial stages of solid acid-catalyzed reactions of adsorbed propane. A mechanistic study by in situ MAS NMR, *J. Am. Chem. Soc.*, 2004, **126**, 599–606.
86. A. G. Stepanov, S. S. Arzumanov, M. V. Luzgin, H. Ernst, D. Freude, V. N. Parmon, In situ ^{1}H and ^{13}C MAS NMR study of the mechanism of H/D exchange for deuterated propane adsorbed on H-ZSM-5, *J. Catal.*, 2005, **235**, 221–228.

87. S. S. Arzumanov, S. I. Reshetnikov, A. G. Stepanov, V. N. Parmon, D. Freude, In Situ ^{1}H and ^{13}C MAS NMR kinetic study of the mechanism of H/D exchange for propane on zeolite H–ZSM-5, *J. Phys. Chem. B*, 2005, **109**, 19748–19757.
88. W. Hua, A. Goeppert, J. Sommer, H/D exchange and isomerization of small alkanes over unpromoted and Al_2O_3–promoted SO_4^{2-}/ZrO_2 catalysts, *J. Catal.*, 2001, **197**, 406–413.
89. B. Schoofs, J. A. Martens, P. A. Jacobs, R. A. Schoonheydt, Kinetics of hydrogen–deuterium exchange reactions of methane and deuterated acid FAU- and MFI-type zeolites, *J. Catal.*, 1999, **183**, 355–367.
90. P. Ahlberg, A. Karlsson, A. Goeppert, S. O. N. Lill, P. Diner, J. Sommer, H/D isotope exchange between methane and magic acid (HSO_3 F–SbF_5): an in situ NMR study, *Chem.-Eur. J.*, 2001, **7**, 1936–1943.
91. S. Walspurger, A. Goeppert, M. Haouas, J. Sommer, H/D isotope exchange between methane and magic acid (HSO_3F/SbF_5): an in situ NMR study, *New. J. Chem.*, 2004, **28**, 266–269.
92. A. Goeppert, J. Sommer, H/D exchange, protolysis and oxidation of isopentane and propane in Superacids. σ-basicity and reactivity of C-H bonds, *New J. Chem.*, 2002, **26**, 1335–1339.
93. J. M. Vollmer, T. N. Truong, Mechanisms of hydrogen exchange of methane with H-zeolite Y: an ab initio embedded cluster study, *J. Phys. Chem. B*, 2000, **104**, 6308–6312.
94. G. J. Kramer, R. A. Van Santen, C. A. Emeis, A. K. Nowak, Understanding the acid behaviour of zeolites from theory and experiment, *Nature*, 1993, **363**, 529–530.
95. P. M. Esteves, M. A. C. Nascimento, C. J. A. Mota, Reactivity of alkanes on zeolites: a theoretical ab initio study of the H/H exchange *J. Phys. Chem. B*, 1999, **103**, 10417–10420.
96. G. A. Olah, Y. Halpern, J. Shen, Y. K. Mo, Electrophilic reactions at single bonds. XII. Hydrogen-deuterium exchange, protolysis (deuterolysis), and oligocondensation of alkanes with superacids, *J. Am. Chem. Soc.*, 1973, **95**, 4960–4969.
97. G. A. Olah, R. H. Schlosberg, Chemistry in super acids. I. Hydrogen exchange and polycondensation of methane and alkanes in FSO_3H–SbF_5 ('magic acid') solution. Protonation of alkanes and the intermediacy of CH_5^+ and related hydrocarbon ions. The high chemical reactivity of 'paraffins' in ionic solution reactions, *J. Am. Chem. Soc.*, 1968, **90**, 2726–2727.
98. G. A. Olah, G. Klopman, R. H. Schlosberg, Super acids. III. Protonation of alkanes and intermediacy of alkanonium ions, pentacoordinated carbon cations of CH_5^+ type. Hydrogen exchange, protolytic cleavage, hydrogen abstraction, polycondensation of methane, ethane, 2,2–dimethylpropane and 2,2,3,3–tetramethylbutane in FSO_3H–SbF_5, *J. Am. Chem. Soc.*, 1969, **91**, 3261–3268.
99. H. Hogeveen, A. F. J. Bickel, Chemistry and spectroscopy in strongly acidic solutions: electrophilic substitution at alkane-carbon by protons, *J. Chem. Soc., Chem. Commun.*, 1967, 635–636.
100. E. A. Lombardo, R. Pierantozzi, W. K. Hall, The mechanism of neopentane cracking over solid acids, *J. Catal.*, 1988, **110**, 171–183.
101. J. Engelhardt, W. K. Hall, Contribution to the understanding of the reaction chemistry of isobutane and neopentane over acid catalysts, *J. Catal.*, 1990, **125**, 472–487.
102. F. Evanics, I. Kiricsi, G. Tasi, H. Förster, P. Fejes, Cracking of neopentane over acidic zeolites: influence of isobutane and isobutene admission, *J. Mol. Catal A*, 1995, **95**, 269–276.
103. H. K. Beyer, J. Horvath, F. Reti, Catalytic cracking and dehydrogenation of butanes and neopentane on reduced copper-exchanged Y-zeolites, *React. Kinet. Catal. Lett.*, 1980, **14**, 395–401.
104. P. V. Menacherry, G. L. Haller, Neopentane conversion over zeolite-supported platinum and palladium catalysts, *J. Catal.*, 1997, **167**, 425–437.
105. T. K. Cheung, J. L. d'Itri, F. C. Lange, B. C. Gates, Neopentane cracking catalyzed by iron- and manganese-promoted sulfated zirconia, *Catal. Lett.*, 1995, **31**, 2–3.
106. S. Walspurger, Y. Y. Sun, A. Sani-Souda-Sido, J. Sommer, Neopentane and solid acids: direct hydron exchange before cracking, *J. Phys. Chem. B*, 2006, **110**, 18368–18373.
107. G. D. Lei, B. T. Carvill, W. M. H. Sachtler, Single file diffusion in mordenite channels: neopentane conversion and H/D exchange as catalytic probes, *Appl. Catal. A*, 1996, **142**, 347–359.

108. E. G. Derouane, J. M. André, A. A. Lucas, Surface curvature effects in physisorption and catalysis by microporous solids and molecular sieves, *J. Catal.*, 1988, **110**, 58–73.
109. M. Hino, K. J. Arata, Synthesis of solid superacid catalyst with acid strength of Ho $\leq -16{,}04$, *J. Chem. Soc., Chem. Commun.*, 1980, 851–852.
110. M. T. Tran, N. S. Gnep, G. Szabo, M. Guisnet, Influence of the calcination temperature on the acidic and catalytic properties of sulphated zirconia, *Appl. Catal. A*, 1998, **171**, 207–217.
111. X. Li, K. Nagaoka, L. J. Simon, R. Olindo, J. A. Lercher, A. Hofmann, J. Sauer, Oxidative activation of *n*-butane on sulfated zirconia *J. Am. Chem. Soc.*, 2005, **127**, 16159–16166.
112. Y. Y. Sun, S. Walspurger, B. Louis, J. Sommer, Investigation of factors influencing catalytic activity for n-butane isomerization in the presence of hydrogen on Al-promoted sulfated zirconia, *Appl. Catal. A*, 2005, **292**, 200–207.
113. X. Li, K. Nagaoka, R. Olindo, J. A. Lercher, Synthesis of highly active sulfated zirconia by sulfation with SO_3, *J. Catal.*, 2006, **238**, 39–45.
114. M. Bevilacqua, A. Gutierrez-Alejandre, C. Resini, M. Casagrande, J. Ramirez, G. Busca, An FTIR study of the accessibility of the protonic sites of H-mordenites, *Phys. Chem. Chem. Phys.*, 2002, **4**, 4575–4583.
115. I. Salla, T. Montanari, P. Salagre, Y. Cesteros, G. Busca, Fourier transform infrared spectroscopic study of the adsorption of CO and nitriles on Na–mordenite: evidence of a new interaction, *J. Phys. Chem. B*, 2005, **109**, 915–922.
116. O. Marie, F. Thibault-Starzyk, Confirmation of the strongest nitriles–hydroxy groups interaction in the side pockets of mordenite zeolites, *Phys. Chem. Chem. Phys.*, 2000, **2**, 5341–5349.
117. N. S. Nesterenko, F. Thibault-Starzyk, V. Montouillout, V. V. Yuschenko, C. Fernandez, J.-P. Gilson, F. Fajula, I. I. Ivanova, Accessibility of the acid sites in dealuminated small-port mordenites studied by FTIR of co-adsorbed alkylpyridines and CO, *Micropor. Mesopor. Mater.*, 2004, **71**, 157–166.
118. A. H. Fuchs, A. K. Cheetham, Adsorption of guest molecules in zeolitic materials: computational aspects *J. Phys. Chem. B.*, 2001, **105**, 7375–7383.
119. C. J. A. Mota, P. M. Esteves, Theoretical studies of the electrostatic effect on carbocations adsorbed over zeolitic sites, *Molec. Eng.*, 1997, **7**, 429–438.
120. M. A. C. Nascimento, Computer simulations of the adsorption process of light alkanes in high-silica zeolites, *J. Mol. Struct. (Theochem)*, 1999, **464**, 239–247.

6

Chemistry in Self-Assembled Nanoreactors

Jarl Ivar van der Vlugt, Tehila S. Koblenz, Jeroen Wassenaar, and Joost N. H. Reek[1]

[1] Supramolecular and Homogeneous Catalysis, Van 't Hoff Institute for Molecular Sciences, University of Amsterdam, 1018 WV Amsterdam, The Netherlands

6.1 Introduction

Supramolecular chemistry, defined by Jean-Marie Lehn as the 'chemistry beyond the molecule', is the application of programmed molecules that assemble into larger molecular architectures via intermolecular non-covalent bonds.[1,2] An important class of supramolecular structures is that of the host–guest assemblies, where the host is a receptor that selectively binds (generally smaller) guest molecules.[3–5] For some time now, a lot of attention has focused on mimicking Nature and controlling non-covalent host–guest interactions in synthetic systems, with the aim to arrive at artificial supramolecular catalysis, i.e. defined substrate orientation and pre-activation at an encapsulated catalytically active centre or in a cavity.[6] It has appeared to be a major challenge to go beyond the 'straightforward' assembly of appropriate known catalysts, functional groups and substrates, e.g. in bioinorganic chemistry, as this in general has so far not resulted in catalytic activities or selectivities that can match those displayed by nature's machinery, although many an astounding (structural) feat has been accomplished.[7] To fully capture all beneficial aspects displayed by enzymes it should be emphasized that several effects contribute to the unique properties exhibited by the (metallo)enzyme active site, such as substrate preorganization, restricted substrate motion, protein dynamics, covalent binding of the transition state and desolvation of the substrate,[8] most of which are induced by the specific micro-environment created inside the enzyme cavity. Especially desolvation, i.e. the exclusion of typical 'organic' solvent or aqueous media in which generally speaking homogeneous catalysis is taking place, and which tend to heavily influence chemical reaction, and the substitute governing role taken on by the specific micro-environment created around the substrates are largely underestimated and two of the challenging features of enzyme-mimicking. The conditions created within the micro-environment depend on precisely constructed favorable interactions as well as a dissimilar dielectric constant. However, such cavity effects are difficult to study and have mainly been put forth on the basis of computational studies. Synthetic approaches to selectively assemble such nanosized reaction chambers, i.e. nanoreactors, are therefore highly desired, since they would enable experimental verification with synthetic analogues. In addition, the ability to control the reaction environment by creating a confined and well-defined nanospace around the substrates could provide new entries and opportunities to develop sustainable catalytic processes.

6.2 Self-Assembled Nanocapsules

Molecular capsules are a special class of host molecules with a very defined three-dimensional structure, including a hollow interior which under judicious conditions could engage in 'binding' of guest molecules by encapsulation within the enclosed internal space, i.e. the cavity.[9] Besides mere physical entrapment, it can be easily envisioned that additional attractive forces, e.g. hydrophobic interactions, π–π interactions, or weak coordinative interactions can aid in the selective encapsulation of particular guest molecules. Two types of these nanometer-sized molecular capsules can be distinguished; the covalent based capsules and the non-covalent based capsules i.e. self-assembled or supramolecular capsules.[10] Guest exchange in-and-out of non-covalent capsules can, in addition to a gating mechanism, i.e. diffusion of molecules through voids in the supramolecular skeletal construction, proceed via (partial) dissociation of the capsule (*vide infra*). Consequently, guest exchange is often, but not always, more facile for non-covalent capsules compared to their covalent counterparts. Another advantage of non-covalent capsules can be the circumvention of tedious multistep syntheses that are necessary for the formation of covalent capsules. In this review we will focus on the chemistry taking place inside self-assembled capsules with closed cavities, however, one should keep in mind that the boundaries between capsules with open- and closed cavities are not always very clear.

Since the nineties of the last century, research groups around the world have explored the application of nanocapsules as nanoreactors, i.e. reaction vessels for chemical transformations, and the influence of different cavity effects.[11,12] In this chapter the focus will mainly be on recent developments concerned with synthetic nanoreactors that can be obtained in a selective and controlled manner through the use of self-assembly principles and rational design, and on their application as catalytically active capsules for respective chemical reactions. For the sake of clarity, each specific type of nanoreactor will be discussed in a separate section. Particular types of nanocapsules to be reviewed include assemblies held together by hydrogen bonding, metal-ligand interactions and hydrophobic effects. At the end of the chapter we will shortly address some recent developments in the application of micellar and vesicular self-assembled structures.

6.3 Encapsulation Effects in Catalysis

It is important to realize that the selectivity of a reaction is a matter of relative reaction rates between competitive pathways. Also the isolation (or trapping) of otherwise instable reaction intermediates by encapsulation and issues such as product inhibition can be explained in terms of reaction rates as this is a matter of changing the relative rate of a sequence of reaction steps. It is of great relevance to understand how the rate equation of a reaction changes when the process takes place inside the capsule.

6.3.1 Encapsulation Effects in Terms of Rate Equation

In order to understand encapsulation effects of chemical processes, the rate equation should be analyzed rather than making a comparison with enzymes. For a simple bimolecular reaction where substrates A and B are giving product C (Equation 6.1) the general rate equation can be used (Equation 6.2). If reactions take place inside a capsule, substrate encapsulation and product release in-and-out of the nanoreactor (NR) should be taken into account, resulting in a more complex rate equation as all three events have to be considered (Equation 6.3). In an ideal case where the reaction between A and B inside the nanoreactor (NR) is the rate-determining step, the rate equation simplifies analogously to Michaelis–Menten kinetics and depends solely on the rate-constant of this step, i.e. k_b, and on the capsule concentration with the encapsulated substrates ($NR \supset A \bullet B$) (Equation 6.4).[13] In terms of energy profiles these equations can be translated to the energy diagrams depicted in Figure 6.1a.

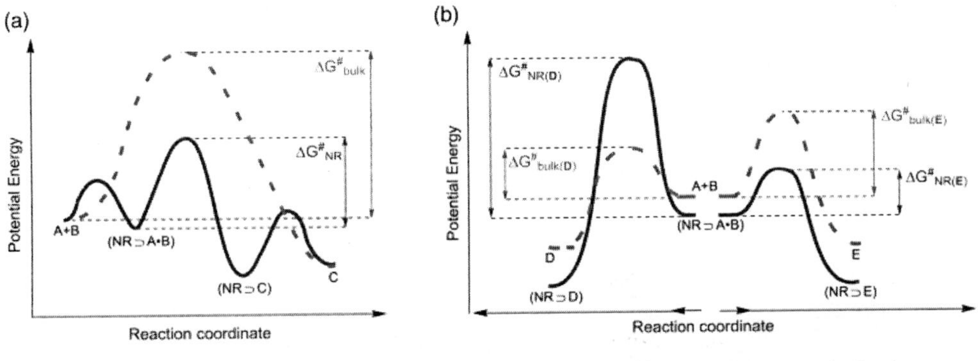

Figure 6.1 a) Simplified reaction profiles of a reaction in the bulk solution (dashed line) and of a reaction within a nanoreactor (blue line). b) Simplified reaction profiles of a reaction leading to product D which is destabilized by the nanoreactor (solid line) compared to the bulk solution (dashed line), and of a reaction leading to product E which is stabilized by the nanoreactor (blue line) compared to the bulk solution (dashed line). [6] Reproduced by permission of The Royal Society of Chemistry (RSC)

$$A + B \xrightarrow{k_a} C \tag{6.1}$$

$$v = d[C]/dt = k_a [A][B] \tag{6.2}$$

$$NR + A + B \rightleftharpoons (NR \supset A \bullet B) \xrightarrow{k_b} (NR \supset C) \rightleftharpoons NR + C \tag{6.3}$$

$$v = d[C]/dt = k_b [(NR \supset A \bullet B)] \tag{6.4}$$

The rate-constant k is a function of the thermodynamic activation parameters i.e. the Gibbs free energy of activation ($\Delta G^{\#}$) and hence the activation enthalpy ($\Delta H^{\#}$) and the activation entropy ($\Delta S^{\#}$) via the Eyring and Arrhenius equations: $\Delta G^{\#} = \Delta H^{\#} - T \bullet \Delta S^{\#} = -RT(\ln k) + c$. (T = temperature, R = gas constant, c = a constant). Two extreme scenarios can be distinguished when reactions take place inside a capsule (a practical situation might be a combination of the two):

1. The nanoreactor does not alter the activation parameters (i.e. $\Delta G^{\#}_{NR} = \Delta G^{\#}_{bulk}$) and the only effect to be expected is the effect of physically bringing substrates together. In essence, the reaction goes from a bimolecular to an intramolecular reaction pathway. Since the activation parameters in equations (2) and (4) are the same, the rate constant remains the same, $ka = kb$, and the difference in rate is due to the fact that $[(NR \supset A \bullet B)] > [A][B]$. In this chapter we have referred to this as *effective concentration.*[14] For a reaction that is carried out at a millimolar concentration, the increase in reaction rate will be a factor 1000 (using a millimolar capsule), whereas the increase in reaction rate at higher concentrations is much smaller. At extremely high substrate concentrations (above 1 molar) other issues will become important and the current simplification will no longer hold.
2. The nanoreactor does change the activation parameters (i.e. $\Delta G^{\#}_{NR} \neq \Delta G^{\#}_{bulk}$). Like enzymes, nanoreactors can *stabilize transition states* ($\Delta G^{\#}$) of the reactions taking place in the capsule, which lowers the activation energy barrier (i.e. $\Delta G^{\#}_{NR} < \Delta G^{\#}_{bulk}$) of the reaction, as is depicted in Figure 1a.[15] Besides the obvious enthalpic stabilization ($\Delta H^{\#}$) via non-covalent interactions between the transition state and the surrounding, entropy ($\Delta S^{\#}$) can also play a key-role in this stabilization. The specific size, shape and chemical environment of the confined nanospace *preorganizes encapsulated substrates* towards the transition state by restricting their translational and rotational degrees of freedom and directing their orientation within the enclosed cavity. Generally, a precise fit between the cavity and the substrates results in a more effective preorganization, reducing the potential negative entropy of a reaction. The cavity can also activate substrates by forcing them to adopt the most reactive conformation of those populated in the bulk, i.e. a statistical effect.[15]

6.3.2 New Reactivities and Selectivities

The specific size, shape and hydrophobic environment of the cavity as well as partly desolvation and isolation of the substrates from the bulk solvent can induce new activities and selectivities.[16] The new microenvironment within the nanoreactor can for example enforce the substrates to adopt conformations which are not or less populated in the bulk or it can stabilize certain reactive intermediates. In bimolecular reactions the cavity can direct the relative spatial orientation of the two substrates, facilitating selective reactions by e.g. blocking bulk solution pathways that require an orientation which is not possible within the cavity.[17] Indeed, manipulation of reaction energetics and environment through encapsulation might provide access to reaction pathways which are otherwise inaccessible because of their high energy-barrier i.e. the energy landscape within the nanoreactor might be different from that outside. The former can be illustrated by considering a kinetic controlled reaction between A and B which can give two products: D and E. The specific interactions between the nanoreactor and the encapsulated substrates can alter the reaction activation energy barrier $\Delta G^{\#}$ of products D and E compared to the bulk. The transition state leading to product E can be stabilized upon encapsulation and thereby favouring its formation. In contrast, the transition state leading to product D can be destabilized upon encapsulation, thereby favouring the formation of product E (Figure 6.1b). The former example describes a situation where the activity and selectivity of a reaction can be reversed upon replacing the bulk solvent with a nanoreactor.

Substrate size- and shape-selective reactions can also be explained in terms of effective concentration and transition state stabilization. 1) In a mixture of substrates a higher concentration can be achieved for those having a complementary size and shape to the nanoreactor portals and cavity (and thus can enter the nanoreactor) compared to those substrates that can not easily enter or do not fit within the cavity. In some cases, when the substrate encapsulation is a slow process (slow diffusion), substrate encapsulation can become the rate-determining step. 2) One can also imagine that substrates of identical size and shape can both enter the cavity, but the transition state of one of the reaction pathways is stabilized to a greater extent than the other. Importantly, for all encapsulated reactions one should keep in mind that the substrate residence time within the nanoreactor and the kinetic rates of the encapsulated reaction should at least have a comparable magnitude.[18]

For nanoreactors with an active site such as transition metal species, similar effects as described above can be expected.[19–21] The rate equation (Equation 6.4) is of course different and it might include the catalyst concentration. However, the principles remain the same: capsules can bring reactant and catalyst together or the activation parameters can be modified. Transition metal-catalyzed reactions consist of several steps, i.e. the catalytic cycle, with one of these steps being the rate-determining step. A transition metal-catalyzed reaction carried out in a restricted environment could change the rate-determining step of the catalytic cycle as a consequence of the change in activation parameters, resulting in new selectivities.[22] The cavity can also modify the structure of the encapsulated active site, giving rise to inherently different properties.[23,24] Similar to the above discussion, the second-coordination-sphere around the active-site can induce substrate size-, shape-, and regio-selectivities. Active-site encapsulation can also result in stabilization of the active site, which strongly interrelates to preventing catalyst decomposition. For transition metal catalysis this stabilization effect might be one of the driving forces towards application.

6.3.3 Product Stabilization

Nanoreactors might give rise to product inhibition when the nanoreactor has a higher affinity for the product compared to the substrate(s). Inhibiting product release prevents a catalytic turnover or at least lowers the reaction rate. In the case of product inhibition, the encapsulated product, i.e. ($NR \supset C$), is very low in energy and the barrier for product release from the nanoreactor cannot be overcome (Figure 6.1a). This higher preference for the product can be enthalpic, e.g. the product has multiple attractive interactions with the host. Entropic disadvantages in bimolecular reactions arise from the need to replace a single product by two substrates, which can also contribute to product inhibition effects. In some cases, product inhibition can be circumvented, for example by implementing a follow-up (domino) reaction with the product to yield a compound which exhibits lower affinity for the nanoreactor.[15] Product inhibition is occasionally observed in metallo-cages when the product is too large to leave the nanoreactor through its portals. This steric barrier can be overcome by using a less constrained, more open shaped nanoreactor or by aiming for products that are too small to cause product inhibition.[17] The reverse has been shown for H-bonded capsules, where formation of compounds with dimensions larger than the 'side-pockets' of the capsule enforce a (partial) disrupture or opening of the structure to allow for guest exchange.[25] Interestingly, product inhibition can also be used for the stabilization of reactive intermediates or labile products that are otherwise difficult to isolate and analyse.[26] The stabilization of reactive intermediates can provide new information on mechanism of the particular chemical reaction.

6.4 Hydrogen Bonded Capsules

Hydrogen bonds can be typified as weak, non-covalent interactions that are highly directional and dynamic, all of which make them ideally suited as a motif for the construction of supramolecular capsules. If properly designed, and even when using a non-stoichiometric mixture, solvation of the respective building blocks will lead to the spontaneous and instantaneous formation of such hydrogen bonded assemblies, albeit in a dynamic fashion, displaying fairly high rates of dissociation and recombination compared to e.g. coordinative interactions. As a result of this, and dependent on the specific multiplicity of the H-bonding motifs present in the supramolecular system, the lifetime of the resulting capsules can range from microseconds to hours.[27–29] Capsule assembly is induced by encapsulation of appropriate guests (templating agent), which can be solvent molecules or specific guests. Rebek and co-workers have shown that optimal guest occupation is approx. 55% of the available volume of the cavity.[30] However, stable self-assembled capsules with guests that occupy much more or less than 55% of the volume exist as well. Most H-bonded capsules do not have sizeable apertures for guest exchange and consequently guest exchange occurs via partial or complete rupture of the capsule structure, depending on the properties of the host, guest and solvent. The residence time of the guest within the capsule is in the order of milliseconds to hours. H-bonded capsules are stable in apolar organic solvents such as dichloromethane and mesitylene while competitive solvents like DMSO and water lead to capsule dissociation.

Rebek and co-workers have developed the cylinder-shaped capsule **A**.[27–29] This capsule consists of two self-complementary vase-shaped cavitands, i.e. resorcinarenes, substituted with four imide-functionalities on their upper rim (Figure 6.2a). The cavitands dimerize in solution into the cylindrical capsule **A** upon the formation of a seam of eight bifurcated hydrogen bonds (i.e. sixteen H-bonds). Capsule **A** has a cylindrical cavity with an internal volume of *ca.* 425 $Å^3$. The cavity contains a gradient of polarity and shape along its length and accommodates elongated guests. The capsule can simultaneously encapsulate two

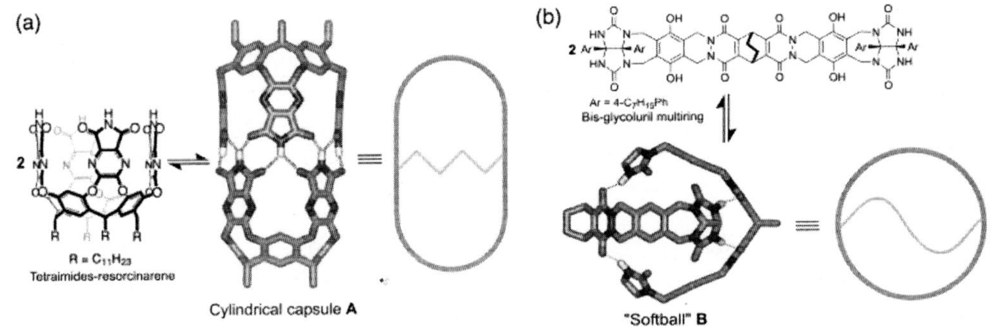

Figure 6.2 a) Cylindrical capsule **A**. b) 'Softball' capsule **B**. In the modelled structures some substituents, hydrogen atoms and hydrogen bonds have been omitted for clarity. Reproduced with permission from [21] Copyright 2002 Wiley-VCH Verlag GmbH & Co. KGaA

different guests, i.e. selective pairwise recognition, such as one molecule of benzene together with one *p*-xylene molecule, which can be of interest for coupling reactions. During guest exchange two flaps of the capsule open, allowing guest exchange without capsule dissociation.

6.4.1 Stoichiometric 1,3-Dipolar Cycloaddition

Rebek and co-workers have applied the cylindrical capsule **A** as a nanoreactor for the 1,3-dipolar cycloaddition between phenylazide **1** and phenylacetylene **2** (Scheme 6.1).[31] Nanoreactor **A** simultaneously encapsulates reagents **1** and **2** to give [**A** ⊃ **1** • **2**] (this notation denotes that **A** encapsulates **1** and **2**) next to the homo combinations [**A** ⊃ **12**] and [**A** ⊃ **22**] which are formed slightly less than the statistical distribution would predict. It was already shown that the cylindrical cavity of **A** constrains the orientation of the guest molecules in an edge-to-edge fashion, so that their substituents are forced to interact. The cycloaddition within nanoreactor **A** results exclusively in the 1,4-triazole after several days at millimolar concentrations. In contrast, the reaction in the absence of **A** produces a 1 : 1 mixture of regioisomeric products of 1,4- and 1,5-triazoles and has a half-life of several years. The observed rate enhancement is explained by an effective concentration of 3.7 M within **A**. The high regioselectivity results from preorganized orientation of the substrates within **A**, imposed by the nanoreactor boundaries. Nanoreactor **A** has also been shown to impose substrate size selectivity; the encapsulated cycloaddition between the larger azides 1-naphthyl azide or 4-biphenyl azide and phenylacetylene was not accelerated. The system suffers, as expected, from product inhibition and therefore stoichiometric amounts of **A** have to be used. The product could be liberated upon addition of DMF resulting in denaturation of the capsule.

The 'softball' capsule **B** (Figure 6.2b), also developed by Rebek and co-workers, is composed of two multi-ring structures having a bridged bicyclic centrepiece and two glycoluril units on both ends of the multi-ring, providing proper rigidity, curvature and functional groups necessary for capsule assembly (Figure 6.2b).[27–29] The self-complementary glycoluril-based building blocks dimerize in solution into capsule **B** due to the formation of a seam of sixteen hydrogen bonds. 'Softball' **B** is a closed-shell capsule of roughly

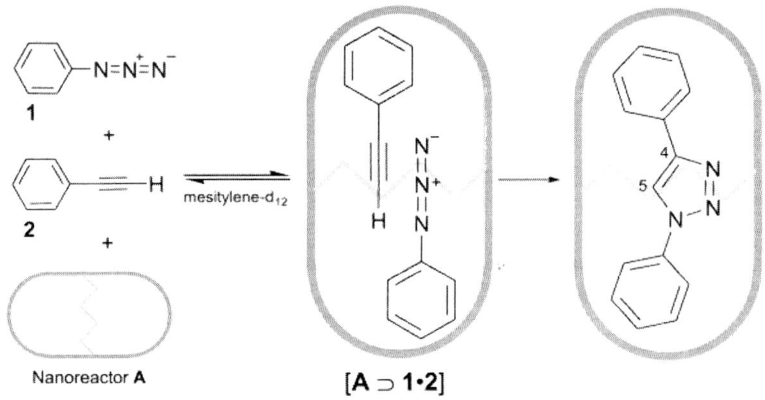

Scheme 6.1 *1,3-Dipolar cycloaddition between phenylazide **1** and phenylacetylene **2** within nanoreactor **A***

spherical shape with an internal volume of *ca.* 400 $Å^3$. The 'softball' can simultaneously encapsulate two different guests such as one molecule of deuterated benzene together with one molecule of deuterated monofluorobenzene. Guests encapsulation occurs by opening of two separate flaps of the 'softball' and subsequently departure of the guest as the incoming guest approaches, i.e. a gating mechanism.

6.4.2 Catalytic Diels–Alder Reaction

Rebek and co-workers have used 'softball' **B** as a nanoreactor for bimolecular Diels–Alder reactions.[25] The Diels–Alder reaction between *p*-benzoquinone (**3**) and cyclohexadiene **4a** within nanoreactor **B**, present in stoichiometric amounts, has been accelerated 170-fold compared to the bulk and resulted in the encapsulated adduct [**B** ⊃ **5a**] (Scheme 6.2a). Even though nanoreactor **B** enhances the rate of the encapsulated reaction, no true catalytic behaviour was observed because of product inhibition by **5a**. After encapsulating two quinones **3** within **B**, one molecule is occasionally replaced by a thiophene dioxide **4b**, which leads to the encapsulated Diels–Alder product **5b**. Product inhibition is partly suppressed if thiophene dioxide derivative **4b** is used as the diene (Scheme 6.2b). This is because two molecules of the *p*-benzoquinone (**3**), i.e. [**B** ⊃ $(\mathbf{3})_2$] have a higher affinity for nanoreactor **B** than derivative **5b**, the corresponding Diels–Alder product of the thiophene dioxide. After each turnover the encapsulated product is released and replaced by two quinone reactants and true catalysis can take place. When catalytic amounts of **B** (10 mol%) are used, a tenfold rate enhancement (at 10 mM substrate concentration) compared to the background reaction was observed, which is still lower than can be expected on the basis of the effective concentration.

6.5 Capsules Based on Metal–Ligand Interactions

Metal–ligand interactions are strong and highly directional and lead to the self-assembly of robust stable coordination cages.[32] Kinetically labile metal-ligand (M–L) interactions

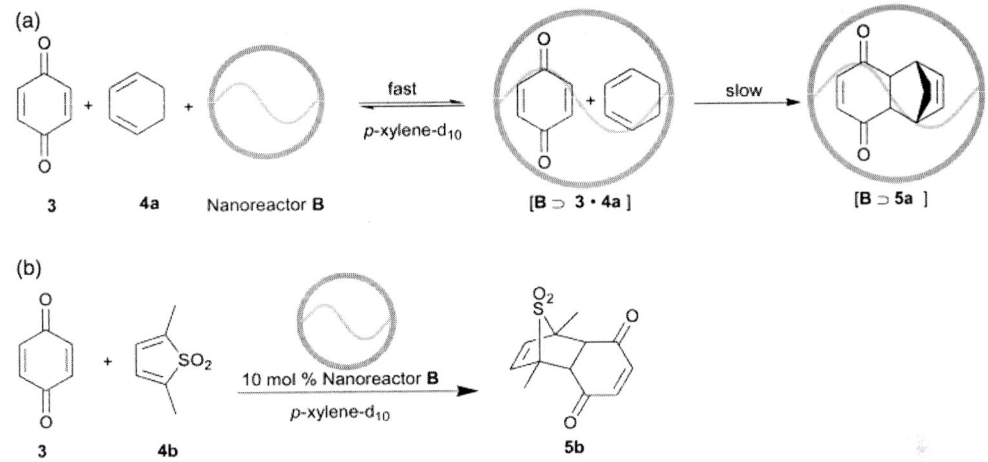

Scheme 6.2 *a) Diels–Alder reaction between p-benzoquinone* ***(3)*** *and cyclohexadiene* ***4a*** *within nanoreactor* ***B****. b) Catalytic Diels–Alder reaction between p-benzoquinone* ***(3)*** *and the thiophene dioxide derivative* ***4b*** *within nanoreactor* ***B***

are essential for converting initially formed kinetic products to the more stable thermodynamic product. Therefore appropriate conditions should be applied for the assembly of M–L based capsules in high yields. The metal coordination geometry in combination with multidentate organic ligands are used as codons for curvature. The preorganized and rigid nature of the ligands results in stable and well defined capsules with cage-like architectures. Unlike H-bonded capsules, guest exchange in and out of the coordination cages occurs by expansion of the cage apertures without M–L bond rupture.[33] Consequently, the size and shape of the cage apertures function as a gate keeper as they dictate the permitted size and shape of the guests. Besides the encapsulation of substrate molecules to facilitate certain conversions, also catalytically active metal centres can thus be encapsulated. Indeed sufficient space should be available to allow the transformation of encapsulated substrates at the active site.

6.6 Tetrahedral Cages Based on Octahedral M^{3+} Ions

Raymond and co-workers have developed the chiral tetrahedral $[M_4L_6]^{12-}$ coordination cage **C** consisting of four metal ions with an octahedral coordination geometry, e.g. Ga^{3+}, and six naphthalene-based bis-bidentate catechol amide ligands (Figure 6.3).[19]

The metal ions are situated at the corners of the tetrahedron and the ligands span the edges of the tetrahedron. The tris-bidentate chelation of the metal atoms renders them chiral (Δ or Λ), and the mechanical coupling between the metals through the rigid ligands results in the exclusive formation of the homochiral assemblies $\Delta,\Delta,\Delta,\Delta$ and $\Lambda,\Lambda,\Lambda,\Lambda$. Resolution of both enantiomers by means of selective precipitation upon association with the chiral cation *S*-nicotinium was recently described.[34] The desired M_4L_6 stoichiometry with a tetrahedral shaped cage is achieved by the presence of an appropriate guest

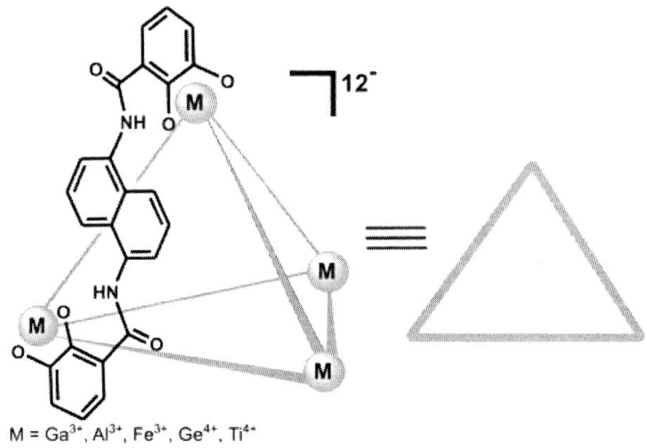

Figure 6.3 *Tetrahedral M_4L_6 coordination cage **C**. [6] Reproduced by permission of The Royal Society of Chemistry (RSC)*

template during the assembly process such as NR_4^+ (R = Me, Et, Pr). The highly negatively charged tetrahedral cage is soluble in water and other polar solvents and contains a hydrophobic cavity of ~300–500 Å^3. The highly anionic character of the cage allows for exclusive encapsulation of monocationic guests such as alkylammonium ions and cationic organometallic complexes such as $[CpRu(C_6H_6)]$ + (Cp = η^5-C_5H_5). Recently, tetrahedral nanocages with even larger interior voids have been assembled by the group of Raymond, through modification of the ligand backbone, using binaphthyl as the central framework instead of the formerly employed mononaphthyl skeleton. Calculated inner volumes in excess of ~700 Å^3 were reported, leading to favourable encapsulation of guests such as tetraphenylphosphonium ions.[35]

6.6.1 Hydrolysis

The tetrahedral coordination cage **C** has been used by Raymond, Bergman and co-workers as a catalyst for the acidic hydrolysis of orthoformates in basic solutions.[36] This is an example where the inner phase facilitates reactions that are not possible in the exterior environment. Nanoreactor **C** has a much higher affinity for monocationic guests over neutral guests. Addition of neutral and even weakly basic compounds such as amines, phosphines and orthoformates, $HC(OR)_3$, to aqueous solutions of **C** resulted in encapsulation of their protonated form. The protonation is thermodynamically driven by stabilization of the protonated species. Hydrolysis of orthoformates involves a protonated intermediate. Indeed catalytic amounts of **C** (1 mol %) in basic solution (pH 11) gave rapid hydrolysis of orthoformates to the corresponding formate esters, HC(O)(OR), which is subsequently hydrolyzed by OH^- to formate, HCO_2^- (Scheme 6.3a). Product inhibition is not taking place and the empty nanoreactor **C** can re-enter the catalytic cycle like a true catalyst. Rate accelerations of up to 890-fold were observed for tris(isopropyl)orthoformate. As expected, nanoreactor **C** exhibits substrate size selectivity and only orthoformates smaller than tripentyl orthoformate are readily hydrolyzed.

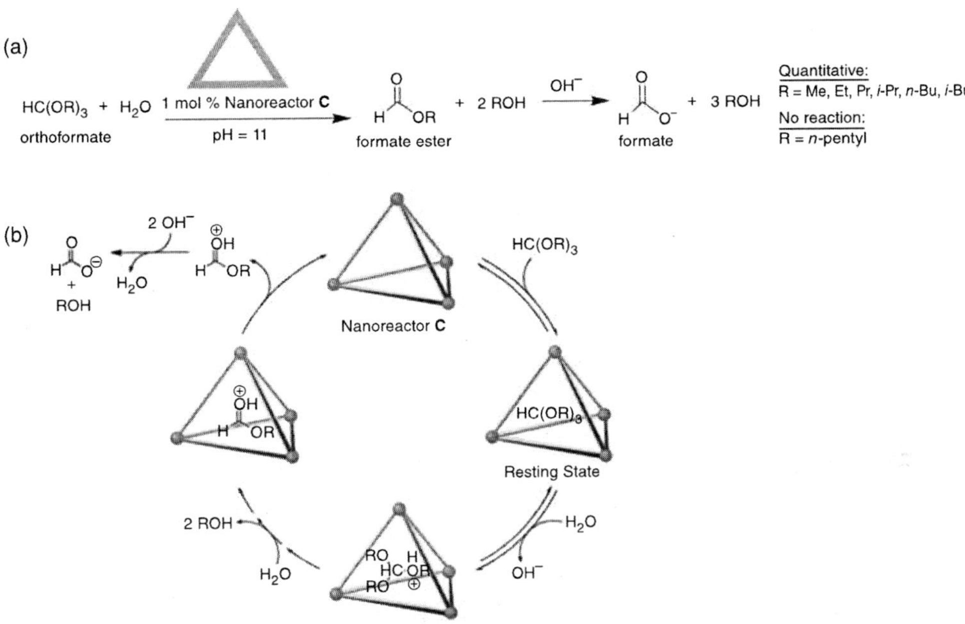

Scheme 6.3 *a) Catalytic hydrolysis of orthoformates within nanoreactor* ***C****. b) Mechanism for catalytic orthoformate hydrolysis within* ***C****. Reproduced in part with permission from Science*[36]

A mechanistic study of the catalytic reaction has revealed a catalytic cycle in which the neutral orthoformate is first encapsulated within nanoreactor **C** (Scheme 6.3b). Protonation of the encapsulated substrate, presumably by deprotonation of water, results in the stabilized monoprotonated orthoformate. Subsequently, two successive hydrolysis steps within **C** liberate two equivalents of the corresponding alcohol. Finally, the protonated formate ester is ejected from **C** and is further hydrolyzed by OH^- in solution. The shift in the effective basicity of the encapsulated guests compared to the free analogue is four orders of magnitude, typically also found for enzymes that modify basic properties of the encapsulated substrates. This example is clearly based on stabilization of the transition state by interactions between the capsule and the encapsulated transition state. The mechanism of the encapsulated reaction involves an initial pre-equilibrium step followed by a first-order rate-limiting step. This Michaelis–Menten kinetics is parallel to enzymatic pathways. An inhibition study with NPr_4^+, a strongly interacting but reversible guest for **C**, has revealed that the encapsulated hydrolysis reaction exhibits competitive inhibition.

The unimolecular Aza-Cope rearrangement of ally[15,19] and propargyl[37] enammonium cations within nanoreactor **C** was also investigated by Raymond, Bergman and co-workers. The former reaction was catalyzed by 13 mol % of **C** and a rate acceleration of up to 850-fold was observed, whilst with propargyl enammonium cations, the substoichiometric [3,3]-sigmatropic rearrangement was found to be enhanced up to a factor of 184 in the presence of nanocage **C**. Since this is an intramolecular reaction, it represents

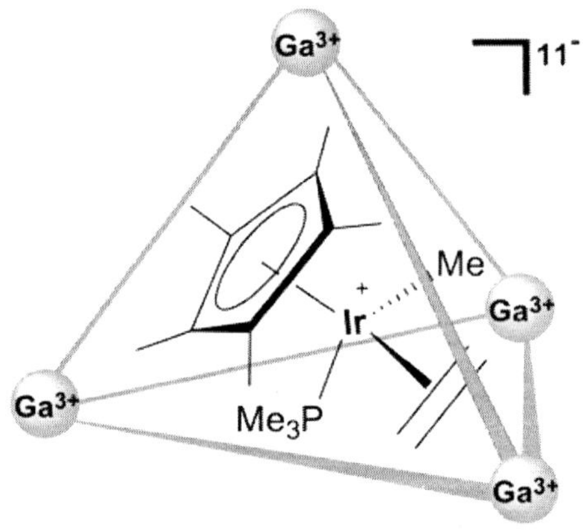

[C ⊃ Ir$^+$-complex]

Figure 6.4 *Cationic Ir-complex encapsulated within **C**, i.e. [**C** ⊃ Ir$^+$-complex] (this notation denotes that C encapsulates an Ir$^+$-complex). [6] Reproduced by permission of The Royal Society of Chemistry (RSC)*

a clear-cut example in which the capsule preorganizes the substrate to reduce the entropic contributions to the activation energy. The nanoreactor induces substrate size- and shape-selectivity and acts as a true catalyst, since release and hydrolysis of the iminium product circumvent product inhibition.

Raymond and co-workers have encapsulated cationic transition metal complexes such as $[Cp^*(PMe_3)Ir(Me)(C_2H_4)]^+$ and $[(PMe_3)_2Rh(COD)]^+$ within their tetrahedral coordination cage **C** via non-directional, non-covalent bonds (Figure 6.4).[38,39] Concerning the former, the group of Raymond in conjunction with Bergman and co-workers have studied the thermal C–H bond activation of aldehydes and ethers by an Ir(III)-complex encapsulated within nanoreactor **C**. This encapsulated iridium complex [**C** ⊃ Ir-complex] induced highly specific substrate size and shape selectivities. However, product inhibition has prevented this system to achieve catalytic turnover.[19,38]

6.6.2 Allylic Alcohol Isomerization

More success was garnered when the cationic rhodium complex $[(PMe_3)_2Rh(COD)]^+$, encapsulated within the tetrahedral coordination cage **C**, was used as an isomerization catalyst for allylic alcohols (Table 6.1).[39] The size selection by nanoreactor **C** only allowed encapsulation of the catalyst precursor $[(PMe_3)_2Rh(COD)]^+$ and not the PEt_3-equivalent. The encapsulated precursor was hydrogenated to give the encapsulated active catalyst $[\mathbf{C} \supset (PMe_3)_2\,Rh(OD_2)_2^+]$, i.e. [**C** ⊃ **6**], which remained encapsulated for 12 h. [**C** ⊃ **6**] is not the thermodynamic product as longer reaction times result in release of **6** from the capsule. Therefore, the application of [**C** ⊃ **6**] is restricted to fast reactions such as allylic alcohol isomerisation, where substrate entrance and product release are rapid and occur prior to active site release. In contrast to the non-encapsulated catalyst **6**, the encapsulated catalyst [**C** ⊃ **6**] only isomerized small allyl alcohols (Table 6.1, entries 1b–3b). Encapsulation within nanoreactor **C** imposes two-fold substrate selectivity, i.e. size- and shape-selectivity, to the active site. The origin of this dual-selectivity lies in the

Table 6.1 *Catalytic isomerization of allylic alcohols by **6**, $[(PMe_3)_2Rh(OD_2)_2^+]$, in bulk-solution or within nanoreactor **C**: [**C** ⊃ **6**]. [6] Reproduced by permission of The Royal Society of Chemistry (RSC)*

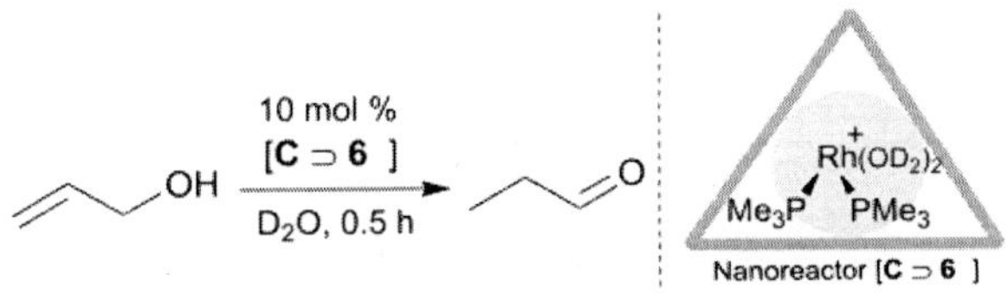

Entry	Substrate	Catalyst	Yield
1a	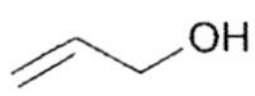	**6**	95%
1b		[**C** ⊃ **6**]	95%
2a	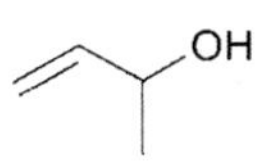	**6**	95%
2b		[**C** ⊃ **6**]	n.r.
3a	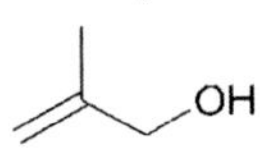	**6**	95%
3b		[**C** ⊃ **6**]	n.r.
4a	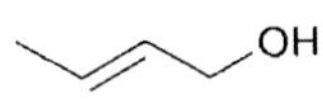	**6**	n.r.
4b		[**C** ⊃ **6**]	n.r.

defined apertures of the cage, which inhibit inclusion of larger branched substrates. Terminal substituted crotyl alcohol inhibits the catalyst (Table 6.1, entry 4a), as is confirmed by the simultaneous addition of both allyl alcohol and crotyl alcohol to the non-encapsulated catalyst **6**, which did not result in isomerization of either substrate. However, addition of both allyl alcohol and crotyl alcohol to [**C** ⊃ **6**] led to selective isomerization of the allyl alcohol. Hence, nanoreactor **C** protects the encapsulated catalyst **6** against decomposition and poisoning of the complex.

Using a slightly different approach, Hupp and co-workers have described the assembly of an open box structure, encapsulating a Mn(III)-porphyrin via direct metal-ligand interactions between the guest molecule and the host structure.[20] The open box is based on four Zn(II)-porphyrins coordinated to four $[Re(CO)_3Cl]$ complexes via their pyridyl ligands. The encapsulated Mn(III)-porphyrin appeared to be a stable and selective epoxidation catalyst. Although the open box type assemblies are somewhat beyond the scope of this chapter, it is important to realize that these more open assemblies can also provide interesting catalysts.

6.7 Octahedral and Square Pyramidal Cages Based on Square-Planar M^{2+} Ions

Upon mixing suitably *cis*-protected square planar Pd^{2+} or Pt^{2+} complexes, e.g. $[Pd(en)]^{2+}$ (en = ethylenediamine) together with planar C3-symmetric, triangular and tripodal ligands such as 2,4,6-tris(4-pyridyl)-1,3,5-triazine ligand in a 6:4 relative stoichiometry, spontaneous self-assembly results in the formation of octahedral nanocage **D**, as described by

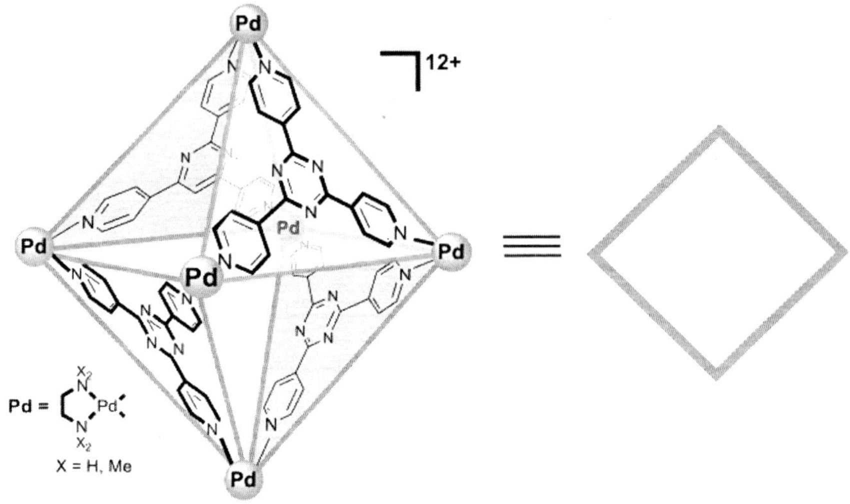

Figure 6.5 *Octahedral M_6L_4 coordination cage* ***D****. [6] Reproduced by permission of The Royal Society of Chemistry (RSC)*

Fujita and co-workers (Figure 6.5).[40,41] By a slight adjustment in the architecture of the linking unit, i.e. employing 2,4,6-tris(3-pyridyl)-1,3,5-triazine, and using the same relative ratio of 6 : 4 (M : L) the related half-open, square-pyramidal cage **E** was formed.

Within these aggregates, the corners of the octahedron are defined by the metal complexes, whereas the faces of the octahedron are occupied by the planar tripodal ligand scaffolds. As a consequence of the use of neutral donor ligands, the resulting hexanuclear complex carries a high overall positive charge and thus shows appreciable solubility in aqueous media, featuring a hydrophobic inner environment. As a result, these cages are very suitable hosts to accommodate a range of neutral and anionic guest molecules, including ferrocene and adamantane. Guest exchange can take place via the relatively large pores of the host. The large inner cavity has a diameter of approx. 2.2 nm and allows encapsulation of one to four guest molecules per cage, depending on the guest size. The coordination cage **D** can selectively recognize two different guests, such as one anthracene-type molecule and one maleimide-type molecule, in a pair-wise manner. Remarkably, no external guests are needed in order to selectively form the Pd-based octahedral cage and the cage can be prepared at large scale (up to 50 g). This indicates that also polar solvents such as water can be included under judicious conditions. The Fujita group has reported many other types of cage structures, but cage **D** has been mostly used as reaction vessel.

6.7.1 Diels–Alder Reaction

The octahedral coordination cage **D** described by Fujita and co-workers has been applied as a nanoreactor for bimolecular Diels–Alder reactions in water.[17] Suspending 9-(hydroxymethyl)anthracene **7** and *N*-cyclohexylmaleimide **8a** in an aqueous solution of

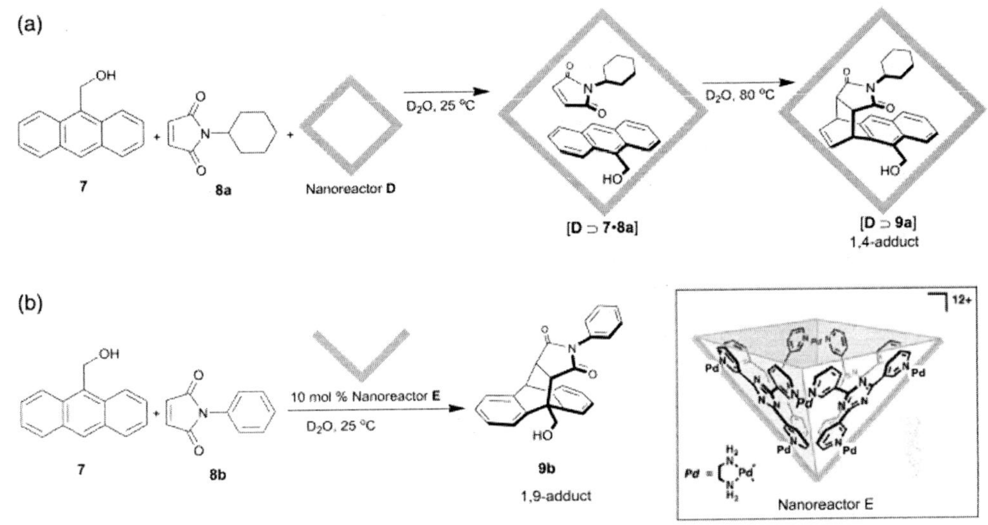

Scheme 6.4 *a) Diels–Alder reaction between the anthracene 7 and N-cyclohexylmaleimide **8a** within nanoreactor **D**. b) Catalytic Diels–Alder reaction between the anthracene 7 and N-phenylmaleimide **8b** within square-pyramidal nanoreactor **E**. Reproduced in part with permission from Science*[17]

near-stoichiometric quantities of **D** resulted in the selective formation of [**D** ⊃ **7** • **8a**] (Scheme 6.4a). Upon warming the reaction mixture, the Diels–Alder product **9a** formed in excess of 98% and could be subsequently isolated by extraction with chloroform. The nanoreactor induces unusual regioselectivity by promoting reaction at the terminal anthracene ring to give the syn-1,4-Diels–Alder adduct **9a**, while reaction in the bulk solvent yields the product bridging at the central anthracene ring, the 1,9-Diels–Alder adduct, in 44%. The unusual stereo- and regioselectivity observed for the encapsulated Diels–Alder reaction are explained by the fixed orientation of the two substrates within **D**, preventing interaction at the 9,10 position of the anthracene **7**. This is an example where preorganization within the nanoreactor prevents the energetically most favoured product. However, product inhibition by strong complexation of **9a** within **D** prevents catalytic turnover.

Interestingly, the Diels–Alder reaction of the anthracene **7** and *N*-phenylmaleimide **8b** could be performed in a catalytic fashion using 10 mol % of a square-pyramidal bowl **E**. In this reaction the conventional regiochemistry of the 1,9-Diels–Alder adduct **9b** (Scheme 6.4b) was obtained, but in excellent yield (99%), whereas without the nanoreactor present, only 3% conversion was observed under the same conditions.[17] Bowl **E** does not suffer from product inhibition because its open structure allows facile guest exchange (no kinetic traps), and the affinity for the substrate and product is based on aromatic stacking, which is more pronounced for the substrate than for product **9b**. It is unknown if the rate acceleration is caused by effective concentration and/or transition-state stabilization. Remarkably, nanoreactor **D** can also efficiently promote the Diels–Alder reaction of highly stable and relative inert aromatic molecules such as triphenylene and perylene with **8a** to give the corresponding *endo* Diels–Alder adduct with *syn* stereochemistry in high yields.[42] The steric demand of the *N*-cyclohexyl group on **8a** is important as it directs the orientation of the substrate within the cavity of **D**.

6.7.2 Olefin Photodimerization

The octahedral coordination cage **D** has been used by Fujita and co-workers as a nanoreactor for the bimolecular [2+2] photodimerization of olefins.[43] Suspending bulky olefins such as acenaphthylenes **10a** or **10b** in an aqueous solution of **D** resulted in complexes [**D** ⊃ (**10a**)$_2$] and [**D** ⊃ (**10b**)$_2$], respectively (Scheme 6.5a). Isolation and irradiation of these complexes yielded the encapsulated *syn*-dimers [**D** ⊃ **11a**] and [**D** ⊃ **11b**] in >98%. The products were isolated by extraction with chloroform. Compared to the non-encapsulated photodimerization of **10a** in benzene, the nanoreactor accelerated the reaction and improved the stereoselectivity for the *syn*-dimer **11a** from 53% to 100%. In addition, the nanoreactor protected the encapsulated product **11a** against photodissociation. Photodimerization of the non-symmetrically substituted 1-methylacenaphthylene **10b** can yield up to four different isomers. Under standard conditions, however, no dimerization occurs due to the steric hindrance of the methyl group, but photodimerization of **10b** within nanoreactor **D** does occur with a high stereo- and regioselectivity, as is demonstrated by the exclusive formation of the head-to-tail *syn*-dimer **11b**.

The cross-photodimerization reaction represents a great challenge because it requires **D** to selectively recognize two different olefins in a pair-wise fashion prior to irradiation, whereas the cross-coupling reaction under standard conditions shows no preference over the homo-coupling. The cross-photodimerization of acenaphthylene with substituted naphthoquinones within **D** resulted in exclusive formation of the cross syn-dimer only

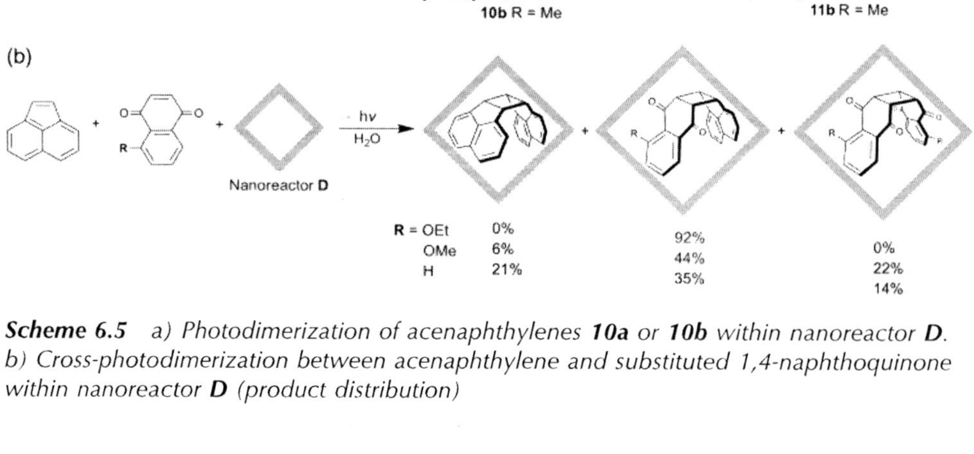

Scheme 6.5 *a) Photodimerization of acenaphthylenes **10a** or **10b** within nanoreactor **D**. b) Cross-photodimerization between acenaphthylene and substituted 1,4-naphthoquinone within nanoreactor **D** (product distribution)*

for 5-ethoxynaphthoquinone, while the less steric olefins, having no substituent or a 5-methoxy group, yielded mixtures of homo- and heterodimers (Scheme 6.5b).[44] In addition, even relatively inert polycyclic aromatic compounds such as pyrene, phenanthrene and fluoranthene were selectively [2+2] cross-photodimerized with *N*-cyclohexylmaleimide within **D**, resulting in high stereo- and regioselectivity.[42] Again, the steric demand of the *N*-cyclohexyl group of the maleimide is essential, as less sterically demanding groups were not reactive in the photodimerization with pyrene.

Subsequent work has established the formation of chiral derivatives of nanoreactor **D** through the use of chiral diamine end-capping groups instead of ethylenediamine as ligand on the Pd building blocks and subsequent self-assembly with functionalized triazine frameworks. These chiral nanocages are able to remotely control the chiral [2+2] photo addition of fluoranthenes and several maleimides, a reaction previously unknown to be performed in an asymmetric fashion, in up to 50% ee.[45]

Substrate desolvation and the hydrophobic environment within the cavity create a *new inner phase* for new applications. Indeed, the water-soluble nanoreactor **D** with a hydrophobic cavity has been used as phase-transfer catalyst for Wacker oxidation of styrene in an aqueous phase.[46] Another interesting feature of nanoreactors is their molecular frame that can play an active role during reaction of encapsulated molecules. For example, the triazine ligands of Fujita octahedral cage **D** can be photochemically excited; subsequent electron transfer from an encapsulated alkane (i.e. adamantane) to the cage induces substrate oxidation by O_2 (or H_2O, depending on reaction conditions), via the intermediacy of an adamantyl radical and the triazine-based radical anion of cage **D**.[47] Also, the photochemical cyclization of diphenylethanedione, concomitant with effective suppression of photolytic cleavage of the substrate was reported to be enhanced by nanoreactor **D**.[48] Photomediated 1,4-radical additions to quinones are another example where the constrained enclathration of substrate molecules in **D** enforces unusual reactivity.[49] No mention was made on the potential interaction or involvement of the photoactive cage-faces (the triazine scaffolds) in the latter two cases.

As mentioned previously, nanoreactors can also stabilize *in-situ* generated reactive intermediates and labile products like in the oligomerization of trialkoxysilanes within **D**, which could not be isolated in the absence of this nanoreactor.[26]

6.8 Hydrophobic Effects as the Driving Force for the Self-Assembly of Nanocapsules

Gibb and co-workers have used water-soluble cavitands that self-assemble into capsule **F**. The assembly process is based on hydrophobic effects (and possibly on non-directional π–π stacking between the two building blocks) and the capsule formation occurred only in the presence of hydrophobic guests that function as template.[50]

The template plays an important role in the formation of capsule **F** because no directional non-covalent bonds such as H-bonding and M–L interactions are involved in the self-assembly process. In addition, an external hydrophobic template is necessary because aqueous solvent does not template the formation of **F**. Capsule **F** consists of two octa-acid, deep-cavity cavitands with a pseudo-conical hydrophobic cavity (Figure 6.6a). The eight carboxylic acid groups are located at the periphery of the cavitand, thereby inducing

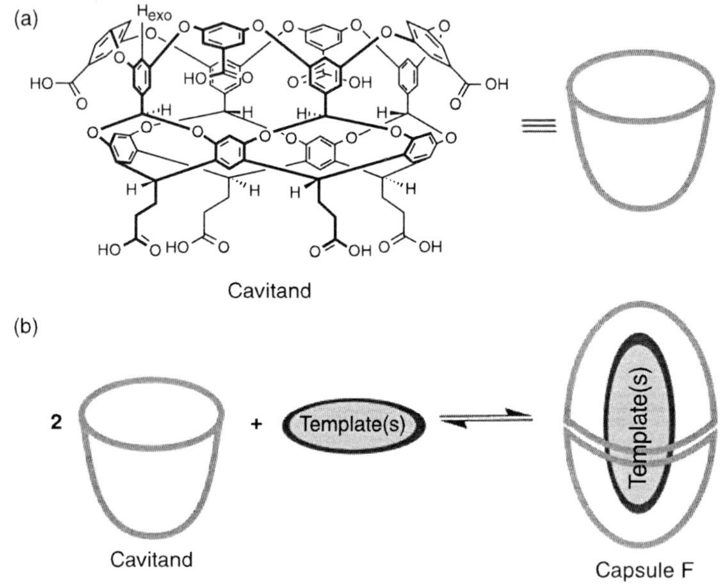

Figure 6.6 a) Deep-cavity cavitand. b) Self-assembly of capsule **F**. [6] Reproduced by permission of The Royal Society of Chemistry (RSC)

high water-solubility under basic conditions. The cavitands dimerize upon encapsulation of complementary hydrophobic templates, e.g. a rigid steroid or two smaller alkanes (Figure 6.6b). As mentioned above, the capsule is also held together by non-directional π–π stacking interactions between the aromatic rings on the wide hydrophobic rim of the two cavitands; therefore the structural integrity of these capsules is not fixed. Capsule **F** has a calculated internal cavity of approx. 500 $Å^3$.

6.8.1 Photooxidation

The cavitand-based capsule **F** has been applied by Gibb and Ramamurthy as a nanoreactor for the oxidation of methyl cycloalkenes by singlet oxygen to give allylic hydroperoxides.[50,51] Addition of a stoichiometric amount of ethyl cyclohexene **12** to the deep-cavity cavitand in a basic aqueous solution (which is necessary for capsule formation) resulted in formation of the host–guest complex, wherein nanoreactor **F** encapsulates two substrate molecules, i.e. [**F** $\supset$ 122] (Scheme 6.4a). In the photooxidation reaction a singlet oxygen ($^1O_2$) is added to the alkene bond and simultaneously one of the allylic hydrogens is abstracted, which can result in three different allylic hydroperoxides **13a**, **13b** and **13c** (Scheme 6.6a). Two different photosensitizers were used to generate singlet oxygen, namely the water soluble Rose Bengal (RB) or the water insoluble dimethyl benzil (DMB). In the later case, DMB was itself encapsulated within nanoreactor **F** [**F** $\supset$ DMB], *vide infra*. Irradiation of [**F** $\supset$ 122] in the presence of photosensitizer RB or [**F** $\supset$ DMB] yielded the allylic hydroperoxides **13a** and **13c** in 60–70% with 90–95% selectivity towards the regioisomer **13c**. The products were isolated by extraction with chloroform. As can be seen in Scheme 6.4a, in the absence of nanoreactor **E** (in acetonitrile with

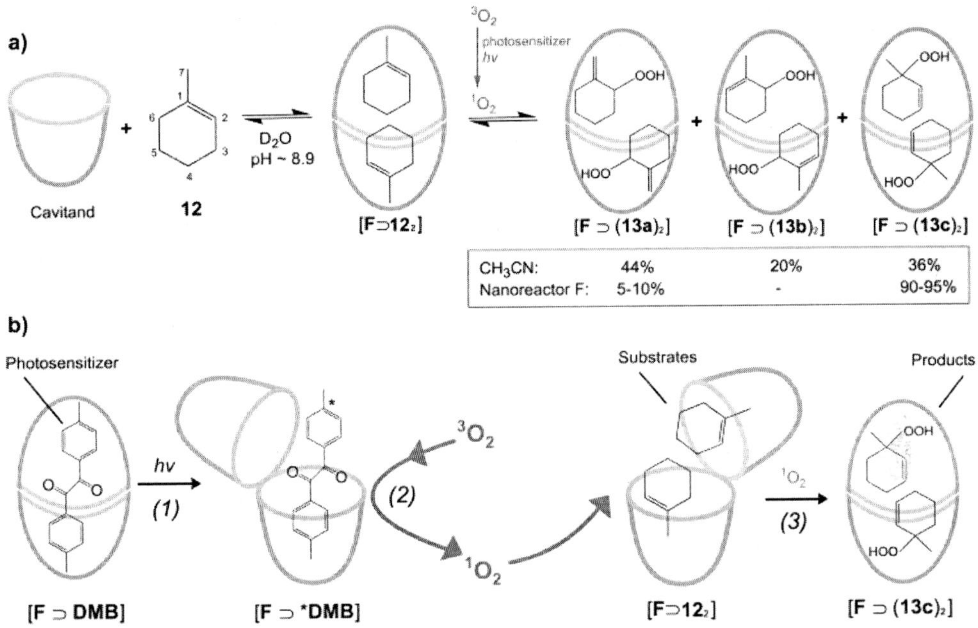

Scheme 6.6 *a) Photooxidation of 1-methyl cyclohexene **12** within nanoreactor **F**. b) 1. Excitation of an encapsulated photosensitizer [**F** ⊃ DMB]. b) Excitation of triplet oxygen by an encapsulated photosensitizer [**F** ⊃ *DMB]. c) Photooxidation of encapsulated **12** [**F** ⊃ **12**$_2$] by singlet oxygen*

photosensitizer RB) three hydroperoxides are formed, **13a** (44%), **13b** (20%) and **13c** (36%). Thus, nanoreactor E exhibits very high regioselectivity in this particular oxidation reaction. NMR studies suggested that the methyl groups of the encapsulated substrates are anchored at the narrowest parts of the nanoreactor (Scheme 6.4a). Hence, the encapsulated substrate is oriented such that singlet oxygen is prevented from approaching the methyl group. It is important to note that it is very difficult to control selectivity in reactions with singlet oxygen. In addition, the NMR studies also support the observations that the allylic hydrogen H_3 of **12** is the most accessible of the three allylic hydrogen sets (H_3, H_6 and H_7). Interestingly, nanoreactor **F** also stabilizes the product because the encapsulated hydroperoxides [**F** ⊃ 132] remained stable for weeks.

A very interesting aspect of this study is the use of the photosensitizer dimethyl benzil (DMB). This photosensitizer does not dissolve in water, unless it is encapsulated within nanoreactor **F**, [**F** ⊃ DMB] (Scheme 6.6b). It is important to notice that when [**F** ⊃ DMB] and [**F** ⊃ 122] were mixed no guest exchange was observed and the capsules remained independent. The oxidation reaction of **12** involving [**F** ⊃ DMB] starts with the generation of excited DMB, i.e. [**F** ⊃ *DMB]. Next, the former nanoreactor opens and allows contact between oxygen and *DMB, resulting in the formation of singlet oxygen ($^1O_2$). In a subsequent step, the singlet oxygen diffuses from [**F** ⊃ DMB] to [**F** ⊃ 122], resulting in regioselective oxidation of **12**. This system has properties related to those important for biological signaling.[52]

6.9 Ligand Template Approach Using Lewis Acid/Base Interactions

Our own work in the area of supramolecular self-assembly and the application of capsular structures for catalytic conversions has dealt mainly with the introduction and application of ligand-template approaches as a new strategy for the encapsulation of transition metal catalysts.[21–24,53] The template ligands have a bifunctional character in that they contain functional groups for capsule assembly as well as a donor-atom site for metal coordination. This results in the encapsulation of the metal within the ligand-template capsule. Tris(pyridyl)phosphines, $P(Py)_3$, are privileged template ligands as the nitrogen atoms of the pyridyl groups selectively and strongly coordinate to Zn(II)-porphyrins or Zn(II)-salphens, [Zn], resulting in the ligand-template capsule **G**, i.e. $P(Py)_3 \bullet [Zn]_3$ (Figure 6.7a).

The zinc building blocks create a hemispherical capsule around the tris(pyridyl)phosphine. Subsequently, the phosphorus atom can coordinate to a transition metal such as Rh or Pd, resulting in selective encapsulation, i.e. [**G** ⊃ Rh or Pd] (this notation denotes that **G** encapsulates Rh or Pd). Addition of the zinc building blocks to the rhodium-bis-tris(*meta*-pyridyl)phosphine-complex, $Rh(P(m\text{-}Py)_3)_2(CO)(acac)$, creates sterical hindrance around the encapsulated metal and results in the facile dissociation of one of the two pyridylphosphine ligands to give $[P(m\text{-}Py)_3 \bullet [Zn]_3 \supset Rh(CO)(acac)]$ i.e. [**G** ⊃ Rh(CO)

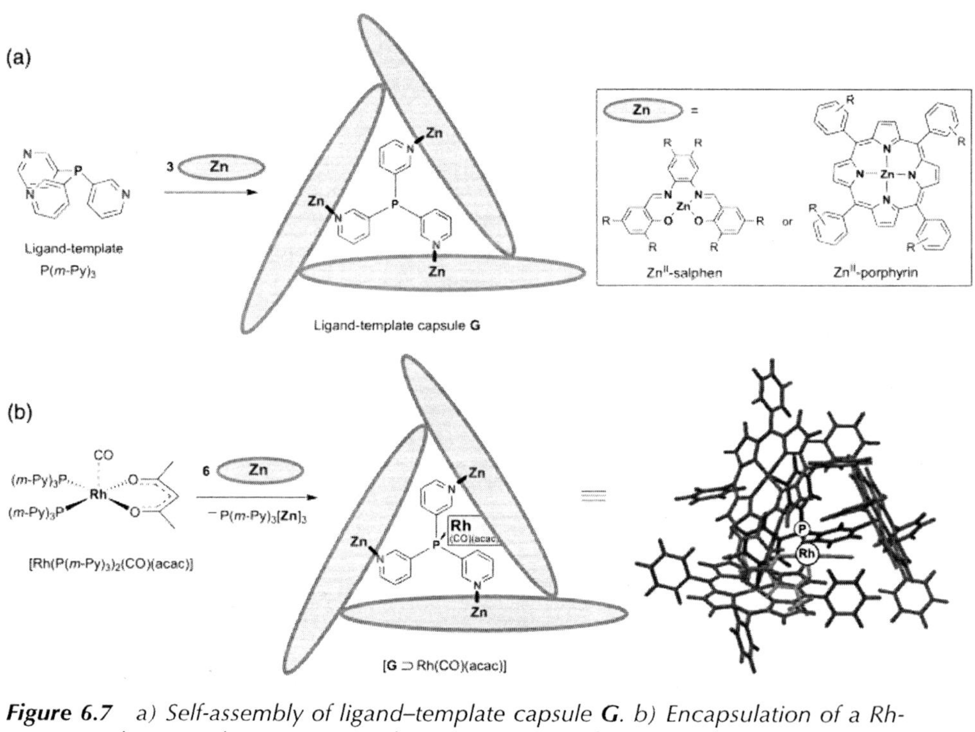

Figure 6.7 *a) Self-assembly of ligand–template capsule **G**. b) Encapsulation of a Rh-species within capsule **G**, i.e. [**G** ⊃ Rh(CO)(acac)] (modelling picture on the right).Reproduced with permission from [21] Copyright 2002 Wiley-VCH Verlag GmbH & Co. KGaA*

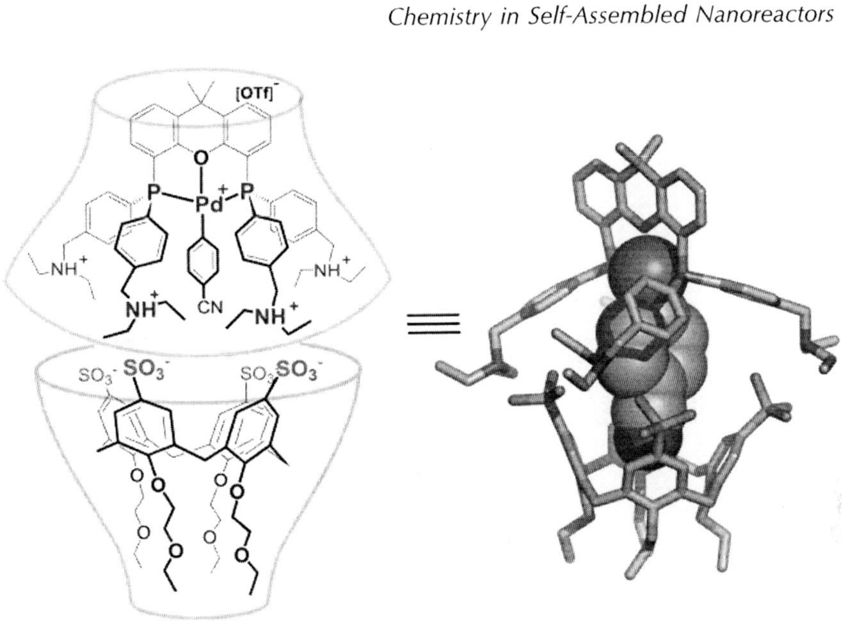

Figure 6.8 Bisphosphine based heterocapsule ***H*** encapsulating a transition metal, [***H*** ⊃ Pd-Ar], (modeling picture on the right). In the modeled structure, some substituents and hydrogen atoms have been omitted for clarity. [6] Reproduced by permission of The Royal Society of Chemistry (RSC)

(acac)] (Figure 6.7b). The encapsulated rhodium catalysts were shown to have unusual reactivity and regioselectivity in the hydroformylation reaction (*vide infra*).

We have also reported a ligand-template approach for metal encapsulation in which the ligand-template is an integrated part of the capsule.[54] This example involves a concave shaped bifunctional diphosphine ligand which can complexate a transition metal and also contains functional groups for capsule formation. Self-assembly of the tetracationic diphosphine ligand with a tetraanionic calix[4]arene leads to the formation of a reversible heterocapsule based on ionic interactions (**H**). Coordination of palladium to the template-ligand results in metal encapsulation [**H** ⊃ Pd–Ar] (Figure 6.8).

6.9.1 Hydroformylation

Transition metal catalysts encapsulated within the ligand-template nanoreactor **G**, $P(Py)_3$ • $[Zn]_3$, have been applied to catalyze industrially relevant processes such as hydroformylation and Heck reaction.[22–24] Nanoreactor [**G** ⊃ Rh(CO)(acac)] encapsulates a Rh-species that contains only one tris(*meta*-pyridyl)phosphine ligand, P(*m*-$Py)_3$, surrounded by three Zn-porphyrins or Zn-salphens. Under syngas pressure (H_2/CO), rhodium species like Rh(CO)(acac)$P(Py)_3$ transform into a complex of type $HRh(CO)_3P(Py)_3$, which is the active species for the hydroformylation reaction. In this reaction terminal alkenes are converted into linear and/or branched aldehydes, and the ratio of these products strongly depends on the specific catalyst applied. Hydroformylation of 1-octene by encapsulated rhodium, [**G** ⊃ $HRh(CO)_3$], resulted in a 10-fold rate enhancement compared

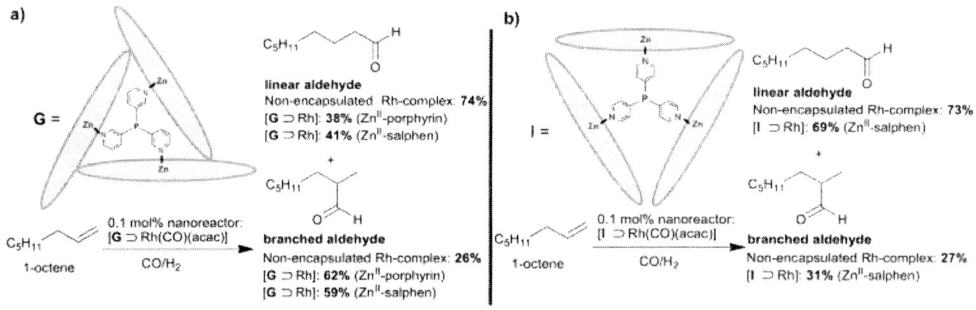

Scheme 6.7 *Hydroformylation of 1-octene a) within nanoreactor [**G** ⊃ Rh] and b) within nanoreactor [**I** ⊃ Rh]. Product distribution of the aldehyde products*

to the non-encapsulated rhodium catalyst (Scheme 6.7a).[23,24] In addition, the selectivity for the product has reversed; nanoreactor [**G** ⊃ $HRh(CO)_3$] containing Zn(II)-porphyrins provides 63% of the branched aldehyde compared to 26% observed for the non-encapsulated species. The unusual selectivity and increased rate can only partly be explained by modification of the catalytically active species upon encapsulation. Indeed, upon encapsulation the rhodium complex goes from bisphosphine to monophosphine, which generally produces more branched aldehydes along with higher rates compared to the bisphosphine species. In addition, a part of the effect was ascribed to complete encapsulation of the catalyst, as open cage analogues containing only two porphyrins instead of three, and which were shown to also have only one phosphine ligand coordinated to the rhodium, proved less active and less selective toward branched aldehyde.

An advantage of the ligand-template approach for capsule assembly is the possibility to modify the shape of the capsule by only minor changes of the template-ligand, as demonstrated by using either tris(*meta*-pyridyl)- or tris(*para*-pyridyl)phosphine. Addition of Zn(II)-salphen scaffolds to these two ligand templates resulted in selective formation of nanoreactors **G** and **I**, respectively (Scheme 6.7).[23] Hydroformylation of 1-octene by the tris(*meta*-pyridyl)phosphine based nanoreactor [**G** ⊃ Rh] gave a preference for the branched aldehyde (Scheme 6.7a), whereas the tris(*para*-pyridyl)phosphine based nanoreactor [**I** ⊃ Rh] predominantly gave the linear aldehyde, as with the non-encapsulated Rh–bisphosphine complex (Scheme 6.7b). These selectivities imply that nanoreactor [**G** ⊃ Rh] corresponds to a Rh–monophosphine complex whereas the more open structure of nanoreactor [**I** ⊃ Rh] allows the formation of Rh–bisphosphine species.

More recently, nanoreactor [**G** ⊃ Rh] has operated as a high-precision catalyst for the regioselective hydroformylation of internal alkenes.[22] The non-encapsulated complex $HRh(P(m\text{-}Py)_3)_2(CO)_2$ yields a near statistical mixture of the two expected internal aldehydes 2-methyloctanal and 2-ethylheptanal. However, nanoreactor [**G** ⊃ Rh], encapsulating the Rh-catalyst, exhibits a strong preference to form 2-ethylheptanal in 88% (Scheme 6.8). This outstanding selectivity is unprecedented in the hydroformylation of internal alkenes. A similar selectivity was found in the hydroformylation of 3-octene. Experiments using various partial H_2 and CO pressures resulted to the proposition that the hydride migration is the selectivity-determining step. This step requires a rotation of the coordinated alkene which is hampered by the steric restrictions imposed by the interior of the

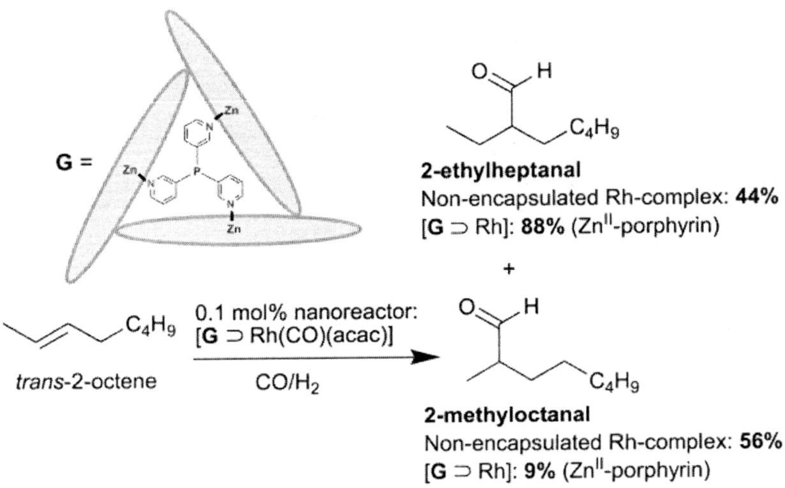

Scheme 6.8 *Hydroformylation of* trans-2-octene *by nanoreactor [**G** ⊃ Rh]. Product distribution of the aldehyde products*

capsule. Apparently, rotation of the intermediate olefin complex leading to the 2-ethylheptanal product is more facile, explaining the observed regioselectivity. Therefore it can be concluded that the nanoreactor determines the regiochemical outcome of the reaction by imposing its sterical restrictions during a specific step in the catalytic sequence. It is interesting to note that there are no other reactions that can distinguish between the two carbon atoms of 3-octene, demonstrating that reactions carried out in nanoreactors can result in unprecedented selective reactions.

In related work, we have employed α-diimine ligands with pyridine functionalities in conjunction with our Zn-porphyrin self-assembly protocol to arrive at supramolecular ligands for the Pd-catalyzed copolymerization of styrene and CO.[55] It was shown that both the reactivity of the catalyst and the stereoregularity of the polymer formed can be controlled in a supramolecular fashion, by subtle changes of the building blocks involved in the catalyst assembly. These supramolecular catalysts among the most active and selective catalysts for this reaction, and the origin of the selectivity is currently being studied. It is hypothesized that the effective encapsulation of the metal center by the porphyrin units has a significant impact on the overall dynamics of the system.

So far we have not used the chromophoric character of the porphyrin building blocks, but we recently studied systems in which we capitalize on these light capturing devices. We reported a catalyst system that is formed by assembly, which is able to use light as primary energy source to generate molecular hydrogen (Scheme 6.9).[56] The primary biomimetic [2Fe2S] hydrogenase catalyst is utilized with supramolecular handles to assemble the light capturing chromophore in close proximity of the active diiron centre to facilitate the photo-induced electron transfer. Importantly, closely related [2Fe2S] complexes with very similar electrocatalytic properties show a distinctively different behavior upon (photo)reduction. For this reason the supramolecular approach has been important as it enables the easy modular variation of complexes. In addition, we found an active

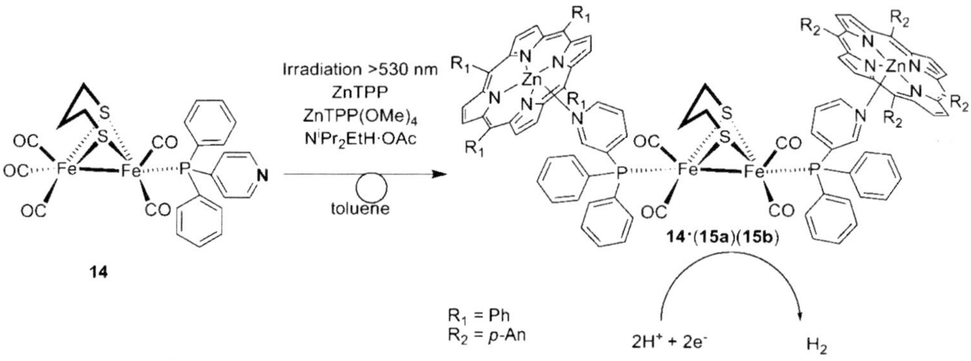

Scheme 6.9 *Photocatalytic reduction of protons using self-assembled, bio-mimetic Fe_2S_2-diphosphine complex **14 • (15a)(15b)***

catalyst that is difficult to prepare by traditional strategies and which we did not predict in advance. Detailed studies show that in both experiments with complexes **14** the assembly **14** • (**15a**)(**15b**) is the active species formed, either directly or via light induced disproportionation. The hydrogen producing system requires two different chromophores (**15a** and **15b**) present in solution, suggesting that **14** • (**15a**)(**15b**) is an active species. This study reconfirms the observation that small changes in the parent [2Fe2S] complex can result in different and unpredictable reaction pathways upon reduction. The supramolecular approach is highly promising for further development in this important area, as the modular nature of the building blocks applied also facilitates rapid optimization.

6.10 Virus Capsids, Proteins and Micellar Systems

Several approaches have recently been developed that directly apply natural architectures for artificial chemical reactions, some of which are detailed in different chapters of this book. Although not classified as homogeneous catalysis, the reduction of metal salts inside nanoreactors could be the first step on the way to reactivity with the corresponding metal colloids or nanoparticles in e.g. hydrogenation reactions. A variety of carrier systems have been studied lately, including virus capsids, polymeric micelles, mini-emulsions and hollow core-shell particles, as nanoreactors and hosts for the synthesis and encapsulation of well-defined, stable nanoparticles.[57]

Viruses and proteins such as ferritin or streptavidin[58] are large self-assembled micellar aggregates of bio-amphiphiles that naturally function as carriers for RNA. Removal of this RNA core yields the corresponding *apo*-capsids that can be further functionalized. Hence, apoferritin has been used as a container for the formation of Cu and Ag nanoparticles, by reduction of corresponding metal salts inside the virus capsid interior.[59] Also Fe-nanoparticles have been synthesized in a controlled manner within the cavity of genetically engineered human H-chain ferritin protein.[60] The groups of Nolte and Cornelissen have published initial results on the use of virus capsids as nanoreactors for single enzyme catalysis.[61]

6.11 Micellar Systems

Catalysis inside micellar systems is an attractive approach that might also find application in recycling and recovery of homogeneous catalysts, which would enable introduction to continuous catalysis. Besides covalent approaches, using crosslinked, unimolecular micelles or latex particles and electrostatic interactions between the surfactant head groups and the cationic metal species,[62] non-covalent, dynamic micelles charged with catalytically active transition metal complexes have also been employed. Recent examples include the work by Weberskirch on the dual-catalyst hydroaminomethylation of 1-octene in aqueous media through the use of phosphane-functionalized macromonomers.[63] Reasonable n/iso ratios of up to 11 and turnover frequencies of up to ~600 h^{-1} were obtained. In a similar fashion, micellar systems derived from 2-oxaline based amphiphilic block-copolymers, partially functionalized with Co–salen complex at the hydrophobic interior, were used for the hydrolytic kinetic resolution of epoxides in water.[64] Excellent ee's for both the epoxide and the diol product were obtained, but catalyst recycling experiments showed loss in selectivity after a few runs. The droplets present in specific inverted (water-in-oil) mini-emulsions have been employed as nanoreactors to perform a polymerase chain reaction (PCR).[65] Micelles built-up in part from amphiphilic dendrons have been exploited as vessels for the aqueous-phase Diels–Alder addition reaction between C_{60} and polycyclic hydrocarbons.[66] Inverted block-copolymeric micelles, featuring a poly(4-vinylbenzoic acid) interior and a polystyrene corona in toluene, were utilized as hosts for Fe-bis(imino)pyridine based single-site alkene polymerization catalysis of ethene.[67] Rod-coil based amphiphiles, wherein the rod-fragment consisted of linear polyphenylene moieties and the coil-tail was represented by poly(ethylene glycol) chains, were shown to aggregate into micellar structures in water. The specific architecture as well as the strict hydrophobic interior were shown to provide a good platform to perform the Suzuki coupling with standard aryl boronic acids and halides in water at room temperature at low catalyst loadings.[68] Our own group reported on functionalized, xanthene based diphosphine ligands that were decorated with surface-active pendant groups (i.e.-4-$C_6H_4O(CH_2)_nC_6H_4(SO_3Na)$), which spontaneously self-assembled into vesicular aggregates in water. These supramolecular structures proved stable even at elevated temperature (90 °C), so upon introduction of a suitable Rh-precursor, the selective aqueous biphasic hydroformylation of higher alkenes (e.g. octene), which typically show very low solubility in water, could be carried out, coupled with repetitive recycling of the catalyst system.[69]

The groups of Rowan and van Hest have jointly reported on the use of polymersomes – enlarged liposomes with a vesicular structure containing a hydrophobic bilayer – as nanoreactor vessels for a three-step cascade reaction.[70] The stepwise, controlled incorporation of horseradish peroxidase (HRP) and glucose oxidase (GO) in separate inner-domains of the polymersome structure was confirmed by fluorescence microscopy. More specifically, GO was selectively located inside the hydrophilic core enclosed by the bilayer, which itself contained HRP, whilst a third enzyme, *Candida antarctica* lipase B (CALB), was not encapsulated but was added to the bulk solvent surrounding the reactors. The following reaction sequence was studied as a proof-of-principle: i) conversion of 1,2,3,4-tetra-*O*-acetyl-*β*-glucopyranoseglucose into glucose by CALB, ii) oxidation of glucose to the corresponding lactone concomitant with formation of H_2O_2 by GO and iii) the conversion of 2,2′-azinobis(3-ethylbenzothiazoline-6-sulfonic acid (ABTS) into $ABTS^{\cdot+}$ with consumption of H_2O_2 by HRP. For earlier results in the area of macromolecular micelles, polymers and biomacromolecules the reader is referred to the recent review literature.[11]

6.12 Conclusions and Outlook

A wide array of self-assembled molecular capsules based on various building blocks and non-covalent interactions has been developed in the last decade. The nanospace within these supramolecular capsules is generally in the range of 300–500 $Å^3$ (although larger examples are known as well), which is sufficient for the selective encapsulation of appropriately sized guest molecule(s). A number of distinctly different capsule architectures have been characterized, highly dependent on the specific type of non-covalent interaction, building block structure and templating agent, and as a result guest shielding and guest exchange rates can strongly vary. A diversity of chemical processes has been carried out within molecular capsules and the effects observed so far are, although academic, very interesting. Reactions can be *accelerated* and the *selectivity* of a chemical process can be changed completely. These observations can be explained by stabilization of the reaction transition state by the capsule (based on enthalpic and entropic contributions) or by concentration effects in the case of bimolecular reactions, such as Diels–Alder reactions. More important are the unique reaction *selectivities* induced by the novel finite microenvironment within the capsule. The size and shape of the nanoreactor's cavity and that of the nanoreactor's gates can control the substrate selectivity by controlling the access to the cavity. Similarly, a nanoreactor can protect an *active site* located in the cavity that otherwise would be poisoned by chemicals present in solution. The *regio*- and chemoselectivities can also be changed by the capsule, through a change in the relative reaction rates of competing pathways. This was for example observed for encapsulated rhodium complexes that were used as hydroformylation catalysts. In addition to these effects, reaction *intermediates*, that otherwise exhibit too short lifetimes for identification, have been observed as stable species in nanoreactors. In these cases the reaction rate of the subsequent step after the formation of the intermediate is slowed down by the nanocapsule. *Product inhibition*, which is a frequently encountered problem in bimolecular coupling reactions carried out within enclosed cavities, is fundamentally related to the former. The coupling product might have a higher affinity for the capsule than the substrates, and consequently product release from the nanoreactor becomes the slowest step in the reaction. Also entropic factors might be important, as a unimolecular product has to exchange with two substrate molecules. Product inhibition can prohibit the utility of nanoreactors as true catalysts.

In addition to the nanoreactors discussed in this review, which are formed by assembly of at least two building blocks, capsules based on covalent bonds have also been applied as nanoreactors.[11] Such covalent nanoreactors were beyond the scope of the current review, but similar effects in catalysis are observed. However, self-assembled capsules also have guest exchange mechanisms via partial disassembly of the capsule, whereas exchange for the covalent analogues is restricted to portal slippage. It is this unique property that enables the combination of complete encapsulation with sufficiently rapid in–out exchange, which might proof to be very advantageous for potential catalytic applications.

Although the research field is still in its infancy, several examples of reactions, wherein self-assembled nanoreactors are applied and are shown to dramatic enhance or alter reactivity, have appeared, thereby demonstrating the power of the supramolecular concept. Detailed studies are required to fully understand the mechanisms behind the effects observed when carrying out reactions in nanoreactors. The results obtained so far sketch a bright prospective, as reactions have been observed that are unique to those carried out in capsules. In this review we have focussed on reactions that take place *inside* the capsules. However, molecular capsules have also been used to control reactions that take place *outside* the capsule for example by controlling the release of reagents, making the nanoreactor applications virtually unlimited.[71]

Acknowledgements

We thank the Netherlands Organization of Scientific Research NWO (VENI grant to J.I.v.d.V.; VICI grant to J.N.H.R.) and the University of Amsterdam for financial support, and all co-workers, whose names appear in the references, for their contributions.

References

1. J. M. Lehn, Supramolecular chemistry: receptors, catalysts, and carriers, *Science*, 1985, **227**, 849.
2. J. M. Lehn, *Supramolecular Chemistry: Concepts and Perspectives*, WILEY-VCH, 1995.
3. D. J. Cram, Cavitands: orgnaic hosts with enforced cavities, *Science*, 1983, **219**, 1177.
4. D. J. Cram, Molecular Container Compounds, *Nature*, 1992, **356**, 29.
5. D. J. Cram, J. M. Cram, *Container Molecules and Their Guests*, (Monographs in Supramolecular Chemistry, No. 4, Series Editor J. F. Stoddart) The Royal Society of Chemistry, Cambridge, U.K., 1997.
6. T. S. Koblenz, J. Wassenaar, J. N. H. Reek, Reactivity within a confined self-assembled nanospace, *Chem. Soc. Rev.*, 2008, **37**, 247.
7 a) J. K. M. Sanders, Supramolecular catalysis in transition, *Chem. Eur. J.*, 1998, **4**, 1378; b) A. J. Kirby, Enzyme mechanisms, models, and mimics, *Angew. Chem. Int. Ed.*, 1996, **35**, 707; c) W. B. Motherwell, M. J. Bingham, Y. Six, Recent progress in the design and synthesis of artificial enzymes, *Tetrahedron*, 2001, **57**, 4663; d) J. I. van der Vlugt, F. Meyer, Homogeneous copper-catalyzed oxidations, *Top. Organomet. Chem.*, 2007, **22**, 191; e) J. I. van der Vlugt, F. Meyer, Synthetic models for the active sites of nickel-containing enzymes, *Met. Ions Life Sci.*, 2007, **2**, 181.
8. J. Gao, S. Ma, D. T. Major, K. Nam, J. Pu, D. G. Truhlar, Mechanisms and free energies of enzymatic reactions, *Chem. Rev.*, 2006, **106**, 3188; M. J. S. Dewar, D. M. Storch, Alternative view of enzyme reactions, *Proc. Natl. Acad. Sci. USA*, 1985, **82**, 2225; X. Zhang, K. N. Houk, Why enzymes are proficient catalysts: beyond the Pauling paradigm, *Acc. Chem. Res.*, 2005, **38**, 379.
9. L. R. MacGillivray, J. L. Atwood, Structural classification and general principles for the design of spherical molecular hosts, *Angew. Chem. Int. Ed.*, 1999, **38**, 1018.
10. In the early days of this research field also covalent based capsules were assigned as 'supramolecular capsules' because of their encapsulation properties. However, because of the developments of the last decade we designate here only non-covalent based capsules as 'supramolecular capsules'.

11. D. M. Vriezema, M. C. Aragones, J. A. A. W. Elemans, J. J. L. M. Cornelissen, A. E. Rowan, R. J. M. Nolte, Self-assembled nanoreactors, *Chem. Rev.*, 2005, **105**, 1445.
12. A. Lützen, Self-assembled molecular capsules - even more than nano-sized reaction vessels, *Angew. Chem. Int. Ed.*, 2005, **44**, 1000.
13. An important difference between enzymes and nanoreactors is product release from the cavity. The Michaelis–Menten kinetics assumes that product release occurs rapidly (this assumption is also done in equation 4). However, this assumption does not always counts for nanoreactors because they can display product inhibition (*vide infra*).
14. For an alternative analysis of supramolecular effects in bimolecular reactions using the '*effective* molarity' approach, see: S. Di Stefano, L. Mandolini, Effective molarities in supramolecular catalysis of two-substrate reactions, *Acc. Chem. Res.*, 2004, **37**, 113.
15. D. Fiedler, H. van Halbeek, R. G. Bergman, K. N. Raymond, Supramolecular catalysis of unimolecular rearrangements: substrate scope and mechanistic insights, *J. Am. Chem. Soc.*, 2006, **128**, 10240 and references therein.
16. K. Takaoka, M. Kawano, T. Ozeki, M. Fujita, Crystallographic observation of an olefin photodimerization reaction that takes place via thermal molecular tumbling within a self-assembled host, *Chem. Commun.*, 2006, 1625.
17. M. Yoshizawa, M. Tamura, M. Fujita, Diels-alder in aqueous molecular host: unusual regioselectivity and efficient catalysis, *Science*, 2006, **312**, 251.
18. A. V. Davis, R. M. Yeh, K. N. Raymond, Supramolecular assembly dynamics, *Proc. Natl. Acad. Sci. USA*, 2002, **99**, 4793.
19. D. Fiedler, D. H. Leung, R. G. Bergman, K. N. Raymond, Selective molecular recognition, C-H bond activation, and catalysis in nanoscale reaction vessels, *Acc. Chem. Res.*, 2005, **38**, 351 and references therein.
20. M. L. Merlau, M. Del Pilar Mejia, S. T. Nguyen, J. T. Hupp, Artificial enzymes formed through directed assembly of molecular square encapsulated epoxidation catalysis, *Angew. Chem. Int. Ed.*, 2001, **40**, 4239.
21. A. W. Kleij, J. N. H. Reek, Ligand-template directed assembly: an efficient approach for the supramolecular encapsulation of transition-metal catalysts, *Chem. Eur. J.*, 2006, **12**, 4218.
22. M. Kuil, T. Soltner, P. W. N. M. Van Leeuwen, J. N. H. Reek, High-precision catalysts: regioselective hydroformylation of internal alkenes by encapsulated rhodium complexes, *J. Am. Chem. Soc.*, 2006, **128**, 11344.
23. A. W. Kleij, M. Lutz, A. L. Spek, P. W. N. M. van Leeuwen, J. N. H. Reek, Encapsulated transition metal catalysts comprising peripheral Zn(II)salen building blocks: template-controlled reactivity and selectivity in hydroformylation catalysis, *Chem. Commun.*, 2005, 3661.
24. V. F. Slagt, P. C. J. Kamer, P. W. N. M. van Leeuwen, J. N. H. Reek, Encapsulation of transition metal catalysts by ligand-template directed assembly, *J. Am. Chem. Soc.*, 2004, **126**, 1526 and references therein.
25. J. Kang, J. Santamaria, G. Hilmersson, J. Rebek Jr., Self-assembled molecular capsule catalyzes a Diels-Alder reaction, *J. Am. Chem. Soc.*, 1998, **120**, 7389.
26. M. Yoshizawa, T. Kusukawa, M. Fujita, S. Sakamoto, K. Yamaguchi, Cavity-directed synthesis of labile silanol oligomers within self-assembled coordination cages, *J. Am. Chem. Soc.*, 2001, **123**, 10454.
27. J. Rebek Jr., Simultaneous encapsulation: molecules held at close range, *Angew. Chem. Int. Ed.*, 2005, **44**, 2068 and references therein.
28. L. C. Palmer, J. Rebek Jr., The ins and outs of molecular encapsulation, *Org. Biomol. Chem.*, 2004, **2**, 3051 and references therein.
29. F. Hof, S. L. Craig, C. Nuckolls, J. Rebek Jr., Molecular encapsulation, *Angew. Chem. Int. Ed.*, 2002, **41**, 1488 and references therein.
30. S. Mecozzi, J. Rebek Jr., The 55% solution: a formula for molecular recognition in the liquid state, *Chem. Eur. J.*, 1998, **4**, 1016.
31. J. Chen, J. Rebek Jr., Selectivity in an encapsulated cycloaddition reaction, *Org. Lett.*, 2002, **4**, 327.
32. C. H. M. Amijs, G. P. M. van Klink, G. van Koten, Metallasupramolecular architectures, an overview of functional properties and applications, *Dalton Trans.*, 2006, 308.
33. M. D. Pluth, K. N. Raymond, Reversible guest exchange mechanisms in supramolecular host-guest assemblies, *Chem. Soc. Rev.*, 2007, **36**, 161 and references therein.

34. A. V. Davis, D. Fiedler, M. Ziegler, A. Terpin, K. N. Raymond, Resolution of chiral, tetrahedral M_4L_6 metal-ligand hosts, *J. Am. Chem. Soc.*, 2007, **129**, 15354.
35. S. M. Biros, R. M. Yeh, K. N. Raymond, Design and formation of a large tetrahedral cluster using 1,1-'binaphthyl ligands', *Angew. Chem. Int. Ed.*, 2008, **47**, 6062.
36. a) M. D. Pluth, R. G. Bergman, K. N. Raymond, Acid catalysis in basic solution: a supramolecular host promotes orthoformate, *Science*, 2007, **316**, 85; b) M. D. Pluth, R. G. Bergman, K. N. Raymond, Catalytic deprotection of acetals in basic solution with a self-assembled supramolecular nanozyme, *Angew. Chem. Int. Ed.*, 2007, **46**, 8587; c) M. D. Pluth, R. G. Bergman, K. N. Raymond, Supramolecular catalysis of orthoformate hydrolysis in basic solution: an enzyme-like mechanism, *J. Am. Chem. Soc.*, 2008, **130**, 11423.
37. C. J. Hastings, D. Fiedler, R. G. Bergman, K. N. Raymond, Aza Cope rearrangement of propargyl enammonium cations catalyzed by a self-assembled 'nanozyme', *J. Am. Chem. Soc.*, 2008, **130**, 10977.
38. D. H. Leung, R. G. Bergman, K. N. Raymond, Scope and mechanism of the C-H bond activation reactivity within a supramolecular host by an iridium guest: a stepwise ion pair guest dissociation mechanism, *J. Am. Chem. Soc.*, 2006, **128**, 9781 and references therein.
39. D. H. Leung, R. G. Bergman, K. N. Raymond, Highly selective supramolecular catalyzed allylic alcohol isomerization, *J. Am. Chem. Soc.*, 2007, **129**, 2746.
40. M. Fujita, M. Tominaga, A. Hori, B. Therrien, Coordination assemblies from a Pd(II)-cornered square complex, *Acc. Chem. Res.*, 2005, **38**, 371.
41. M. Fujita, K. Umemoto, M. Yoshizawa, N. Fujita, T. Kusukawa, K. Biradha, Molecular paneling via coordination, *Chem. Commun.*, 2001, 509.
42. Y. Nishioka, T. Yamaguchi, M. Yoshizawa, M. Fujita, Unusual [2+4] and [2+2] cycloadditions of arenes in the confined cavity of self-assembled cages, *J. Am. Chem. Soc.*, 2007, **129**, 7000.
43. M. Yoshizawa, Y. Takeyama, T. Kusukawa, M. Fujita, Cavity-directed, highly stereoselective [2+2] photodimerization of olefins within self-assembled coordination cages, *Angew. Chem., Int. Ed.*, 2002, **41**, 1347.
44. M. Yoshizawa, Y. Takeyama, T. Okano, M. Fujita, Cavity-directed synthesis within a self-assembled coordination cage: highly selective [2+2] cross-photodimerization of olefins, *J. Am. Chem. Soc.*, 2003, **125**, 3243.
45. Y. Nishioka, T. Yamaguchi, M. Kawano, M. Fujita, Asymmetric [2+2] olefin cross photoaddition in a self-assembled host with remote chiral auxiliaries, *J. Am. Chem. Soc.*, 2008, **130**, 8160.
46. H. Ito, T. Kusukawa, M. Fujita, Wacker oxidation in an aqueous phase through the reverse phase-transfer catalysis of a self-assembled nanocage, *Chem. Lett.*, 2000, 598.
47. M. Yoshizawa, S. Miyagi, M. Kawano, K. Ishiguro, M. Fujita, Alkane oxidation via photochemical excitation of a self-assembled molecular cage, *J. Am. Chem. Soc.*, 2004, **126**, 9172.
48. T. Furusawa, M. Kawano, M. Fujita, The confined cavity of a coordination cage suppresses the photocleavage of α-diketones to give cyclization products through kinetically unfavorable pathways, *Angew. Chem. Int. Ed.*, 2007, **46**, 5717.
49. T. Yamaguchi, M. Fujita, High selective photomediated 1,4-radical addition to *o*-quinones controlled by a self-assembled cage, *Angew. Chem. Int. Ed.*, 2008, **47**, 2067.
50. C. L. D. Gibb, B. C. Gibb, Well-defined, organic nanoenvironments in water: the hydrophobic effect drives a capsular assembly, *J. Am. Chem. Soc.*, 2004, **126**, 11408.
51. A. Natarajan, L. S. Kaanumalle, S. Jockusch, C. L. D. Gibb, B. C. Gibb, N. J. Turro, V. Ramamurthy, Controlling photoreactions with restricted spaces and weak intermolecular forces: exquisite selectivity during oxidation of olefins by singlet oxygen, *J. Am. Chem. Soc.*, 2007, **129**, 4132. For a second application of nanocapsule E, see: L. S. Kaanumalle, C. L. D. Gibb, B. C. Gibb, V. Ramamurthy, Photo-Fries reaction in water made selective with a capsule, *Org. Biomol. Chem.*, 2007, **5**, 236.
52. A. Greer, Organic chemistry: molecular cross-talk, *Nature*, 2007, **447**, 273.
53. M. J. Wilkinson, P. W. N. M. van Leeuwen, J. N. H. Reek, New directions in supramolecular transition metal catalysis, *Org. Biomol. Chem.*, 2005, **3**, 2371.
54. T. S. Koblenz, H. L. Dekker, C. G. De Koster, P. W. N. M. Van Leeuwen, J. N. H. Reek, Bisphosphine base hetero-capsules for the encapsulation of transition metals, *Chem. Commun.*, 2006, 1700.

55. J. Flapper, J. N. H. Reek, Templated encapsulation of pyridyl-bian palladium complexes: tunable catalysts for CO/4-*tert*-butylstyrene copolymerization, *Angew. Chem. Int. Ed.*, 2007, **46**, 8590.
56. A. M. Kluwer, R. Kapre, F. Hartl, M. Lutz, A. L. Spek, A. M. Brouwer, P. W. N. M. van Leeuwen, J. N. H. Reek, Self-assembled biomimetic [2Fe2S]-hydrogenase-based photocatalyst for molecular hydrogen evolution, *Proc. Natl. Acad. Sci. USA*, 2009, **106**, 10460.
57. A. D. Crespy, M. Stark, C. Hoffmann-Richter, U. Ziener, K. Landfester, Polymeric nanoreactors for hydrophilic reagents synthesized by interfacial polycondensation on miniemulsion droplets, *Macromolecules*, 2007, **40**, 3122; b) O. Gazit, R. Khalfin, Y. Cohen, R. Tannenbaum, Self-assembled diblock copolymer 'nanoreactors' as 'catalysts' for metal nanoparticle synthesis, *J. Phys. Chem. C*, 2009, **113**, 576.
58. We will not discuss the recent applications of the coupled streptavidin–biotin motif in bio-inspired asymmetric catalysis, as an entire chapter is dedicated to 'Artificial Metalloenzymes', written by Prof. Thomas Ward.
59. N. Gálvez, B. Fernandez, E. Valero, P. Sánchez, R. Cuesta, J. M. Domínguez-Vera, Aproferritin as a nanoreactor for preparing metallic nanoparticles, *C. R. Chimie*, 2008, **11**, 1207.
60. M. Uchida, M. L. Flenniken, M. Allen, D. A. Willits, B. E. Crowley, S. Brumfield, A. F. Willis, L. Jackiw, M. Jutila, M. J. Young, T. Douglas, Targeting of cancer cells with ferrimagnetic ferritin cage nanoparticles, *J. Am. Chem. Soc.*, 2006, **128**, 16626.
61. M. Comellas-Aragonès, H. Engelkamp, V. I. Claessen, N. A. J. M. Sommerdijk, A. E. Rowan, P. C. M. Christianen, J. C. Maan, B. J. M. Verduin, J. J. L. M. Cornelissen, R. J. M. Nolte, A virus-based single-enzyme nanoreactor, *Nature Nanotech.*, 2007, **2**, 635.
62. R. Sablong, J. I. van der Vlugt, R. Thomann, S. Mecking, D. Vogt, Disperse amphiphilic submicron particles as non-covalent supports for cationic homogeneous catalysts, *Adv. Synth. Catal.*, 2005, **347**, 633; K. Kunna, J. Loos, C. Müller, D. Vogt, Aqueous-phase hydroformylation of 1-octene: styrene latices a phase-transfer agents, *Angew. Chem. Int. Ed.*, 2006, **45**, 7289.
63. B. Gall, M. Bortenschlager, O. Nuyken, R. Weberskirch, Cascade reactions in polymeric nanoreactors: mono (Rh)- and bimetallic (Rh/Ir) micellar catalysis in the hydroaminomethylation of 1-octene, *Macromol. Chem. Phys.*, 2008, **209**, 1152.
64. B. M. Rossbach, K. Leopold, R. Weberskirch, Self-assembled nanoreactors as highly active catalysts in the hydrolytic kinetic resolution (HKR) of epoxides in water, *Angew. Chem. Int. Ed.*, 2006, **45**, 1309.
65. A. Musyanovych, V. Mailänder, K. Landfester, Miniemulsion droplets as single molecule nanoreactors for polymerase chain reaction, *Biomacromolecules*, 2005, **6**, 1824.
66. A. Simonyan, I. Gitsov, Linear-dendritic supramolecular complexes as nanoscale reaction vessels for 'green' chemistry. Diels-Alder reactions between fullerene C_{60} and polycycilic aromatic hydrocarbons in aqueous medium, *Langmuir*, 2008, **24**, 11431.
67. C. Bouilhac, E. Cloutet, D. Taton, A. Deffieux, R. Bosali, H. Cramail, Block copolymer micelles as nanoreactors for single-site polymerization catalysts, *J. Polym. Sci. A: Polym. Chem.*, 2009, **47**, 197.
68. J.-H. Ryu, C.-J. Jang, Y.-S. Yoo, S.-G. Lim, M. Lee, Supramolecular reactor in an aqueous environment: aromatic cross Suzuki coupling reaction at room temperature, *J. Org. Chem.*, 2005, **70**, 8956.
69. M. Schreuder Goedheijt, B. E. Hanson, J. N. H. Reek, P. C. J. Kamer, P. W. N. M. van Leeuwen, Accelerated biphasic hydroformylation by vesicle formation of amphiphilic diphosphines, *J. Am. Chem. Soc.*, 2000, **122**, 1650.
70 a) D. M. Vriezema, P. M. L. Garcia, N. Sancho Oltra, N. S. Hatzakis, S. M. Kuiper, R. J. M. Nolte, A. E. Rowan, J. C. M. van Hest, Positional assembly of enzymes in polymersome nanoreactors for cascade reactions, *Angew. Chem. Int. Ed.*, 2007, **46**, 7378; b) S. M. Kuiper, M. Nallani, D. M. Vriezema, J. J. L. M. Cornelissen, J. C. M. van Hest, R. J. M. Nolte, A. E. Rowan, Enzymes containing porous polymersomes as nano reaction vessels for cascade reactions, *Org. Biomol. Chem.*, 2008, **6**, 4315.
71. J. Chen, S. Körner, S. L. Craig, D. M. Rudkevich, J. Rebek Jr., Chemistry: amplification by compartmentalization, *Nature*, 2002, **415**, 385.

7

Concave Reagents

Ulrich Lüning

Otto-Diels-Institut für Organische Chemie, D-24098 Kiel, Germany

7.1 Introduction

7.1.1 Supramolecular Chemistry and Enzymes

Supramolecular Chemistry has been defined as the chemistry beyond the molecule.[1] If you wish, any reaction, and especially catalysis, is a supramolecular phenomenon. The term supramolecular catalysis,[2] however, is used in a much more narrow sense. A good approach to avoid endless discussions on definitions is to observe Mother Nature and look for supramolecular chemistry there. The molecular recognition of enzymes and their ability to catalyse numerous reactions with outstanding selectivities is a challenge for the (supramolecular) chemist.

What is so special about the enzymes? Which elements have to be copied if one wanted to construct an enzyme-like molecule? And what are the reasons to attempt a construction of an enzyme-like molecule?

Enzymes have been brought to perfection by billions of years of evolution. Nevertheless, enzymes for every conceivable reaction do not exist. A Diels–Alderase for instance has long been sought after, and even today there is dispute about its existence.[3] In addition, enzymes have been perfected in aqueous media, and usually only little variation in the solvent composition is possible. It is therefore attractive to copy the features of enzymes and to construct artificial molecules which may operate in any solvent and which may carry out reactions Nature has not taken into her reservoir of techniques.

For such enzyme-like molecules, two central features of an enzyme have to be imitated. First, there is a reactive site – the active site. This active site consists of one or more functional groups which, in many cases, may also carry out the same type of reaction if they are not part of an enzyme. But the selectivity, and also often the reactivity, will not be the same. Responsible for this aspect is the second important feature of enzymes, the embedding of the active site in the three-dimensional structure of the protein. The polypeptide backbone forms helices, sheets and other suprastructures which usually build up clefts, caves or niches in which the active site is located. This sterical shielding usually is responsible for the fascinating selectivities of enzymatic reactions.

The active site itself may just consist of the functional groups of the proteinogenic amino acids, or additional co-enzymes can be embedded. Many of them contain metal ions, and thus the enzymes can be divided into metallo-enzymes and metal-free ones. In the laboratory, bio-inorganic chemistry and organocatalysis try to understand and reproduce these two classes.

In this chapter, the geometric features will be the guide to enzyme-inspired artificial systems. The vast majority of substrates have a shape which resembles that of a potato. In other words, their overall shape is convex and irregular. For a selective recognition of a surface, a multipoint interaction is necessary. Because the substrate is convex, the recognition region of the enzyme has to be concave which is ensured by a cleft or cave type geometry (see above). Host–guest recognition has been discussed since Emil Fischer[4,5] established the lock-and-key principle. More elaborate models discuss flexible fit and induced fit, and conformational flexibility is essential in this context. But adaptability does not result in a loss of the concave shape. One way to ensure a general concave structure is the use of macrocycles, and another approach is to employ stiff subunits. The latter is realized in cleft molecules, also termed pincers, and within the group of macrocycles, especially bimacrocycles ensure concavity because they enclose the void in three dimensions.

For a reagent or a catalyst, a reactive functional group is needed. Inspired by the structure of enzymes, the functional group must be located in the concave region. The challenge is an *endo*-functionalization. In 1987,[6] we presented the concept of concave reagents

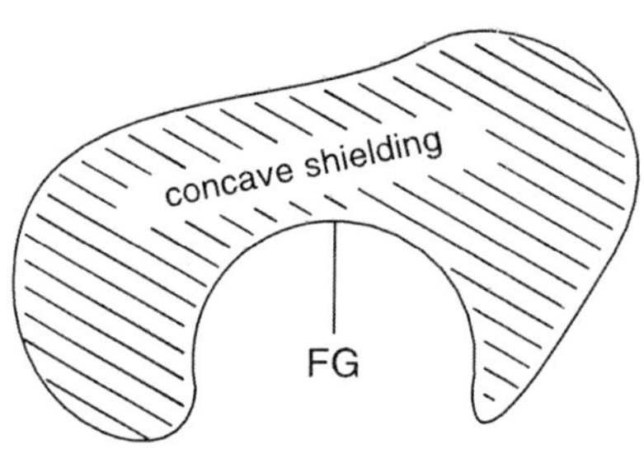

Figure 7.1 *Schematic drawing of a molecule with a functional group FG in a concave environment. The functional group FG is placed in an* endo-*position the same way the active site is located in an enzyme*

Figure 7.2 The geometry of a concave reagent resembles that of a light bulb in a lamp shade. The shade is the concave shielding. The (re)active site, the light bulb, is located on its inside and can only be touched by a molecule with matching size and shape

in which we compared the geometry of a concave reagent with that of a light bulb in a lamp shade (Figure 7.2). The lamp shade represents the concave shielding, the light bulb the functional reactive group within this concave confinement. Next, the macroscopic features have to be translated into the nano-world. The rim of the lamp shade is a cycle, and in order to let molecules or part of a molecule pass, it must be macrocyclic. Then, the functional group must be placed above this entry which means that the macrocycle must be spanned by an additional bridge carrying the functional group. The last task is to ensure that the functional group remains in an *endo*-position, and a lamp-like molecule is completed.

By turning the bimacrocycle upside-down, our concave concept was termed some years later as reaction bowls.[7] Functional groups including less stable ones such as a sulfenic acid, a selenenic acid, and a monosubstituted simple enol have been incorporated in cyclophanes and bridged calixarenes[8] (for an example see Figure 7.15).

7.1.2 Reagents and Catalysts

Over the past decades, numerous reactions have been carried out inside supramolecular complexes. In a large number of examples, no additional functional group was present in the cavity but the confinement and often the electrostatic potential was responsible for reactivity and selectivity. In contrast, *endo*-functionalized reagents and catalysts are active due to the embedded functional group.

Early examples for the first group of reactions are those in which a charge is developing or a charge is redistributed. Schmidtchen's[9] tetrahedral trimacrocyclic tetracationic receptors for instance accelerate the decarboxylation and ring opening of a benzisoxazole

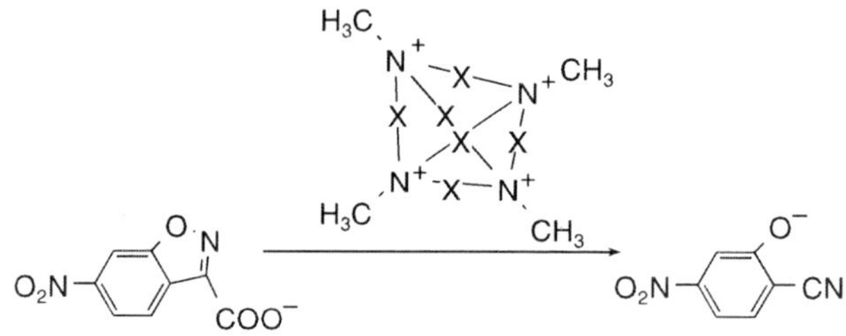

Figure 7.3 Ring opening and decarboxylation of a benzisoxazole, catalysed by a macrotricyclic tetracationic host molecule $[X = (CH_2)_{6,8}]$[9]

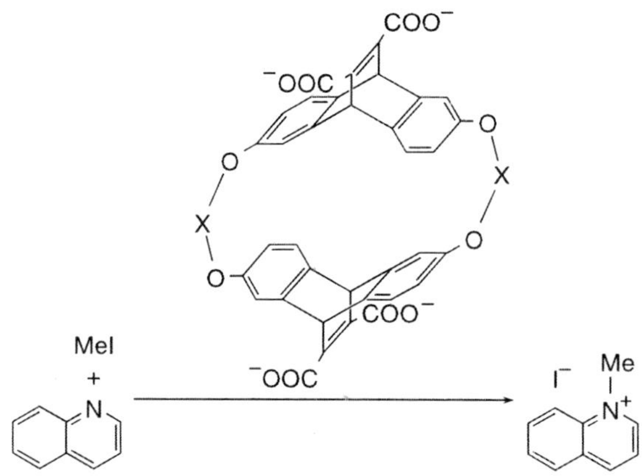

Figure 7.4 π-Rich tetraanionic macrocycles (X = p-phenylene, 1,4-cyclohexanebismethylene) catalyse the quaternization of quinoline with methyl iodide[10]

(Figure 7.3) while Dougherty's[10] anionic, π-rich macrocycles catalyse the alkylation of quinoline by stabilizing the charge in the transition state (Figure 7.4).

In the past decade, self-assembled cages have been made accessible for instance in the groups of Raymond,[11] Fujita[12] or Stang[13]. The resulting host molecules have now been used to catalyse reactions related to the ones mentioned above (Diels–Alder,[12] see Figure 7.5, 3,3-sigmatropic rearrangement,[11] see Figure 7.6).

In other host molecules, binding sites are located in such a way that the substrates or reagent and substrate are bound in proper proximity within a host molecule. The proximity is entropically favorable for a reaction and thus rate accelerations have been observed. In addition, the proper placement of the reactants leads to altered selectivities. For instance Sanders[14] described a macrocyclic trimeric porphyrin in which two substrates are preorganized in such a way that the stereochemistry of a Diels–Alder reaction is governed by the inclusion (see Figure 7.7).

But most related to enzymes are those concave entities in which reactive functional groups are pointing inwards. Several subclasses can be identified: there are reagents and catalysts, and these can be divided into metal-containing and metal-free compounds.

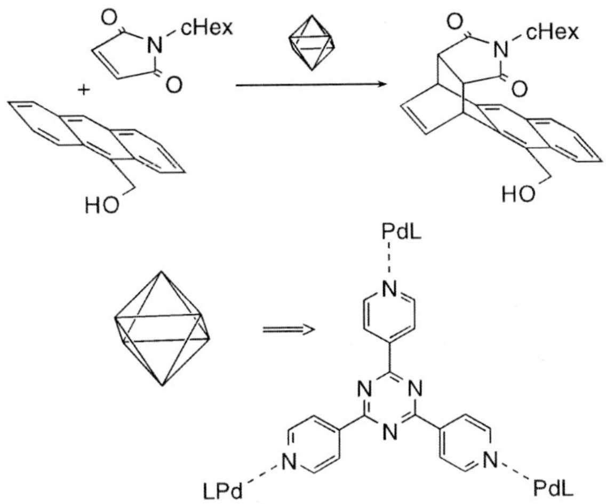

Figure 7.5 An octahedral self-assembled hexapalladium cluster (abbreviated as an octahedron) catalyzes a Diels–Alder reaction with unusual selectivity. The octahedron consists of tripyridinetriazine units which are connected by palladium ions bearing tetramethylethylene-1,2-diamine ligands L[12]

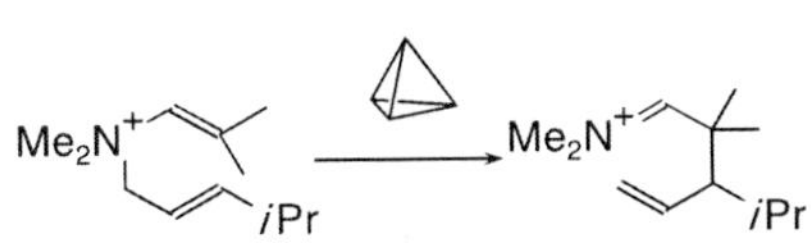

Figure 7.6 A tetrahedral self-assembled cluster consisting of four gallium ions and six 1,5-bis(2′,3′-dihydroxybenzamido)naphthalene bridges (abbreviated above as a tetrahedron) accelerates a 3,3-sigmatropic rearrangement by almost three orders of magnitude[11]

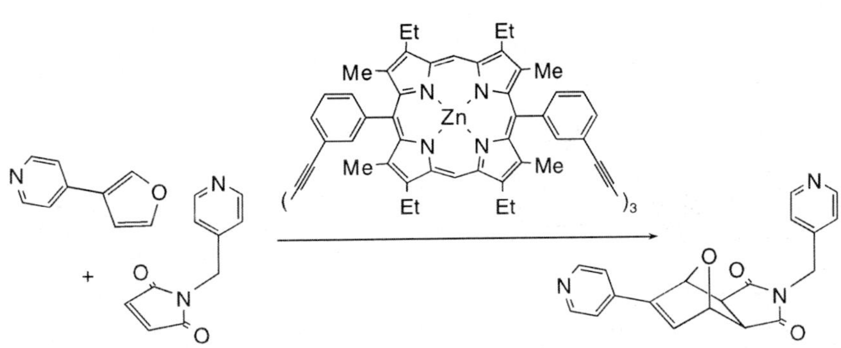

Figure 7.7 A trimeric zinc-porphyrin macrocycle catalyzes a Diels-Alder reaction with exo-selectivity[14]

In the case of a reagent, a functional group is altered in the course of the reaction. An acid for instance can carry out a protonation which results in the corresponding base at the end of the reaction. Or a redox-reagent oxidizes or reduces a substrate and alters its redox state. This is a problem for expensive reagents like concave ones, and therefore recycling is essential in these cases (but easy to carry out in case of protonations/deprotonations or redox-reactions).

In catalysis, however, the catalyst is unaltered after each catalytic cycle. This means that much less material is needed if the catalysis has a good turn-over and that a recovery is possible, at least in principle.

Concave reagents have been employed in many reactions but for the reasons mentioned above, most of the reactions were catalyses. Besides acid or base catalyses, especially transition metal ion catalysed reactions are of interest and have been investigated. The following chapters will first present major classes of concave reagents and will then discuss some reactions and catalyses and the influence of the concave shielding on rate and especially selectivity.

7.2 Classes of Concave Reagents

The first[6] examples for concave reagents were bimacrocyclic concave pyridines with the general structure shown in Figure 7.8. A 2,6-disubstitution of the pyridine was chosen in order to ensure the *endo*-position of the pyridine nitrogen atom's lone pair. Later, the framework of the bimacrocycle has been changed,[15] and in principle, several approaches to concave pyridines, and to concave reagents in general, can be envisioned.

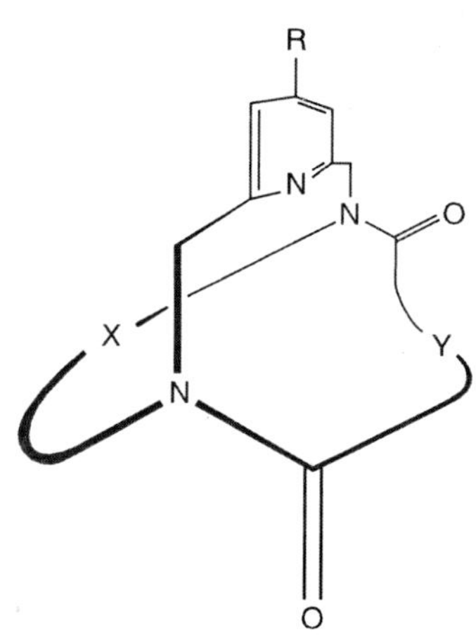

Figure 7.8 *The first class of concave reagents[6] were bimacrocyclic 2,6-disubstituted pyridines with amide bridgeheads building up the bimacrocycle. The bridges X and Y were polymethylene but also polyethylene glycol chains, the basicity of the pyridine could be tuned by respective substituents R in 4-position[16]*

7.2.1 Strategies for the Construction of Bimacrocycles

For a concave reagent with a lamp-like geometry, at least a bimacrocycle has to be constructed. Let us discuss a pyridine-containing bimacrocycle (see Figure 7.9). The synthetic task can be accomplished by two subsequent macrocyclizations with a monomacrocyclic intermediate (see Figure 7.9, left and right), or in a double macrocyclization which gives the desired bimacrocycle from a non-macrocyclic starting material in one reaction step (see Figure 7.9, middle). The two step reaction can be subdivided into two approaches. Either the pyridine is part of the starting material from the very beginning (Figure 7.9, right), or a pyridine-free macrocycle is spanned by a pyridine-containing bridge (Figure 7.9, left).

Although at first glance, the one-pot reaction may seem to be advantageous, all three methods have their pros and cons. In a double macrocyclization, two reactive ends are needed for the closure of each bridge, i.e. a total of four reactive ends exist in the precursor. This implies a selectivity problem in the macrocyclization, because not only the desired bimacrocycle may be formed, also a *bis*-monomacrocycle whose structure resembles that of an earmuff is conceivable (see Figure 7.10). For several different classes of concave reagents, this problem has been investigated, but fortunately, only in the case of very long chains, the formation of an 'earmuff' is a problem.[17] Therefore, many concave reagents may efficiently be built up using this strategy.

If the two-step syntheses for bimacrocyclic concave reagents are compared with one another, the use of a pyridine-free macrocycle as the starting material seems to be attractive because numerous macrocycles are easily available, some of which can even be purchased. Consequently, for instance calixarenes[18] have been spanned with pyridine (and other) bridges.[19] Although a very powerful ligand for a copper(I) catalyst has been found

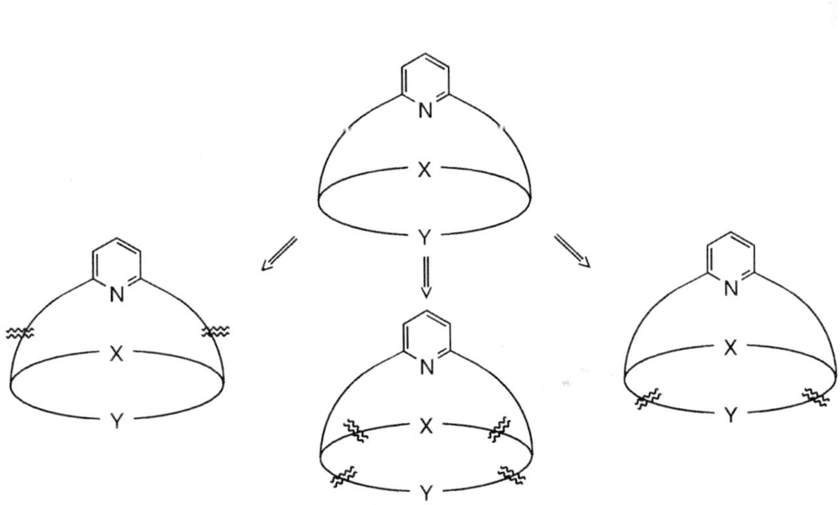

Figure 7.9 *A bimacrocyclic pyridine can be constructed following three different approaches: (a) a non-macrocyclic precursor is doubly cyclized to give the bimacrocycle, shown in the middle, or (b) it can be built up stepwise. The left retrosynthesis represents the bridging of a X and Y containing macrocycle by a pyridine bridge while the right retro-synthesis starts from a pyridine-containing macrocycle*

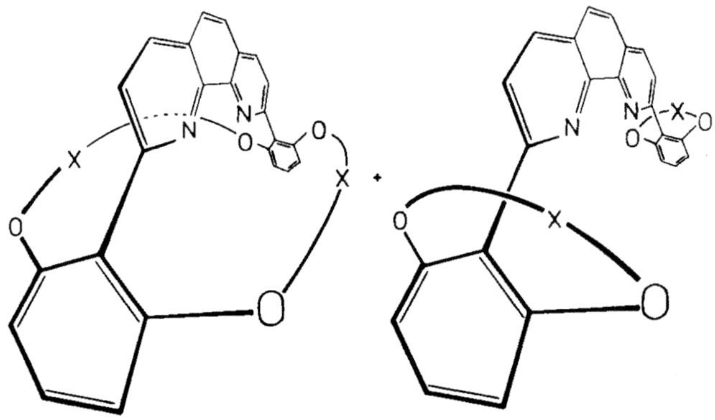

Figure 7.10 The first concave 1,10-phenanthrolines with aryl bridgeheads were synthesized from a tetraphenolic precursor applying a quadruple Williamson ether synthesis (X = polymethylene or polyethylene glycol). The alternative product, a bis-metacyclophane (sometimes called 'earmuff', shown on the right), was not formed. The bimacrocyclic structure of the concave 1,10-phenanthroline was confirmed by X-ray analyses[17]

using this approach (see below),[20] the strategy is limited because the size of the starting macrocycle is preassigned in cases such as calixarenes or cyclodextrins. In the case of the bridged calixarenes, only the calix[6]arene derivative could be synthesized in good yield.[19] The respective calix[4] derivative did not form at all,[19] the calix[5] one was formed in low yield,[21] and with calix[8]arene, regioisomers were formed.[21] Furthermore, a fine-tuning of the size of the macrocycle is not possible.

Therefore, also the third approach is useful. By synthesizing macrocyclic diamines first which were then bridged by a pyridine (or other) bridge, a large variety of bimacrocycles were formed and minute changes in the length and composition of the bridges were possible (see Figure 7.8).[6,16]

Thus a pool of concave reagents has been made available using one or another of these three general strategies. Some of these molecules have been attached to dendrimers or polymers to allow easier recovery.[22,23]

7.2.2 Ring-closure

The formation of macrocycles is well established but not as straightforward as the synthesis of a five- or six-membered ring.[24] The latter ones possess almost no strain and are therefore formed much more easily than strained three- or four-membered rings or medium size rings. However, macrocycles which contain more than twelve atoms usually are unstrained, too, still their synthesis is more difficult. The major competition to macrocyclization is oligo- and polymerization, and thus a number of methods had to be elaborated to produce macrocycles in good yields. Two classes of syntheses must be distinguished: syntheses under thermodynamic control, and kinetically controlled ones. In a kinetically controlled macrocyclization, the rate of polymerization and cyclization differ in the reaction order. While (macro)cyclization is a unimolecular, first-order reac-

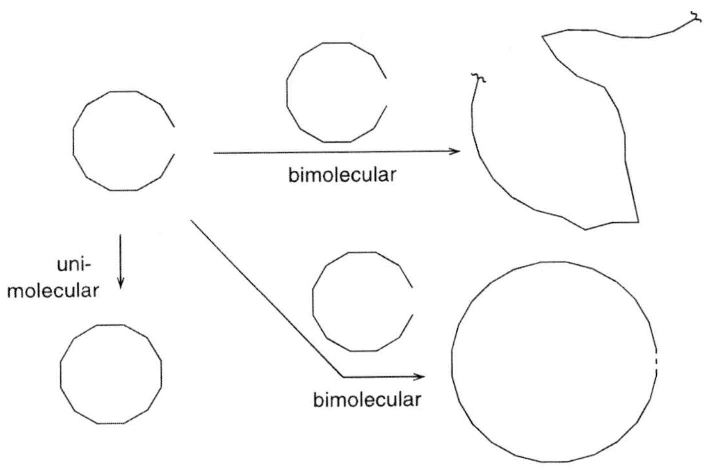

Figure 7.11 *The cyclization of a non-macrocyclic precursor is a unimolecular process while the competing processes, the elongation to bi-, oligo- or polymeric material is bimolecular. Please note that the final macrocyclization to an oligomeric macrocycle (indicated by a dashed line in the drawing) is again unimolecular*

tion, oligo- or polymerization take place between *two* molecules and thus these reactions are second-order in concentration of the precursors (see Figure 7.11). The latter reactions are therefore slowed down proportional to the square of the dilution while the macrocyclization of a single molecule is dilution independent. The respective rate only decreases linearly with dilution. This has been exploited by the high-dilution principles,[25] and several modifications thereof have been worked out [e.g. one-component, two-component dilution, syringe pump technique]. The strong point of the kinetic control is the fact that a macrocycle, once it is formed, remains a macrocycle.

This is not the case in a thermodynamically controlled reaction. Thermodynamic control means that all compounds in the reaction mixture are in equilibrium, they constantly are interconverted into one another. The relative ratio of the products depends on the relative thermodynamic stability of each compound. If the desired macrocycle is the most stable compound, fine, but how can the concentration of a macrocycle be increased when it is not the global most stable product? It has to be forced to be a more stable compound. This can be done by adding an additional component which interacts with the desired macrocycle in such a way that the resulting complex is the most stable compound.

Such an approach uses the template effect, and by carefully adjusting the template to the macrocycle, the number of stabilizing interactions between the template and the desired product can be maximized while the number of interactions with starting materials or oligo- and polymers is smaller due to a lesser amount of complementarity (see Figures 7.12 and 7.13).

The field of dynamic combinatorial chemistry[26] takes advantage of the template effect,[27] and numerous macrocycles have been stabilized by a proper template.[28,29] Also for the construction of macrocycles for concave reagents, the template effect is very valuable.

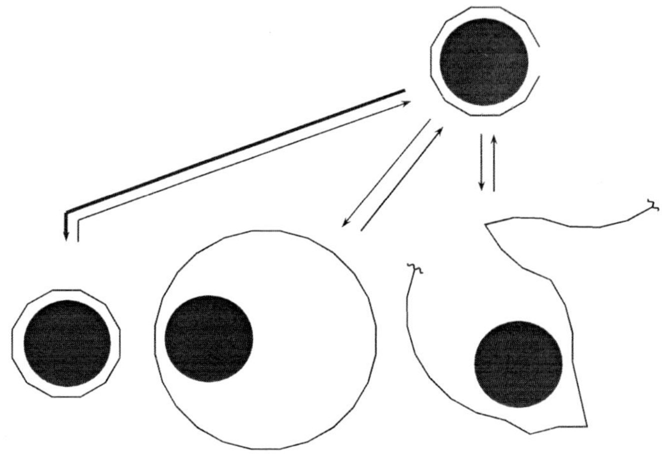

Figure 7.12 *In a thermodynamically controlled macrocyclization, all products are in equilibrium: desired macrocycle (left), oligomers (middle) and polymers (right). By adding a template which is complementary to the desired macrocycle, the interactions between these two particles are maximized and the equilibrium will shift towards the complex formed from the template and the macrocycle*

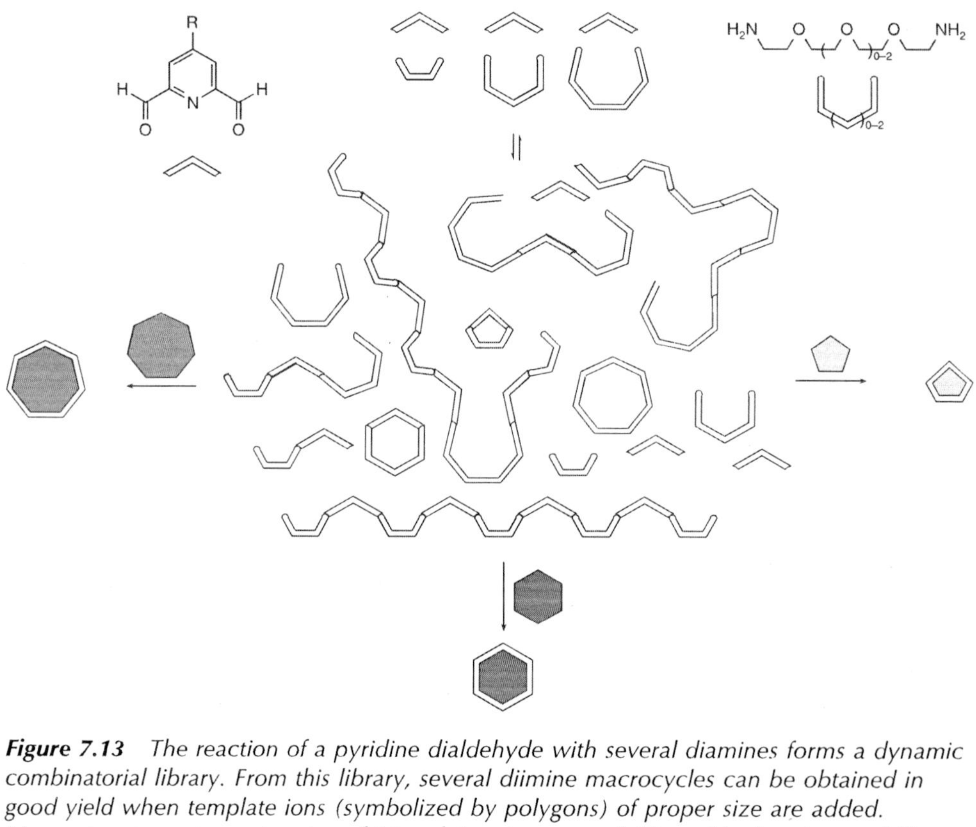

Figure 7.13 *The reaction of a pyridine dialdehyde with several diamines forms a dynamic combinatorial library. From this library, several diimine macrocycles can be obtained in good yield when template ions (symbolized by polygons) of proper size are added. Magnesium ions give a ring size of 15, calcium ions one of 18, and barium ions stabilize the 21-membered ring*[29]

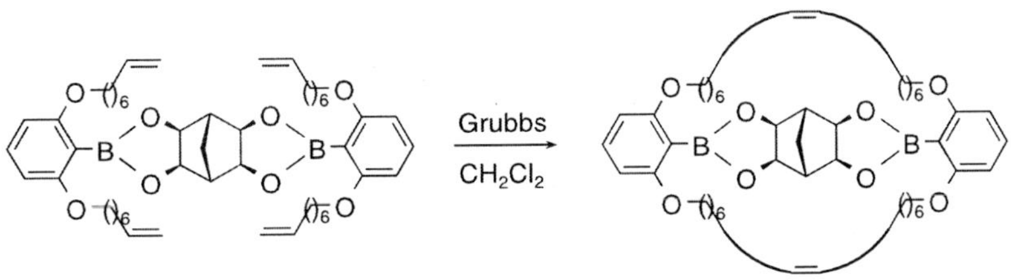

Figure 7.14 The boronic ester formation can be used to prearrange appropriate building blocks. By ring closing metathesis, the non-macrocyclic precursor is turned into a bimacrocycle. When the norbornane template is removed (by ester hydrolysis or carbon-boron bond cleavage), endo-functionalized macrocycles are produced[31]

Already in the beginning of the history of concave reagents, the template effect of metal cations was utilized,[6] but also organic molecules can be applied. Sanders[30] for instance used the interaction of pyridines with zinc porphyrins in the construction of macrocyclic oligomeric porphyrins, and recently, the interaction between polyols and boronic acids has been used in order to synthesize new *endo*-functionalized macrocycles (Figure 7.14).[31]

Which reactions are useful for a ring-closing reaction? In principle, all reactions can be used. All irreversible reactions may be used in kinetically controlled cyclizations. However, it is advisable to use reactions which possess a decent reaction rate because the dilution and the potentially slow addition of one or more starting material to the reaction mixture lead to long reaction times anyway. Prominent reactions used for kinetically controlled macrocyclizations are nucleophilic substitutions as for instance found in Williamson ether syntheses (for an example see the product shown in Figure 7.10) or the formation of amides from acid chlorides and amines (see Figure 7.8).

For a dynamic combinatorial templated reaction, a reversible reaction is needed. Preferentially, a reaction is chosen whose reversibility can be switched on and off at wish, for instance by temperature or pH change or by the addition (and later removal or inhibition) of a catalyst. Also subsequent fast reactions can scavenge the macrocyclic reaction products. Prominent examples[26] for reversible reactions in macrocycle syntheses are disulfide exchange, ester exchange, both pH controlled, ruthenium catalysed ring-closing metatheses or the formation of hydrazones and imines, the latter one usually being scavenged by reduction of the imines to respective amines.

7.2.3 Concave Acids, Bases and Ligands

Bimacrocycles synthesized by one of the strategies described above may contain functional groups of different nature. First, basic centres were introduced,[6,15–17,19,32,33] next, complementary concave acids[34,35] were realized, and third, the lone pairs of the bases were utilized to coordinate a catalytically active metal ion.[17,36,37] But in principle, any kind of reactive group can be placed into the concave environment.[38] Just recently, the first concave N-heterocyclic carbenes have been realized.[39,40] Figures 7.15–7.21 summarize several classes of concave reagents and catalysts. As can be seen easily from the final structures, the different strategies yield products of different complexity. When a calixarene is spanned by a bridged or a tetrafunctionalized non-macrocyclic precursor is doubly cyclized, products with higher symmetry are formed than in the case of a stepwise

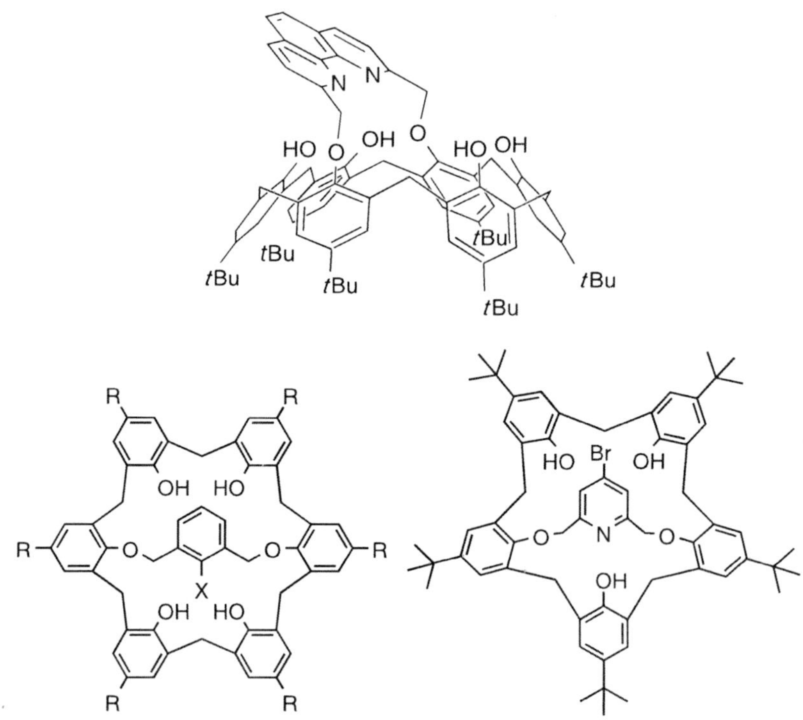

Figure 7.15 *Three selected concave reagents obtained by spanning calixarenes with a bridge which contains a functional group.*[8,19,21,33,42,43] *For the 1,10-phenanthroline bridged calix[6]arene, the predominant conformation is shown in which the bridge is tilted towards one side of the calixarene rim.*[42] *For its importance, see the catalysis chapter below. Calix[6]arene has been A,D-bridged with numerous bridges, e.g. with an aniline*[33] *(R = tBu, X = NH_2) or with a sulfenic acid*[7,8] *(R = H, X = SOH). Also pyridines*[19] *may be used as bridges as shown here with a calix[5]arene*[21] *example*

synthesis which starts from the building block containing the reagent. In addition, isomers may be formed such as conformers (in the case of carboxamides[6,16]), *cis/trans*-mixtures (when ring-closing metathesis[31,41] is used) or even regioisomers (when a larger calixarene such as calix[8]arene[21] is bridged).

As mentioned above, the objective of the concave shielding of a functional group in a concave reagent is selectivity: chemo-, regio-, diastereo- and especially enantioselectivity have to be controlled. To influence the latter, chiral concave reagents are necessary. The usual way to turn a non-chiral compound into a chiral one is to add a stereogenic center, typically an asymmetrically substituted carbon atom. If applied to concave molecules there are two possible regions to attach it to. If attached to the inside of the cavity, the void becomes chiral but is filled to a large part by this chiral substituent. But if attached to the outside, the concave inside remains mostly unaltered. Therefore axial chirality has been envisioned because it desymmetrizes the cavity without filling it up. Biaryls are configurationally stable if they are unsymmetrically substituted with

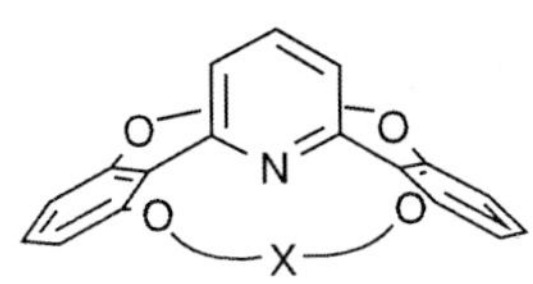

Figure 7.16 *Concave pyridines can be obtained in one-step or two-step macrocyclizations starting from appropriately 2,6-disubstituted pyridines. This figure displays a concave pyridine obtained by a one-step macrocyclization [ring-closing metathesis, followed by hydrogenation: X = $(CH_2)_{10}$]. For the products of the two-step macrocyclizations see Figure 7.8 and Figure 7.15*

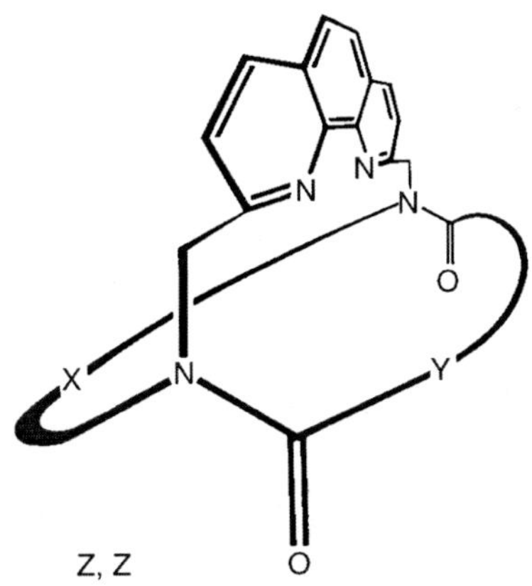

Figure 7.17 *Concave 1,10-phenanthrolines can also be obtained in one-step or two-step macrocyclizations starting from appropriately 2,9-disubstituted 1,10-phenanthrolines. Here the ZZ conformer (X, Y = polyethylene glycol or Y = polymethylene) of a two step macrocyclization starting from a 1,10-phenanthroline containing monomacrocycle is shown. For a one-step macrocyclization see Figure 7.10, and for a bridged calixarene see Figure 7.15*

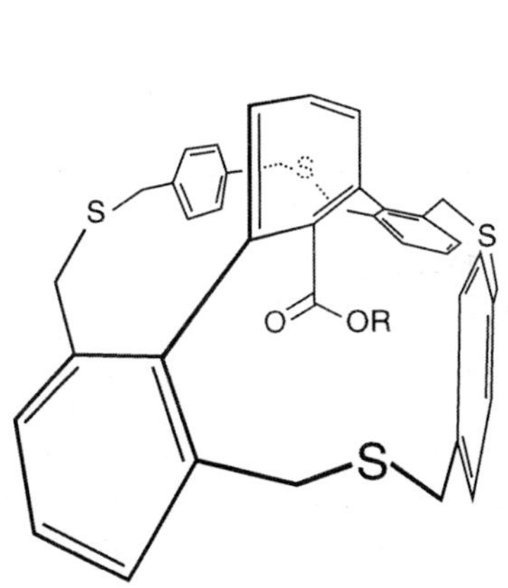

Figure 7.18 *Concave ester and acid obtained from an appropriately tetrafunctionalized* m-*terphenyl (R = Me, H)*[34,35]

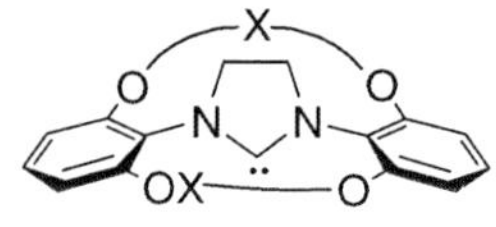

Figure 7.19 *Concave N-heterocyclic carbene (X = polymethylene)*[39]

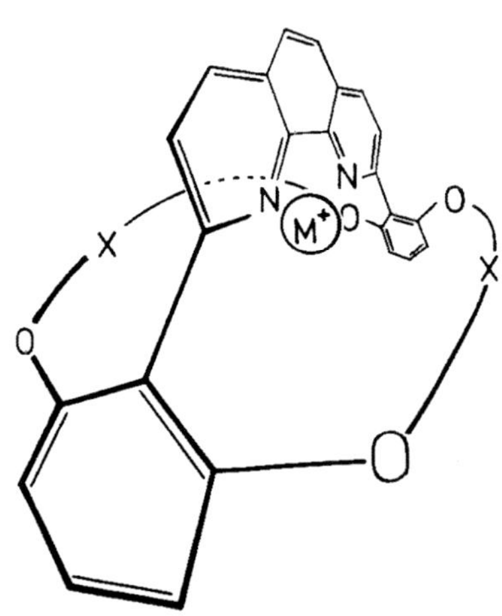

Figure 7.20 *Concave 1,10-phenanthrolines (X = polymethylene or polyether) are bidentate ligands and form stable complexes with transition metal ions*[17,36]

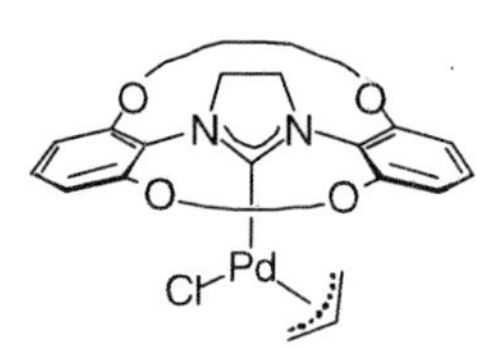

Figure 7.21 *The allyl palladium complex of a concave N-heterocyclic carbene is one example for concave NHC-metal complexes (the bimacrocycle forming decamethylene chains are abbreviated as arc of a circle).*

appropriate substituents in their *ortho*-positions, and additional stability will be introduced by the bimacrocyclic structure. In many of the concave reagents, biaryl substructures exist (see Figure 7.16, Figures 7.18–7.21), usually two. If both were axially chiral, diastereomers would be formed, one of which is chiral but the other one a *meso*-compound.[44,45] Therefore, only one unsymmetrical bridgehead has been used (Figure 7.22).[40,44] The resulting bimacrocyclic concave imidazolinium ion was then reacted with a chiral counter ion (TRISPHAT or BINPHAT[46]) and resulting diastereomeric ion pairs could be observed by NMR.[40] Unfortunately, the barrier of rotation is yet too low to separate the enantiomers.

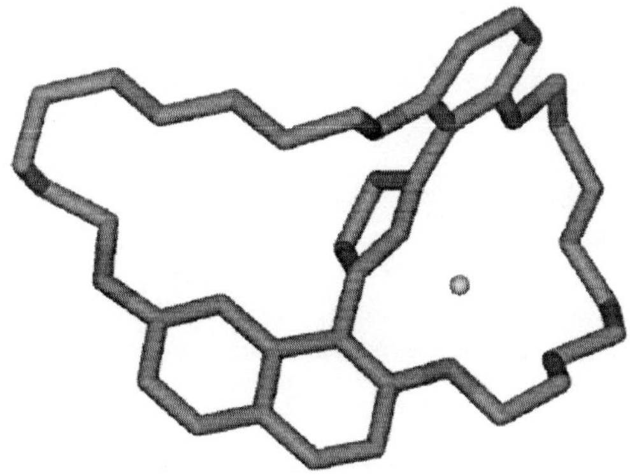

Figure 7.22 X-ray structure of a chiral version (only one enantiomer shown) of a concave imidazolinium ion,[39] precursor to a chiral concave NHC. Hydrogen atoms are omitted for clarity, the single atom represents the chloride counter ion

7.3 Reactions and Catalyses

As presented in the introduction, concave reagents can be used as reagents or as catalysts. As reagents they will be 'used up' with the need of recycling, as catalysts they can carry out numerous catalytic cycles and can be recovered from the mixture.

7.3.1 Reagents

Due to the simple recycling, acid-base reactions, i.e. protonations or deprotonations, have been carried out successfully with concave reagents. There is no need to differentiate between acids and bases because for instance a protonated pyridine can transfer a proton in the same way as for example a carboxylic acid can. Much more important is the acidity of the corresponding acids. Respective pK_a values for the deprotonation have been determined in water (per definition) but many reactions are carried out in various organic solvents (which is a strong point of concave reagents, see introduction).[47] The influence of the solvent on acidity very often is enormous but at the same time this difference is frequently underestimated. If for instance the solvent water is exchanged by ethanol, the pK_a-analogous values of pyridines in ethanol hardly differ from the pK_a values in water.[16] But for carboxylic acids, a drastic drop of several orders of magnitude can be observed when exchanging water by ethanol.[35] Consequently, less polar or even non-protic solvents may have an even more drastic effect on the relative acidities. In the pyridinium-carboxylic acid comparison,[16,35] the discrepancy is extreme because for a pyridinium ion a charge is lost upon deprotonation while a carboxylic acid develops a charge when the anion is formed. And the less polar a solvent is, the more is the equilibrium shifted towards the uncharged particles, the base in the case of pyridine, the acid for carboxylic acids.

In protonations and deprotonations there are two fundamentally different pathways by which the proton can be transferred: the general and the specific protonation.[48,49] The terms general and specific have been coined for acid-catalysis[48] but they can be used for proton-transfers in the same sense.[49] *General* means that generally (i.e. always) the

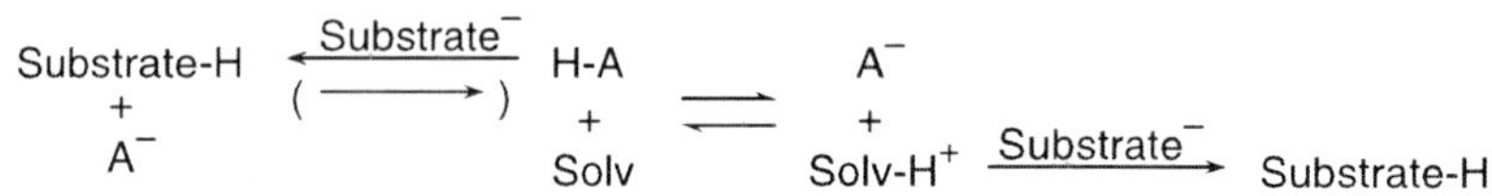

Figure 7.23 *The proton delivering agent for the protonation of a substrate can either be the acid H-A (specific protonation) or the protonated solvent Solv-H⁺ (general protonation). Only in the case of a specific protonation can the acid influence selectivity in the protonation step. But even in this case, a reversibility of the protonation, due to the basicity of the conjugate base of the acid, can ruin the controlled specific protonation. In this latter case, the thermodynamically most stable protonation products will be observed*

proton is transferred by the same particle, explicitly the protonated solvent. In contrast, a *specific* protonation is carried out by a specific proton donor, for instance a pyridinium ion or a carboxylic acid. A change in the proton source will therefore only have an influence on the outcome of a protonation if the general protonation pathway is avoided. At first glance, the solution to this problem is the use of a hardly basic solvent and an acid of moderate acidity. However, there is a second problem in selective protonations. If the proton source shall have an influence on the selectivity, the protonation must be kinetically controlled. A thermodynamic control would produce the more stable protonation product and the protonating acid will finally have no influence on the product distribution. Unfortunately, the chance for a thermodynamically controlled protonation, an equilibrium, rises with increasing basicity of the corresponding base, and therefore a fine balance must be found between acidities and basicities of all involved compounds. This may be the reason why only few examples exist for selective protonations, including enantioselective ones.[50] In the case of concave reagents, concave proton donors have been used to carry out the Nef reaction under mild conditions[51] and for some stereoselective protonations.[52]

7.3.2 Catalysts

Besides the use as a reagent, concave reagents can be used in a catalytic fashion. From the vast possibility of potential reactions, so far nucleophilic catalyses and metal catalyses have been investigated with several types of concave catalysts.

7.3.2.1 Organocatalysis

For nucleophilic catalyses, concave pyridines, concave 1,10-phenanthrolines and concave NHC can be used. However, the substitution pattern which ensures the *endo*-placement of the nucleophilic lone pair does not allow all kinds of nucleophilic catalyses. Pyridine and especially its more basic sibling DMAP [4-(dimethylamino)pyridine] are excellent catalysts for acyl transfers. However the pyridine's catalytic power drops enormously when substituents are added to the 2- and 6-position.[53] A reaction which demands less space is the formation of a hydrogen bond to the lone pair of the nitrogen, and thus catalyses via hydrogen-bond formation are easily possible with concave pyridines. Specifically, the activation of an alcohol by the formation of a hydrogen bond to the lone pair of a pyridine nitrogen atom in a concave environment has proven to be very valuable for selective acylations of these OH groups. An activated alcohol adds much faster to a ketene than a free alcohol. Thus the use of pyridine, or related bases, can speed up the addition of alcohols to for instance diphenylketene. This ketene has been chosen because it is reactive enough in this addition reaction but still stable enough to be handled in normal laboratory practice.[54] This reaction allows to distinguish numerous alcohols,[17b,23,54–56] and in the case of a glucose derivative with two very similar OH groups, one of them may be acylated in high yield exclusively. The concave shielding of the pyridine allows to discriminate between two secondary axial OH-groups (Figure 7.24).[56]

More recently, concave NHC have been investigated as nucleophilic catalysts.[57] The catalytic potential of NHC is well known since Breslow[58] elucidated the mechanism of thiamine dependent enzymes.[59] With aldehydes, NHC may catalyse the benzoin condensation in the same way, a cyanide ion can. But when additional substrates are added, the nature of the NHC becomes important.[60] Thus, in a mixture of aldehydes and enals, two competing pathways may be catalysed and either so called Stetter products (for instance 1,4-diones) or γ-lactones can be formed (Figure 7.25). Using one particular concave NHC,

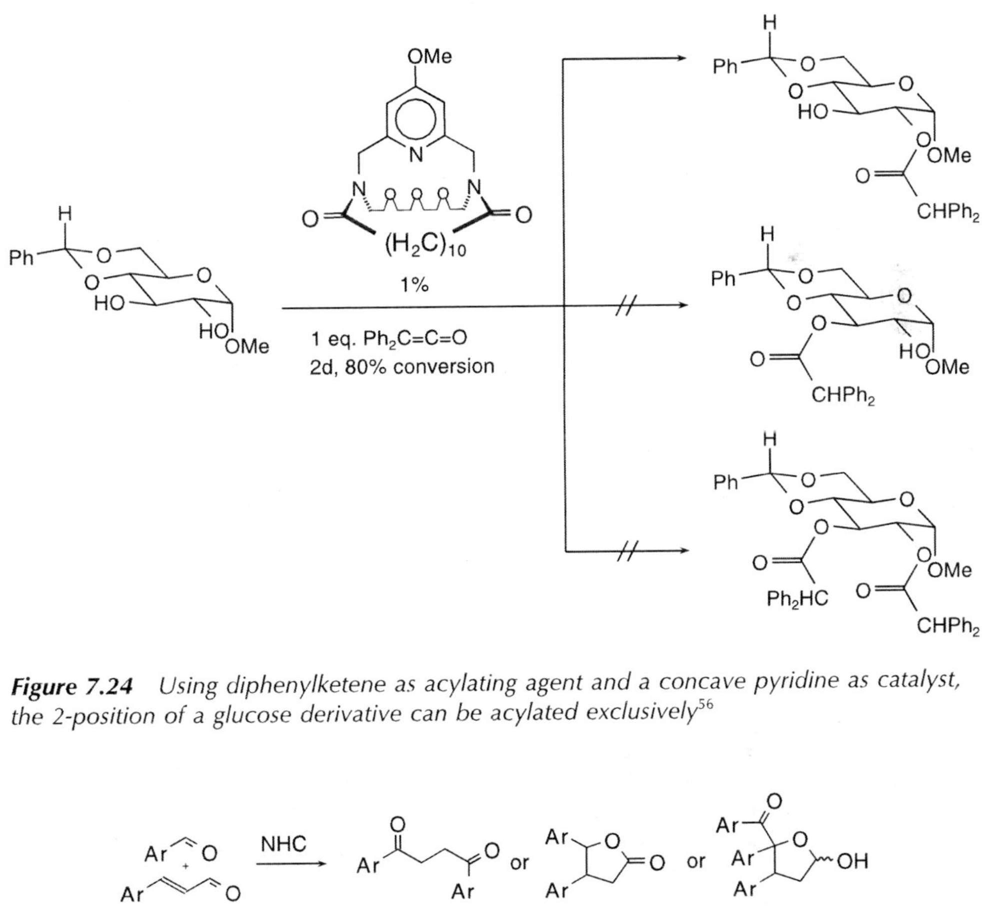

Figure 7.24 *Using diphenylketene as acylating agent and a concave pyridine as catalyst, the 2-position of a glucose derivative can be acylated exclusively*[56]

Figure 7.25 *N-Heterocyclic carbenes catalyze reactions between aryl aldehydes and enals. Depending on the nature of the catalysts,*[57,60] *1,4-diones (Stetter products), γ-lactones or even a 2+1 adduct (a hemiacetal) can be formed*

even a fourth product, a hemiacetal from a 2+1 addition was found[57] but its formation is not yet mechanistically completely understood.

7.3.2.2 Metal Catalysis

Probably the largest catalytic potential lies in the metal complexes of concave reagents. Two sub-structures are well established as metal ligands: 1,10-phenanthroline and NHC. Numerous transition metal complexes have been synthesized[36,37,61,62] and some of them have been investigated in metal catalysed reactions such as Lewis-acid catalysed Diels–Alder reactions,[36] palladium-catalysed allylations,[63] and copper(I)-catalysed cyclopropanations.[20,41,64] In the latter reaction, the full potential of concave reagents and the importance of the fine structure of the concave shielding has become obvious. With two different types of concave ligands, two complementary stereoselectivities have been found.

When copper(I) catalysed, indene for instance reacts with diazoacetates to give either *exo*- or *endo*-cyclopropanes (see Figure 7.26).[20] The distinct difference in the selectivities achieved by using different concave 1,10-phenanthrolines lies in the different geometry of the ligands.

In 2,9-diaryl-1,10-phenanthrolines (see Figures 7.10 and 7.20), bridgeheads and 1,10-phenanthroline unit are connected directly, leading to a rather stiff entity. The active centre, the copper ion bound to the nitrogen atoms of the 1,10-phenanthroline can only be reached from 'below' through the bimacrocycle. In the cyclopropanation, the active species is a copper-carbenoid. When the alkene approaches it, the substituents of the alkene and the carboxylate of the carbenoid try to avoid sterical compression within the cavity and take up maximal distance to one another. Thus, the *exo*-product is formed with high selectivity.

In contrast in the bridged calixarene (see Figure 7.15), the methyleneoxy units between the 1,10-phenanthroline and the calixarene act as flexible joints and this leads to a different geometry in which the 1,10-phenanthroline bridge is tilted towards one chain of the calixarene rim. The calixarene macrocycle is too small to accommodate the carboxylate or to let an alkene approach the carbenoid. This results in an orientation in which both, the alkene and the carboxylate residue of the carbene moiety, have to be squeezed in the gap between the 1,10-phenanthroline and part of calixarene rim. By this sterical shielding of the calixarene, the carboxylate and the residues of the alkene are oriented next to each other, and thus the predominant formation of the *endo*-product is observed.[64b]

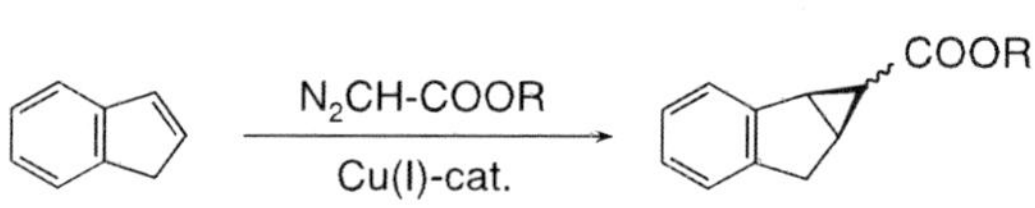

Figure 7.26 *When concave 1,10-phenanthrolines are used as ligands for copper(I) ions in the cyclopropanation of indene by diazoacetates, the* exo/endo-*selectivities can be controlled by the choice of the ligand. The concave 1,10-phenanthroline from Figure 7.10 [X = $(CH_2)_{10}$] in combination with* tert-butyl *diazoacetate is highly* exo-*selective* (exo/endo = *140 : 1) while the 1,10-phenanthroline bridged calix[6]arene from Figure 7.15 is* endo-*selective (best results for $N_2CHCOOMe$,* exo/endo = *14 : 86)*[20]

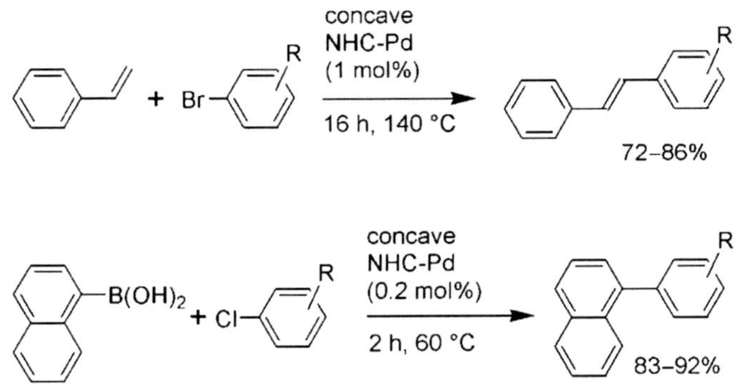

Figure 7.27 *Palladium complexes of concave NHC (see Figure 7.19, Figure 7.21) are good catalysts in Heck and Suzuki couplings*[62]

The last decade has shown the increasing importance of N-heterocyclic carbenes as ligands for transition metals. In many complexes, the stabilization of special oxidation states of a transition metal is even better by a NHC than by phosphines,[65] the second generation of Grubbs' catalyst being a prominent example.[66] Also with concave NHC, many complexes can easily be prepared,[37,62] and some of them have been applied to catalyses.[62] Copper complexes may be used for cyclopropanations, too, but yet with smaller selectivities than 1,10-phenanthrolines exhibit. With palladium however (see Figure 7.21), cross-coupling reactions such as Heck or Suzuki couplings can be carried out with good efficiency (Figure 7.27).[62]

7.4 Summary and Outlook

The features of enzymes – active centre and concave shielding – have been combined in (usually) bimacrocyclic synthetic molecules. The resulting concave reagents can be used as reagents or preferentially as catalysts. In several reactions as for instance acylations or cyclopropanations, the concave reagents have demonstrated their selectivity-determining power. In current research, first chiral concave reagents have been synthesized and it is the task for the future to optimize the structures and to establish enantioselective reactions and catalyses. The similarity between concave reagents and enzymes, besides all advantages, produces one drawback. Indeed, by optimizing the concave geometry, selectivities can be tailored, but as with enzymes, a perfect reagent-substrate match leads to the problem that the optimized catalyst is only perfect for the substrate it was optimized for. In enzyme chemistry there exists an exaggerated saying: 'one enzyme, one substrate'. In this respect, concave reagents cannot surpass enzymes. The higher the selectivity, the smaller is the substrate tolerance. But concave reagents can be made from any starting material. The backbone of the bimacrocyclic structure is not restricted to amino acids. And thus concave reagents can be made more robust, for any solvent, for higher temperatures. In addition, also the functional groups are not restricted to those found in enzymes. The guideline for the future is therefore: try to

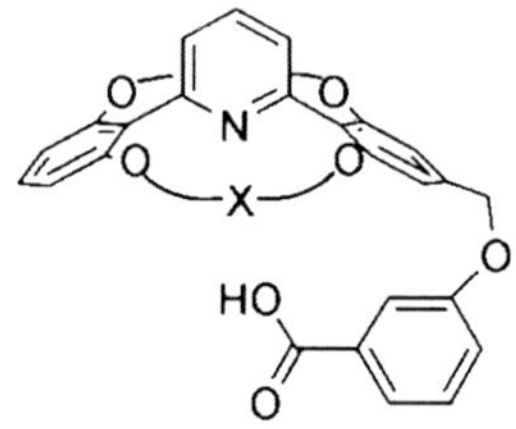

Figure 7.28 *To allow bifunctional acid-base catalysis with concave reagents, an acid function has been attached to a concave pyridine [X = $(CH_2)_{10}$]*[41e]

copy the excellency of enzymes as much as possible but try to use optimized backbones and any functional group.

So far, most approaches to enzyme-like activity have used just one of the functional groups which are present in enzymes. However, many enzymes only operate by a cooperation of functional groups (see for instance the catalytic triade in peptidases). There, the enzyme's functional groups perform a multifunctional catalysis. Therefore in (organo) catalysis, bifunctional catalysis has been developed, too.[67] In the field of concave reagents, first bifunctional catalysts have been constructed (Figure 7.28),[41e] and future will tell how capable they are to catalyse reactions with their acidic *and* basic functionalities.

References

1. a) J.-M. Lehn, *Supramolecular Chemistry: Concepts and Perspectives*, VCH, Weinheim, New York, Basel, Cambridge, Tokyo, 1995; b) G. R. Desiraju, Chemistry beyond the molecule, *Nature*, 2001, **412**, 397–400.
2. P. W. N. M. van Leeuwen (Ed.), *Supramolecular Catalysis*, Wiley-VCH, Weinheim, 2008.
3. a) Suggestion of a Diels–Alderase: E. M. Stocking, R. M. Williams, Chemie und Biologie natürlicher Diels–Alder-Reaktionen, *Angew. Chem.*, 2003, **115**, 3186–3223; *Angew. Chem. Int. Ed.*, 2003, **42**, 3078–3115; b) Put into question: J. M. Serafimov, D. Gillingham, S. Kuster, D. Hilvert, The putative Diels–Alderase macrophomate synthase is an efficient aldolase, *J. Am. Chem. Soc.*, 2008, **130**, 7798–7799.
4. E. Fischer, Einfluss der Configuration auf die Wirkung der Enzyme, *Ber. Dtsch. Chem. Ges.*, 1894, **27**, 2985–2993.
5. 100 years of lock and key principle: a) A. Eschenmoser, Hundert Jahre Schlüssel-Schloß-Prinzip, *Angew. Chem.*, 1994, **106**, 2455; *Angew. Chem. Int. Ed. Engl.*, 1994, **33**, 2363; b) F. W. Lichtenthaler, Hundert Jahre Schlüssel-Schloß-Prinzip: Was führte Emil Fischer zu dieser Analogie? *Angew. Chem.*, 1994, **106**, 2456–2467; *Angew. Chem. Int. Ed. Engl.*, 1994, **33**, 2364–2374; c) D. E. Koshland Jr., Das Schüssel-Schloß-Prinzip und die Induced-fit-Theorie, *Angew. Chem.*, 1994, **106**, 2468–2472; *Angew. Chem. Int. Ed. Engl.*, 1994, **33**, 2408–2412.
6. U. Lüning, Concave reagents: synthesis of macrobicyclic pyridines, *Liebigs Ann. Chem.*, 1987, 949–955.
7. K. Goto, N. Tokitoh, R. Okazaki, Synthese einer stabilen Arensulfensäure mit einem schüsselförmigen makrobicyclischen Cyclophangerüst, *Angew. Chem.*, 1995, **107**, 1202–1203; *Angew. Chem. Int. Ed. Engl.*, 1995, **34**, 1124–1126.
8. a) K. Goto, M. Holler, R. Okazaki, Synthesis, structure, and reactions of a sulfenic acid bearing a novel bowl-type substituent: the first synthesis of a stable sulfenic acid by direct oxidation of a thiol, *J. Am. Chem. Soc.*, 1997, **119**, 1460–1461; b) K. Goto, R. Okazaki, Molecular bowls

and capsules with an endohedral functionality. The stabilization of highly reactive species in their inner phase, *Liebigs Annalen/Recueil*, 1997, 2393–2407.

9. F. P. Schmidtchen, Host–guest interactions. The binding mode of 6–nitrobenzisoxazole-3–carboxylate to quaternary ammonium macrocycles, *J. Chem. Soc., Perkin Trans. 2*, 1986, 135–141.
10. a) D. A. Stauffer, A. David, R. E. Barrans Jr., D. A. Dougherty, Biomimetische Katalyse einer S_N2–Reaktion als Folge einer neuartigen Übergangszustandsstabilisierung, *Angew. Chem.*, 1990, **102**, 953–956; *Angew. Chem. Int. Ed.*, 1990, **29**, 915–918; b) A. McCurdy, L. Jimenez, D. A. Stauffer, D. A. Dougherty, Biomimetic catalysis of S_N2 reactions through cation–π interactions. The role of polarizability in catalysis, *J. Am. Chem. Soc.*, 1992, **114**, 10314–10321.
11. D. Fiedler, H. van Halbeek, R. G. Bergman, K. N. Raymond, Supramolecular catalysis of unimolecular rearrangements: substrate scope and mechanistic insights, *J. Am. Chem. Soc.*, 2006, **128**, 10240–10252.
12. M. Yoshizawa, M. Tamura, M. Fujita, Diels–Alder in aqueous molecular hosts: unusual regioselectivity and efficient catalysis, *Science*, 2006, **312**, 251–254.
13. S. R. Seidel, P. J. Stang, High symmetry coordination cages via self-assembly, *Acc. Chem. Res.*, 2002, **35**, 972–983.
14. C. J. Walker, H. L. Anderson, J. K. M. Sanders, *exo*-Selective acceleration of an intermolecular Diels–Alder reaction by a trimeric porphyrin host, *Chem. Commun.*, 1993, 458–460.
15. U. Lüning, R. Baumstark, W. Schyja, Concave reagents, 14: concave pyridines with a 2,6–diarylpyridine core, *Tetrahedron Lett.*, 1993, **34**, 5063–5066.
16. a) U. Lüning, R. Baumstark, K. Peters, H. G. v. Schnering, Concave reagents, 3: synthesis, basicity and conformation of new concave pyridines, *Liebigs Ann. Chem.*, 1990, 129–143; b) U. Lüning, R. Baumstark, M. Müller, Concave reagents, 8: concave pyridines and 1,10–phenanthrolines with sulfonamide bridgeheads. Increased basicity by 4–diethylamino substitution of the pyridine unit, *Liebigs Ann. Chem.*, 1991, 987–998; c) O. Storm, U. Lüning, Concave reagents, 38: concave pyridines with extended π-systems, *Eur. J. Org. Chem.*, 2002, 3680–3685; d) O. Storm, U. Lüning, Concave reagents, 39: basicity and solvatochromism of concave pyridines with extended π-systems in protic and nonprotic solvents, *Eur. J. Org. Chem.*, 2003, 3109–3116.
17. The bimacrocycle/bismetacyclophane problem is discussed in: a) U. Lüning, M. Müller, Concave reagents, 5: bimacrocyclic 1,10–phenanthroline cyclophanes, *Chem. Ber.*, 1990, **123**, 643–645. Since then, numerous X-ray structures prove the lamp geometry e.g.; b) U. Lüning, M. Müller, M. Gelbert, K. Peters, H. G. von Schnering, M. Keller, Concave reagents, 15: new concave 1,10–phenanthrolines: catalysts for the alcohol addition to ketenes, and ligands in transition metal complexes, *Chem. Ber.*, 1994, **127**, 2297–2306. Only in one case has a biscyclophane been found when extremely long chains were used: c) O. Winkelmann, *diploma thesis*, Kiel, 2005.
18. a) C. D. Gutsche, *Calixarenes*, The Royal Society of Chemistry, Cambridge, 1989 and 1992; b) C. D. Gutsche, *Calixarenes Revisited*, The Royal Society of Chemistry, Cambridge, 1998.
19. a) H. Ross, U. Lüning, Konkave Reagenzien, 16: Konkave Reagenzien auf Calixarenbasis, *Angew. Chem.*, 1995, **107**, 2723–2725; *Angew. Chem. Int. Ed. Engl.*, 1995, **34**, 2555–2557; b) H. Ross, U. Lüning, Concave reagents, 18: synthesis, conformational behaviour, and basicity of new pyridine-bridged calix[6]arenes, *Liebigs Ann.*, 1996, 1367–1373.
20. F. Löffler, M. Hagen, U. Lüning, Concave reagents, 32: *Syn*- and *anti*-selective cyclopropanation of alkenes with diazoacetates catalysed by copper(I) complexes of concave 1,10–phenanthrolines, *Synlett*, 1999, 1826–1828.
21. S. Konrad, C. Näther, U. Lüning, Concave reagents, 44: calix[5]- and calix[8]arenes bridged with heterocycles, *Eur. J. Org. Chem.*, 2005, 2330–2337.
22. a) U. Lüning, M. Gerst, Concave reagents, 12: polymerfixation of concave pyridines, *J. Prakt. Chem./Chem.-Ztg.*, 1992, **334**, 656–660; b) T. Marquardt, U. Lüning, Towards golf ball shaped reagents: dendrimer fixed concave pyridines (Concave reagents 24), *Chem. Commun.*, 1997, 1681–1682; c) U. Lüning, T. Marquardt, Concave reagents, 29: dendrimer fixed concave pyridines, *J. Prakt. Chem.*, 1999, **341**, 222–227.

23. U. Lüning, W. Hacker, Concave reagents, 31: a Merrifield bound concave Pyridine for the selective acylation of polyols, *J. Prakt. Chem.*, 1999, **341**, 662–667.
24. Third Schiemenz 'law' of organic chemistry: 'whenever five- or six-membered rings may be formed they will'. For an illustration of the three 'laws' see: U. Lüning, *Organische Reaktionen. Eine Einführung in Reaktionswege und Mechanismen*, Elsevier – Spektrum, Spektrum Akademischer Verlag, 2007 and 2010.
25. See e.g. F. Vögtle, Verdünnungsprinzip-Reaktionen – Einteilung und Durchführung, *Chem.-Ztg.*, 1972, **96**, 396–403.
26. a) I. Huc, J.-M. Lehn, Virtual combinatorial libraries: dynamic generation of molecular and supramolecular diversity by self-assembly, *Proc. Natl. Acad. Sci. USA*, 1997, **94**, 2106–2110; b) P. A. Brady, J. K. M. Sanders, Thermodynamically-controlled cyclisation and interconversion of oligocholates: metal ion templated 'living' macrolactonisation, *J. Chem. Soc., Perkin Trans. 1*, 1997, 3237–3253; c) S. J. Rowan, S. J. Cantrill, G. R. L. Cousins, J. K. M. Sanders, J. F. Stoddart, Dynamische kovalente Chemie, *Angew. Chem.*, 2002, **114**, 938–993; *Angew. Chem. Int. Ed.*, 2002, **41**, 898–952; d) P. T. Corbett, J. Leclaire, L. Vial, K. R. West, J.-L. Wietor, J. K. M. Sanders, S. Otto, Dynamic combinatorial chemistry, *Chem. Rev.*, 2006, **106**, 3652–3711.
27. Although the term dynamical combinatorial chemistry has only been coined at the end of the last century, the template effect has been used for a much longer time. Some of the best known macrocyclic ligands such as cyclam probably would not have been known so well if their synthesis had not implied the template effect. Probably the first example of a metal ion template for the formation of imino-macrocycles: N. F. Curtis, D. A. House, Structure of some aliphatic Schiff-base complexes of nickel (II) and copper (II), *Chem. and Ind.*, 1961, **42**, 1708–1709.
28. F. Diederich, P. J. Stang (eds.), *Templated Organic Synthesis*, Wiley-VCH, Weinheim, New York, Chichester, Brisbane, Singapore, Toronto, 2000.
29. O. Storm, U. Lüning, How to synthesize macrocycles efficiently using virtual combinatorial libraries, *Chem. Eur. J.*, 2002, **8**, 793–798.
30. H. L. Anderson, J. K. M. Sanders, Synthesis of cyclic porphyrin oligomers with amines as templates, *Angew. Chem.*, 1990, **102**, 1478–1480; *Angew. Chem., Int. Ed. Engl.*, 1990, **29**, 1400–1403.
31. S. Lüthje, C. Bornholt, U. Lüning, Concave reagents, 46: polyols as templates for the synthesis of macrocycles from boronic acid building blocks, *Eur. J. Org. Chem.*, 2006, 909–915.
32. U. Lüning, M. Müller, Concave reagents, 2: macrobicyclic 1,10–phenanthrolines, *Liebigs Ann. Chem.*, 1989, 367–374.
33. H. Ross, U. Lüning, Concave reagents, 19: synthesis, structure and basicity of a calix[6]arene bridged by an aniline, *Tetrahedron*, 1996, **52**, 10879–10882.
34. a) U. Lüning, C. Wangnick, K. Peters, H. G. v. Schnering, Concave reagents, 7: concave benzoic acids, *Chem. Ber.*, 1991, **124**, 397–402; b) U. Lüning, C. Wangnick, Concave reagents, 11: concave benzoates, sterically extremely shielded esters, *Liebigs Ann. Chem.*, 1992, 481–484.
35. U. Lüning, H. Baumgartner, C. Wangnick, Concave reagents, 17: steric effects on the acidity of concave sulfinic acids and concave benzoic acids, *Tetrahedron*, 1996, **52**, 599–604.
36. a) M. Gelbert, U. Lüning, Concave reagents, 25: transition metal complexes of concave 1,10–phenanthrolines as catalysts for [4+2]-cycloadditions. Ligand effects on the *exo/endo*-selectivity, *Supramolecular Chem.*, 2001, **12**, 435–444; b) M. Gelbert, C. Körber, O. Friedrich, F. Fahrenkrug, M. Keller, U. Lüning, Concave reagents, 36: transition metal complexes with concave 1,10–phenanthrolines, *Supramolecular Chem.*, 2002, **14**, 199–210; c) R. Cacciapaglia, S. Di Stefano, F. Fahrenkrug, U. Lüning, L. Mandolini, Concave reagents, 40: the copper(II) complex of a concave reagent as a selective catalyst for ester methanolysis, *J. Phys. Org. Chem.*, 2004, **17**, 350–355.
37. O. Winkelmann, C. Näther, U. Lüning, Concave reagents, 55: bimacrocyclic NHC transition metal complexes, *J. Organomet. Chem.*, 2008, **693**, 923–932.
38. a) U. Lüning, H. Baumgartner, Concave reagents, 13: 2′-Substituted *m*-terphenyls as building blocks. Synthesis of a concave thiol acetate, *Synlett*, **1993**, 571–572; b) M. Abbass, C. Kühl,

C. Manthey, A. Müller, U. Lüning, Concave reagents, 42: new 2′-substituted *m*-terphenyls, *Collect. Czech. Chem. Commun.*, 2004, **69**, 1325–1344.

39. O. Winkelmann, C. Näther, U. Lüning, Concave reagents, 52: concave imidazolinium salts as precursors to concave N-heterocyclic carbenes, *Eur. J. Org. Chem.*, 2007, 981–987.
40. O. Winkelmann, D. Linder, J. Lacour, C. Näther, U. Lüning, Concave reagents, 53: chiral concave imidazolinium salts as precursors to chiral concave N-heterocyclic carbenes, *Eur. J. Org. Chem.*, 2007, 3687–3697.
41. a) U. Lüning, F. Fahrenkrug, M. Hagen, Concave reagents, 35: concave 1,10–phenanthrolines via ring-closing metathesis, *Eur. J. Org. Chem.*, 2001, 2161–2163; b) U. Lüning, F. Fahrenkrug, Concave reagents, 47: concave 1,10–phenanthrolines as tailored ligands for copper(I)-catalysed diastereoselective cyclopropanations, *Eur. J. Org. Chem.*, **2006**, 916–923; c) U. Lüning, M. Abbass, F. Fahrenkrug, Concave reagents, 37: a facile route to aryl-substituted 1,10–phenanthrolines by means of Suzuki coupling reactions between substituted areneboronic acids and halogeno-1,10–phenanthrolines, *Eur. J. Org. Chem.*, 2002, 3294–3303; d) U. Lüning, F. Fahrenkrug, Concave reagents, 41: synthesis of concave 1,10–phenanthrolines by a combination of Suzuki coupling, ring closing metathesis and hydrogenation, *Eur. J. Org. Chem.*, 2004, 3119–3127; e) T. Liebig, M. Abbass, U. Lüning, Concave reagents, 51: concave pyridines for bifunctional acid-base catalysis, *Eur. J. Org. Chem.*, 2007, 972–980.
42. a) H. Ross, U. Lüning, Concave reagents, 23: Synthesis of calix[6]arene bridged by a 1,10-phenanthroline, *Tetrahedron Lett.*, 1997, **38**, 4539–4542; b) J. P. W. Eggert, J. Harrowfield, U. Lüning, B. W. Skelton, A. H. White, F. Löffler, S. Konrad, Concave reagents, 43: Improved synthesis and conformational analysis of an A,D-1,10-phenanthroline bridged calix[6]arene, *Eur. J. Org. Chem.*, 2005, 1348–1353.
43. a) U. Lüning, H. Ross, I. Thondorf, Concave reagents, 27: Steric requirements of intraannular subtituents in A,D-bridged calix[6]arene, *J. Chem. Soc. Perkin Trans. 2*, 1998, 1313–1317; b) S. Konrad, C. Näther, U. Lüning, Concave reagents, 49: Calix[6]arenes carrying camphorsulfonyl substituents, *Synthesis*, 2006, 2097–2099; c) S. Konrad, M. Bolte, C. Näther, U. Lüning, Concave reagents, 50: Chiral 1,10-phenanthroline-bridged calix[6]arene dendrimers, *Eur. J. Org. Chem.*, 2006, 4717–4730; d) J. P. W. Eggert, U. Lüning, Concave reagents, 54: Towards A,D-(1,10-phenanthroline)-bridged calix[6]arene dendrimers, *Eur. J. Org. Chem.*, 2007, 6046–6052.
44. Obeying ‘Murphy’s law’, the *meso*-compound was prefered in preliminary experiments: M. Hagen, *dissertation*, Christian-Albrechts-Universität zu Kiel, 1999.
45. The problem of diastereomers of dinaphthylsubstituted NHC has been investigated: X. Luan, R. Mariz, M. Gatti, C. Costabile, A. Poater, L. Cavallo, A. Linden, R. Dorta, Identification and characterization of a new family of catalytically highly active imidazolin-2–ylidenes, *J. Am. Chem. Soc.*, 2008, **130**, 6848–6858.
46. Tris(tetrachlorobenzenediolato)phosphate(V) anion: a) J. Lacour, C. Ginglinger, C. Grivet, G. Bernardinelli, Synthese und Trennung der Enantiomere des konfigurationsstabilen Tris(tetrachlorbenzo1diolato)phosphat(V)-Ions, *Angew. Chem.*, 1997, **109**, 660–662; *Angew. Chem. Int. Ed. Engl.*, 1997, **36**, 608–609; b) Bis(tetrachlorobenzenediolato)mono([1,1′]binaphthalenyl-2,2′-diolato) phosphate(V) anion: J. Lacour, A. Londez, C. Goujon-Ginglinger, V. Buß, G. Bernardinelli, Configurational ordering of cationic chiral dyes using a novel C2–symmetric hexacoordinated phosphate anion, *Org. Lett.*, 2000, **2**, 4185–4188.
47. Although by definition, a pK_a value is determined in water, analogous ‘pK_a’ values have been determined in other solvents, too. See for instance: K. Izutsu (Ed.), *Acid-base dissociation constants in dipolar aprotic solvents*, Chemical Data Series No. 35, International Union of Pure and Applied Chemistry, published by Blackwell Scientific Publications, Oxford, London, Edinburgh, Boston, Melbourne, 1990.
48. For the definition of general and specific acid catalysis see: H. Maskill, *The physical basis of organic chemistry*, Oxford University Press, New York, 1993.
49. General and specific protonations are discussed in: U. Lüning, C. Wangnick, M. Kümmerlin, Regioselective protonation of allyllithium compounds, *Chem. Ber.*, 1994, **127**, 2431–2433.

50. B. Schäfer, Enantioselektive Protonierungen: Kinetische Racematspaltung, *Chem. i. uns. Zeit*, 2002, **36**, 382–389.
51. a) U. Lüning, R. Baumstark, M. Müller, C. Wangnick, F. Schillinger, Concave reagents, 4: protonation of nitronate anions via concave pyridines. Stereoselective C-protonation versus Nef reaction, *Chem. Ber.*, 1990, **123**, 221–223; b) U. Lüning, F. Schillinger, Concave reagents, 6: can the structure of concave reagents determine selectivities? *Chem. Ber.*, 1990, **123**, 2073–2075.
52. a) U. Lüning, M. Müller, Konkave Reagenzien, 10: Konkav verpackte Protonen: Reagenzien für kontra-thermodynamische Protonierungen, *Angew. Chem.*, 1992, **104**, 99–102; *Angew. Chem. Int. Ed. Engl.*, 1992, **31**, 80–82; b) U. Lüning, H. Baumgartner, C. Manthey, B. Meynhardt, Concave reagents, 20: sterically shielded *m*-terphenyls as selective agents in general protonations, *J. Org. Chem.*, 1996, **61**, 7922–7926.
53. a) A. R. Butler, V. Gold, The hydrolysis of acetic anhydride. Part VII. Catalysis by pyridine and methylpyridines in acetate buffers, *J. Chem. Soc.*, 1961, 4362–4367; b) L. W. Deady, W. L. Finlayson, Catalysis by α-substituted pyridines in the hydrolysis of aryl acetates and acetic anhydride, *Aust. J. Chem.*, 1983, **36**, 1951–1956.
54. U. Lüning, R. Baumstark, W. Schyja, Concave reagents, 9: alcoholysis of diphenylketene catalysed by concave pyridines and open-chain analogs, *Liebigs Ann. Chem.*, 1991, 999–1002.
55. a) W. Schyja, S. Petersen, U. Lüning, Concave reagents, 21: selective acylations of primary and secondary alcohols by ketenes, *Liebigs Ann.*, 1996, 2099–2105; b) U. Lüning, S. Petersen, W. Schyja, W. Hacker, T. Marquardt, K. Wagner, M. Bolte, Concave reagents, 26: concave pyridines for selective acylations of polyols, *Eur. J. Org. Chem.*, 1998, 1077–1084.
56. S. Petersen, U. Lüning, Concave reagents, 28: comparison of bimacrocyclic, monomacrocyclic and non-macrocyclic bis(amidomethyl)pyridines as catalysts in the base catalysed addition of alcohols to ketenes, *Eur. J. Org. Chem.*, 1999, 847–854.
57. O. Winkelmann, C. Näther, U. Lüning, Organocatalysis by bimacrocyclic NHC: unexpected formation of a cyclic hemiacetal instead of a γ-butyrolactone, *Org. Biomol. Chem.*, 2009, 553–556.
58. R. Breslow, On the mechanism of thiamine action. Evidence from studies on model systems, *J. Am. Chem. Soc.*, 1958, **80**, 3719–3726.
59. Reviews on thiamine-dependent enzymes: a) G. A. Sprenger, M. Pohl, Synthetic potential of thiamin diphosphate-dependent enzymes, *J. Mol. Catal. B*, 1999, **6**, 145–159; b) R. Duggleby, Domain relationships in thiamine diphosphate-dependent enzymes, *Acc. Chem. Res.*, 2006, **39**, 550–557.
60. a) H. Stetter, Die katalysierte Addition von Aldehyden an aktivierte Doppelbindungen – Ein neues Syntheseprinzip, *Angew. Chem.*, 1976, **88**, 695–704; *Angew. Chem. Int. Ed. Engl.*, 1976, **15**, 639–647; b) C. Burstein, F. Glorius, Organokatalysierte konjugierte Umpolung von α,β-ungesättigten Aldehyden zur Synthese von γ-Butyrolactonen, *Angew. Chem.*, 2004, **116**, 6331–6334; *Angew. Chem., Int. Ed.*, 2004, **43**, 6205–6208; c) S. S. Sohn, E. L. Rosen, J. W. Bode, N-Heterocyclic carbene-catalysed generation of homoenolates: γ-butyrolactones by direct annulations of enals and aldehydes, *J. Am. Chem. Soc.*, 2004, **126**, 14370–14371; d) D. Enders, O. Niemeier, A. Henseler, Organocatalysis by N-heterocyclic carbenes, *Chem. Rev.*, 2007, **107**, 5606–5655; e) N. Marion, S. Diez-Gonzáles, S. P. Nolan, N-heterocyclische Carbene als Organokatalysatoren, *Angew. Chem.*, 2007, **119**, 3046–3058; *Angew. Chem. Int. Ed.*, 2007, **46**, 2988–3000; f) K. Hirano, I. Piel, F. Glorius, Diastereoselective synthesis of trifluoromethylated γ-butyrolactones via N-heterocyclic carbene-catalysed conjugated umpolung of α,β-unsaturated aldehydes, *Adv. Synth. Catal.*, 2008, **350**, 984–988.
61. a) F. Löffler, U. Lüning, G. Gohar, Concave reagents, 33: transition metal complexes of 1,10–phenanthroline bridged calix[6]arenes, *New J. Chem.*, 2000, **24**, 935–938; b) J. P. W. Eggert, J. M. Harrowfield, U. Lüning, B. W. Skelton, A. H. White, Lanthanide(III) ion coordination by a concave reagent, *Polyhedron*, 2006, 910–914.
62. O. Winkelmann, C. Näther, U. Lüning, Concave reagents, 56: synthesis, structure and catalytic activity of a bimacrocylic NHC palladium allyl complex, *J. Organomet. Chem.*, 2008, **693**, 2784–2788.

63. B. Meynhardt, U. Lüning, C. Wolff, C. Näther, Concave reagents, 30: diastereoselective generation of quarternary stereocenters by ligand controlled palladium catalysed allylations, *Eur. J. Org. Chem.*, 1999, 2327–2335.
64. a) M. Hagen, U. Lüning, Concave reagents, 22: cyclopropanation of alkenes with ethyl diazoacetate: copper (I) complexes of concave 1,10–phenanthrolines as diastereoselective catalysts, *Chem. Ber./Recueil*, 1997, **130**, 231–234; b) M. Bühl, F. Terstegen, F. Löffler, B. Meynhardt, S. Kierse, M. Müller, C. Näther, U. Lüning, Concave reagents, 34: on the mechanism and stereoselectivity of the copper(I) catalysed cyclopropanation of olefins. A combined experimental and density-functional study, *Eur. J. Org. Chem.*, 2001, 2151–2160.
65. As ligands, the nucleophilic NHC are strong two-electron σ-donors, displaying similar ligand properties as trialkylphosphines: K. Öfele, W. A. Herrmann, D. Mihalios, M. Elison, E. Herdtweck, W. Scherer, J. Mink, Mehrfachbindungen zwischen Hauptgruppenelementen und Übergangsmetallen CXXVI. Heterocyclen-Carbene als phosphananaloge Liganden in Metallkomplexen, *J. Organomet. Chem.*, 1993, **459**, 177–184.
66. a) T. Weskamp, F. J. Kohl, W. Hieringer, D. Gleich, W. A. Herrmann, Hochaktive Rutheniumkatalysatoren für die Olefinmetathese: die Synergie N-heterocyclischer Carbene und koordinativ labiler Liganden, *Angew. Chem.*, 1999, **111**, 2573–2576, *Angew. Chem. Int. Ed.*, 1999, **38**, 2416–2419; b) M. Scholl, S. Ding, C. W. Lee, R. H. Grubbs, Synthesis and activity of a new generation of ruthenium-based olefin metathesis catalysts coordinated with 1,3–dimesityl-4,5–dihydroimidazol-2–ylidene ligands, *Org. Lett.*, 1999, **1**, 953–956; c) J. Huang, E. D. Stevens, S. P. Nolan, J. L. Petersen, Olefin metathesis-active ruthenium complexes bearing a nucleophilic carbene ligand, *J. Am. Chem. Soc.*, 1999, **121**, 2674–2678.
67. Just a few reviews on this quickly developping field: a) D. H. Paull, C. J. Abraham, M. T. Scerba, E. Alden-Danforth, T. Lectka, Bifunctional asymmetric catalysis: cooperative lewis acid/base systems, *Acc. Chem. Res.*, 2008, **41**, 655–663; b) M. Kanai, N. Katob, E. Ichikawab, M. Shibasaki, Power of cooperativity: Lewis acid-Lewis base bifunctional asymmetric catalysis, *Synlett*, 2005, 1491–1508,; c) M. Shibasaki, M. Kanai K. Funabashi, Recent progress in asymmetric two-center catalysis, *Chem. Commun.*, 2002, 1989–1999.

8

Reactivity Control by Calixarenes

Luigi Mandolini, Roberta Cacciapaglia, and Stefano Di Stefano
IMC CNR Sezione Meccanismi di Reazione, c/o Dipartimento di Chimica,
Università La Sapienza, 00185 Roma, Italy

8.1 Introduction

Calixarenes are oligomeric cyclophanes derived from the condensation of phenols with formaldehyde. They became popular in the late seventies and have played a very important role in Supramolecular Chemistry ever since, as witnessed by numerous books[1] and review articles[2] on the subject. Calixarenes are important building blocks for the preparation of a large variety of supramolecular complexes and materials, thanks to their ability to act as hosts, and the possibility of introducing a large number of functions at either rim by means of selective derivatization.

This chapter focuses on reactivity control in stoichiometric and catalytic reactions taking place in the confines of supramolecular complexes of reactants with calixarene receptors. Earlier work on the subject was reviewed by us in 2000.[3] Quite recently, Homden and Redshaw have published an extensive review on the use of calixarenes in metal-based catalysis.[2i] Because of space limitations and, even more importantly, to avoid extensive overlap with the already reviewed material, the scope of this chapter is restricted to works treated only marginally, or not treated at all, in the above review article. The first section deals with examples in which reactivity control takes place via substrate inclusion into the calixarene cavity. The other section illustrates the use of the calix[4]arene upper rim in the construction of di- and trimetallic complexes capable of esterase and nuclease activity.

8.2 Calixarenes as Hosts

The limited size of the calix[4]arene backbone allows deep penetration within its cavity of only small guests. Notably, Rathore, Kochi and co-workers[4] reported that the tetra-*O*-methylated *p-tert*-butylcalix[4]arene **1** (R—Me), upon oxidation in CH_2Cl_2 followed by exposure to gaseous NO, or upon treatment with $NOSbCl_6$, forms a dark purple, highly stable 1 : 1 complex ($K > 5 \times 10^8 M^{-1}$), that shows no sign of decomposition for several months, both in solution and in the solid state. The X-ray structure showed that NO^+ is completely entombed in the cylindrical cavity of the 1,3-*alternate* conformation. ^1H NMR spectroscopic evidence was obtained that in solution all four conformers (*cone*, *partial cone*, 1,2-*alternate*, 1,3-*alternate*) of the calix[4]arene host form complexes of definite stability with NO^+. The preference of 1,3-*alternate* in the solid state indicates a crystallization induced drainage of the equilibria in solution. Rudkevich *et al.*[5,6] subsequently discovered that caged NO^+ complexes of the same kind are reversibly formed upon bubbling NO_2/N_2O_4 through a chloroform solution of a tetra-*O*-alkylated calix[4]arene (**1**, R—Hex) fixed in the 1,3-*alternate* conformation, in the presence of $SnCl_4$ (Equation 8.1). Quite remarkably, encapsulated NO^+ was found to act as a highly selective nitrosating agent towards secondary amides. The reaction mechanism outlined in Scheme 8.1 is consistent with the observed size-shape selectivity and with the finding that the chemical properties of encapsulated NO^+ are different from those in bulk solution, and are controlled by the calix[4]arene cavity. The reaction proved to be quite sensitive to the size of –Alk, but not to that of –Alk′. Fair to good yields (50–95%) of *N*-nitrosated products were obtained only with *N*-methylated amides, independent of the size of –Alk′ from ethyl to heptyl. No reaction occurred with amides alkylated with –Alk bulkier than methyl, on account of the hindered approach to the encapsulated NO^+ species.

$$\mathbf{1} \xrightarrow[SnCl_4, CHCl_3]{NO_2/N_2O_4} [\mathbf{1} \cdot NO]^+ NO_3SnCl_4^- \quad (8.1)$$

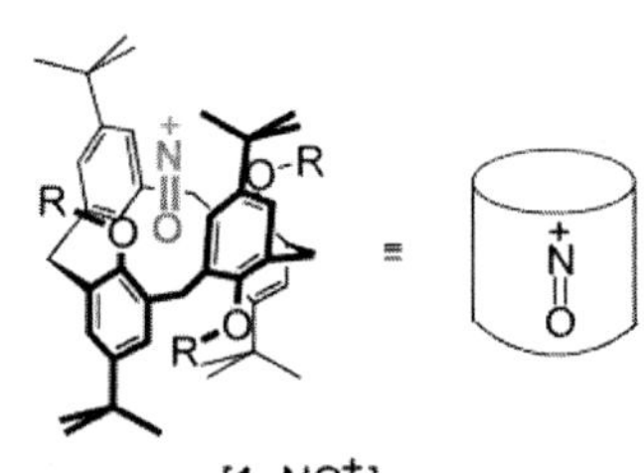

[1·NO⁺]

The wider cavity of calix[6]arene has been masterfully exploited by the group of Reinaud in the design of ligands that mimic the coordination sphere of the metal ion of copper enzymes.[7] A calix[6]arene functionalized at the smaller rim by three independent *N*-donor groups, upon binding to copper is constrained in a *cone* conformation. The free coordination site of the metal situated inside the calix[6]arene cavity is available for coordination to small donor molecules to afford the so-called 'funnel complexes'.[8] Structure **2** shows a funnel complex featuring a tris(pyridine) coordination site, a perfectly tetrahedral geometry for cuprous ion, and a CH_3CN ligand embedded in the calix[6]arene cavity. A preliminary investigation of the electrochemical behaviour of a tris(pyridine)–calix[6]arene–Cu complex[9a] was extended to additional families of Cu–N_3–calix[6]arene

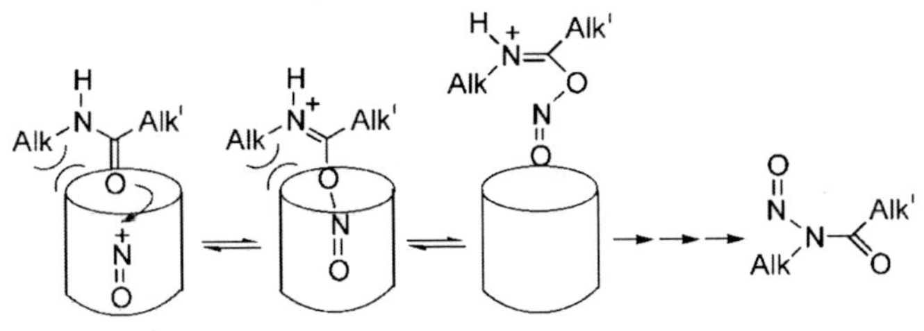

Scheme 8.1 *Proposed mechanism of* N*-nitrosation of secondary amides with encapsulated* NO^+

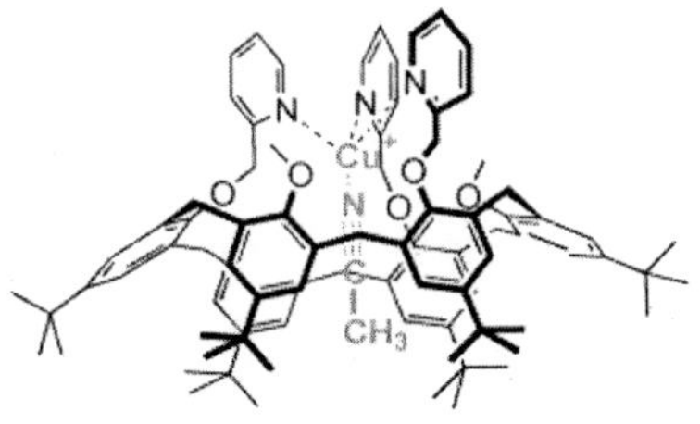

2

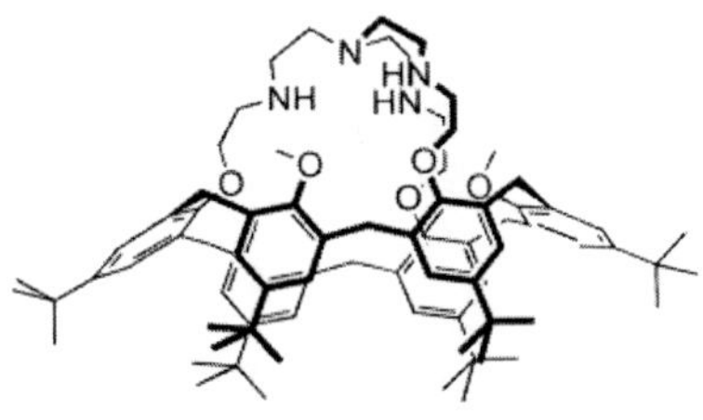

3

complexes including tertiary amine and *N*-alkylimidazole ligands.[9b] The important discovery was made that the electrochemical behaviour of the complexes could be resolved and analyzed in CH_3CN, but not in non-coordinating solvents such as CH_2Cl_2 and THF. The beneficial action of CH_3CN was obviously ascribed to its ability to form funnel complexes. The supramolecular concept of the 'shoetree' molecule guest was proposed. The 'shoetree' CH_3CN guest in the calix[6]arene cavity ensures the overall fixation of the coordination at every stage of the Cu(II)/Cu(I) reorganization process. As such, the 'shoetree' guest is a prerequisite for a dynamic redox behaviour.

Another family of calix[6]arene funnel complexes is provided by the calixtren ligand **3**.[10] Here a calix[6]arene is capped by a tren unit, thereby presenting a N_4 coordination site at the end of a concave calix[6]arene core. Calixtren–Cu complexes are endowed with unique properties, when compared to all previously described tren–Cu complexes. The **3**–Cu(I) complex does not disproportionate in concentrated solution, and is stable in chlorinated solvents for days. Furthermore the **3**–Cu(II) complex is a strong receptor for neutral ligands, such as nitriles, alcohols and amides, but not for anionic ligands. This was ascribed to the strong electrostatic repulsion exerted by the lone pairs of the six oxygen atoms of the calix[6]arene core. Unlike the Cu(I) complexes of the calix[6]arene-based ligands equipped with tris(imidazole) or tris(pyridine) coordination sites, the calixtren-complexed Cu(I) centre in **3**–Cu(I), was found to be very sensitive towards O_2 (Scheme 8.2), leading to a transient $(CuO_2)^+$ species, whose nature was assigned as a reactive Cu(II)–superoxide adduct. In the presence of CH_3CN, well suited for penetrating the calix[6]arene cavity and for stabilizing the Cu(II) centre, O_2^- is ejected from the complex into the reaction medium. Even more importantly, in a non coordinating solvent such as CH_2Cl_2 the metal-bound superoxide species directly oxidizes a CH bond of the tren backbone, a result that has important implications for the modes of action of copper monooxygenases.

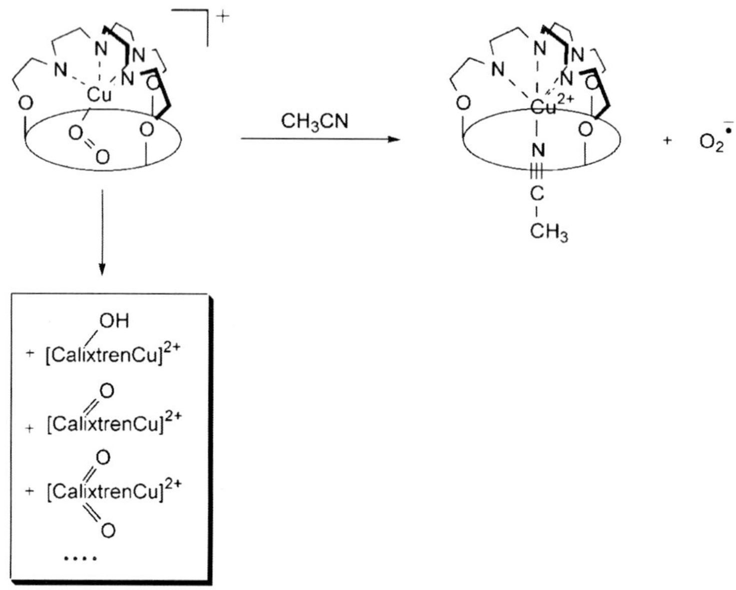

Scheme 8.2 *Reactions of the proposed superoxide transient adduct [**3**·Cu·O$_2$]$^+$ in the presence and absence of acetonitrile*

The *p*-sulfonatocalix[n]arenes **4**, originally synthesized by Shinkai *et al.*[11] as water soluble calixarenes for catalytic studies in water solution, were more recently used by Ueoka *et al.*[12] as catalysts in the specific acid catalyzed methanolysis of *N*-Ac-L-amino acids (Phe, Tyr, Trp, His, Lys, Arg). Rates of methanolysis in the presence of the calix[n]arene catalysts, normalized per sulfonic group, were compared with rates obtained in the presence of the noncyclic analogue *p*-hydroxybenzenesulfonic acid. Rate enhancements – ranging from 12- to 86-fold – relative to control were recorded only in the methanolysis of basic amino acid derivatives (His, Lys, Arg), but neutral amino acid derivatives (Phe, Tyr, Trp) responded virtually in the same way to the presence of cyclic and non-cyclic catalyst. Michaelis–Menten kinetics and ^{1}H NMR spectral evidence pointed to the intermediacy of inclusion complexes of **4** with the protonated form of basic amino acid derivatives, as shown in **5** for the His–**4** (n = 4) combination.

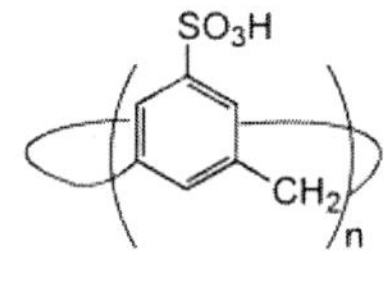

4 (n = 4, 6, 8)

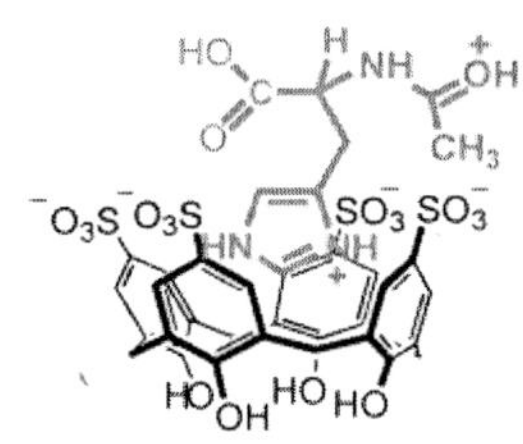

5

Ditopic receptors **6** were designed and synthesized by de Mendoza *et al.*[13a] as mimics of the phosphocholine binding site of McPC603 antibody. Receptors **6** bind strongly to dioctanoyl-L-α-phosphatidylcholine ($K = 73000\,M^{-1}$ for **6a** and $95000\,M^{-1}$ for **6b** in chlo-

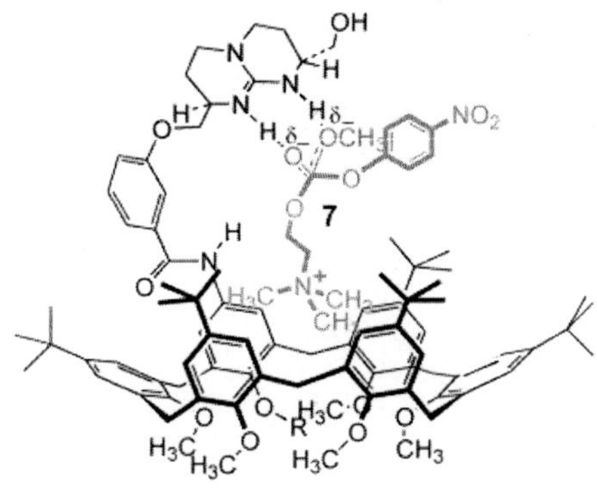

6a, R = H
6b, R = H_2C–

roform, 25 °C), as a result of a combined binding of the choline trimethylammonium head to the calix[6]arene subunit through cation–π interactions, and of the anionic phosphate to the guanidinium subunit through ion pairing and hydrogen bonding. Since phosphate diesters are well known transition state analogues for the basic hydrolysis of esters, it was reasoned that receptors **6** would catalyze a $B_{Ac}2$-type cleavage of choline esters. Indeed, the methanolysis of choline *p*-nitrophenyl carbonate (PNPCC), in $CHCl_3$/MeOH 99 : 1 in the presence of diisopropylethylamine–perchlorate salt buffer at 25 °C (Equation 8.2) was effectively catalyzed with turnover by **6**.[13b] A rate enhancement of 76-fold relative to background was recorded in the presence of 1 mM **6a**, and one of 149-fold in the presence of 1 mM **6b**. Kinetic analysis yielded maximum rate accelerations at saturation of 600 and 1000 for **6a** and **6b**, respectively. Much lower rate enhancements were brought about by equimolar mixtures of the disconnected subunits, showing that the two subunits in **6** bind to the transition state **7** with considerable degree of synergism.

$$\underset{\text{PNPCC}}{\text{PNPO-}\overset{\text{O}}{\overset{\|}{\text{C}}}\text{-OCH}_2\text{CH}_2\overset{+}{\text{N}}(\text{CH}_3)_3} + \text{CH}_3\text{O}^- \xrightarrow[25\ ^\circ\text{C}]{\text{CHCl}_3\,/\,\text{CH}_3\text{OH}\ \ 99:1} \text{PNPO}^- + \text{H}_3\text{CO-}\overset{\text{O}}{\overset{\|}{\text{C}}}\text{-OCH}_2\text{CH}_2\overset{+}{\text{N}}(\text{CH}_3)_3 \quad (8.2)$$

A similar strategy of catalyst design based on ditopic receptors composed of distinct recognition elements for the reacting and non-reacting parts of the substrates, was adopted by the group of Rebek in a number of works involving reactions of choline or choline derivatives. In these studies, the recognition subunit for the choline trimethylammonium head is a cavitand derived from resorcinol.[14] Receptor **8** features a zinc(II) salophen unit fused to a resorcinarene. As suggested by an energy minimized structure, the trimethylammonium group of a PNPCC guest is well positioned deep within the cavity, whereas the ester carbonyl is simultaneously coordinated to the metal electrophile. In accordance with this picture, PNPCC has a very high affinity for **8**, much higher than that of choline. The complexed PNPCC undergoes basic hydrolysis in buffered CH_2Cl_2 containing 0.01% water (Equation 8.3) no less than 50 times more rapidly than uncomplexed PNPCC. Consistent with a complexation catalysis scheme, the much less reactive acetylcholine acts as a competitive inhibitor in the hydrolysis of PNPCC. Complex cavitand **8** is also an effective catalyst for the acylation of choline.[15] Here the salophen-complexed metal

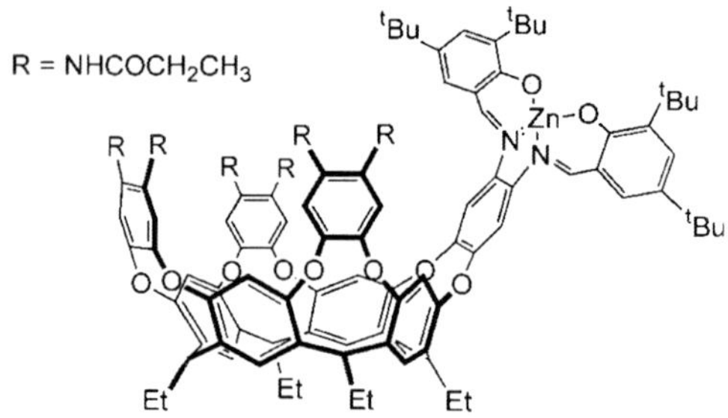

8

ion is believed to activate the incoming acylating agent toward nucleophilic attack from the cavitand-complexed choline hydroxyl. In the presence of 2 mol% of **8**, the reaction with acetic anhydride (Equation 8.4) takes place 1900 times faster than the background reaction. Acetylation is an order of magnitude slower when the zinc(II)–salophen complex and the cavitand unit are not covalently connected.

$$PNPCC + HO^- \xrightarrow[RT]{CH_2Cl_2\,(0.01\%\,H_2O)} PNPO^- + HOCH_2CH_2\overset{+}{N}(CH_3)_3 + CO_2 \quad (8.3)$$

$$HOCH_2CH_2\overset{+}{N}(CH_3)_3 + Ac_2O \xrightarrow[25°C]{DMSO\text{-}d_6} AcOCH_2CH_2\overset{+}{N}(CH_3)_3 + AcOH \quad (8.4)$$

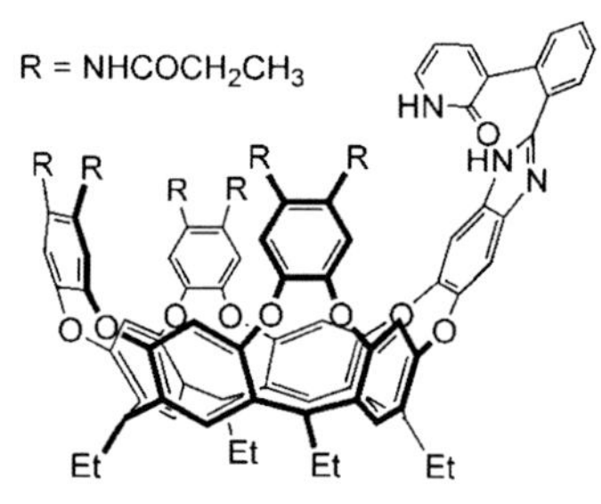

9

As a variation on the theme, instead of a zinc(II)–salophen unit, receptor **9**[16] features a 2-pyridone unit, that is a known bifunctional catalyst for the rate limiting breakdown of the tetrahedral intermediate involved in the aminolysis of active esters in aprotic solvents. Turnover catalysis was indeed observed when the reaction between propylamine and PNPCC (Equation 8.5) was carried out in CH_2Cl_2/CH_3CN 99 : 1 in the presence of receptor **9**. It was estimated that the reaction inside **9** is *ca.* 6000 times faster than the reaction in the bulk solvent.

$$PNPCC + PrNH_2 + CH_3O^- \xrightarrow[RT]{CH_2Cl_2\,/\,CH_3CN\ \ 99:1} PNPO^- + PrNH\text{-}\overset{O}{\overset{\|}{C}}\text{-}OCH_2CH_2\overset{+}{N}(CH_3)_3 + CH_3OH \quad (8.5)$$

In mesitylene-d_{12} cavitand **10** exhibits moderate to high binding affinity for maleimides *N*-substituted with medium sized alkyl groups, such as adamantyl and cyclooctyl.[17] These

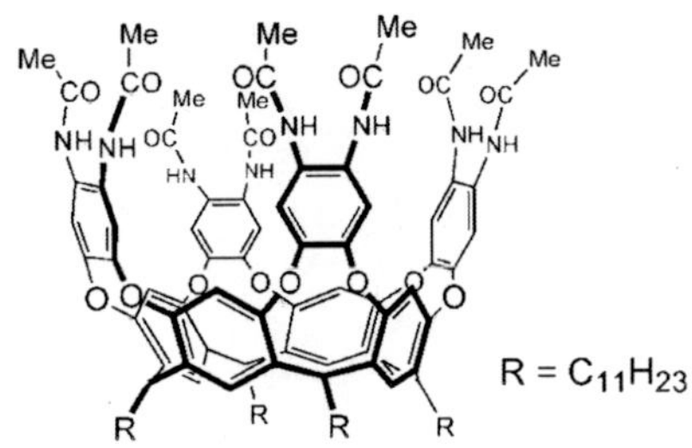

10

groups, when deeply hosted in the cavity of **10**, position the maleimide carbonyl group in close proximity to the organized *H*-bonding network, whereas the double bond remains exposed above the receptor. The Diels–Alder reaction of the complexed maleimide with 9-anthracenemethanol (Equation 8.6) is strongly accelerated relative to the reaction of the uncomplexed reactant. This is believed to arise from interaction of the dienophile with the hydrogen bonding network at the cavitand rim. The hydroxyl of 9-anthracenemethanol plays a role, because anthracenes lacking a hydroxyl group do not react. The increase in steric bulk of the reaction product causes fast release from its complex with **10**, thus ensuring high turnover efficiency with little or no product inhibition.

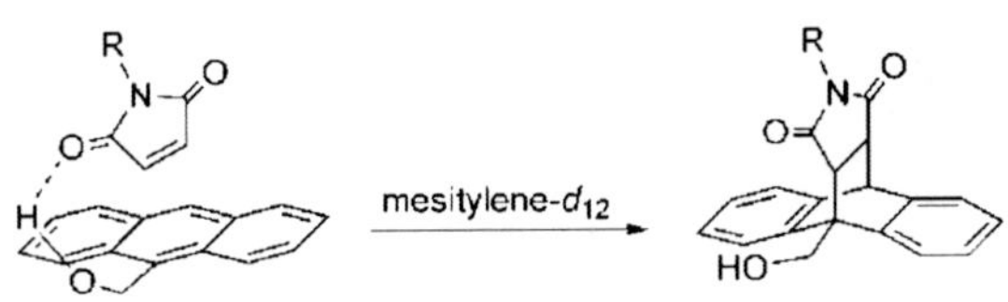

(8.6)

Cavitand **10** was also found to stabilize the Meisenheimer complexes of S_NAr reactions of 2-nitro-4-trifluoromethylchlorobenzene with piperidine and its next higher homologues azepane (7-ring) and azocane (8-ring) in the non-polar solvent *p*-xylene-d_{10} (Equation 8.7).[18] The affinity for the host is low with piperidine ($K = 14\,M^{-1}$) and is somewhat higher with the larger rings ($K = 40\,M^{-1}$). When carried out in the presence of stoichiometric **10**, the reaction of piperidine with the activated aryl chloride shows a 12-fold rate enhancement relative to the control reaction. Seven- and eight-membered cyclic amines, whose affinity for the host is somewhat higher, show accelerations of 67- and 35-fold, respectively. These were proposed to arise from the ability of the polar amide groups at the upper rim to stabilize the anionic portion of the dipolar transition state, possibly combined with cation–π interactions between the aromatic walls of **10** and the developing positive charge of the nucleophilic nitrogen.

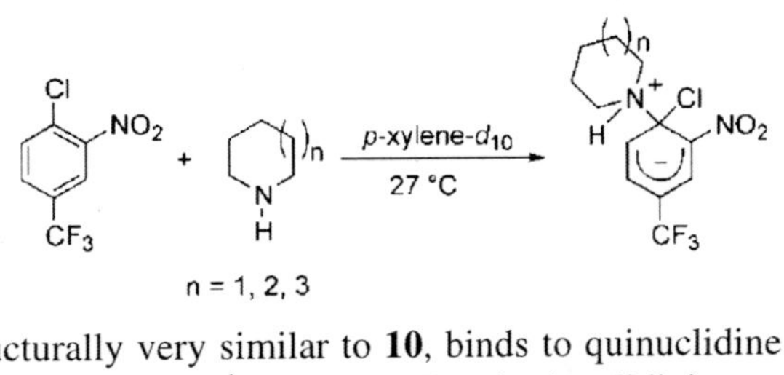

(8.7)

Cavitand **11**, structurally very similar to **10**, binds to quinuclidine in acetone solution with moderate affinity ($K = 40\,M^{-1}$), but the *N*-butylquinuclidinium cation is bound much

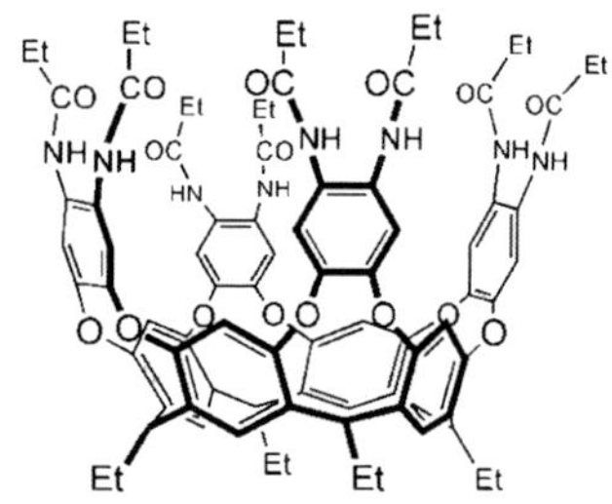

11

more strongly ($K > 4000\,M^{-1}$), on account of cation–π interactions between the hosted cation and the inner aromatic walls of the container molecule.[19] When hosted in the cavity of **11**, quinuclidine undergoes the Mentschutkin reaction with butyl halides and sulfonates $CH_3CH_2CH_2CH_2X$ in acetone solution (Equation 8.8) much more rapidly than uncomplexed quinuclidine, with rate enhancements of 1600, 1300, 150, and 100 for X = Br, Cl, Mes, Tos, in the given order. Product inhibition prevents turnover, and the reactions are not catalytic.

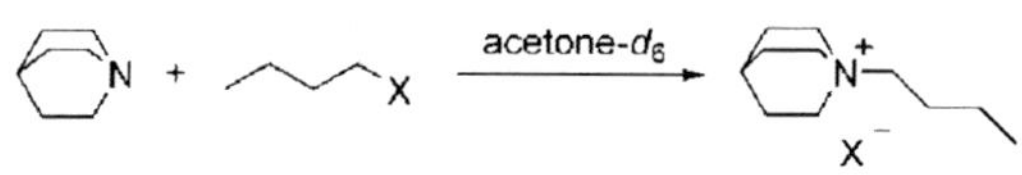

(8.8)

Resorcinarene-based deep cavitands were also used by Rebek and coworkers to stabilize and detect otherwise elusive reaction intermediates. The entropically favoured reaction of primary amines hosted in the cavity of **12** with the introverted aldehyde functionality (Scheme 8.3) leads to previously undetectable hemiaminal intermediates, whose half-lives vary from 30 min to 100 h.[20] The nature of the stabilization arises from a combination of mechanical isolation from the medium, hydrogen bonding to the organized amide seam at the cavitand rim, and steric repulsion between the alkyl group and the cavitand wall occurring for the repositioning of the alkyl group when going from hemiaminal to imine, as schematically shown in Scheme 8.3 for the reaction of *i*-propylamine. The hemiaminal derived from *i*-propylamine has a half-life of 135 minutes, whereas half-lives lower by an order of magnitude are observed for hemiaminals derived from unbranched propylamine and butylamine. The exceptionally long lifetimes of the hemiaminals derived from ethanolamine ($t_{1/2}$ = 3520 h) and ethylenediamine ($t_{1/2}$ = 5640 h) are due to further stabilization arising from intramolecular hydrogen bond.

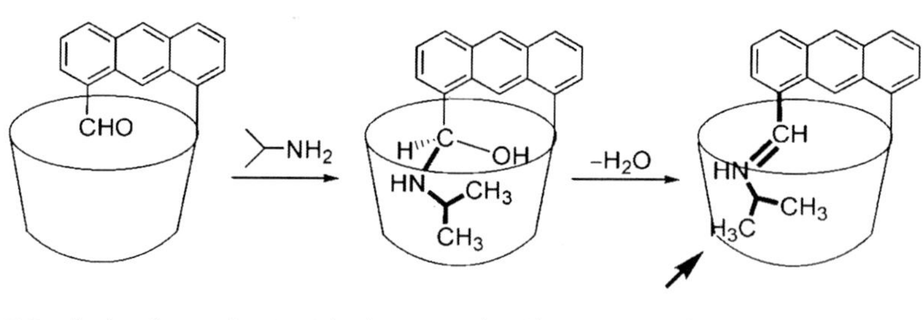

Scheme 8.3 *Imine formation inside the capsule. The arrow indicates a possible steric mismatch increasing the energy barrier for water elimination*

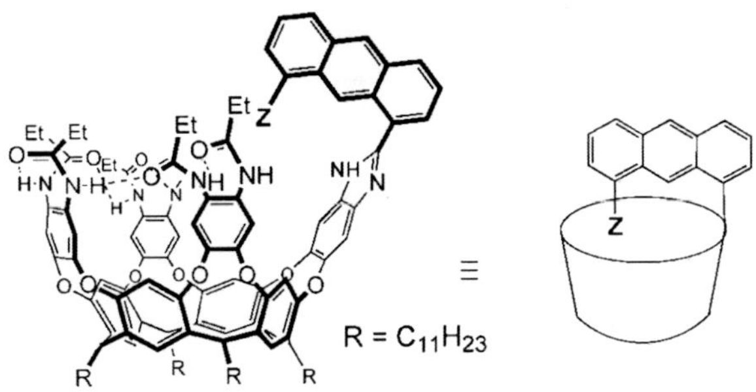

12 Z = CHO
13 Z = CO_2H

A similar strategy was adopted for the observation of elusive isoimide intermediates.[21] The reaction between a carboxylic acid and an aliphatic isonitrile in mesitylene-d_{12} (Equation 8.9) proceeds through an isoimide intermediate that rearranges to a *N*-acylformamide (Mumm rearrangement). When cavitand **13** was reacted at millimolar concentrations with isopropyl isonitrile in mesitylene-d_{12}, ^{1}H NMR peaks and IR bands arising from the isoimide intermediate were detected. Over the course of a few hours, the signals arising from the intermediate disappeared and were replaced by signals from the *N*-acylformamide product. Again, stabilization by supramolecular interactions and steric barriers to Mumm rearrangement due to restricted molecular motions in the interior of the cavity, explain the slow rate of decomposition of the isoimide intermediate.

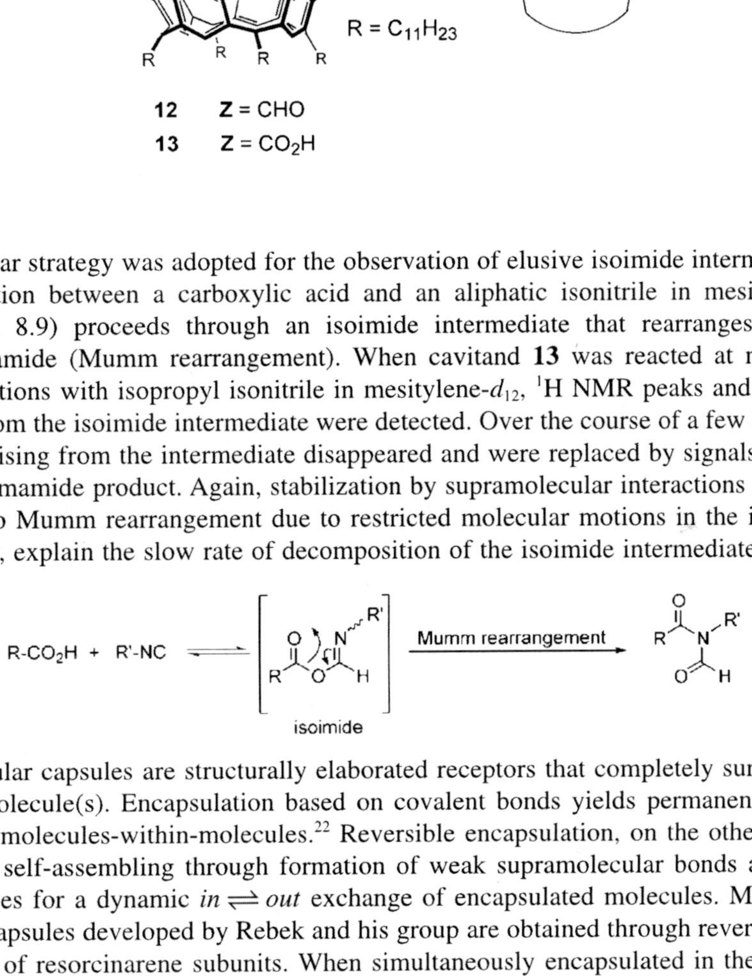

(8.9)

Molecular capsules are structurally elaborated receptors that completely surround the hosted molecule(s). Encapsulation based on covalent bonds yields permanent arrangements of molecules-within-molecules.[22] Reversible encapsulation, on the other hand, is based on self-assembling through formation of weak supramolecular bonds and offers possibilities for a dynamic *in* $\rightleftharpoons$ *out* exchange of encapsulated molecules. Most of the dimeric capsules developed by Rebek and his group are obtained through reversible self-assembly of resorcinarene subunits. When simultaneously encapsulated in the cylindrically shaped inner space of these capsules, two reactant molecules are temporarily isolated from others in solution and display reactivity features different from those in bulk solution. The matter has been extensively reviewed,[23] and will not be discussed here.

8.3 Calixarenes as Molecular Platforms

Many hydrolytic enzymes possess two metal ions in their active site.[24] Consequently, there has been a wide interest in the biomimetic hydrolytic activity of dinuclear metal complexes. In many instances such dinuclear complexes have been obtained from homoditopic receptors composed of two identical metal complexing units connected by a suitable spacer. The catalytic efficiency of these complexes is influenced by a variety of factors, among which the nature of the spacer itself plays an important role. The use of either rim of calix[4]arenes as suitable spacers is well precedented.[3] Particularly successful has been the use of the upper rim of calix[4]arenes blocked in the *cone* conformation, that was pioneered by the group of Reinhoudt.[25] Dinuclear zinc(II) and copper(II) complexes of ditopic receptors composed of two nitrogen ligand units at the diagonal position of a calix[4]arene scaffold, showed a high degree of catalytic cooperation of the two metal centres in the cleavage of phosphodiester bonds, as a result of a good compromise between preorganization and flexibility (dynamic preorganization).[25] Most recent investigations in this area, resulting from joint efforts involving the authors' group in Roma, Reinhoudt's group in Enschede, and Ungaro's group in Parma, are summarized in this section. These studies were aimed at widening the scope of the calix[4]arene scaffold in the construction of di- and trinuclear catalysts capable of esterase and nuclease activity.

8.3.1 Artificial Esterases: Barium(II) Complexes

The basic ethanolysis of esters functionalized with a distal carboxylate anchoring group is catalyzed with turnover by dinuclear barium complexes according to the mechanism depicted in Scheme 8.4.[26] The high affinity of carboxylate for the ligated barium ion ensures efficient binding of the ester substrate to the metal catalyst. The function of the anchoring step is manifold: it selects the substrate, provides a moderate electronic activation to the ester undergoing nucleophilic attack, and transforms an otherwise intermolecular ethoxide delivery to the ester carbonyl into an intramolecular one.

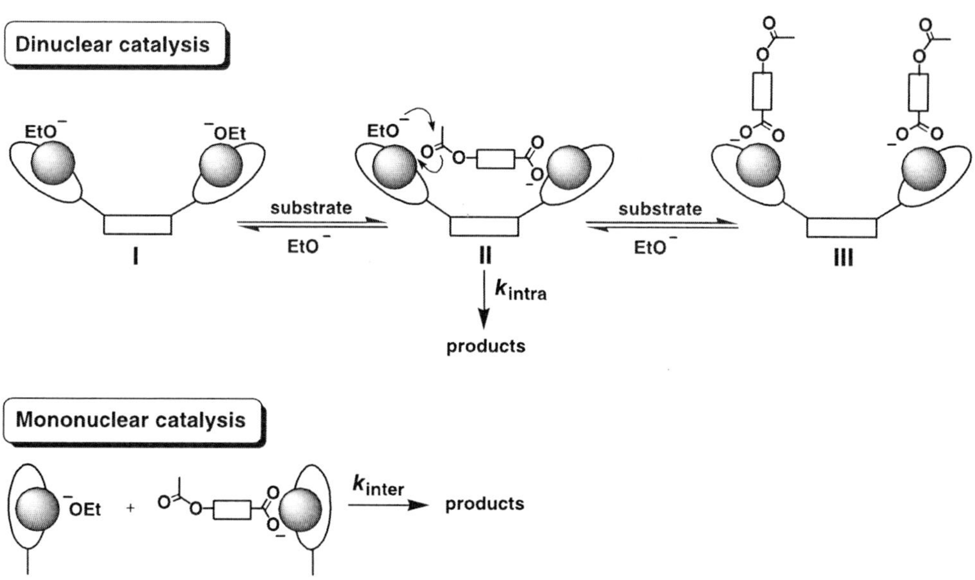

Scheme 8.4

Table 8.1 *Basic ethanolysis of esters* ***18–21*** *catalysed by* Ba^{2+} *complexes of ligands* ***14–17***[a]

Catalyst	k_{obs}/k_o[b]			
	18	**19**	**20**	**21**
17–Ba	300	20	52	12
14–Ba_2 (1,2-vicinal)	35100	22000	20000	170
15–Ba_2 (1,3-distal)	1340	420	570	24
16–Ba_2	2760	1220	4240	300

[a] Data from ref. 27. Runs carried out in EtOH, 25 °C, 0.025 mM substrate, 1 mM $EtONMe_4$, 0.2 mM monotopic or 0.1 mM ditopic ligand, 0.2 mM $Ba(SCN)_2$.
[b] k_o is the pseudo-first-order rate constant obtained in the presence of 1.00 mM $EtONMe_4$ alone. Ester **18**, $k_o = 1.34 \times 10^{-4}\,s^{-1}$; ester **19**, $k_o = 9.10 \times 10^{-4}\,s^{-1}$; ester **20**, $k_o = 1.51 \times 10^{-4}\,s^{-1}$; ester **21**, $k_o = 1.32 \times 10^{-3}\,s^{-1}$.

The catalytic efficiencies of regioisomeric complexes **14**–Ba_2 and **15**–Ba_2 in the basic ethanolysis of esters **18–21** are compared in Table 8.1 with that of the dinuclear barium complex of the closely related ligand **16**, as well as with that of the mononuclear control **17**–Ba.[27] Rate measurements refer to conditions where no less than 90% of the ester is bound to the catalyst in the productive (Michaelis) complex **II** (Scheme 8.4), while the concentration of the unproductive 2:1 complex **III** is negligibly low. The superiority of dinuclear catalysts over the mononuclear counterpart demonstrates that the two metal ions work in a cooperative fashion, in accordance with the bifunctional catalytic mechanism in Scheme 8.4, but the extent of cooperation is strongly dependent on the catalyst–substrate combination. This is not really surprising, as the catalytic mechanism implies that the substrate and the catalyst must form a well matched pair in terms of size and geometrical features. A convenient comparison of catalytic performances of regioisomeric

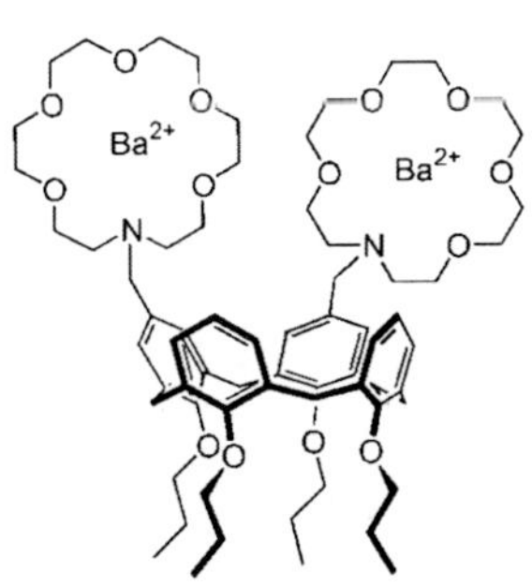

14 – Ba_2

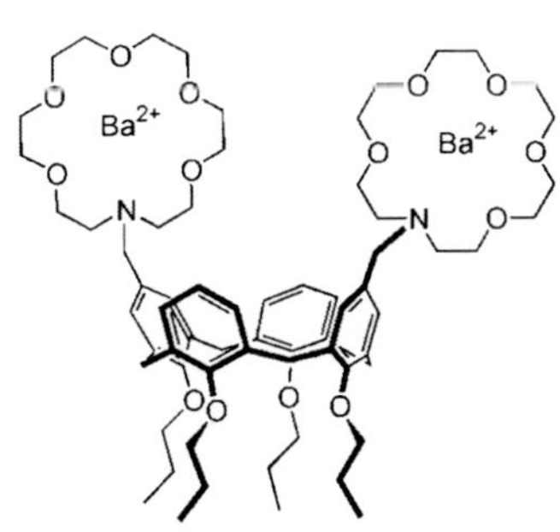

15 – Ba_2

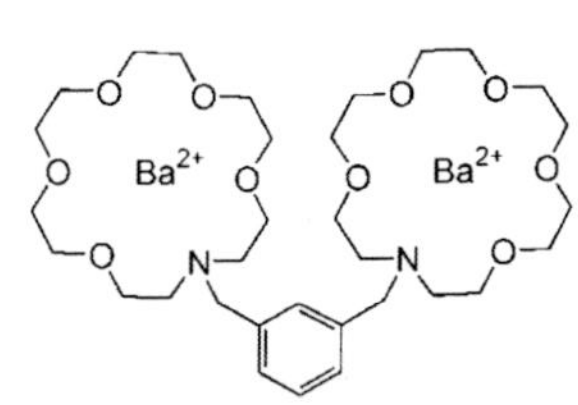

16 – Ba_2

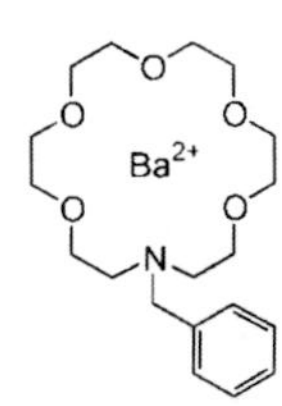

17 – Ba

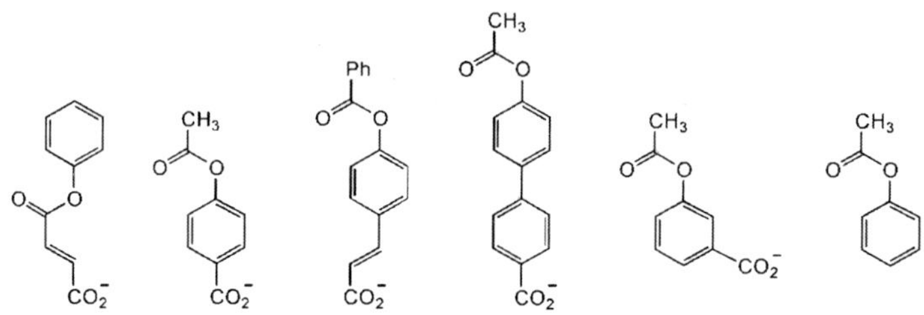

18 19 20 21 22 23

14–Ba_2 and **15**–Ba_2 complexes, and of the analogous *m*-xylylene derivative **16**–Ba_2 is carried out in terms of effective molarities (*EM*s).[23b] Since the bimetallic catalysts transform an otherwise intermolecular reaction between ester and ethoxide ion into an intramolecular one, the *EM* notion, defined by the ratio k_{intra}/k_{inter} (Scheme 8.4), strictly applies to the catalytic processes at hand. The *EM* provides an absolute measure of the efficiency of the intramolecular catalytic step, that is independent of the intrinsic reactivity of end groups. Unlike the catalytic rate enhancement and the advantage of dinuclear over mononuclear catalyst, the *EM* is independent of reactant concentrations, and its numerical value is solely determined by the usual choice of molarity as concentration unit.

Plots of *EM vs.* the carboxylate-carbonyl distances in the ester substrates, taken as a gross measure of their size, are shown in Figure 8.1. It is apparent that the dinuclear catalyst **14**–Ba_2, in which the azacrown units are linked to vicinal positions of the calix[4]arene scaffold, is not only far superior to its diagonal regioisomer **15**–Ba_2 in all cases, but it is also superior to **16**–Ba_2 in the reactions of esters **18–20**. The *EM* profile shows that the catalytic efficiency of **14**–Ba_2 reaches its maximum value in the reaction of ester **19**, and drops to a very low value in the reaction of the 'longest' ester **21**. This indicates that **14**–Ba_2 cannot expand its intermetal distance to fit the long carboxylate-carbonyl distance in **21**. In conclusion, reactivity data obtained for the various catalyst–substrate combinations indeed show that a close fit of ester size to metal-to-metal distance is an important prerequisite for catalysis. However, the marked superiority of **14**–Ba_2 over **15**–Ba_2 with all of the esters can hardly be ascribed to a more suitable intermetal distance in the former. This indicates that in addition to intermetal distance, still poorly understood effects may contribute significantly to catalytic efficiency of dinuclear metal catalysts.

8.3.2 Artificial Esterases: Zinc(II) Complexes

The zinc(II) complex of 2,6-bis[(dimethylamino)methyl]pyridine (BAMP) is catalytically active in the methanolysis of esters at pH 10.4.[28] Rate enhancements (Table 8.2) brought about by BAMP–Zn[28,29] are much higher for the carboxylate decorated esters **19–22** than for the parent phenyl acetate **23**, because binding to the metal ion transforms the moderately electron-releasing (rate-retarding) carboxylate into an electron-withdrawing (rate-enhancing) one. In the reactions of **23**, moderate differences in rate-enhancements among mono-, bi-, and trimetallic complexes indicate that the various metal centres act essentially as independent non-cooperative catalytic units. In marked contrast to the behaviour of phenyl acetate **23**, the reactions of the carboxylate-decorated esters are in many cases catalyzed by dinuclear complexes **24**–Zn_2 and **25**–Zn_2 much more strongly – 4 orders of

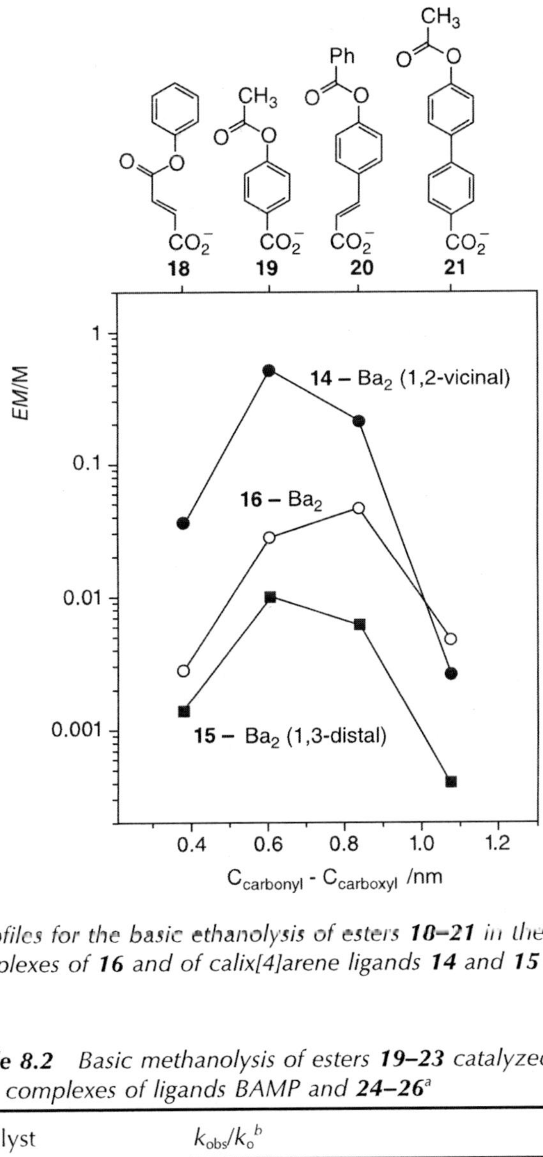

Figure 8.1 *EM profiles for the basic ethanolysis of esters **18–21** in the presence of the dinuclear Ba^{2+} complexes of **16** and of calix[4]arene ligands **14** and **15** (data from ref. 27)*

Table 8.2 *Basic methanolysis of esters **19–23** catalyzed by Zn^{2+} complexes of ligands BAMP and **24–26***[a]

Catalyst	k_{obs}/k_o[b]				
	23	**22**	**19**	**20**	**21**
BAMP–Zn	5.6	120	38	81	40
24–Zn_2 (1,2-vicinal)	10	15000	6200	1400	110
25–Zn_2 (1,3-distal)	4.4	1000	240	100	42
26–Zn_3	13	13000	24000	3400	180

[a] Data from refs. 28 and 29. Runs carried out in MeOH, pH 10.4 (*N,N*-diisopropyl-*N*-(2-methoxyethyl)amine buffer), 25 °C; 1 mM catalyst; substrate concentration in the range of 0.1–0.5 mM.

[b] k_o is the pseudo-first-order rate constant for reactions carried out in the presence of buffer alone. Ester **23**, $k_o = 9.4 \times 10^{-7}\ s^{-1}$; ester **22**, $k_o = 2.2 \times 10^{-7}\ s^{-1}$; ester **19**, $k_o = 5.5 \times 10^{-7}\ s^{-1}$; ester **20**, $k_o = 1.6 \times 10^{-7}\ s^{-1}$; ester **21**, $k_o = 8.9 \times 10^{-7}\ s^{-1}$.

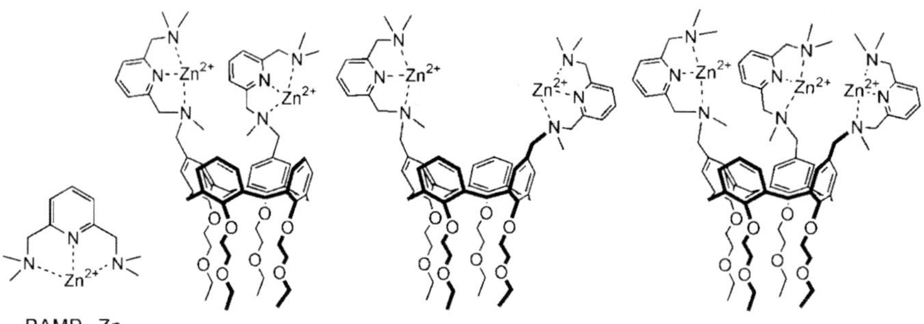

BAMP–Zn **24** – Zn_2 **25** – Zn_2 **26** – Zn_3

magnitude in the most favourable cases – than by mononuclear BAMP–Zn.[28,29] This is consistent with a bifunctional catalytic mechanism (Figure 8.2) in which one of the metal ions serves as a docking site for the carboxylate, and the other delivers an activated methoxide to the ester function. In the reactions catalyzed by the 1,2-vicinal complex **24**–Zn_2, the extent of cooperation of the two metal centres is quite high for esters **22** and **19**, significantly lower for ester **20** and even lower for ester **21**. Clearly, catalyst **24**–Zn_2 experiences an increasing difficulty to fit the altered substrate in the transition state when the ester size exceeds that of **19**. A similar trend of reactivity is seen in the reactions carried out in the presence of the less effective catalyst **25**–Zn_2. In this case, however, there is no sign of cooperation between metal centres in the reactions of **20** and **21**.

The marked superiority of the 1,2-vicinal complex **24**–Zn_2 over its distal regioisomer **25**–Zn_2 closely parallels that observed in the basic ethanolysis catalyzed by dinuclear complexes **14**–Ba_2 and **15**–Ba_2 (Table 8.1). The results obtained in the basic methanolysis of esters catalyzed by the zinc(II) complexes of calix[4]arenes **28–31** decorated with 1,5,9-triazacyclododecane ([12]aneN_3) ligands fully confirm the superiority of 1,2-vicinal bimetallic catalysts over their distal regioisomers.[30] Thus, the order of catalytic efficiency 1,2-vicinal ≫ 1,3-distal appears to be a substrate-independent feature of upper rim calix[4]arene-based bimetallic catalysts for the cleavage of carboxylate-functionalized esters, which is unaffected by the nature of the metal ion and of the corresponding ligating unit.

Kinetic data obtained in the reactions catalyzed by the BAMP ligated zinc complexes permit reliable estimates of the k_{intra} and k_{inter} quantities needed for the calculation of *EM*s.

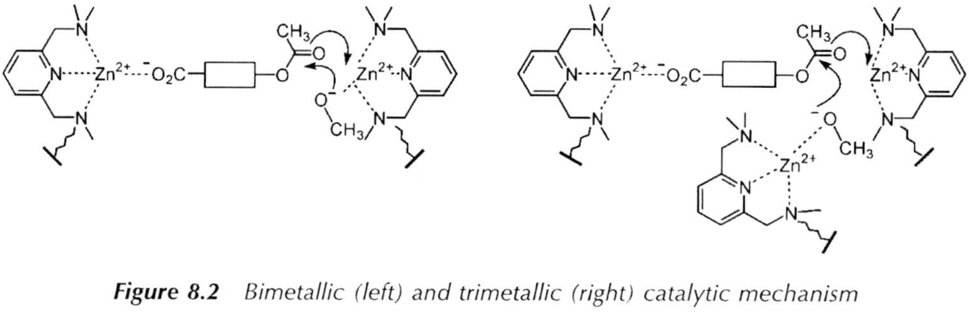

Figure 8.2 *Bimetallic (left) and trimetallic (right) catalytic mechanism*

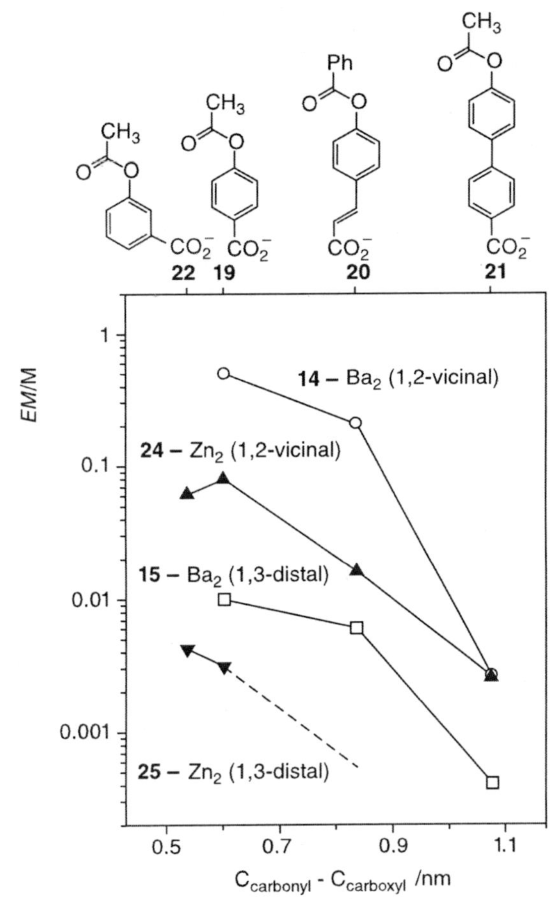

Figure 8.3 *EM profiles for the basic methanolysis of esters* ***19–22*** *in the presence of the dinuclear* Ba^{2+} *complexes of* ***14*** *and* ***15*** *(data from ref. 27) and of the dinuclear* Zn^{2+} *complexes of* ***24*** *and* ***25*** *(data from ref. 29)*

EM data are plotted in Figure 8.3 together with available data related to the basic ethanolysis catalyzed by dinuclear complexes **14**–Ba_2 and **15**–Ba_2. The superiority of both **14**–Ba_2 and **15**–Ba_2 to the structurally related **24**–Zn_2 and **25**–Zn_2 indicates a lower adaptability of the zinc(II) complexes to the altered substrates in the transition state, which is believed to arise from more stringent requirements of the coordinative interactions of a *d*-block metal ion compared with an *s*-block metal ion.

8.3.3 Artificial Esterases: Trimetallic Complexes

The highest catalytic activity in the methanolysis of esters **18–22** is exhibited in most cases by trinuclear complexes. In principle, a trinuclear complex could work as a dinuclear complex, with the third metal ion acting as a spectator. In **26**–Zn_3 and **31**–Zn_3 there are two 1,2-vicinal and one 1,3-distal bimetallic arrangements. Since the efficiency of 1,3-

distal bimetallic complexes is very low, the activity of trimetallic complexes may be thought as the sum of contributions arising from two 1,2-vicinal bimetallic arrangements. The statistical advantage of **26**–Zn_3 (**31**–Zn_3) over **24**–Zn_2 (**29**–Zn_2) is 2 under subsaturating conditions, but reduces to 4/3 under saturating conditions.[30] Thus, the statistical advantage of a trimetallic complex over its 1,2-vicinal bimetallic analogue should lie somewhere between 2 and 1.3, most likely closer to the lower limit because either set of catalytic experiments was carried out under conditions not far from saturation. The finding that **26**–Zn_3 is about four times as effective as **24**–Zn_2 in the cleavage of **19** (Table 8.2) provides a strong indication of the cooperation of three metal ions in the catalytic mechanism. Efficient cooperation of three metal ions (Zn, Cu) in the cleavage of phosphodiesters was reported in a number of studies[25,31] but, to the best of our knowledge, trimetallic catalysis in ester cleavage is unprecedented.

The proposed mechanism depicted in Figure 8.2 emphasizes the three different functions performed by the metal ions: (*i*) substrate recognition through binding to carboxylate, (*ii*) Lewis-acid activation of the ester carbonyl, and (*iii*) nucleophile delivery. We further note that k_{obs}(**26**–Zn_3)/k_{obs}(**24**–Zn_2) = 2.4 in the cleavage of **20** (Table 2), and that k_{obs}(**31**–Zn_3)/k_{obs}(**29**–Zn_2) = 2.4 also in the cleavage of **22**.[30] It appears therefore that there are small, but higher than statistical advantages of trinuclear over 1,2-vicinal dinuclear catalysts, which might be taken as an indication of a modest contribution from the third metal ion.

8.3.4 Artificial Nucleases: Zinc(II) Complexes

Efficient catalysis of HPNP transesterification (Equation 8.10) arising from synergic action of the two metal centres was reported in previous studies[25] for the dinuclear zinc(II) complex of BAMP functionalized calix[4]arene **25**. In marked contrast to what observed on the catalytic cleavage of esters (Table 8.2), the 1,2-vicinal complex **24**–Zn_2 was subsequently found[28] to be significantly less effective than its 1,3-distal regioisomer **25**–Zn_2.

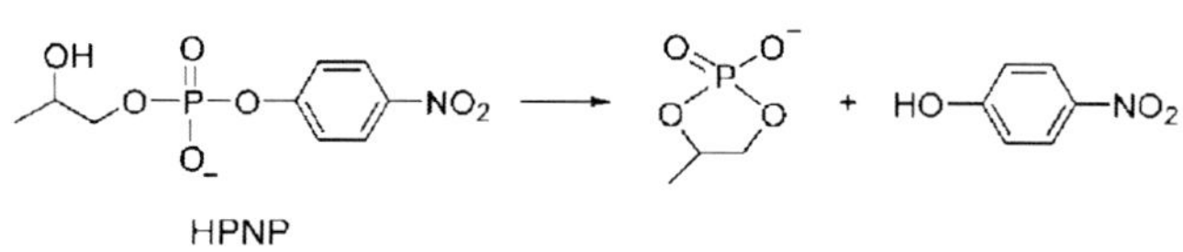

(8.10)

A major limitation of the BAMP functionalized catalysts is the limited solubility in water, that requires the use of an organic cosolvent. Another limitation is the affinity of BAMP for zinc(II), that is large enough for catalytic studies run at millimolar catalyst concentrations in 50% aqueous CH_3CN at pH 7.0 ($K \approx 1 \times 10^5 M^{-1}$), but drops to $1 \times 10^3 M^{-1}$ in water at pH 7.0. In the aim at developing metal catalysts which are soluble enough in water to avoid the use of organic cosolvents, the [12]aneN_3 ligating unit seemed appropriate, because its high affinity for zinc(II) in water (Log K = 8.4 at 25 °C) ensures extensive binding of the metal ion to the ligand even at submillimolar concentrations.[32] Unfortunately, the zinc(II) complexes of ligands **28–31** were not soluble enough in water for catalytic studies. Furthermore, catalytic studies carried out in 50% aqueous CH_3CN (Table 8.3) revealed a disappointing lack of synergism between metal ions, which is at variance with the results obtained with the corresponding BAMP complexes.[25,28] The moral of this story is that even promising arrangements of catalytically effective ligated

Table 8.3 *Transesterification of HPNP catalyzed by Zn^{2+} and Cu^{2+} complexes of [12]aneN_3-based ligands*[a]

Catalyst	k_{obs}/k_o	
	M = Zn[b]	M = Cu[c]
[12]aneN_3–M	450	15
27–M	190	26
29–M_2 (1,2-vicinal)	330	1090
30–M_2 (1,3-distal)	320	40
31–M_3	1100	590

[a] Data from ref. 32.
[b] Reaction conditions: 50% CH_3CN/H_2O, 20 mM HEPES, pH 7, 25 °C, 0.2 mM catalyst, 0.15 mM HPNP. In the presence of buffer alone: $k_o = 1.9 \times 10^{-8}\,s^{-1}$.
[c] Reaction conditions: water, 20 mM HEPES, pH 7, 25 °C, 0.2 mM catalyst, 0.1 mM HPNP. In the presence of buffer alone: $k_o = 2.2 \times 10^{-7}\,s^{-1}$.

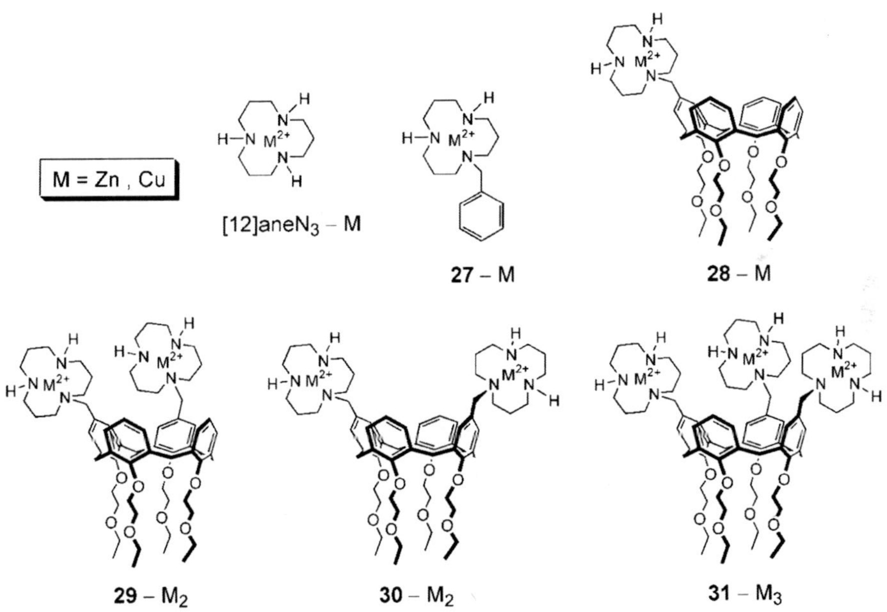

metal ions on a molecular platform by no means guarantee effective cooperation between catalytic units.

8.3.5 Artificial Nucleases: Copper(II) Complexes

The affinity of [12]aneN_3 for copper(II) in water (Log K = 12.6 at 25 °C) is still higher than that of zinc(II) and the copper(II) complexes of ligands **28–31** are soluble enough in water for kinetic experiments, the highest solubility being experienced by **31**–Cu_3. Although mononuclear copper(II) complexes are an order of magnitude less effective than the corresponding zinc(II) complexes in the cleavage of HPNP (Table 3), cooperation

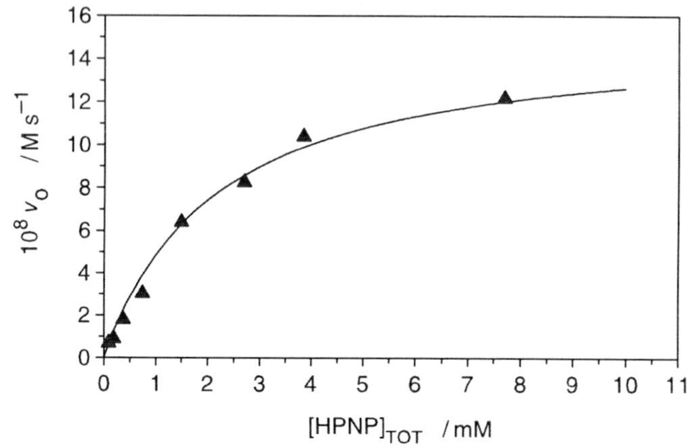

Figure 8.4 Initial rate of HPNP transesterification vs. substrate concentration for the reaction catalyzed by 0.2 mM **31**–Cu_3 (water, pH 7 (20 mM HEPES), 25 °C)

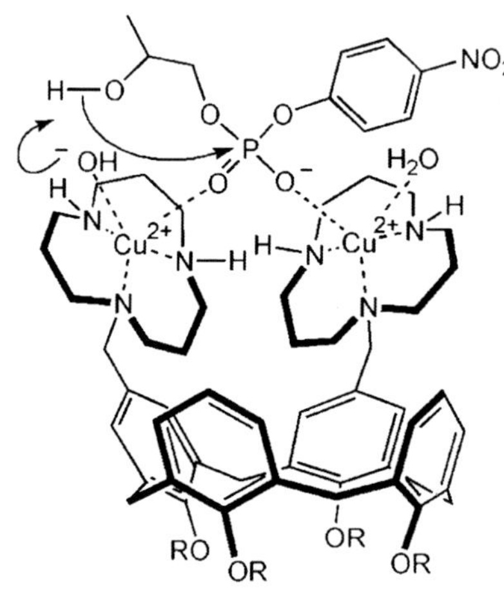

Figure 8.5 HPNP transesterification catalyzed by **29**–Cu_2. Proposed mode of substrate binding and activation

between metal centres is significant in the 1,2-vicinal dinuclear complex **29**–Cu_2.[32] In contrast, the catalytic efficiency of **30**–Cu_2 is about twice as great as that of **27**–Cu, showing that there is no rate acceleration per metal centre. It also appears that the trimetallic complex **31**–Cu_3 behaves as a 1,2-vicinal dinuclear catalyst, whose efficiency is somewhat lowered by the third metal centre. The close adherence to the Michaelis–Menten equation (Figure 8.4) indicates the intermediacy of a reversibly formed, moderately stable HPNP-catalyst complex ($K = 1/K_M = 500\,M^{-1}$). A likely mode of binding of HPNP to the catalyst involves interaction of the phosphate moiety to the copper(II) ions in the 1,2-vicinal position (Figure 8.5), in what amounts to double Lewis-acid activation, with the putative involvement of the metal hydroxide as a general base.

Table 8.4 *Cleavage of diribonucleotide 3′-5′-monophosphates N′pN″ in the presence of trinuclear catalyst* **31**–Cu_3[a]

N′pN″	k_{obs} (s^{-1})	k_{rel}	k_{obs}/k_o[b]
CpA	$<2 \times 10^{-7}$[c]	<0.14	$<3 \times 10^2$
GpA	1.4×10^{-6}	1.0	
CpG	2.1×10^{-6}	1.5	0.6×10^4
CpC	2.5×10^{-6}	1.8	0.6×10^4
ApG	2.7×10^{-6}	1.9	1.3×10^4
GpG	2.8×10^{-6}	2.0	1.3×10^4
GpU	3.7×10^{-6}	2.6	
UpG	4.2×10^{-5}	30	
UpU	5.5×10^{-5}	39	1.4×10^5

[a] Data from ref. 32. Reaction conditions: 1 mM **31**–Cu_3, 0.1 mM *N′pN″*, water, pH 7 (HEPES 20 mM), 50 °C.
[b] $10^{10} k_o$ (pH 7, 50 °C) (s^{-1}) : CpA, 6.1; CpC, 4.3; UpU, 3.9; CpG, 3.5; GpG, 2.2; ApG, 2.1.
[c] No reaction after 24 h.

It is well known that species that cleave effectively *p*-nitrophenyl phosphates are not necessarily good catalysts for the cleavage of unactivated phosphate esters. It was therefore quite gratifying to find that **31**–Cu_3 cleaves with high efficiency eight diribonucleoside monophosphates (Equation 8.11) out of nine in the tested lot (Table 8.4).[32] Catalytic rate accelerations are in the order of 10^4-fold in most cases, to be compared with the 10^3-fold rate accelerations in the cleavage of HPNP (Table 8.3). The rate enhancement rises to 10^5-fold in the cleavage of UpU and this was suggested to be caused by the additional binding site arising from the copper-assisted deprotonation of the uracyl moiety at the 5′-hydroxyl terminus.

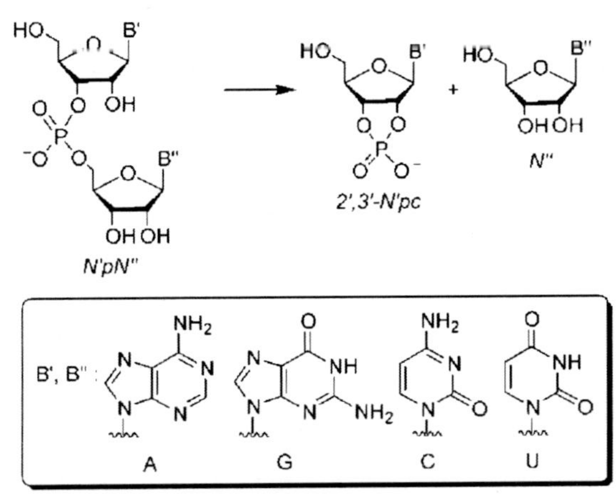

(8.11)

The inertness of CpA in the presence of **31**–Cu_3 (Table 8.4) is not due to the inherently low reactivity of the substrate, nor to its insensitivity to copper(II) catalysis, as shown by the finding that CpA is cleaved by [12]aneN_3–Cu and **30**–Cu_2 about 3 times more rapidly than UpU.[32] Furthermore, the efficiency of **31**–Cu_3 in the cleavage of UpU is similar to that of **29**–Cu_2, and this reinforces the view that **31**–Cu_3 uses only two vicinal metal centres in the catalysis, with the third metal ion acting as a more or less innocent spectator.

The phosphodiesterase activity of the copper(II) complexes was further explored in the cleavage of oligomeric ribonucleotides.[33] The kinetics was investigated using gel electrophoresis to separate and analyze reactant and products having a radioactive phosphate label in the 5′-terminal position (Equation 8.12). Rate constants for the cleavage of all of the scissile bonds of oligoribonucleotides **32–37**, obtained either from time-course kinetics (e.g. Figure 8.6) or from initial rates, are listed in Table 8.5, where the fatness of the arrows pointing to the scissile bonds gives a rough indication of the relative reactivity. There is an undeniable tendency for the catalytic efficiency to increase with the number of metal units, revealing variable extents of cooperation between metal centres of trimetallic and bimetallic complexes. In a number of cases the 1,2-vicinal complex **29**–Cu_2 is superior to its distal regioisomer **30**–Cu_2, in line with what observed in the reactions of HPNP and UpU, but in other cases the reverse holds. A similar, not clearly defined situation, holds for the relative efficiency of trinuclear *vs.* dinuclear complexes, because the observed reactivity order strongly depends on oligoribonucleotide identity. In any event, whenever the trinuclear complex is the best cleaving agent in the lot, the advantage due to the third metal ion hardly exceeds a factor of 2, which indicates that a compelling evidence of the operation of trimetallic catalysis is lacking. It is worth noting that the rate enhancement of *ca.* 5×10^5-fold relative to background in the cleavage of CpA bound in **35** by **31**–Cu_3 is one of the highest values recorded for a synthetic metallonuclease in the cleavage of ribonucleotide dimers or higher oligomers.[25,34]

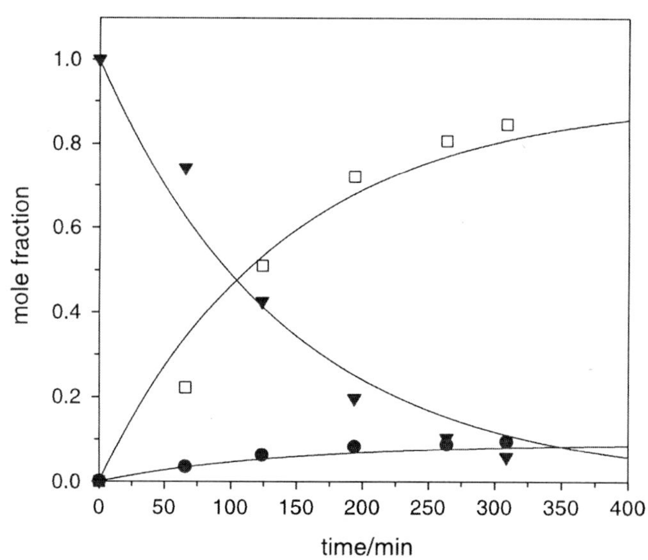

Figure 8.6 *Reaction of 5′-p*ACCAUC (**32**) promoted by 50 μm **31**-Cu_3 (water, pH 7, 25 °C). Mole fraction of reactant (▼), and of fragments 5′-p*ACC (□; CpA cleavage), and of 5′-p*AC (●; CpC cleavage), vs time*

Table 8.5 *Cleavage of oligoribonucleotides **32–37** promoted by Cu^{2+}-calixarene complexes. Pseudo-first order rate constants (10^6k, s^{-1}) in water, pH 7.4, 50 °C*[a]

<table>
<tr><th>Catalyst</th><th colspan="6">Substrate</th></tr>
<tr><td></td><td>32</td><td>33</td><td>34</td><td>35</td><td>36</td><td>37</td></tr>
<tr><td>28–Cu</td><td>ACCAUC</td><td>CGCUGA</td><td>AGGUUAA</td><td>CAGGCC</td><td>CCGGCA</td><td>ACUAUC</td></tr>
<tr><td></td><td>C|C 0.3
C|A 5.8</td><td>b)</td><td>b)</td><td>C|A 7.1</td><td>b)</td><td>U|A 2.1</td></tr>
<tr><td>29–Cu_2
(1,2-vicinal)</td><td>ACCAUC</td><td>CGCUGA</td><td>AGGUUAA</td><td>CAGGCC</td><td>CCGGCA</td><td>ACUAUC</td></tr>
<tr><td></td><td>C|C 4.6
C|A 35</td><td>b)</td><td>U|A 13</td><td>C|A 130</td><td>C|C 76
C|G 17
C|A 47</td><td>U|A 25</td></tr>
<tr><td>30–Cu_2
(1,3-distal)</td><td>ACCAUC</td><td>CGCUGA</td><td>AGGUUAA</td><td>CAGGCC</td><td>CCGGCA</td><td>ACUAUC</td></tr>
<tr><td></td><td>C|C 2.8
C|A 13</td><td>C|G 3.7
C|U 1.3
U|G 0.4</td><td>U|A 5.2</td><td>C|A 110</td><td>C|C 15
C|G 3.2
C|A 10</td><td>U|A 50</td></tr>
<tr><td>31–Cu_3</td><td>ACCAUC</td><td>CGCUGA</td><td>AGGUUAA</td><td>CAGGCC</td><td>CCGGCA</td><td>ACUAUC</td></tr>
<tr><td></td><td>C|C 11
C|A 110</td><td>C|G 6.4
C|U 1.1
U|G 1.1</td><td>U|A 9.5</td><td>C|A 300</td><td>C|C 17
C|G 2.1
C|A 11</td><td>U|A 51</td></tr>
</table>

[a]Data from ref. 33. Reaction conditions: <0.2 nM substrate, 50 μM catalyst.
[b]No reaction in 5 h.

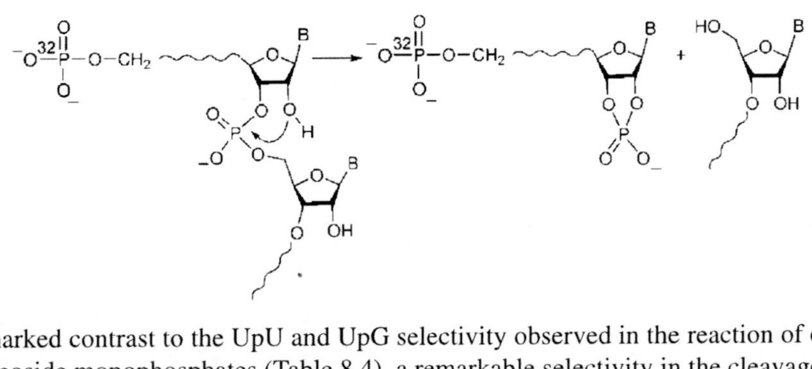

(8.12)

In marked contrast to the UpU and UpG selectivity observed in the reaction of oligoribonucleoside monophosphates (Table 8.4), a remarkable selectivity in the cleavage of the CpA bond of oligoribonucleotides **32** and **35** is observed for all metal complexes (Table 8.5). This is fully confirmed by the behaviour of heptadecamer **38**, in which all CpA bonds are cleaved by **29**–Cu_2 (Table 8.6), whereas other bonds do not undergo appreciable scission. A closely similar behaviour was experienced in the presence of **30**–Cu_2. CpA

Table 8.6 *Cleavage of **38** promoted by **29**-Cu_2 complex. Pseudo-first order rate constants (10^6k, s^{-1}) in water, pH 7.4, 50 °C*[a]

Catalyst	Substrate	
29-Cu_2 (1,2-vicinal)	**38** 5'-GCAAGCACAGACAUCAG-3' α β γ δ ε	
	C|A (α)	42
	C|A (β)	41
	C|A (γ)	38

[a]Data from ref. 33. Reaction conditions: 0.1 nM substrate, 10 μM catalyst.

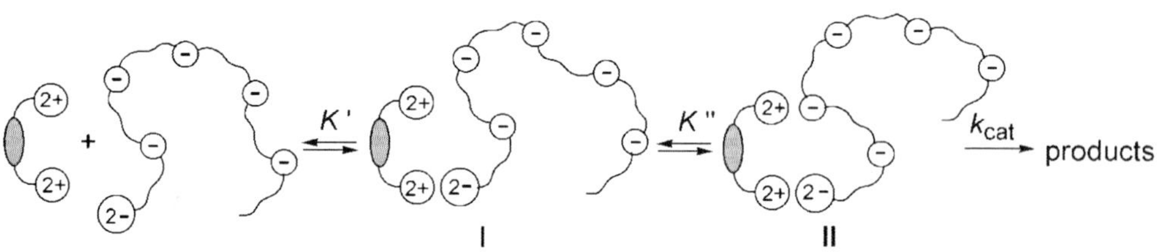

Figure 8.7 *Catalysis by a dinuclear Cu^{2+}-calixarene complex in the cleavage of an oligoribonucleotide labelled at the 5′-terminus with a ^{32}P phosphate dianion*

bonds labelled as α, β and γ are cleaved much more rapidly than bonds labelled as δ and ε, which suggests that far away CpA bonds are cleaved more slowly than bonds closer to the 5′-terminal position. This view is corroborated by the reactivity ratios of 27 : 10 : 1 calculated from the data in Table 8.5 for the **31**–Cu_3 catalyzed cleavage of the CpA bond in **35**, **32** and **36**, respectively. A mechanism consistent with the above findings is depicted in Figure 8.7. The primary binding interaction between catalyst and substrate involves one of the metal centres and the ^{32}P-radiolabelled 5′-terminal phosphate dianion. One of the possible productive complexes (**II**) is depicted as a dinuclear metallomacrocycle arising from intramolecular binding of one of the singly charged phosphates to the other metal centre. The fact that K^{II} is expected to decrease on increasing the ring size is well in keeping with the finding that scissile bonds close to the 5′-terminus are cleaved more easily.

8.4 Concluding Remarks

In this chapter we have shown that calixarenes have undergone a limited, but highly significant amount of research on their ability to include into their cavities reactant guests or part-structures of reactants, thereby exerting a strong reactivity control of hosted species in stoichiometric and catalytic reactions.

The versatility of calixarenes as platforms on which to build mono- and multitopic ligands has been known for long. In the second part of the chapter we have focussed on the dynamic preorganization of the calix[4]arene platform, that induces high levels of cooperation between two and, possibly, three ligated metal ions in the cleavage of carboxylate esters and phosphodiesters, including RNA oligonucleotides.

It is felt that calixarenes have not exhausted their potential yet, and that future research in the area of reactivity control and catalysis by calixarenes looks promising of exciting new results.

References

1. a) C. D. Gutsche, *Calixarenes*, The Royal Society of Chemistry, Cambridge, 1989; b) C. D. Gutsche, *Calixarenes Revisited*, The Royal Society of Chemistry, Cambridge, 1998; c) J. Vicens, V. Böhmer (Eds.), *Calixarenes: A Versatile Class of Macrocyclic Compounds*, Kluwer Academic Publishers, Dordrecht, 1991; d) L. Mandolini, R. Ungaro (Eds.), *Calixarenes in Action*, Imperial College Press, London, 2000; e) Z. Asfari, V. Böhmer, J. Harrowfield, J. Vicens (Eds.), *Calixarenes 2001*, Kluwer, Dordrecht, 2001; f) J. Vicens, J. Harrowfield (Eds.) *Calixarenes in the Nanoworld*, Springer, Dordrecht, 2006. g) W. Sliwa, C. Kozlowski, *Calixarenes and Resorcinarenes: Synthesis Properties and Applications*, Wiley-VCH, Weinheim, 2009.
2. a) L. Wang, X. F. Shi, X. J. Hu, Y. Liu, Research and application of the calixarene-containing polymer materials, *Prog. Chem.*, 2002, **14**, 217–224; b) A. Casnati, F. Sansone, R. Ungaro, Peptido- and glycocalixarenes: playing with hydrogen bonds around hydrophobic cavities, *Acc. Chem. Res.*, 2003, **36**, 246–254; c) S. E. Matthews, P. D. Beer, Calixarene-based anion receptors, *Supramol. Chem.*, 2005, **17**, 411–435; d) L. Baldini, A. Casnati, F. Sansone, R. Ungaro, Calixarene-based multivalent ligands, *Chem. Soc. Rev.*, 2007, **36**, 254–266; e) J. S. Kim, D. T. Quang, Calixarene-derived fluorescent probes, *Chem. Rev.*, 2007, **107**, 3780–3799; f) S. M. Biros, J. Rebek Jr., Structure and binding properties of water-soluble cavitands and capsules, *Chem. Soc. Rev.*, 2007, **36**, 93–104; g) W. Maes, W. Dehaen, Oxacalix[*n*](het)arenes, *Chem. Soc. Rev.*, 2008, **37**, 2393–2402; h) A. W. Coleman, S. Jebors, P. Shahgaldian, G. S. Ananchenko, J. A. Ripmeester, *para*-Acylcalix[*n*]arenes: from molecular to macroscopic assemblies, *Chem. Commun.*, 2008, 2291–2303; i) D. M. Homden, C. Redshaw, The use of calixarenes in metal-based catalysis, *Chem Rev.*, 2008, **108**, 5086–5130.
3. R. Cacciapaglia, L. Mandolini, Calixarene based catalytic systems, in *Calixarenes in Action*, L. Mandolini, R. Ungaro (Eds.), Imperial College Press, London, 2000, pp. 241–264.
4. R. Rathore, S. V. Lindeman, K. S. S. P. Rao, D. Sun, J. K. Kochi, Guest penetration deep within the cavity of calix[4]arene hosts: the tight binding of nitric oxide to distal (cofacial) aromatic groups, *Angew. Chem. Int. Ed.*, 2000, **39**, 2123–2127.
5. a) G. V. Zyryanov, Y. Kang, D. M. Rudkevich, Sensing and fixation of NO_2/N_2O_4 by calix[4]arenes, *J. Am. Chem. Soc.*, 2003, **125**, 2997–3007; b) Y. Kang, G. V. Zyryanov, D. Rudkevich, Towards supramolecular fixation of NO_x gases: encapsulated reagents for nitrosation, *Chem. Eur. J.*, 2005, **11**, 1924–1932.
6. Our good friend and distinguished colleague Dmitry Rudkevich passed away prematurely in 2007 at the age of 44. The authors wish to dedicate this chapter to his memory.
7. O. Reinaud, Y. Le Mest, I. Jabin, Models of metallo-enzyme active sites, in *Calixarenes in the Nanoworld*, J. Vicens, J. Harrowfield (Eds.), Springer, Dordrecht, 2006.
8. Y. Rondelez, M.-N. Rager, A. Duprat, O. Reinaud, Calix[6]arene-based cuprous 'funnel complexes': a mimic for the substrate access channel to metalloenzyme active sites, *J. Am. Chem. Soc.*, 2002, **124**, 1334–1340.
9. a) N. Le Poul, M. Campion, G. Izzet, B. Douziech, O. Reinaud, Y. Le Mest, Electrochemical behavior of the tris(pyridine)–Cu funnel complexes: an overall induced-fit process involving

an entatic state through a supramolecular stress, *J. Am. Chem. Soc.*, 2005, **127**, 5280–5281; b) N. Le Poul, M. Campion, B. Douziech, Y. Rondelez, L. Le Clainche, O. Reinaud, Y. Le Mest, Monocopper center embedded in a biomimetic cavity: from supramolecular control of copper coordination to redox regulation, *J. Am. Chem. Soc.*, 2007, **129**, 8801–8810.

10. G. Izzet, J. Zeitouny, H. Akdas-Killig, Y. Frapart, S. Ménage, B. Douziech, I. Jabin, Y. Le Mest, O. Reinaud, Dioxygen activation at a mononuclear Cu(I) center embedded in the calix[6]arene-tren core, *J. Am. Chem. Soc.*, 2008, **130**, 9514–9523.
11. S. Shinkai, S. Mori, H. Koreishi, T. Tsubaki, O. Manabe, Hexasulfonated calix[6]arene derivatives: a new class of catalysts, surfactants, and host molecules, *J. Am. Chem. Soc.*, 1986, **108**, 2409–2416.
12. K. Goto, Y. Yano, E. Okada, C.-W. Liu, K. Yamamoto, R. Ueoka, Catalytic specificity exhibited by *p*-sulfonatocalix[*n*]arenes in the methanolysis of *N*-acetyl-L-amino acids, *J. Org. Chem.*, 2003, **68**, 865–870.
13. a) J. O. Magrans, A. R. Ortiz, A. Molins, P. H. P. Lebouille, J. Sánchez-Quesada, P. Prados, M. Pons, J. de Mendoza, A designed non-peptidic receptor that mimics the phosphocholine binding site of the McPC603 antibody, *Angew Chem. Int. Ed.*, 1996, **35**, 1712–1715; b) F. Cuevas, S. Di Stefano, J. O. Magrans, P. Prados, L. Mandolini, J. de Mendoza, Toward an artificial acetylcholinesterase, *Chem. Eur. J.*, 2000, **6**, 3228–3234.
14. S. Richeter, J. Rebek Jr., Catalysis by a synthetic receptor sealed at one end and functionalized at the other, *J. Am. Chem. Soc.*, 2004, **126**, 16280–16281.
15. F. H. Zelder, J. Rebek Jr., Cavitand templated catalysis of acetylcholine, *Chem Comm.*, 2006, 753–754.
16. A. Gissot, J. Rebek Jr., A functionalized, deep cavitand catalyzes the aminolysis of a choline derivative, *J. Am. Chem. Soc.*, 2004, **126**, 7424–7425.
17. R. J. Hooley, J. Rebek Jr., A deep cavitand catalyzes the Diels–Alder reaction of bound maleimides, *Org. Biomol. Chem.*, 2007, **5**, 3631–3636.
18. S. Butterfield, J. Rebek Jr., A cavitand stabilizes the Meisenheimer complex of S_NAr reactions, *Chem. Comm.*, 2007, 1605–1607.
19. B. W. Purse, A. Gissot, J. Rebek Jr., A deep cavitand provides a structured environment for the Menschutkin reaction, *J. Am. Chem. Soc.*, 2005, **127**, 11222–11223.
20. a) R. J. Hooley, T. Iwasawa, J. Rebek Jr., Detection of reactive tetrahedral intermediates in a deep cavitand with an introverted functionality, *J. Am. Chem. Soc.*, 2007, **129**, 15330–15339; b) T. Iwasawa, R. J. Hooley, J. Rebek Jr., Stabilization of labile carbonyl addition intermediates by a synthetic receptor, *Science*, 2007, **317**, 493–496.
21. P. Restorp, J. Rebek Jr., Reaction of isonitriles with carboxylic acids in a cavitand: observation of elusive isoimide intermediates, *J. Am. Chem. Soc.*, 2008, **130**, 11850–11851.
22. D. J. Cram, J. M. Cram, *Container Molecules and Their Guests*, The Royal Society of Chemistry, Cambridge, 1994.
23. J. Rebek Jr., Simultaneous encapsulation: molecules held at close range, *Angew. Chem. Int. Ed.*, 2005, **44**, 2068–2078; b) R. Cacciapaglia, S. Di Stefano, L. Mandolini, Effective molarity in supramolecular catalysis of two-substrate reactions, *Acc. Chem. Res.*, 2004, **37**, 113–122.
24. See, for example: a) J. Weston, Mode of action of bi- and trinuclear zinc hydrolases and their synthetic analogues, *Chem. Rev.*, 2005, **105**, 2151–2174; b) J. A. Cowan, Metal activation of enzymes in nucleic acid biochemistry, *Chem. Rev.*, 1998, **98**, 1067–1087; c) D. E. Wilcox, Binuclear metallohydrolases, *Chem. Rev.*, 1996, **96**, 2435–2458.
25. P. Molenveld, J. F. J. Engbersen, D. N. Reinhoudt, Dinuclear metallo-phosphodiesterase models: application of calix[4]arenes as molecular scaffolds, *Chem. Soc. Rev.*, 2000, **29**, 75–86.
26. R. Cacciapaglia, S. Di Stefano, E. Kelderman, L. Mandolini, Supramolecular catalysis of ester and amide cleavage by a dinuclear barium(II) complex, *Angew Chem. Int. Ed.*, 1999, **38**, 348–351.
27. R. Cacciapaglia, A. Casnati, S. Di Stefano, L. Mandolini, D. Paolemili, D. N. Reinhoudt, A. Sartori, R. Ungaro, Dinuclear barium(II) complexes based on a calix[4]arene scaffold as catalysts of acyl transfer, *Chem Eur. J.*, 2004, **10**, 4436–4442.

28. R. Cacciapaglia, A. Casnati, L. Mandolini, D. N. Reinhoudt, R. Salvio, A. Sartori, R. Ungaro, Di- and trinuclear Zn^{2+} complexes of calix[4]arene based ligands as catalysts of acyl and phosphoryl transfer reactions, *J. Org. Chem.*, 2005, **70**, 624–630.
29. R. Cacciapaglia, A. Casnati, L. Mandolini, D. N. Reinhoudt, R. Salvio, A. Sartori, R. Ungaro, Calix[4]arene-based Zn^{2+} complexes as shape and size-selective catalysts of ester cleavage, *J. Org. Chem.*, 2005, **70**, 5398–5402.
30. R. Cacciapaglia, A. Casnati, L. Mandolini, D. N. Reinhoudt, R. Salvio, A. Sartori, R. Ungaro, Di- and trinuclear arrangements of zinc(II)-1,5,9-triazacyclododecane units on the calix[4]arene scaffold: efficiency and substrate selectivity in the catalysis of ester cleavage, *Inorg. Chim. Acta*, 2007, **360**, 981–986.
31. M. Komiyama, S. Kina, K. Matsumura, J. Sumaoka, S. Tobey, V. M. Lynch, E. Anslyn, Trinuclear copper(II) complex showing high selectivity for the hydrolysis of 2′-5′ over 3′-5′ for UpU and for 3′-5′ over 2′-5′ for ApA ribonucleotides, *J. Am. Chem. Soc.*, 2002, **124**, 13731–13736.
32. R. Cacciapaglia, A. Casnati, L. Mandolini, D. N. Reinhoudt, R. Salvio, A. Sartori, R. Ungaro, Catalysis of diribonucleoside monophosphate cleavage by water soluble copper(II) complexes of calix[4]arene based nitrogen ligands, *J. Am. Chem. Soc.*, 2006, **128**, 12322–12330.
33. R. Cacciapaglia, A. Casnati, L. Mandolini, A. Peracchi, D. N. Reinhoudt, R. Salvio, A. Sartori, R. Ungaro, Efficient and selective cleavage of RNA oligonucleotides by calix[4]arene-based synthetic metallonucleases, *J. Am. Chem. Soc.*, 2007, **129**, 12512–12520.
34. K. Nwe, C. M. Andolina, J. R. Morrow, Tethered dinuclear europium(III) macrocyclic catalysts for the cleavage of RNA, *J. Am. Chem. Soc.*, 2008, **130**, 14861–14871.

9

Reactions Inside Carcerands

Ralf Warmuth

Rutgers. The State University of New Jersey, Department of Chemistry and Chemical Biology, Piscataway, NJ 08854, USA

9.1 Introduction

Molecular container molecules are spherical, hollow hosts with inner cavities that are large enough to accommodate one or more guest molecules.[1] They were developed by Donald J. Cram in the mid 1980s and have been used extensively as molecular reaction flask in order to study the effect of the confinement on reactions involving the encapsulated guest.[2]

In 1985, Cram and co-workers synthesized carcerand **1** by multiply linking cavitand **2** and **3** (Figure 9.1).[3] The name carcerand is derived from the Latin word *carcer*, which means 'prison'. During the shell-closure, **1** trapped almost every component present in the reaction flask. In carcerands, the incarcerated guest cannot leave the container even at high temperature. Complexes with permanently imprisoned guests are termed carceplexes. In contrast, hemicarcerands form stable hemicarceplexes at ambient temperature, but incarcerate and release guests at elevated temperature, whereby the guest passes through one of the size-restricted equatorial openings in the host shell (Figure 9.2).

Hemicarceplexes are stabilized by intrinsic and constrictive binding energy.[4] The latter, a physical barrier, is the activation free energy required for a guest to enter the inner cavity and typically amounts to 25–30 kcal mol^{-1}.[4b] In carceplexes, this barrier is so high that guest escape is impossible without host destruction. Isolated from the bulk phase, the

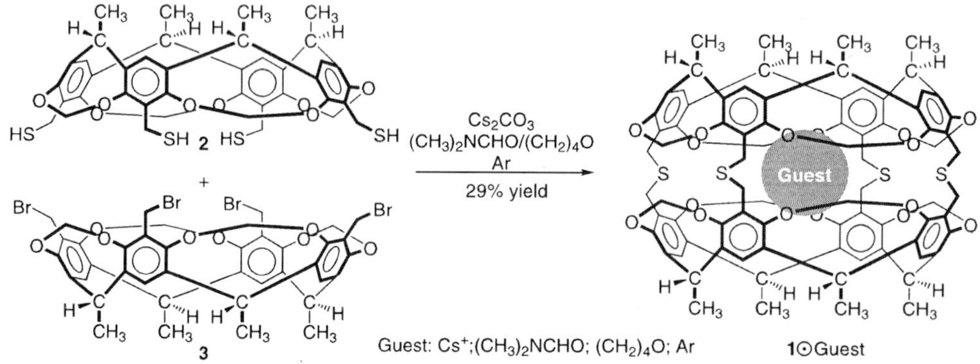

Figure 9.1 *Templated synthesis of carceplex **1**⊙Guest by four-fold connecting of cavitands **2** and **3***[3]

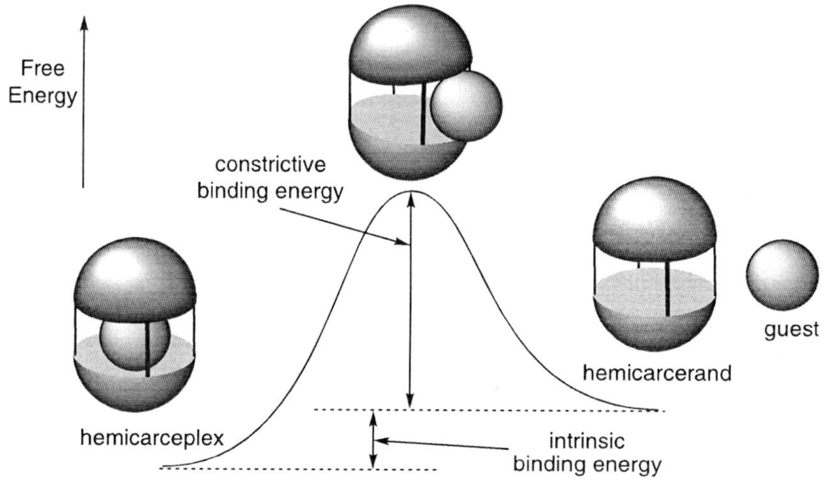

Figure 9.2 *Free energy profile of hemicarceplex dissociation*

guest can freely rotate inside its molecular prison. Cram coined the name 'inner phase' for the interior of a container compound since the properties of an incarcerated guest molecule are different from those in the bulk phase.

A large variety of hemicarcerands have been synthesized by connecting two cavitands with four appropriate linkers (Figure 9.3).[4b,5] Among those, perhaps, the most versatile container molecule with respect to binding properties is hemicarcerand **4**, which forms stable hemicarceplexes with guests ranging in size from benzene to naphthalene. Owing to its excellent binding properties and the ease of preparing hemicarceplexes, **4** has been the host of choice for applications as molecular reaction flask to study reactions of encapsulated guests.

The synthesis of hemicarcerands and carceplexes is a templated reaction and the proper choice of template molecule, which complements the shape and size of the inner cavity

(a)

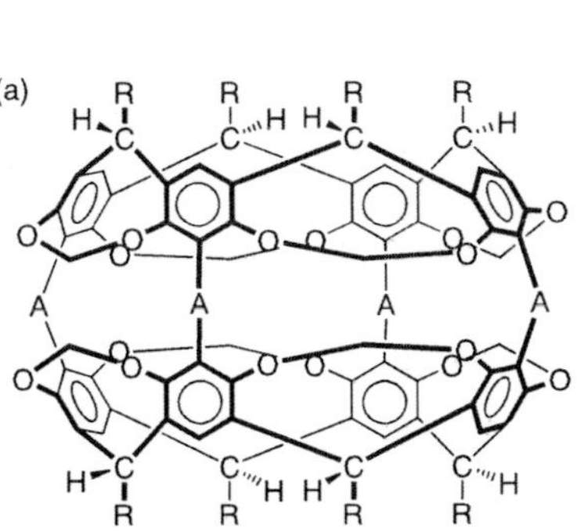

R = CH_3; $(CH_2)_4CH_3$; CH_2CH_2Ph

Compound #	A	Compound #	A
4	O-$(CH_2)_4$-O	**26**	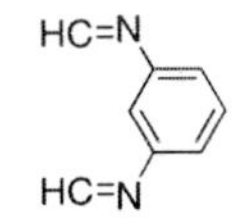
5	O-CH_2-O		
14	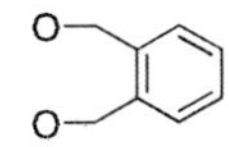	**109**	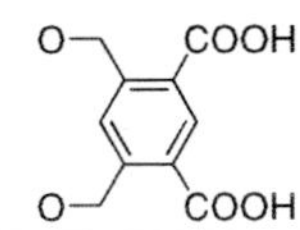

(b)

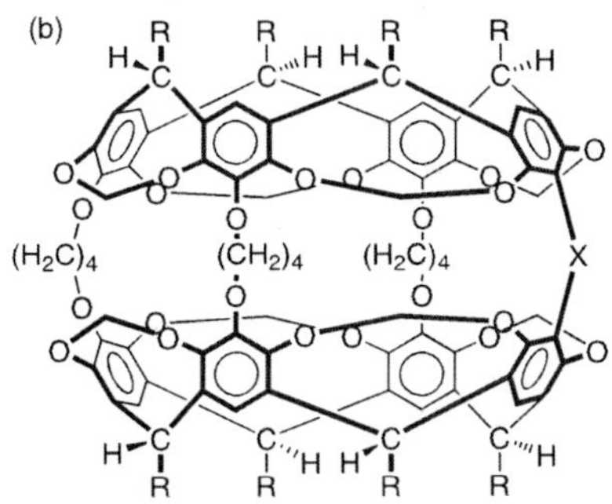

Compound #	X	Compound #	X
7	OH HO	**12**	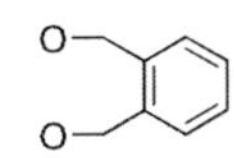
8	O-CH_2-O		
9	O-$(CH_2)_2$-O	**13**	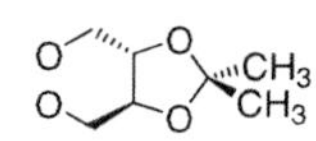
10	O-$(CH_2)_3$-O		
11	O-$(CH_2)_5$-O		

Figure 9.3 *Structures of symmetrical (A) and asymmetrical hemicarcerands (B)*

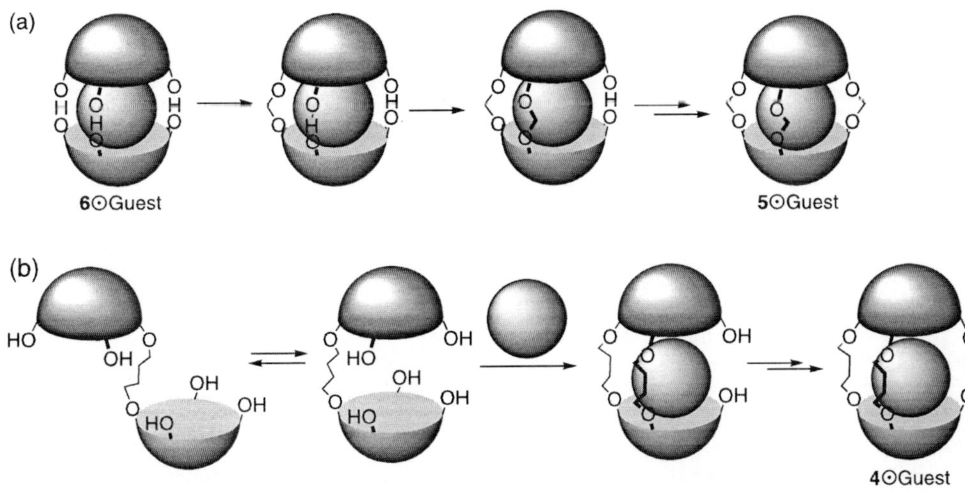

Figure 9.4 *Proposed mechanism for the formation of hemicarceplexes* ***5****⊙Guest (A) and* ***4****⊙Guest (B) and relative templation efficiencies in the formation of* ***5****⊙Guest (C)*

of the desired hemicarcerand, is vitally important.[6–7] For carceplex **5**⊙Guest, yields ranged from zero to 89% and template efficiency correlates nicely with the size and shape complementarity of inner cavity and guest.[8] Yields are highest, when the cavitands and template can be preassembled in a trimeric hydrogen-bonded complex **6**.[8b–c,9] The H-bonding complex **6** is unimportant for the formation of hemicarceplexes **4**⊙Guest and of other larger hemicarceplexes.[10] For the larger host **4**, linking the two cavitands with the first bridge disrupts the H-bonds of **6** allowing for subsequent intermolecular and intramolecular reactions. The template does not affect the course of the reaction until after the second linker is introduced and has determined the fate of the product. In *N*-methylpyrrolidinone as solvent, closing the host shell with the fourth bridge is rate determining and requires heating.[5e,f] At room temperature, diol **7** accumulates and becomes available. Diol **7** provides an easy entry to unsymmetrical hemicarcerands with three unique and one different linker (e.g. **8–13** in Figure 9.3).[5e,f,11] Furthermore, using the so-called 'seal-in' procedure, hemicarceplexes **4**⊙Guest or **8–13**⊙Guest can be synthesized by reacting **7** with a linker unit in the presence of the guest in the solvent hexamethylphosphoramide, which is too large to serve as template.[5e,f] The seal-in procedure allows synthesis of hemicarceplexes with guests that are thermally too unstable to be 'heated' into empty **4** or **8–13**.[12]

The novelty of hemicarcerand structures has lured molecular container chemists to study properties and reactivity of incarcerated guests. Interesting questions that seek answers are: Could one conduct reactions inside hemicarcerands and how would they differ from their bulk phase counterparts? How do small reactants such as protons, electrons or photons pass through the host skeleton in order to reach the incarcerated guest? Can we even generate and protect highly reactive molecules inside the inner phase and prevent their self-destruction via dimerization or the reaction with bulk phase reactants that are too large to pass through the protective host skin? Is catalysis possible in such novel reaction chambers? In the following, I will summarize recent efforts in exploring inner phase chemistry and in stabilizing reactive intermediates by incarceration.

9.2 Types of Inner Phase Reactions

Conceptually, one can divide inner phase reactions into four groups (Figure 9.5):

(1) Intermolecular inner phase reactions
(2) Intramolecular inner phase reactions
(3) Mother molecule–daughter molecule reactions
(4) Innermolecular inner phase reactions.

In intermolecular inner phase reactions, the incarcerated guest reacts with a bulk phase reactant. This may require full or partial passage of the bulk phase reactant through one of the openings in the host shell. In most cases, it is difficult to discriminate between both models. Many different intermolecular inner phase reactions have been carried out with sometimes very surprising outcomes and are summarized in section 9.4.

Light or heat triggers intramolecular inner phase reactions and causes rearrangement of the incarcerated guest or fragmentation. Such reactions have lead to the generation of incarcerated reactive intermediates and are discussed in sections 9.5–6.

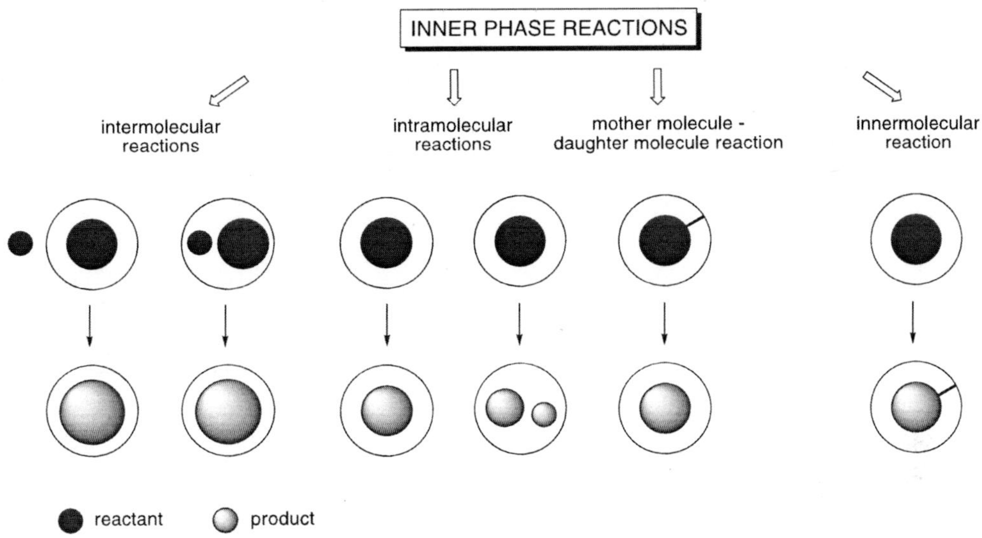

Figure 9.5 *Different types of inner phase reactions*

Two further new inner phase reaction types have been introduced recently. In Okazaki's mother molecule–daughter molecule-reactions, an external stimulus leads to the release of a guest (daughter), which was covalently linked to the inner surface of the surrounding container (mother) to form a new mother molecule–daughter molecule complex.[13] The opposite are innermolecular reactions.[14] The incarcerated guest reacts with the inner surface of the surrounding molecular container leading to a covalent adduct. The latter two reaction types are particularly interesting. The mother molecule–daughter molecule complex formation allows the generation of an encapsulated reactive intermediate or of an endohedral fixed reactive functionality, which both are protected by the surrounding capsule shell. Innermolecular reactions are interesting since these reactions take part at the concave inner surface of one reactant (the host), which is rather uncommon.

9.3 Probing the Properties of the Inner Phase

9.3.1 Amide C–N Bond Rotation and Ring-Flip of Cyclohexanes

The precise mix of free and occupied space inside the reactant filled container,[15] the shape of the inner phase, the rigidity of the container and the electronic nature of its inner surface may all or in part contribute to the reaction dynamics and selectivity of an inner phase reaction. Reversible conformational changes of guest molecules are easily tractable spectroscopically and ideal to study the effect of confinement on transition states. Cram and co-workers studied the *cis-trans* isomerization of $(CH_3)_2NCHO$ and $(CH_3)_2NCOCH_3$ inside **5**.[5a] For $(CH_3)_2NCHO$, the C–N rotational barrier decreased in the order liquid phase > inner phase > vacuum and was ~1 kcal mol^{-1} lower inside **5** than in nitrobenzene. For $(CH_3)_2NCOCH_3$, the order was inner phase > liquid phase > vacuum and the barrier

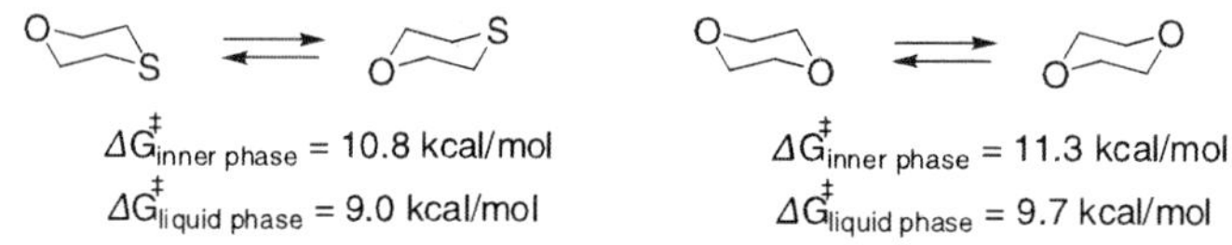

Figure 9.6 *Ring-flip dynamics of 1,4-thioxane and 1,4-dioxane inside 5*

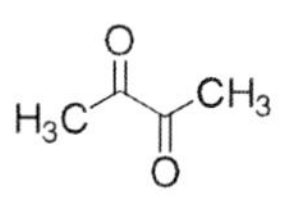

Figure 9.7 *Structure of biacetyl*

~2 kcal mol^{-1} higher inside **5** than in solution. Cram explained these trends with the different ratio of free and occupied space in the inner phase. In CPK models, $(CH_3)_2NCHO$ is loosely held inside the container, whereas $(CH_3)_2NCOCH_3$ is strongly compressed against the host walls in the ground state and even more so in the transition state for bond rotation. Thus, the rigid container resists being deformed more than the solvent cage resists being moved to accommodate the transition state. Depending on the mix of free and occupied space, the inner phase may be more like vacuum, liquid or even solid. Likewise ring inversion of 1,4-thioxane and 1,4-dioxane has a 1.8 kcal mol^{-1} and 1.6 kcal mol^{-1} higher barrier inside **5** than in the liquid phase (Figure 9.6).[16] The origin for the increased barriers is not fully clear and may result from ground state effects, such as stabilizing host-guest contacts that are lost in the transition state, or steric constraints in the transition state. Nevertheless, one can conclude that host rigidity translates into high sensitivity towards small structural perturbations of the guest.

9.3.2 Spectroscopic Probes

Molecules that change their photophysical characteristics in response to changes in the environment are well suited to probe the electronic and spatial properties of confined spaces.[17] Pina *et al.* measured large bathochromic shifts for absorption, fluorescence and phosphorescence maxima upon incarcerating biacetyl inside **14** (Figures 9.3 and 9.7).[18] These shifts are far outside the range of values of free biacetyl in any solvent. Nau interpreted these red-shifts with an unusually high polarizability inside **14**.[19] Absorption maxima and inverse oscillator strength correlated linearly with solvent polarizability. From these solvatochromic shifts, Nau estimated the polarizability of the inner phase, which is substantially higher than that of the highly polarizable diodomethane. Compared to the cavity of cyclodextrins or cucurbit[7]urils, which have polarizabilities that are closer to the other extreme,[19] the inner phase constitutes an environment of unusually high polarizability, which has consequences for the dynamics of inner phase reactions as will be discussed later.

Nau's interpretation was later challenged by Romanova *et al.* who studied the fluorescence and phosphorescence properties of biacetyl inside three hemicarcerands with different cavity sizes.[20] Emission maxima experienced bathochromic shifts and increased in the order **4** < **14** < **15**. However, the singlet-triplet excited state energy gap $\Delta S_1–T_1$ remained nearly constant for all three hemicarceplexes. Since the $S_1 \rightarrow S_0$ transition moment is orders of magnitude larger than that of the forbidden $T_1 \rightarrow S_0$ transition, a coupling between the transition dipole and polarizability dipole cannot be the origin for the emission shifts. Romanova *et al.* explained these shifts with smaller amounts of free space inside the inner cavity as compared to liquids and postulated a higher 'effective local pressure' inside hemicarcerands. With decreasing cavity size biacetyl may twist out of planarity, which would explain the observed trend.

9.4 Through-Shell Reactions

9.4.1 Proton Transfer Reactions

The investigation of proton transfer between incarcerated bases and bulk phase acids provides insight into the effect of incarceration on the guest's acidity or basicity. It is expected that the hydrophobicity of the inner phase will alter the pK_a of the incarcerated guest, which is reminiscent to many enzyme-catalyzed reactions.[21] Cram and co-workers studied proton transfers between a strong bulk phase acid and incarcerated amines **15**⊙pyridine, **15**⊙$(CH_3CH_2)_2NH$ and **15**⊙$CH_3(CH_2)_3NH_2$.[4a] Despite a large enough opening in the shell of **15**, attempts to protonate incarcerated pyridine with CF_3COOD in $CDCl_3$ failed. Cram suggested that the reduced basicity is a result of ineffective solvation of the pyridinium ion by the rigid host, the inability to form a contact ion pair in the inner phase and the larger size of the pyridinium ion compared to pyridine. Inaccessibility of the unshared electron pair of incarcerated pyridine would also be an explanation, but seems very unlikely (Figure 9.8A).

For **15**⊙$(CH_3CH_2)_2NH$, instantaneous decomplexation of **15**⊙$(CH_3CH_2)_2ND_2^+$ accompanied through-shell proton transfer. The ability to protonate **15**⊙$(CH_3CH_2)_2NH$ with CF_3COOD in $CDCl_3$ results from the location of the nitrogen of $(CH_3CH_2)_2NH$ in the equatorial region close to the portals (Figure 9.8B). After protonation, the counter ion

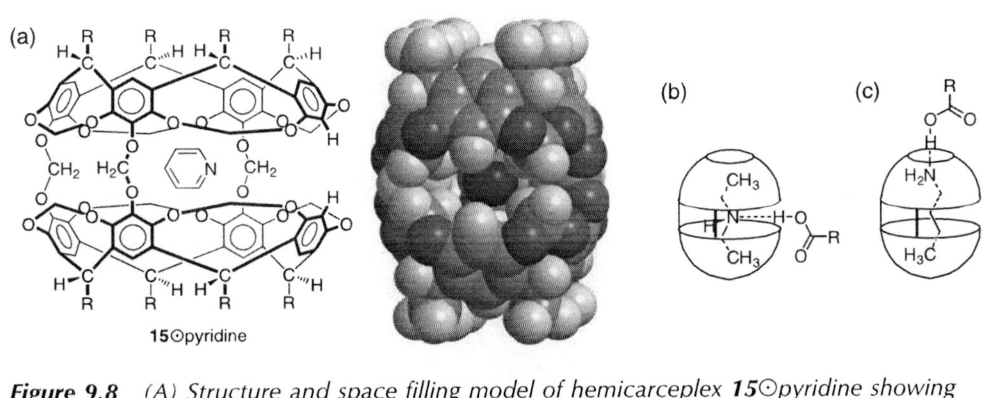

Figure 9.8 *(A) Structure and space filling model of hemicarceplex **15**⊙pyridine showing the pyridine nitrogen. (B) Proposed proton transfer mechanism for **15**⊙$(CH_3CH_2)_2NH$ and **15**⊙$CH_3(CH_2)_3NH_2$ (C).*

pulled the guest out of the inner phase. Addition of excess CF_3COOD to **15**⊙$CH_3(CH_2)_3NH_2$ lead to a 2:1 mixture of **15**⊙$CH_3(CH_2)_3ND_3^+$ and **15**⊙$CH_3(CH_2)_3ND_2$, which remained constant over time although slow decomplexation took place. Complete protonation of **15**⊙$CH_3(CH_2)_3NH_2$ required 100 eq. of CF_3COOD. Excess CD_3COOD only H/D exchanged the amine protons. These results show that the acidity of incarcerated $CH_3(CH_2)_3NH_2$ is comparable to that of CF_3COOH in $CDCl_3$. Furthermore, the strong upfield-shifted amine protons of **15**⊙$CH_3(CH_2)_3NH_2$ imply guest alignment along the polar axis of **15**. In this orientation, through-shell protonation most likely occurs through the holes in the polar caps of **15** (Figure 9.8C).

9.4.2 Electron Transfer Reactions

Electron transfer (ET) plays an important role in many organic reactions.[22] Although ET processes are most efficient in contact donor acceptor complexes, electron tunneling over long distance is possible.[23] Thus, ET reactions are well suited to be studied between an incarcerated guest and a bulk phase reducing or oxidizing agent. An oxidation-reduction cycle for hydroquinones **16–19** could be carried out in the interior of **4** (Figure 9.9).[24] Oxidation with $Ce(NH_4)_2(NO_2)_6$–silica gel—$CDCl_3$ or $Tl(O_2CCF_3)_3$—CCl_4 led to the parent incarcerated quinones **20–23** in essentially quantitative yields.

In the absence of light, the incarcerated quinones were stable below 100 °C, despite their usual instability in solution. Reduction back to the hydroquinones was possible with SmI_2/MeOH. The same reagent reduced nitrobenzene **24** to *N*-hydroxyl-aniline **25**. Surprisingly, aniline, which is the product in the liquid phase, is not formed. The latter result, the high yields, and the instability of free *o*-quinones, suggests that all reduction/oxidation took place inside **4**, rather than by a dissociation – bulk phase reaction – association mechanism. It also demonstrates that electrons are transferred readily through the host shell in and out of the inner phase. The exact mechanism of the ET and the role of the intervening medium (the host) are not clear yet. To get more insight, Kaifer compared the electrochemical behavior of free and incarcerated ferrocene inside **26** (Figures 9.3 and 9.9).[25] In the inner phase, ET was strongly hindered kinetically and thermodynamically. A more positive half-way potential for the oxidation, due to the hydrophobicity of the inner phase, as well as 10-fold rate retardation was measured. To support the observation

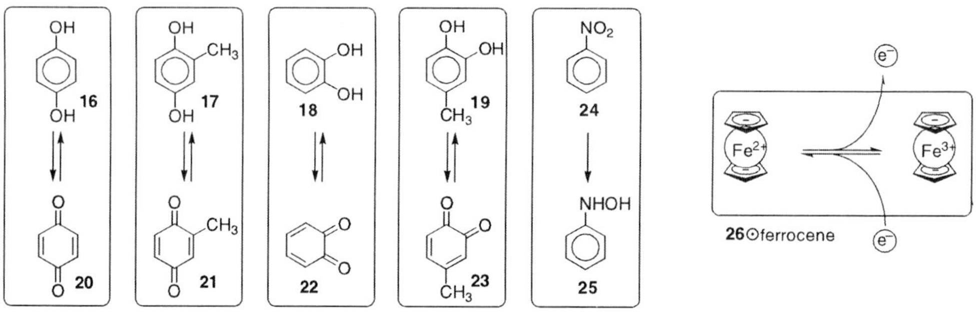

*Figure 9.9 Through-shell oxidation-reduction cycles of hydroquinones **16–19** and nitrobenzene **24** inside **4** and electrochemical oxidation of **26**⊙ferrocene*

of an inner phase oxidation/reduction cycle, Kaifer isolated oxidized ferrocene hemicarceplex **26**⊙ferrocene$^+$. The NMR spectrum showed large spectral shifts and line-broadening compared to that of **26**⊙ferrocene, which is consistent with the formation of paramagnetic ferrocene$^+$. The blue color of ferrocene$^+$ and its NMR spectrum persisted at room temperature for days until **26**⊙ferrocene$^+$ slowly reverted back to **26**⊙ferrocene, revealing that guest dissociation/association can be excluded within the time frame of the electrochemical experiments. Kaifer suggested that the slower ET between the incarcerated ferrocene and the electrode surface could result partially from the higher mass of **26**⊙ferrocene compared to ferrocene and also from a reduction of the electronic coupling between the ferrocene center and the electrode which is affected by the increase in distance from 3.5 Å to about 9 Å. Whether the hemicarcerand's aromatic structure mediates the electron coupling is not clear.

9.4.3 Nucleophilic Substitutions and Isotopic Exchanges

The alkylation studies of Kurdistani *et al.* provide much insight into the relationship between reactivity and guest orientation and bulk phase reagent size.[5e] Different phenols were alkylated in the inner phase of **4**. Two factors determined the observed reactivity: (a) portal size and (b) preferred guest orientation relative to the equatorial-located portals. Alkylation with NaH/MeI in THF of 4-$HOC_6H_4CH_3$ (*p*-cresol) or 4-HOC_6H_4OH (*p*-hydroquinone) was impossible. Under the same conditions, 2-$HOC_6H_4CH_3$ (*o*-cresol), 3-$HOC_6H_4CH_3$ (*m*-cresol), and 3-HOC_6H_4OH (resorcinol) were quantitatively methylated. 2-HOC_6H_4OH (catechol) gave a mixture of mono- and dimethylated carceplexes. An examination of typical crystal structures of hemicarceplexes with 1,4-disubstituted benzene guest suggests that the OH-group of 4-$HOC_6H_4CH_3$ is located in a protected polar cap of the host.[5b] In *ortho*- or *meta*-disubstituted benzenes, one substituent resides inside a shielded polar cap, whereas the second substituent is located near a portal. This suggests that these reactions must occur in the entryways through a linear transition state, which is partially 'solvated' by the alkoxy-units that align the host's portals (Figure 9.10). Since this 'pseudo solvent cage' has limited flexibility, alkylation with larger alkylating agents failed.

Likewise in D_2O-saturated $CDCl_3$, no D-for-H exchange of OH groups was possible when guests were 4-$HOC_6H_4CH_3$, 2-HOC_6H_4OH or 4-HOC_6H_4OH.[5e] In the presence of diazobicyclo[5.4.0]undec-7-ene, 4-HOC_6H_4OH exchanged its hydroxyl-protons, but not

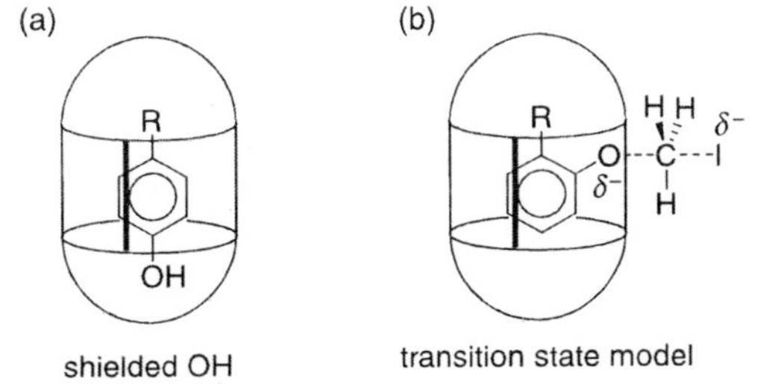

Figure 9.10 *Preferred orientation of* para-*substituted phenols and transition state model for the methylation of* ortho-*substituted phenolates inside* ***4***

the conformationally more fixed 4-$HOC_6H_4CH_3$. In THF-NaH at 25 °C followed by D_2O-quench, the hydroxyl-protons of 2-HOC_6H_4OH, which are more exposed to the equatorial-located portals, exchange, but not the protected hydroxyl-protons of 4-HOC_6H_4OH and 4-$HOC_6H_4CH_3$.

Such steric interactions in the transition state, that serve as selectivity criteria in these inner phase reactions, also contribute to the highly structural recognition in enzyme-catalyzed reactions. The understanding of these interactions provides valuable guidelines for the rational design of novel highly selective catalysts in the future.

9.4.4 Nucleophilic Additions

Inner phase guest orientation and mobility also control reactivity in through-shell borane and methyllithium additions to benzaldehyde **27**, benzocyclobutenone **28** and benzocyclobutadione **29** inside hemicarcerand **4**.[26] BH_3-THF reduced all three incarcerated guests to the corresponding alcohols **30–32** (Figure 9.11A–C). Guest reactivity differed from that in the liquid phase and increased in the order **29** ≈ **28** > **27**. Furthermore, incarcerated **29** added only one equivalent of BH_3-THF. Reduction of the second carbonyl group required aqueous work-up followed by addition of BH_3-THF to **4⊙33**. Hydrolysis of **4⊙33** gave incarcerated *cis*-benzocyclobutenediol **34**. X-ray structures helped to rationalize these observations (Figure 9.11D–E).

In **4⊙27** and **4⊙28**, the guest's carbonyl groups are located inside a host's cavitand and reduction requires reorientation of the guest. The additional conformational energy adds to the activation energy and is higher for **27** as compared to **28**. In **4⊙29**, one carbonyl is shielded. The other is perfectly positioned for through-shell reaction inside an entryway. After addition to the exposed C=O, coordination of the boron of **35** to a host's ether oxygen hinders guest rotation and prevents exposure of the second C=O until **35** is hydrolyzed.

Guest orientations also explain outcomes of CH_3Li additions to incarcerated **27–29** (Figure 9.12). Again, guest reactivity decreased in the order **29** ≫ **28** > **27**. **29** added one equivalent of CH_3Li already at −78 °C to yield **36** and Moore rearrangement product **37**.[27] No double-addition took place. Hemicarceplex **4⊙28** required room temperature for a complete reaction. Under the same conditions, **4⊙27** reacted sluggishly and incompletely.

Very interesting are the formation of host cleavage products **39** and **7** in these reactions, which resulted from innermolecular reactions of the incarcerated lithium alcoholates. In the CH_3Li addition to **4⊙28**, lithium alcoholate **40** partially underwent an innermolecular nucleophilic displacement at one of the host's acetals yielding hemicarcerand **39** and **39⊙41** (Figures 9.12–9.13). On the other hand, at 0 °C incarcerated lithium alcoholates **43** and/or **44** cleaved one of the O—$(CH_2)_4$—O linker of **4** via *β*-elimination (Figure 9.13B). Bulk phase lithium alcoholates are not basic enough to induce this reaction. The incarcerated counterparts must be several orders of magnitude more reactive. Three factors contribute to this rate-acceleration. (1) The absence of aggregation of R′OLi in the inner phase.[28] (2) The poor ability of **4** to 'solvate' R′OLi, which increases its basicity.[4a] (3) Lithium coordination to an oxygen lone pair of the cleaved C—O bond positions the alkoxide O in close proximity to the *β*-H of the bridge and provides charge compensation during the concerted *syn*-elimination (Figure 9.13B).[29] These examples and those discussed in the previous section show that through-shell and inner phase chemistry clearly

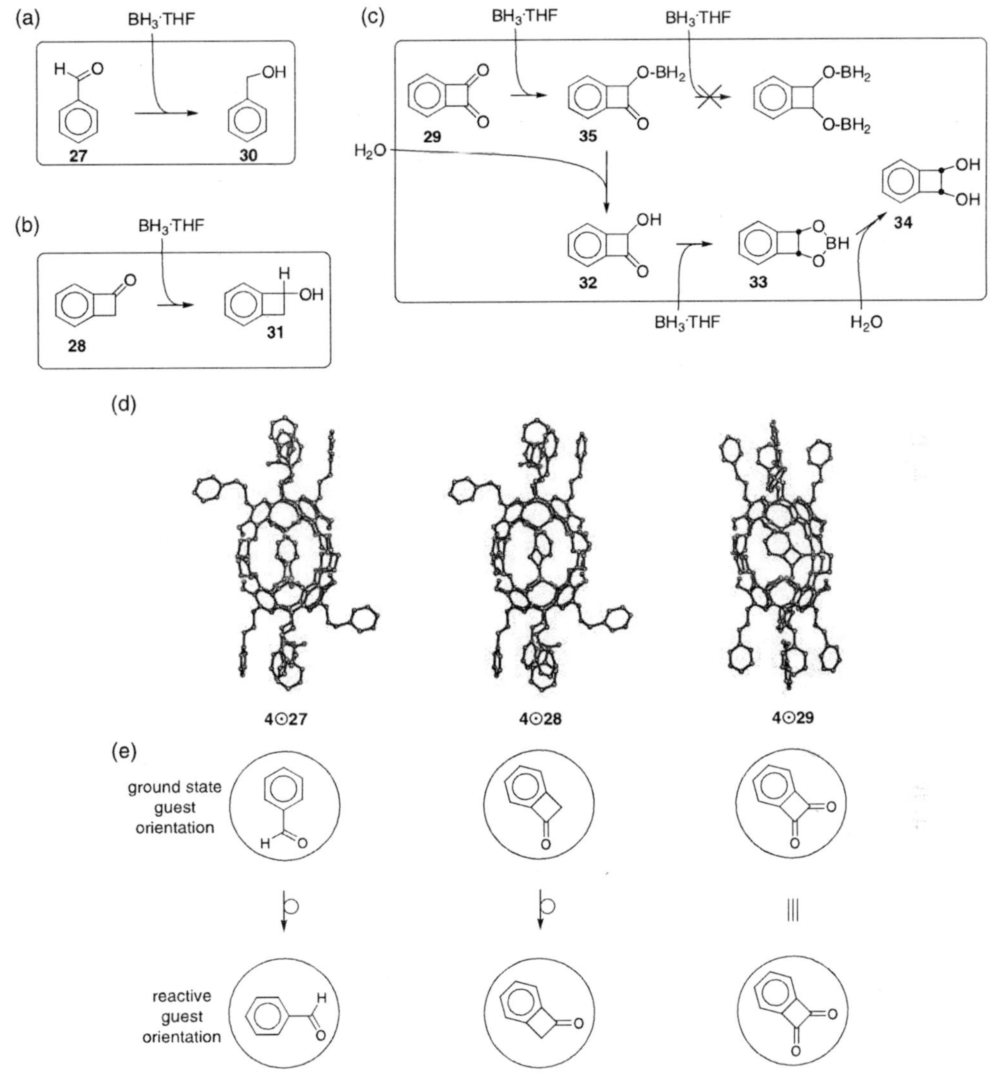

Figure 9.11 *(A-C) Borane reductions of **27–29** inside **4**. (D) X-ray structures of hemicarceplexes **4⊙27**, **4⊙28** and **4⊙29**. Reprinted with permission from [26]. Copyright 2003 American Chemical Society. (E) Ground state and 'reactive' guest orientations*

differs from 'conventional' chemistry in the bulk phase with respect to reactivity and selectivity. Inner phase and through-shell reactions show the following characteristic features:

(1) Guest functional groups that reside inside a host's polar cap are less reactive than those exposed to an equatorial portal, which have the potential for high through-shell reactivity.
(2) The reactivity of bulk phase reactants is largely influenced by their size and shape relative to that of the host's equatorial portals.

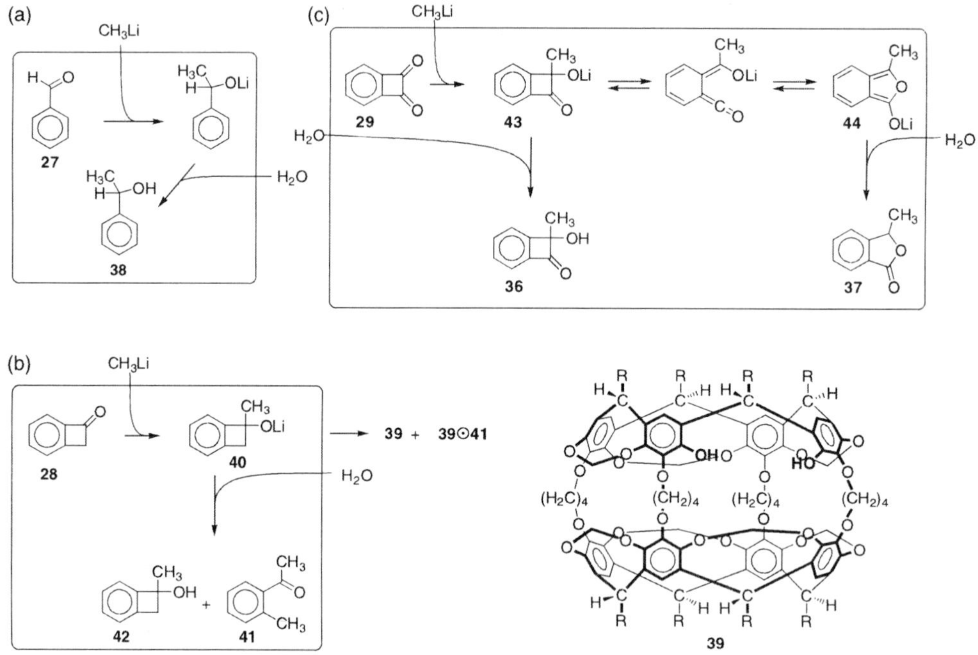

Figure 9.12 *Through-shell CH_3Li additions to incarcerated **27–29** (A–C)*

(3) If functional groups are protected in the guest's most favorable orientation, reactivity depends on the inner phase rotational mobility of the guest.

(4) The basicity and nucleophilicity of incarcerated lithium alcoholates exceeds those of bulk phase alcoholates by several orders of magnitude resulting in efficient inner-molecular elimination or nucleophilic transacetalisation and formation of hemicarcerands with one extended portal. In these innermolecular reactions, small structural changes of the guest have a sound effect on the reaction mode.

9.5 Intramolecular Thermal Reactions

In the previous section we have seen that the nucleophilicity and basicity of incarcerated lithium alcoholates strongly exceeds those of bulk phase alcoholates mainly as a result of the lack of aggregation and poor solvation by the host.[26] As a consequence, innermolecular reaction involving lithium alcoholates are strongly accelerated. Several extrusion reactions have been studied inside container molecules and have highlighted additional ways how encapsulation inside a hemicarcerand may accelerate or decelerate inner phase reactions.

9.5.1 Diazirine Fragmentation

An investigation of the thermal fragmentation of aryldiazirines inside hemicarcerands showed that the walls of hemicarcerands stabilize transition states, if bond breaking

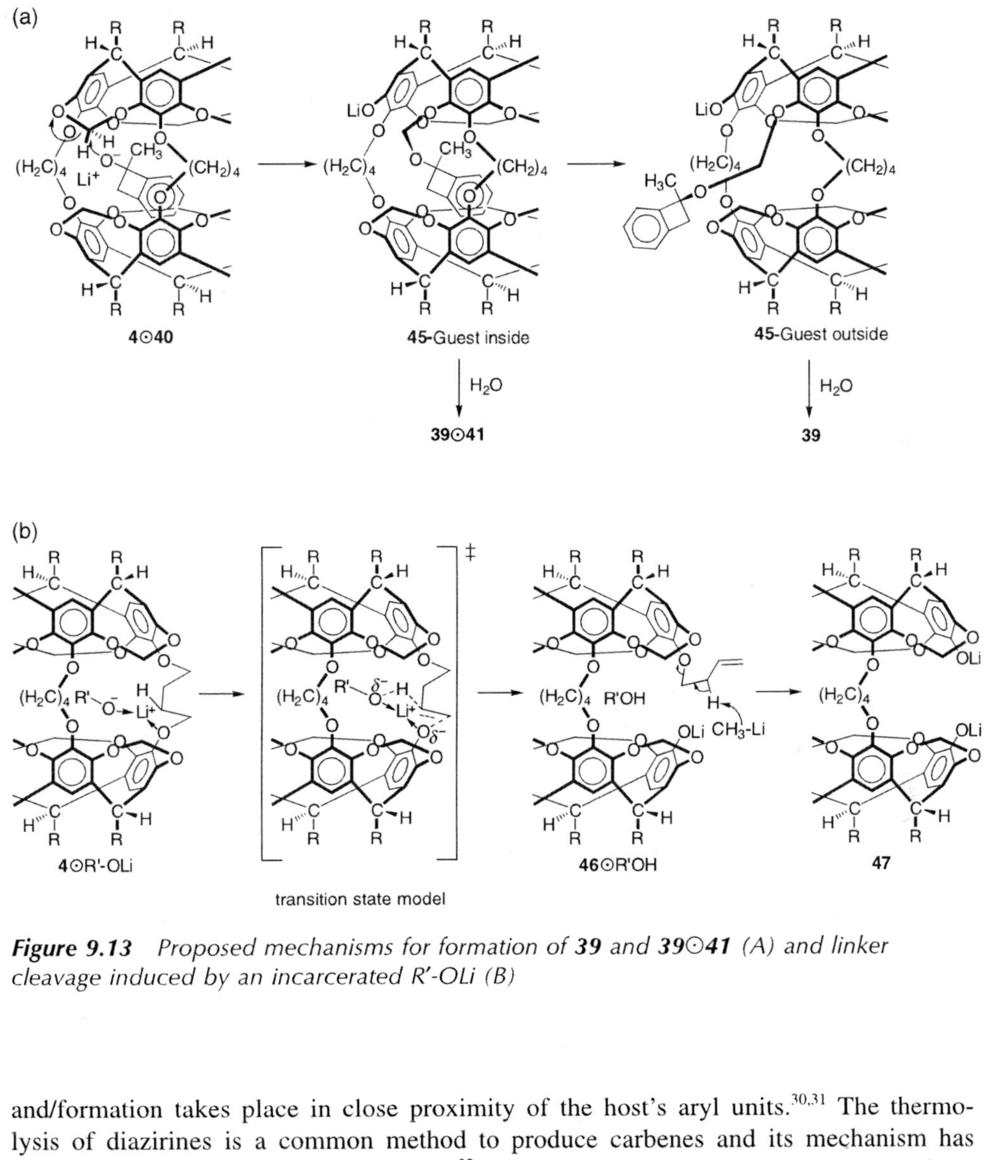

Figure 9.13 *Proposed mechanisms for formation of **39** and **39⊙41** (A) and linker cleavage induced by an incarcerated R′-OLi (B)*

and/formation takes place in close proximity of the host's aryl units.[30,31] The thermolysis of diazirines is a common method to produce carbenes and its mechanism has been studied in detail (Figure 9.14A).[32] Compared to the bulk phase, inner phase fragmentation of **48** is 15-fold accelerated, that of **49** slightly faster (1.2 fold) and that of **50** 2.4-fold slower.[30] Furthermore, all inner phase transition states are stabilized enthalpically by 2–3 kcal mol^{-1}, which, in the case of **49** and **50**, is partially or fully compensated by unfavorable entropic contributions to $\Delta G^\ddagger$. The unfavorable $\Delta T\Delta S^\ddagger$ term likely results from loss of vibrational degrees of freedom as the guest expands upon reaching the transition state leading to a tighter hemicarceplex. The favorable enthalpic stabilization is interesting and was explained with the high polarizability of the inner phase.[19,31] The stretched C—N bonds of the transition state are more polarizable than those of the ground state. Thus, the transition state will be stronger stabilized

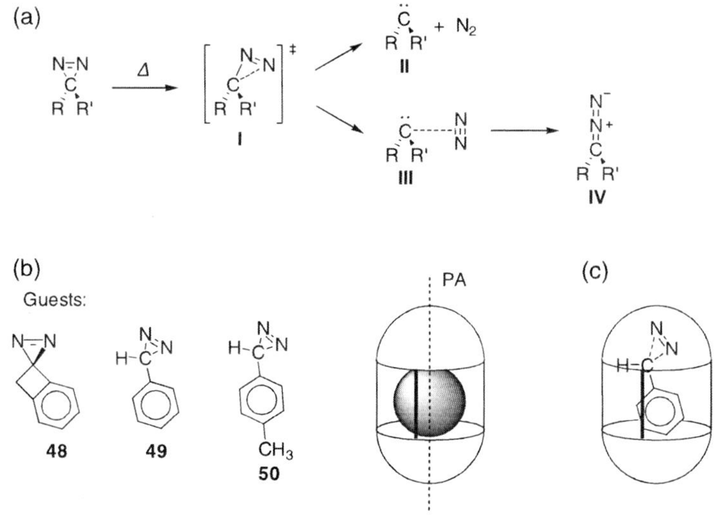

Figure 9.14 (A) Products and mechanism of diazirine fragmentation. (B) Inner phase orientation of aryldiazirines ***48–50*** positioning the diazirine group in van der Waals contact to the highly polarizable aryl units of the upper cavitand. (C) Transition state model for the inner phase phenyldiazirine fragmentation

through dispersion interactions,[33] especially since bond breaking takes place in close proximity of the highly polarizable aryl units of a cavitand (Figure 9.14B–C).

In an extended study, Sanchez Carrera *et al.* explored the influence of hemicarcerand shape on the thermolysis of phenyldiazirines.[31] All inner phase transition states were slightly accelerated. Rate constants increase in the order: **9⊙49** > **12⊙49** > **10⊙49** > **4⊙49** > **13⊙49** and vary overall by a factor of 3.2. Again, inner phase transition states where stabilized enthalpically by about 2 kcal mol^{-1} and moderately destabilized entropically. Furthermore, faster rates were observed for hemicarceplexes whose fourth shorter bridge induces a slight kink in the host structure. This was interpreted with an induced fit model and a better ability of the latter hemicarcerands to accommodate the slightly bend transition state.

9.5.2 Fragmentation of 3-Sulfolene

A very interesting thermal pericyclic reaction inside the asymmetric carcerand **51** was recently reported by Reinhoudt and co-workers.[34] They studied the extrusion of SO_2 and butadiene from incarcerated 3-sulfolene by mass spectrometry (Figure 9.15).[34b]

The extrusion of SO_2 and butadiene from free 3-sulfolene readily takes place at 100–130 °C. Substantially higher temperatures were required for carceplex **51**⊙3-sulfolene. SO_2 was only detected above 170–180 °C, and butadiene at 215 °C. Up to 180 °C only the intact carceplex was observed. Above 180 °C, also empty **51** but neither a SO_2 carceplex, nor a butadiene carceplex was detected. Since this host is stable at such high temperatures, guest escape due to the thermal destruction of **51** can be excluded. Hence, the detected SO_2 and butadiene must result from **51**⊙3-sulfolene and must escape the

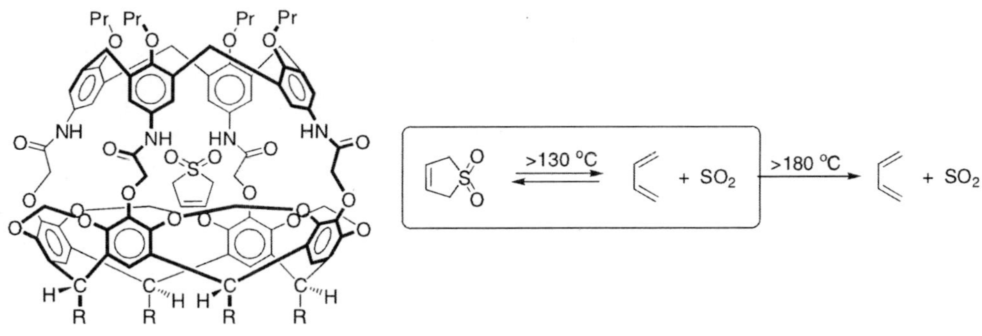

Figure 9.15 Structure of 51⊙3-sulfolene and thermal extrusion of SO_2 from 3-sulfolene

inner phase through one of the larger side portals. Reinhoudt explained the unusually high thermal stability of incarcerated 3-sulfolene with a fast recombination in the inner phase (Figure 9.15). Below 180 °C, a thermal equilibrium between 3-sulfolene, SO_2 and butadiene is established, which is pulled towards the extrusion products via their escape from the inner phase above 180 °C. This shows impressively how the confinement changes the rates of bimolecular reactions by providing a very high local concentration of both reactants.[35]

9.6 Inner Phase Photochemistry

9.6.1 Inner Phase Stabilization of Reactive Intermediates: Concept

The possibility to photolyze incarcerated guest molecules opens up the opportunity to generate and protect highly strained and reactive molecules inside carcerands.[2] This allows NMR spectroscopic characterization of otherwise fleeting species, which complements matrix isolation spectroscopy, ultra-fast spectroscopy or flow techniques. The concept of reactive intermediate stabilization by incarceration was introduced by Cram, Tanner and Thomas with 'the taming of cyclobutadiene' and is outlined in Figure 9.16.[36]

Photolysis of a suitable, stable photochemical precursor yields the reactive intermediate in the inner phase. Once generated, the surrounding host prevents destructive reactions, such as dimerization or trapping with bulk phase reactants that are too large to pass through an opening in the host shell. Difficult to prevent are innermolecular reactions with the surrounding host, which may take place with incarcerated carbenes, nitrenes, radicals and arynes, thus limiting their lifetime. In the following sections, several examples will be discussed.

9.6.2 Cyclobutadiene

'The taming of cyclobutadiene' inside **15** is the first example of an inner phase stabilization of a reactive intermediate and nicely demonstrates the power of this approach (Figure 9.17).[36] Cyclobutadiene **52** is the prototypical example to verify theory of aromaticity.[37,38]

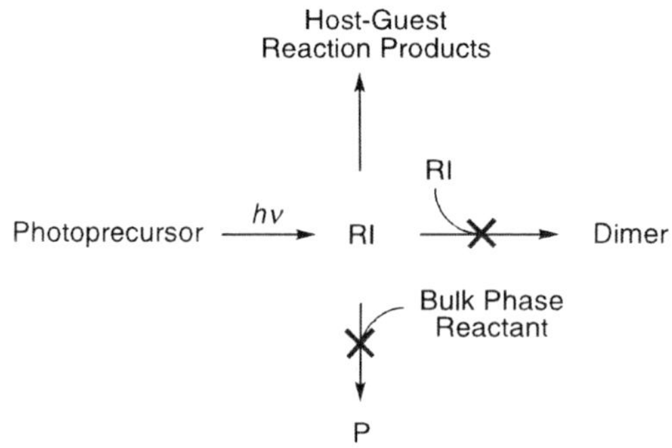

Figure 9.16 *Reactive intermediate stabilization by incarceration*

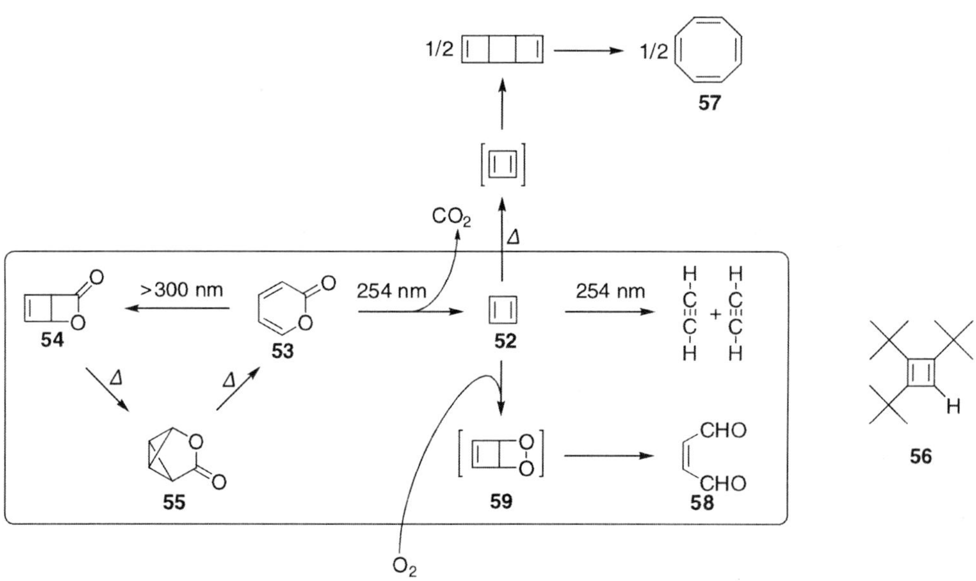

Figure 9.17 *Photochemical generation of **52** reactions in **15**. Structure of **56***

It is severely angle-strained in addition to being antiaromatic.[38,39] Cram and co-workers generated **52** inside **15** by irradiating α-pyrone hemicarceplex **15⊙53**.

Hemicarcerand **15** has a larger, slot-shaped opening that permits passage of **53** at >120 °C, but retains the guest at ambient temperature. Irradiation above 300 nm converted **15⊙53** to photopyrone **15⊙54**, which as a solid, rearranged to **15⊙55** at 90 °C. At higher temperature, **15⊙55** reverted quantitatively back to **15⊙53**. Controlled irradiation of **15⊙53** with unfiltered UV light produced nearly quantitatively cyclobutadiene. Prolonged photolysis split encapsulated cyclobutadiene into acetylenes. The latter escaped the inner phase and could be precipitated as red cuprous acetylide. In the absence of oxygen, cyclobutadiene was stable up to 60 °C! and was characterized for the first time by ^{1}H NMR spectroscopy. The four protons of **52** resonated at δ 2.27, which is 3.03 ppm upfield from the ring proton of **56**,[40] due to the shielding effect by the aryl units of the surrounding host. The lifetime of incarcerated cyclobutadiene is controlled by the barrier of its passage through the larger opening inside **15**. If a solution of **15⊙52** was heated in a sealed tube at high temperatures, the guest escaped the protective shelter and dimerized. Also, oxygen, which easily passes through the host shell, trapped the guest as malealdehyde **58**, presumably via an intermediate dioxetane **59**.

9.6.3 Anti-Bredt Bridgehead Olefins

The successful inner phase stabilization of cyclobutadiene suggests that this approach may allow stabilization of other molecules with highly strained multiple bonds. An interesting class of hydrocarbons with twisted C=C bonds are *anti*-Bredt bridgehead olefins.[41] These bicycloalkenes have a *trans*-cycloalkene and are unstable if their olefinic strain (OS) is $OS \geq 21\,\text{kcal mol}^{-1}$.[42] Bicyclo[2.2.2]oct-1-ene **60** and (Z)-bicyclo[3.2.1]oct-1-ene **61** have $OS = 46.4$ and $21.9\,\text{kcal mol}^{-1}$,[42b] respectively and are fleeting in solution, in part due to their high tendency to dimerize or rearrange,[43,44] but were recently room temperature stabilized inside hemicarcerand **4**.[45] For the inner phase synthesis of **60** and **61**, Jones' carbene route was chosen (Figure 9.18). Contrary to pyrolysis of **64**,[43] inner phase photolysis of diazirine **66** gave a complex product mixture composed of hemicarceplexes **4⊙60**, **4⊙61**, and **4⊙67** and of small amounts of carbene-hemicarcerand insertion products. Mechanistic studies suggest, that photochemically excited **66*** directly rearranges to **60** and **61** without participation of carbene **63** (Figure 9.19).[46] Both incarcerated *anti*-Bredt olefins are stable at room temperature in the absence of oxygen. In aerated solution, they add O_2 to yield ketoaldehydes **68** and **69**. Oxidation of **4⊙60** is rapid at room temperature, but requires days at 60 °C in the case of **61**, which reflects their different OS. Trapping studies also suggested formation of small amounts of the highly strained **62** in the photolysis of **4⊙66**. **62** escaped spectroscopic detection due to its fast isomerization to **61**, but could be trapped with O_2, if photolysis was carried out in aerated solution.

A thermal *retro*-Diels–Alder reaction of **60**, which in Jones' seminal pyrolysis studies had served as indirect proof for formation of **60**,[43] could also be induced inside **4**. At 62 °C, **60** slowly rearranged to triene **65**, which escaped the inner phase and was identified in the bulk by its characteristic ^{1}H NMR spectrum.

9.6.4 *o*-Benzyne

The inner phase stabilization of *o*-benzyne **70** inside hemicarcerand **4** uncovered some of the limits of the inner phase approach.[14,47] Since its discovery,[48–49] *o*-benzyne has received

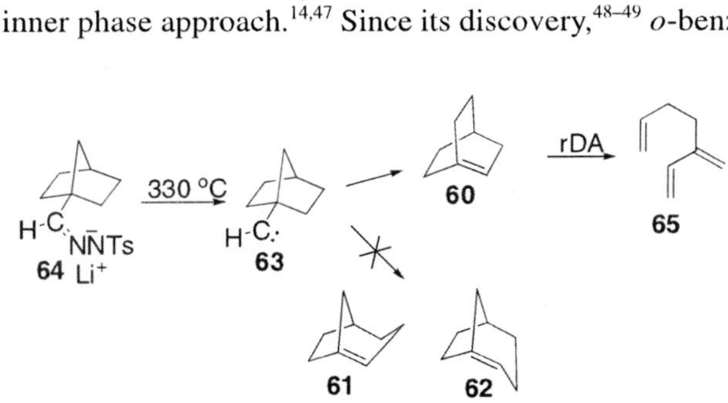

Figure 9.18 *Carbene route to* ***60***[43]

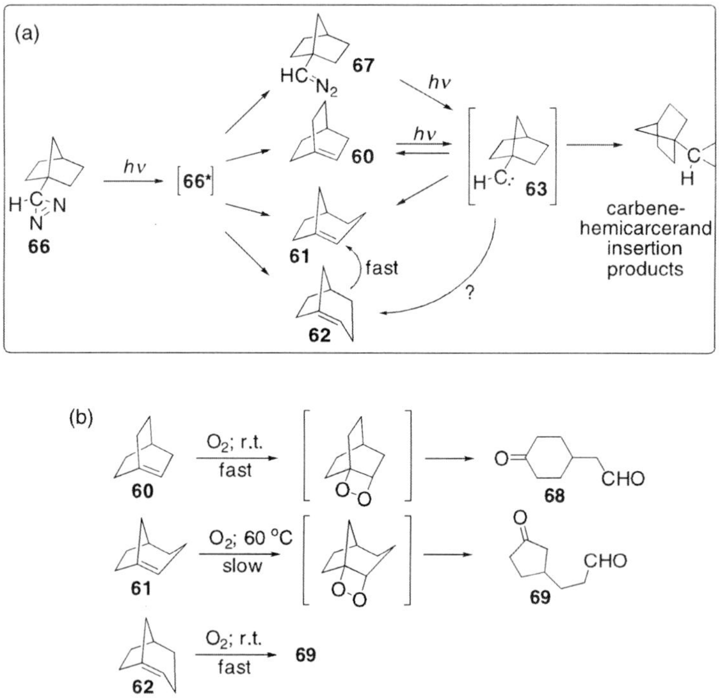

Figure 9.19 *(A) Photochemistry of incarcerated **66** and (B) oxygen trapping of incarcerated* anti-*Bredt olefins **60–62***

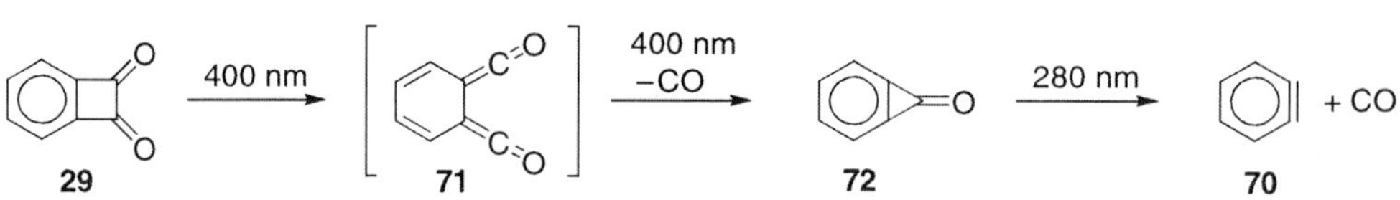

Figure 9.20 *Photochemistry of **29** in argon at 8 K and inside hemicarcerand **4**[47,54]*

much attention by the scientific community as intermediate in organic synthesis[50] and due to its interesting structural and electronic properties.[51,52] Benzyne has been investigated in cryogenic matrices by UV/Vis-,[53] FT-IR-,[54,55] and very recently solid state NMR spectroscopy.[56] Chapman first matrix-isolated *o*-benzyne by photolyzing benzocyclobutenedione **29** at 8 K (Figure 9.20).[54] The same route led to the successful inner phase synthesis of *o*-benzyne.[47]

Photolysis of hemicarceplex **4**⊙**29** yielded hemicarceplex **4**⊙**72**. Benzocyclopropenone **72** had previously been studied in solution below −78 °C, where it rapidly hydrolyzed in the presence of moisture.[57] However, protected inside the hemicarcerand, it was stable at room temperature in water-saturated $CDCl_3$. The proton chemical shifts of incarcerated **72** suggest co-alignment of its C2 axis with the C4 axis of **4** (Figure 9.21) such that the reactive C=O resides in the shielded region of the inner phase, which explains the high stability of the incarcerated guest.

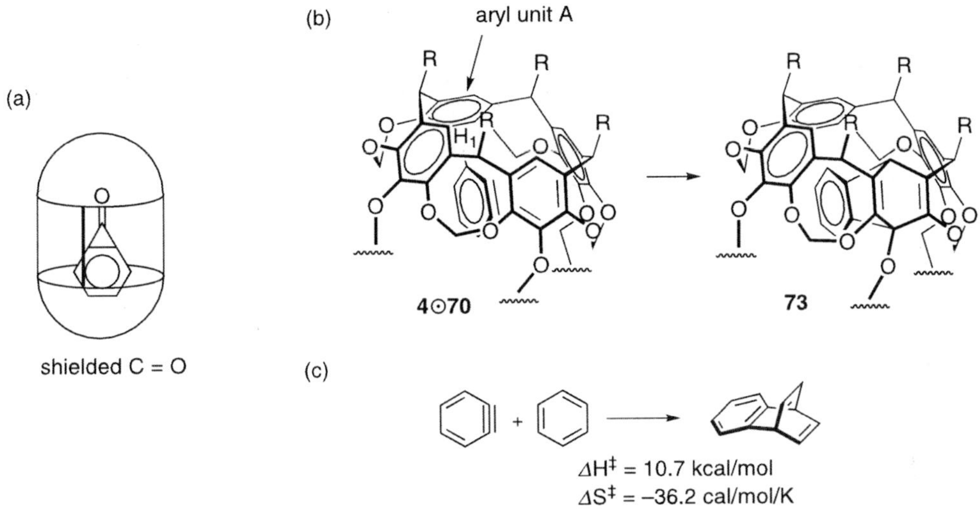

Figure 9.21 *Inner phase orientation of **72** (A) and Diels-Alder reactions of **70** (B–C)*

Further photolysis of **4⊙72** gave **4⊙70**. Due to the high π-bond strain of 50 kcal mol^{-1},[58] *o*-benzyne underwent an innermolecular Diels–Alder reaction with the surrounding **4** that was fast above −75 °C (Figure 9.21).[14] This reaction is very selective and **70** adds exclusively across the 1,4-position of an aryl unit of **4**. The MM3* minimum energy conformer of **4⊙70** shows strong preorganization of the reactive triple bond for the observed Diels–Alder reaction with distances of 4.53 Å and 4.05 Å between the reacting carbons of host and guest.[59] This high preorganization is reflected in the moderately negative activation entropy $\Delta S^{\ddagger}_{innermolecular}(298\,K) = -10.7\ cal\,mol^{-1}\,K^{-1}$.[14] Interestingly, the measured $\Delta H^{\ddagger}_{innermolecular}$ for the innermolecular addition is slightly higher than the calculated $\Delta H^{\ddagger}_{calc}$ for the addition of **70** to benzene (Figure 9.21C).[59] Thus, the increased reactivity of **4** must be compensated by steric interactions in the Diels–Alder-transition state originating from a repulsion between H(1) and aryl unit A (Figure 9.21B). This suggests that an incarcerated 3,6-disubstituted *o*-benzyne may not be able to react with the host and may be stable at room temperature.

At −75 °C, the Diels–Alder reaction was slow enough to record ^{1}H-NMR spectra of **4⊙70**.[47] Protons of **70** resonate at δ 4.99 and δ 4.31. Under the assumption that they feel the same shielding by the surrounding host as the protons of benzene, the chemical shifts of 'free' *o*-benzyne were estimated at δ 7.0 and δ 7.6 in excellent agreement to the calculated shifts.[52a] Much less relatively upfield shifted are the guest ^{13}C-signals and provide more insight into the electronic properties of *o*-benzyne. The measured chemical shift for the quaternary carbon of **70** at δ 181.33 is within the experimental error of the average of the three chemical shift tensor principle values δ 193 ± 15 of matrix isolated ^{13}C-enriched **70** at 20 K in argon.[56] The ^{13}C-NMR spectrum of incarcerated *o*-benzyne also provided direct ^{13}C–^{13}C-coupling constants. Comparison of the experimental ^{13}C–^{13}C-coupling constants with the coupling constants of model compounds suggested a cumulenic character of *o*-benzyne (Figure 9.22), which, however, contradicts most recent results of *ab initio* calculations.[52a,60] These calculations predict that *o*-benzyne is

70a 70b

Figure 9.22 Resonance structures of o-benzyne

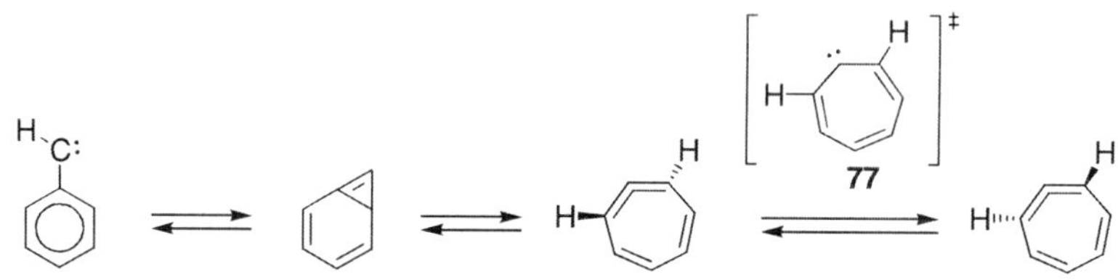

74 75 76

Figure 9.23 The Baron mechanism of the phenylcarbene rearrangement[62]

aromatic according to its geometric, energetic and magnetic properties and that the in-plane π-bond induces a small amount of bond localization resulting in an acetylenic character.

9.6.5 Phenylcarbene Rearrangement

The phenylcarbene rearrangement is one of the most important and fascinating carbene rearrangements and was studied inside hemicarcerands **4** and **13**. In the gas phase, phenylcarbene (PC) **74** ring-expands to cyclohepta-1,2,4,6-tetraene (CHTE) **76** involving bicyclo[4.1.0]hepta-1,3,5-triene **75** as intermediate (Figure 9.23).[61–64] CHTE, a bend and twisted allene with 40 kcal mol^{-1} strain energy,[64] is the local minimum on this part of the potential energy surface and enantiomerizes via the planar cyclohepta-1,3,5-trienylidene **77** as transition state.[63,64]

Ring-expansion can also be triggered photochemically via excitation of triplet phenylcarbene ^{3}**74** generated from phenyldiazomethane or phenyldiazirine (Figure 9.24).[12,65–69]

However, even at 77 K, photolysis of **4⊙49** produced only insertion products **78** (85% yield) and **79** (4.5%). Both formed via insertion of transient **74** into an inward pointing acetal C—H and linker α-C—H bonds of **4**, respectively. The magnitude and temperature dependence of the kinetic isotope effects of the acetal insertion, suggested considerable tunnelling contributions.[70] After partial deuteration of **4**, the kinetic isotope effect of k_H/k_D = 9.8 inside **80** slowed innermolecular insertions and sufficiently increased the lifetime of ^{3}**74**, such that photochemical rearrangement to CHTE was possible in 17% yield and 30% yield at 77 K and 15.5 K, respectively. In the related inner phase *p*-tolylcarbene rearrangements, the yields of 5-methylcyclohepta-1,2,4.6-tetraene were even higher. Photolysis of **4⊙50** and perdeuterated **81⊙50** at 77 K afforded **4⊙82** and **81⊙82** in 41% and 67% yield, respectively. The higher yields likely have an extrinsic origin (confinement). Any carbene-host reaction, be it arylcyclopropanation, acetal C—H insertion or linker C—H or C—O insertion, requires tilting of the carbene out of its preferred orientation, in which the long axis of the carbene is aligned with the polar axis of the host. The greater steric demand of the methyl group

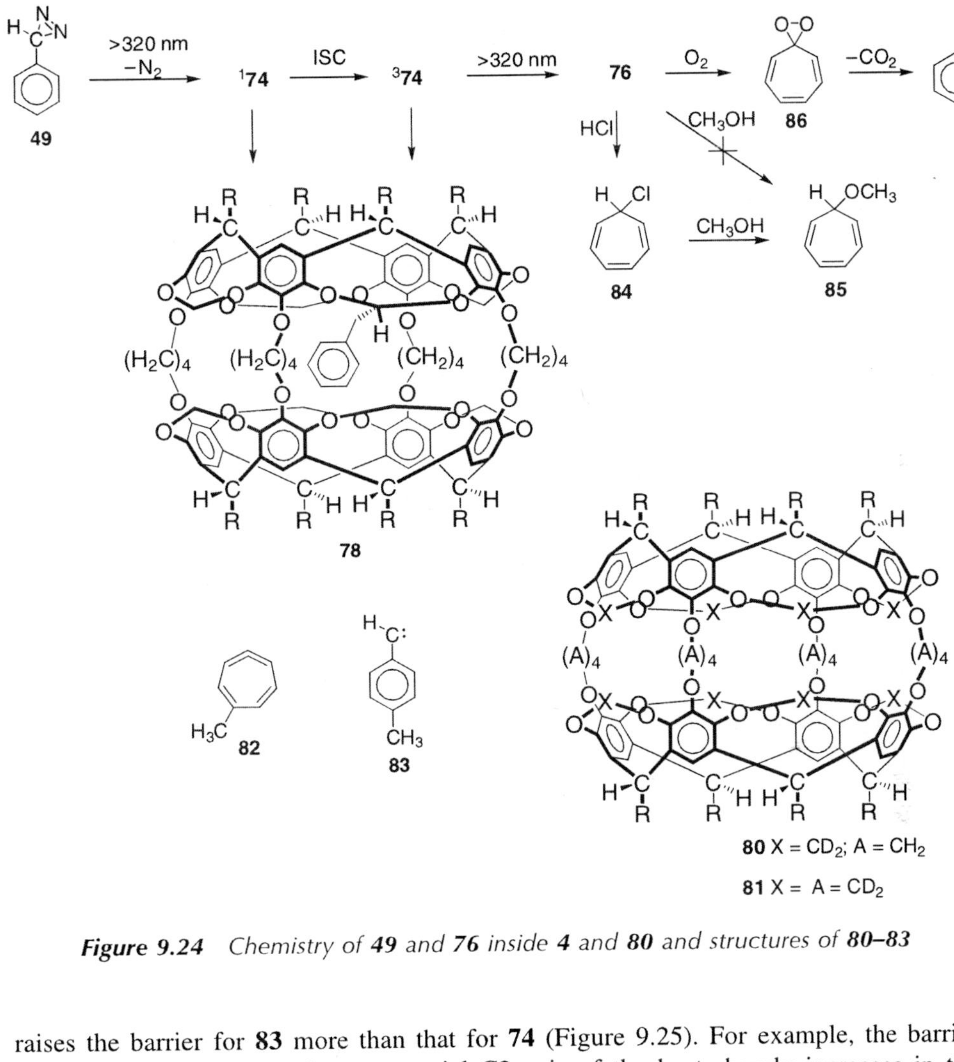

Figure 9.24 *Chemistry of **49** and **76** inside **4** and **80** and structures of **80–83***

raises the barrier for **83** more than that for **74** (Figure 9.25). For example, the barrier for guest tumbling around an equatorial C2 axis of the host sharply increases in the order benzene ≪ toluene ≪ *para*-xylene.[5b,10,71]

Both incarcerated allenes **76** and **82**, persisted for months at room temperature in the absence of oxygen and could be characterized by ^{1}H NMR spectroscopically. An investigation of their inner phase chemistry showed several interesting phenomena. Exposing hemicarceplex **80**⊙CHTE to HCl gas produced instantaneously hemicarceplex **80**⊙C_7H_7Cl. Subsequent addition of small amounts of CH_3OH to the bulk phase, trapped incarcerated C_7H_7Cl (**84**) and gave quantitatively $C_7H_7OCH_3$ (**85**) (Figure 9.24). Addition of CH_3OH to **80**⊙CHTE resulted in no reaction even at elevated temperatures. This seems at first surprising since free CHTE yields C_7H_7OR instantaneously, if generated in the presence of an alcohol ROH.[72] Labeling studies support a mechanism, in which protonation of cycloheptatrienylidene (**77**) yields an intermediate tropylium ion ($C_7H_7^+$). In the inner phase, **77** lies more than 20 kcal mol^{-1} above ground state CHTE, which explains

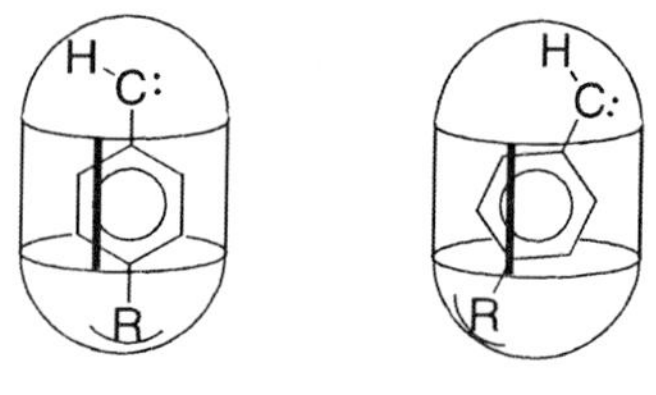

Figure 9.25 Confinement raises barrier for inner phase arylcarbene – hemicarcerand reactions

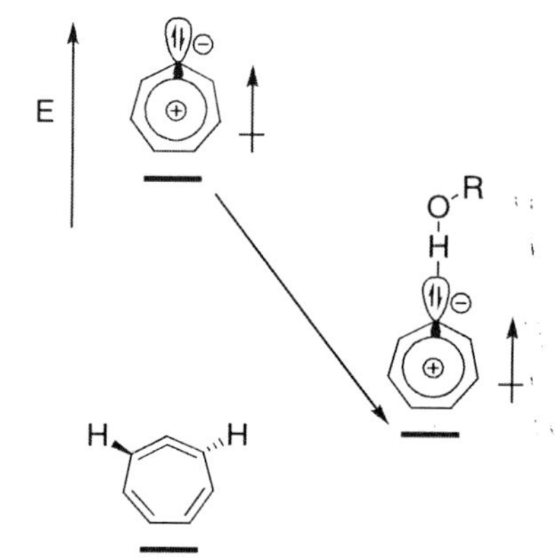

Figure 9.26 Relative energy of CHTE and cycloheptatrienylidene 77 as function of the medium polarity

the inertness of **76** inside **4**.[12] However, in a polar protic solvent, the larger dipole moment of **77** and stabilizing carbene ... H—OR hydrogen bonds bring **77** closer in energy to CHTE (Figure 9.26).[12]

Also interesting is the autoxidation of incarcerated CHTE in aerated solution. Under these conditions, oxygen diffuses into the hemicarcerand and adds to the central allene C to produce a spirocyclic dioxirane **86**. The latter **86** rapidly decarboxylates to benzene at room temperature.

The constrictively stabilized CHTE and MeCHTE allowed the measurement of several barriers of phenyl- and tolylcarbene rearrangements, which up to then were only available from high level calculations and allow comparison between experiment and theory.[63,64,73,74] In an attempt to measure the enantiomerization barrier of CHTE, **49** was photolyzed inside hemicarcerand **13** and produced diastereomeric hemicarceplexes **13**⊙(+)-CHTE and **13**⊙(–)-CHTE in a 2 : 3 ratio.[12] In the asymmetric host environment guest protons H2 experienced different host-induced shielding allowing differentiation by ^{1}H NMR spectroscopy. The absence of coalescence at 100 °C gave a lower limit of 19.6 kcal mol^{-1}

for the enantiomerization barrier, which agrees with all current calculations.[63,64] For the corresponding MeCHTE hemicarceplexes, exchange rate constants could be extracted from line shape analysis of high temperature NMR spectra.[68] Furthermore, since photolysis of **13⊙50** produced **13**⊙(+)-MeCHTE and **13**⊙(−)-MeCHTE in a ratio = 1 : 1.15 (de = 7%), which substantially differed from the thermodynamic ratio = 1 : 1.8 at room temperature, exchange rate constants could also be measured just below ambient temperature. The enantiomerization free energy of the minor and major enantiomer $\Delta G^{\ddagger}$(minor) = 20.3 kcal mol^{-1} and $\Delta G^{\ddagger}$(minor) = 20.6 kcal mol^{-1} agreed with estimates from DFT calculations ($\Delta G^{\ddagger}$ = 20.0 kcal mol^{-1}; B3LYP/6-311G**).

At elevated temperature, the lifetime of incarcerated MeCHTE strongly decreased due to ring-contraction resulting in *p*- and *m*-tolylcarbene (TC) formation (Figure 9.27). The latter react instantaneously with the surrounding hemicarcerand. A careful analysis of rates and products of these ring-contractions gave activation parameters for the MeCHTE to *m*- and MeCHTE to *p*-TC rearrangements. Again, both activation free energies agreed very well with those calculated at the B3LYP/6-311G** level of theory.[73] Also, the inner phase and computed activation enthalpy and entropy for the MeCHTE to *p*-TC matched very well. Very interesting, however, is the large enthalpy-entropy compensation in the inner phase MeCHTE to *m*-TC rearrangement. This enthalpy-entropy compensation was interpreted with a hemicarcerand and solvent reorganization as a consequence of a dramatic change in the guest shape along the reaction coordinate and is taken as indirect evidence for the intermediacy of **87** in this rearrangement (Figure 9.27).

This example illustrates nicely how kinetic experiments in confined space allow mapping of potential energy surfaces of important organic chemical processes involving highly reactive intermediates, which is difficult to achieve with other techniques such as laser flash photolysis, CID or matrix isolation.

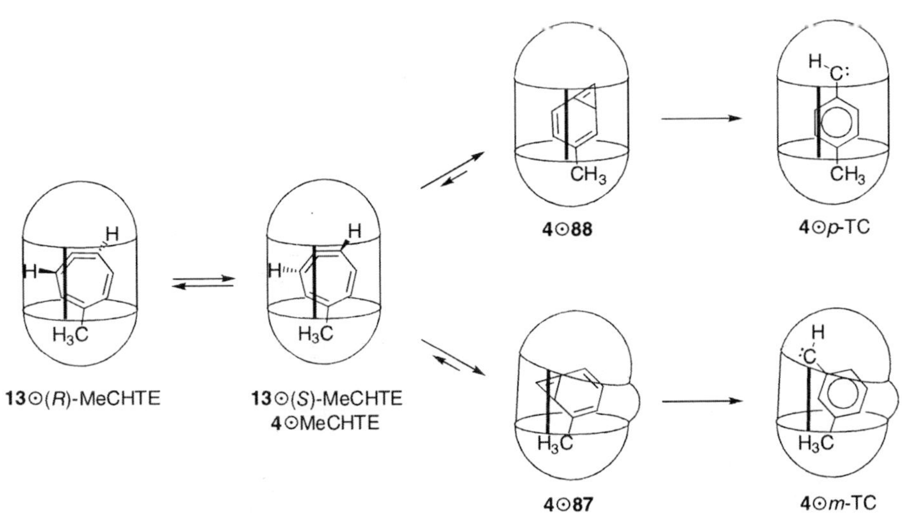

Figure 9.27 *Thermal chemistry of incarcerated MeCHTE*

9.6.6 Carbenes

In the previous chapter, we have seen that incarcerated phenylcarbene rapidly inserts into C–H bonds of hemicarcerands even at 77 K.[12,67] Such high reactivity is expected and typical for alkyl or arylcarbenes.[75] However, the stability and reactivity of carbenes R–C–R′ depends on the nature of the substituent R and R′ and can be tailored.[76,77] Heteroatoms strongly stabilize the singlet state through electron donation and many diaminocarbenes are stable and isolable at room temperature.[78] In cases, where intrinsic stabilization (e-donation) is not sufficient, extrinsic effects (incarceration), may render an otherwise fleeting singlet carbene stable under normal conditions. Fluorophenoxycarbene **89** is such specie and was recently room temperature stabilized by incarceration (Figure 9.28).[79,80] Both substituents stabilize **89** by ~60 kcal mol^{-1} compared to methylene.[77] Yet, **89**, if photochemically generated from diazirine **90**, is fleeting at room temperature and instantaneously dimerizes, reacts with moisture or is trapped in the presence of alkenes.[81]

Liu *et al.* generated incarcerated **89** by irradiation of fluorophenoxydiazirine hemicarceplex **4⊙90** at low temperature.[79] Incarcerated **89** persisted for weeks at room temperature. The ^{13}C and ^{19}F NMR spectra of **4⊙89** provided interesting insight into the electronic properties of **9**. The carbenic carbon resonated at δ 285.7 ppm, which compares well to chemical shifts of other persistent heteroatom substituted carbenes.[78] The strongly downfield shifted fluorine, the unusually large ^{19}F–^{13}C coupling constant and the considerable upfield shift of the *ipso* carbon of **89** relative to that of **90** point towards strong participation of both O and F atoms in the carbene stabilization through push–push effects.

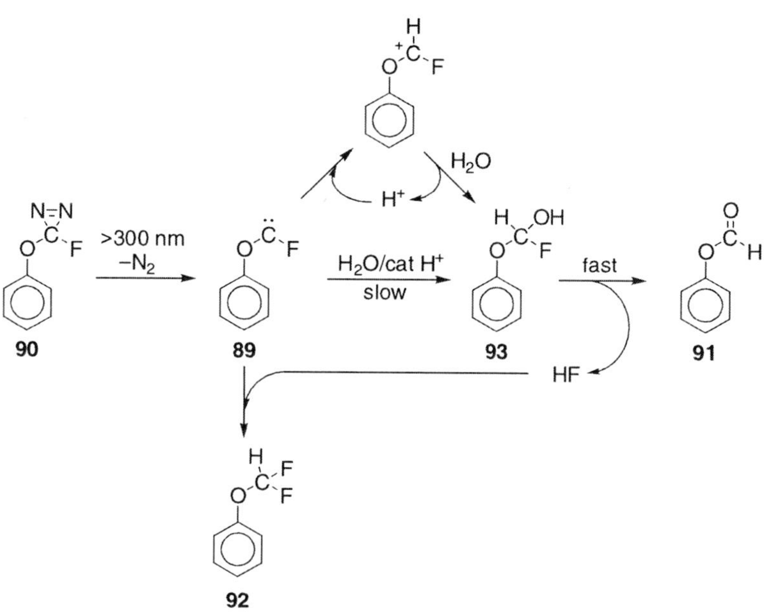

*Figure 9.28 Photochemistry of **90** inside hemicarcerand **4** and mechanism of the acid catalyzed trapping of carbene **89** with water*

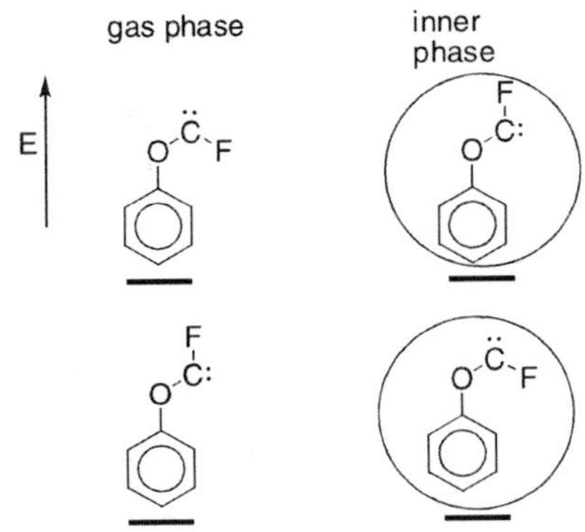

Figure 9.29 *Relative energy of* cis- *and* trans-**89** *in the gas and inner phase*

The NMR spectroscopic investigations also showed that incarceration had a strong effect on the conformational preference of **89** (Figure 9.29). Contrary to gas phase *ab initio* calculations, which predict *trans*-**89** as the lower energy conformation, the conformational equilibrium is reversed in the inner phase. Thus, confinement favors the coiled and more compact *cis*- over the extended *trans* conformation, which is consistent with the observations of coiled linear alkane conformations in self-assembled molecular capsules.[82]

In the presence of trace amounts of acid, incarcerated **89** slowly reacted with water in the bulk phase to yield phenylformate hemicarceplex **4⊙91** and phenyl difluoromethyl ether hemicarceplex **4⊙92** (Figure 9.29). The requirement of acid catalysis in the inner phase water trapping reaction is surprising since catalysis is not required for free **89**. This suggests that the water-trapping of **89** is initiated by protonation and that water is not acidic enough in the inner phase contrary to the bulk phase. The hydrophobicity of the inner phase and lack of solvation of the hypothetical ion pair [**89**H]$^+$[OH]$^-$ are likely reasons for the absence of this acid base reaction similar to the examples discussed in section 9.4.1.This shows that, incarceration not only prevents dimerization of **89**, but also slows trapping reactions with water by many orders of magnitude.

9.6.7 Phenylnitrene

Very recently, phenylnitrene (PN) and its intramolecular rearrangement have been investigated inside hemicarcerand **4**.[83] PN is an important reactive intermediate for organic synthesis and photoaffinity labeling of biomacromolecules.[84] It is isoelectronic with phenylcarbene (PC) and undergoes very similar intramolecular chemistry.[61c,85,86] Above −100 °C, ^{1}PN rapidly ring-expands to the highly strained cyclic ketenimine **95**, which can be trapped with amines or other nucleophiles. Below −100 °C, ^{1}PN intersystem crosses to triplet ^{3}PN (Figure 9.30).

Though at first glance PN and PC show great similarity in their chemistry, their reactivity differs dramatically, which has been subject of extensive investigations over the past decades and reflects itself in the inner phase chemistry of both species. For example, in solution ^{1}PN ring-expands rapidly at room temperature to **95** in the sub-nanosecond time scale, whereas ring-expansion of ^{1}PC can only be observed at elevated temperatures in

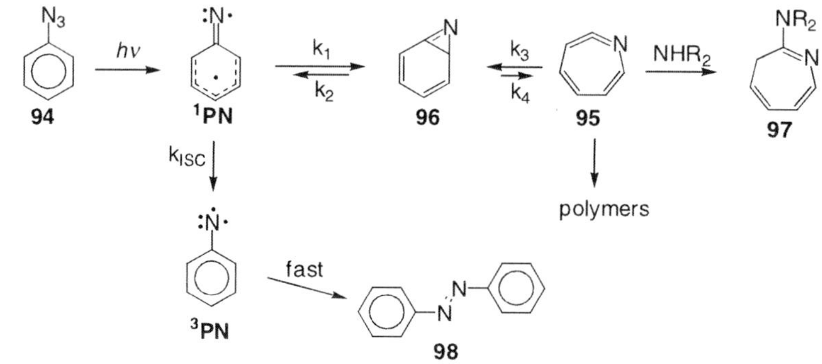

Figure 9.30 The mechanism of the phenylnitrene rearrangement

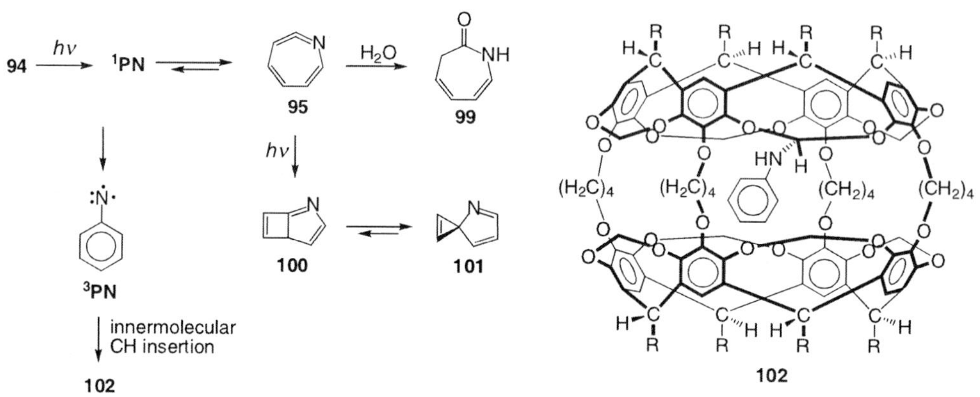

Figure 9.31 Inner phase photochemistry of phenylazide **94** inside hemicarcerand **4**

the gas phase due to the substantially higher activation energy and the higher intermolecular reactivity of ^{1}PC.[73,87]

As a consequence, ^{1}PC does not rearrange to CHTE, if generated inside **4**.[12,67] Ring-expansion cannot compete with the much faster insertion into hemicarcerand bonds. The situation is different for ^{1}PN, in which case intramolecular pathways (intersystem crossing and ring-expansion) are much faster than reactions with the hemicarcerand. Thus, photolysis of incarcerated phenylazide at −86 °C, at which temperature ring-expansion is faster than intersystem crossing, produced **4⊙95**, whose ^{13}C NMR spectrum could be recorded for the first time (Figure 9.31).[83a] At this temperature, incarcerated **95** slowly decayed within 5 hours to triplet ^{3}PN by ring-contraction and intersystem crossing, allowing for a precise determination of the activation parameters of this process. If the photolysis was carried out in THF/water 8 : 1, **95** could be trapped with water as lactam **99**.

The inner phase photolysis studies of **4⊙94** uncovered a novel photochemical reaction of **95**.[83b] Under the photolysis conditions, **95** underwent efficient photochemical electrocyclization to the *anti*-Bredt imine **100**, which at −5 °C thermally rearranged via a 1,5-shift to the slightly more stable 1-azaspiro[2.4]hepta-1,4,6-triene **101**.

The reactivity and life-time of triplet ^{3}PN and triplet PC differed remarkably in the inner phase of **4**. Whereas PC reacts with **4** already at 15 K and persist probably only a few minutes at this temperature,[12,67] the lifetime of incarcerated ^{3}PN is 13.6 min at −3 °C.[83b] Both ^{3}PC and ^{3}PN preferentially insert into inward pointing acetal C-H bonds of **4** to produce **78** and **102**, respectively (Figures 9.24 and 9.31). The difference in reactivity towards C—H insertion between triplet carbenes and an isoelectronic triplet nitrenes is well known and has been explained with a nitrogen rehybridization in the rate limiting H-abstraction step of the nitrene. Rehybridization is not needed in the carbene reaction.[61c,86]

C—H Insertion reactions involving free phenylnitrene are essentially impossible to study in solution by laser flash photolysis since C—H insertion cannot compete with dimerization, which is orders of magnitude faster.[88] Thus, confining ^{3}PN inside the molecular container, which eliminates dimerization, provides an elegant way to explore this important type of chemistry and allowed for the first time an accurate measurement of the activation parameters of a C—H insertion involving ^{3}PN.[89]

9.6.8 Norrish Type II Photochemistry

Sherman and co-workers recently investigated the photochemistry of butyrophenone **103** in the inner phase of carcerand **104** (Figure 9.32).[90]

The photochemistry of arylalkylketones is rich and important to organic synthesis.[91] Photochemical reactions of arylalkylketones typically arise from the triplet $^3(n, \pi^*)$ state. If a γ-hydrogen is present, as for **103**, the excited state abstracts the γ-hydrogen through a cyclic six-membered transition state (Norrish Type II photochemistry). The resulting 1,4-biradical disproportionates back to **103**, cyclizes to cyclobutanol **107** (Yang product) and/or β-fragments to ene **106** and enol **105** (Norrish Type II cleavage), which ketonizes in solution. The modulation of both photochemical processes through confinement has always been of great interests to supramolecular photochemists.[92]

Irradiation of **103** inside carcerand **104** gave a 5.6 : 1 mixture of Norrish Type II cleavage products **105** and **106** and Yang product **107**, which is very similar to the ratio found in solution. On the other hand, incarcerated hexyrophenone was completely stable to irradiation even though it shows similar photoreactivity as **103** in solution. The inertness of **105** is a result of its inner phase conformation, in which the distance between γ-protons and carbonyl oxygen precludes abstraction.

Both Norrish Type II photoproducts remained incarcerated. Furthermore, the surrounding host prevented ketonization of **105**, which requires acid or base catalysis. In fact, the resistance of incarcerated photoenol to ketonize is remarkable. Incarcerated **105** required

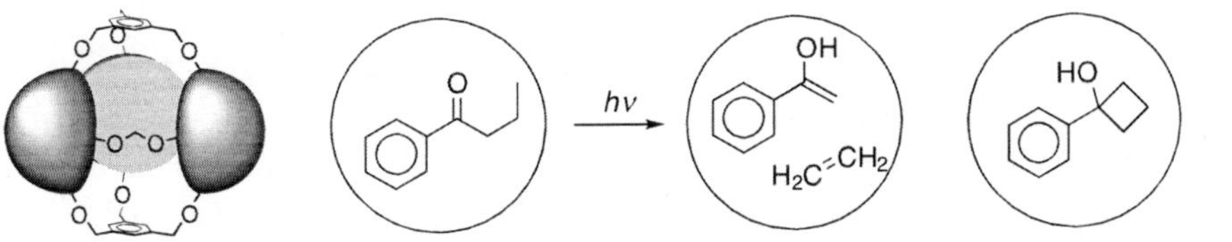

104 **104⊙103** **104⊙105+106** **104⊙107**

Figure 9.32 Photochemistry of butyrophenone **103** inside carcerand **104**

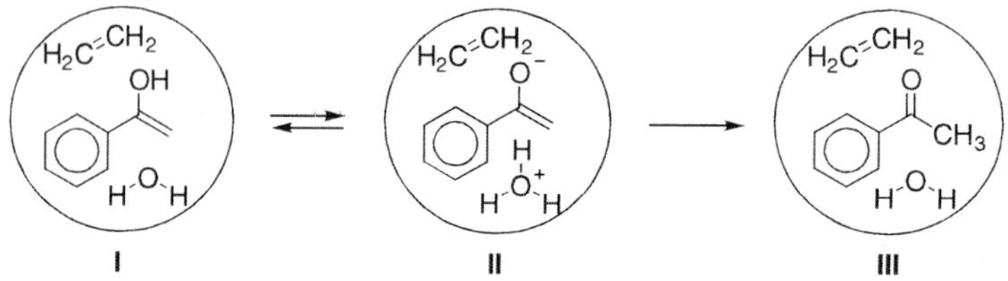

Figure 9.33 Mechanism of the inner phase ketonization of **104⊙105** + **106** in D_2O-saturated d_5-$PhNO_2$ at 100 °C

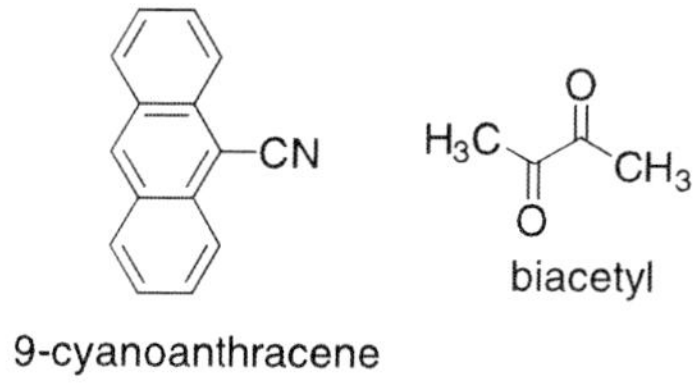

Figure 9.34 Structures of 9-cyanoanthracene and biacetyl

heating to 100 °C in water-saturated nitrobenzene to yield acetophenone **108**. The extrapolated life time of **15⊙105** + **106** at room temperature was more than seven orders of magnitude higher than that of free **105**.

An analysis of the kinetics and product distribution of ketonization in D_2O saturated nitrobenzene gave interesting insight into the mechanism of inner phase ketonization (Figure 9.33). Ketonization showed biphasic kinetics and is initiated by deprotonation of **105** (**I** → **II**). Ion pair **II** either collapses back to **105** (**I**) or forward to acetophenone (**III**).

The slow rate of inner phase ketonization is likely a result of a very high energy barrier associated with the formation of the ion pair in the hydrophobic and only weekly 'solvating' inner cavity of **104**. The inner phase mechanism is consistent with that proposed by Kresge for the uncatalyzed ketonization of free **105**.[93] However, in the inner phase competing proton transfer to the enolate O (**II** → **I**) or C (**II** → **III**) have similar rates, whereas the latter step is rate determining in the ketonization of free **105**.Thus, the inner phase perturbs the mechanism by substantially raising both barriers and by making them comparable in magnitude.

9.6.9 Incarcerated Excited States

In section 9.3.2, we have seen that spectral shifts of incarcerated guests can provide insight into the electronic and spatial characteristics of the inner phase. Of special interest to the photophysical community are also excited state lifetimes of molecules enclosed in constrained media.[94] Investigation of these systems may lead to the development of novel photophysical probes for immunoassays and devices for information storage or solar-energy conversion. Parola *et al.* investigated the photophysical properties of 9-cyanoanthracene (CA) inside hemicarcerand **26** (Figures 9.3 and 9.34).[95]

They observed major differences in the photophysical properties of CA upon incarceration. The surrounding host lead to a red-shift of the absorption spectrum and reduced the

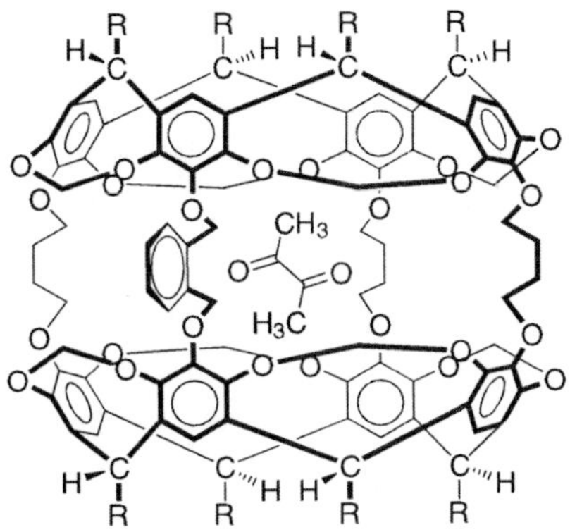

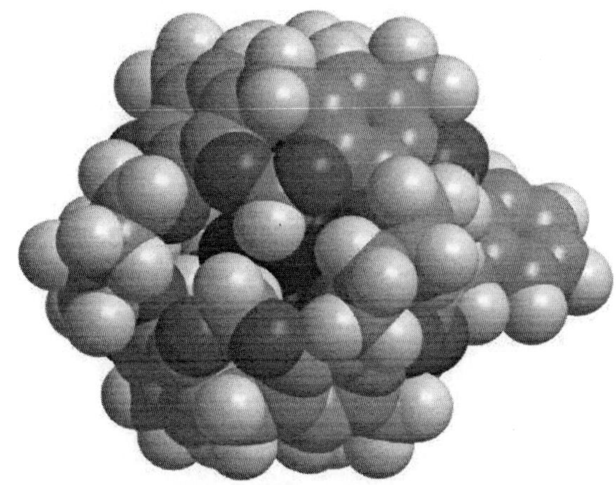

12⊙biacetyl

*Figure 9.35 Structure and space filling model of hemicarceplex **12**⊙biacetyl showing shielding of biacetyl (black) by **12***

lifetime of the singlet exited state ^1La by a factor of 40. In addition, the fluorescence quantum yield was 50 times smaller for incarcerated CA compared to free CA. Since guest emission bands and host absorption bands didn't overlap, energy transfer quenching from the host to the guest can be ruled out. Parola *et al.* concluded that quenching most likely takes place by electron transfer from the methoxybenzene units of **26** and that such electron transfer would be exothermic. Further studies from this group and independently by Farrán and Deshayes with triplet sensitizer biacetyl incarcerated in **14** and **12** provided insight into the role of host and solvent on the lifetime of triplet excited states.[18,96] Typically, the life-time of triplet excited states depends on the rates of non-radiative intersystem crossing (ISC), phosphorescence and of intermolecular quenching via electron transfer, energy transfer or H-abstraction. In the absence of intermolecular quenching processes, the T_1 excited state lifetime of biacetyl increases in the order gas phase, inner phase of **14**, liquid phase. This was interpreted with incarcerated biacetyl being in a not too tight cavity, which provides more free space with less specific cage–guest interactions than a solvent cage, but causes still a higher number of excited state–host collisions compared to the gas phase.

The lifetime of incarcerated biacetyl is unaffected by the solvent as long as it doesn't contain heavy atoms.[18] Interestingly, halogenated solvents enhance intersystem crossing via a remote external heavy atom effect. This was demonstrated by Romanova *et al.* for hemicarceplex **12**⊙biacetyl (Figure 9.35).[97,98]

It is well known, that ISC is strongly enhanced in molecules with covalently linked atoms of high atomic number (internal heavy atom effect).[99,100] Solvents containing heavy atoms, such as halides, also enhance ISC. Such external solvent heavy atom effects are believed to require direct contact between chromophore and perturbing molecule.[99,101] Contrary to this believe, Romanova *et al.* measured a pronounced decrease in the phosphorescence lifetime of incarcerated triplet biacetyl upon changing the solvent from benzene to CH_2Cl_2 to $CHBr_3$ to CH_2I_2.[97] In CH_2I_2, the phosphorescence lifetime was 7-fold shorter than in benzene. The shielding of the guest from direct contact with bulk components shows that long-distance spin–orbit coupling is possible through the walls of the hemicarcerand (Figure 9.35). This conclusion is further supported by the excellent agreement between the measured increase in the ISC rate induced by the external heavy atom effect upon switching from $CHBr_3$ to CH_2I_2 and that computed based on the current theory of ISC induced by intermolecular spin–orbit coupling.[102,103]

9.6.10 Photoelectron and Triplet Energy Transfer

Triplet excited state quenching by photoelectron transfer (PET) and energy transfer (ET) play a central role in biological photosynthesis,[104] visual transduction,[105] organic photochemistry,[106] semiconductor photocatalysis and imaging.[107–109] PET and ET across the insulating walls of a hemicarcerand have recently been studied in order to improve our understanding of these photophysical processes.[98]

Equations 9.1 and 9.2 schematically describe energy and electron transfer quenching of an incarcerated triplet excited state Host⊙G(T1) with a bulk phase quencher Q:[110]

$$\begin{aligned}\text{Host}\odot\text{G}(\text{T}_1)+\text{Q}(\text{S}_0) &\rightleftarrows \text{Host}\odot\text{G}(\text{T}_1)\cdots\text{Q}(\text{S}_0)\rightarrow\text{Host}\odot\text{G}(\text{S}_0)\cdots\text{Q}(\text{T}_1)\\ &\rightleftarrows \text{Host}\odot\text{G}(\text{S}_0)+\text{Q}(\text{T}_1)\end{aligned} \tag{9.1}$$

$$\text{Host}\odot\text{G}(\text{T}_1)+\text{Q}\rightleftarrows\text{Host}\odot\text{G}(\text{T}_1)\cdots\text{Q}\rightarrow\text{Host}\odot\text{G}^-\cdots\text{Q}^+\rightleftarrows\text{Host}\odot\text{G}^-+\text{Q}^+ \tag{9.2}$$

ET proceeds by a Dexter electron exchange mechanism and is a weakly coupled non-adiabatic process. Its rate constant k_{ET} can be approximated by the Golden Rule:[111–113]

$$k_{ET}=(2\pi/\hbar)\times|v|^2\times FCWDS \tag{9.3}$$

where *FCWDS* are the Franck–Condon weighted density of states and v is the electronic coupling matrix element. In a semiclassical treatment, this equation can be separated into a preexponential factor A and an exponential term that relates k_{ET} to the driving force ΔG and nuclear reorganization energies of reactant λv and solvent λs:

$$k_{ET}=A\times\exp\left\{-(\lambda_s+\Delta G+\lambda_v)^2/4\lambda_s k_B T\right\} \tag{9.4}$$

The dependence of the rate constant for photoinduced electron transfer k_{PET} on the driving force and reorganization energies is similar. Equation 4 predicts a parabolic dependence of log k_{ET} on the driving force. At $-\Delta G=(\lambda s+\lambda v)$, k_{ET} is largest and decreases at smaller (normal region) and more exothermic driving force (inverted region). The experimental observation of the Marcus inverted region for electron or energy transfer between non-covalently linked triplet excited state/quencher pairs has been very difficult mainly because ET and PET transfer at high driving forces is much faster than the rate of diffusional encounter. In the encounter complexes of the hemicarceplex/quencher systems (Host⊙G(T_1)....Q(S_0) and Host⊙G(T_1)....Q) excited state and quencher are separated by about 7 Å. As a consequence of a strong distance dependence,[111b,114] ET and PET are substantially slower than diffusion, which has made observation of the inverted region possible in these systems[96,110,115] In their seminal work on through space triplet ET, Deshayes and co-workers studied acetophenone (AC) hemicarceplex **14**⊙Ac and probed through-shell triplet ET chemically via the isomerization of *cis*-piperylene to *trans*-piperylene (Figure 9.36).[116]

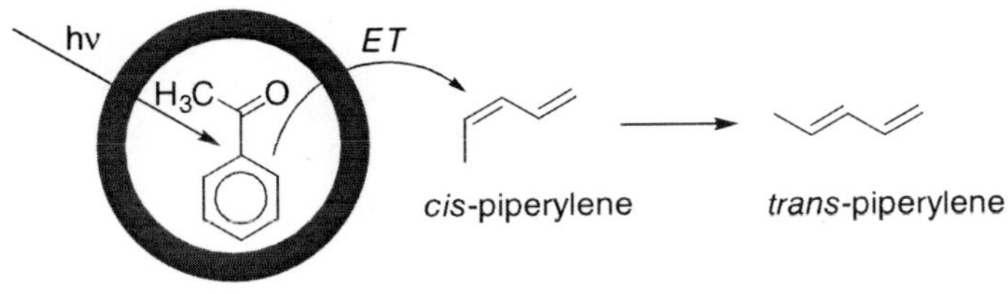

Figure 9.36 *Photosensitized isomerization of* cis-*piperylene catalyzed by* ***14*⊙*Ac***

From a Stern–Volmer analysis and the lifetimes of free and incarcerated triplet acetophenone, Deshayes determined a 2.7-fold slower triplet ET rate for **14**⊙Ac compared to free acetophenone. This corresponds to an almost diffusion controlled rate for **14**⊙Ac. Since triplet energy is transferred through an electron exchange mechanism, which is believed to require a close contact between donor and acceptor, sufficient overlap of the HOMO and LUMO of the donor-acceptor pair through the host-shell must exist. Whether the intervening hemicarcerand plays a role in the through-space ET is not clear. As an extension of this work, Farrán and Deshayes studied triplet ET between incarcerated biacetyl (**14**⊙biacetyl) and various bulk phase acceptors.[96] In all cases, hemicarcerand **14** retarded triplet ET, which suggests a reduced electron-coupling between donor and acceptor as a result of their larger separation. Also, log k_{ET} and ΔG showed a hyperbolic relationship as predicted by the Golden Rule.

Also interesting is the extremely slow triplet ET rate to O_2, which is typically a highly efficient quencher. Farrán and Deshayes concluded that when oxygen is prevented from coming into direct contact with the donor, the quenching rate drops off drastically. Parola *et al.* independently measured triplet ET rates from **14**⊙biacetyl to those quenchers used by Deshayes and several others and agreed with Deshayes that the difference between k_{ET} of free and incarcerated biacetyl results from different electronic exchange matrix elements v.[110] However, they were careful in taking the observed parabolic like relationship as firm evidence for inverted behavior especially since their data were strongly scattered. A parabolic-like correlation may simply reflect non-homogeneity of the quenchers, as a consequence of their different sizes, which leads to different donor-acceptor distances and/or orientations and hence to different values for v. Most likely for the same reason, evidence for the inverted region could not be found for quenching of **14**⊙biacetyl by electron transfer from bulk phase aromatic amine donors.

In a subsequent investigation, Deshayes and Piotrowiak provided clear support for the parabolic Marcus relationship and explained the original data by Farrán and Deshayes and Parola *et al.* in a quantitative manner by taking into account the different internal nuclear reorganization energies λv of the acceptors.[115] According to MO calculations, λv varies by more than 20 kcal mol^{-1} among the different acceptors. As a consequence, two acceptors with nearly identical driving forces and transfer rates may belong to different regions of Marcus parabola. This is the case for dibromoanthracene (DBA) and diphenylbutadiene (DPB) (Figure 9.37). Both were assigned to the correct region of a Marcus parabola based on their activation energy of transfer, which is negative for the former

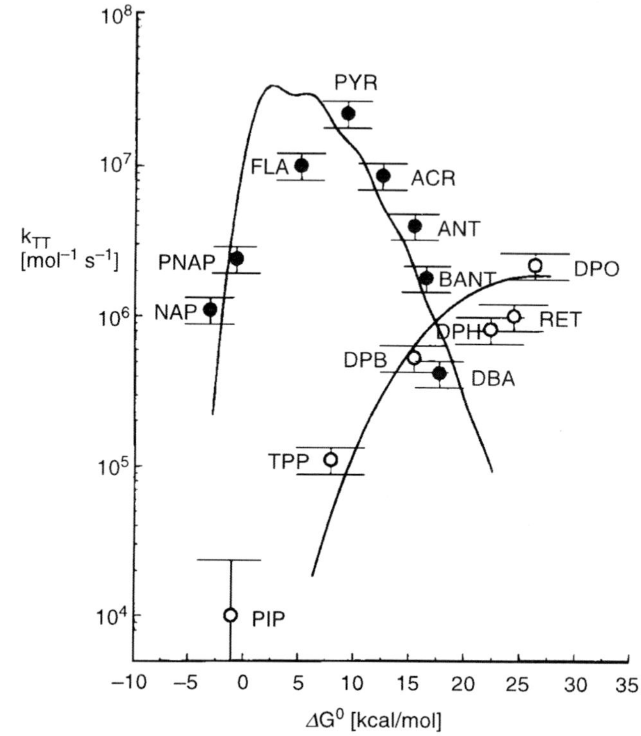

Figure 9.37 *Rate constant versus driving force −ΔG of triplet ET from **14**⊙biacetyl to aryl (●) and alkene (○) acceptors and theoretical curves generated using the semiclassical Marcus–Jortner formalism of triplet energy transfer. Reprinted with permission from [115]. Copyright 1998 American Chemical Society*

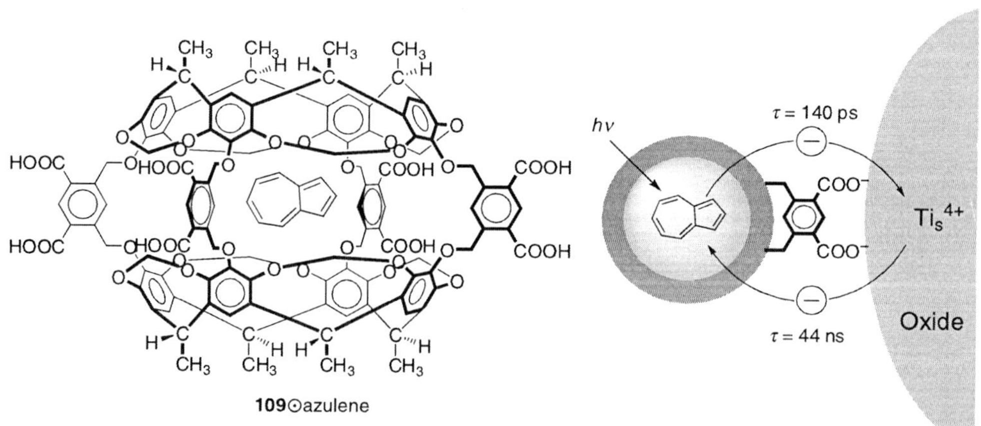

Figure 9.38 *Structure of hemicarceplex **109**⊙azulene (left). Schematic representation of **109**⊙azulene/TiO_2 assembly and PET between azulene and the TiO_2 nanoparticle*

(typical for the inverted region), but positive for the latter (normal region behavior). Deshayes and Piotrowiak identified four groups of acceptors. (1) Rigid aromatics that display small λv. (2) Acyclic olefins that twist around the double bond upon triplet excitation and therefore have large λv. (3) Cyclic olefins with even larger λv and (4) O_2, which has essentially no λv. Each acceptor group has its own $\log k_{ET}$ versus ΔG correlation (Figure 9.37). The remaining scattering in the experimental data may result from differences in size and shape of the acceptors, leading to different effective electronic couplings in the corresponding encounter complexes and possibly also to different encounter frequencies.

The investigation detailed above were mainly concerned with the dependence of triplet ET rate on ΔG and λv and provided clear support for the inverted region. As pointed out earlier, electronic coupling between donor and acceptor is strongly reduced upon separating them by an intervening hemicarcerand, leading to a slowing of energy and electron transfer. In a more recent investigation, Deshayes and Piotrowiak addressed dependence of electronic coupling between incarcerated biacetyl and the bulk phase quencher on the hemicarcerand size.[20] The electronic coupling between incarcerated biacetyl and the acceptor can be described by a simple superexchange mechanism and viewed as a sequence of guest–hemicarcerand and hemicarcerand–solute interactions. Thus, the total electronic coupling matrix element v_{total} is the product of that for the guest–hemicarcerand v_{GH} and the hemicarcerand–acceptor interaction v_{HA}:

$$v_{total} \propto v_{GH} \times v_{HA} \tag{9.5}$$

Hemicarceplexes **4**⊙biacetyl, **15**⊙biacetyl and **14**⊙biacetyl, which vary in size and linker characteristics, were included in this study. k_{ET} increased with decreasing host size in the order $k_{ET}(\mathbf{4}) < k_{ET}(\mathbf{14}) \ll k_{ET}(\mathbf{15})$. The average electronic coupling matrix elements showed the same trend and are $|v(\mathbf{4})| = 0.20\ cm^{-1}$, $|v(\mathbf{14})| = 0.26\ cm^{-1}$, and $|v(\mathbf{15})| = 0.66\ cm^{-1}$. Since v_{HA} should be independent of the hemicarcerand size, the differences in v_{total} reflect differences in the guest–hemicarcerand electronic couplings v_{HA} among the hosts. v_{total} was slightly higher for ET to neat acceptors. However, the trends were the same and v_{HA} differences measured in neat acceptors were identical to those measured in acceptor solutions. These trends are consistent with predictions. The time-averaged guest–hemicarcerand interaction should depend on the size of the cavity and should increase with decreasing cavity. One can also conclude, that the *o*-xylylene linkers in **14** don't cause special enhancement of electronic coupling.

These investigations not only unraveled the influence of the hemicarcerand on the mechanism of through-shell triplet ET, but also contribute to the understanding of solvent-mediated electron transfer,[117,118] in which a solvent molecule separates donor and acceptor and provides the pathway for electronic coupling. Since the thickness of a hemicarcerand is comparable to that of common organic solvents, the measured electronic coupling matrix elements are good estimates for the magnitude of solvent-mediated contributions to electronic coupling in triplet excitation transfer.

9.6.11 Hemicarcerand-based Photoactive Assemblies

The investigations of triplet ET and PET across the walls of a hemicarcerand described in the previous section have improved the understanding of nonradiative electronic ET processes and have allowed experimental verification of theoretical models. These studies also form the basis for application of hemicarceplexes in the fabrication of nanodevices for solar energy conversion.[108] Deshayes and Piotrowiak recently studied the photophysical properties of hemicarceplex **109**⊙azulene immobilized onto the surface of TiO_2 nanoparticles via its eight peripheral carboxylic groups.[119] Immobilization led to nearly complete quenching of the azulene fluorescence emission, which suggests efficient ET from the excited state to the conducting band of TiO_2. The time constant for electron injection was about 1000-times slower than that for the 1- or 2-carboxyazulene immobilized on TiO_2. Also, back-electron transfer **109**⊙azulene$^+$/TiO_2^- → **109**⊙azulene/TiO_2 was strongly retarded by about the same factor (Figure 9.38). Interestingly, the latter kinetics could be described with a single exponential in contrast to the 1-carboxyazulene/ TiO_2 and other dye/TiO_2 systems,[120] for which multi-exponential behavior is observed as a consequence of inhomogeneity of binding and presence of surface traps. The homogeneous recombination kinetics indicates that tunnelling across the hemicarcerand walls is the rate-limiting step of back-electron transfer and that the slow tunneling averages out the large distribution of rates that are normally seen. Thus, non-covalent immobilization of dye molecules through encapsulation inside a TiO_2 bound hemicarcerand, not only increases the lifetime of the charge separated state but also leads to a more homogeneous system with respect to the kinetic of charge recombination. It remains to be seen how these photophysical properties affect the properties of photovoltaic devices synthesized from dye-hemicarceplex/TiO_2 hybrid assemblies.

9.7 Conclusions and Outlook

This review shows that the inner phase of molecular container compounds are interesting new confined environments, in which a large variety of different reactions have been carried out and new once have been discovered. Moreover, molecular container compounds have become real tools for physical organic chemists to study reaction mechanism and long-distance phenomena. Molecular containers made possible the investigation of highly strained and reactive molecules under normal working conditions by generating them in the protective inner phase. They also allowed the investigation of electronic interactions between encapsulated and bulk phase molecules through the intervening hemicarcerand and have provided experimental support for theoretical models of long distance spin-orbit coupling, electron and energy transfer. It is anticipated that this field of research will further grow and that the recent development of multi-cavitand nanocapsules[7b,121] will make possible the investigation of chemical reactivity of macromolecular guests that are of interest to material and biological sciences.

Acknowledgements

The author would like to thank his current and former co-workers for their efforts to advance this exciting research field. He warmly thanks the National Science Foundation (Grant CHE-0518351) for financial support of his research.

References

1. a) D. J. Cram, Molecular container compounds, *Nature*, 1992, **356**, 29–36; b) D. J. Cram, J. M. Cram, *Container Molecules and Their Guests*, The Royal Society of Chemistry, Cambridge, U. K., 131–216, 1994; c) E. Maverick, D. J. Cram, Carcerands and hemicarcerands: host that imprison molecular guests, in *Comprehensive Supramolecular Chemistry*, F. Vögtle (Ed.), Pergamon, Oxford, U.K., 367–418, 1996; d) A. Jasat, J. C. Sherman, Carceplexes and hemicarceplexes, *Chem. Rev.*, 1999, **99**, 931–967; e) R. Warmuth, J. Yoon, Recent highlights in hemicarcerand chemistry, *Acc. Chem. Res.*, 2001, **34**, 95–105.
2. a) R. Warmuth, Inner phase stabilization of reactive intermediates, *Eur. J. Org. Chem.*, 2001, 423–437; b) R. Warmuth, The inner phase of molecular container compounds as a novel reaction environment, *J. Incl. Phenom.*, 2000, **37**, 1–38.
3. D. J. Cram, S. Karbach, Y. H. Kim, L. Baczynskyj, G. W. Kalleymeyn, Shell closure of two cavitands forms carcerand complexes with components of the medium as permanent guests, *J. Am. Chem. Soc.*, 1985, **107**, 2575–2576.
4. a) D. J. Cram, M. E. Tanner, C. B. Knobler, Guest release and capture by hemicarcerands introduces the phenomenon of constrictive binding, *J. Am. Chem. Soc.*, 1991, **113**, 7717–7727; b) D. J. Cram, M. T. Blanda, K. Paek, C. B. Knobler, Constrictive and intrinsic binding in a carcerand containing four portals, *J. Am. Chem. Soc.*, 1992, **114**, 7765–7773.
5. a) J. C. Sherman, C. B. Knobler, D. J. Cram, Syntheses and properties of soluble carceplexes, *J. Am. Chem. Soc.*, 1991, **113**, 2194–2204; b) T. Robbins, C. B. Knobler, D. Bellew, D. J. Cram, A highly adaptive and strongly binding hemicarcerand, *J. Am. Chem. Soc.*, 1994, **116**, 111–121; c) M. L. C. Quan, D. J. Cram, Constrictive binding of large guests by a hemicarcerand containing four portals, *J. Am. Chem. Soc.*, 1991, **113**, 2754–2755; d) J. Yoon, D. J. Cram, The first water-soluble hemicarceplexes, *Chem. Commun.*, 1998, 497–498; e) S. K. Kurdistani, R. C. Helgeson, D. J. Cram, Stepwise shell closures provide hosts that expose or protect guests from outer-phase reactants, *J. Am. Chem. Soc.*, 1995, **117**, 1659–1660; f) J. Yoon, C. Sheu, K. N. Houk, C. B. Knobler, D. J. Cram, Synthesis, binding properties, and structures of seven new hemicarcerands each composed of two bowls bridged by three tetramethylenedioxy groups and a fourth unique linkage, *J. Org. Chem.*, 1996, **61**, 9323–9339; g) J. Yoon, D. J. Cram, Chiral recognition properties in complexation of two asymmetric hemicarcerands, *J. Am. Chem. Soc.*, 1997, **119**, 11796–11806.
6. J. C. Sherman, Molecules that can't resist templation, *Chem. Commun.*, 2003, 1617–1623.
7. For thermodynamically controlled hemicarceplex formations that do not require a template see: a) S. Ro, S. J. Rowan, A. R. Pease, D. J. Cram, J. F. Stoddart, Dynamic hemicarcerands and hemicarceplexes, *Org. Lett.*, 2000, **2**, 2411–2414; b) X. Liu, Y. Liu, G. Li, R. Warmuth, One-pot 18 component synthesis of an octahedral nanocontainer molecule, *Angew. Chem. Int. Ed.*, 2006, **45**, 901–904.
8. a) R. G. Chapman, N. Chopra, E. D. Cochien, J. C. Sherman, Carceplex formation: scope of a remarkably efficient encapsulation reaction, *J. Am. Chem. Soc.*, 1994, **116**, 369–370; b) R. G. Chapman, J. C. Sherman, Study of templation and molecular encapsulation using highly stable and guest-selective self-assembling structures, *J. Am. Chem. Soc.*, 1995, **117**, 9081–9082; c) R. G. Chapman, J. C. Sherman, Templation in the formation of carceplexes, *J. Org. Chem.*, 1998, **63**, 4103–4110.
9. R. G. Chapman, G. Olovsson, J. Trotter, J. C. Sherman, Crystal structure and thermodynamics of reversible molecular capsules, *J. Am. Chem. Soc.*, 1998, **120**, 6252–6260.
10. D. A. Makeiff, D. J. Pope, J. C. Sherman, Template effects in the formation of a tetramethylene-bridged hemicarceplex, *J. Am. Chem. Soc.*, 2000, **122**, 1337–1342.
11. J. Yoon, D. J. Cram, Decomplexation rate comparisons of hemicarceplexes whose single unique host bridge is changed in length and blocking power, *Chem. Commun.*, 1997, 1505–1506.
12. R. Warmuth, M. A. Marvel, Chemistry and properties of cycloheptatetraene inside a hemicarcerand, *Chem. Eur. J.*, 2001, **7**, 1209–1220.

13. a) S. Watanabe, K. Goto, T. Kawashima, R. Okazaki, An endohedral simple enol: the first isolation of a β-unsubstituted simple enol utilizing a lantern-shaped molecular framework, *J. Am. Chem. Soc.*, 1997, **119**, 3195–3196; b) K. Goto, R. Okazaki, Molecular bowls and capsules with an endohedral functionality. The stabilization of highly reactive species in their inner phase, *Liebigs Ann./Recueil.*, 1997, 2393–2407.
14. R. Warmuth, First *innermolecular* Diels–Alder reaction of *o*-benzyne inside a molecular container compound, *Chem. Comm.*, 1998, 59–60.
15. S. Mecozzi, J. Rebek Jr., The 55% solution: a formula for molecular recognition in the liquid state, *Chem. Eur. J.*, 1998, **4**, 1016–1022.
16. R. G. Chapman, J. C. Sherman, Restricted motion of guests confined in carceplexes and capsules, *J. Org. Chem.*, 2000, **65**, 513–516.
17. a) C. Reichardt, *Solvents and Solvent Effects in Organic Chemistry*, 2nd Ed., VCH, Weinheim, 1988; b) P. Suppan, N. Ghoneim, *Solvatochromism*, The Royal Society of Chemistry, Cambridge, 1997.
18. F. Pina, A. J. Parola, E. Ferreira, M. Maestri, N. Armaroli, R. Ballardini, V. Balzani, Supramolecular photochemistry and photophysics. Biacetyl imprisoned in a hemicarcerand, *J. Phys. Chem.*, 1995, **99**, 12701–12703.
19. C. Marquez, W. M. Nau, Polarizabilities inside molecular containers, *Angew. Chem. Int. Ed.*, 2001, **40**, 4387–4390.
20. Z. S. Romanova, K. Deshayes, P. Piotrowiak, Triplet excitation transfer through the walls of hemicarcerands: dependence of the electronic coupling on the size of the molecular cage, *J. Am. Chem. Soc.*, 2001, **123**, 11029–11036.
21. A. Warshel, P. K. Sharma, M. Kato, Y. Xiang, H. Liu, M. H. M. Olsson, Electrostatic basis of enzyme catalysis, *Chem. Rev.*, 2006, **106**, 3210–3235. A. Warshel, G. Naray-Szabo, F. Sussman, J.-K. Hwang, How do serine proteases really work? *Biochemistry*, 1989, **28**, 3629–3637.
22. L. E. Eberson, *Electron Transfer Reactions in Organic Chemistry*, Springer Verlag, New York, 1987.
23. H. B. Gray, J. R. Winkler, Long-range electron transfer, *PNAS*, 2005, **102**, 3534–3539.
24. T. A. Robbins, D. J. Cram, Through-shell oxidation and reduction reactions of guests in a hollow container single molecule, *J. Am. Chem. Soc.*,1993, **115**, 12199.
25. S. Mendoza, P. D. Davidov, A. E. Kaifer, Electrochemistry of encapsulated guests: ferrocene inside Cram's hemicarcerands, *Chem. Eur. J.*, 1998, **4**, 864–870.
26. R. Warmuth, E. F. Maverick, C. B. Knobler, D. J. Cram, Through-shell alkyl lithium additions and borane reductions, *J. Org. Chem.*, 2003, **68**, 2077–2088.
27. a) S. T. Perri, L. D. Foland, O. H. W. Decker, H. W. Moore, Synthesis of benzoquinones and annulated derivatives from conjugated ketenes, *J. Org. Chem.*, 1986, **51**, 3067–3068; b) L. D. Foland, J. O. Karlsson, S. T. Perri, R. Schwabe, S. J. Xu, S. Patil, H. W. Moore, Rearrangement of 4–alkynylcyclobutenones. A new synthesis of 1,4–benzoquinones, *J. Am. Chem. Soc.*, 1989, **111**, 975–989.
28. J. H. Exner, E. C. Steiner, Solvation and ion pairing of alkali-metal alkoxides in dimethyl sulfoxide. Conductometric studies, *J. Am. Chem. Soc.*, 1974, **96**, 1782–1787.
29. R. A. Bartsch, Ionic association in base-promoted β-elimination reactions, *Acc. Chem. Res.*, 1975, **8**, 239–245.
30. R. Warmuth, J.-L. Kerdelhué, S. Sánchez Carrera, K. J. Langenwalter, N. Brown, Rate acceleration through dispersion interactions: insight into the effect of a hemicarcerand on the transition state of inner phase diazirine decompositions, *Angew. Chem. Int. Ed.*, 2002, **41**, 96–99.
31. S. Sánchez Carrera, N. Brown, J.-L. Kerdelhué, K. J. Langenwalter, R. Warmuth, Inner phase reaction dynamics: the influence of hemicarcerand polarizability and shape on the potential energy surface of an inner phase reaction, *Eur. J. Org. Chem.*, 2005, 2239–2249.
32. a) M. T. H. Liu, The thermolysis and photolysis of diazirines, *Chem. Soc. Rev.*, 1982, **11**, 127–140; b) I. D. R. Stevens, M. T. H. Liu, N. Soundararajan, N. Paike, The thermal decomposition of diazirines: 3–(3–methyldiazirin-3–yl)propan-1–ol and 3–(3–methyldiazirin-3–yl) propanoic acid, *J. Chem. Soc. Perkin Trans.*, 1990, **2**, 661–667; c) M. T. H. Liu, Y.-K. Choe, M. Kimura, K. Kobayashi, S. Nagase, T. Wakahara, Y. Niino, M. O. Ishitsuka, Y. Maeda, T.

Akasaka, Effect of substituents on the thermal decomposition of diazirines: experimental and computational studies, *J. Org. Chem.*, 2003, **68**, 7471–7478.

33. a) D. A. Stauffer, R. E. Barrans Jr., D. A. Dougherty, Biomimetic catalysis of an S_N2 reaction as a consequence of a novel transition-state stabilization, *Angew. Chem. Int. Ed.*, 1990, **29**, 915–918; b) S. M. Ngola, D. A. Dougherty, Evidence for the importance of polarizability in biomimetic catalysis involving cyclophane receptors, *J. Org. Chem.*, 1996, **61**, 4355–4360.
34. a) P. Timmerman, W. Verboom, F. C. J. M. van Veggel, J. P. M. van Duynhoven, D. N. Reinhoudt, A new type of stereoisomerism in carceplexes formed from calix[4]arene units, *Angew. Chem., Int. Ed.*, 1994, **33**, 2345–2348; b) A. M. A. van Wageningen, P. Timmerman, J. P. M. van Duynhoven, W. Verboom, F. C. J. M. van Veggel, D. N. Reinhoudt, Calix[4] arene-based (hemi)carcerands and carceplexes: synthesis, functionalization, and molecular modeling study, *Chem. Eur. J.*, 1997, **3**, 639–654.
35. a) J. Kang, J. Rebek Jr., Acceleration of a Diels–Alder reaction by a self-assembled molecular capsule, *Nature*, 1997, **385**, 50–52; b) M. Yoshizawa, M. Tamura, M. Fujita, Diels–alder in aqueous molecular hosts: unusual regioselectivity and efficient catalysis, *Science*, 2006, **312**, 251–254.
36. D. J. Cram, M. E. Tanner, R. Thomas, The taming of cyclobutadiene, *Angew. Chem Int. Ed.*, 1991, **30**, 1024–1027.
37. a) M. P. Cava, M. J. Mitchell, *Cyclobutadiene and Related Compound*, Academic Press, New York, 1967; b) G. Maier, Tetrahedrane and cyclobutadiene, *Angew. Chem., Int. Ed.*, 1988, **27**, 309–332.
38. A. A. Deniz, K. S. Peters, G. J. Snyder, Experimental determination of the antiaromaticity of cyclobutadiene, *Science*, 1999, **286**, 1119–1122.
39. O. L. Chapman, C. L. McIntosh, J. Pacansky, Photochemical transformations. XLVIII. Cyclobutadiene, *J. Am. Chem. Soc.*, 1973, **95**, 614–617.
40. G. Maier, A. Alzérreca, Small rings. 15. Tri-*tert*-butylcyclobutadiene, *Angew. Chem., Int. Ed.*, 1973, **12**, 1015–1016.
41. a) P. M. Warner, Strained bridgehead double bonds, *Chem. Rev.*, **1989**, 89, 1067–1093. b) H. Hopf, *Classics in Hydrocarbon Chemistry*, VCH, Weinheim, 122–137, 2000; c) M. Jones Jr., The carbene route to bridgehead double bonds, in *Advances in Carbene Chemistry*, U. H. Brinker (Ed.), Jai Press, Stamford, CT, Vol. 2, 77–96, 1998.
42. a) W. F. Maier, P. V. R. Schleyer, Evaluation and prediction of the stability of bridgehead olefins, *J. Am. Chem. Soc.*, 1981, **103**, 1891–1900; b) I. Novak, Molecular modeling of anti-bredt compounds, *J. Chem. Inf. Model.*, 2005, **45**, 334–338.
43. A. D. Wolf, M. Jones Jr., Carbene route to bridgehead olefins. Bicyclo[2.2.2]oct 1 ene, *J. Am. Chem. Soc.*, 1973, **95**, 8209–8210.
44. a) N. Bian, M. Jones Jr., The stereochemistry of addition of an alkylcarbene, *Tetrahedron Lett.*, 1993, **34**, 3967–3969; b) R. T. Ruck, M. Jones Jr., Bicyclo[2.2.1]heptan-1-ylcarbene can ring expand, *Tetrahedron Lett.*, 1998, **39**, 4433–4436.
45. P. Roach, R. Warmuth, The room-temperature stabilization of bicyclo[2.2.2]oct-1–ene and bicyclo[3.2.1]oct-1–ene, *Angew. Chem., Int. Ed.*, 2003, **42**, 3039–3042.
46. D. C. Merrer, R. A. Moss, Kinetics of intramolecular carbene reactions, in *Advances in Carbene Chemistry*, Brinker U. H. (Ed.) Elsevier, Amsterdam, Vol. 3, 53–114, 2001.
47. R. Warmuth, *o*-benzyne: strained alkyne or a cumulene? – NMR characterization in a molecular container, *Angew. Chem., Int. Ed.*, 1997, **37**, 1347–1350.
48. G. Wittig, *Naturwissenschaften*, 1942, **30**, 696–703.
49. J. D. Roberts, H. E. Simmons Jr., J. A. Carlsmith, C. W. Vaughan, Rearrangement in the reaction of chlorobenzene-1–C14 with potassium amide, *J. Am. Chem. Soc.*, 1953, **75**, 3290–9391.
50. R. W. Hoffmann, *Didehydrobenzene and Cycloalkynes*, Academic Press, New York, Vol. 11, 1967.
51. H. H. Wenk, M. Winkler, W. Sander, One century of aryne chemistry, *Angew. Chem., Int. Ed.*, 2003, **42**, 502–528.
52. a) H. Jiao, P. von Ragué Schleyer, B. R. Beno, K. N. Houk, R. Warmuth, Theoretical studies of the structure, aromaticity, and magnetic properties of *o*-benzyne, *Angew. Chem., Int. Ed.*,

1997, **37**, 2761 and cited references; b) F. De Proft, P. v. R. Schleyer, J. H. Van Lenthe, F. Stahl, P. Geerlings, Magnetic properties and aromaticity of *o*-, *m*-, and *p*-benzyne, *Chem. Eur. J.*, 2002, **8**, 3402–3410.

53. J. G. G. Simon, N. Muenzel, A. Schweig, Matrix isolation photochemistry: photoequilibrium between transient *o*-benzyne, carbon monoxide, and transient cyclopentadienylideneketene in an argon matrix, *Chem. Phys. Lett.*, 1990, **170**, 187–192.
54. O. L. Chapman, K. Mattes, C. L. McIntosh, J. Pacansky, G. V. Calder, G. Orr, Photochemical transformations. LII. Benzyne, *J. Am. Chem. Soc.*, 1973, **95**, 6134–6135.
55. a) O. L. Chapman, C.-C. Chang, J. Kolc, N. R. Rosenquist, H. Tomioka, Photochemical method for the introduction of strained multiple bonds. Benzyne C≡C stretch, *J. Am. Chem. Soc.*, 1975, **97**, 6586–6588; b) J. G. Radziszewski, B. A. Hess Jr., R. Zahradnik, Infrared spectrum of *o*-benzyne: experiment and theory, *J. Am. Chem. Soc.*, 1992, **114**, 52–57.
56. A. M. Orendt, J. C. Facelli, J. G. Radziszewski, W. J. Horton, D. M. Grant, J. Michl, ^{13}C dipolar NMR spectrum of matrix-isolated *o*-benzyne-1,2–$^{13}C_2$, *J. Am. Chem. Soc.*, 1996, **118**, 846–852.
57. J. G. G. Simon, N. Münzel, A. Schweig, ^{1}H and ^{13}C NMR spectra of benzocyclopropenone in liquid solutions at 193 K, *Chem. Phys. Lett.*, 1993, **201**, 377–382.
58. R. P. Johnson, K. J. Daoust, Interconversions of cyclobutyne, cyclopentyne, cyclohexyne, and their corresponding cycloalkylidenecarbenes, *J. Am. Chem. Soc.*, 1995, **117**, 362–367.
59. B. R. Beno, C. Sheu, K. N. Houk, R. Warmuth, D. J. Cram, Cycloaddition of *o*-benzyne to benzene and the inner phase of a hemicarcerand, *Chem. Comm.*, 1998, 301–302.
60. S. G. Kukolich, M. C. McCarthy, P. Thaddeus, Molecular structure of *o*-benzyne from microwave measurements, *J. Phys. Chem. A*, 2004, **108**, 2645–2651.
61. a) P. P. Gasper, J.-P. Hsu, S. Chari, M. Jones Jr., The phenylcarbene rearrangement revisited, *Tetrahedron*, 1985, **41**, 1479–1507; b) R. P. Johnson, Strained cyclic cumulenes, *Chem. Rev.*, 1989, **89**, 1111–1124; c) M. S. Platz, Comparison of phenylcarbene and phenylnitrene, *Acc. Chem. Res.*, 1995, **28**, 487–492.
62. W. J. Baron, M. Jones Jr., P. P. Gaspar, Interconversion of *o*-, *m*-, and *p*-tolylcarbene, *J. Am. Chem. Soc.*, 1970, **92**, 4739–4740.
63. S. Matzinger, T. Bally, E. V. Patterson, R. J. McMahon, The C_7H_6 potential energy surface revisited: relative energies and IR assignment, *J. Am. Chem. Soc.*, 1996, **118**, 1535–1542.
64. a) M. W. Wong, C. Wentrup, Interconversions of phenylcarbene, cycloheptatetraene, fulvenallene, and benzocyclopropene. A theoretical study of the C_7H_6 energy surface, *J. Org. Chem.*, 1996, **61**, 7022–7029; b) P. R. Schreiner, W. L. Karney, P. v. R. Schleyer, W. T. Borden, T. P. Hamilton, H. F. Schaefer III, Carbene rearrangements unsurpassed: details of the C_7H_6 potential energy surface revealed, *J. Org. Chem.*, 1996, **61**, 7030–7039; c) W. L. Karney, W. T. Borden, *Ab Initio* study of the ring expansion of phenylnitrene and comparison with the ring expansion of phenylcarbene, *J. Am. Chem. Soc.*, 1997, **119**, 1378–1387.
65. a) P. R. West, O. L. Chapman, J.-P. LeRoux, 1,2,4,6–Cycloheptatetraene, *J. Am. Chem. Soc.*, 1982, **104**, 1779–1782; b) R. J. McMahon, C. J. Abelt, O. L. Chapman, J. W. Johnson, C. L. Kreil, J.-P. LeRoux, A. M. Mooring, P. R. West, 1,2,4,6–Cycloheptatetraene: the key intermediate in arylcarbene interconversions and related C_7H_6 rearrangements, *J. Am. Chem. Soc.*, 1987, **109**, 2459–2469.
66. O. L. Chapman, C. J. Abelt, Synthesis and photochemistry of 1–diazo-2–cyclopentene and 2–diazobicyclo[3.2.0]hepta-3,6–diene, *J. Org. Chem.*, 1987, **52**, 1218–1221.
67. R. Warmuth, M. A. Marvel, 1,2,4,6–Cycloheptatetraene: room temperature stabilization inside a hemicarcerand, *Angew. Chem., Int. Ed.*, 2000, **39**, 1117–1119.
68. R. Warmuth, The enantiomerization barrier of 5-methylcyclohepta-1,2,4,6-tetraene, *J. Am. Chem. Soc.*, 2001, **123**, 6955–6956.
69. J.-L. Kerdelhué, Kevin J. Langenwalter, R. Warmuth, Mapping the potential energy surface of the tolylcarbene rearrangement in the inner phase of a hemicarcerand, *J. Am. Chem. Soc.*, 2003, **125**, 973–986.
70. C. Kemmis, R. Warmuth, Temperature dependence of the kinetic isotope effect and tunneling contribution in the innermolecular reaction of phenylcarbene inside a hemicarcerand, *J. Supramol. Chem.*, 2003, 253–267.

71. Y. Liu, R. Warmuth, A 'through-shell' binding isotope effect, *Angew. Chem. Int. Ed.*, 2005, **44**, 7107–7110.
72. W. Kirmse, K. Loosen, H.-D. Sluma, Carbenes and the oxygen-hydrogen bond: cyclopentadienylidene and cycloheptatrienylidene, *J. Am. Chem. Soc.*, 1981, **103**, 5935–5937.
73. C. M. Geise, C. M. Hadad, Substituent effects in the interconversion of phenylcarbene, bicyclo[4.1.0]hepta-2,4,6–triene, and 1,2,4,6–cycloheptatetraene, *J. Org. Chem.*, 2002, **67**, 2532–2540.
74. X. Wang, Z. Yang, J. Wang, J. Zhang, W. Cao, Computational study of the rearrangement reaction mechanism of phenylcarbene in a molecular container: Cram's hemicarcerand, *Theochem*, 2006, **766**, 169–175.
75. M. Jones Jr., R. A. Moss, Singlet carbenes, in *Reactive Intermediate Chemistry*, R. A. Moss, M. S. Platz, M. Jones Jr. (Eds.), Wiley-Interscience, Hohoken, NJ, 273–328, 2004.
76. R. A. Moss, Carbenic philicity, in *Carbene Chemistry: From Fleeting Intermediates to Powerful Reagents*, G. Bertrand (Ed.), Fontis Media-Marcel Dekker, Lausanne, 57–90, 2002; b) R. A. Moss, Carbenic selectivity in cyclopropanation reactions, *Acc. Chem. Res.*, 1980, **13**, 58–64; c) R. A. Moss, Carbenic reactivity revisited, *Acc. Chem. Res.*, 1989, **22**, 15–21.
77. N. G. Rondan, K. N. Houk, R. A. Moss, Transition states and selectivities of singlet carbene cycloadditions, *J. Am. Chem. Soc.*, 1980, **102**, 1770–1776.
78. D. Bourissou, O. Guerret, F. P. Gabbai, G. Bertrand, Stable carbenes, *Chem. Rev.*, 2000, **100**, 39–92.
79. X. Liu, G. Chu, R. A. Moss, R. R. Sauers, R. Warmuth, Fluorophenoxycarbene inside a hemicarcerand: a bottled singlet carbene, *Angew. Chem. Int. Ed.*, 2005, **44**, 1994–1997.
80. W. Kirmse, Incarcerated carbenes, *Angew. Chem. Int. Ed.*, 2005, **44**, 2476–2479.
81. R. A. Moss, G. Kmiecik-Lawrynowicz, K. Krogh-Jespersen, The philicity of fluorophenoxycarbene, *J. Org. Chem.*, 1986, **51**, 2168–2172.
82. A. Scarso, L. Trembleau, J. Rebek Jr., Hydrocarbon conformation in self-assembled nanocapsules, *Angew. Chem. Int. Ed.*, 2003, **42**, 5499–5502.
83. a) R. Warmuth, S. Makowiec, The phenylnitrene rearrangement in the inner phase of a hemicarcerand, *J. Am. Chem. Soc.*, 2005, **127**, 1084–1085; b) R. Warmuth, S. Makowiec, Photochemical and thermal reactions of intermediates in the phenylnitrene rearrangement inside a hemicarcerand, *J. Am. Chem. Soc.*, 2007, **129**, 1233–1241.
84. H. Bayley, *Photogenerated Reagents in Biochemistry and Molecular Biology*, Elsevier, Amsterdam, 1983.
85. a) N. P. Gritsan, M. S. Platz, Kinetics, spectroscopy, and computational chemistry of arylnitrenes, *Chem. Rev.*, 2006, **106**, 3844–3867; b) M. S. Platz, Nitrenes, in *Reactive Intermediate Chemistry*, R. A. Moss, M. S. Platz, M. Jones Jr. (Eds.), Wiley-Interscience, Hoboken, NJ, 501–559, 2004.
86. W. T. Borden, N. P. Gritsan, C. M. Hadad, W. L. Karney, C. R. Kemnitz, M. S. Platz, The interplay of theory and experiment in the study of phenylnitrene, *Acc. Chem. Res.*, 2000, **33**, 765–771.
87. a) N. P. Gritsan, T. Yuzawa, M. S. Platz, Direct observation of singlet phenyl nitrene and measurement of its rate of rearrangement, *J. Am. Chem. Soc.*, 1997, **119**, 5059–5060; b) N. P. Gritsan, Z. Zhu, C. M. Hadad, M. S. Platz, Laser flash photolysis and computational study of singlet phenylnitrene, *J. Am. Chem. Soc.*, 1999, **121**, 1202–1207.
88. T.-Y. Liang, G. B. Schuster, Photochemistry of 3- and 4-nitrophenyl azide: detection and characterization of reactive intermediates, *J. Am. Chem. Soc.*, 1987, **109**, 7803–7810.
89. A. Reiser, L. Leyshon, Correlation between negative charge on nitrogen and the reactivity of aromatic nitrenes, *J. Am. Chem. Soc.*, 1970, **92**, 7487.
90. D. A. Makeiff, K. Vishnumurthy, J. C. Sherman, Ketonization of incarcerated acetophenone enol, *J. Am. Chem. Soc.*, 2003, **125**, 9558–9559.
91. a) P. J. Wagner, Type II photoelimination and photocyclization of ketones, *Acc. Chem. Res.*, 1971, **4**, 168–177; b) P. J. Wagner, P. Klan, Norrish type II photoelimination of ketones: cleavage of 1,4-biradicals formed by γ-hydrogen abstraction, in *CRC Handbook of Organic Photochemistry and Photobiology*, W. Horspool, F. Lenci (Eds.) CRC Press LLC, Boca Raton,

Fla, 2nd Ed., 52/1–52/31, 2004; c) P. J. Wagner, Yang photocyclization: coupling of biradicals formed by intramolecular hydrogen abstraction of ketones, in *CRC Handbook of Organic Photochemistry and Photobiology*, W. Horspool, F. Lenci (Eds.) CRC Press LLC, Boca Raton, Fla, 2nd Ed., 58/1–58/70, 2004.

92. a) V. Ramamurthy, D. F. Eaton, J. V. Caspar, Photochemical and photophysical studies of organic molecules included within zeolites, *Acc. Chem. Res.*, 1992, **25**, 299–307; b) C. L. D. Gibb, A. K. Sundaresan, V. Ramamurthy, B. C. Gibb, Templation of the excited-state chemistry of α-(n-alkyl) dibenzyl ketones: how guest packing within a nanoscale supramolecular capsule influences photochemistry, *J. Am. Chem. Soc.*, 2008, **130**, 4069–4080; c) A. Natarajan, L. S. Kaanumalle, S. Jockusch, C. L. D. Gibb, B. C. Gibb, N. J. Turro, V. Ramamurthy, Controlling photoreactions with restricted spaces and weak intermolecular forces: exquisite selectivity during oxidation of olefins by singlet oxygen, *J. Am. Chem. Soc.*, 2007, **129**, 4132–4133; d) M. Yoshizawa, Y. Takeyama, T. Kusukawa, M. Fujita, Cavity-directed, highly stereoselective [2+2] photodimerization of olefins within self-assembled coordination cages, *Angew. Chem. Int. Ed.*, 2002, **41**, 1347–1347.

93. a) Y. Chiang, A. J. Kresge, J. A. Santaballa, J. Wirz, Ketonization of acetophenone enol in aqueous buffer solutions. Rate-equilibrium relations and mechanism of the uncatalyzed reaction, *J. Am. Chem. Soc.*, 1988, **110**, 5506–5510; b) J. Andraos, A. J. Kresge, P. A. Obraztsov, Solvent and secondary substrate isotope effects on the acid-catalyzed ketonization of acetophenone enol in aqueous solution, *J. Phys. Org. Chem.*, 1992, **5**, 322–326.

94. a) H.-J. Schneider, H. Dürr, *Frontiers in Supramolecular Organic Chemistry and Photochemistry*, VCH, Weinheim, 1991; b) V. Ramamurthy, R. G. Weiss, G. S. Hammond, A model for the influence of organized media on photochemical reactions, in *Advances in Photochemistry*, Vol. 18, D. H. Volman, G. S. Hammond, D. C. Neckers (Eds.), John Wiley & Sons, 67–234, 1993; c) V. Ramamurthy, D. F. Eaton, Photochemistry and photophysics within cyclodextrin cavities, *Acc. Chem. Res.*, 1988, **21**, 300–306; d) V. Ramamurthy, *Photochemistry in Organized and Constrained Media*, VCH, New York, 1991.

95. A. J. Parola, F. Pina, M. Maestri, N. Armaroli, V. Balzani, Supramolecular photochemistry and photophysics. 9-Cyanoanthracene imprisoned in a hemicarcerand, *New J. Chem.*, 1994, **18**, 659–661.

96. A. Farrán, K. Deshayes, Free energy dependence of intermolecular triplet energy transfer: observation of the inverted region, *J. Phys. Chem.*, 1996, **100**, 3305–3307.

97. Z. S. Romanova, K. Deshayes, P. Piotrowiak, Remote intermolecular 'heavy-atom effect': spin-orbit coupling across the wall of a hemicarcerand, *J. Am. Chem. Soc.*, 2001, **123**, 2444–2445.

98. Electron and excitation transfer in hetero-supramolecular assemblies and at molecule-nanoparticle interfaces, P. Piotrowiak, K. Deshayes, Z. S. Romanova, C. Pagba, S. Hore, G. Zordan, I. Place, A. Farran, *Pure Appl. Chem.*, 2003, **75**, 1061–1068.

99. a) M. Kasha, Collisional perturbation of spin-orbital coupling and the mechanism of fluorescence quenching. A visual demonstration of the perturbation, *J. Chem. Phys.*, 1952, **20**, 71–74; b) G. W. Robinson, Intensity enhancement of forbidden electronic transitions by weak intermolecular interactions, *J. Chem. Phys.*, 1967, **46**, 572–585.

100. G. Kavarnos, T. Cole, P. Scribe, J. C. Dalton, N. J. Turro, Molecular photochemistry. XXXIX. External heavy-atom-induced spin-orbital coupling. Spectroscopic study of naphthonorbornanes, *J. Am. Chem. Soc.*, 1971, **93**, 1032–1034.

101. a) D. R. Kearns, W. A. Case, Investigation of singlet $\rightarrow$ triplet transitions by the phosphorescence excitation method. III. Aromatic ketones and aldehydes, *J. Am. Chem. Soc.*, 1966, **88**, 5087–5097; b) A. P. Marchetti, D. R. Kearns, Investigation of singlet-triplet transitions by the phosphorescence excitation method. IV. The singlet-triplet absorption spectra of aromatic hydrocarbons, *J. Am. Chem. Soc.*, 1967, **89**, 768–777; c) T. H. Cheng, N. Hirota, S. W. Mao, PMDR [phosphorescence-microwave double resonance] studies in heavy atom containing hosts, *Chem. Phys. Lett.*, 1972, **15**, 274–278.

102. G. Basu, M. Kubasik, D. Anglos, A. Kuki, Spin-forbidden excitation transfer and heavy-atom induced intersystem crossing in linear and cyclic peptides, *J. Phys. Chem.*, 1993, **97**, 3956–3967.

103. P. Petelenz, J. Fuenfschilling, Resonantly enhanced external heavy atom effect in organic glasses, *Chem. Phys.*, 1990, **145**, 333–342
104. E. Kohen, R. Santus, J. G. Hirschberg, *Photobiology*, Academic Press, San Diego, 1995, 177–225.
105. E. Kohen, R. Santus, J. G. Hirschberg, *Photobiology*, Academic Press, San Diego, 1995, 261–274.
106. G. J. Kavarnos, *Fundamentals of Photoinduced Electron Transfer*, VCH Publishers, New York, 1993.
107. B. O' Regan, M. Graetzel, A low-cost, high-efficiency solar cell based on dye-sensitized colloidal titanium dioxide films, *Nature*, 1991, **353**, 737–740.
108. a) M. Grätzel, Conversion of sunlight to electric power by nanocrystalline dye-sensitized solar cells, *J. Photochem. Photobiol. A*, 2004, **164**, 3–14; b) A. Hagfeldt, M. Graetzel, Light-induced redox reactions in nanocrystalline systems, *Chem. Rev.*, 1995, **95**, 49–68.
109. D. F. Eaton, Electron transfer processes in imaging, *Top. Curr. Chem.*, 1990, **156**, 199–225.
110. A. J. Parola, F. Pina, E. Ferreira, M. Maestri, V. Balzani, Photoinduced electron- and energy-transfer processes of biacetyl imprisoned in a hemicarcerand, *J. Am. Chem. Soc.*, 1996, **118**, 11610–11616.
111. a) G. L. Closs, M. D. Johnson, J. R. Miller, P. Piotrowiak, A connection between intramolecular long-range electron, hole, and triplet energy transfers, *J. Am. Chem. Soc.*, 1989, **111**, 3751–3753; b) G. L. Closs, P. Piotrowiak, J. M. MacInnis, G. R. Fleming, Determination of long-distance intramolecular triplet energy-transfer rates. Quantitative comparison with electron transfer, *J. Am. Chem. Soc.*, 1988, **110**, 2652–2653; c) M. A. Sigman, G. L. Closs, Free energy and structure dependence of intramolecular triplet energy transfer in organic model compounds, *J. Phys. Chem.*, 1991, **95**, 5012–5017. d) D. B. MacQueen, J. R. Eyler, K. S. Schanze, Intramolecular energy transfer in the inverted region, *J. Am. Chem. Soc.*, 1992, **114**, 1897–1878.
112. J. Jortner, Temperature dependent activation energy for electron transfer between biological molecules, *J. Chem. Phys.*, 1976, **64**, 4860–4867.
113. J. Ulstrup, J. Jortner, Effect of intramolecular quantum modes on free energy relations for electron transfer reactions, *J. Chem. Phys.*, 1975, **63**, 4358–4368.
114. N. Koga, K. Sameshima, K. Morokuma, Ab initio MO calculations of electronic coupling matrix elements on model systems for intramolecular electron transfer, hole transfer, and triplet energy transfer: distance dependence and pathway in electron transfer and relationship of triplet energy transfer with electron and hole transfer, *J. Phys. Chem.*, 1993, **97**, 13117–13125.
115. I. Place, A. Farran, K. Deshayes, P. Piotrowiak, Triplet energy transfer through the walls of hemicarcerands: temperature dependence and the role of internal reorganization energy, *J. Am. Chem. Soc*, 1998, **120**, 12626–12633.
116. A. Farrán, K. Deshayes, C. Matthews, I. Balanescu, 'Through space' triplet energy transfer: movement of electrons through the hemicarcerand skeleton, *J. Am. Chem. Soc.*, 1995, **117**, 9614–9615.
117. N. E. Miller, M. C. Wander, R. J. Cave, A Theoretical study of the electronic coupling element for electron transfer in water, *J. Phys. Chem. A*, 1999, **103**, 1084–1093; b) E. W. Castner Jr., D. Kennedy, R. J. Cave, Solvent as electron donor: donor/acceptor electronic coupling is a dynamical variable, *J. Phys. Chem. A*, 2000, **104**, 2869–2885.
118. a) R. J. Cave, M. D. Newton, K. Kumar, M. B. Zimmt, Theoretical study of solvent effects on the electronic coupling matrix element in rigidly linked donor-acceptor systems, *J. Phys. Chem.*, 1995, **99**, 17501–17504; b) K. Kumar, Z. Lin, D. H. Waldeck, M. B. Zimmt, Electronic coupling in C-clamp-shaped molecules: solvent-mediated superexchange pathways, *J. Am. Chem. Soc.*, 1996, **118**, 243–244; c) H. Han, M. B. Zimmt, *J. Am. Chem. Soc.*, 1998, **120**, 8001; d) R. W. Kaplan, A. M. Napper, D. H. Waldeck, M. B. Zimmt, Solvent mediated coupling across 1 nm: not a π bond in sight, *J. Am. Chem. Soc.*, 2000, **122**, 12039–12040; e) A. M. Napper, I. Read, R. Kaplan, M. B. Zimmt, D. H. Waldeck, Solvent mediated superexchange in a C-clamp shaped donor-bridge-acceptor molecule: the correlation between solvent electron affinity and electronic coupling, *J. Phys. Chem. A*, 2002, **106**, 5288–5296.

119. C. Pagba, G. Zordan, E. Galoppini, E. L. Piatnitski, S. Hore, K. Deshayes, P. Piotrowiak, Hybrid photoactive assemblies: electron injection from host-guest complexes into semiconductor nanoparticles, *J. Am. Chem. Soc.*, 2004, **126**, 9888–9889.
120. a) O. Enea, J. Moser, M. Gratzel, Achievement of incident photon to electric current conversion yields exceeding 80% in the spectral sensitization of titanium dioxide by coumarin, *J. Electroanal. Chem.*, 1989, **259**, 59–65; b) J. Moser, M. Gratzel, Observation of temperature independent heterogeneous electron transfer reactions in the inverted Marcus region, *Chem. Phys.*, 1993, **176**, 493–500.
121. a) D. A. Makeiff, J. C. Sherman, A six-bowl carceplex that entraps seven guest molecules, *J. Am. Chem. Soc.*, **2005**, 127, 12363–12367; b) E. S. Barrett, J. L., Irwin, A. J. Edwards, M. S. Sherburn, Superbowl container molecules, *J. Am. Chem. Soc.*, 2004, **126**, 16747–16749; c) X. Liu, R. Warmuth, Solvent effects in thermodynamically controlled multi-component nanocapsule syntheses, *J. Am. Chem. Soc.*, 2006, **128**, 14120–14127; d) Y. Liu, X. Liu, R. Warmuth, Multi-component dynamic covalent assembly of a rhombicuboctahedral nanocapsule, *Chem. Eur. J.*, 2007, **13**, 8953–8959; e) D. Xu, R. Warmuth, Edge-directed dynamic covalent synthesis of a chiral nanocube, *J. Am. Chem. Soc.*, 2008, **130**, 7520–7521.

10

Encapsulation of Reactive Intermediates

Jean-Luc Mieusset and Udo H. Brinker
Institut für Organische Chemie, Universität Wien, A–1090 Wien, Austria

10.1 Introduction

The inclusion of molecules is a very common technique in the pharmaceutical sciences to protect sensitive compounds and increase their usability and their storage time. For example, drugs were enclosed in host molecules like cyclodextrins. This topic has been previously reviewed.[1]

The same approach can be applied to organic chemistry where very labile species can be encapsulated. In this way, species that usually would rapidly decompose or polymerize even at low temperatures can be stored at room temperature, especially when the guest has no chance to escape from its host. To achieve this goal, hemicarcerands have been mostly used, leading to the preparation of cages containing highly reactive species such as 1,3-cyclobutadiene, cycloheptatetraene, or benzyne. This subject is presented in detail in a specially dedicated chapter (Reactions inside Carcerands, Ralf Warmuth).

Similarly, very useful results can be obtained if the labile species is not incarcerated but can easily be released from its cage. Obviously, this property is significant in order to have a chance to use the guest or its products. In this section, we will first present what can be done to increase the thermodynamic and kinetic stability of a labile species with the aim to increase its relative concentration during the reaction or to make it more persistent. Second, we will show a few examples to demonstrate how to stabilize some specific noncovalent assemblies and to enforce a change in the conformation of the guest. Finally, the focus will be put on highly reactive intermediates (radicals, radical cations, carbenes, and nitrenes) showing how to protect them from the surroundings in order to prevent their trapping but also to modify their reactivity to be able to obtain products that would not be found otherwise. Taking examples from the supramolecular chemistry of carbenes and nitrenes, more details will be given about the methodology used in these studies.

10.2 Encapsulation of Labile Species

The following results show some recent improvements toward the possibility to carry out water sensitive reactions in aqueous solution using hydrophobic catalysts.

The first example concerns the attempt to isolate a common reactive intermediate. An ordinary easily occurring reaction is the addition of primary amines to carbonyl groups to form imines. The main intermediate of this condensation reaction is a hemiaminal which can be observed only in special cases because the release of water proceeds very rapidly. A cavitand could be built which is able to increase the concentration of hemiaminal during the reaction and allows its observation by NMR spectroscopy (Scheme 10.1).[2] Cavitand **1a** is based on a resorcinarene-based bowl, a macrocycle easily available

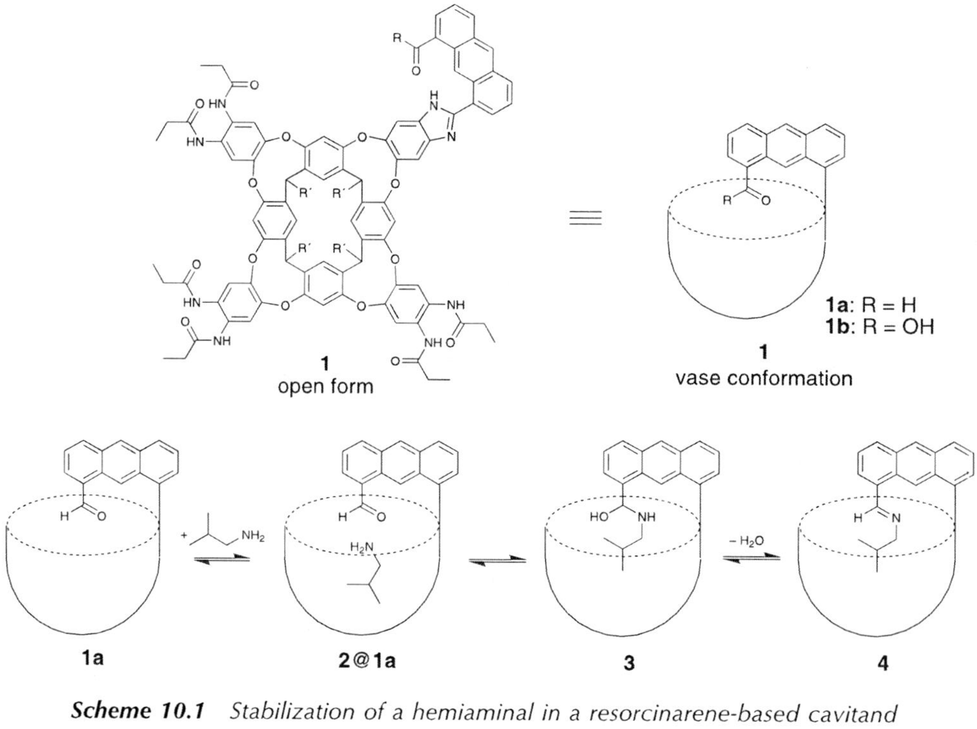

Scheme 10.1 *Stabilization of a hemiaminal in a resorcinarene-based cavitand*

by condensation of resorcinol with aldehydes under acidic conditions. Usually, long chain aldehydes are used to improve the solubility of the container. The hydroxy groups on the upper rim offer the possibility to deepen the cavity and to introduce a large variety of functionalities. In the case of **la**, the upper rim has been functionalized with an anthracene arm bearing an aldehyde group and with amido groups to obtain a rigid vase conformation firmly maintained by intramolecular hydrogen bonds. In mesitylene, a solvent too large to be accommodated in the cavity and therefore not competiting for inclusion, various small amines **2** can be complexed. Formation of the hemiaminal **3** is relatively fast but the elimination of water to afford imine **4** is slowed. In solution, dehydration is induced by a second amine molecule. Due to the confinement in a limited space, this step is hampered. Thus, the hemiaminal accumulates and its concentration declines only slowly over time with a half life of 30 to 90 min.

This strategy has been expanded to the observation of isoimide intermediates **5**. Isoimides are obtained by a formal insertion of isonitriles into the OH bond of carboxylic acids. Usually, high temperatures are required and the addition product is rapidly stabilized by a 1,3-*O*→*N* acyl transfer to form *N*-acylformamide **6**.[3] Under these conditions, no significant amounts of isoimide **5** are present in the solution during the reaction. However, the previously described cavitand **1a**, after oxidation to carboxylic acid **1b**, can be used to reveal the intermediacy of the isoimide (Scheme 10.2).[4] Addition of isopropyl isonitrile to a solution of cavitand **1b** in mesitylene-d_{12} induces the folding of the host to its vase conformation and inclusion of the guest in direct proximity to the acidic group. In this confined space, addition occurs at room temperature and the isoimide intermediate **5** can be observed by NMR and IR spectroscopy at millimolar concentrations before completion of the rearrangement to acylformamide **6**.

One of the most remarkable contributions is the generation of iminium ions in water in good yields combined with the ability to conserve these ions for months in this solution. Iminium ions are easily formed by the condensation of secondary amines with carbonyl groups in an equilibrium reaction. However, they are very sensitive to hydrolyses and are not viable in water except if they are suitably encapsulated. The water soluble self-assembled $Ga_4L_6^{12-}$ complex[5] is a suitable container for this purpose (Scheme 10.3).[6] The four gallium ions are bridged by six bis-bidentate catecholamide ligands L forming a 12-fold negatively charged tetrahedron. This assembly is a perfect host for organic cations. Indeed, stirring a mixture of ketone, pyrrolidine (**7**), and $Ga_4L_6^{12-}$ leads to the inclusion of the corresponding iminium ions **8**. Obviously, the generation of a suitable guest and its inclusion constitute a sufficient driving force to shift the equilibrium

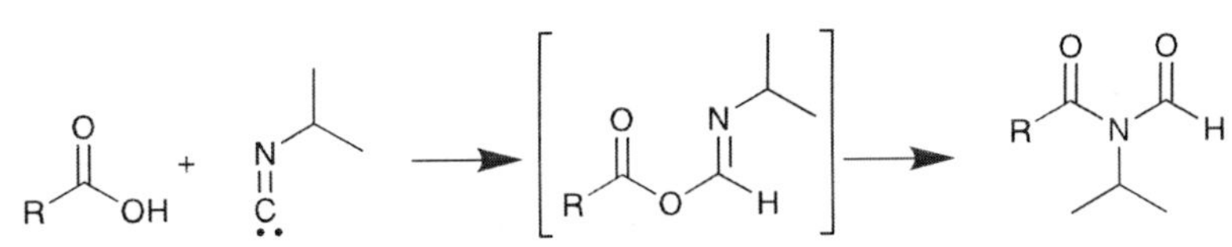

1b
R = cavitand

5

6

Scheme 10.2 *Stabilization of the isoimide intermediate **5** generated by the reaction of an isonitrile with a carboxylic acid*

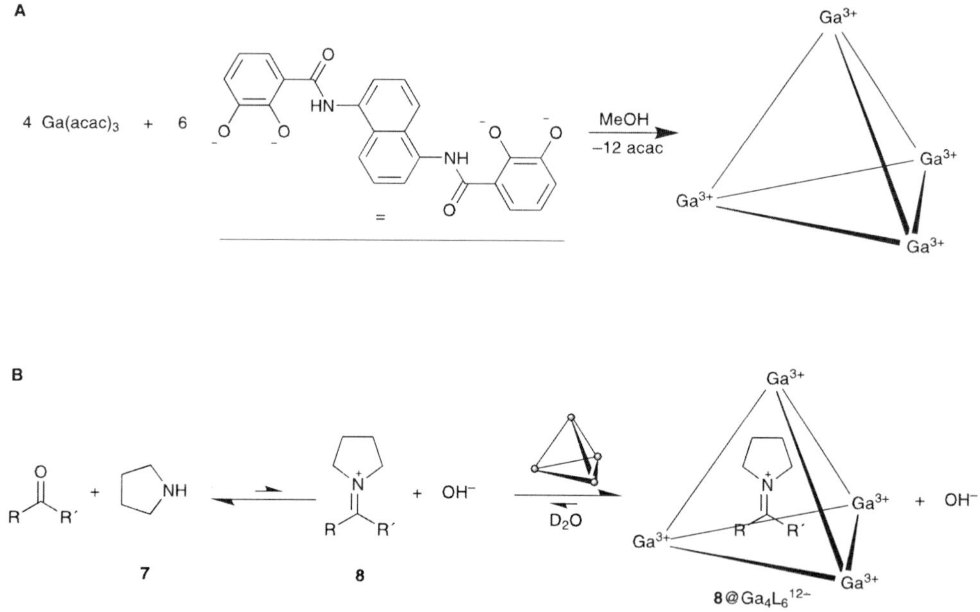

Scheme 10.3 A) Self-assembly of the $Ga_4L_6^{12-}$ tetrahedron. B) Encapsulation of iminium ions in water

toward the iminium product. The best yields are obtained with pentanones and hexanones (80–90%) due to an optimal fit of the size of the guest in the hydrophobic cavity.

An analogous approach is the stabilization of phosphonium ions by addition of phosphanes (PMe_3, PEt_3, $PPhMe_2$, and PPh_2Me) to methyl ketones (acetone, methyl ethyl ketone, 1,1,1-trifluoroacetone, and fluoroacetone) in an aqueous solution of $Ga_4L_6^{12-}$ (Scheme 10.4).[7] These cations decompose in water but they are persistent for weeks in the nanovessel. However, this stability is pH-dependant. The pH should be low because the guest should be protonated and is regularly exposed to the bulk of water as a consequence of the dynamic behavior of the assembly, but the pH should not be too low because in that case the host would disassemble. Therefore, the optimal value is 5.2. The exact mechanism is still unknown; it has been ascertained that the protonated phosphane can be encapsulated but it is not possible to determine where the addition to the ketone occurs, inside or outside the cavity. Due to the chirality of the $Ga_4L_6^{12-}$ tetrahedron, a kinetic diastereoselectivity is also obtained with chiral ions leading to diastereomeric excesses of 30–50%.

The effect of encapsulation can be quite impressive. White phosphorus (P_4 tetrahedrons), a strongly pyrophoric compound, spontaneously burns when exposed to air releasing considerable amounts of energy deriving from the release of strain and the replacement of the weak P—P bonds by strong P—O bonds. Under inclusion in the self-assembled tetrahedral cage $Fe_4L_6^{4-}$, the hydrophobic P_4 becomes air-stable and can be handled in water in the presence of oxygen (Scheme 10.5).[8] The cage is formed by stirring 2-formylpyridine (**9**) with 4,4′-diaminobiphenyl-2,2′-disulfonic acid (**10**) in the presence of iron(II) sulfate; the bis-bidentate imine ligand is formed *in situ*.[9] The water-soluble $Fe_4L_6^{4-}$ cage

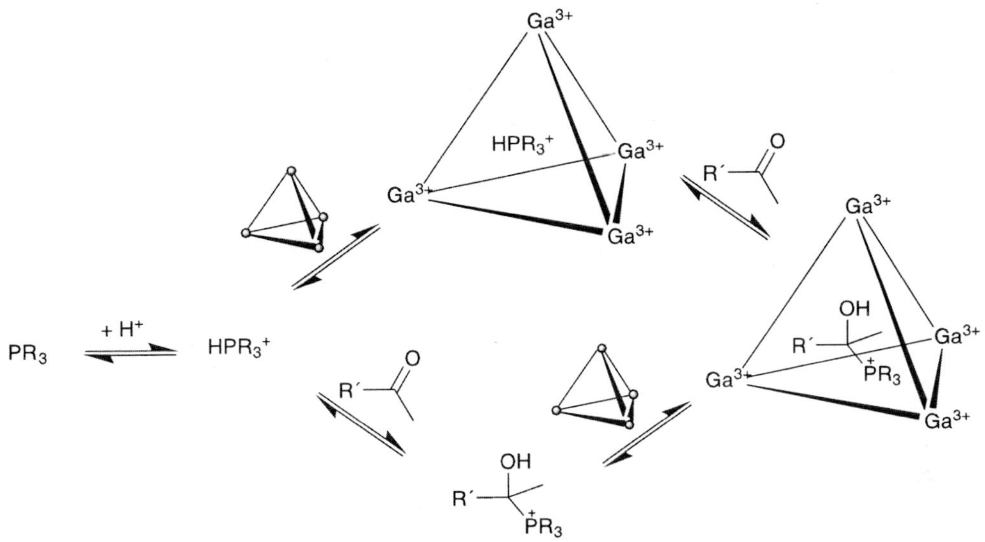

Scheme 10.4 *Stabilization of phosphonium ions in water by encapsulation*

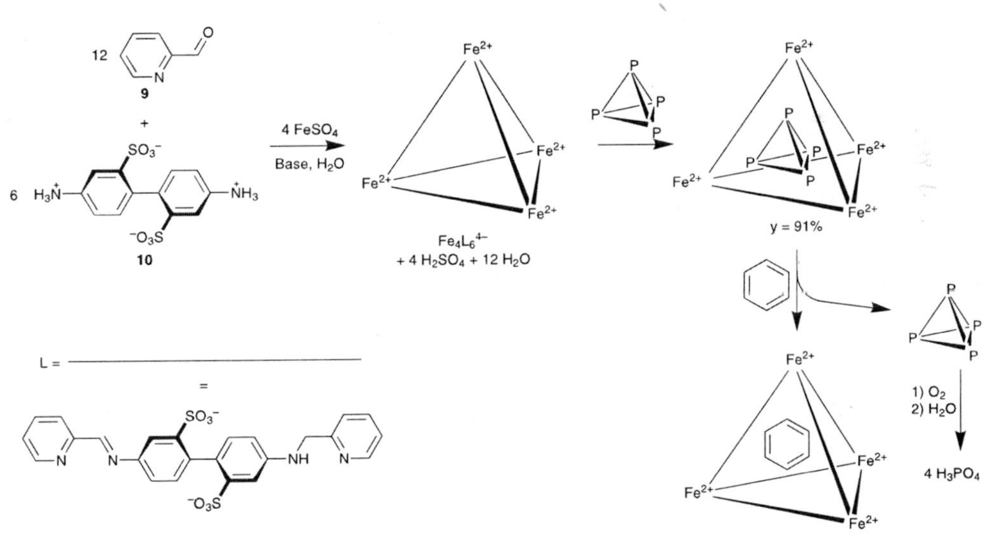

Scheme 10.5 *Encapsulation of P_4 in $Fe_4L_6^{4-}$*

has a purple color and is diamagnetic because iron(II) is in its low spin state. With a cavity volume of 140 Å^3, it encloses preferably small cyclic hydrocarbons but no organic cations, probably because of its small total charge. The cage dissociates back to its building block in a reversible manner at low pH as a consequence of protonation of the ligands. It can also be irreversibly decomposed by addition of a chelating amine resulting in an imine exchange reaction.[10] The preparation of the complex is achieved by mixing an aqueous solution of $Fe_4L_6^{4-}$ with solid white phosphorus; the crystallization is induced by vapor diffusion of acetone yielding 91% of P_4@$Fe_4L_6^{4-}$. The decrease in reactivity is not mainly due to exclusion of dioxygen from P_4 as the dynamic nature of the host and the small size of dioxygen do not allow obtaining a strict separation of the reactant. The main reason is that the reaction products and intermediates are too large to fit into the cavity. However, this dynamic inclusion has one important advantage: the reactive guest can easily be released by extraction with benzene or cyclohexane where P_4 can exhibit its reactivity and be oxidized to finally give phosphoric acid H_3PO_4. In this case, the removal of P_4 is performed by using a competing guest for the cavity; with *n*-heptane, the extraction is not successful.

A further example concerns the isolation of cyclic trimers of siloxanes.[11] These cyclic trimers are usually short-lived kinetic products generated during the first phase of the polycondensation of trisilanols.[12] In a bulk, they usually further condense to higher oligomers which are thermodynamically more stable and therefore, these trimers cannot be isolated. However, when the condensation of phenyltrimethoxysilane $PhSi(OMe)_3$ is performed in the $M_6L_4^{12+}$ nanocage **12**, only the cyclic trimer **14** is obtained under stereochemistry control giving exclusively the all-*cis* isomer (Scheme 10.6).

Nanocage **12** is a stable complex which can easily be prepared in boiling water by self-assembly of 6 units of a platinum nitrate salt with 4 molecules of a tris(4-pyridyl) triazine **11** panel as ligand. The holes between the panels are sufficiently large to allow solvents and small molecules like $PhSi(OMe)_3$ to enter or exit the cage. At 100 °C in an aqueous solution of capsule **12**, $PhSi(OMe)_3$ is hydrolyzed within five minutes to trisilanol **13**. The large cavity of **12** encloses three or four molecules of **13** at once forming the complex $(\mathbf{13})_{3-4}$@**12** which can be monitored by NMR. The condensation takes place in the cavity until trimer **14** is formed. However, **14** is too bulky to escape or to condense further. The reaction is then stopped at this level and **14** imprisoned in cage **12** can be isolated in 92% yield by filtration and recrystallization. The complex **14**@**12** is now stable in an aqueous solution and even under acidic conditions (pH < 1).

Using this methodology, the size and shape of the cavity determines which oligomer will be isolated (Scheme 10.7). Indeed, bowl **16** produces dimer **19** whereas the small coordination tube **21**, which allows the complexation of only one molecule of the reactant, affords monomer **18**, a silanetriol.[13] In this case, starting from 4 equiv of 3,5-bis(3-pyridyl) pyridine **20** and 6 equiv of $Pd(en)(NO_3)_2$, tube **21** can only be prepared *in situ* by means

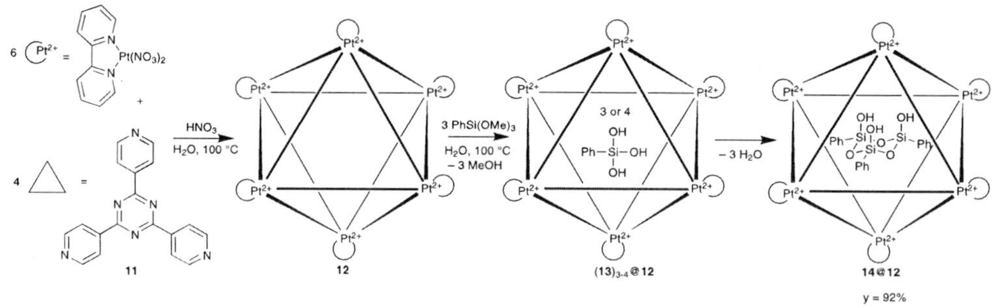

Scheme 10.6 *Isolation of the cyclic siloxane trimer **14***

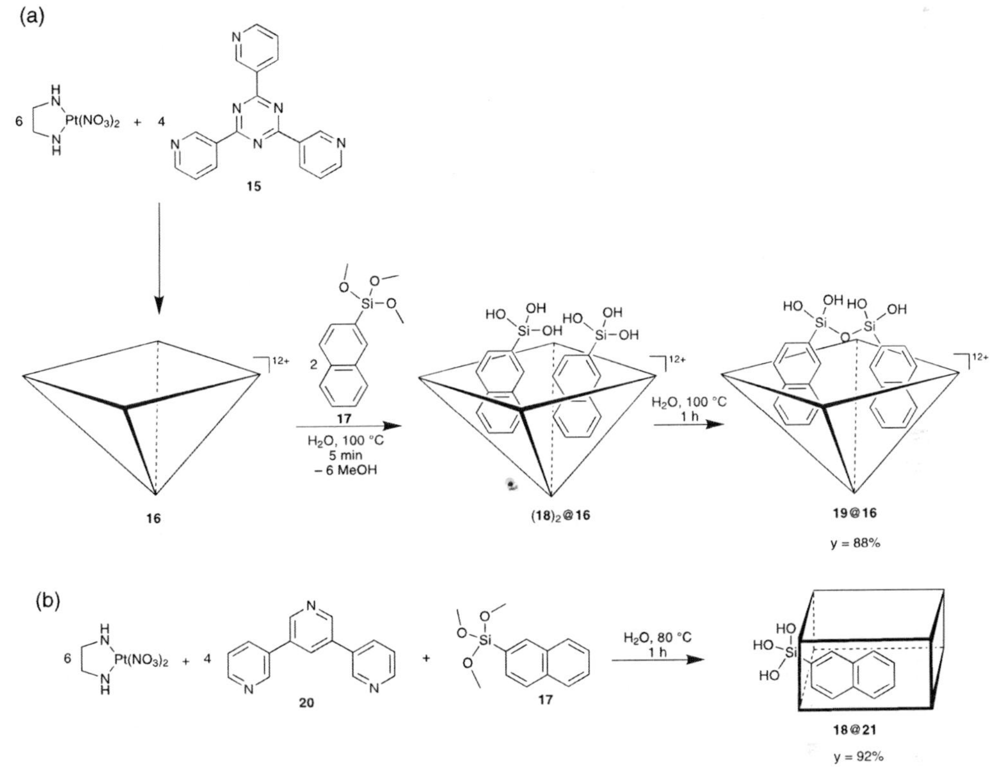

Scheme 10.7 *Cavity-directed synthesis of labile silanols: A) Dimer* ***19*** *in bowl* ***16****. B) Monomer* ***18*** *in tube* ***21***

of the template effect of 2-naphthyltrimethoxylsilane **17** forming complex **17@21**. Aromatic compound **17** is a very suitable guest due to strong π–π and CH–π interaction between its naphthyl group and the pyridyl panels. At this temperature, **17** slowly hydrolyzes to **18** as revealed by monitoring of the reaction by NMR. Prolonged heating gives **18@21**, a complex in which only the hydrophobic naphthyl group is deeply included in the tube. The hydrophilic $Si(OH)_3$ group is still exposed outside the cavity but the sterically demanding tube framework prevents further polycondensation. Moreover, these experiments demonstrate that full enclosure of the reactive center in a cage is not always necessary to stabilize an intermediate and partial inclusion can be sufficient. **18@21** was isolated in very good yield (92%), recrystallized from water, and submitted to X-ray analysis.

10.3 Isolation of Non-covalently Bonded Aggregates

In absence of a suitable guest molecule, in general the cavities are filled with solvent molecules. Usually, several of these small compounds are enclosed in the inner space and are relatively loosely bonded and disordered. Using as a host the coordination cage

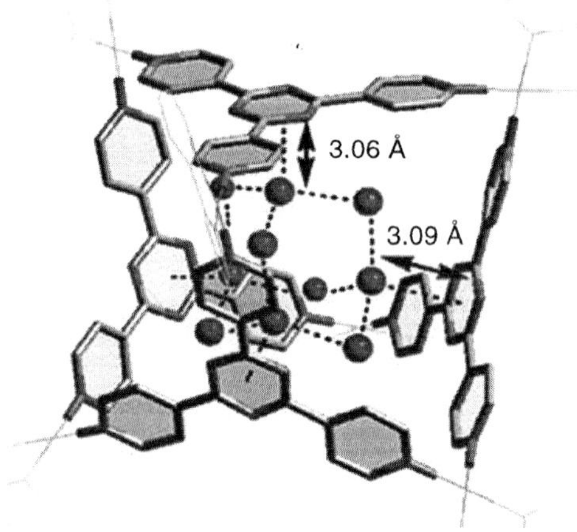

Figure 10.1 X-ray crystal structure of **22b** along with oxygen atoms (water molecules) around the cage. Reprinted with permission from [14]. Copyright 2005 American Chemical Society

$Pd_6L_4^{12+}$ **22b** self-assembled from four tris-pyridyl ligands **11** and six $Pd(bipy)^{2+}$, a highly ordered adamantanoid $(H_2O)_{10}$ cluster has been obtained; this structure corresponds to the basic unit of I_6-type ice (Figure 10.1).[14] The structure could be revealed by an X-ray diffraction study at −180 °C. The position of the hydrogen atoms was determined by single-crystal neutron diffraction at room temperature with crystals prepared from D_2O. In this case, the recognition of the molecular ice by the cage is quite peculiar because it does not occur through OH–π hydrogen bonding but through interaction of the lone pair of the 4 bridgehead water molecules with the aromatic π-system of the 4 triazine walls of the cationic $Pd_6L_4^{12+}$ cage **22b**. This highly ordered structure of the water cluster is quite stable: it is still not molten at room temperature.

Similarly, encapsulation favors the formation of stable hydrogen-bonded dimers. In aqueous solution, the hydrophobic cavity protects the hydrogen bonds between the pair from the concurrence of the water molecules. For example, four base pairs are required for the stable association of two DNA strands in water. In a cage, only one pair of nucleotide is necessary to achieve a selective binding as demonstrated by the formation of a hydrogen-bonded aggregate between 5′-adenosine monophosphate **24** and 5′-uridine monophosphate **25** resulting in the preferred formation of complex **24.25@23** over $(\mathbf{24})_2$**@23** and $(\mathbf{25})_2$**@23** (Scheme 10.8).[15] This selectivity is due to the fact that the nucleobases cannot efficiently form hydrogen bridges with themselves. The pillared cage **23** is an appropriate host because it possesses a flat hydrophobic pocket perfectly suited for the binding of planar aromatic guests as a result of an ideal interplanary distance between the panels (6.6 Å). **23** is obtained by self-assembly of two tris(4-pyridyl)triazine and three 1,4-pyrazine ligands with six $Pt(en)(NO_3)_2$ edges in the presence of coronene as an aromatic template.

Structures held together by metal-metal *d–d* interaction can also be prepared using encapsulation. Infinite stacking of metal complexes is quite common in the solid or in the liquid crystalline state and these structures possess interesting electroconductive and

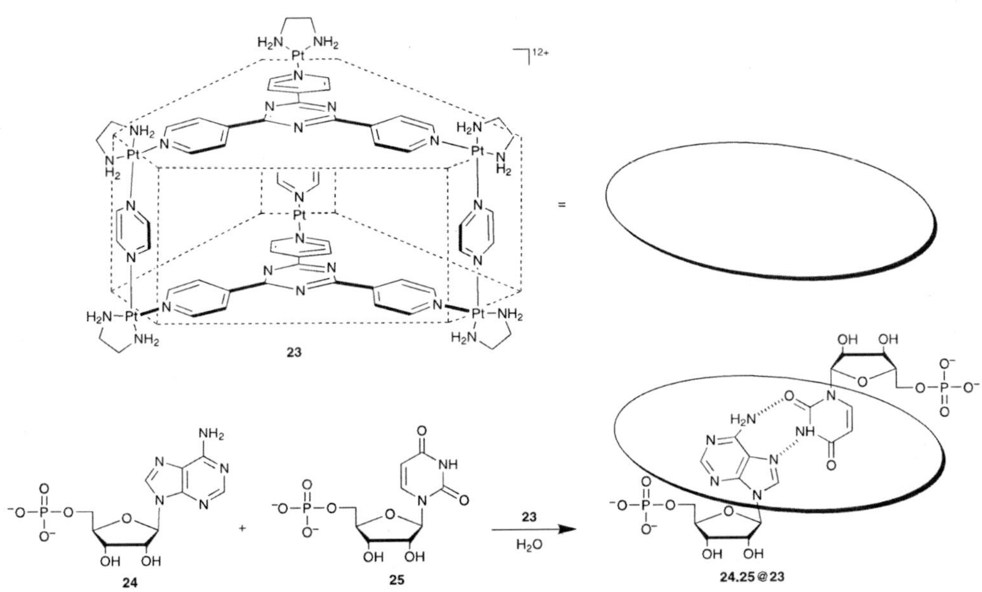

Scheme 10.8 *Structure of the enclathrated nucleotide base pair* ***24.25@23***

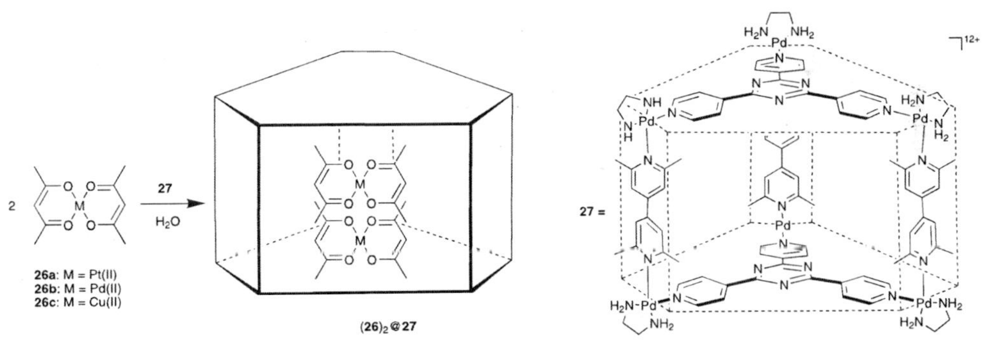

Scheme 10.9 *Discrete stacking of two planar metal complexes* ***26*** *within the pillared cage* ***27*** *causing metal-metal interactions between the guests*

magnetic properties. Guest inclusion into a molecular cage allows the stacking of a finite number of metal complexes and the study of their properties. For example, two molecules of Pt(acac)$_2$ (**26a**) can be included into the cage $Pd_6L^1_3L^2_2{}^{12+}$ **27** by stirring the mixture of the two components in water at room temperature (Scheme 10.9).[16] The quantitative formation of the complex (**26a**)$_2$@**27** is revealed by NMR spectroscopy on account of the upfield-shift of the guest signals and by CSI-MS. Single crystals obtained by slow evaporation of water provide more details about the structure of the assembly: inclusion of the guests induces conformational change of the cage which is now twisted by 23 ° to maximize the interactions with the twisted dimer of Pt(acac)$_2$. Evidence for the *d–d* interaction is given by the short Pt(II)–Pt(II) distance of 3.32 Å, an impressive downfield shift of the

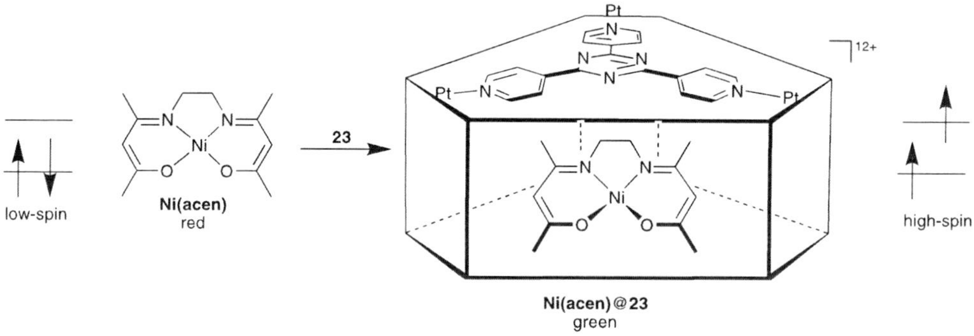

Scheme 10.10 *Encapsulation-driven spin crossover of Ni(II)(acen)*

^{195}Pt NMR signal upon inclusion, and by UV spectra of the powdered orange crystals showing a new absorption band at 500 nm. The interactions are weaker in aqueous solution and the color disappears. Similar stacked dimers can be obtained with $Pd(acac)_2$ (**2b**) and $Cu(acac)_2$ (**2c**). As a monomer, $Cu(acac)_2$ is a radical with one unpaired electron ($S = ½$). ESR spectroscopy reveals the presence of a triplet species ($S = 1$) and spin–spin interaction between the copper atoms in the complex $(\mathbf{26c})_2$@**27**. Using a similar approach, higher definite stacked towers can be synthesized consisting of seven or nine large aromatic molecules.[17]

The spin state of a transition metal complex can even be changed by encapsulation. Spin crossover are usually induced by temperature or pressure changes, by light or by magnetic stimuli.[18] Using cage **23** as a host for the red, planar, low spin complex Ni(II)(acen) ($S = 0$) a green 1 : 1 complex is obtained (Scheme 10.10).[19] SQUID measurements show that the ground state is then coupled with the high-spin state ($S = 1$) via a spin-orbit interaction. The spin crossover results from interactions between the d_{z^2} orbital of Ni(II) with the triazine ring of the panel forming a pseudo octahedral ligand field.

10.4 Inclusion of Reactive Intermediates

10.4.1 Viologen Radical Cations

Viologens (4,4′-bipyridinium) are widely used because of their electron-transfer properties. However, application is often hindered by the poor stability of the corresponding cation radical. Inclusion in a macromolecule should largely prevent intermolecular reactions or dimerization of this reactive species. Cucurbiturils, which easily can be synthesized from glycoluril and formaldehyde, are perfect hosts for organic cations due to their rigidity and the combination of a hydrophobic cavity with a very polar rim of urea groups leading to high association constants. The main drawback is their poor solubility in water and in organic solvents.[20] Thus, the dication methyl viologen (**28**) is tightly bound by cucurbit[7]uril ($k = 1 \times 10^5\,M^{-1}$ for **28**@CB[7]).[21] This value is significantly higher than those determined, for the previously studied complexes with the more flexible cyclodextrins (10^3–$10^4\,M^{-1}$). Upon inclusion, **28** indeed keeps its electrochemical properties and

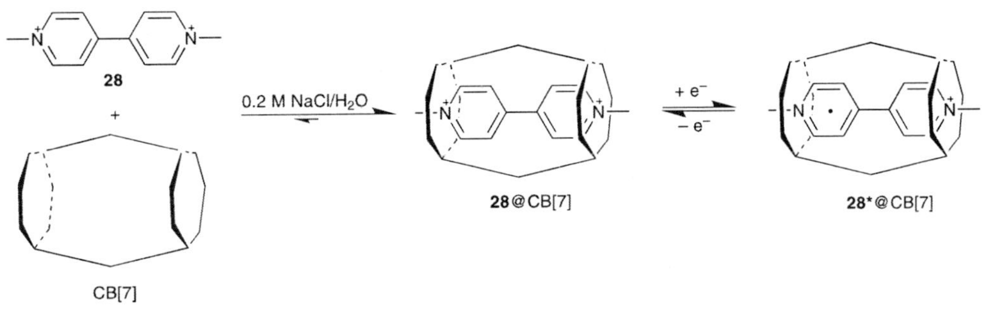

Scheme 10.11 *Protection of a viologen radical cation by inclusion in a cucurbituril*

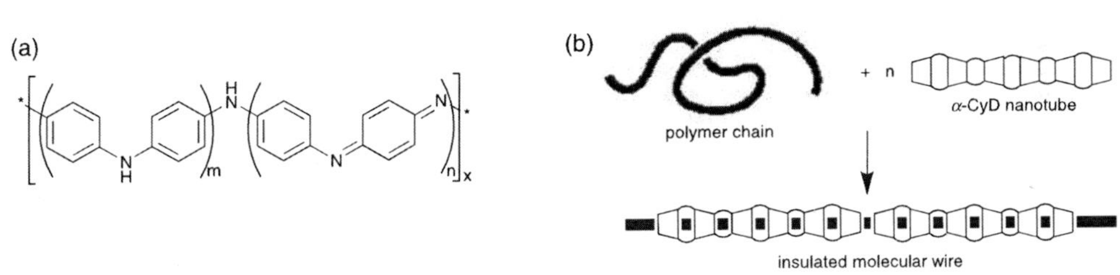

Scheme 10.12 *A) Structure of the three different states of polyaniline: leucoemeraldine (m = 1, n = 0), emeraldine (m = 0.5, n = 0.5), and pernigraniline (m = 0, n = 1). B) Conformation change from coil to rod of a polyaniline chain by threading into a cyclodextrin nanotube*

still undergoes a reversible one-electron reduction to radical cation **28*** without changes in the kinetic. The reduced complex **28***@CB[7] with the radical cation is still firmly bonded because the guest is still charged. Further reduction leads to a neutral viologen which for the most part is released from the host.[22]

This approach can also be applied for example to the protection of polyaniline (PANI), one of the most useful conducting polymer due to its stability and the simple control of the conductivity through doping/dedoping strategies with protons or oxidizing agents. Polyaniline is obtained by oxidation of aniline, mostly with ammonium persulfate. The reduced form is the colorless leucoemeraldine with phenyl rings connected by amine bonds (Scheme 10.12A). In the fully oxidized violet pernigraniline state, the links consist of imine groups. The most interesting form is the half-oxidized emeraldine, a blue-green paramagnetic polyradical cation which possesses a high conductivity of $1000\,S\,cm^{-1}$. This conjugated polymer is therefore very interesting for application in molecular electronics. However, interchain interactions, coiling and the chemical stability of the molecule are common problems. A widely used approach is the encapsulation of the chain to form an insulated molecular wire.[23] In the case of emeraldine, it has been shown that *β*-cyclodextrin can be threaded into the polymer forming a molecular necklace, a pseudopolyrotaxane. In *N*-methyl-2-pyrrolidone, complexation by *β*-cyclodextrin forces the polyaniline chain to change its conformation from coil to rod because the close packing of host molecules with a narrow cavity strongly favors the linear all-*trans* configuration of the chain, the

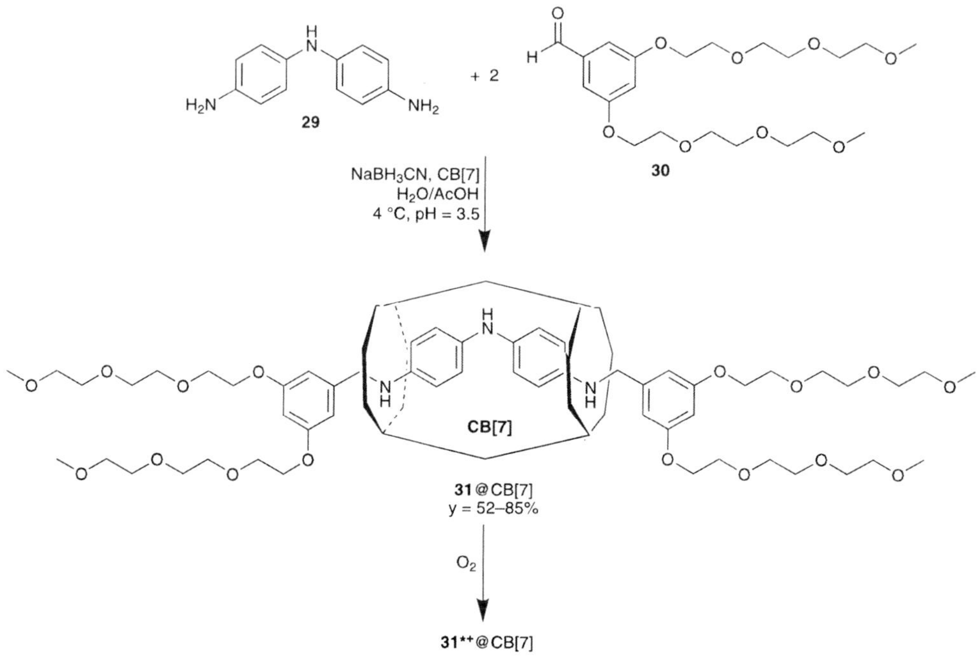

Scheme 10.13 *Kinetic and thermodynamic stabilization of a radical cation as a cucurbituril rotaxane*

configuration required to realize high conductivity.[24] This process is disfavored by entropy and therefore, it is only observed at temperatures below 2 °C. Moreover, this insulated emeraldine is protected from chemical oxidation by iodine. Similarly, polyaniline has been threaded into a β-1,3-glucan[25] and also into a more rigid molecular nanotube (Scheme 10.12B) consisting of α-CD macrocycles which have been crosslinked with epichlorhydrin using polyethylene glycol as a template.[26] This linearized molecular wire could be observed by atomic force microscopy.[27]

The strength of this approach is best illustrated by the stabilization of the radical cation **31**$^{*+}$ of the dimer of polyaniline **29** confined as a rotaxane in the cavity of cucurbit[7]uril, an assembly soluble in a wide range of polar organic solvents (Scheme 10.13).[28] Whereas in water the neutral rotaxane **31**@CB[7] is rapidly oxidized by oxygen to form the blue radical cation **31**$^{*+}$@CB[7] which is relatively persistent toward further oxidation, the free dimer **31** requires ammonium persulfate oxidation yielding the benzoquinoid dication **31**$^{2+}$ almost directly. As shown by square wave voltammetry, these findings can be explained by a thermodynamic stabilization of the radical cation **31**$^{*+}$ in comparison to the neutral **31** as a consequence of the strong affinity of the cucurbiturils for cations (The first oxidation potential is lowered by 570 mV upon rotaxanation). The resistance against further oxidation is the result of a kinetic effect (The second oxidation potential remains practically unchanged). The rotaxane **31**@CB[7] is synthesized in good yields by reductive amination of the complexed and protonated aniline **29**@CB[7] with the aldehyde stoppers **30**.

10.4.2 Radicals

The supramolecular chemistry of radicals has been widely investigated, especially with stable radicals like TEMPO or benzyl *t*-butyl nitroxide.[29] This is because radicals can be observed by EPR spectroscopy, a method particularly suitable to obtain kinetic information about the association and dissociation processes in the time range between 10^{-9} and 10^{-5} seconds, i.e., for complexes displaying fast exchange on the NMR time scale.

More reactive radicals have been generated by photochemical cleavage of esters and ketones in dendrimers. Under these conditions, the radical pair remains encapsulated in a cage and the two moieties can better recombine. Furthermore, their mobility is reduced. One of the most impressive results is obtained by the cleavage of 1-naphthyl phenylacyl ester **32** (Scheme 10.14).[30] Whereas the photolysis in hexane solution leads to a mixture of seven products, the reaction in an aqueous solution of **G2** or **G3** poly(alkyl aryl ether) dendrimers of the second (respectively, the third) generation at pH 9 is highly selective. This increase in selectivity is mainly caused by the short life-time of the singlet radical pair **33** and **34** in the confined environment of the dendritic pocket. In hexane, the high persistency of the radicals allows the dimethyl phenylacyl radical **34** to split off CO to form the stable tertiary radical **35**. The rate of this decarbonylation reaction has been determined to be $1.5 \times 10^8\,s^{-1}$. The high persistency of the radicals is also the reason why they have enough time to diffuse away in hexane and finally recombine to the symmetrical diphenylbutane **41** and naphthol **40**. In the dendrimer, recombination occurs in less than 6 ns. On this time scale, as a consequence of the high viscosity of the medium and the restricted mobility of the radical pair, the recombination remains highly selective and formation of the *ortho*-ketone **36** is strongly preferred (93% in **G2** and 96% in **G3**). The sole side product is the *para*-isomer **37**, the other pathways are still not open. Moreover, during this short time period, the radicals do not leak from the hydrophobic dendritic cavity to the water bulk. Thus, no products result from hydrolysis. Especially this absence of hydrolysis products and the high selectivity of the reaction in the relatively open dendrimer of the second generation are remarkable. Experiments based on the generation of long-lived triplet radical pairs which can only recombine if they undergo intersystem crossing have shown that the dendrimer cavities are leak-proof even on the microsecond time scale. Performing this photolysis in a SDS (sodium dodecyl sulfate) micelle leads to similar results but the reaction is less selective. Additional examples are described in the first chapter of this book.

The particular steric conditions prevailing in guest–host systems allow the observation of some unusual reactions which are rarely observed in solution (Scheme 10.15). This is the case for the photooxidation of adamantane **43** in the $Pd_6L_4^{12+}$ cage **22b** which leads to the generation of an adamantyl radical in an anion radical (**43**$_3$.**43***@**22b***$^-$).[31] This reaction is caused by the narrow confinement of four adamantane molecules in **43**$_4$@**22b** which are pressed against the triazine core of the ligand (host–guest distance = 2.6 Å). This triazine being coordinated to three pyridines and being part of a cationic cage is a particularly efficient electron acceptor which can remove an electron from the guest. The freshly *in situ* generated adamantyl radical cation then loses a proton and selectively forms the tertiary 1-adamantyl radical **43*** because it is more stable than the 2-adamantyl radical. Finally, reaction with water or oxygen leads to the formation of 1-adamantanol **44** and

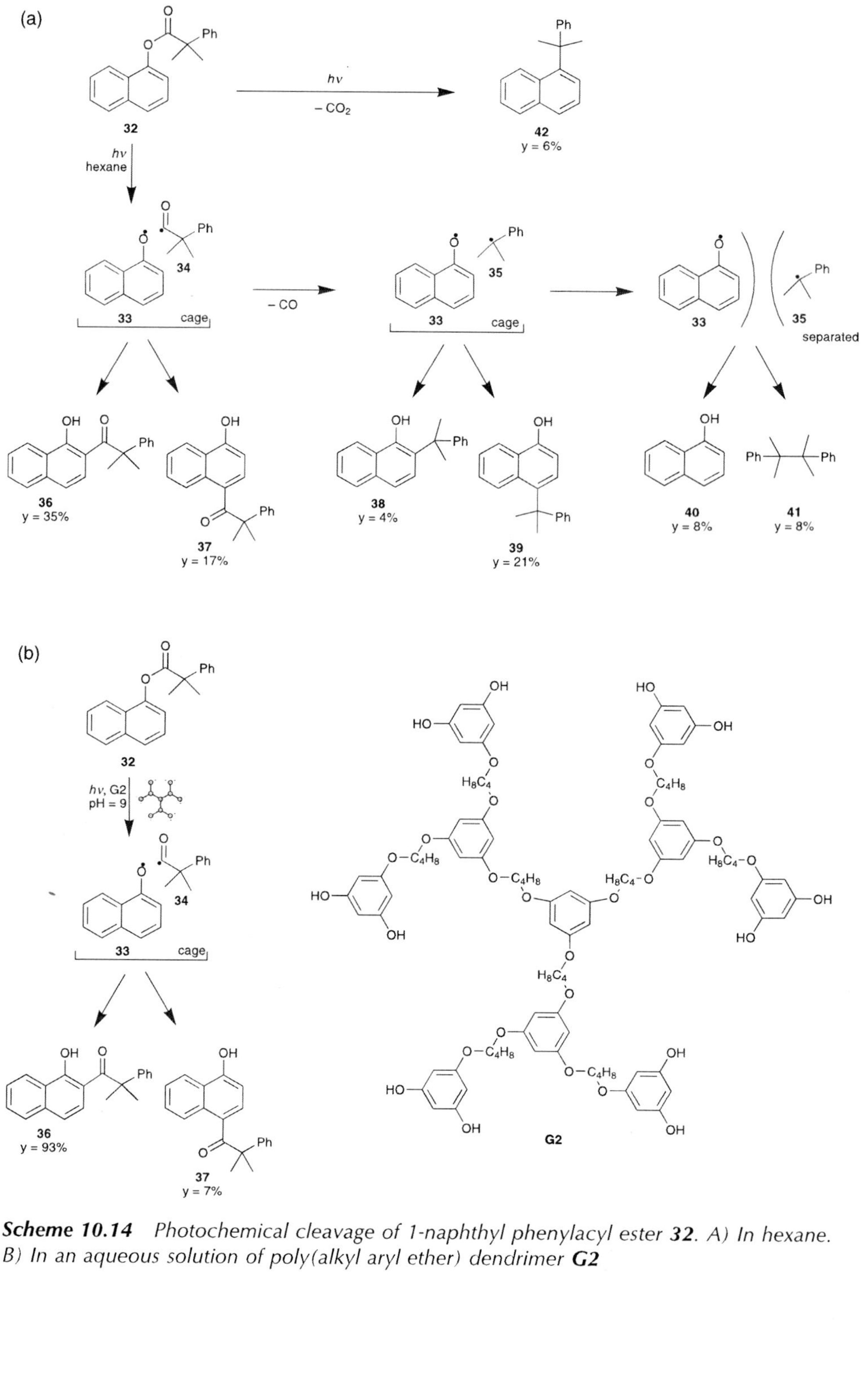

Scheme 10.14 *Photochemical cleavage of 1-naphthyl phenylacyl ester* ***32****. A) In hexane. B) In an aqueous solution of poly(alkyl aryl ether) dendrimer* ***G2***

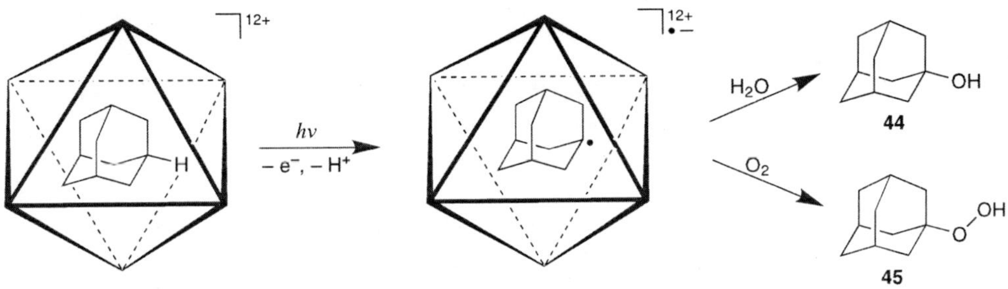

43$_4$**@22b** **43**$_3$**.43*@22b**$^{\bullet-}$ **44** **45**

Scheme 10.15 *Photooxidation of an alkane blocked in a* $Pd_6L_4^{12+}$ *cage built with triazine panels*

1-adamantyl hydroperoxide **45** in an overall yield of 24% (only one of the four adamantanes reacts). The blue radical anion could be detected by EPR and differential IR spectroscopy.

10.4.3 Carbenes

Carbenes are species comprising a divalent carbon atom. Because this carbon has only six valence electrons, it is highly electron-deficient and reactive.[32] However, depending on the groups attached to the carbenic carbon, the stability and the philicity of these species vary greatly.[33] For example, the *N*-heterocyclic carbenes (NHC) which have been recognized as powerful organocatalysts and ligands are nucleophilic and can even be isolated. However, they remain air and moisture-sensitive. These compounds illustrate why a supramolecular approach is useful in this field. It allows the broadening of the reaction conditions in which sensitive intermediates can be applied and manipulated. Classical approaches include complexation of the carbene with Lewis-acids like CO_2 or the highly electrophilic $SiCl_4$. However, these compounds tend to be too stable. Good results are also obtained with the use of a polydimethylsiloxane matrix which compensates the weaker acidity of the silicon atoms in silicone by its macromolecularity. The stabilization interaction is low (3–5 kcal mol^{-1}), preserving the reactivity but nevertheless allows for storing and handling in the presence of air.[34]

Most of the work concerning the modification of carbene reactivity has been performed on cyclodextrins (CD) with reactive alkyl carbenes. Diazirines have proven to be the most convenient precursors due to the small size of the three-membered ring and the volatility of the leaving group, molecular nitrogen. Diazirines are usually obtained from the corresponding ketone in two steps: in a methanolic solution of ammonia, hydroxylamine-*O*-sulfonic acid (HOSA) is added yielding a diaziridine which then is oxidized to the diazirine, most conveniently with iodine.

On a laboratory scale, the complexes are prepared by the coprecipitation method:[35] the guest is added to an aqueous solution of the hydrophilic cyclodextrin. The complex is usually less soluble, precipitates and can be isolated by filtration or centrifugation. For water-insoluble guests, diethyl ether can be used to facilitate the phase transfer. The ether then can be blown away by bubbling argon through the solution. The main driving forces for the complex formation are hydrophobic and van-der-Waals interactions. These are the reasons why the most important criteria for a high association constant are a good fit between the dimensions of the guest and the cavity of cyclodextrins and a high polarizability of the guest molecules. In water, the driving force for complex formation is governed by the hydrophobicity of the inner surface of the cyclodextrins. Indeed, any water molecules residing inside the cavity are higher in energy because they cannot build an extensive network of hydrogen bonds. They are then readily displaced by the hydrophobic guest in order to minimize the surface where water molecules are in contact with apolar units.

To understand the reactivity of a carbene entrapped inside a host molecule, the properties of the inclusion complex should be investigated first. This occurs by measurement of the changes in the physical properties of the guest upon complexation. Whereas spectrophotometric and fluorescence spectroscopy are most useful for dyes and conductometric studies for ionic compounds, diazirines in cyclodextrins are best studied applying circular dichroism[36] and NMR spectroscopy. Indeed, circular dichroism which measures the difference in absorbance between left and right circularly polarized light is a very powerful method because only the complex gives rise to an absorption spectrum, an effect which is called induced circular dichroism (ICD). In fact, two conditions should be met to obtain a signal: first, a chromophore absorbing UV or visible light, a condition fulfilled only by the diazirine and second, a center of chirality, e.g., as observed in cyclodextrins. Titration experiments allow the determination of the strength of the complex and the stoichiometry is obtained by the method of continuous variation. Moreover, information can be obtained about the orientation of the chromophore in relation to the cyclodextrin depending on the sign of the ICD effect (Figure 10.2). According to Harata's rule[37] which is based on the Kirkwood–Tinoco[38] theory of polarizabilities, a positive sign of the ICD effect means that the transition dipole moment vector of the chromophore is parallel to the symmetry axis of the cyclodextrin, whereas a negative sign is caused by a perpendicular arrangement. Harata's rule is valid for an aromatic chromophore located inside the cavity; in case where the chromophore is situated outside, the signs are reversed, according to Kodaka's rule.[39]

During the past twenty years, significant work has been performed. Several carbenes have been studied in order to observe changes in their reactivity. Especially the supra-

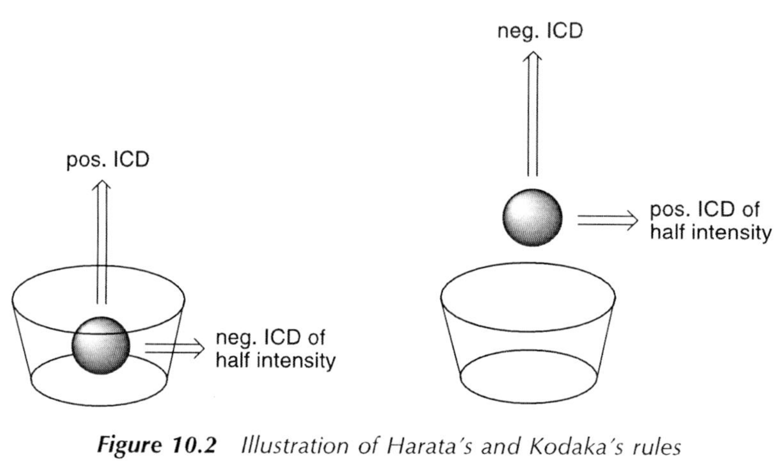

Figure 10.2 Illustration of Harata's and Kodaka's rules

molecular chemistry of 4-oxocyclohexa-2,5-dienylidene[40], 2-methylcyclohexylidene, 3-nortricyclanylidene, cyclooctylidene, chlorophenylcarbene, and adamantanylidene was investigated.[41]

10.4.3.1 4-Oxocyclohexa-2,5-dienylidene

4-Oxocyclohexa-2,5-dienylidene (**48**) is known as an intermediate in the degradation of 4-chlorophenol (**46**), a widespread organic pollutant. Carbene **48** is generated by photoinduced dehalogenation (λ = 280 nm) of **46** and is relatively long-lived (*ca.* 1 μs) in water because this triplet carbene is reluctant to abstract hydrogen from OH groups (Scheme 10.16).[40] Depending on the reaction conditions, carbene **48** can be trapped efficiently either by oxygen to form *p*-benzoquinone (**50**) through the intermediacy of benzoquinone-*O*-oxide (**49**). It can also be scavenged as biphenyl **54** by reaction with the starting material, or as hydroquinone (**53**) by reaction with water, or can be trapped by an alkene. In an aqueous cyclodextrin (CD) solution, however, the usual reaction pathway to benzoquinone **50** is hampered; instead, phenol (**52**) is obtained. In the presence of CDs, although the association constants are rather small (140 M^{-1} for α-CD and 300 M^{-1} for β-CD), carbene **48** becomes difficult to trap. Instead, hydrogen abstraction from the host strongly dominates to form phenol (**52**). This is especially true for α-CD: a 0.03 M solution of α-CD (83% complexation) yields 91% of phenol even in the presence of air. Indeed, the rate constant for the quenching of triplet **48** by α-CD ($k_q = 5.7 \times 10^8\,M^{-1}\,s^{-1}$) is very high. This is not only due to the good hydrogen donor property of secondary alcohols but also to the confinement in the narrow cavity of α-CD as revealed by a comparison with the significantly slower hydrogen abstraction from alcohols or glucose ($k_q = 8.5 \times 10^6\,M^{-1}\,s^{-1}$). In fact, the life-time of **48** is impressively reduced to a few tens of nanoseconds in the presence of cyclodextrin. Photolysis in the solid state also leads to phenol (**52**).[42]

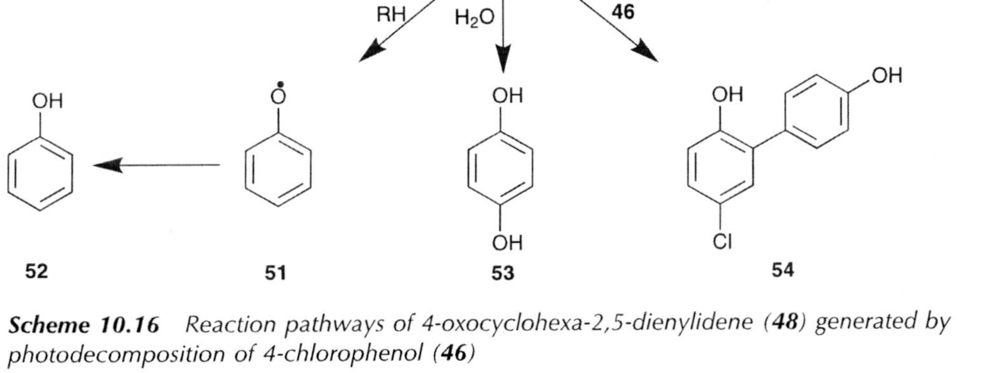

Scheme 10.16 *Reaction pathways of 4-oxocyclohexa-2,5-dienylidene (**48**) generated by photodecomposition of 4-chlorophenol (**46**)*

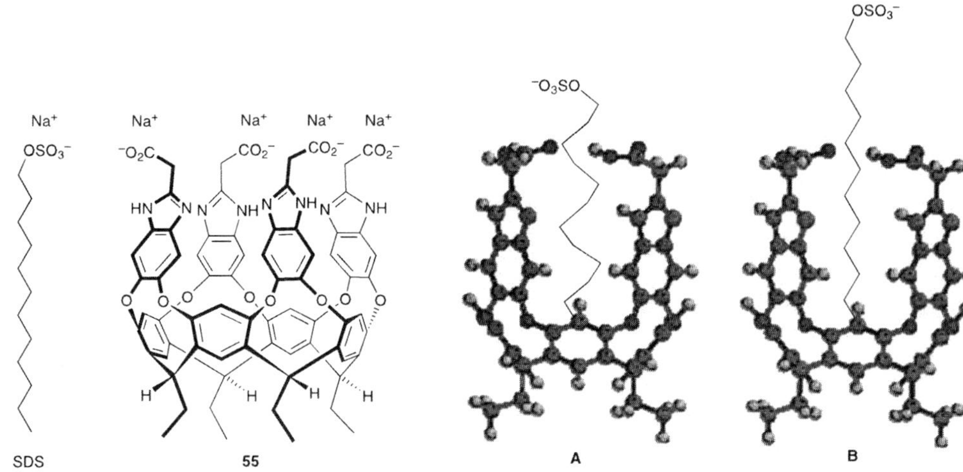

*Figure 10.3 Structure of the complex between SDS and cavitand 55. (**A**) Scheme illustrating the coiling of the alkyl chain within the cavity. (**B**) Extended conformation of SDS in the cavity*

The selectivity of an intramolecular carbene reaction is very strongly dependent on the conformation of the molecule, due to the high reactivity of the divalent carbon and its poor selectivity. If a carbene precursor or the carbene entrapped within a cavity is forced to take on a different conformation than the one prevailing in the liquid state, new products can be expected. For molecules with functional groups, a change in the geometry can be enforced by using noncovalent interactions like hydrogen bonds, dipole–dipole, or ion–ion interactions. For alkanes only few strategies remain; a convex structure can be built up to press the alkyl groups into a specific pocket[43] or the hydrophobic effect can be employed to compress apolar units into the cavity of a host (Figure 10.3).[44] These approaches have proven to be successful for the packing of alkyl chains into reduced space. Alkyl chains which are normally fully extended in solution (all *anti*) can be forced to coil to a helix in order to fit into the available restricted space. Use of polarizable side walls slightly helps the change in conformation by increasing van-der-Waals interactions and C—H/π interactions. An example is the coiling of the alkyl chain of the anionic surfactant sodium dodecyl sulfate (SDS) in the anionic cavitand **55** in water. The coiling is monitored best by NMR spectroscopy as the CH_2 protons are strongly shifted upfield when they are placed in proximity of an aromatic ring and by NOE measurements which allow determining the size of the helix in the most representative conformation. In the case of **SDS@55**, an eight-carbon helix seems to be present together with a few percent of a six-carbon helix. Of course, due to the flexibility of the alkyl chain and the fact that an optimally filled cavity should have a packing coefficient of only 55%,[45] rapid changes in the conformation still occur. The energetic cost for the coiling process has been estimated at circa 0.5 kcal·mol^{-1} for each *gauche* interaction.

10.4.3.2 2-Methylcyclohexylidene

One of the first attempts to follow this concept was performed with 2-methylcyclohexylidene (**57**). Carbene **57** is known to yield 1-methylcyclohexene (**59**) and 3-methylcy-

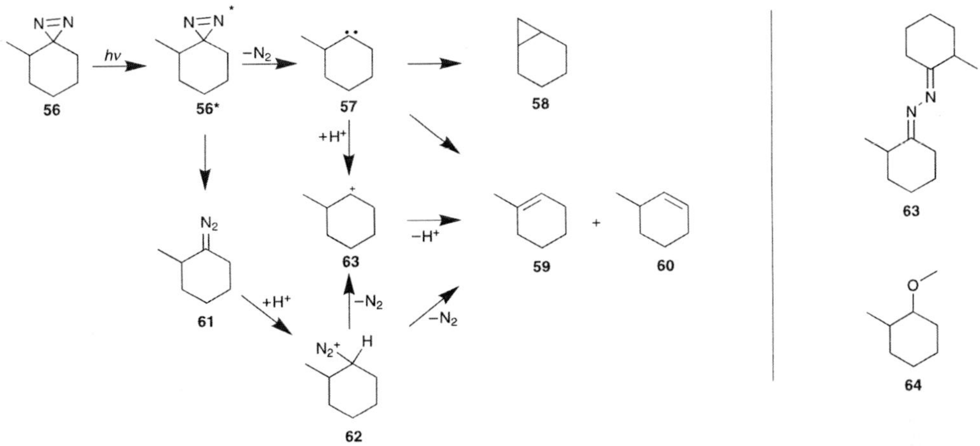

Scheme 10.17 Generation of 2-metylcyclohexylidene (*57*) and its subsequent reactions

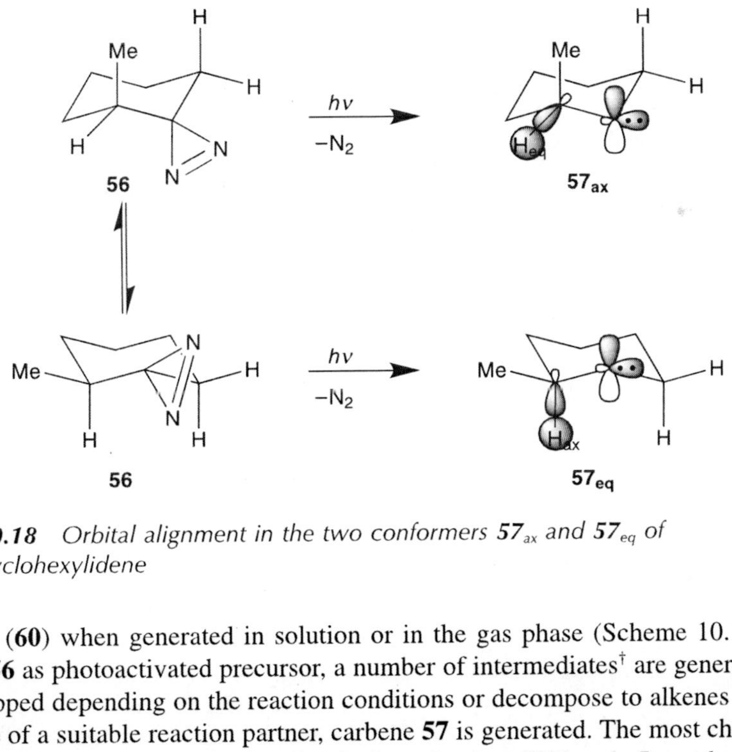

Scheme 10.18 Orbital alignment in the two conformers 57_{ax} and 57_{eq} of 2-methylcyclohexylidene

clohexene (**60**) when generated in solution or in the gas phase (Scheme 10.17).[46] With diazirine **56** as photoactivated precursor, a number of intermediates[†] are generated which can be trapped depending on the reaction conditions or decompose to alkenes **59** and **60**. In absence of a suitable reaction partner, carbene **57** is generated. The most characteristic reaction for its presence is the intramolecular insertion into CH bonds. It was hypothesized that the steric constraints in the cavity of a cyclodextrin may favor the conformer $\mathbf{57}_{ax}$ with an axial methyl group over conformer $\mathbf{57}_{eq}$ which is the most stable conformer under normal conditions (Scheme 10.18).[47] As a consequence of the high reactivity of carbenes

[†]This concerns essentially the photoactivated diazirine **56*** which may react directly by a rearrangement in the excited state (RIES) or ring-open to the labile diazo compound **61**. Under slightly acidic conditions, protonation of the carbene to carbenium ion **63** and the intermediacy of diazonium ion **62** should also be considered.

Table 10.1 *Product distribution from the decomposition of diazirine **56***

	ratio **59**:**60**	alkene **59**	alkene **60**	azine **63**	ether **64**
0.1 M **56** in MeOH	1.8	43	24		33
56@β-CD	1.9	66	34		
0.1 M **56** in *n*-C_5H_{12}	2.1	45	21	34	
56@γ-CD	2.1	62	30	8	
56@α-CD	2.6	72	28		

and their propensity to react with a wide range of bonds that are suitably aligned, the presence of a different conformer of the reactive intermediate should lead to a different product distribution. For an efficient hydrogen shift, the filled molecular orbital of the breaking CH bond (HOMO) should be aligned with the empty orbital at the carbenic center (LUMO). In $\mathbf{57}_{ax}$, the arrangement of the orbitals is much more favorable for a 1,3-CH insertion which would produce norcarane **58**. Moreover, formation of alkene **59** should be hampered. Unfortunately, starting from diazirine **56**, no significant change is obtained; **58** is not formed and the alkene ratio does not significantly vary (Table 10.1).

The intermediates being electron-deficient, formation of alkene **59** is always favored in accordance with Saytzeff's rule. The highest yield of **59** is obtained from the solid state photolysis of **56**@α-CD owing to a relative high **59**:**60** ratio and the absence of intermolecular products.

10.4.3.3 3-Nortricyclanylidene

Similarly, 3-nortricyclanylidene (**66**) was studied to obtain some evidence for an operating corset effect imposed by the macrocyclic host.[48] In the gas phase and also in solution, **66** ring-opens to 4-ethynylcyclopentene (**67**) via cyclopropylcarbene fragmentation, a reaction which causes a significant change in the geometry of the guest (Scheme 10.19). It was expected that generating carbene **66** in the nanopore of a cyclodextrin would favor the formation of quadricyclane (**69**) through a 1,3-CH insertion in analogy to the reactivity of 2-norbornanylidene (**73**) affording **68** in a nearly quantitative yield.[49] However, **69** was never observed from any reaction of the cyclodextrin complexes. Instead and in contrast to the results obtained with methylcyclohexylidene **57**, a very significant amount of the carbene **66** reacted with the OH groups at the rims of cyclodextrins (**71**: 52% with β-CD and 80% with α-CD, Table 10.2). This finding can be explained by the difference in the life-time of carbenes **57** and **66**, respectively. Whereas **57** can rearrange very easily to an alkene by a 1,2-H-shift, this pathway does not exist for **66**. As an alternative, the cleavage of the three-membered ring requires to overcome a considerable energy barrier. This leaves more time for the guest to react innermolecularly with the host. An intriguing result is the formation of the significant amount (14%) of nortricyclane (**68**) in the narrow and rigid cavity of α-CD. Product **68** is caused by a twofold abstraction of hydrogen. However, insertion of the carbene into a CH bond of the host does not occur, a reaction that up to now has never been observed with cyclodextrins. Formation of azine **70** in β-CD suggests that **65**@β-CD crystallizes in a head-to-head arrangement with two diazirine groups facing each other.

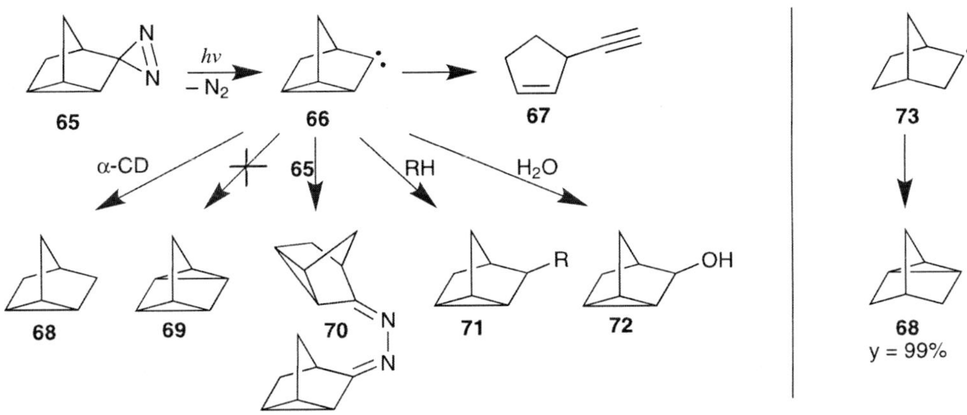

Scheme 10.19 3-Nortricyclanylidene (**66**) and its subsequent reactions

Table 10.2 Main products (%) resulting from the photolysis of 3-azinortricyclane (**65**)

	enyne **67**	nortricyclane **68**	azine **70**	insertion into the medium **71**	alcohol **72**
cyclohexane	19	1	49	20	1
65@β-CD	11	4	10	52	19
65@α-CD	6	14	–	80	–

10.4.3.4 Adamantanylidene

Photolyses of the α-, β-, and γ-cyclodextrin complexes **74**@CD of 2-aziadamantane (**74**) in the solid state indeed afforded markedly different product distributions and also different inclusion geometries (Figure 10.4).[51,52] The complexes **74**@CD have been investigated in detail using ICD.[53] This study performed in aqueous solutions shows that whereas the diazirine group in **74**@β-CD and **60**@γ-CD is placed in the vicinity of polar OH groups, the diazirine function in **74**@(α-CD)$_2$ is located in an apolar microenvironment as is revealed by the fine structure of its CD and UV spectrum. Moreover, for **74**@(α-CD)$_2$, a very large ICD ($\Delta\varepsilon_\infty$ = −0.732 M^{-1} cm^{-1} at 380 nm) is obtained due to strong interaction of the guest with the chiral cyclodextrin. This is in contrast to the weak absorption measured for **74**@γ-CDs which can be explained by the increased mobility of the guest **74** in the large and more flexible cavity of γ-CD.[52] The validity of computational calculations of host–guest complexes has been questioned.[54] Titration experiments have revealed the thermodynamic parameters of the 2-aziadamantane β-CD complex; it has also been shown that the results obtained are in good agreement with molecular dynamics simulations of this complex under explicit consideration of water molecules.[55] These differences in the strength of the complex are found again chemically: the reactivities of the enclosed adamantanylidene diverge significantly (Scheme 10.20 and Table 10.3).[51,52]

Especially the 1 : 2 complex formed with α-cyclodextrin leads to a drastic modification of the product distribution. Due to a complete encapsulation of the guest, formation of azine **78** is totally inhibited, thus inner- and intramolecular reaction pathways are favored.

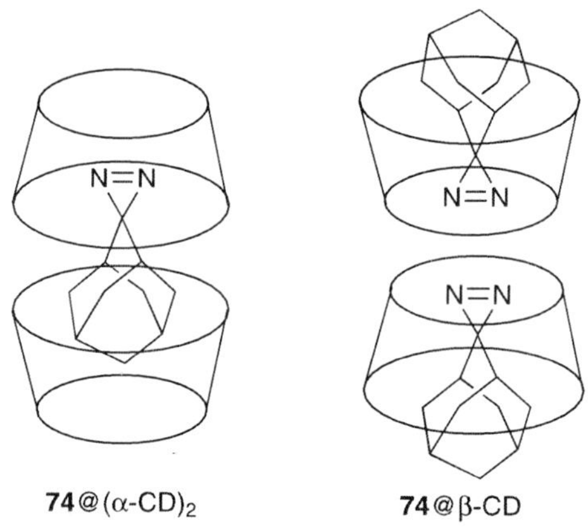

Figure 10.4 Proposed structures for complexes **74**@(α-CD)$_2$ and **74**@β-CD

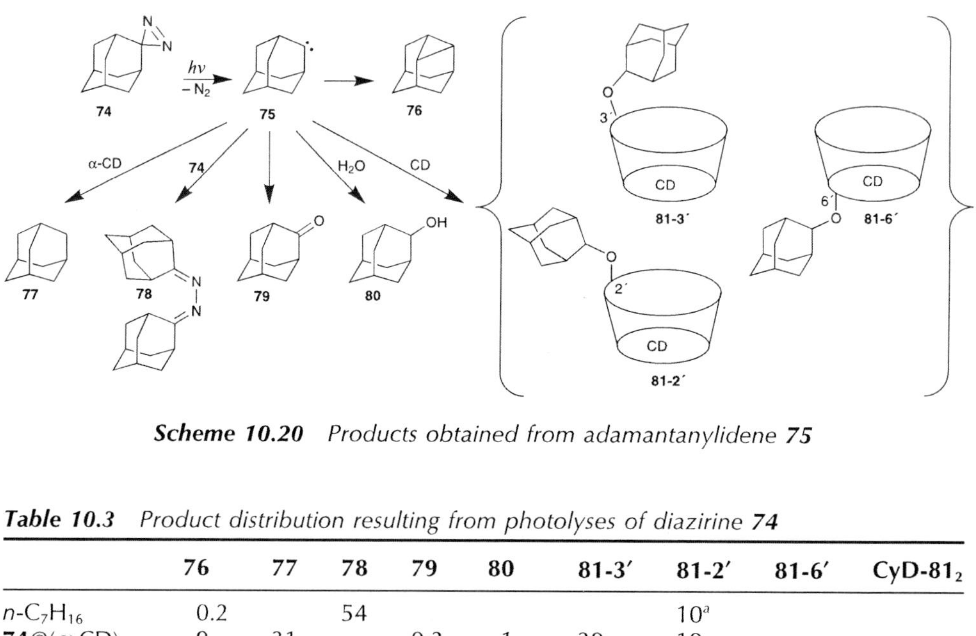

Scheme 10.20 Products obtained from adamantanylidene **75**

Table 10.3 Product distribution resulting from photolyses of diazirine **74**

	76	77	78	79	80	81-3′	81-2′	81-6′	CyD-81$_2$
n-C_7H_{16}	0.2		54				10[a]		
74@(α-CD)$_2$	9	31	–	0.3	1	39	19	–	–
74@β-CD	11	4	33	2	0.3	–	7	17	10
74@γ-CD	9	6	29	9	13	9	4	5	–

[a] Insertion into the solvent.

However, within the confined environment, the 1,3 C—H insertion reaction of carbene **75** is not favored; the yield of didehydroadamantane (**76**) remains low (9%) and CH-insertions into the host do not occur as well. Instead, the production of adamantane (**77**) is strongly enhanced and reaches 31%. The formation of **77** is particularly facilitated by the fact that the singlet-triplet gap in adamantanylidene **75** is low (3.1 to 4.8 kcal mol^{-1}).[56] Although

75 has been generated as a singlet and generally reacts as a singlet, the specific steric constraints prevailing in the rigid cavity of two α-CDs hamper concerted insertion reactions which have high requirements on the alignment of the orbitals involved in the reaction. Instead, intersystem crossing and hydrogen abstraction are favored. Accordingly, oxidized α-CD is also recovered. This pathway is more significant for adamantanylidene **75** than for nortricyclanylidene **66** because **66** has a larger singlet-triplet gap due to a higher ring strain. In fact, the main reaction of **75** corresponds to an etherification of the secondary rim of the cyclodextrin (39% for **81-3′** and 19% for **81-2′**). The predilection for reaction at position 3 corresponds to the geometrical requirement of the most stable conformation of the alkylated cyclodextrins **81**. In **81-3′**, the adamantyl group preferably remains directly on the top of the cavity, whereas in **81–2′**, it points outwards. Reaction with the primary hydroxy group at C6 does not occur. The photolyses of the 1 : 1 complexes with β- and γ-CD afford a wider range of products.[52] First of all, with β- and γ-CD, etherification at C6 is geometrically possible and does happen. With **74**@β-CD, even two adamantyl groups can react with the hydroxy functions of the same cyclodextrin molecule. Second, the main product is azine **78** along with adamantanone (**79**) which probably arises from the hydrolysis of an intermediate on the way to azine **78** or directly from **78**.[52] And finally, it is worth noticing that the yield of adamantanol **80** very strongly depends on the dryness of the hygroscopic cyclodextrin complexes. Overall, the ICD spectra and the products obtained from photolysis of 2-aziadamantane (**74**) in CD allow to draw conclusions about the supramolecular structure. Indeed, the results are mostly compatible with a head-to-head arrangement of the cyclodextrins in the **74**@$(\alpha\text{-CD})_2$ complex (Figure 10.4). For the β-CD complex, the major product being azine **78**, a 2 : 2 structure is more likely (Figure 10.4). Such a combination of CDs would be the analogue of the structure determined by X-ray analysis for the complex of 5-hydroxy-2-aziadamantane with β-CD[50] (Figure 10.5).

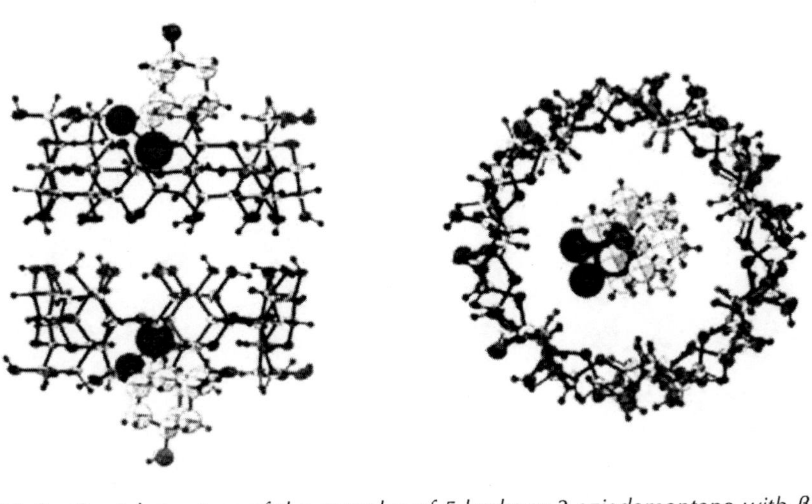

Figure 10.5 *Crystal structure of the complex of 5-hydroxy-2-aziadamantane with β-CD obtained by X-ray diffraction. Reprinted with permission from [50]. Copyright 2000 Elsevier*

10.4.3.5 Arylcarbenes

The following example emphasizes the fact that to observe drastic changes in the reactivity, encapsulation in a host molecule is not compulsory. It is often sufficient to perform the reaction in a constrained system under the condition that the adequate conformation is present. Thus, striking results have been obtained starting from 1-(1,1′-biphenyl-4-yl)-1-diazo-2-phenylpropane (**82**). Although photolysis of **82** in solution yields a complex mixture of products resulting from hydrogen and phenyl shifts, photolysis in the solid state is highly stereoselective and affords almost exclusively the (*Z*)-alkene **84** (Scheme 10.21) as a consequence of the conformational constraints of the crystal packing which are nearly the same in the reactant and in the (*Z*)-product.[57] Moreover, the migrating hydrogen is already perfectly aligned with the empty p orbital of the carbene intermediate **83** and no rotation is required for the 1,2 hydrogen shift.

10.4.3.6 Chloro(phenyl)carbene

Attempts were also made to observe the rearrangement of chloro(phenyl)carbene (**86**) to 1-chloro-1,2,4,6-cycloheptatetraene (**87**) in the cavity of cyclodextrins.[58] Starting from diazirine **85**, the rearrangement to **87** can indeed be observed in an argon matrix at 10 K.[59] Under preparative conditions, 100% of azine **88** is obtained. What would happen in a molecular capsule? Reactive intermediate **86** belongs to the category of carbenes that are particularly suitable for intermolecular reactions as a result of the absence of a reaction pathway like the common hydrogen shifts leading to the formation of alkenes. As a consequence, carbene **86** possesses a relatively long life-time (3.6 μs in isooctane)[60] and prefers to react with the OH groups of CDs yielding almost exclusively benzaldehyde (**89**) (Scheme 10.22).[58]

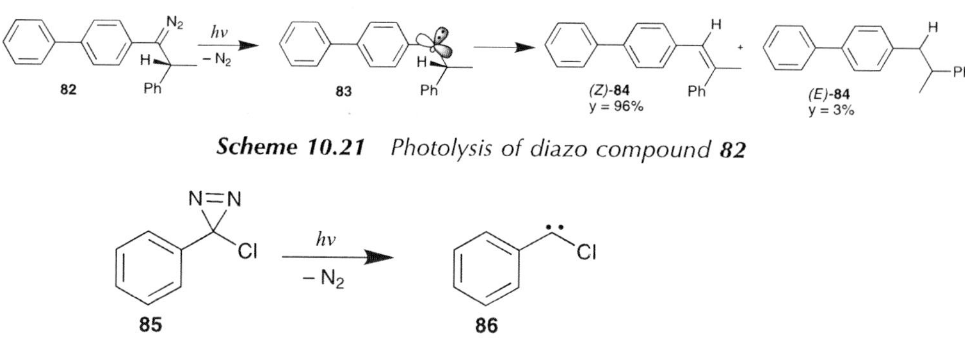

Scheme 10.21 *Photolysis of diazo compound **82***

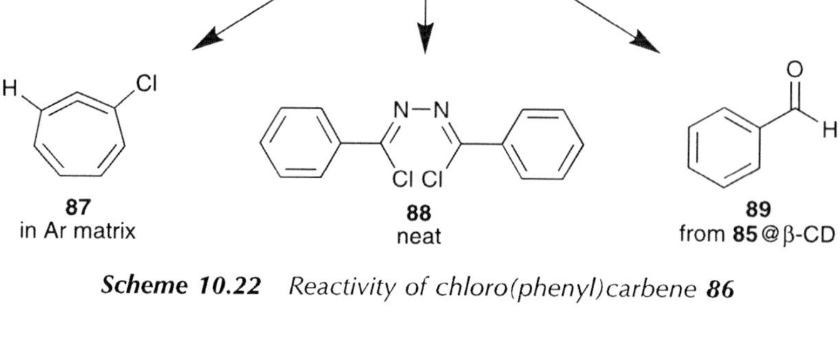

Scheme 10.22 *Reactivity of chloro(phenyl)carbene **86***

This high propensity of entrapped carbenes to react with native α- and β-cyclodextrin motivated us to search for more inert hosts. Therefore, after a few experiments with hydrophilic cyclodextrins derivatives, i.e., permethylated cyclodextrins (TRIMEB in the case of β-CD, TRIMEA for α-CD) were performed successfully as molecular reactors for supramolecular carbene chemistry.[61] This is due to their ready availability,[62] their high solubility in water and in organic solvents and their good crystallization properties. The absence of OH groups at the rims of TRIMEB makes this host inert toward reactions of carbenes. However, owing to the lack of intramolecular hydrogen bridges, TRIMEB is more flexible and forms less stable complexes with lower association constants in comparison to native β-cyclodextrin.[61]

10.4.3.7 Bicyclo[3.2.1]octan-8-ylidenes

The most recent experiments have been performed with 3-substituted bicyclo[3.2.1]octan-8-ylidenes **91**. These carbenes were chosen because of their symmetry. In the presence of a chiral host these achiral intermediates could give rise to chiral products **92** and **93** (Scheme 10.23). The aim of this study was to determine whether the intrinsically chiral cyclodextrin hosts are able to cause the formation of one enantiomer over the other.[63] The unsubstituted bicyclo[3.2.1]octan-8-ylidene (**91**) reacts exclusively through 1,3-CH insertion to tricyclooctane **92**.[64] For the 3-*endo*-substituted carbenes **91**, a second reaction pathway becomes competitive yielding bicyclo[3.3.0]octene **93** by a 1,2 alkyl shift. It was found that not only the host is able to induce a modest enantiomeric excess but the corset effect of the host can also significantly alter the bicyclooctene :tricyclooctane ratio **93**:**92** (Table 10.4). Once again, the rigid and narrow bowl of α-CD exerts the strongest influence on the guest. **90b**@(α-CD)$_2$ is the complex for which the highest ee (8%) is obtained. Also the formation of **93** is preferred over **92**.[63]

For **90b**@TRIMEB, a single crystal suitable for X-ray analysis could be grown (Figure 10.6).[61] The structure obtained ascertains the complete separation of one guest molecule from another and explains why azine **94** is lacking in the product mixture resulting from the photolysis of **90b**@TRIMEB. **90b**@TRIMEB is a good example for a complex for

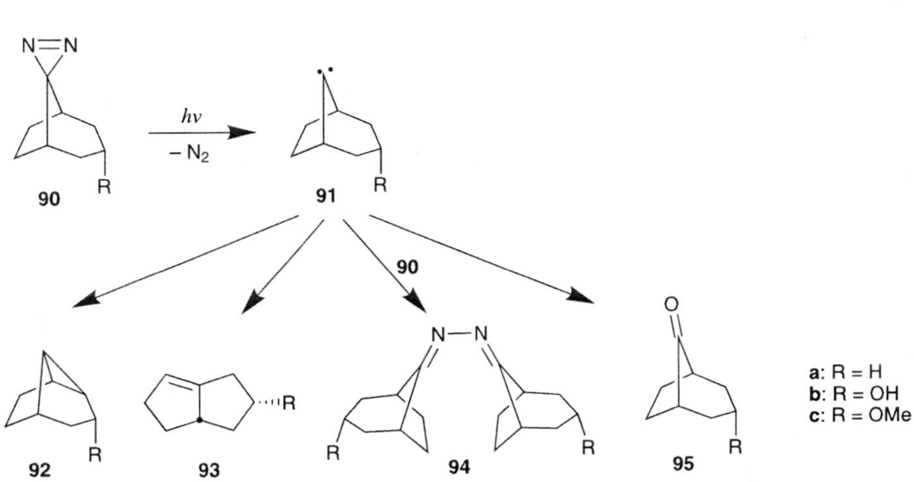

Scheme 10.23 *Generation of bicyclo[3.2.1]octanylidene **91** and its subsequent reactions*

Table 10.4 Product distribution resulting from the decomposition of diazirine **90**

	94:**95**	**93**:**92**	ee **92**
90b@β-CD		45:55	4%
90b@TRIMEB		21:79	4%
90b@(α-CD)$_2$		55:45	8%
90b		34:66	
90c@β-CD	89:11		
90c@TRIMEB		**92c** excl.	
90c		36:64	
90c in benzene	38:5	15:42	

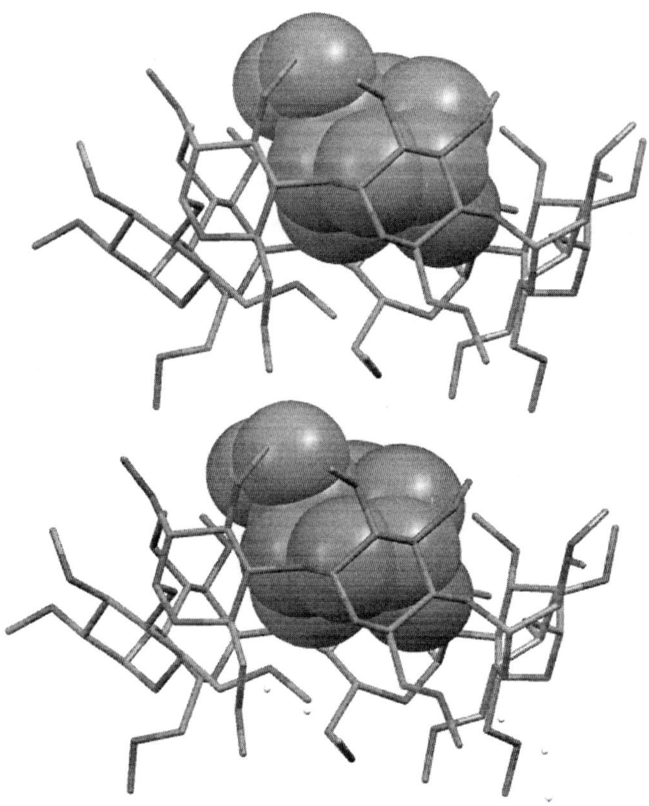

Figure 10.6 Crystal structure of **90b**@TRIMEB

which the orientation of the guest depends on its environment. In the solid state, the hydroxy group of the guest is positioned in the middle of the cyclodextrin cavity forming a hydrogen bridge with a glucosidic oxygen atom. In contrast, in an aqueous solution, the orientation is reversed: the hydroxy group is pointing toward the larger rim and the bulk of the solvent whereas the diazirine unit is deeply enclosed within the hydrophobic cavity as demonstrated from ICD spectra and 2D ROESY experiments. The association constant determined by the curve fitting method is $550\,M^{-1}$ in water; this is still a fair value but already considerably less then the association constant of the corresponding **90b**@β-CD ($11200\,M^{-1}$).[61]

10.4.4 Nitrenes

Our knowledge about supramolecular nitrene chemistry and the reactivity of these intermediates in a constrained system is still in its infancy: even though nitrenes are widely used for photoaffinity labelling.[65] However, the exact structures of the products formed after reaction with the active sites are often unknown. Therefore, a better understanding of the binding properties of a nitrene precursor within the host molecule is necessary. Moreover, it is essential to learn which reactions still do occur inside a supramolecular structure.

Azides **96** are the most popular precursors for nitrenes **97**, which can be generated either thermally or photochemically.[66] In a nitrene, the nitrogen atom is monovalent and possesses a sextet of electrons in its outer shell. Therefore, nitrenes are highly reactive intermediates. Usually, they are generated as singlets with paired electrons. However, intersystem crossing to the diradical triplet ground state is quite fast. (Scheme 10.24)

The most characteristic reaction of nitrenes consists of secondary amine **98** formation by insertion into C—H bonds either in a concerted process for singlets or through hydrogen abstraction followed by recombination of the radical couple for triplets, a process called pseudoinsertion. An even more efficient reaction is the addition to alkenes to form aziridines. For singlets, this addition is concerted and thus stereoselective. Whereas for triplets, the reaction proceeds through a stepwise mechanism involving radicals and therefore is not stereoselective. Further typical reactions are double hydrogen abstraction of triplet nitrenes to afford primary amines **100** and formation of azo compound **99** by reaction with the azide precursor or by dimerization when the nitrene concentration is high.

Our group has investigated the behavior of nitrenes enclosed in host molecules. For this purpose, we have obtained initial results with adamantyl nitrenes which can be generated either thermally or photochemically from readily available adamantane azides.[67]

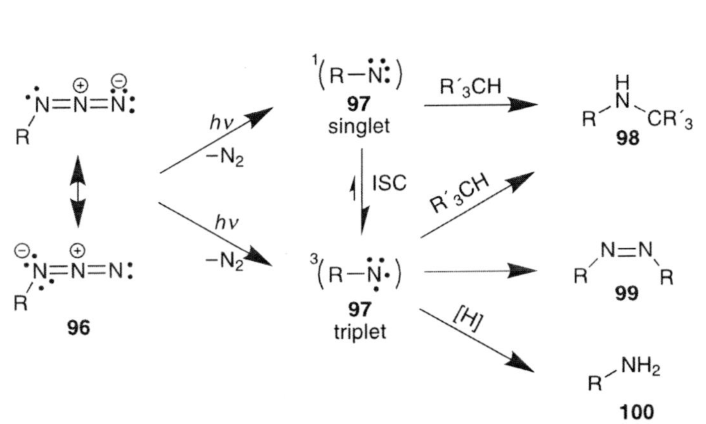

Scheme 10.24 *Nitrene generation*

Two main reasons guided this choice. On one hand, the reactions of these compounds under classical conditions are already known. This facilitates to find out, how nitrene reactivity has been modified by inclusion. On the other hand, as has been shown earlier in this chapter, the adamantane skeleton is a good fit for the cavity of α- and β-cyclodextrins leading to high association constants upon complexation.

10.4.4.1 1-Adamantanyl nitrene

1-Adamantanyl nitrene (**102**) is generated by photolysis of 1-adamantane azide (**101**). Nitrene **102** rearranges by an alkyl shift to the unstable imine **103** (Scheme 10.25). The C=N double bond is quite reactive because it is located at a bridge-head. Therefore, **103** dimerizes to azetidine **105** in a head-to-tail fashion in an alkane solution or in the solid state.[68]

In α- and also in β-cyclodextrin, the outcome of the reaction is totally different because dimerization to **105** cannot occur. Instead, during the aqueous work-up the hemiaminal 4-azahomoadamantan-3-ol (**104**) is formed by addition of water to the bridge-head imine (Table 10.5).

10.4.4.2 2-Adamantanyl nitrene

2-Adamantane azide (**106**) behaves similarly (Scheme 10.26): after generation of nitrene **107**, the main reaction consists of a 1,2-alkyl shift to form 4-azahomoadamant-4-ene (**108**) which can be isolated.[69] As a side product, adamantane imine (**109**) is produced through

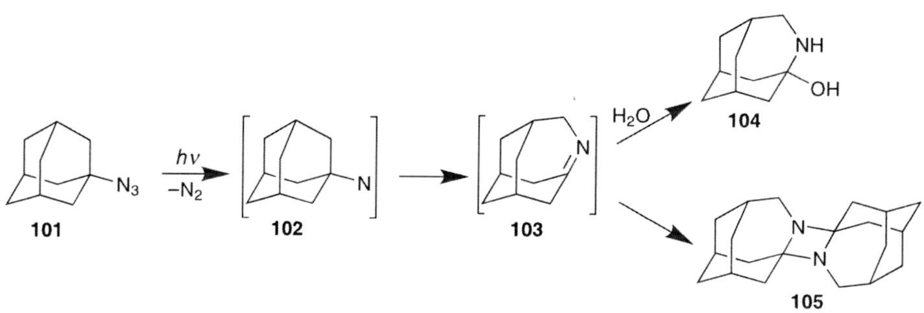

Scheme 10.25 *Generation of 1-adamantyl nitrene **102** and its subsequent reactions*

Table 10.5 *Product distribution resulting from the decomposition of 1-adamantane azide **101***

	Yield (%)	
	104	**105**
101 (solid state)		84
101 (alkane solution)		88
101@(α-**CD**)$_2$	94[a]	
101@β-**CD**	91	traces

[a] Traces of 1-adamantane amine and adamantan-1-ol.

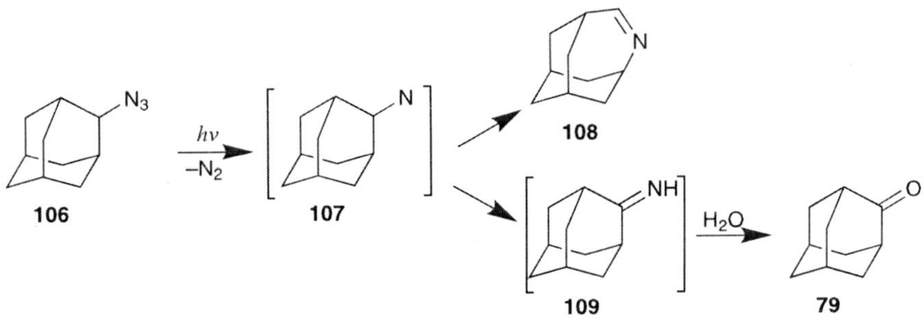

Scheme 10.26 *Generation of 2-adamantanyl nitrene **107** and its subsequent reactions*

Table 10.6 *Product distribution resulting from the decomposition of 2-adamantane azide **106***

	Yield (%)	
	108	**79**
106 (solid state)	84	15
106 (alkane solution)	87	5
106@(α-CD)$_2$	81	10
106@β-CD	77	23

a 1,2-H shift in **107**. Compound **109** easily hydrolyzes to afford adamantanone (**79**). When nitrene **107** is generated inside the cyclodextrin cavity, very similar results are obtained (Table 10.6). In any case, both adamantanyl nitrenes did not form any isolable innermolecular products with α- or β-cyclodextrin.

10.4.4.3 Ferrocenyl nitrene

The third case study concerns ferrocenyl nitrene. The geometry of the inclusion complexes of its precursor inside α- and β-cyclodextrins was thoroughly investigated using induced circular dichroism, quantum mechanical and NMR studies.[70] Because of the poor solubility of these complexes in water, alternative solvents had to be employed. Aprotic polar solvents like DMSO and DMF possess the best solubilizing properties but at the cost of low association constants. Ethylene glycol is a better compromise between binding and solubilizing properties.

Indeed, the behavior of ferrocenyl azide inside β-cyclodextrin is striking, since the alignment of the guest changes depending on the solvent and the temperature (Figure 10.7). In ethylene glycol and in DMSO/H_2O 50/50, a negative ICD is obtained which results from an axial orientation according to Harata's rule (Figure 10.8, Structure A). With DMSO contents higher than 62%, the ICD band is positive, ferrocenyl azide adopts an equatorial arrangement (Structure B). Structure B is also favored at higher temperatures as can be shown by the sign reversal of the ICD spectra between 10 and 40 °C (Figure 10.7B). The proposed geometries of inclusion are also corroborated by 2D ROESY spectroscopy (Table 10.7).

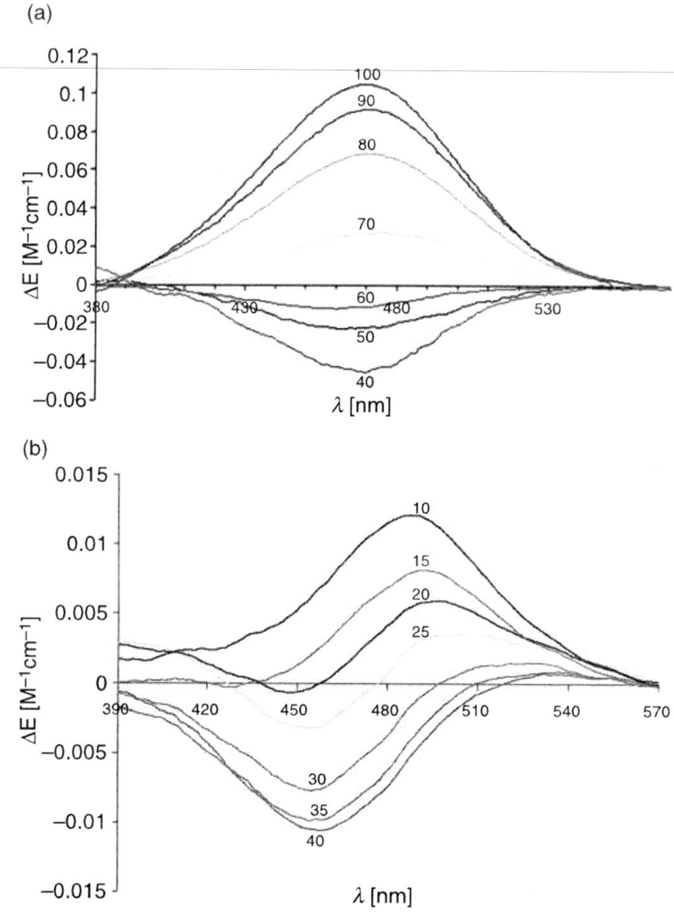

Figure 10.7 *A) ICD signal dependence of $FcN_3@\beta$-CD on the DMSO content (% v/v) of the aqueous solution, $T = 25\,°C$.* ***B)*** *Temperature effect (°C) on the ICD spectra sign at a constant solvent composition, ($DMSO/H_2O$: 62/38 (v/v), [FcN_3]: 0.0049 M, [β-CD]: 0.0496 M). Reprinted with permission from [70]. Copyright 2006 American Chemical Society*

Finally, the ferrocenyl complexes were decomposed photochemically and thermally. Thermolyses can be performed in this case because the decomposition temperature of the azide (**96**) (80 °C) is much lower than those employed for other diazirines and azides. The results obtained from photolyses and thermolyses do not differ significantly. Here again, α-CD causes the most drastic changes because of the complete encapsulation of the guest in a 1 : 2 complex (Scheme 10.27). In accord with the other reactions performed in α-CD, the main reaction pathway is hydrogen abstraction from the host. Upon thermolysis ferrocenyl amine (**112**) is obtained in a yield up to 60%. More remarkably, ferrocenyl nitrene (**111**) seems to react in very low yields with α-CD. However, the structure of the reaction product **113** has not been fully established yet and is quite unexpected because **113** is the result of a glucopyranose-furanose conversion. In contrast, the products obtained by thermolysis of ferrocenyl nitrene in the solid state, namely

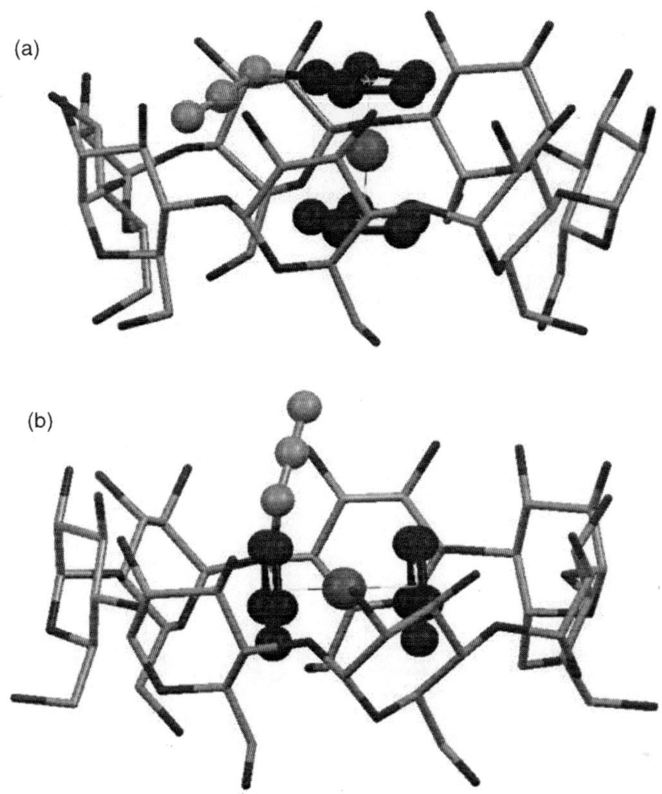

Figure 10.8 *Proposed orientation of ferrocenyl azide inside the β-cyclodextrin cavity according to ICD. (A) axial, (B) equatorial arrangement. Reprinted with permission from [70]. Copyright 2006 American Chemical Society*

Table 10.7 *Yields of isolated products from decomposition of ferrocenyl azide (FcN_3, **110**) complexes*

Complex		FcH (**115**)	FcN=NFc (**114**)	$FcNH_2$ (**112**)	FcN-CD
FcN_3 (solid)	Δ	9	47	–	–
FcN_3@$(\alpha\text{-CD})_2$	Δ	<0.5	–	60	9
FcN_3@$(\alpha\text{-CD})_2$	$h\nu$	<0.5	–	53	3
FcN_3@β-CD	Δ	<0.5	5	22	no
FcN_3@β-CD	$h\nu$	<0.5	8	27	no
FcN_3@γ-CD	Δ	4	4	19	yes
FcN_3@γ-CD	$h\nu$	1	2	9	yes
FcN_3@TRIMEB	Δ	7	2	–	yes

azoferrocene (**114**) (47%) and ferrocene (**115**) (9%) are not formed from FcN_3@$(\alpha\text{-CD})_2$. With FcN_3@β-CD, ferrocenyl amine (**112**) remains the main product (22%) but there is no insertion into the host and azoferrocene formation (**114**: 5%) cannot be totally avoided. FcN_3@TRIMEB yields ferrocene and azoferrocene. The crystal structure of FcN_3@TRIMEB is quite uncommon: the guest adopts a bimodal arrangement. In one half of the

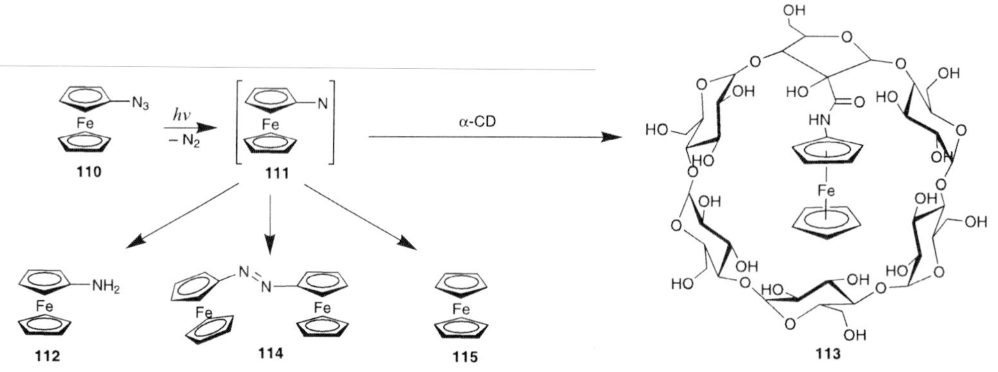

Scheme 10.27 *Generation of ferrocenyl nitrene (**111**) and its subsequent reactions*

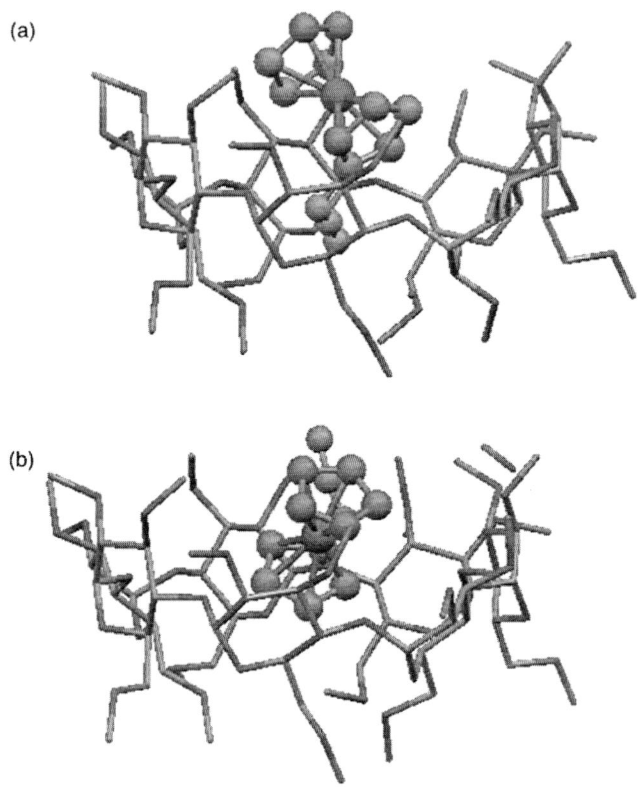

Figure 10.9 *Crystal structure of FcN_3@TRIMEB showing the two orientations of the guest*

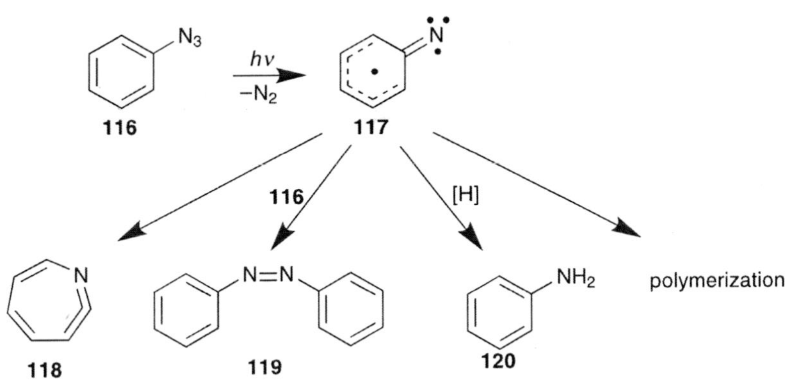

Scheme 10.28 *Solution chemistry of phenyl azide*

ferrocenyl azide molecules the azido group is oriented toward the bottom of the cavity (Figure 10.9, orientation A) whereas in the other half this group is pointing toward the secondary rim (orientation B).[67]

10.4.4.4 Phenylnitrene

Phenylnitrene (**117**) can be generated as a singlet from phenyl azide (**116**) under UV irradiation. Below 165 K, intersystem crossing to the triplet state occurs. Otherwise, an alkyl shift leads to the formation of the easily polymerizing 1-azacyclohepta-1,2,4,6-tetraene (**118**), (Scheme 10.28) a compound which is in equilibrium with singlet phenylnitrene.[71] Finally, the nitrene may be trapped if suitable compounds are present, e.g. formation of azobenzene (**119**) by reaction with phenyl azide. Otherwise, hydrogen abstraction affords aniline (**120**) and polymerization products are often obtained.[72] Interestingly, the chemical behavior of phenylnitrene is strongly altered when this intermediate is generated inside the resorcin[4]arene-based cavitand **121**.[72] Especially, **121** can be functionalized regioselectively depending on the structure of the complex.

The complex **116@121** is obtained in a 1.5 : 1 stoichiometry by crystallization from a saturated solution of cavitand **121** in phenyl azide (**116**) and drying at 25 °C (18 h at 0.4 Torr). Irradiation of the 1.5 : 1 complex of **116@121** leads to the formation of *N*-alkylaniline **122** as the major product (35%) through a regioselective C–H insertion of phenylnitrene into one of the alkyl chains (Scheme 10.29). The formation of product **122** is a strong hint that in **116@121** excess of phenyl azide is located between the alkyl chains. The major secondary product is azobenzene (**119**) (22%). This relatively high yield of dimerization product can be explained by the specific packing found in the crystal structure of **116@121**, namely a head-to-head arrangement in which two azido groups are in close proximity.

DSC (Differential Scanning Calorimetry) analysis reveals how a 1 : 1 complex can be obtained (Figure 10.10). Indeed, two dissociation processes are observed. First, excess phenyl azide is lost between 70 and 125 °C ($\Delta H = 4.0\,\text{kcal mol}^{-1}$). Second, the remaining 1 : 1 complex dissociates at 142 °C with a fairly high ΔH of $13.1\,\text{kcal mol}^{-1}$. It is worth noticing that MPWB1K/6-31+G(d,p) calculations ($\Delta H = 11.1\,\text{kcal mol}^{-1}$) are in good agreement with the experimental results concerning the orientation of the guest and the strength of the interactions.

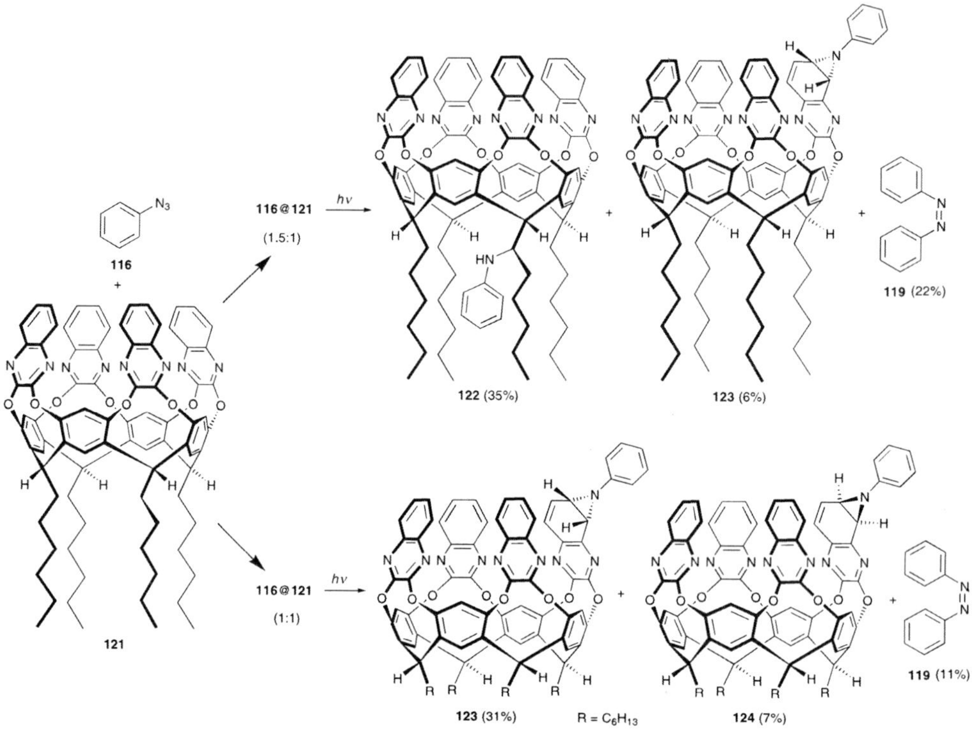

Scheme 10.29 *Irradiation of* ***116@121***

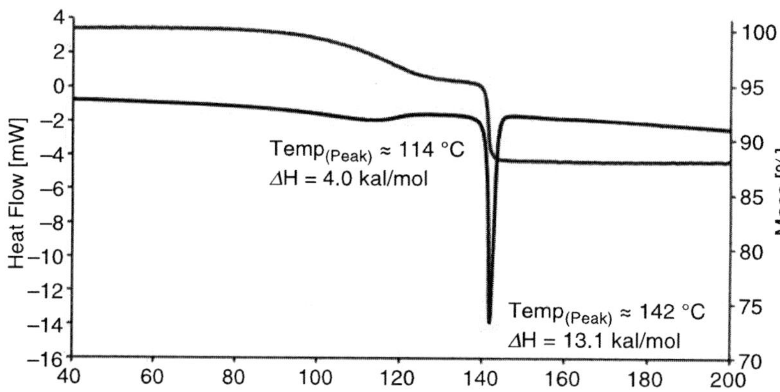

Figure 10.10 *DSC diagram and thermogravimetric analysis (starting on the top) of* ***116@121****. Reprinted with permission from [72]. Copyright 2009 American Chemical Society*

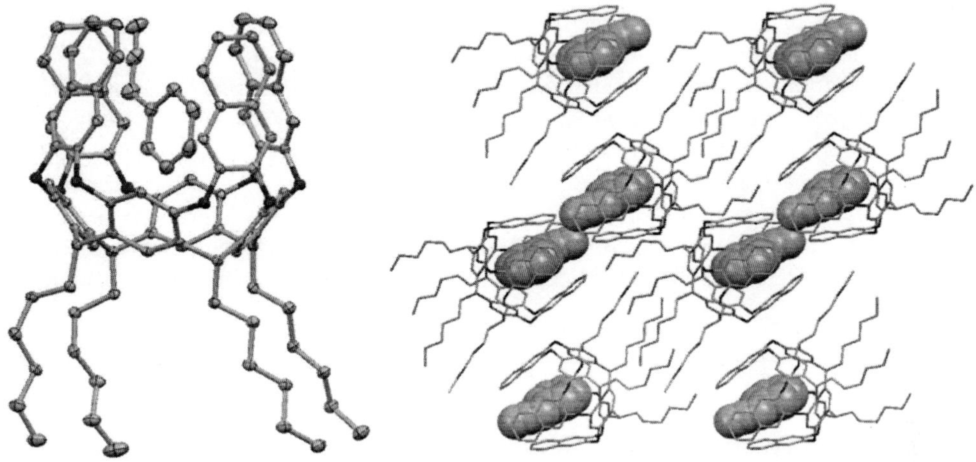

Figure 10.11 Crystal structure of **116@121**: side view and detail of crystal lattice. Reprinted with permission from [72]. Copyright 2009 American Chemical Society

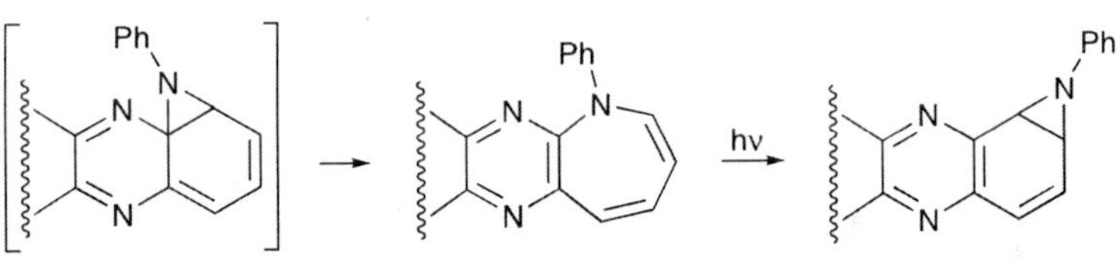

Scheme 10.30 Isomerization of aziridines **123** and/or **124** with ratio reversal

For synthetic purposes, the 1.5 : 1 complex can be treated thermally at 112 °C for 35 min at 1 atm to furnish **116@121** as a 1 : 1 complex (Figure 10.11). Its photolysis does not afford any C—H insertion product **122**. Instead, phenylnitrene adds to a quinoxaline ring and yields the two diastereomeric aziridines **123** (31%) and **124** (7%) together with azobenzene (**119**) (11%). This high regioselectivity and even the aziridine formation can only be obtained using solid state photolysis and are the result of the restricted motion of the nitrene in the cavitand. Indeed, photolysis of phenyl azide in presence of 10 equiv of quinoxaline in benzene affords only aniline (**120**) (17%) and azobenzene (**119**) (9%). The azanorcaradienes **123** and/or **124** can be converted into azepine **126** upon heating via a 1,5-sigmatropic nitrogen migration to aziridine **125** followed by valence isomerization (Scheme 10.30). Irradiation of azepine **126** affords back the aziridines **123** and **124** in a ratio of 1 : 5; this reaction chain allows reversing the ratio of **123 : 124**.

In conclusion, the generation of reactive intermediates in the cavity of a supramolecular structure offers a straightforward approach to the preparation of functionalized hosts as can be seen for example from the results obtained from the different regioselective modifications of a resorcin[4]arene cavitand by phenylnitrene.[72] The properties of the guest can be significantly changed: intersystem-crossing can be enforced and therefore a carbene can be coerced to react as a triplet. Accordingly, the reactivity may vary greatly. Moreover, as a trend, the encapsulated intermediates tend to be less reactive, especially in intermolecular reactions and toward the inner side of the host. This is partly due to the steric requirement for the reaction and also due to the need to have an optimal alignment of the orbitals of the reactant. As a consequence of the confinement in a constrained system, these reactions are slowed down. However, especially in rigidly and densely packed complexes, some reactions which do not have strong requirements on the geometry of the transition state are strongly accelerated, e.g., electron transfer or hydrogen abstraction. Indeed, the narrow cavity of α-CD has proven to be a very efficient hydrogen donor toward triplet intermediates and radicals favoring the reduction of the guest.

References

1. a) R. Arun, C. K. Ashok Kumar, V. V. N. S. S. Sravanthi, Cyclodextrins as drug carrier molecule: a review, *Sci. Pharm.*, 2008, **76**, 567–598; b) R. Gaudana, J. Jwala, S. H. S. Boddu, A. K. Mitra, Recent perspectives in ocular drug delivery, *Pharm. Res.*, 2009, **26**, 1197–1216; c) A. Vyas, S. Saraf, S. Saraf, Cyclodextrin based novel drug delivery systems, *J. Incl. Phenom. Macrocycl. Chem.*, 2008, **62**, 23–42.
2. T. Iwasawa, R. J. Hooley, J. Rebek Jr., Stabilization of labile carbonyl addition intermediates by a synthetic receptor, *Science*, 2007, **317**, 493–496.
3. X. Li, S. Danishefsky, New chemistry with old functional groups: on the reaction of isonitriles with carboxylic acids – a route to various amide types, *J. Am. Chem. Soc.*, 2008, **130**, 5446–5447.
4. P. Restorp, J. Rebek Jr., Reaction of isonitriles with carboxylic acids in a cavitand: observation of elusive isoimide intermediates, *J. Am. Chem. Soc.*, 2008, **130**, 11850–11851.
5. D. L. Caulder, C. Brückner, R. E. Powers, S. König, T. N. Parac, J. A. Leary, K. N. Raymond, Design, formation and properties of tetrahedral M_4L_4 and M_4L_6 supramolecular clusters, *J. Am. Chem. Soc.*, 2001, **123**, 8923–8938.
6. V. M. Dong, D. Fiedler, B. Carl, R. G. Bergman, K. N. Raymond, Molecular recognition and stabilization of iminium ions in water, *J. Am. Chem. Soc.*, 2006, **128**, 14464–14465.
7. J. L. Brumaghim, M. Michels, K. N. Raymond, Hydrophobic chemistry in aqueous solution: stabilization and stereoselective encapsulation of phosphonium guests in a supramolecular host, *Eur. J. Org. Chem.*, 2004, 4552–4559.
8. P. Mal, B. Breiner, K. Rissanen, J. R. Nitschke, White phosphorus is air-stable within a self–assembled tetrahedral capsule, *Science*, 2009, **324**, 1697–1699.
9. P. Mal, D. Schultz, K. Beyeh, K. Rissanen, J. R. Nitschke, An unlockable–relockable iron cage by subcomponent self–assembly, *Angew. Chem. Int. Ed.*, 2008, **47**, 8297–8301.
10. C. D. Meyer, C. S. Joiner, J. F. Stoddart, Template–directed synthesis employing reversible imine bond formation, *Chem. Soc. Rev.*, 2007, **36**, 1705–1723.
11. M. Yoshizawa, T. Kusukawa, M. Fujita, K. Yamaguchi, Ship–in–a–bottle synthesis of otherwise labile cyclic trimers of siloxanes in a self–assembled coordination cage, *J. Am. Chem. Soc.*, 2000, **122**, 6311–6312.
12. R. H. Baney, M. Itoh, A. Sakakibara, T. Suzuki, Silsesquioxanes, *Chem. Rev.*, 1995, **95**, 1409–1430.
13. M. Yoshizawa, T. Kusukawa, M. Fujita, S. Sakamoto, K. Yamaguchi, Cavity–directed synthesis of labile silanol oligomers within self–assembled coordination cages, *J. Am. Chem. Soc.*, 2001, **123**, 10454–10459.
14. M. Yoshizawa, T. Kusukawa, M. Kawano, T. Ohhara, I. Tanaka, K. Kurihara, N. Niimura, M. Fujita, Endohedral clusterization of ten water molecules into a 'molecular ice' within the hydrophobic pocket of a self–assembled cage, *J. Am. Chem. Soc.*, 2005, **127**, 2798–2799.
15. T. Sawada, M. Yoshizawa, S. Sato, M. Fujita, Minimal nucleotide duplex formation in water through enclathration in self–assembled hosts, *Nature Chemistry*, 2009, **1**, 53–56.

16. M. Yoshizawa, K. Ono, K. Kumazawa, T. Kato, M. Fujita, Metal–metal *d–d* interaction through the discrete stacking of mononuclear M(II) complexes (M = Pt, Pd, and Cu) within an organic-pillared coordination cage, *J. Am. Chem. Soc.*, 2005, **127**, 10800–10801.
17. Y. Yamauchi, M. Yoshizawa, M. Fujita, Engineering stacks of aromatic rings by the interpenetration of self-assembled coordination cages, *J. Am. Chem. Soc.*, 2008, **130**, 5832–5833.
18. P. Gütlich, H. A. Goodwin, *Topics in Current Chemistry, Spin Crossover in Transition Metal Compounds I–III*; Springer, Berlin, 2004.
19. K. Ono, M. Yoshizawa, M. Akita, T. Kato, Y. Tsunobuchi, S. Ohkoshi, M. Fujita, Spin crossover by encapsulation, *J. Am. Chem. Soc.*, 2009, **131**, 2782–2783.
20. K. Kim, N. Selvapalam, Y. H. Ko, K. M. Park, D. Kim, J. Kim, Functionalized cucurbiturils and their applications, *Chem. Soc. Rev.*, 2007, **36**, 267–279.
21. W. Ong, M. Gomez–Kaifer, A. E. Kaifer, Cucurbit[7]uril: a very effective host for viologens and their cation radicals, *Org. Lett.*, 2002, **4**, 1791–1794.
22. A. Mirzoian, A. E. Kaifer, Reactive pseudorotaxanes: inclusion complexation of reduced viologens by the hosts *β*–cyclodextrin and heptakis(2,6–di–*O*–methyl)–*β*–cyclodextrin, *Chem. Eur. J.*, 1997, **3**, 1052–1058.
23. M. J. Frampton, H. L. Anderson, Insulated molecular wires, *Angew. Chem. Int. Ed.*, 2007, **46**, 1028–1064.
24. K. Yoshida, T. Shimomura, K. Ito, R. Hayakawa, Inclusion complex formation of cyclodextrin and polyaniline, *Langmuir*, 1999, **15**, 910–913.
25. M. Numata, T. Hasegawa, T. Fujisawa, K. Sakurai, S. Shinkai, *β*–1,3–Glucan (schizophyllan) can act as a one–dimensional host for creation of novel poly(aniline) nanofiber structures, *Org. Lett.*, 2004, **6**, 4447–4450.
26. A. Harada, J. Li, M. Kamachi, Synthesis of a tubular polymer from threaded cyclodextrins, *Nature*, 1993, **364**, 516–518.
27. T. Shimomura, T. Akai, T. Abe, K. Ito, Atomic force microscopy observation of insulated molecular wire formed by conducting polymer and molecular nanotube, *J. Chem. Phys.*, 2002, **116**, 1753–1756.
28. R. Eelkema, K. Maeda, B. Odell, H. L. Anderson, Radical cation stabilization in a cucurbituril oligoaniline rotaxane, *J. Am. Chem. Soc.*, 2007, **129**, 12384–12385.
29. P. Franchi, M. Lucarini, G. F. Pedulli, Use of nitroxide radicals to investigate supramolecular entities, *Curr. Org. Chem.*, 2004, **8**, 1831–1849.
30. L. S. Kaanumalle, J. Nithyanandhan, M. Pattabiraman, N. Jayaraman, V. Ramamurthy, Water–soluble dendrimers as photochemical reaction media: chemical behavior of singlet and triplet radical pairs inside dendritic reaction cavities, *J. Am. Chem. Soc.*, 2004, **126**, 8999–9006.
31. Y. Furutani, H. Kandori, M. Kawano, K. Nakabayashi, M. Yoshizawa, M. Fujita, In situ spectroscopic, electrochemical, and theoretical studies of the photoinduced host–guest electron transfer that precedes unusual host–mediated alkane photooxidation, *J. Am. Chem. Soc.*, 2009, **131**, 4764–4768.
32. R. A. Moss, M. S. Platz, M. Jones Jr., *Reactive Intermediate Chemistry*, Wiley, Hoboken, 2004.
33. J.–L. Mieusset, U. H. Brinker, The carbene reactivity surface: a classification, *J. Org. Chem.*, 2008, **73**, 1553–1558.
34. F. Bonnette, T. Kato, M. Destarac, G. Mignani, F. P. Cossio, A. Baceiredo, Encapsulated *N*–heterocyclic carbenes in silicones without reactivity modification, *Angew. Chem. Int. Ed.*, 2007, **46**, 8632–8635.
35. A. R. Hedges, Industrial applications of cyclodextrins, *Chem. Rev.*, 1998, **98**, 2035–2044.
36. a) Y. A. Zhdanov, Y. E. Alekseev, E. V. Kompantseva, E. N. Vergeichik, Induced optical activity in cyclodextrin complexes, *Russ. Chem. Rev.*, 1992, **61**, 563–575; b) D. Krois, U. H. Brinker, Circular dichroism of cyclodextrin complexes, in *Handbook of Cyclodextrins and Their Complexes*; H. Dodziuk (Ed.), Wiley, New York, 289–298, 2006.
37. K. Harata, H. Uedeira, The circular dichroism of the *β*–cyclodextrin complex with naphthalene derivatives, *Bull. Chem. Soc. Jpn.*, 1975, **48**, 375–378.
38. I. Tinoco Jr., Theoretical aspects of optical activity part two: polymers, *Adv. Chem. Phys.*, 1962, **4**, 113–160.

39. a) M. Kodaka, Sign of circular dichroism induced by β–cyclodextrin, *J. Phys. Chem.*, 1991, **95**, 2110–2112; b) M. Kodaka, Application of a general rule to induced circular dichroism of naphthalene derivatives complexed with cyclodextrins, *J. Phys. Chem. A*, 1998, **102**, 8101–8103.
40. a) M. G. Rosenberg, U. H. Brinker, Carbenes generated within cyclodextrins and zeolites, *Adv. Phys. Org. Chem.*, 2005, **40**, 1–47; b) M. G. Rosenberg, U. H. Brinker, Constrained carbenes, *Eur. J. Org. Chem.*, 2006, 5423–5440.
41. I. Manet, S. Monti, P. Bortolus, M. Fagnoni, A. Albini, The photochemistry of 4–chlorophenol in water revisited: the effect of cyclodextrins on cation and carbene reactions, *Chem. Eur. J.*, 2005, **11**, 4274–4282.
42. J. P. Da Silva, L. F. Vieira Ferreira, A. M. Da Silva, A. S. Oliveira, A comparative study of the photophysics and photochemistry of 4–chlorophenol adsorbed on silicalite and β–cyclodextrin, *J. Photochem. Photobiol. A*, 2002, **151**, 157–164.
43. B. W. Purse, J. Rebek Jr., Self–fulfilling cavitands: packing alkyl chains into small spaces, *PNAS*, 2006, **103**, 2530–2534.
44. L. Trembleau, J. Rebek Jr., Helical conformation of alkanes in a hydrophobic cavitand, *Science*, 2003, **301**, 1219–1220.
45. S. Mecozzi, J. Rebek Jr., The 55% solution: a formula for molecular recognition in the liquid state, *Chem. Eur. J.*, 1998, **4**, 1016–1022.
46. J. W. Wilt, W. J. Wagner, The rearrangement of 2–methylcyclohexylidene, *J. Org. Chem.*, 1964, **29**, 2788–2789.
47. M. G. Rosenberg, S. M. Kam, U. H. Brinker, Encapsulation of an asymmetric diazirine: reactivity of 2–methylcyclohexanylidene, *Tetrahedron Lett.*, 1996, **37**, 3235–3238.
48. M. G. Rosenberg, U. H. Brinker, Effect of supramolecular constraint on the selectivity of 3–nortricyclanylidene, *J. Org. Chem.*, 2001, **66**, 1517–1522.
49. L. Friedman, H. Shechter, Transannular and hydrogen–rearrangement reactions in carbenoid decomposition of diazocycloalkanes, *J. Am. Chem. Soc.*, 1961, **83**, 3159–3160.
50. M. M. Bobek, G. Giester, H. Kählig, U. H. Brinker, A 'sugar–coated' carbene precursor: a single crystal X–ray diffraction and NMR study, *Tetrahedron Lett.*, 2000, **41**, 5663–5667.
51. D. Krois, M. M. Bobek, A. Werner, H. Kählig, U. H. Brinker, Chemospecific monofunctionalization of α–cyclodextrin in the solid state, *Org. Lett.*, 2000, **2**, 315–318.
52. a) U. H. Brinker, R. Buchkremer, M. Kolodziejczyk, R. Kupfer, M. G. Rosenberg, M. D. Poliks, M. Orlando, M. L. Gross, Carbenes in constrained systems I: 1,3-C—H, insertion reaction of adamantylidene within the β-cyclodextrin cavity, *Angew. Chem., Int. Ed. Engl.*, 1993, **32**, 1344–1346; b) D. Krois, L. Brecker, A. Werner, U. H. Brinker, Supramolecular structure–reactivity relationships: photolysis of a series of aziadamantane@cyclodextrin inclusion complexes in the solid state, *Adv. Synth. Catal.*, 2004, **346**, 1367–1374.
53. D. Krois, U. H. Brinker, Induced circular dichroism and UV–*vis* absorption spectroscopy of cyclodextrin inclusion complexes: structural elucidation of supramolecular aziadamantane (spiro[adamantane–2,3′–diazirine]), *J. Am. Chem. Soc.*, 1998, **120**, 11627–11632.
54. a) H. Dodziuk, Modeling of CDs and their complexes, in *Cyclodextrins and Their Complexes*; H. Dodziuk (Ed.), Wiley, Weinheim, 2006; b) E. A. Castro, D. A. J. Barbiric, Molecular modeling and cyclodextrins: a relationship strengthened by complexes, *Curr. Org. Chem.*, 2006, **10**, 715–729.
55. a) B. Sellner, G. Zifferer, A. Kornherr, D. Krois, U. H. Brinker, Molecular dynamics simulations of β–cyclodextrin–aziadamantane complexes in water: simulation versus experiment, *J. Phys. Chem. B*, 2008, **112**, 710–714; b) G. Zifferer, A. Kornherr, U. H. Brinker, Molecular dynamics simulation of configurational properties of complexes between β–cyclodextrin and 2,6–diaziadamantane in water, *Mol. Simul.*, 2008, **34**, 1177–1183.
56. a) T. Bally, S. Matzinger, L. Truttmann, M. S. Platz, S. Morgan, Matrix spectroscopy of 2–adamantylidene, a dialkylcarbene with singlet ground state, *Angew. Chem. Int. Ed. Engl.*, 1994, **33**, 1964–1966; b) G. V. Shustov, M. T. H. Liu, On the possibility of conversion of strained bridgehead alkenes into carbenes via 1,2 hydrogen and 1,2 carbon migrations. A theoretical study of the rearrangements in the adamantene and protoadamantene systems, *Can. J. Chem.*, 1998, **76**, 851–861.

57. a) S. H. Shin, A. E. Keating, M. A. Garcia–Garibay, Transforming a nonselective carbene rearrangement into a highly selective process by using crystalline media, *J. Am. Chem. Soc.*, 1996, **118**, 7626–7627; b) S. H. Shin, D. Cizmeciyan, A. E. Keating, S. I. Khan, M. A. Garcia–Garibay, Control of carbene reactivity by crystals. A highly selective 1,2–H shift in the solid–to–solid reaction of 1–(4′–biphenylyl)–2–phenyldiazopropane to (*Z*)–1–(4′–biphenylyl)–2–phenylpropene, *J. Am. Chem. Soc.*, 1997, **119**, 1859–1868; c) A. E. Keating, S. H. Shin, K. N. Houk, M. A. Garcia–Garibay, Combining quantum mechanical reaction pathways with force field lattice interactions to model a solid–state phototransformation, *J. Am. Chem. Soc.*, 1997, **119**, 1474–1475; d) M. A. Garcia–Garibay, S. Shin, C. N. Sanrame, Engineering reactions in crystalline solids: prediction of intramolecular carbene rearrangements, *Tetrahedron*, 2000, **56**, 6729–6737; e) C. N. Sanrame, C. P. Suhrada, H. Dang, M. A. Garcia–Garibay, Photochemistry of crystalline chlorodiazirines: the influence of conformational disorder and intermolecular Cl···N=N interactions on the solid–state reactivity of singlet chlorocarbenes, *J. Phys. Chem. A*, 2003, **107**, 3287–3294.
58. M. G. Rosenberg, U. H. Brinker, Inter– and innermolecular reactions of chloro(phenyl)carbene, *J. Org. Chem.*, 2003, **68**, 4819–4832.
59. W. W. Sander, Chemiluminescence from arylcarbene oxidation: phenylchlorocarbene and (2-chlorophenyl)carbene, *Spectrochim. Acta A*, 1987, **43A**, 637–46.
60. a) N. J. Turro, J. A. Butcher Jr., R. A. Moss, W. Guo, R. C. Munjal, M. Fedorynski, Absolute rate constants for additions of phenylchlorocarbene to alkenes, *J. Am. Chem. Soc.*, 1980, **102**, 7576–7578; b) I. Naito, A. Oku, N. Otani, Y. Fujiwara, Y. Tanimoto, Formation of oxonium ylide evidenced by the laser flash photolysis of (biphenyl–4–yl)chlorodiazirine in ethers, *J. Chem. Soc., Perkin Trans 2*, 1996, 725–729.
61. J.–L. Mieusset, D. Krois, M. Pacar, L. Brecker, G. Giester, U. H Brinker, Supramolecular recognition and structural elucidation of inclusion complexes of an achiral carbene precursor in *β*– and permethylated *β*–cyclodextrin, *Org. Lett.*, 2004, **6**, 1967–1970.
62. V. Schurig, M. Jung, D. Schmalzing, M. Schleimer, J. Duvekot, J. C. Buyten, J. A. Peene, P. Mussche, CGC enantiomer separation on diluted cyclodextrin derivatives coated on fused silica columns, *J. High Resolut. Chromatogr.*, 1990, **13**, 470–474.
63. J.–L. Mieusset, G. Wagner, K.–J. Su, M. Steurer, M. Pacar, M. Abraham, U. H. Brinker, Supramolecular photochirogenesis with carbenes entrapped in cyclodextrins, *Eur. J. Org. Chem.*, 2009, 5907–5912.
64. G. N. Fickes, C. B. Rose, Reactions of bicyclo[3.2.1]octan–8–ylidene and bicyclo[3.2.1]oct–2–en–8–ylidene, *J. Org. Chem.*, 1972, **37**, 2898–2899.
65. a) W. Lwowski, Nitrenes in photoaffinity labeling: speculations of an organic chemist, *Ann N Y Acad Sci*, 1980, **346**, 491–500; b) E. L. Vodovozova, Photoaffinity labeling and its application in structural biology, *Biochemistry (Moscow)*, 2007, **72**, 1–20; c) S. Bräse, C. Gil, K. Knepper, V. Zimmermann, Organic azides: an exploding diversity of a unique class of compounds, *Angew. Chem. Int. Ed.*, 2005, **44**, 5188–5240.
66. a) M. S. Platz, Nitrenes, in *Reactive Intermediate Chemistry*, R. A. Moss, M. S. Platz, M. Jones Jr. (Eds.), Wiley, Hoboken, 501–560, 2004; b) M. P. Doyle, Synthetic carbene and nitrene chemistry, in *Reactive Intermediate Chemistry*, R. A. Moss, M. S. Platz, M. Jones Jr. (Eds.), Wiley, Hoboken, 561–592, 2004.
67. P. Walla, Supramolecular chemistry of organic azides, Ph.D Thesis, University of Vienna, Austria, 2009.
68. a) H. Quast, P. Eckert, Brückenkopfazide, Die Photolyse von 1–Azidoadamantan, ein einfacher Weg zu Azahomoadamantanen, *Liebigs Ann. Chem.*, 1974, 1727–1741; b) I. R. Dunkin, C. J. Shields, H. Quast, B. Seiferling, The photolysis of 1–azido–4–methylbicyclo[2.2.2]octane and 1–azidoadamantane in low–temperature matrices, *Tetrahedron Lett.*, 1983, **24**, 3887–3890; c) J. Michl, G. J. Radziszewski, J. W. Downing, K. B. Wiberg, F. H. Walker, D. R. Miller, P. Kovacic, M. Jawdosiuk, V. Bonacic–Koutecky, Highly strained single and double bonds, *Pure Appl. Chem.*, 1983, **55**, 315–322.
69. T. Sasaki, S. Eguchi, N. Toi, Photochemical synthesis of 4–azahomoadamant–4–enes and further studies on their reactivity in some cycloadditions, *J. Org. Chem.*, 1979, **44**, 3711–3715.

70. P. Walla, V. B. Arion, U. H. Brinker, Solvent– and temperature–tuned orientation of ferrocenyl azide inside β–cyclodextrin, *J. Org. Chem.*, 2006, **71**, 3274–3277.
71. a) R. Warmuth, S. Makowiec, The phenyl nitrene rearrangement in the inner phase of a hemicarcerand, *J. Am. Chem. Soc.*, 2005, **127**, 1084–1085; b) E. Leyva, M. S. Platz, G. Persy, J. Wirz, Photochemistry of phenyl azide: the role of singlet and triplet phenylnitrene as transient intermediates, *J. Am. Chem. Soc.*, 1986, **108**, 3783–3790; c) W. T. Borden, N. P. Gritsan, C. M. Hadad, W. L. Karney, C. R. Kemnitz, M. S. Platz, The interplay of theory and experiment in the study of phenylnitrene, *Acc. Chem. Res.*, 2000, **33**, 765–771.
72. G. Wagner, V. B. Arion, L. Brecker, C. Krantz, J.–L. Mieusset, U. H. Brinker, Controllable selective functionalization of a cavitand via solid state photolysis of an encapsulated phenyl azide, *Org. Lett.*, 2009, **11**, 3056–3058.

11

Dye Encapsulation

Jeremiah J. Gassensmith[1], *Easwaran Arunkumar*[2], *and Bradley D. Smith*[1]

[1]Department of Chemistry and Biochemistry, 251 Nieuwland Science Hall, University of Notre Dame, Notre Dame, IN 46556, USA

[2]Molecular Targeting Technologies Inc, 833 Lincoln Ave, West Chester, PA 19380, USA

11.1 Introduction

Organic dyes have been used throughout the ages as pigments in everyday items like paints, clothing, and printed matter. More recently, fluorescent dyes have emerged as crucial components in many high technology devices such as secure banknotes, lasing media, optical data storage, biomedical probes, and environmental sensors. A critical issue in all of these applications is dye performance, which can be judged by several parameters such as chemical stability, photochemical stability, colour intensity, fluorescence quantum yield, solubility, toxicity, etc. The systematic development of organic dyes has been pursued since the first reported synthesis of Indigo by von Baeyer in 1882,[1] an accomplishment which set in motion the German chemical industry.[2] In recent times, interest in fluorescent dyes has increased dramatically due to their growing impact in biotechnology and nanotechnology. It is worth noting that fluorescent dyes were a central component in the large scale sequencing methods that mapped the human genome, one of mankind's most impressive scientific achievements.[3]

The most obvious way to alter dye performance is to synthetically modify the covalent structure. However, despite the continued advances in synthetic organic chemistry, the process of dye synthesis is still a tedious task that consumes materials and human resources. As a way of circumventing this problem, supramolecular chemists have started

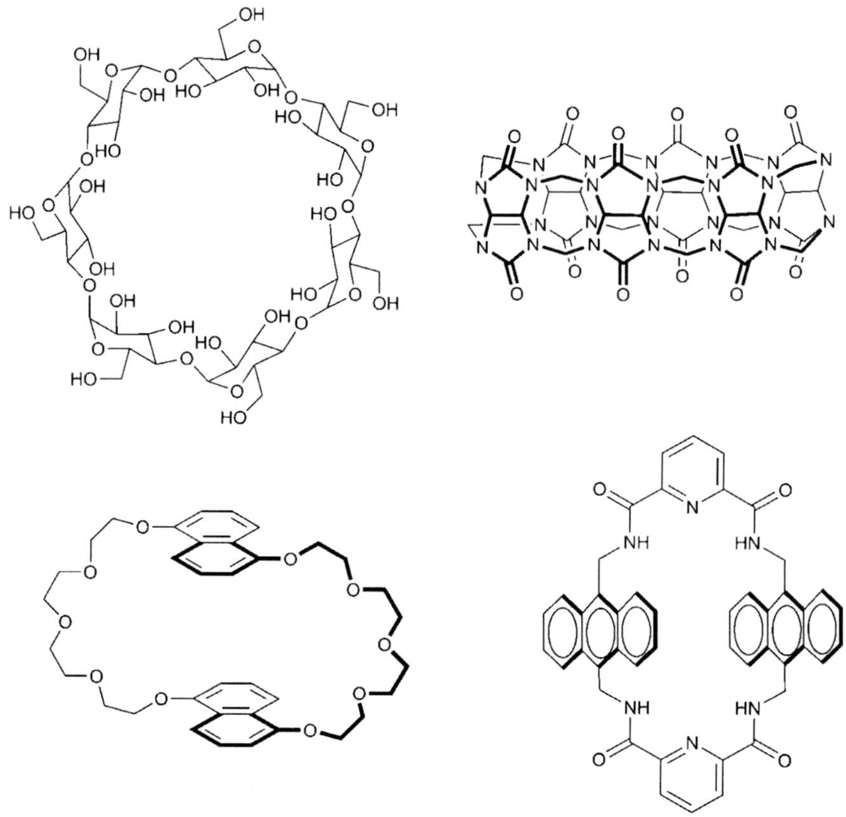

Scheme 11.1 Common container molecules: (top left) Cyclodextrin (CD), (top right) Cucurbituril (CB), (bottom left) crown ether cyclophane, (bottom right) tetralactam cyclophane

to explore non-covalent methods for altering dye properties. The focus of this chapter is on mechanical encapsulation of dyes inside: (a) protective organic container molecules such as cyclodextrins, cucurbiturils, cyclophanes, and crown ethers (Scheme 11.1), (b) biological receptors such as antibodies, aptamers and peptides, and (c) inorganic cages such as zeolites and silica nanoparticles. In some cases the dye binding process is reversible and produces a host–guest complex that is in equilibrium with the free species; in other cases the mechanical bonding is so strong that the host–guest complex can be considered a permanent molecule. Upon encapsulation, the microenvironment around the dye is changed, which often alters the absorption wavelength or the fluorescence quantum yield. Further, the dye is sterically protected, thus inhibiting intermolecular processes that often degrade dye performance like self-aggregation which reduces brightness and colour purity and photooxidation which leads to bleaching. Steric protection of long wavelength dyes is expected to be particularly helpful, because these dyes inherently have narrow HOMO–LUMO gaps and thus tend to be highly reactive.

There are various practical reasons to develop supramolecular methods of dye encapsulation. Reversible dye binding systems can be incorporated into dynamic association schemes that produce changes in colour or fluorescence emission. Thus, they can act as analytical sensing processes for environmental assays or as molecular probes for cell biology. Permanent encapsulation inside a container molecule can be achieved by several rotaxane synthesis methods that have been developed over the last twenty years.[4] These dye rotaxanes can act as bright and highly stable fluorescent probes for diagnostics and bioimaging applications. Another way to make a bioimaging agent is to use a genetically expressible bioreceptor such as an antibody or aptamer with a recognition motif that selectively binds a dye with high affinity.[5] These tagged bioreceptors can be visualized by fluorescence microscopy and used as imaging agents in cell biology research. Finally, permanent encapsulation of dyes inside porous inorganic matrices produces probes and advanced materials with many potential applications in biotechnology and nanotechnology.[6]

11.2 Reversible Dye Encapsulation Inside Organic Container Molecules

Organic container molecules are large macrocycles whose internal cavities can accommodate guest molecules. Historically, the most studied family of water-soluble container molecules are the cyclodextrins (CDs) which have non-polar cavities that can accommodate organic dyes of appropriate size and insulate them from an aqueous solvent. The three most common CDs are α-CD, β-CD and γ-CD with cavity diameters of 5.7, 7.8 and 9.5 Å respectively.[7] Encapsulation of fluorescent dyes inside a CD typically improves chemical and photochemical stability and also raises the fluorescence quantum yield. However, a drawback with CD encapsulation is the relatively weak association constant ($K_a \sim 10^3$–$10^4\,M^{-1}$) which can be readily disrupted. This has prompted a move in recent years to the cucurbit[n]uril (CB) family of container molecules which have much higher dye affinities in water and, like CDs, possess different diameters depending upon the number of repeating subunits in the macrocycle.[8] For example, the association constant for the dye Rhodamine in water is $50{,}000\,M^{-1}$ with cucurbit[7]uril (CB[7]) but only $210\,M^{-1}$ with β-CD.[9] The Rhodamine inclusion complex, in addition to possessing greater chemical stability and reduced dye aggregation, also exhibits properties distinct from uncomplexed Rhodamine dyes including longer fluorescence lifetimes and decreased nonspecific adsorption to surfaces.[10] The inclusion complex appears quite suitable for dye based lasers, which operate most efficiently with aqueous lasing media. An aqueous solution of Rhodamine 6G complexed inside CB[7] produces better beam shape profiles when compared to free Rhodamine 6G in ethanol. CB[7] can also be used in stoichiometric ratios and thus it negligibly impacts the refractive qualities of the water, unlike detergents and emulsifiers, which also produce a dye deaggregating effect, but adversely affects performance due to bubble and foam formation.[11]

Related studies have shown that CB[7] binds fused tricyclic dyes such as Proflavine, Pyronine Y, and Thionine with higher association constants than the analogous CD, and that dye encapsulation increases the fluorescence quantum yield.[12] In contrast to CB[7], the larger CB[8] encapsulates two dye molecules while the smaller CB[5] does not effectively encapsulate any tricyclic dye. In Figure 11.1, the perceived colour of the dye

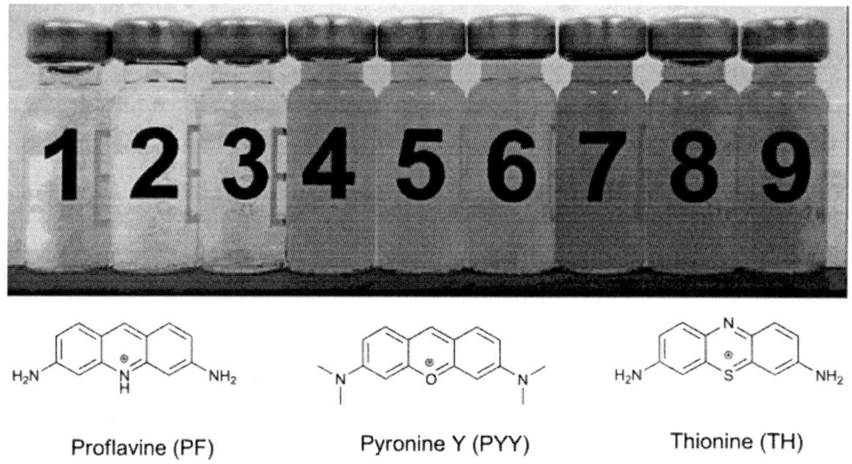

Figure 11.1 Vials containing dyes in the absence or presence of CBs. From left to right: 1) PF; 2) PF-CB[7]; 3) $(PF)_2$-CB[8]; 4) PYY; 5) PYY-CB[7]; 6) $(PYY)_3$-CB[8]; 7) TH; 8) TH-CB[7]; 9) $(TH)_2$-CB[8]. Reprinted with permission from [12]. Copyright 2008 Wiley-VCH Verlag GmbH & Co. KGaA

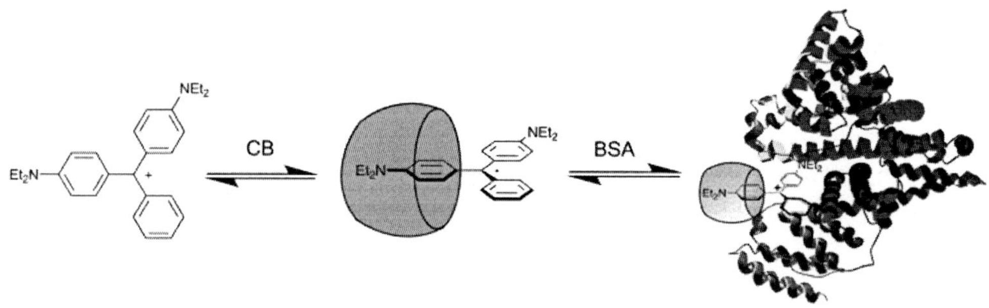

Figure 11.2 Complexation of Brilliant Green inside CB[7] causes a fluorescence increase which is further enhanced when the complex binds to the protein, bovine serum albumin. Reprinted with permission from [13]. Copyright Wiley-VCH Verlag GmbH & Co. KGaA

solution is altered by these encapsulation processes. This colour difference is attributed to changes in polarity induced by encapsulation and, when CB[8] is used, changes in dye aggregation. Separate studies have shown that CB[7] can also encapsulate Brilliant Green, a fluorescent antimicrobial trimethane dye, and form a 1 : 1 association complex with a binding constant of 17,000 M^{-1}. Intriguingly, the fluorescence quantum yield for the inclusion complex is further enhanced when it associates with bovine serum albumin (Figure 11.2).[13]

Reversible dye binding systems can be developed into colourimetric and fluorescent assays for analytical detection. As shown in Figure 11.3, two different binding schemes can be envisioned, (a) competitive dye/analyte inclusion, and (b) cooperative dye/analyte inclusion. The competitive inclusion process is the basis for the 'dye displacement assay' where the analyte displaces the dye from the container molecule. For this process to be visualized, the properties of complexed and uncomplexed dyes must be markedly dif-

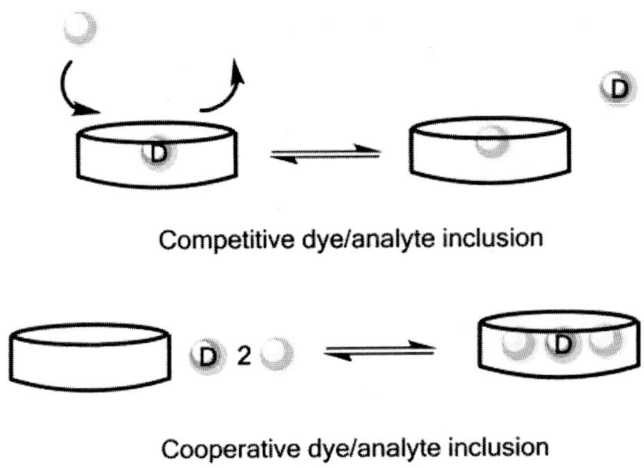

Figure 11.3 *(top) Competitive dye/analyte inclusion; (bottom) cooperative dye/analyte inclusion*

ferent. For example, CB[7] and the fluorescent dye Dapoxyl have been incorporated into an assay that monitors the activity of amino acid decarboxylase enzymes.[14] At pH 6, Dapoxyl has a relatively high affinity for CB[7] (K_a = 20,000 M^{-1}) and it is over 200 times more fluorescent when inside the CB[7] cavity. Zwitterionic amino acids have much lower binding affinities for CB[7] (K_a < 1,000 M^{-1}) and cannot displace the dye, but the presence of amino acid decarboxylase leads to decarboxylated cationic products that have substantially stronger affinities for CB[7] (~30,000 M^{-1}). The overall scheme is a 'switch off' assay, where the product of enzyme action displaces the dye from the container molecule which decreases the fluorescence intensity. In general, 'switch on' fluorescence assays are technically easier to monitor but they are harder to devise because most dyes are more fluorescent when they are encapsulated. A rare example uses the dye 2,3-diazabicyclo[2.2.2]oct-2-ene (DBO), which is less fluorescent when it is bound to a *p*-sulfonatocalix[4]arene macrocycle.[14] The action of amino acid decarboxylase leads to displacement of the DBO dye from the macrocycle and an eight-fold increase in fluorescence intensity.

The other binding scheme in Figure 11.3 is cooperative dye/analyte inclusion, where the analyte promotes association of the dye inside the container molecule. Compared to dye displacement, this is a more complicated supramolecular system because it requires the container molecule to have multi-guest recognition properties. Proof-of-concept studies have been reported using crown ether cyclophanes as container molecules that can simultaneously bind two types of guests: flat aromatic dyes and metal cations. In Figure 11.4 is a crown ether derived naphthalene macrocycle whose weak affinity for a pyromellitic diimide is increased by an order of magnitude by the presence of lithium salts.[15] The crown ether oxygens coordinate to the lithium cations which in turn promotes simultaneous dye inclusion. The result is a stunning colour change from a pale yellow to bright red due to formation of a charge transfer complex. In principle, this multi-component assembly can be employed as a colorimetric sensor for lithium cations.

A related cooperative binding system is shown in Figure 11.5. A crown ether derived macrocycle only binds a squaraine dye when sodium cations are present in the solution.[16] The sodium cations bridge the dye and the crown ether oxygens inside the macrocyclic

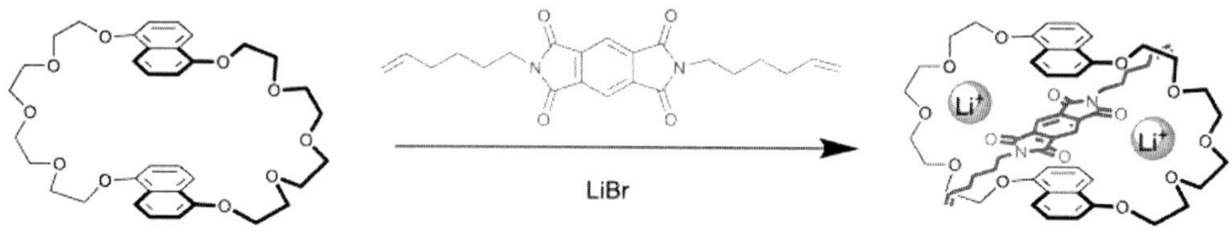

Figure 11.4 Crown ether cyclophane binds pyromellitic diimide dye only in the presence of lithium cations

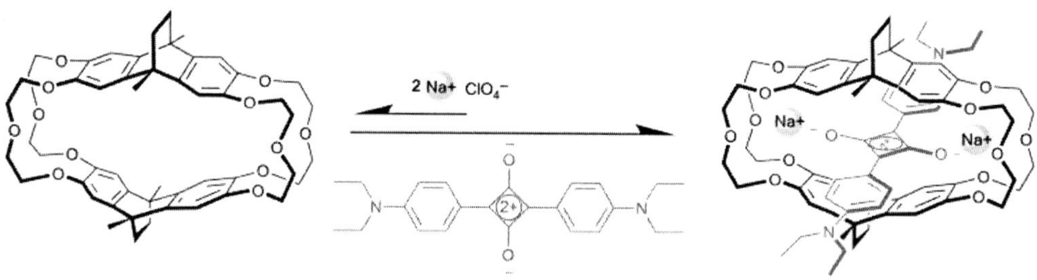

Figure 11.5 Crown ether cyclophane binds squaraine dye only in the presence of sodium cations

cavity, an inclusion process that enhances the dye's fluorescence. Addition of potassium cations leads to displacement of the sodium and consequent dye ejection from the cavity because the larger potassium ions bind to the cyclophane exterior and thus dye inclusion is no longer favoured. This multicomponent assembly system exhibits rudimentary characteristics of a combination NOT/AND logic device where the presence of sodium alone causes fluorescence signal increase, but a mixture of sodium and excess potassium or potassium alone produces no change in fluorescence.

11.3 Reversible Dye Encapsulation by Biological Receptors

There is growing interest in biological receptors for fluorescent dyes, in particular, genetically expressed receptors such as proteins or oligonucleotides.[17] A major goal of this research is to discover high affinity dye recognition motifs that can be utilized for bioimaging applications. The recognition motif can be encoded into the sequence of a larger protein or oligonucleotide of interest where the dye can be used to non-covalently label the genetically expressed target in a living cell or animal. The initial challenge is to find the appropriate dye recognition motifs using standard biotechnology screening methods such as phage display,[18] yeast display,[19] or SELEX (systematic evolution of ligands by exponential enrichment).[20] A recent example employed phage display technology to identify a 7-mer peptide sequence (IQSPHFF) with subnanomolar affinity for a fluorescent near-IR benzindolium dye (Dye A in Figure 11.6).[21] Molecular modelling indicates that the peptide sequence forms a tweezer structure with two aromatic arms surrounding the flat dye. Incorporating the IQSPHFF sequence into larger protein structures allows

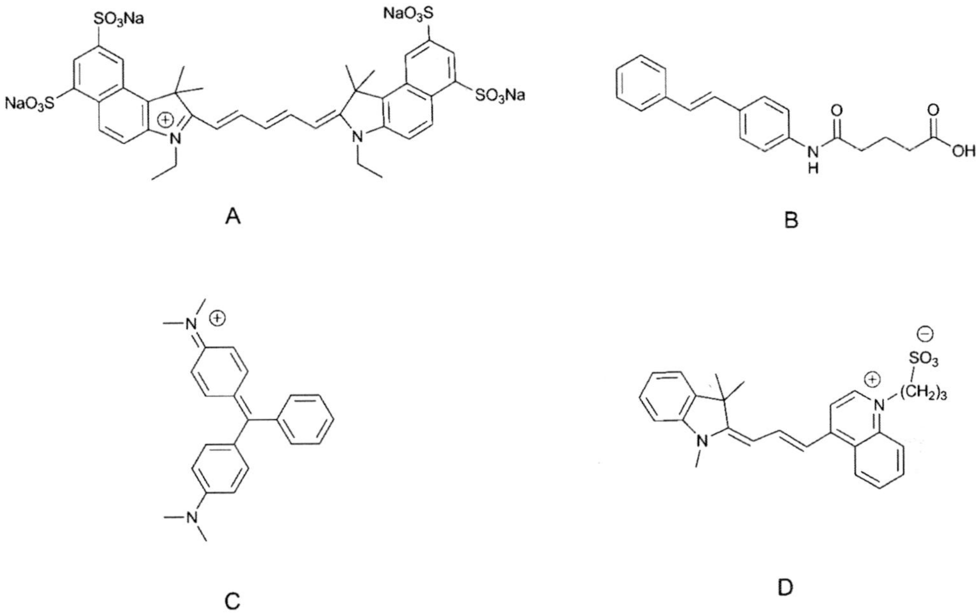

Figure 11.6 Dyes A, B, C and D

selective dye labelling in a range of assays including ELISA, flow cytometry, high throughput screens, microscopy and *in vivo* imaging. However, a drawback with this protein labelling system is the small change in the dye emission wavelength and intensity that occurs upon binding which makes it difficult to distinguish between bound and unbound dye. This means there may be an undesirably high background signal which would lower the contrast. This problem can be minimized by using a dye whose fluorescence in free solution is weak but increases substantially upon association with the bioreceptor. Several antibodies have been identified that exhibit this fluorescence activating phenomenon. One example binds a stilbene derivative (dye B in Figure 11.6) producing a blue-emitting exciplex band due to interaction of the excited state stilbene with a ground-state tryptophan residue buried in the antibody.[22] Another research programme has used yeast display technology to identify single chain antibodies as fluorescence activating proteins.[23] The antibodies selectively bind target dyes such as dye C and increase the emission intensity by factors of several thousand. There has also been progress finding RNA and DNA aptamers that bind dyes and raise quantum yields.[24] For example, a double-stranded RNA aptamer binds the cyanine dye D with $K_a = 10^7\,M^{-1}$ and enhances the fluorescence quantum yield by 60-fold.[25]

11.4 Permanent Dye Encapsulation Inside Rotaxanes

Permanent encapsulation eliminates any ambiguity due to partial dissociation of the dye from the host. The most common strategy for trapping a dye inside a container molecule

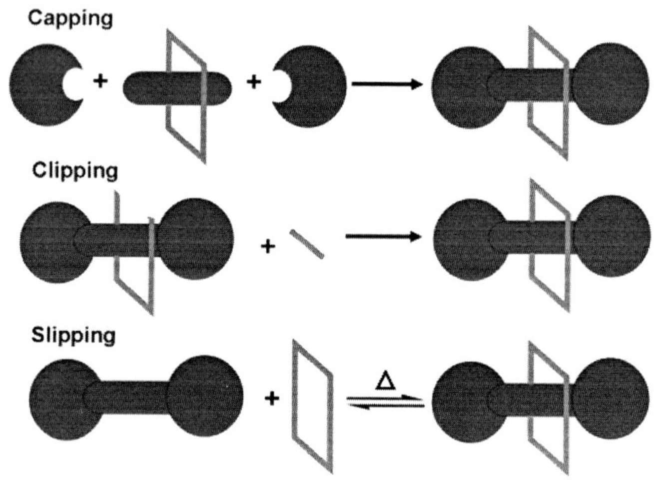

Figure 11.7 *Three strategies for rotaxane synthesis: capping, clipping and slipping*

is to form an interlocked structure known as a rotaxane. Rotaxane synthesis has improved greatly over the last few decades with the discovery and refinement of a number of templated reactions. The three most common methods for rotaxane formation are clipping, capping, and slipping (Figure 11.7), and each has its strengths and weaknesses. To date, much of the research effort on dye rotaxanes has focused on developing synthetic methods that are compatible with the reactive functional groups in organic dyes. Yields for the rotaxane formation step are often low (5–30%) and eventually they will have to be raised if dye rotaxanes are to be implemented in commercial applications. A promising approach that warrants further attention is to use solid-state synthesis methods (Figure 11.7).[26]

The pioneering dye rotaxane synthesis, reported by Anderson and coworkers in 1996, utilized Glaser coupling under aqueous conditions in the presence of a water soluble cyclophane macrocycle to produce a mixture of [2] and [3]rotaxanes with a conjugated phenylacetylene fluorophore as the axle component (Figure 11.8).[27] The purpose of the encapsulation was to insulate the conjugated π-system from quenching processes and both rotaxanes were found to be six fold more fluorescent.

Many of the subsequent preparations of dye rotaxanes have employed CD as the protective macrocycle. In each case, a threaded complex is assembled in aqueous solution and a covalent capping reaction produces the permanently interlocked rotaxane. In Figures 11.9 and 11.10 are two capping reactions that encapsulate dyes inside CD to produce acene[28] and cyanine rotaxanes.[29] Similar methods have also been used to make phenylacetylene,[30] and oligo(thiophene) rotaxanes.[31] With the larger γ-CD it is possible to simultaneously encapsulate two dyes and make homo- and hetero-[3]rotaxanes that can achieve extremely efficient energy transfer between the chromophores.[32] While water is a useful solvent for driving hydrophobic dyes into the CD cavity to make the precursor threaded complexes, it does limit the scope of the capping reactions in that they must be water compatible.

The acene [2]rotaxane in Figure 11.9 was prepared by mixing an anthracene bisboronic acid with β-CD in water and conducting a Suzuki capping reaction to produce the permanent rotaxane. Unlike the free acene axle, the rotaxane does not exhibit an excimer

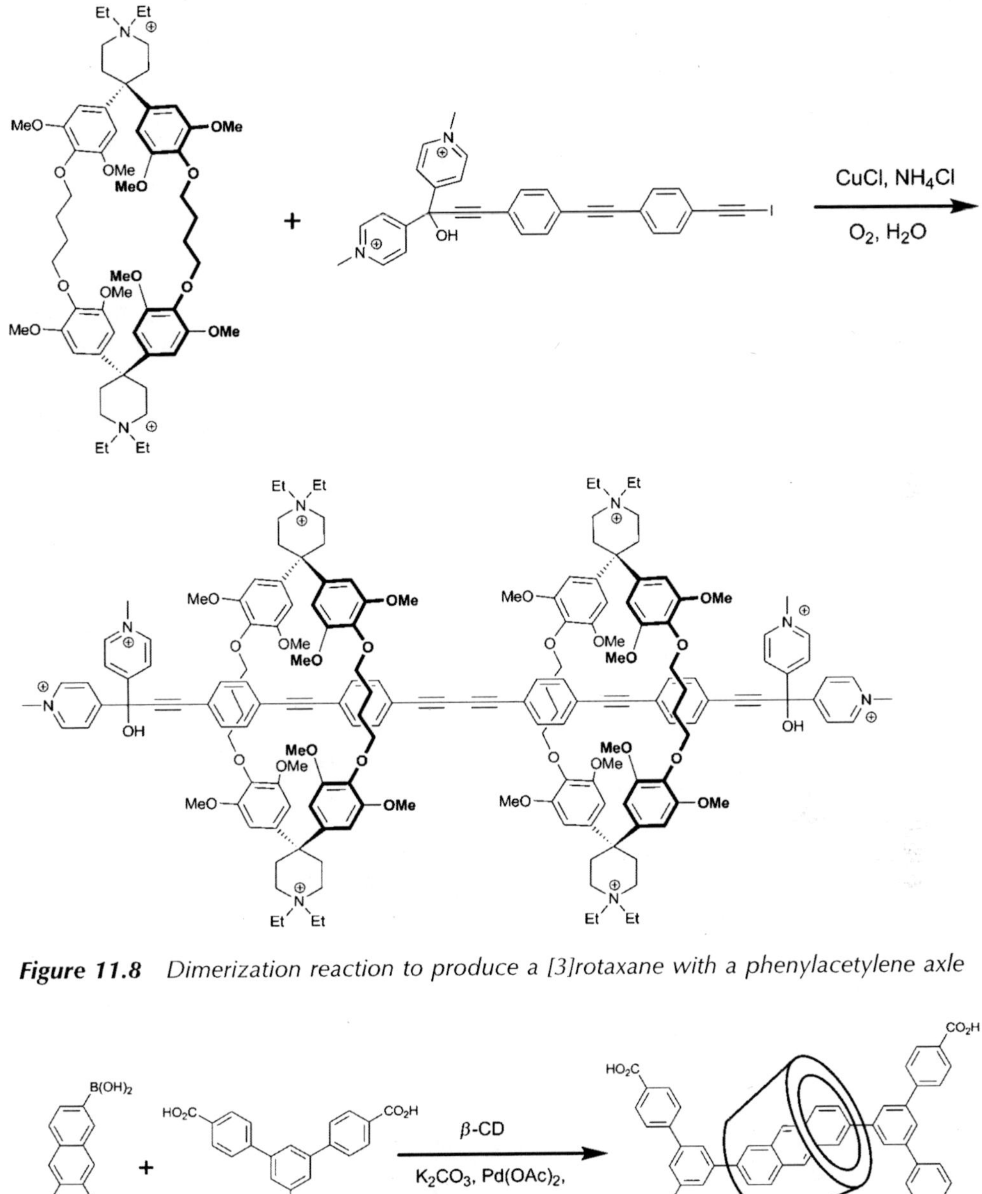

Figure 11.8 Dimerization reaction to produce a [3]rotaxane with a phenylacetylene axle

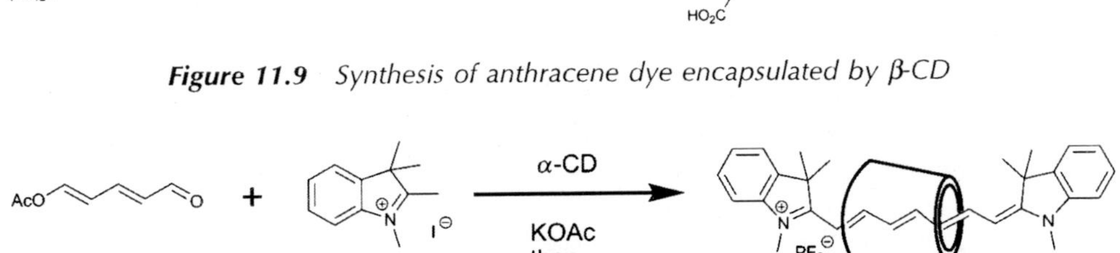

Figure 11.9 Synthesis of anthracene dye encapsulated by β-CD

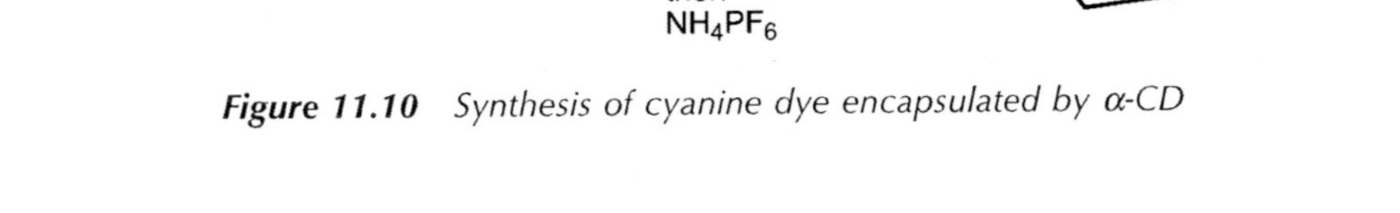

Figure 11.10 Synthesis of cyanine dye encapsulated by α-CD

emission band at elevated concentration, indicating that the surrounding CD prevents self-aggregation of the acene chromophore. However, a dye rotaxane structure does not automatically prevent all types of bimolecular reactions with the encapsulated dye. For instance, anthracene is well known to undergo a [4+4] photodimerization process as well as a [4+2] cycloaddition with singlet oxygen to produce a bridged 9,10-endoperoxide. A comparison of the reactivities of the acene axle component and the rotaxane in Figure 11.9 showed that the surrounding β-CD completely prevented the photodimerization process, but it only slowed the rate of photooxidation by about an order of magnitude. Thus, the chemical protection provided by the surrounding CD is diminished when the reactant is sterically small like molecular oxygen.

The difficulty in preventing dye photooxidation is further illustrated by the cyanine rotaxane in Figure 11.10. Cyanine dyes are a very popular class of red and near-IR fluorescent dyes but they are susceptible to photobleaching. It is known that photogenerated singlet oxygen can decompose cyanine dyes by oxidative cleavage of the conjugated double bonds.[33] The capping reaction shown in Figure 11.10 has been optimized for gram scale production of the cyanine rotaxane which has similar photophysical properties as the free dye.[34] In terms of photostability, air saturated solutions of free dye decompose 3.9 times faster than encapsulated dye, a modest amount of oxidative protection. The above two rotaxane examples demonstrate the difficulty in sterically blocking dye photooxidation. Future efforts to design high stability dye rotaxanes should be based on a clear understanding of the chemical and photochemical degradation processes and where they occur on the chromophore. By knowing what decomposition reactions to prevent, more effective encapsulation strategies can be developed.

The encapsulation of squaraine dyes as axle components inside rotaxanes has lead to some dramatic improvements in dye performance. Squaraines exhibit sharp, strong absorption and emission bands in the red and near-IR wavelengths and they are attractive dyes for bioimaging.[35] However, there are two serious limitations; they tend to form non-fluorescent aggregates and the dye's electron deficient C_4O_2 core is susceptible to chemical attack by nucleophiles.[36] Both of these drawbacks are overcome by dye inclusion inside a tetralactam macrocycle, a process that is driven by hydrogen bonding interactions with the squaraine oxygens. First-generation squaraine rotaxanes were prepared in 20–30% yield by a templated clipping reaction in weakly polar organic solvent.[37] The rotaxanes exhibit very similar photophysical properties as the precursor squaraines, but the encapsulating macrocycle protects the dye from nucleophilic attack. For example, the unprotected squaraine loses its blue colour within minutes in serum solution while squaraine rotaxanes retain their colour for days. In addition, the surrounding tetralactam macrocycle diminishes squaraine absorption band broadening by inhibiting interchromophoric interactions both in solution and solid state.

It is straightforward to conjugate cell targeting ligands to squaraine rotaxanes and produce fluorescence probes for bioimaging. For example, attachment of zinc coordinated dipicolylamine groups, which are known to target bacterial cell surfaces, produces a bright, photostable, and non-toxic probe for bacteria imaging.[38] This probe is over 20 times more photostable than an analogous probe composed of the popular cyanine dye, Cy5. The enhanced photostability allows acquisition of fluorescent movies of bacterial cell division; these movies were previously unavailable due to photobleaching of the fluorescent probe.[39]

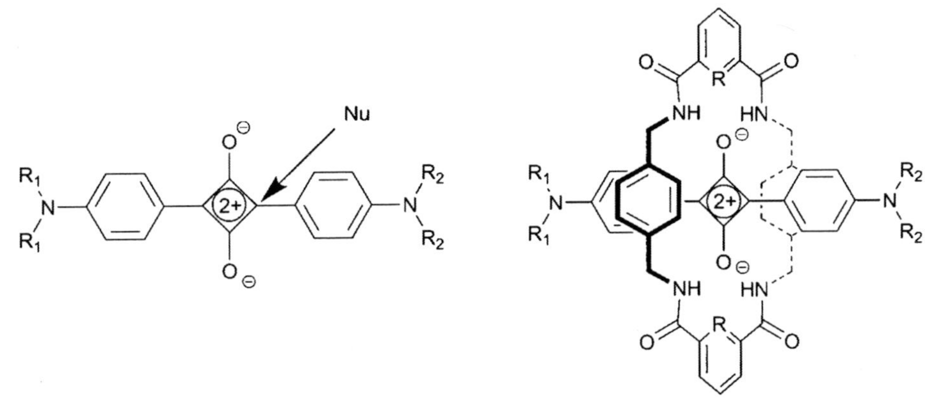

Figure 11.11 Squaraine dye is susceptible to nucleophilic attack (left), whereas the squaraine rotaxane is essentially inert (right)

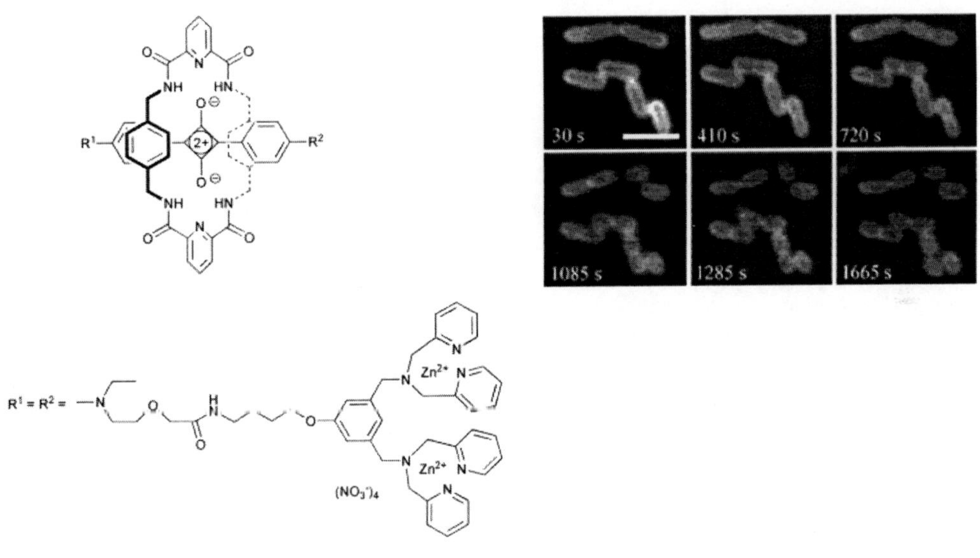

Figure 11.12 Montage from fluorescent movie showing division of bacteria cells labeled with squaraine rotaxane. Reprinted with permission from [39]. Copyright Wiley-VCH Verlag GmbH & Co. KGaA

A series of second-generation squaraine rotaxanes can be produced using the anthrylene tetralactam macrocycle shown in Scheme 11.1. The high solubility of this macrocycle in organic solvents allows squaraine rotaxanes to be produced in essentially quantitative yield by either a thermally promoted slipping process that forces the dye through the macrocyclic cavity,[40] or capping chemistry that covalently adds stopper groups to a self-assembled pseudo-rotaxane complex.[41] The tetralactam macrocycle wraps tightly around the squaraine dye and there is about a 40 nm red-shift in the emission wavelength, a phenomenon that has been observed with a related perylene diimide rotaxane system.[42]

Thus, the structure of the surrounding macrocycle is a molecular design parameter that can be used to non-covalently fine-tune the photophysical properties of the encapsulated dye.

11.5 Permanent Encapsulation Inside Inorganic Matrices

A conceptually different way to permanently encapsulate dyes is to trap them inside inorganic cages.[43] The most straightforward approach is to employ preformed molecular sieves or zeolites. If the dye is smaller than the diameter of the pores then solution and gas phase diffusion techniques can be used to load the dye molecules inside the zeolite (Figure 11.13). Shown in Figure 11.14 is a more sophisticated approach, which assembles the dye inside the zeolite (i.e. 'ship in a bottle synthesis') using a precursor molecule that is grafted to the cavity walls.[44] These dye doped inorganic materials can act as solar energy capture devices and as optical sensors. An example of the latter is a fluorescent humidity sensor based on Nile Red encapsulated inside NaY zeolite supercages.[45] The fluorescence of Nile Red is sensitive to solvent polarity and incorporation within the zeolitic pores further enhances this effect, partly because the zeolite cavities are hydrophilic.

Clay materials, like zeolites, have also been explored as naturally occurring microporous hosts for dyes.[46] Trapping the dyes within these three dimensional scaffolds provides protection from external quenchers but inhibits aggregation behaviour by immobilizing and segregating them within the framework of the material. Interestingly, dye encapsulation within micro and mesoporous materials was performed by the ancient Mayas several thousand years ago. Indigo, a naturally derived blue dye extracted from many plants of the genus *Indigofera*, has modest stability but encapsulation inside palygorskite clay produces a dye of exceptional stability, one that is inert to high and low pH solutions, photobleaching and oxidation. The packing arrangement of the dye within the clay is still a matter of investigation but computational and spectroscopic evidence suggests the zeolitic water in the clay is released upon inclusion of indigo and H-bond contacts

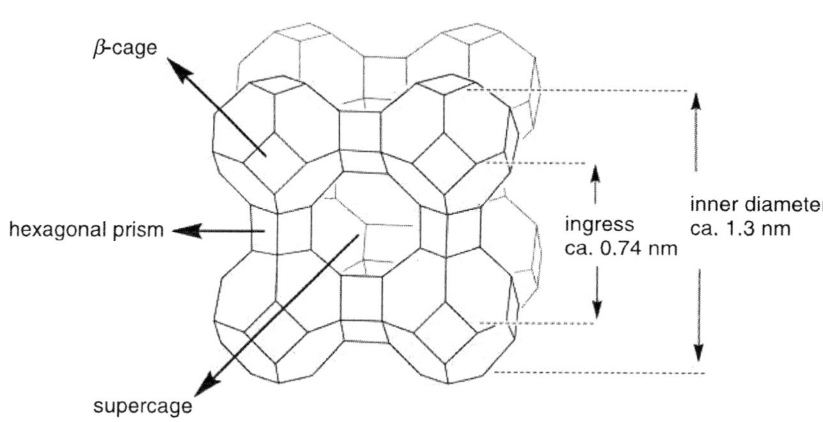

Figure 11.13 *Conceptualization of a microporous zeolite. Adapted from Reference 46*

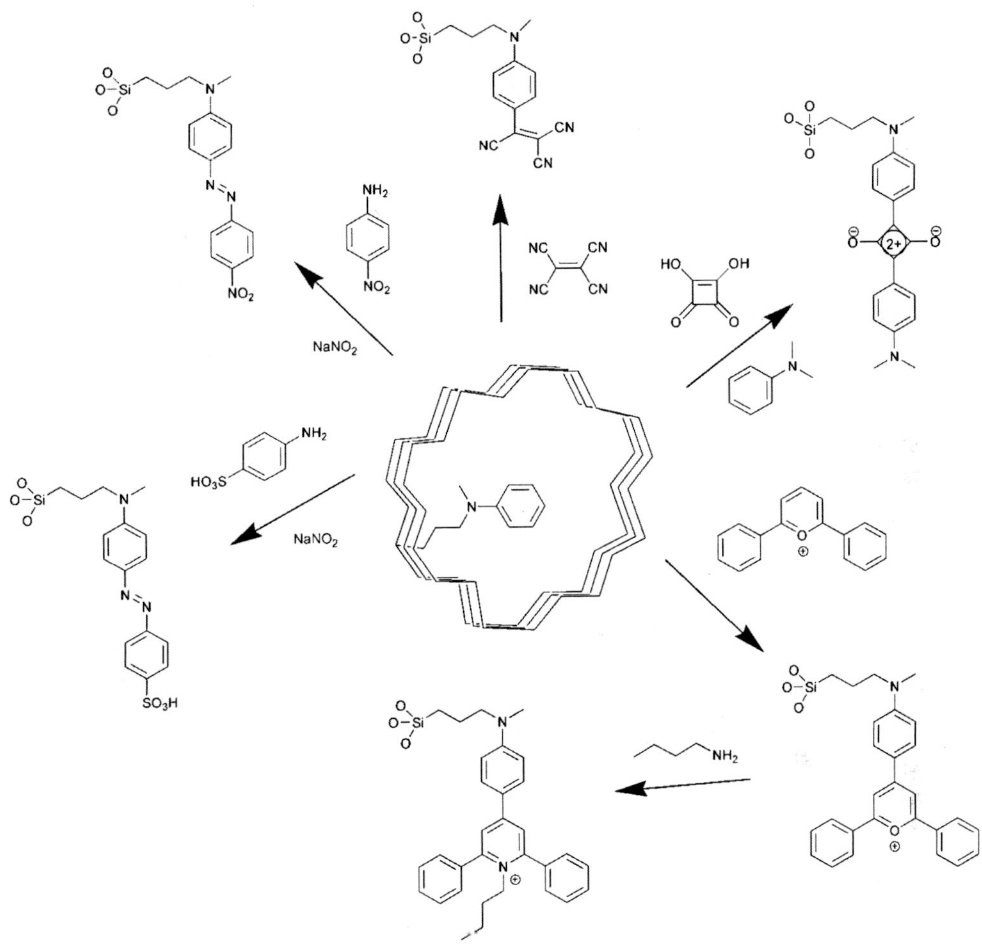

Figure 11.14 Dye synthesis inside silica cage. Adapted from Reference 44

between the dye walls of the clay and structural water molecules (those molecules that make up the scaffold of the clay) form stabilizing interactions.[47]

The desire to produce non-toxic fluorescent nanoparticles for bioimaging and diagnostic applications has lead to new methods of doping dyes within the lattice structures of silica nanoparticles.[48] The trapped dyes exhibit enhanced stability, increased fluorescence, and decreased self-aggregation. A range of related microemulsion methods have been reported to produce dye-doped and core-shell silica nanoparticles.[49] In some cases, the dyes are covalently conjugated to the silicate precursor which leads to greater dye dispersal within the nanoparticle.[50] An alternative method employs a cationic surfactant to template the formation of mesoporous silica containing the molecularly dispersed dyes. Fluorescent silica nanoparticles are potentially useful as sensors for lab-on-chip diagnostics and as probes for cell and animal imaging. Nanoparticle probes that contain photosensitizing dyes are potential candidates for photodynamic therapy.[51] Most photosensitizing dyes are hydrophobic, which leads to formulation problems and poor *in vivo* distribution. Covalently attaching photosensitizers to the nanoparticles enables convenient delivery to tumor cells without the problem of dye diffusion away from the target site. Molecular oxygen can penetrate the pores of the nanoparticles, thus irradiation of the target site after nanoparticle delivery leads to singlet oxygen release within the tumor. Additional cell targeting properties can be acquired by decorating the nanoparticle surface with affinity ligands or cell penetrating peptides.[52]

11.6 Conclusion

Most of the organic dyes that are currently employed in colourimetric and biosensing assays have been around for many decades and although they do not always have optimal performance properties they remain useful because they have low toxicity and relatively low cost. The emergence of new technologies in frontier industries like biomedicine, materials science, and nanotechnology creates a growing demand for dyes that have improved performance. One of the major attractions of molecular encapsulation is that it uses established, familiar dyes as central chromophores; however, a potential drawback with some applications is that the encapsulation matrix will add considerable size and mass to the final product. In most cases, the crucial factor that will determine the eventual scientific and societal impact of encapsulated dye products is the economic cost of fabrication, and there is a continuing need to develop high-yielding templated synthesis methods. The remarkable synthetic success that has been achieved in recent years is reason to expect further major progress. Intellectually, dye encapsulation is likely to expand as an interdisciplinary field, connecting fundamental scientists and engineers who aim to invent next-generation nanoscale devices and molecular machines.

Acknowledgements

We are grateful for financial support from the NIH and the NSF (USA).

References

1. A. Baeyer, V. Drewson, Darstellung von Indigoblau aus Orthonitrobenzaldehyd, *Ber. Chem.*, 1882, **14**, 2856–2864.
2. H. G. Schröter, A. S. Travis, An issue of different mentalities: national approaches to the development of the chemical industry in Britain and Germany before 1914, in *The Chemical Industry in Europe, 1850–1914: Industrial Growth, Pollution, and Professionalization*, E. Homburg, A. S. Travis, H. G. Schröter (Eds.), Chemists and Chemistry, Vol. **17**, Kluwer Academic Publishers, Boston, 1998.
3. S. Thangadurai, The human genome project: the role of analytical chemists., *Anal. Sci.*, 2004, **20**, 595–601.
4. E. Arunkumar, C. C. Forbes, B. D. Smith, Improving the properties of organic dyes by molecular encapsulation, *Eur. J. Org. Chem.*, 2005, **19**, 4051–4059.
5. B. N. G. Giepmans, S. R. Adams, M. H. Ellisman, R. Y. Tsien, The fluorescent toolbox for assessing protein location and function, *Science*, 2006, **312**, 217–224.

6. A. Burns, H. Ow, U. Wiesner, Fluorescent core-shell silica nanoparticles: towards 'Lab on a particle' architectures for nanobiotechnology, *Chem. Soc. Rev.*, 2006, **35**, 1028–1042.
7. J. Szejtli, Introduction and general overview of cyclodextrin chemistry, *Chem. Rev.*, 1998, **98**, 1743–1754.
8. W. M. Nau, J. Mohanty, Taming fluorescent dyes with cucurbituril, *Int. J. Photoenergy*, 2005, **7**, 133–141.
9. S. Hamai, K. Sasaki, Investigation of the formation of inclusion complexes between Rhodamine 6G and cyclodextrins by means of capillary electrophoresis using poly(vinylsulfate), *J. Inclusion Phenom. Mol. Rec. Chem.*, 2003, **45**, 19–25.
10. J. Mohanty, W. M. Nau, Ultrastable Rhodamine with cucurbituril, *Angew. Chem. Int. Ed.*, 2005, **44**, 3750–3754.
11. J. Mohanty, H. Pal, A. K. Ray, S. Kumar, W. M. Nau, Supramolecular dye laser with cucurbit[7]uril in water, *ChemPhysChem*, 2007, **8**, 54–56.
12. P. M.-Navajas, A. Corma, H. Garcia, Complexation and fluorescence of tricyclic basic dyes encapsulated in cucurbiturils, *ChemPhysChem*, 2008, **9**, 713–720.
13. A. C. Bhasikuttan, J. Mohanty, W. M. Nau, H. Pal, Efficient fluorescence enhancement and cooperative binding of an organic dye in a supra-bimolecular host-protein assembly, *Angew. Chem. Int. Ed.*, 2007, **46**, 4120–4122.
14. A. Henning, H. Bakirci, W. M. Nau, Label-free continuous enzyme assays with macrocycle-fluorescent dye complexes, *Nat. Methods*, 2007, **4**, 629–632.
15. G. Kaiser, T. Jarrosson, S. Otto, Y.-F. Ng, A. D. Bond, J. K. M. Sanders, Lithium-templated synthesis of a donor acceptor pseudorotaxane and catenane, *Angew. Chem. Int. Ed.*, 2004, **43**, 1993–1996.
16. a) S.-Y. Hsueh, C.-C. Lai, Y.-H. Liu, S.-M. Peng, S.-H. Chiu, Highly selective Na^+-templated formation of [2]pseudorotaxanes exhibiting significant optical outputs, *Angew. Chem. Int. Ed.*, 2007, **46**, 2013–2017,; b) S.-Y. Hsueh, C.-C. Lai, Y.-H. Liu, Y. Wang, S.-M. Peng, S.-H. Chiu, Protecting a squaraine near-IR dye through its incorporation in a slippage-derived [2]rotaxane, *Org. Lett.*, 2007, **9**, 4523–4526.
17. S. Achilefu, Lighting up tumors with receptor specific optical molecular probes, *Technol. Canc. Res. and Treat.*, 2004, **3**, 393–409.
18. D. R. Buckler, A. Park, M. Viswanathan, R. M. Hoet, R. C. Ladner, Screening isolates from antibody phage-display libraries, *Drug Disc. Today*, 2008, **13**, 318–324.
19. S. A. Gai, K. D. Wittrup, Yeast surface display for protein engineering and characterization, *Curr. Opin. Struct. Biol.*, 2007, **17**, 467–473.
20. S. C. B. Gopinath, Methods developed for SELEX, *Anal. Bioanal. Chem.*, 2007, **387**, 171–182.
21. K. A. Kelly, J. Carson, J. R. McCarthy, R. Weissleder, Novel peptide sequence ('IQ-tag') with high affinity for NIR fluorochromes allows protein and cell specific labeling for in vivo imaging, *PLoS ONE*, 2007, **7**, e665.
22. E. W. Debler, G. F. Kaufmann, M. M. Meijler, A. Heine, J. M. Mee, G. Pljevaljčić, A. J. Di Bilio, P. G. Schultz, D. P. Millar, K. D. Janda, I. A. Wilson, H. B. Gray, R. A. Lerner, Deeply inverted electron-hole recombination in a luminescent antibody-stilbene complex, *Science*, 2008, **319**, 1232–1235.
23. C. S.-Gyorgyi, B. F. Schmidt, Y. Creeger, G. W. Fisher, K. L. Zakel, S. Adler, J. A. J Fitzpatrick, C. A. Woolford, Q. Yan, K. V. Vasilev, P. B. Berget, M. P. Bruchez, J. W Jarvik, A. Waggoner, Fluorogen-activating single-chain antibodies for imaging cell surface proteins, *Nat. Biotechnol.*, 2008, **26**, 235–240.
24. a) B. A. Sparano, K. Koide, Fluorescent sensors for specific RNA: a general paradigm using chemistry and combinatorial biology, *J. Am. Chem. Soc.*, 2007, **129**, 4785–4794,; b) S. Sando, A. Narita, Y. Aoyama, Light-up hoechst–DNA aptamer pair : generation of an aptamer-selective fluorophore from a conventional DNA-staining dye, *ChemBioChem*, 2007, **8**, 1795–1803.
25. T. P. Constantin, G. L. Silva, K. L. Robertson, T. P. Hamilton, K. Fague, A. S. Waggoner, B. A. Armitage, Synthesis of new fluorogenic cyanine dyes and incorporation into RNA fluoromodules, *Org. Lett.*, 2008, **10**, 1561–1564.

26. H. W. Daniell, E. J. F. Klotz, B. Odell, T. D. W. Claridge, H. L. Anderson, Solid-phase synthesis of oligo(phenylene ethynylene) rotaxanes, *Angew. Chem. Int. Ed.*, 2007, **46**, 6845–6848.
27. S. Anderson, H. L. Anderson, Synthesis of a water-soluble conjugated [3]rotaxane, *Angew Chem. Int. Ed.*, 1996, **35**, 1956–1959.
28. M. T. Stone, H. L. Anderson, A cyclodextrin-insulated anthracene rotaxane with enhanced fluorescence and photostability, *Chem. Commun.*, 2007, **23**, 2387–2389.
29. J. E. H. Buston, J. R. Young, H. L. Anderson, Rotaxane-encapsulated cyanine dyes: enhanced fluorescence efficiency and photostability, *Chem. Commun.*, 2000, 905–906.
30. J. S. Park, J. N. Wilson, K. I. Hardcastle, U. H. F. Bunz, M. Srinivasarao, Reduced fluorescence quenching of cyclodextrin-acetylene dye rotaxanes, *J. Am. Chem. Soc.*, 2006, **128**, 7714–7715.
31. K. Sakamoto, Y. Takashima, H. Yamaguchi, A. Harada, Preparation and properties of rotaxanes formed by dimethyl-β-cyclodextrin and oligo(thiophene)s with β-cyclodextrin stoppers, *J. Org. Chem.*, 2007, **72**, 459–465.
32. E. J. F. Klotz, T. D. W. Claridge, H. L. Anderson, Homo-and hetero-[3]rotaxanes (with two π-systems clasped in a single macrocycle, *J. Am. Chem. Soc.*, 2006, **128**, 12432–12433.
33. A. Toutchkine, D.-V. Nguyen, K. M. Hahn, Merocyanine dyes with improved photostability, *Org. Lett.*, 2007, **9**, 2775–2777.
34. C. M. S. Yau, S. I. Pascu, S. A. Odom, J. E. Warren, E. J. F. Clotz, M. J. Frampton, C. C. Williams, V. Coropceanu, M. K. Kuimova, D. Philips, S. Barlow, J. L. Brédas, S. R. Marder, V. Millar, H. L. Anderson, Stabilization of a hetptamethine cyanine dye by rotaxane encapsulation, *Chem. Commun.*, 2008, **25**, 2897–2899.
35. B. Oswald, L. Patsenker, J. Duschl, H. Szmacinski, O. H. Wolfbeis, E. Terpetschnig, Synthesis, spectral properties, and detection limits of reactive squaraine dyes, a new class of diode laser compatible fluorescent protein labels, *Bioconj. Chem.*, 1999, **10**, 925–931.
36. a) E. Arunkumar, J. Daub, A. Ajayaghosh, Selective calcium ion sensing with a bichromophoric squaraine foldamer, *J. Am. Chem. Soc.*, 2005, **127**, 3156–3164,; b) J. V. Ros-Lis, B. Garcia, D. Jimenez, R. Martinez-Manez, F. Sancenon, J. Soto, F. Gonzalvo, M. C. Valldecabres, Squaraines as fluoro-chromogenic probes for thiol containing compounds and their application to the detection of biorelevant thiols, *J. Am. Chem. Soc.*, 2004, **126**, 4064–4065.
37. E. Arunkumar, C. C. Forbes, B. C. Noll, B. D. Smith, Squaraine-derived rotaxanes: sterically protected fluorescent near-IR dyes, *J. Am. Chem. Soc.*, 2005, **127**, 3288–3289.
38. W. M. Leevy, J. R. Johnson, C. Lakshmi, J. Morris, M. Marquez, B. D. Smith, Selective recognition of bacterial membranes by zinc(II)-coordination complexes, *Chem. Commun.*, 2006, **15**, 1595–1597.
39. J. R. Johnson, N. Fu, E. Arunkumar, W. M. Leevy, S. T. Gammon, D. Piwinica-Worms, B. D. Smith, Squaraine rotaxanes: superior substitutes for Cy–5 in molecular probes for near-infrared fluorescence cell imaging, *Angew. Chem. Int. Ed.*, 2007, **46**, 5528–5531.
40. J. J. Gassensmith, E. Arunkumar, L. Barr, J. M. Baums, K. M. DiVittorio, J. R. Johnson, B. C. Noll, B. D. Smith, Self-assembly of fluorescent inclusion complexes in competitive media including the interior of living cells, *J. Am. Chem. Soc.*, 2007, **129**, 15054–15059.
41. J. J. Gassensmith, L. Barr, J. M. Baumes, A. Paek, A. Nguyen, B. D. Smith, Synthesis and photophysical investigation of squaraine rotaxanes by clicked capping, *Org. Lett.*, 2008, **10**, 3343–3346.
42. J. Baggerman, D. C. Jagesar, R. A. L. Vallée, J. Hofkens, F. C. De Schryyer, F. Schelhase, F. Vögtle, A. M. Brouwer, Fluorescent perylene diimide rotaxanes: spectroscopic signatures of wheel-chromophore interactions, *Chem.-Eur. J.*, 2007, **13**, 1291–1299.
43. D. Wöhrle, G. Schulz-ekloff, Molecular sieve encapsulated organic dyes and metal chelates, *Adv. Mat.*, 1994, **6**, 875–880.
44. M. Comes, M. D. Marcos, R. Martinez-Manez, M. C. Millan, J. V. Ros-Lis, F. Sancenon, J. Soto, L. A. Villaescusa, Anchoring dyes into multidimensional large-pore zeolites: a prospective use as chromogenic sensing materials, *Chem.-Eur. J.*, 2006, **12**, 2162–2170.
45. I. Pellejero, M. Urbiztondo, D. Izquierdo, S. Irusta, I. Salinas, M. P. Pina, An optochemical humidity sensor based on immobilized Nile Red in Y Zeolite, *Ind. Eng. Chem. Res.*, 2007, **46**, 2335–2341.

46. G. S.-Ekloff, D. Wöhrle, B. van Duffel, R. A. Schoonheydt, Chromophores in porous silicas and minerals: preparation and optical properties, *Micropor. Mesopor. Mater.*, 2002, **51**, 91–138.
47. R. Giustetto, F. X. L. I. Xamena, G. Ricchiardi, S. Bordiga, A. Damin, R. Gobetto, M. R. Chierotti, Maya Blue: a computational and spectroscopic study, *J. Phys. Chem. B*, 2005, **109**, 19360–19368.
48. L. Wang, K. Wang, S. Santra, X. Zhao, L. R. Hilliard, J. E. Smith, Y. Wu, W. Tan, Watching silica nanoparticles glow in the biological world, *Anal. Chem. A-Pages*, 2006, **78**, 646–654.
49. X. Fang, X. Liu, S. Schuster, W. Tan, Designing a novel molecular beacon for surface-immobilized DNA hybridization studies, *J. Am. Chem. Soc.*, 1999, **121**, 2921–2922.
50. Z. Yang, S. Zheng, W. J. Harrison, J. Harder, X. Wen, J. G. Gelovani, A. Qiao, C. Li, Long-circulating near-infrared fluorescence core-cross-linked polymeric micelles: synthesis, characterization, and dual nuclear/optical imaging, *Biomacromolecules*, 2007, **8**, 3422–3428.
51. T. Y. Ohulchanskyy, I. Roy, L. N. Goswami, Y. Chen, E. J. Bergey, R. K. Pandey, A. R. Oseroff, P. N. Prasad, Organically modified silica nanoparticles with covalently incorporated photosensitizer for photodynamic therapy of cancer, *Nano Lett.*, 2007, **7**, 2835–2842.
52. S. Santra, H. Yang, D. Dutta, J. T. Stanley, P. H. Holloway, W. Tan, B. M. Moudgil, R. A. Mericle, TAT conjugated, FITC doped silica nanoparticles for bioimaging applications, *Chem. Commun.*, 2004, **24**, 2810–2811.

12

Organic Cations in Constrained Systems

Werner Abraham and Lutz Grubert
Institut für Chemie, Humboldt-Universität zu Berlin, D-12489 Berlin, Germany

12.1 Introduction

There is growing interest in artificial receptors that serve as sensors of charged species. Studies of the interaction between hosts and organic cations can also advance our understanding of the recognition properties of biological systems. Because many bioactive molecules such as the neurotransmitters acetylcholine, serotonin, nicotine, or protonated γ-aminobutyric acid are ammonium ions, the complexation of these organic cations is of particular interest. Aliphatic biogenic amines are being monitored in several foodstuffs like fish, meat cheese and wine during ageing and storage.[1]

The design of selectively bound host–guest complexes is a challenge for supramolecular chemistry. This chapter deals with host–guest systems including cyclophanes, calixarenes, homocalixarenes, and cucurbiturils as hosts. Such molecular containers (cavitands) should provide a cavity surrounded by rigid walls that enables close proximity of the guest to the wall of the host. In order to take advantage of cation–π interactions, aromatic subunits are necessary. The influences of shape, rigidity, and the chemical structure of the host molecules as well as the nature of the organic cations will be reviewed. More complex systems such as sensors and switchable host–guest complexes are applications which will play a pronounced role in the development of new receptors developed from base structures of the hosts.

12.1.1 Hosts

For binding to occur, a host must possess suitable binding sites with favourable electronic properties such as polarity and hydrogen bond donor/acceptor abilities that complement those of the guest. Cyclophanes such as **I**, calixarenes **II**, homocalixarenes **III**, and resorcinarenes **IV** fulfil these requirements (Scheme 12.1). Cucurbiturils **V** are water soluble container molecules without aromatic subunits. Therefore, both electrostatic interaction other than cation–π interaction and hydrophobic interaction play a dominant role.

The host molecules considered here possess a more or less rigid structure surrounding a cavity that can form complexes with organic cations having various shapes. The strength of binding depends on several parameters such as:

- The type of interaction between host and guest.
- The solvent used.
- The rigidity of the host.
- The anion.

1. While cation–π interaction[2] plays an important role not only in biological systems as illustrated by acetylcholine esterase,[3] artificial receptors with sidewalls containing aromatic groups also exhibit this type of interaction with organic cations. Hydrogen bonds ($^+$N—H ... X) between the guest ammonium ions, formed by protonation of amines, and the functional groups (X = O, N) of the host molecule may be stronger than other electrostatic forces such as cationic–π interaction.

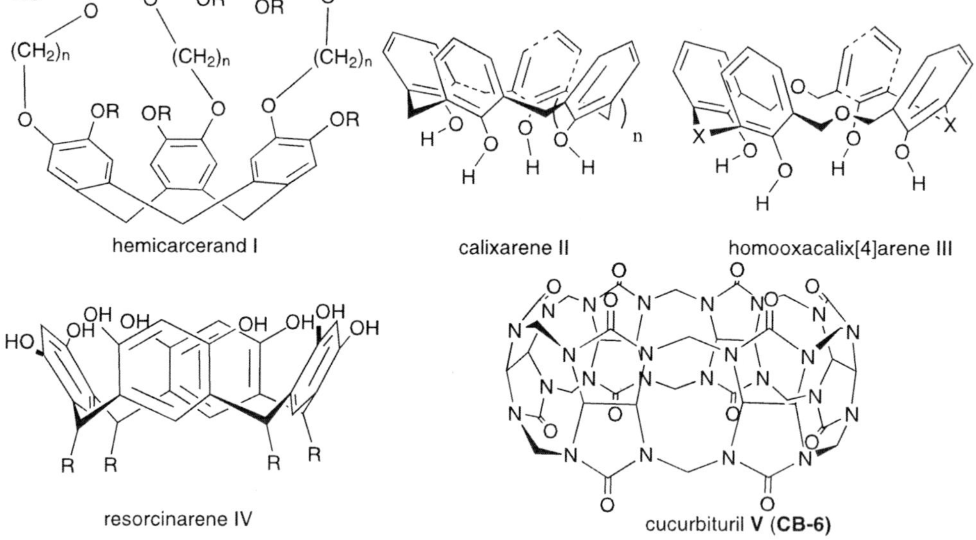

Scheme 12.1 *Host molecules*

Charged groups such as sulfanyl substituents may serve as secondary binding sites operating through salt bridges.

2. The solvent strongly influences the complexation of organic cations. The more fully the cations are solvated, the lower will be the free energy gained by the interaction between the host and the organic cation. Cationic–π interaction has the greatest influence in low polarity solvents such as chloroform. In more polar solvents, such as acetonitrile, cation–π interactions are distinctly diminished as compared with chloroform or tetrachloroethane. Above all, hydrophobic interactions play the main role in aqueous solution. Furthermore, CH–π interaction may contribute to the overall binding strength.
3. To bind guests effectively, a host must have binding sites which are complementary to the properties of the guest molecule. The electronic character of the host must match that of the guest. The conformation of the host must supply strong interactions at the proper distances and angles. If a host does not undergo a significant conformational change upon complexation, it is preorganized. On the other hand, a high flexibility of the host consumes energy which is then not available for the binding process. The overall free energy of complexation represents the difference between the unfavourable reorganisation energy and the favourable binding energy. However, strongly preorganized hosts may hinder the formation of the complexation transition state leading to slower guest binding kinetics (slow exchange results). In contrast, conformationally flexible hosts are able to adjust rapidly to complexation conditions, and both complexation and decomplexation are rapid (fast exchange).
4. Because ion pairs are formed in low polarity aprotic solvents, the anion of organic cations also contributes to the complexation. It has been found that the stability of complexes between aromatic receptors and ammonium and iminium ions is decreased in the order picrate>trifluoroacetate>I^->Br^->Cl^->tosylate>acetate.[4] Thus, the lower the cation-anion attraction, the stronger is the cation–π interaction.[5]

Hosts must provide not only a sufficient interior concave surface, but also an opening for accepting the guest. A large opening leads to a rapid in-out exchange and consequently to small binding constants. A small opening may not provide a portal large enough for the guest molecules to enter. Accordingly, gated (constrictive binding) and non-gated hosts may result.[6]

12.1.2 Guests

The organic cations of the types **A** – **D** are considered in this chapter (Scheme 12.2).

Both the charge density and the shape of the cation will mainly govern the interaction with the π-basic cavity of the hosts of types **I** – **IV** (Scheme 12.1). The accessibility of the cavity of a particular host is controlled both by the size of the cavity and by the conformational mobility of the host.

Ammonium ions **A** represent a special guest class because the cation can be formed by proton transfer in acidic solutions or by acidic functionalities such as the OH-groups of the calixarenes. The N^+H-bond allows formation of ionic hydrogen bonds with H-bond acceptors such as oxygen (e.g., at the narrower rim of calixarenes **II** or at the wider rim of resorcinarenes **IV**). In spherical tetraalkylammonium ions **B**, the positive charge is dispersed over the four alkyl groups. In contrast, the positive charge is non-symmetrically

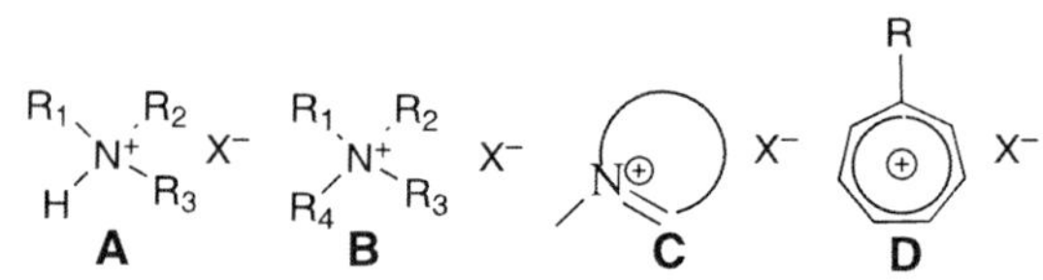

Scheme 12.2 *Types of guest molecules*

distributed in the flat aromatic iminium ions of the type **C**. The charge of aryltropylium ions **D** (R = Ar) is strongly delocalized between the tropylium moiety and the aryl substituent. Therefore, the question arises as to whether both parts of these guests are able to orient toward the interior. Additionally, both **C** and **D** may interact with aromatic units of the hosts by π-stacking.

12.1.3 Structure–Interaction Relationships

The quality of synthetic receptors toward binding of organic cations may be evaluated by thermodynamic parameters such as binding constants, enthalpy (ΔH), entropy (ΔS) and Gibbs free energy (ΔG) and also by structural parameters like complexation-induced shifts of proton resonances (CIS) as determined from NMR-spectroscopy and crystal structures. Determination of binding constants and binding energies can be determined with spectroscopic methods (UV-Vis-, NMR, and Mass spectroscopy). In most cases, K is determined with titration experiments followed by regression plot analysis of the observed data based on suitable equations.[7] UV-Vis and NMR spectroscopy are complementary methods, functioning in different concentration ranges.

Compared with spectroscopic techniques, calorimetry is a universal methodology; it functions independently of probe peculiarities, and has the advantage of yielding enthalpy as a directly observable experimental result.[8] Depending on the extent of interaction between the components from a single calorimetric experiment, the main state functions of the binding process (ΔH, ΔS and ΔG) are accessible. Because in most cases the complete set of parameters is not available, different hosts related to the binding strength and selectivity of various guests may be compared on the basis of binding constants. The CIS values taken from ^{1}H NMR titration experiments provide valuable information about the arrangement of the guest within the host. Protons deep inside the cavity of the host often are strongly shifted up-field due to the shielding effect of the aromatic sidewalls. All methods mentioned above also provide the stoichiometry of complexes.

12.2 Host–guest Complexes with Organic Cations

12.2.1 Cyclophanes

The host **I-1** (Scheme 12.3) possesses an opening that allows guests such as acetylcholine variants **B-1** and **B-2** to enter the hydrophobic cavity. The receptor has a high affinity for these ammonium ions (ΔG = −5.3 and −5.2 kcal mol^{-1}, respectively) in aqueous solution. Probably hydrophobic interactions play the main role. Small changes in the size of the host cavity significantly affect host–guest interaction.[9] CIS values revealed that a loose association is required to achieve a strong binding of quaternary ammonium ions.

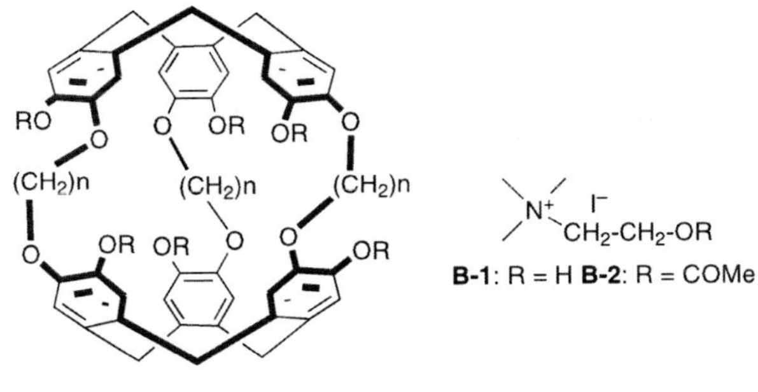

I-1 n = 5, R = CH_2COOH

Scheme 12.3 *Complexes between the cyclophane **I-1** and various organic cations*

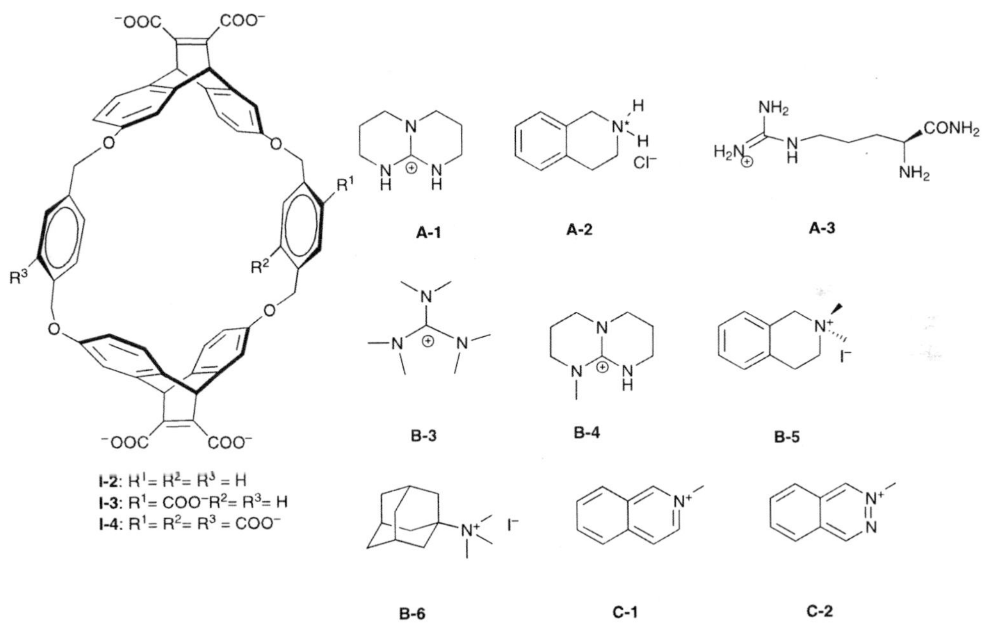

Scheme 12.4 *Different types of organic cations forming complexes with the cyclophanes **I-2–I-4***

The cyclophanes **I-2–I-4** (Scheme 12.4) were synthesized to improve the affinity for protonated amines.[10] While tetraalkylated ammonium ions are strongly complexed by **I-2**, unmodified protonated amines are not well bound. Simple protonated amines are more strongly solvated by water than alkylated ammonium ions.[10] Through the combination of interactions between host and guest, including cation–π, hydrophobic, and salt bridge interactions, the host **I-4** (Scheme 12.4) can bind even well-solvated guests such as **A-1** – **A3**, **B-3–B-5**, and **C-1**, **C-2**. Only the binding of **B-6** is not improved by salt bridge formation. Arginine-NH_2 (**A-3**) and arginine-containing dipeptides were strongly bound in water solution ($\Delta G = -6\,\text{kcal mol}^{-1}$). The effect of salt bridges in addition to cation–π interactions can be estimated as ~1 kcal mol^{-1}. The flat, delocalized π systems such as in **A-3** and **C-1** are well suited for stacking of aromatics of **I-4**. Furthermore, the cation must have a region of intense positive electrostatic potential for an electrostatic interaction to occur with the carboxylate groups in an exposed aqueous environment.

12.2.2 Calix[n]arenes

The calix[n]arenes, perhaps the major class of organic receptors in supramolecular chemistry, are simple to synthesize and can be widely modified by substitutions at the aromatic cores and the phenolic OH-groups. Therefore, they can easily be adapted for specific uses.[11–15] The vase-like conformation of calix[4] and, to a greater extent, the higher homologues calix[5], calix[6], or calix[8]arene is characterized by a high mobility due to flipping of the aromatic units. However, the mobility of the calix[4] and calix[5]arenes in particular can be restricted by substituents at the narrower rim of the cone conformation (Scheme 12.5). Other conformations, such as the 1,3-alternate arrangement of the aryl groups, do not play a significant role in binding of organic cations.

Calixarenes without ionisable substituents at the wider rim are not water soluble. Their complexation behaviour towards organic cations, has, therefore, been studied mainly in lipophilic solvents such as $CDCl_3$ and $(CDCl_2)_2$. The cation–π interaction is the main contribution to the stability of complexes and the association constants are rather low, ranging from 10 to 1000 M^{-1} (see Table 12.1).

The role of the cavity size of calixarenes is heavily predictable. For example, there is no significant hole-selectivity of hosts **II-1–II-3** (Scheme 12.6) in complexes both with the tetraalkylammonium ion **B-7** and the iminium ion **C-3** (Table 12.1). Surprisingly, the flexible methylether **II-1**, which exhibits interconversion between different conformers (cone-partial cone, 1,3 alternate), binds **B-7** whilst the fixed cone with propyl groups at the narrower rim does not.[16] Also, the different cavity sizes of the calix[4]arene **II-4** and calix[5]arene **II-5** fixed in their cone-conformation exhibit no selectivity with the iminium

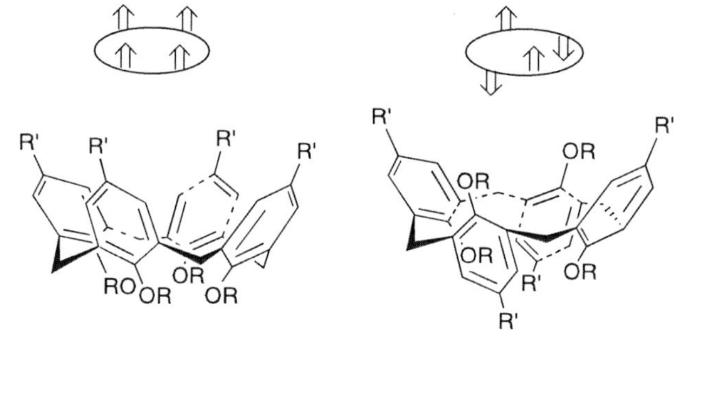

Calix[4]arene II (n = 0)
(cone conformation)

Calix[4]arene II (n = 0)
(1,3 alternate conformation)

Scheme 12.5 *Conformations of calixarenes*

Table 12.1 *Association constants (M^{-1}) of host–guest complexes in $CDCl_3$ solution*

Host	B-2	B-7	B-8	B-9	C-3	C-4	D-1	D-2	D-3
II-1[16]	–	870^a	–	–	120^a	–	–	–	–
II-2[16]	–	870^a	–	–	79^a	363	–	–	–
II-3[16]	–	650^a	–	–	200^a	–	–	–	–
II-4[18]	–	–	–	–	–	277	–	360	92
II-5[21]	210	–	–	210	–	200	–	–	–
II-6[21]	52	–	–	39	–	19	–	–	–
II-8[18]	52	–	3	28	82	190	120	213	190

– not determined, ^{a}solvent $CDCl_3/CD_3CN$ 8/2.

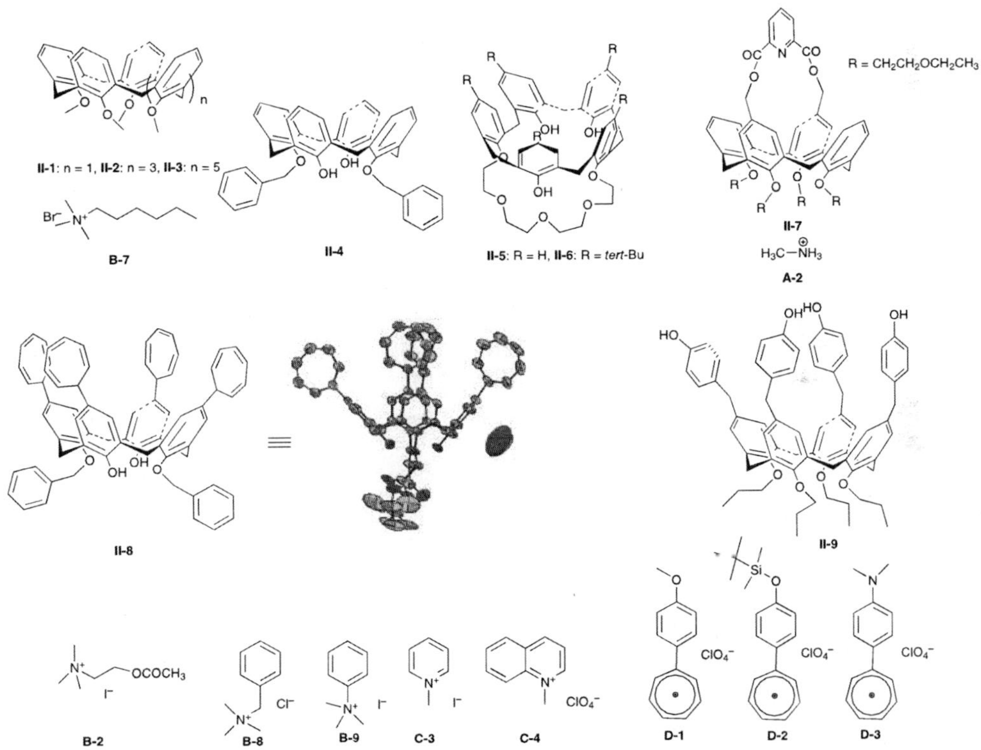

Scheme 12.6 *Calixarene as hosts for different guests*

guest (compare Table 12.1, entries 4 and 5, row **C-4**). In contrast, by restricting the accessibility of the cavity by a bridge such as in host **II-7** (Scheme 12.6) only monomethylammonium ion and not the larger di-, tri- and tetramethylammonium ions are bound.[17] Substituents at the wider rim generally have a considerable influence on the binding of organic cations. The simplest and most often found influence is the size of substituents which hinder guest binding due to steric interference. The bulky *tert*-butyl groups used in the synthesis of calixarenes diminish the association constants (compare entries 5 and 6 of Table 12.1). On the other hand, substituents may enlarge the cavity, thus supporting

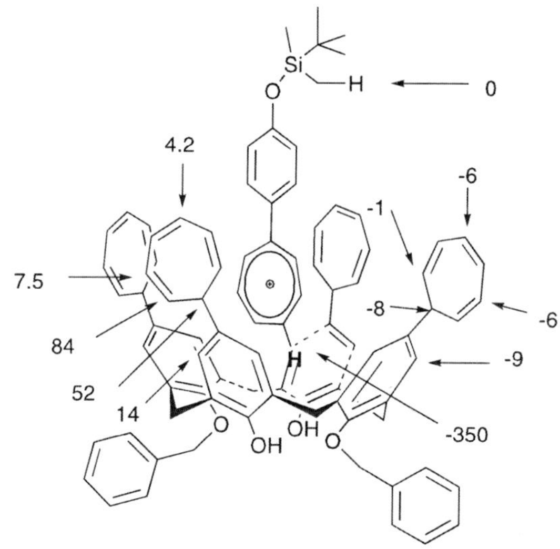

Scheme 12.7 *CIS values (in Hz) observed for different protons in the complex **II-8/D-2***[18]

binding. The cycloheptatrienyl substituents of host **II-8** (Scheme 12.6 and Table 12.1) seem to support the binding of flat iminium and aryl tropylium ions but hinder the complexation of spherical ammonium ions (see Table 12.1, entry 7). The conformation of host **II-8** exhibits a flattened cone-conformation (see crystal structure in Scheme 12.6) which is suitable in order to bind flat aryltropylium ions. According to the CIS values of different protons in both the host and the guest, the guest is sandwiched between two cycloheptatrienyl substituents of the host. The tropylium moiety is closely bound to the cavity, whereas the aryl group is pointed towards the exterior (see Scheme 12.7).[18] A comparison of the CIS values (Hz) observed for the cycloheptatrienyl substituents with those of the protons of the aromatic rings indicates that the cationic guest is located in the upper part of the calixarene cavity. Free horizontal rotation of the aryltropylium ion is slow on the NMR time scale.

The cone conformation of tetraalkoxy calix[4]arenes is not completely blocked in solution but experiences residual mobility between two C_{2v} (flattened cone) structures.[19] The residual mobility can be diminished by functionalization at the narrower rim such as the introduction of crown bridges.[19] Also the alkylation of only two OH-groups at the narrower rim seems to immobilize the cone conformation due to hydrogen bonds. Accordingly, tetraalkoxy derivatives of the host **II-8** exhibit no or very low binding properties towards ammonium and iminium ions.[18] However, a particularly high association constant of $K = 3526\ \mathrm{l\,mol^{-1}}$ was measured for the complex of the host **II-9** (Scheme 12.6) with tetramethylammonium chloride.[20] The anion was proposed to act as a hydrogen bonding acceptor towards the phenolic OH-group of the substituents.

12.2.3 Charged Calixarenes

Studies of water soluble calixarenes such as **II-10** (Scheme 12.8) in complex with organic cations in aqueous media have been most useful because this system represents a typical

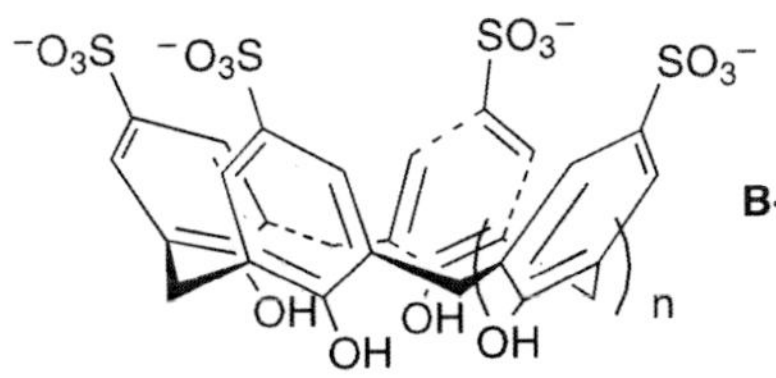

B-2, B-8, B-9

II-10: n = 1, **II-11**: n = 3, **II-12**: n = 5

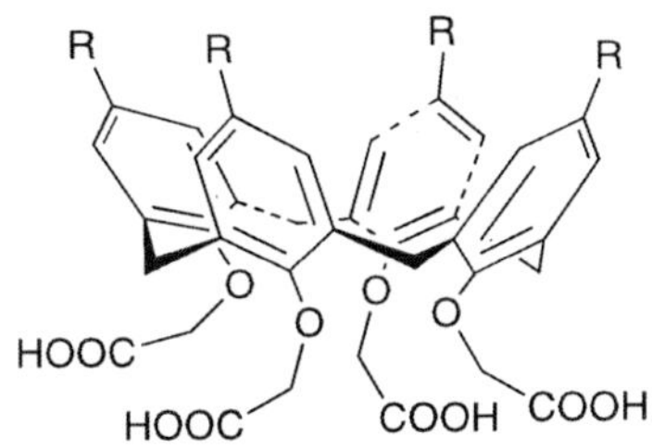

B-8, B-9

II-13: R = SO_3^-, **II-14**: R = H

Scheme 12.8 *Water soluble calix[4]arenes used for complexation studies with guests of the type* ***B***

environment where most biological processes occur.[22] *para*-Sulfonated-calix[4]arene **II-10** is cone shaped with hydrophilic wide and narrow rims separated by a hydrophobic region. This compound binds the basic amino acids lysine and arginine with association constants of ~ $10^3\,M^{-1}$ at pH 8.[23] Compounds **II-11** and **II-12**, in contrast, form only weak complexes. Acetylcholine **B-2** formed very stable complexes with association constants of up to $4 \times 10^5\,l\,mol^{-1}$ with **II-10**.[24] Notably, this affinity constant is comparable to the values observed for natural biological receptors.

A microcalorimetry study of the complexation of **II-10** showed that the mobile host is able to bind tetraalkylammonium and alkylammonium ions in water with association constants up to $47000\,M^{-1}$.[25] The binding of all organic cations is enthalpy driven, but the role of hydrophobic interaction is difficult to evaluate. The entropic term is small and positive (except for the tetraethylammonium ion) and does not correspond to the expectations of the classical hydrophobic effect. But recent studies have indicated that the 'non-classical' hydrophobic effect is controlled by the enthalpy.[26] The positive sign of the entropic term indicates that the solvation of the alkyl chain plays a role. In comparison to **II-10**, the host **II-13**, locked in the cone conformation, binds the tetraethylammonium ion with an affinity that is one order of magnitude weaker (3160 *vs.* $25100\,l\,mol^{-1}$).[27] The binding is assisted by the charged sulfanyl groups at the wider rim. The host **II-14**, which lacks these charged groups, forms a complex with the tetraethylammonium ion with an association constant of only $126\,l\,mol^{-1}$. The ditopic guest **B-9** (Scheme 12.6) is selectively complexed with the aromatic ring pointing toward the interior of the cavity while the **II-14** variant is able to recognise only the polar alkylammonium head group of **B-9** (Scheme 12.9). The complex geometry was discerned from the CIS values observed for protons of the aromatic ring and the protons of the methyl groups,[28] and indicates the

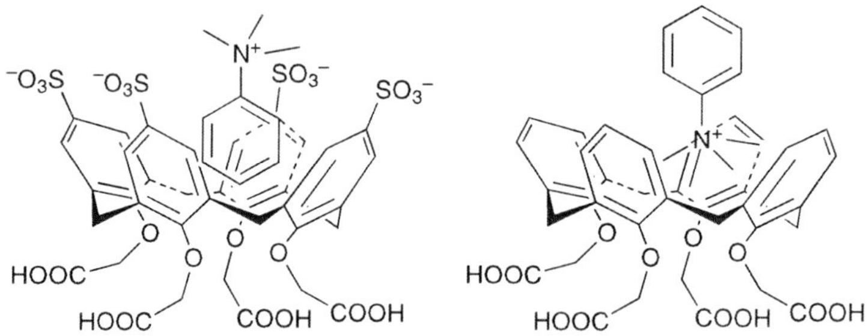

Scheme 12.9 *Structure of complexes between calixarenes and the anilinium ion*

relative importance of electrostatic cation–π interaction. For both guests, the inclusion process is enthalpy driven, but is entropically disfavoured. The negative entropic term results from a favourable term due to the desolvation of host and guest upon complexation and the unfavourable term is due to increased rigidity of the system upon inclusion of the guest. The higher stability of the hosts bearing the sulfonyl groups results from a greater desolvation of the host–guest complex. Surprisingly, the receptor **II-10** is more efficient than the rigidified hosts, which may be explained by the assumption that the host is more adaptable to the shape of the guest due to its conformational mobility.

The calix[4]arenes **II-15** and **II-16** (Scheme 12.10) are only soluble in lipophilic media such as chloroform. Accordingly, the cation–π interaction is expected to mainly control the strength of complexation of acetylcholine iodide **B-2**, chinolinium perchlorate **C-2**, and the tropylium hexafluorophosphates **D-3** and **D-4** (Scheme 12.10). Association constants determined with the help of NMR titration have revealed that the charged host **II-16** binds organic cations more tightly than does the neutral host **II-15** (compare column 3 in Table 12.2).[29] This outcome is surprising and probably the result of complementary cation–π interaction, charge transfer interaction, and π-stacking.[29] If the cationic host **II-16** binds organic cations, the question arises as to whether the cationic host interacts with itself. Dimers have been formed with an association constant of $230\,M^{-1}$ (Scheme 12.11), which demonstrates that the attractive cation–π interaction between each of the tropylium groups with the π-basic walls of the calixarene cavity is stronger than the repulsive interaction between the two positive charges.

Other calix[4]arenes combine both a host and a guest unit, such as the pyridinium-conjugated derivative **II-17** (Scheme 12.12) which forms oligomers by cation–π interaction both in solution, in the solid state, and also in the gas phase, as revealed by ^1H NMR spectroscopy, crystal structure determination, and crystallographic studies.[30]

A general problem in the determination of association constants with the help of NMR titration appeared for the complexes **II-16/B-2** and **II-16/D-3** (Table 12.2). Different association constants were obtained by using signals from different protons (e.g., protons of the tropylium moiety and the dimethylamino group of guest **D-3**). This finding can be interpreted as the formation of isomeric complexes in which different parts of the guest molecule are oriented into the interior of the host molecule (see Scheme 12.13). In such cases, the determination of the microscopic association constants of both complexes is not possible. Obviously the formation of isomeric

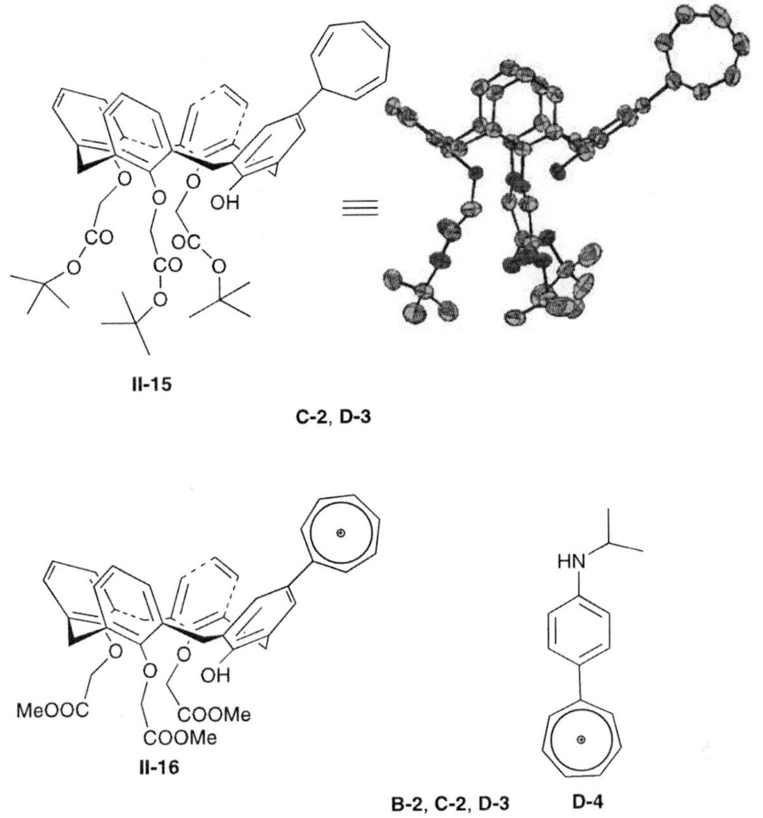

Scheme 12.10 *Modified calix[4]arene hosts and various guests studied*

Table 12.2 *Association constants (M^{-1}) of **II-15** and **II-16** with different guests in $CDCl_3$ solution*

Host	Guests			
	B-2	**C-4**	**D-3**	**D-4**
II-15	–	41 ± 3	21 ± 3 24 ± 1	–
II-16	1090 ± 76[a] 293 ± 24	540 ± 133	163 ± 20[a] 446 ± 43	122 ± 6

– not determined.
[a] association constants obtained with different proton signals of the guest

complexes is related to guests with a more elongated shape, such as acetylcholine and the aryltropylium ions.

With the help of mass spectroscopy (ESI), self-complexation of the charged host **II-16** is detectable even in the gas phase.[29] Also the complexation between the hosts **II-15** and **II-16** is monitored by the related mass peaks.

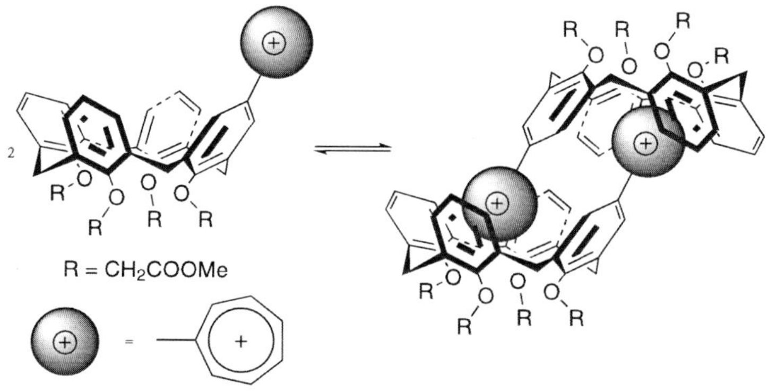

Scheme 12.11 Self complexation of the host ***II-16***

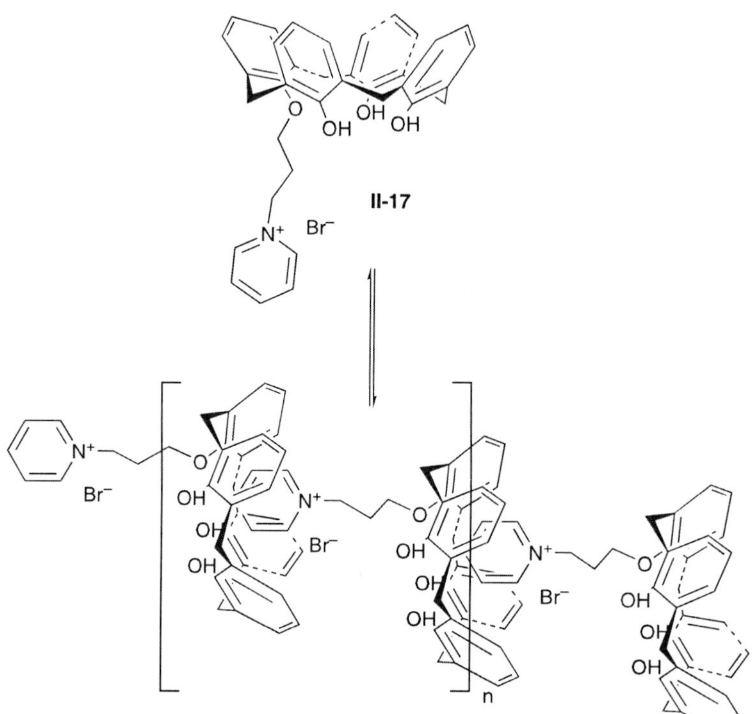

Scheme 12.12 *Oligomeric complexes of a calix[4]arene bearing an organic cation at the narrow rim*

The binding of iminium ions by neutral calixarene hosts in the gas phase has also been detected with the help of SIMS-mass spectroscopy,[31] and even binding constants have been estimated. The peak intensities observed in mass spectrometry reflect the association in a matrix and the stability of the complexes in the gas phase. Thus, mass spectrometry is a suitable tool to detect charged complexes based on cation–π interaction. Because of

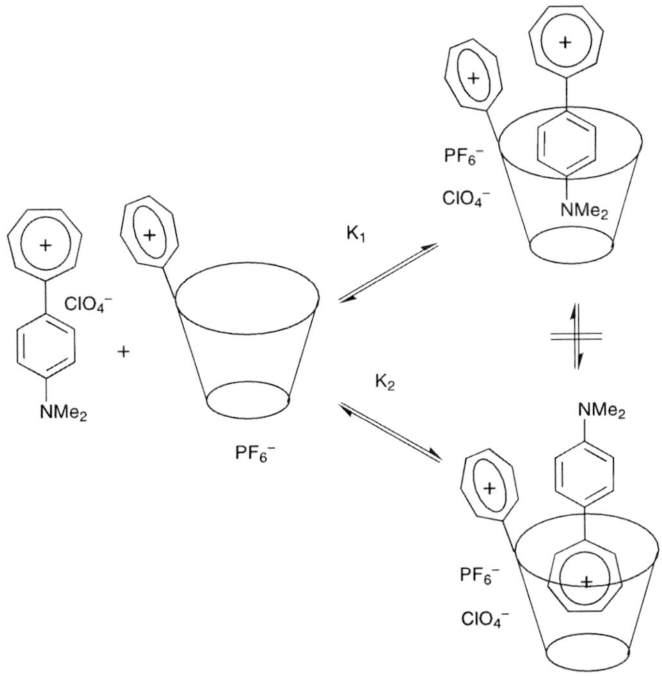

Scheme 12.13 *Complex isomers*

the absence of solvents, the selectivity of binding can be expected to be different from those observed in solution.

12.2.4 Homooxacalixarenes

The introduction of an additional atom such as oxygen into the methylene bridges has two effects: 1) The variety of cyclophanes is considerably increased (two examples are shown in Scheme 12.14). 2) Depending on the number and the distribution of the bridging oxygens, the variety of isomers and conformers is strongly increased. 3) The size of the cavity can be increased relative to the calix[4]arenes; enlarging the cavity may improve the binding strength. 4) The mobility of the cyclophanes is considerably increased.[32] Two effects may result – the higher mobility expends energy and the binding energy is diminished or the higher mobility enables the cyclophane to form the most suitable geometry of the host in the complex.

The whole set of compounds methylated at the narrower rim **III-1–III-7** (Scheme 12.14) has been synthesized and the binding of picrates of organic cations **B-2**, **B-10** **B**-11and **C-3** (Scheme 12.6, Scheme 12.14) was studied.[33] Despite having the same macrocyclic structure of the hosts, there are remarkable differences in both the measured association constants with different guests (see Table 12.3) and the CIS values extrapolated to indicate complete complexation of the guest. These results indicate that the cavity must be somewhat differently arranged in the complexes with different cations due to the balance between cation–π interaction and the strain energy of the host.

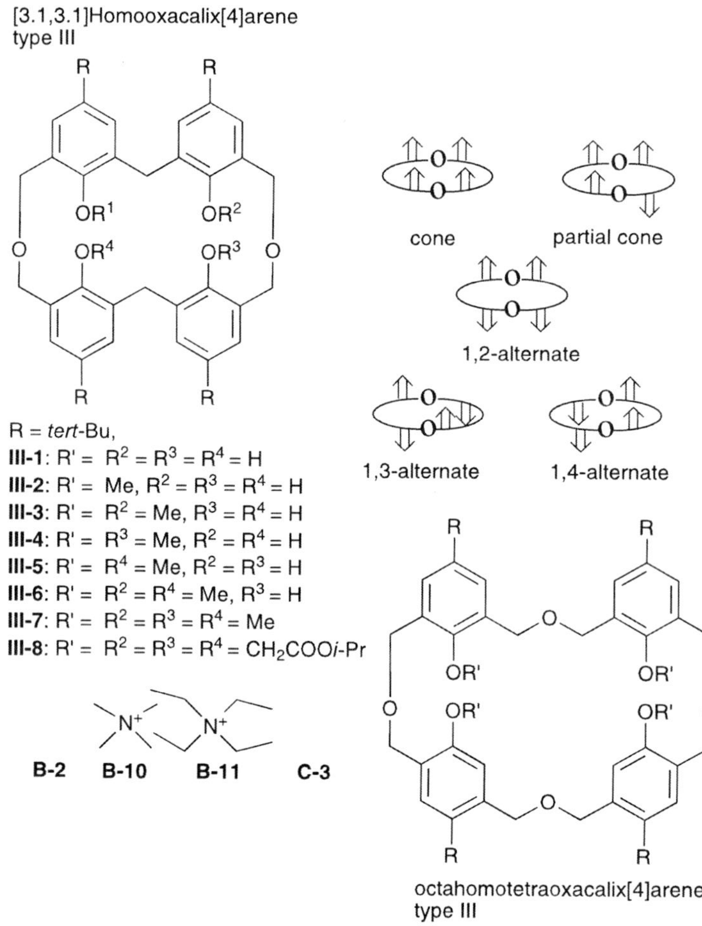

Scheme 12.14 *Homocalixarenes as guests for organic cations of the type B and C*

Table 12.3 *Association constants of complexes in $CDCl_3$ solution (M^{-1})*

host	Guests			
	B-2	**B-10**	**B-11**	**C-3**
III-1	10	15	15	30
III-2	60	110	40	120
III-3	55	130	11	90
III-4	190	740	15	130
III-5	110	410	7	80
III-6	140	490	4	70
III-7	75	220	3	25
III-8	1000	3200	–	650

However, the differences in host geometry are much smaller in complexes than in free hosts.[33,34]

If one assume that similar changes in solvation and in cation-anion attraction occur through complexation of a given guest by **III-1–III-7**, then the association constants reflect the contributions of the reorganisation energy of the mobile host and the cation–π interactions. The hydrogen bonds at the narrower rim, and therefore both the position and the number of OH-substituents, help determine the shape of the host cavity in the host–guest complex.

12.2.5 Resorcinarenes

Resorcinarenes have some degrees of freedom for rotation around the C—C bonds of the connecting CH_2-groups, which results in five conformations, namely crown, boat, chair, diamond and saddle.[35] The hydroxy groups at the wider rim must be linked in order to fix the receptor in the specific cavitand conformation and to improve binding properties.

Phenolate resorcin[4]arenes of the type **IV-1** (Scheme 12.15) are efficient hosts for organic ammonium ions in protic solvents in strongly alkaline organic media by simultaneous charge–charge and cation–π interaction.[36] A drawback of receptors such as **IV-2** is the low selectivity towards alkyltrimethylammonium and bulkier tetraalkylammonium ions.[37]

Increasing the wall of the resorcinarenes, as in **IV-3**, could improve the selectivity.[38] In protic solvents, **IV-3** could combine the selectivity of cavitands with the high affinity of the resorcin[4]arene. The cavitand is built up if the three diaminophenyl groups adopt an axial conformation. The cavity with an open side binds selectively alkylammonium ions complementary in size and shape (see Figure 12.1 and Table 12.4).

The association constants were determined with the help of a competitive displacement assay using **B-17** as a fluorescence indicator[38] (see Figure 12.2). Small changes in size have large consequences because the guest is inserted deeply in the cavity. The long chains present in **B-13** can protrude due to the top openings of the host **IV-3**, resulting in high affinity (Table 12.4). In contrast, tetraalkylated ammonium ions, and NH_4^+ itself, exhibit dramatically weaker complexation behaviour, indicating a bad fit with the cavity of **IV-3**. The host **IV-4** exhibits selectivity in binding naturally occurring ammonium ions in D_2O-saturated $CDCl_3$ solution.[39] Choline **B-1** ($K = 420\,M^{-1}$) and carnitine **B-16** ($K = 2050\,M^{-1}$), compounds which are poorly soluble in water-saturated chloroform, are taken up by this host, but acetylcholine **B-2** is not.

The four-wall water-soluble cavitand **IV-5** (Scheme 12.16) forms a stable deep cavity that has a high affinity towards choline **B-1** and acetylcholine **B-2** (association constants$>10^4\ M^{-1}$) in D_2O.[40] Interestingly, the trimethylammonium head of **B-1** is pointed inward into the cavity of the host. In contrast, the long chain trialkylammonium ions, such as dodecyltrimethylammonium ion, are bound with the (helical folded) alkyl chain in the cavity, but with an association constant of $>10^4\ M^{-1}$. In addition to the favourable electrostatic interaction between the tetracarboxylate wider rim and the tetraalkylammonium core, hydrophobic interactions dominate the binding in this case and reveal the bitopic character of both the host and the guest. Isothermal titration calorimetric data have shown that the complex formation is both enthalpically and entropically driven.

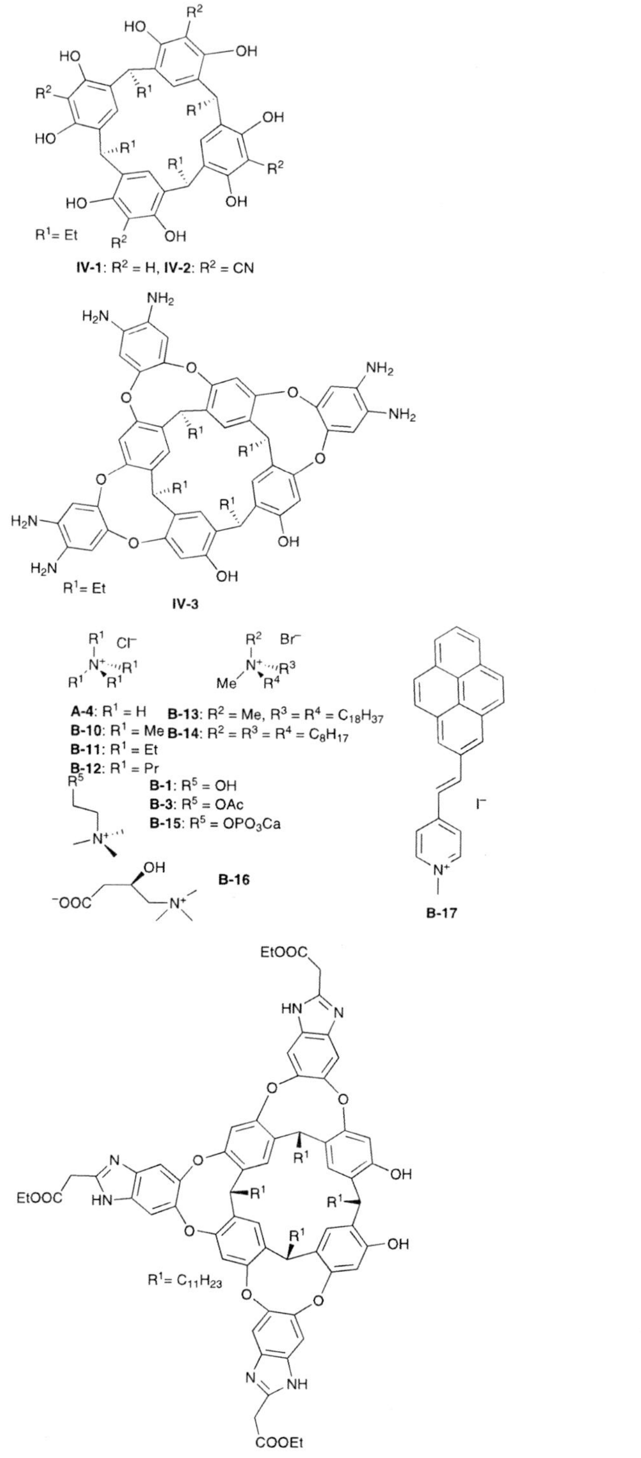

Scheme 12.15 *Resorcinarenes and organic cations studied in complexes*

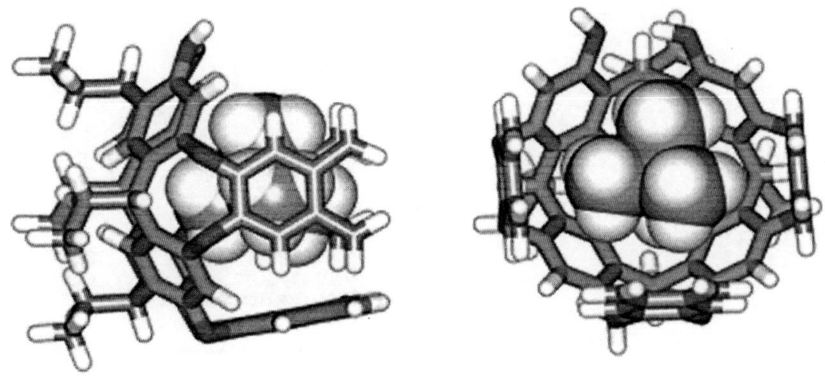

Figure 12.1 Front and side view of the optimized model of the ***IV-3/B-10*** complex. Reprinted with permission from [38]. Copyright 2006 American Chemical Society

Table 12.4 Association constants of complexes in $CDCl_3$ solution (M^{-1})

Guest	K [10^5 M^{-1}]	Guest	K [10^5 M^{-1}]
A-4	<0.001	**B-12**	<0.001
B-1	3.2 ± 0.9	**B-13**	0.1 ± 0.001
B-3	1.0 ± 0.1	**B-14**	<0.001
B-10	5.1 ± 0.6	**B-15**	3.7 ± 0.2
B-11	1.4 ± 0.4	**B-16**	3.3 ± 0.8

Chiral discrimination of the host–guest interaction can be measured in the gas phase. Resorcin[4]arenes bearing chiral substituents are able to select chiral quaternary ammonium ions. In these cases, complexation experiments were conducted by means of electrospray ionization (ESI) mass spectrometry.[41]

12.3 Extended Hosts and Capsules

Complexation properties of calixarene hosts may be improved not only by enlarging the wall of the host but also by connecting two host molecules by bridges (see Scheme 12.17). Compared with the 'mono'-calix[4]arene **II-18**, the bridged hosts **II-19** – **II-22** exhibit a clear increase of the association constants determined with the help of NMR titration in $CDCl_3$ solution (see Figure 12.3).[42] In lipophilic media, ion pairs are complexed, and thus, different salts of the guest molecules were studied in order to compare complexation properties. As expected, both the charge distribution within the cation and the steric requirements of the ion pair, in which the anion is tightly bound to the cation, control the binding process. The spherical tetramethylammonium ion **B-10** can bind with only minor steric adjustments and no such large differences of the association constants of different ion pairs are observed (Table 12.5). From CIS values, the structures given in Scheme 12.18 are proposed.

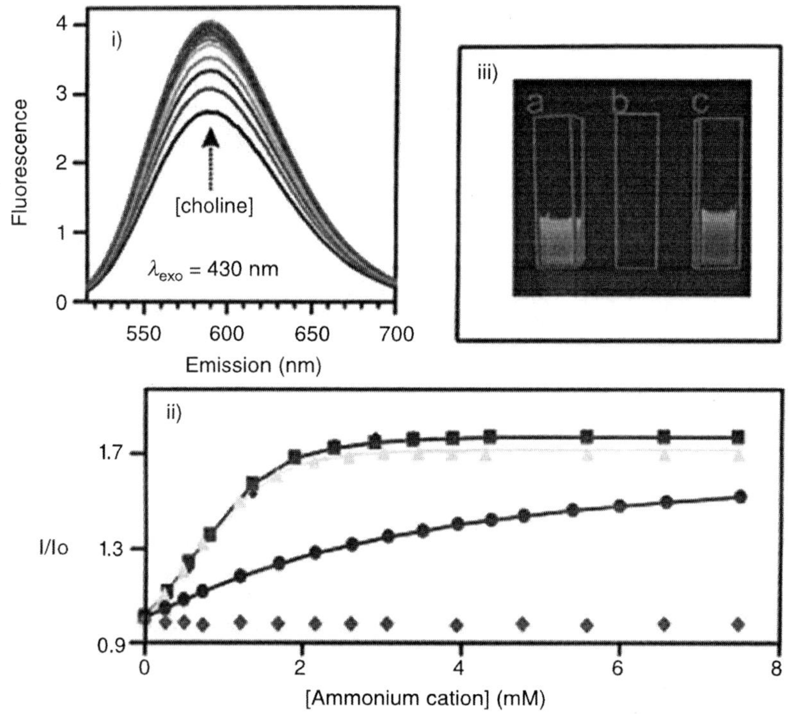

Figure 12.2 *(i) Fluorescence spectra in MeOH of **B-17** (0.184 mM) in the presence of **IV-3** (0.17 mM) upon addition of incremental amounts of **B-1** (0-2.5 mM). (ii) Plots of the increase of the relative fluorescence intensity of **B-17** in competitive binding experiments with **B-15** (circles), **B-16** (squares), **B-13** (triangles), and **B-14** (diamonds) fitted to a competitive binding scheme of two 1 : 1 complexes. (iii)Visual fluorescence regeneration of **B-17**: (a) **B-17** (0.184 mM), (b) mixture of **B-17** (0.17 mM) and **IV-3** (2.5 mM), and (c) mixture of **IV-3** (0.15 mM), **B-17** (0.17 mM), and **B-15** (0.43 mM). Reprinted with permission from [38]. Copyright 2006 American Chemical Society*

The examples shown above demonstrate the improved binding properties caused by connecting two calixarene cavities by bridges at the wider rim. A different feature of binding of type A ammonium ions was reported for calix[5]arenes connected by bridges at the narrower rim (Scheme 12.19).[43] The ^{1}H NMR titration experiments with long chain alkylammonium ions such as **A-5** reveal the nearly exclusive presence of polycapsular assemblies in $CDCl_3/CD_3OD$ (9 : 1) solution. Another type of encapsulation of long-chain dialkylammonium salts was observed with the heterotetratopic host **II-24** (Scheme 12.19).[44]

Very high association constants ($>10^6$ M^{-1}) for the formation of the capsular 1 : 1 complex in CD_2Cl_2/CD_3OD (2 : 1) solution were estimated for alkyl chains of C_{12}–C_{16}. The ureido functionality of the bridge at the wider rim of host **II-24** supports the complexation in nonprotic solvents by binding with anions such as chloride, picrate, or carboxylate (see Figure 12.4). Shorter alkyl chains are encapsulated by only a single cavity.

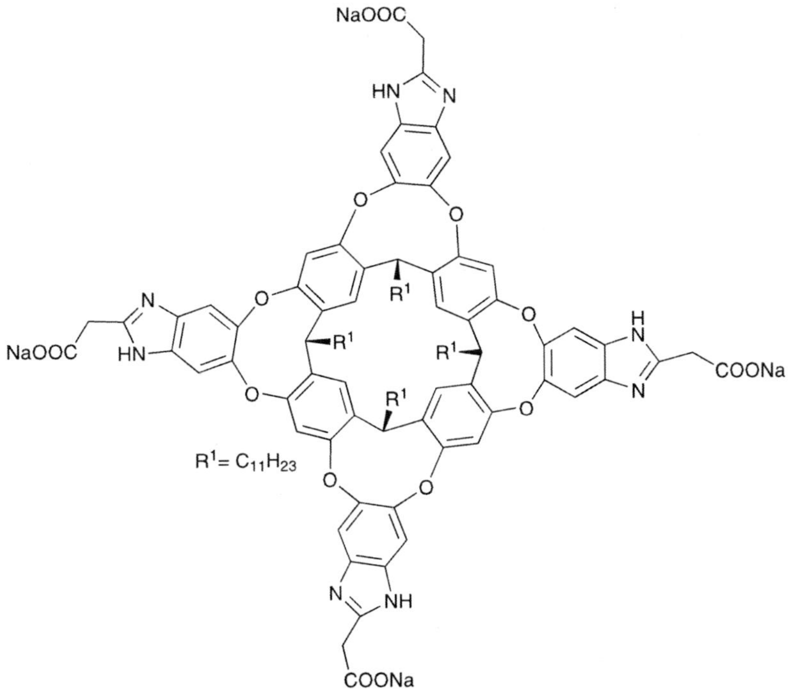

IV-5

Scheme 12.16 *The four-wall water-soluble cavitand* ***IV-5***

The formation of capsular complexes is further corroborated by ESI mass spectra of equimolar host–guest mixtures.

While the capsules described above were formed by, at least, bitopic hosts with covalently connected calixarene moieties, capsules are also formed through noncovalent bonds such as hydrogen bonds if suitable guests are present. The encapsulation of a guest such as the tropylium ion **D-5** is necessary since an empty cavity would be thermodynamically disadvantageous. Dimeric hydrogen-bonded capsules are formed from resorcinarenes bearing hydroxyl substituents at the wider rim both in the solid phase and in solution in the presence **D-5** (Scheme 12.20). The high thermodynamic and kinetic stability of the complex **IV-6/D-5/IV-6** arises from the cooperation of eight CO ... HO hydrogen bonds and the cation–π interaction between the tropylium ion and the resorcinol units (Scheme 12.20).[45]

While the capsule **IV-6/D-5/IV-6** is stabilized by hydrogen bonds, the cocrystallization of resorcinarene **IV-7** with the rigid diquaternary alkylammonium salt **B-18** yields a nanotubular structure held together by intratubular $\pi \ldots \pi$-interaction (see Figure 12.5).[46] Use of the bromide anion, a weaker hydrogen bond acceptor (compared with chloride), drives the assembly process through weaker $\pi \ldots \pi$ and cation–π interactions instead of hydrogen bonding. In contrast, capsules can be formed from resorcinarenes with the diethylammonium ion **B-11**.[47]

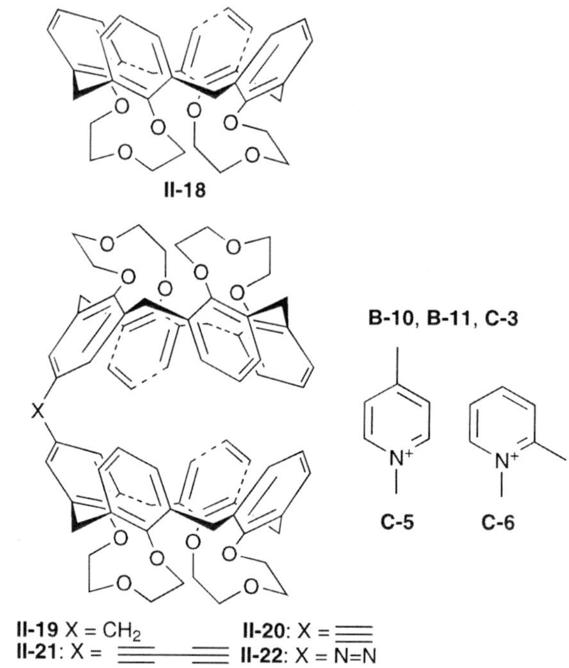

Scheme 12.17 *The rigid calix[4]arene **II-18** and its extended variants*

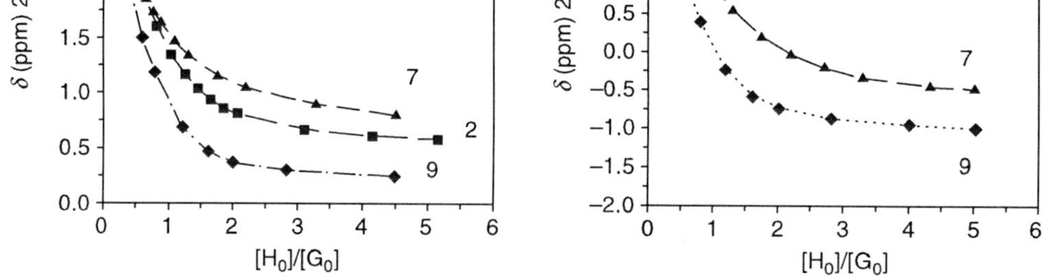

Figure 12.3 *^{1}H NMR titrations of (a) **C-6** iodide and (b) **C-5** iodide with host **II-18** (1), **II-19** (2), **II-20** (7), **II-21**(8), and **II-22** (9) at T = 300 K in $CDCl_3$; symbols denote experimental data points relative to the 4-Me (a) and 2-Me (b) signals, respectively, of the guest; lines are best-fit curves calculated by non-linear regression. Reprinted with permission from [42]. Copyright 2000 Wiley-VCH Verlag GmbH & Co. KGaA*

Table 12.5 Association constants K (M^{-1}) of 1 : 1 complexes ($CDCl_3$)

Host	Guests	K
II-18	**B-10** tosylate	33
II-18	**B-10** chloride	80
II-18	**C-3** iodide	418
II-18	**C-3** tosylate	113
II-18	**C-5** iodide	98
II-19	**B-10** tosylate	101
II-19	**B-10** chloride	340
II-19	**C-3** iodide	>10000
II-19	**C-3** tosylate	1405
II-20	**B-10** tosylate	1407
II-20	**C-3** tosylate	1232
II-20	**C-3** iodide	7260
II-22	**B-10** tosylate	54
II-22	**C-3** iodide	797
II-22	**C-5** iodide	2450

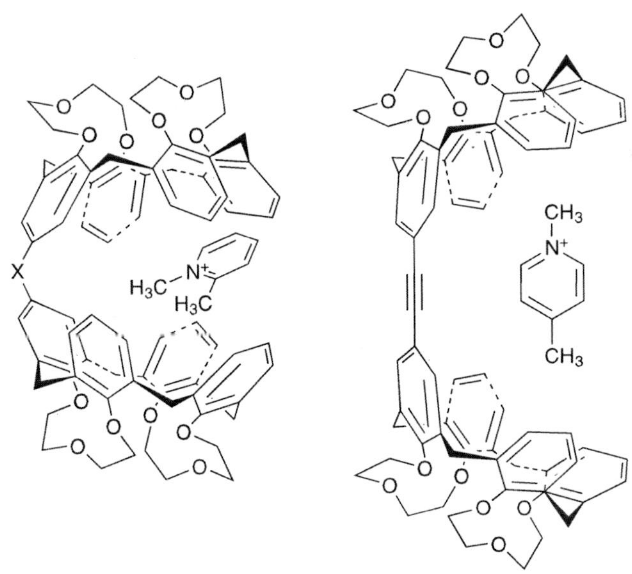

Scheme 12.18 *Proposed structures of the complexes between **II-19** and **C-6** iodide and **II-20** and **C-5***

The calix[4]arene **II-25** (Scheme 12.21) is able to form capsules because of ureidoyl substituents at the wider rim. Tropylium salts form charge transfer complexes with both the monomer and dimers of **II-25** in nonpolar solvents.[48] Due to hydrogen bonds formed in the equatorial region, the affinity of the tropylium ion towards the capsule formed by the dimer is much greater than that of benzene.

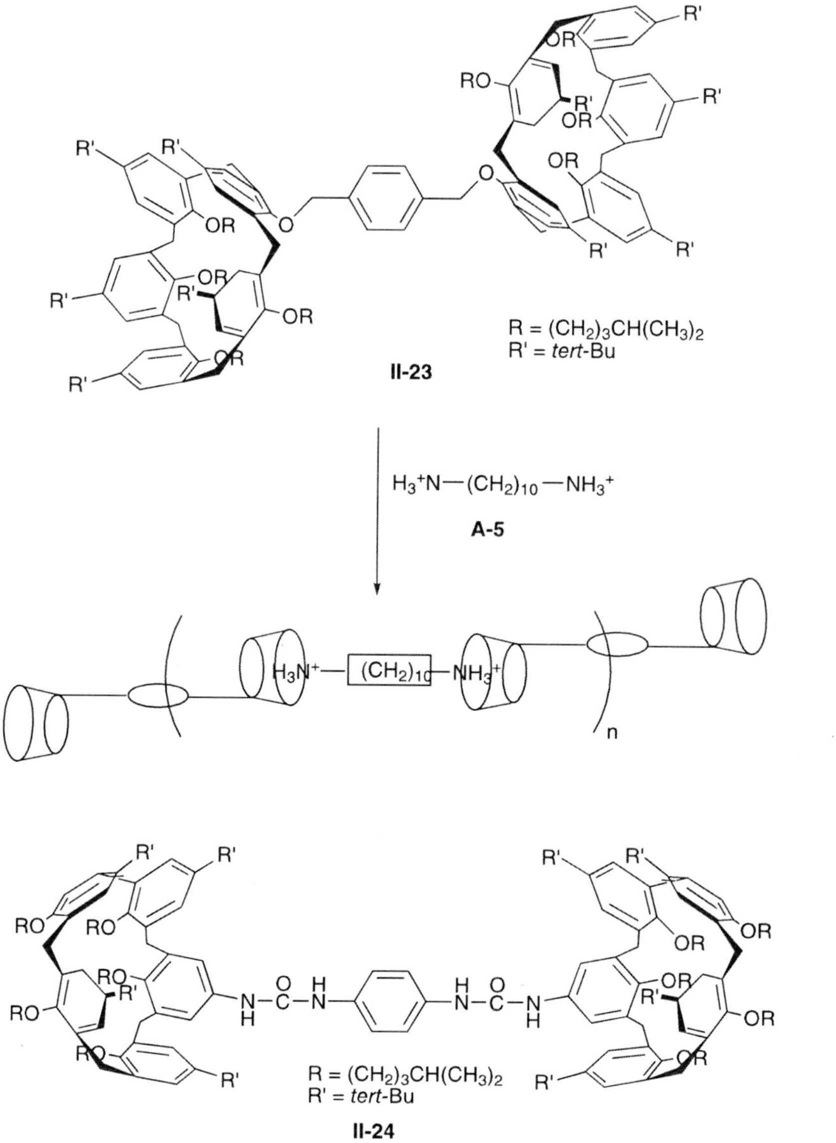

Scheme 12.19 *Extended calixarene hosts which form complexes with ammonium ions*

12.4 Cucurbiturils

Cucurbiturils (CBs) are relatively poorly soluble in water, with only **CB-5** and **CB-7** possessing modest solubility.[49] However, the solubility of **CB-5** and **CB-8** increases strongly in concentrated aqueous acid. Because of the negative electrostatic potential both at the portals and within the cavity, interaction with cationic guests is preferred. Accordingly, alkylammonium ions bind tightly in HCOOH/H_2O 1 : 1 with association

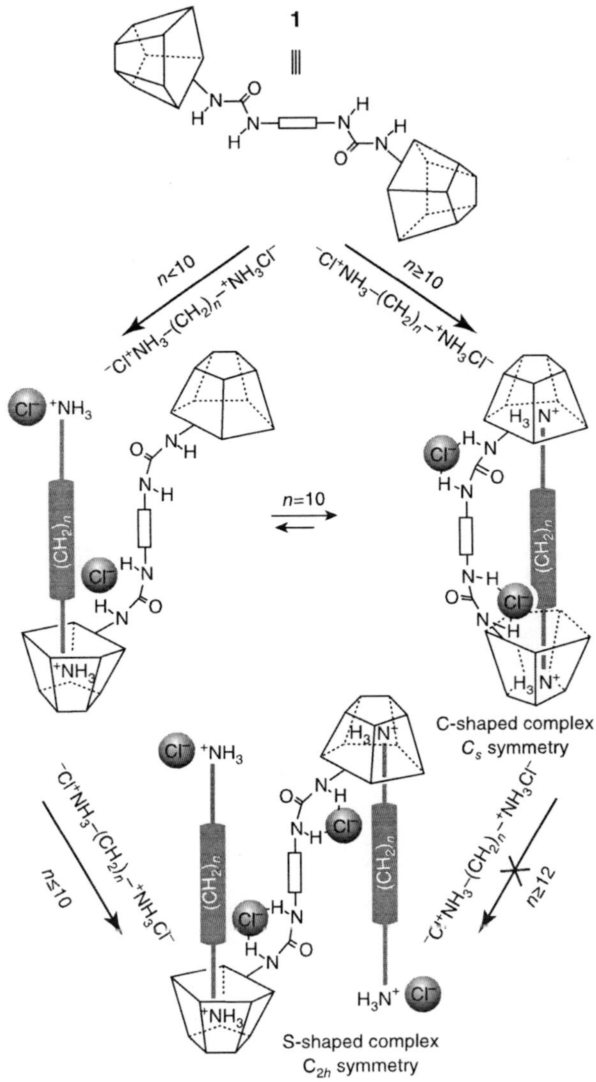

Figure 12.4. *Schematic representation of the influence of the spanning of the guest α,ω-alkanediyldiammonium ions on the complexation pathways leading to C-shaped and/or S-shaped inclusion complexes with host **II-24** (= 1). Reprinted with permission from [44]. Copyright 2005 Wiley-VCH Verlag GmbH & Co. KGaA*

constants up to 10^6 M^{-1} for the complex **CB-6/A-6** (Scheme 12.22).[50] The tight portals of CBs can lead to a constrictive binding,[51] and, accordingly, the exchange of bound and free guests is slow on the ^1H NMR timescale. The narrow portal also causes size selectivity for alkylammonium guests. While **A-7** ($K = 80\,M^{-1}$) and **A-9** ($K = 320\,M^{-1}$) are readily bound, the larger guest **A-8** is not.[51]

The complexation of **A-7** is enthalpy driven. The complexation entropy is negative, indicating a tight complex with a well defined geometry. According to the CIS values

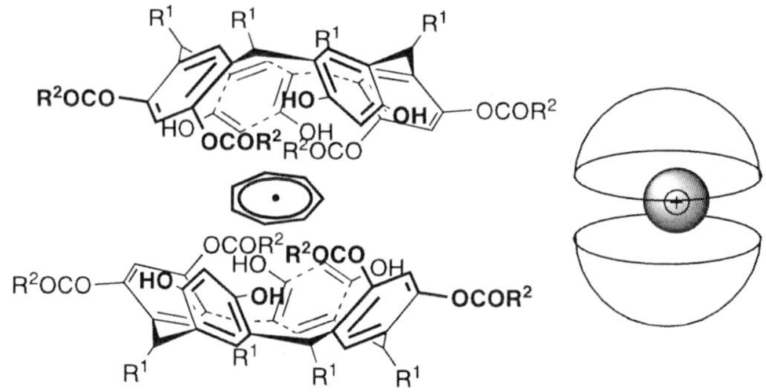

Scheme 12.20 Schematic picture of the complex ***IV-6/D-5/IV-6***

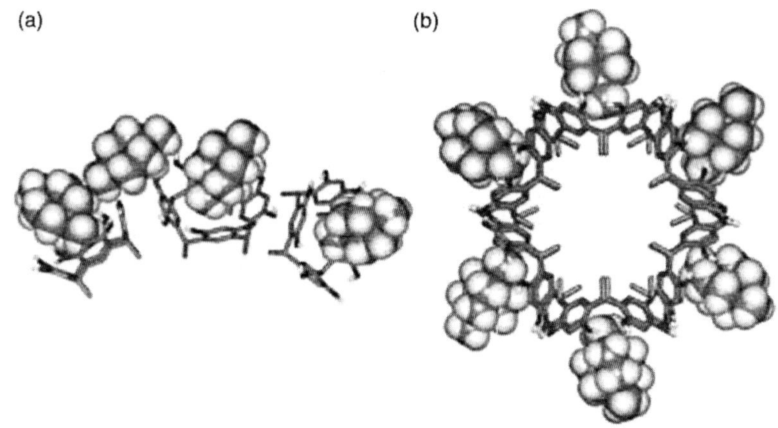

Figure 12.5 *The structure of the nanotube showing resorcinarenes,* ***IV-7****, in stick representation and diquaternary cations* ***B-18****, shown with Van der Waals radii. Anions, water molecules, non-hydrogen bonding hydrogens of resorcinarenes, and disorder of the cations are omitted for clarity. a) The asymmetric unit. b) A view of the tube along the crystallographic a axis. Reprinted with permission from [46]. Copyright 2004 Wiley-VCH Verlag GmbH & Co. KGaA*

observed with **A-7**, the ammonium group interacts with the carbonyl oxygens of **CB-6**, while the cyclohexyl part points towards the interior of the cavity (see Scheme 12.23).[51]

CB-7 is also able to include bipyridinium ions **C-7** (Scheme 12.24) in aqueous solution with association constants of 10^5 M^{-1}.[52] Even the radical cation formed after reduction at the electrode binds strongly to **CB-7** with only a 2-fold decrease as compared with **C-7**.

12.5 Complex Systems and Applications

At present, the association constants of single host–guest complexes may be measured with the help of atomic force microscopy (AFM) force spectroscopy. For this purpose, the host must be immobilized on a gold surface, and the guest must carry one additional

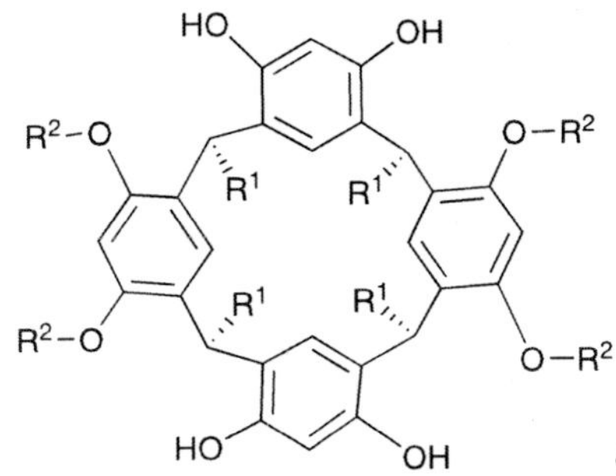

IV-6: $R^1 = C_5H_{11}$, $R^2 =$ 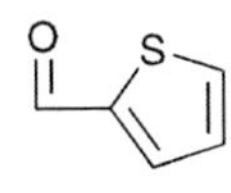
IV-7: $R^1 = Me$, $R^2 = H$

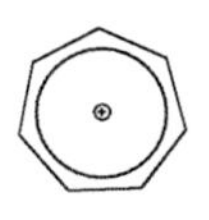

BF_4

D-5

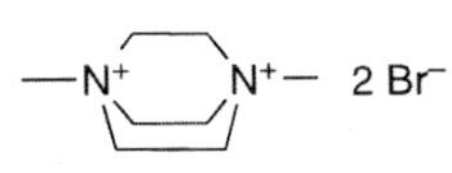

B-18

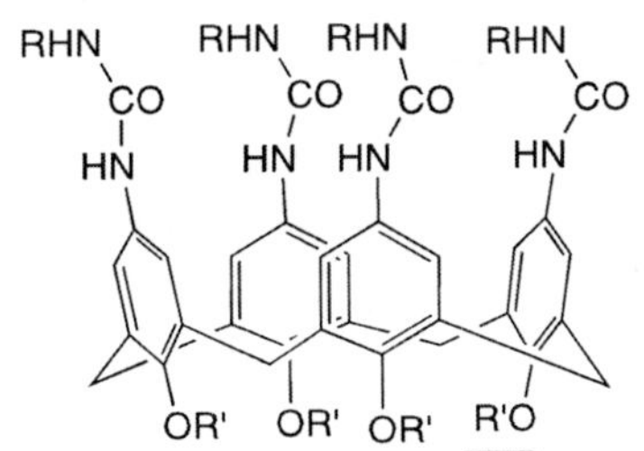

II-25: R= 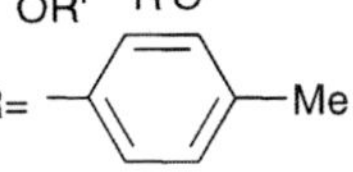

R' = C_5H_{11}

Scheme 12.21 *Guests which support the formation of capsules*

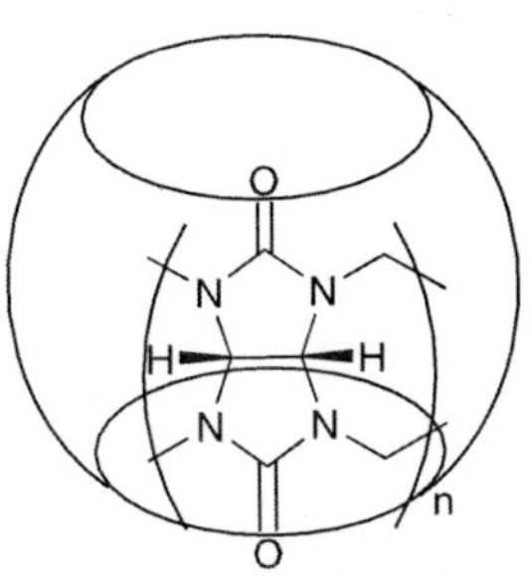

n = 6: **CB-6**
n = 7: **CB-7**

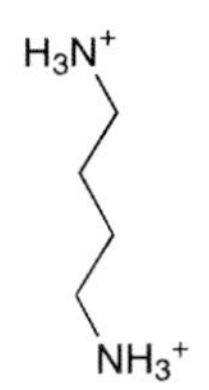

A-6

A-7

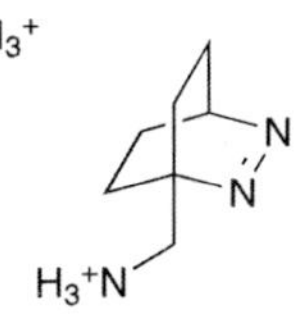

A-8

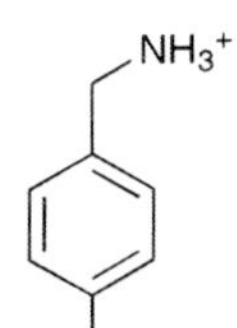

A-9

Scheme 12.22 *Studied complexes between cucurbituril host and ammonium ions*

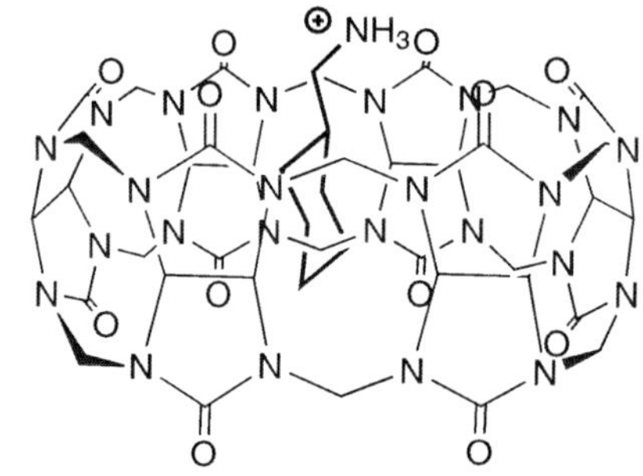

Scheme 12.23 Structure of the complex **CB-7/A-7**

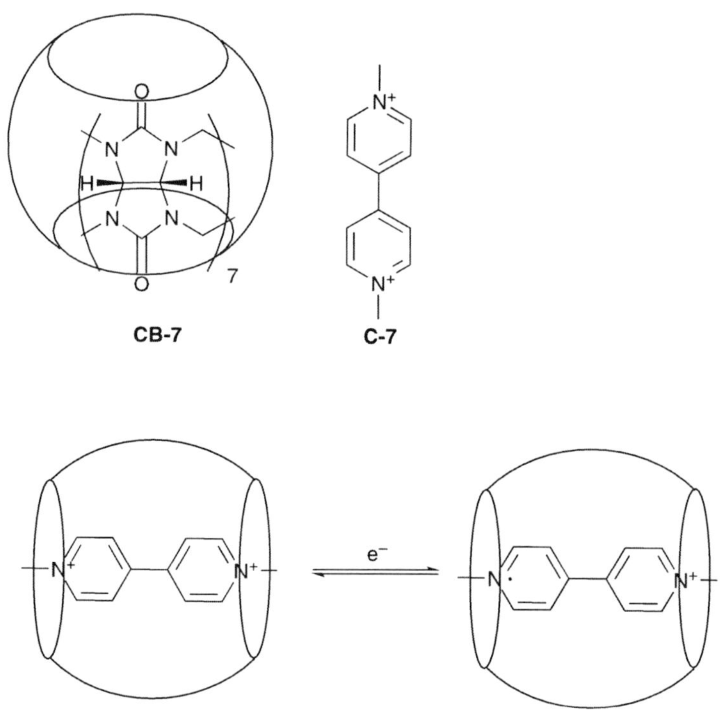

Scheme 12.24 Electrochemical reduction of the complex between **CB-7** and the bipyridinium ion **C-7**

functional group to be covalently attached to the AFM tip (see Figure 12.6).[53] Using this technology, the ammonium ion was shown to be more weakly bound than the trimethylammonium ion.

The binding of **C-3** by a simple calix[4]arene host in nonpolar media can be improved by a factor of 25 by the formation of host clusters on the surfaces of gold nanoparticles.[54]

Applications of host–guest complexes are mainly directed towards the development of sensors. For example, acetylcholine detection at micromolar concentration is based on inclusion of a fluorophore containing the pyridinium unit (*trans*-4[4-(dimethylamino)

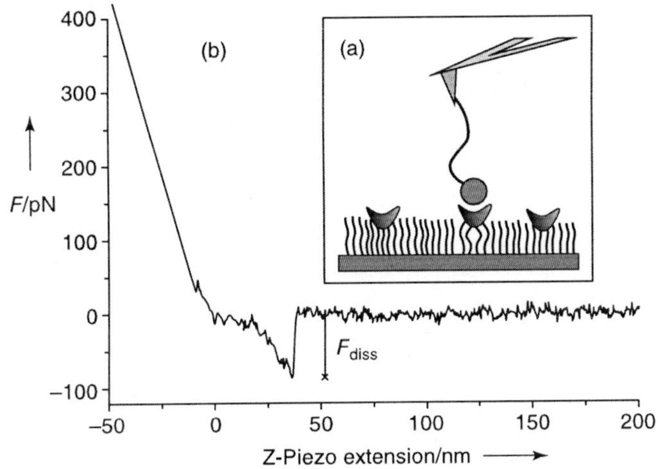

Figure 12.6 *Force spectroscopy: a) schematic setup. The cavitand is immobilized together with didecylsulfide in a 1 : 40 mixture on a gold substrate. The (tetraorganyl) ammonium residue (shaded circle), is attached to a Si_3N_4 AFM tip by a flexible polymer linker. b) Typical force–distance curve (only retractive trace shown). The stretching of the PEG linker over a certain distance prior to bond rupture (tip detachment and relaxation of the cantilever) indicates an unbinding event. Reprinted with permission from [53]. Copyright 2005 Wiley-VCH Verlag GmbH & Co. KGaA*

styryl]-1-methylpyridinium tosylate) in water soluble calixarenes **II-1** – **II-3** (Scheme 12.6). The fluorescence of the guest is both hypsochromically shifted and drastically increased upon complexation. The presence of acetylcholine completely eliminates the new emission band of the bound fluorophore because acetylcholine replaces the fluorophore within the cavity of the calixarene.[55] A sensor was developed by immobilizing the receptor **II-1** on an oxide-containing silicon surface. An analogous system working in alcoholic solution was presented above in Figure 12.2.

As a second example, the coupling of an enzymatic transformation with a dynamic host–guest exchange is also based on fluorescence switching. Combining the decarboxylation of amino acids with a decarboxylase allows the unselective binding of **CB-7** to be used for highly selective amino acid sensing.[56] **CB-7** binds aliphatic amines and diamines in their ammonium form (see above), but the selectivity is low. The neutral fluorophore dapoxyl (Scheme 12.25) and the protonated amine, the product of the enzymatic reaction under physiological conditions, compete for encapsulation. The emission of the fluorophore is strongly enhanced upon complexation. The replacement of the fluorophore by the ammonium ion quenches or 'switches off' the fluorescence (see Scheme 12.26). The enantiospecifity of the enzymatic reaction allows the enantioselective recognition of analytes with an achiral host.

12.5.1 Photoresponsive Hosts

Nature uses chemical signalling in sensory systems such as taste and smell. The most important means for information processing in nature is based on information exchange between DNA and RNA; our brain use signals relying on the behaviour of ions, etc. This

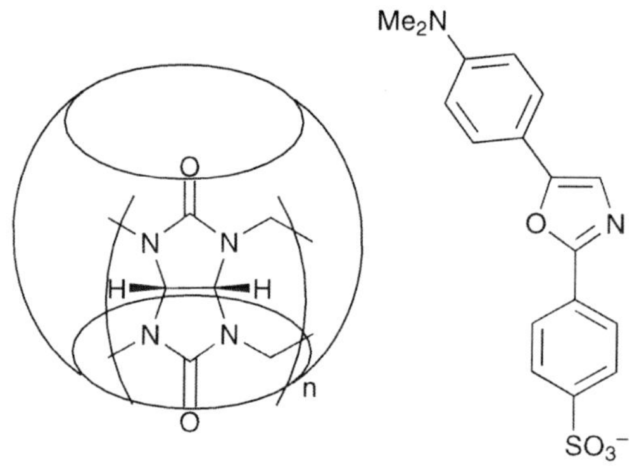

Scheme 12.25 *The host **CB-7** and its guest dapoxyl*

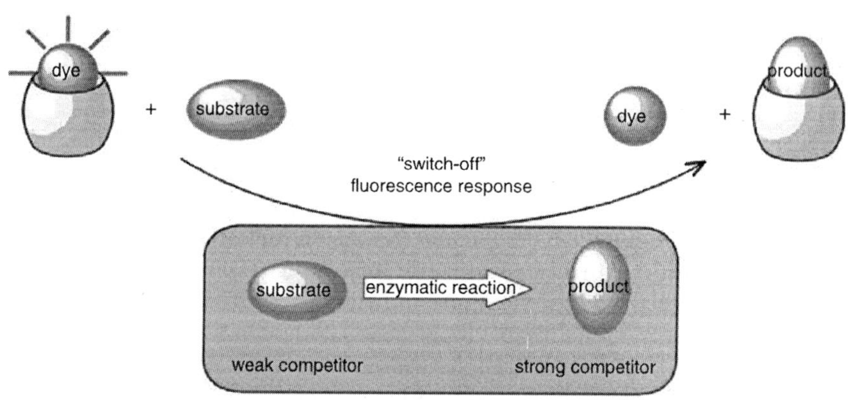

Scheme 12.26 *Working principle of a 'switch-off' product-coupled tandem assay (reproduced by permission from [56]; copyright (2004) Wiley-VCH)*

molecular chemionics is based on molecular recognition.[57] Therefore, hosts with binding properties that can be tuned by an outer stimulus are of particular interest. The advantage of light driven changes is that no chemical fuels are necessary and no waste products are formed.

The host cavity of a resorcin[4]arene bearing two anthracene moieties at the wider rim can be switched by the known photodimerization of the two anthracene substituents resulting in a closed cavity. The process can be reversed by heating of the solution. Whereas the cavity of the open state is well suited to bind ammonium ions such as 2-mercaptoethylamine hydrochloride, the closed form does not bind ammonium ions. The affinity modulation has also been studied by single-molecule force spectroscopy (Figure 12.7). For this purpose, the host has been modified with four didecylsulfide linkers in order to immobilize the host on gold surfaces.[58]

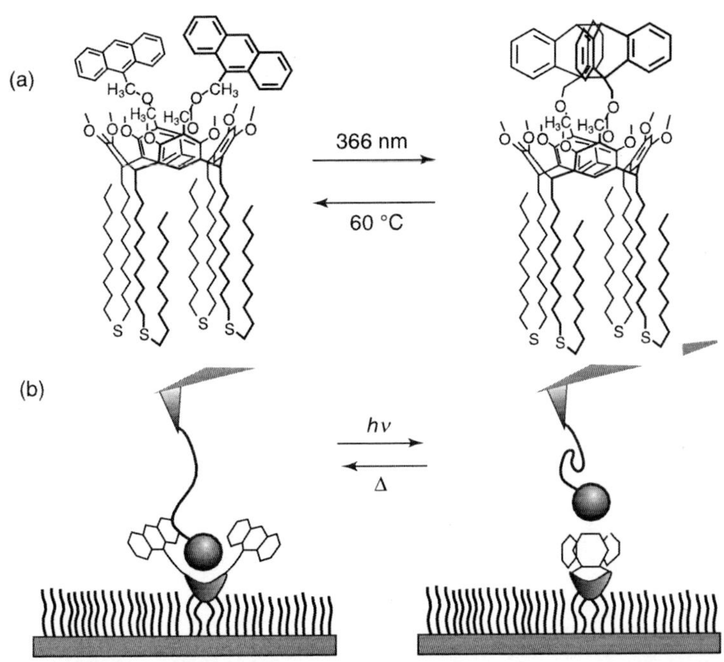

Figure 12.7 *Resorcin[4]arene photoswitch for self-assembly (a) and a pictogram visualizing the diluted self-assembled monolayer of this photoswitch for single-molecule affinity studies by AFM (b). Reprinted with permission from [58]. Copyright 2007 American Chemical Society*

The cavity of the calix[4]arene **II-26** bearing two acridane substituents is open and suitable to bind guests such as **C-3** (K = 1461 ± 458 l mol^{-1} in $CDCl_3$ solution) and **C-4** (K = 117 ± 16 l mol^{-1}) (Scheme 12.27 and Figure 12.8).[59] In contrast, the cavity is blocked by two acridinium substituents formed by photoheterolysis.[60] Therefore, the calixarene in its acridinium state does not bind guests **C-2** or **C-4**. Thus, organic cations can be bound and released depending on the state of the host.

12.6 Conclusions

This chapter has covered some of the developments in the field of supramolecular host–guest complexes with organic cations. Cation–π interaction may play a significant role, while charge-charge attraction often supports the binding process. The negative electrostatic potential at the portals and within the cavity of cucurbiturils has important consequences for their recognition of cationic guests. However, a detailed understanding of cause/effect relations is difficult to derive. We have seen the effects of host structure on the stability and strength of the complexes through the examples given in this chapter. A host should provide a suitable surface and chemical complementarity to the organic cation to enable optimal interaction energy. Rigidified calixarenes and resorcinarenes as well as

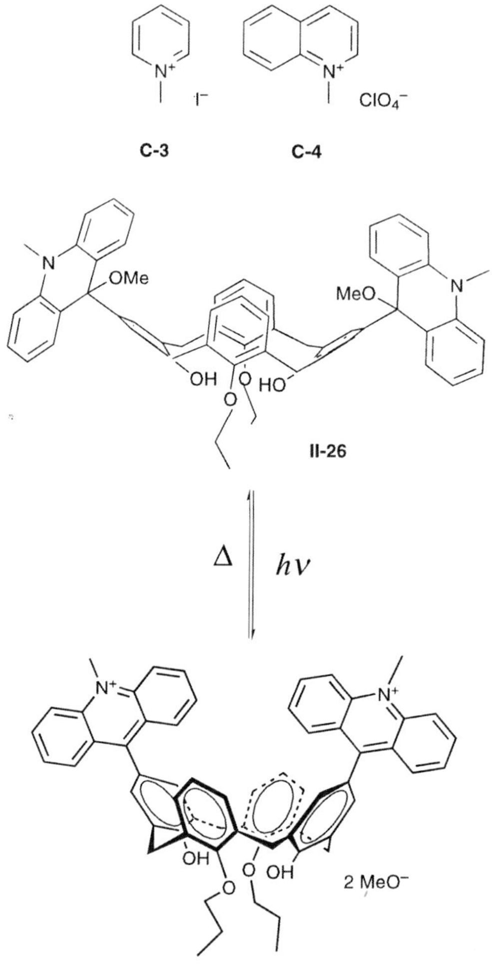

II-27

Scheme 12.27 *Photochemical and thermal switching between **II-26** and the corresponding **II-27***

extended structures may overcome the entropic losses that occur upon complexation and may increase the enthalpic driving force through increased contacts between host and guest. In recent years, the complexity of host systems has been considerably increased. Modified calixarene and resorcinarene receptors for distinct functionalities are the focus of much interest. Switchable receptors and host–guest complexes as sensory material hold great promise and will emerge in the near future. In many cases, the physical properties of guests (pK-values, fluorescence efficiency, chemical reactivity) are changed by complexation which can afford an increasing number of applications. Nevertheless, a greater emphasis should be given to extending the understanding of the cause-effect relationships of the binding properties in order to facilitate more rational approaches to receptor design.

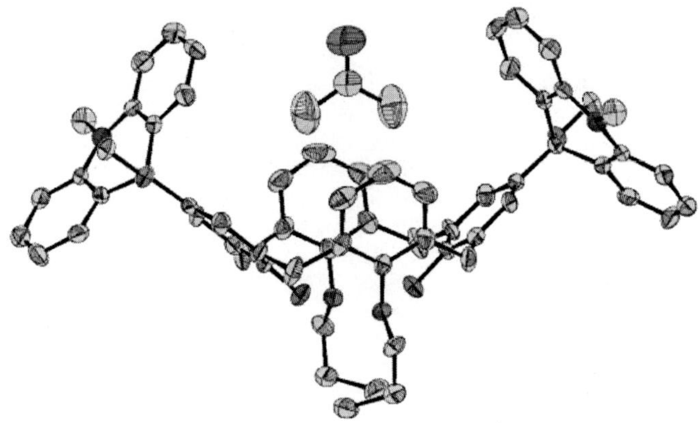

Figure 12.8 Molecular structure of the calixarene ***II-26*** containing acetone as guest (H atoms are omitted for clarity)

References

1. A. R. Shalaby, Significance of biogenic amines to food safety and human health, *Food Res. Int.*, 1996, **29**, 675–690.
2. J. C. Ma, D. A. Dougherty, The cation–π Interaction, *Chem. Rev.*, 1997, **97**, 1303–1324.
3. J. L. Sussman, M. Harel, F. Frolow, C. Oefner, A. Goldman, L. Toker, I. Silman, Atomic structure of acetylcholinesterase from torpedo californica: a prototypic acetylcholine-binding protein, *Science*, 1991, **253**, 872–879.
4. a) V. Böhmer, A. Dalla Cort, L. Mandolini, Counteranion effect on complexation of quats by a neutral calix[5]arene receptor, *J. Org. Chem.*, 2001, **66**, 1990–1902; b) A. Arduini, E. Brindani, G. Giorgi, A. Pochini, A. Secchi, Ainion effects on the recognition of ion pairs by calix[4]arene-based heteroditopic receptors, *J. Org. Chem.*, **67**, 2002, 6188–6194.
5. S. Bartoli, S. Roelens, Electrostatic attraction of counterion dominates the cation–π interaction of acetylcholine and tetramethylammonium with aromatics in chloroform, *J. Am. Chem. Soc.*, 1999, **121**, 11908–11909.
6. a) M. L. C. Quan, D. J. Cram, Constrictive binding of large guests by a hemicarcerand containing four portals, *J. Am. Chem. Soc.*, 1991, **113**, 2754–2755; b) K. N. Houk, K. Nakamura, C. Sheu, A. E. Keating, Gating as a control element in constrictive binding and guest release by hemicarcerands, *Science*, **1996, 273**, 627–629.
7. K. Hirose, Determination of binding constants, *in Analytical methods in Supramolecular Chemistry*, C. A. Schalley (Ed.), Wiley-VCH, Weinheim, 2007.
8. F. P. Schmidtchen, Isothermal titration calorimetry in supramolecular chemistry, in *Analytical methods in Supramolecular Chemistry*, C. A. Schalley (Ed.), Wiley-VCH, Weinheim, 2007.
9. L. G. Garel, B. Lozach, J.-P. Dutasta, A. Collet, Remarkable effect of receptor size on the binding of acetylcholione and related ammonium ions to water-soluble cryptophanes, *J. Am. Chem. Soc.*, 1993, **115**, 11652–11653.
10. S. M. Ngola, P. C. Kearney, S. Mecozzi, K. Russell, D. A. Doughrty, A selective receptor for arginine derivatives in aqueous media. Energetic consequences of salt bridges that are highly exposed to water, *J. Am. Chem. Soc.*, **121**, 1999, 1192–1201.
11. A. Ikeda, S. Shinkai, Novel cavity design using calix[n]arenes skeletons: toward molecular recognition and metal binding, *Chem. Rev.*, 1997, **97**, 1713–1734.
12. C. D. Gutsche, *Calixarenes Revisited*, The Royal Society of Chemistry, Cambridge, 1998.
13. Z. Asfari, V. Böhmer, J. Harrowfield, J. Vicens, *Calixarenes 2001*, Kluwer Academic Publisher, Dordrecht, 2001.

14. W. Abraham, Inclusion of organic cations by calix[n]arenes, *J. Incl. Phenom. Macrocycl. Chem.*, 2002, **43**, 159–174.
15. A. W. Coleman, S. Jebors, P. Shahgaldian, G. S. Ananchenko, J. A. Ripmeester, para-Acylcalix[n]arenes : from molecular to macroscopic assemblies, *Chem. Commun.*, 2008, 2291–2303.
16. F. Inokuchi, K. Araki, S. Shinkai, Facile detection of cation–π interactions in calix[n]arenes by mass spectrometry, *Chem. Lett.*, 1994, 1383–1386.
17. A. Arduni, W. M. McGregor, A. Pochini, A. Secchi, F. Ugozzoli, R. Ungaro, New upper rim pyridine-bridged calix[4]arenes: synthesis and complexation properties toward neutral molecules and ammonium ions in organic media, *J. Org. Chem.*, 1996, **61**, 6881–6887.
18. M. Orda-Zgadzaj, V. Wendel, M. Fehlinger, B. Ziemer, W. Abraham, Inclusion of organic cations by calix[4]arenes bearing cyclohepta-2,4,6-trienyl substituents, *Eur. J. Org. Chem.*, 2001, 1549–1561.
19. A. Arduini, W. M. McGregor, D. Paganuzzi, A. Pochini, A. Secchi, F. Ugozzoli, R. Ungaro, Rigid cone calix[4]arenes as π-donor systems: complexation of organic molecules and ammonium ions in organic media, *J. Chem. Soc., Perkin Trans. 2*, 1996, 839–846.
20. A. Arduini, G. Giorgi, A. Pochini, A. Secchi, F. Ugozzoli, Anion allosteric effect in the recognition of tetramethylammonium salts by calix[4]arene *Cone* conformers, *J. Org. Chem.*, 2001, **66**, 8302–8308.
21. R. Arnecke, V. Böhmer, R. Cacciapaglia, A. Dalla Cort, L. Mandolini, cation–π Interactions between neutral calix[5]arene hosts and cationic organic guests, *Tetrahedron*, 1997, **53**, 4901–4908.
22. F. Perret, A. N. Lazar, A. W. Coleman, Biochemistry of the *para*-sulfonatocalix[n]arenes, *Chem. Commun.*, 2006, 2425–2438.
23. N. D. Douteau-Guével, A. W. Coleman, J.-P. Morel, N. Morel-Desrosiers, Complexation of the basic amino acids lysine and arginine by three sulfonatocalix[*n*]arenes (n = 4, 6 and 8) in water: microcalorimetric determination of the Gibbs energies, enthalpies and entropies of complexation, *J. Chem. Soc., Perkin Trans. 2*, 1999, 629–634.
24. J. M. Lehn, R. Meric, J.-P. Vigneron, M. Cesario, J. Guilhem, C. Pascard, Z. Asfari, J. Vicens, Binding of acetylcholine and other quaternary ammonium cations by sulfonated calixarenes. Crystal structure of a [cholin-tetrasulfonated calix[4]arene] complex, *Supramol. Chem.*, 1995, **5**, 97–103.
25. C. Bonal, Y. Israëli, J.-P. Morel, N. Morel-Desrosiers, Binding of inorganic and organic cations by *p*-sulfonatocalix[4]arene in water: a thermodynamic study, *J. Chem. Soc., Perkin Trans. 2*, 2001, 1075–1078.
26. E. A. Meyer, R. K. Castellano, F. Diederich, Interactions with aromatic rings in chemical and biological recognition, *Angew. Chem. Int. Ed.*, 2003, **42**, 1210–1250.
27. G. Arena, A. Casnati, A. Contino, F. G. Gulino, D. Sciotto, R. Ungaro, Entropic origin of the sulfonate groups' electrostatic assistance in the complexation of quaternary ammonium cations by water soluble calix[4]arenes, *J. Chem. Soc., Perkin Trans. 2*, 2000, 419–423.
28. G. Arena, A. Casnati, A. Contino, G. G. Lombardo, D. Sciotto, R. Ungaro, Water-soluble calixarene hosts that specifically recognize the trimethylammonium group or the benzene ring of aromatic ammonium cations: a combined ^{1}H NMR, calorimetric, and molecular mechanics investigation, *Chem. Eur. J.*, 1999, **5**, 738–744.
29. M. Fehlinger, W. Abraham, Calix[4]arenes bearing a tropylium substituent as hosts for organic cations, *J. Incl. Phenom. Macrocycl. Chem.*, 2007, **58**, 263–274.
30. S. Ishihara, S. Takeoka, Host–guest assembly of pyridinium-conjugated calix[4]arene via cation–π interaction, *Tetrahedron Lett.*, 2006, **47**, 181–184.
31. F. Inokuchi, K. Araki, S. Shinkai, Facile detection of cation–π interactions in calix[n]arenes by mass spectroscopy, *Chem. Lett.*, 1994, 1383–1386.
32. B. Masci, M. Finelli, M. Varrone, Fine tuning of the cavity size in calixarene-like cyclophanes: a complete series of homooxacalix[4]arene ligands for quaternary ammonium ions, *Chem. Eur. J.*, 1998, **4**, 2018–2030.
33. B. Masci, S. L. Mortera, D. Persiani, P. Thuéry, Methyl ether derivatives of *p-tert*-butyl[3.1.3.1] homooxacalixarene. Formation, structure, and complexes with quaternary ammonium ions, *J. Org. Chem.*, 2006, **71**, 504–511.

34. B. Masci, D. Persiani, P. Thuéry, Complexation of quaternary ammonium ions by tetraester derivatives of [3.1.3.1]homooxacalixarene in mobile and in fixed conformation, *J. Org. Chem.*, 2006, **71**, 9784–9790.
35. P. Timmerman, W. Verboom, D. N. Reinhoudt, Resorcinarenes, *Tetrahedron*, 1996, **52**, 2663–2704.
36. H. J. Schneider, D. Guettes, U. Schneider, Host–guest complexes with water-soluble macrocyclic polyphenolates including induced fit and simple elements of a proton pump, *J. Am. Chem. Soc.*, 1988, **110**, 6449–6454.
37. W.-H. Chen, Y. Wei, S.-D. Tan, B. Wang, Z.-L. Xu, Spectrometric study of the size discrimination of quaternary ammonium cations by tetracyanoresorcin[4]arene, *Supramol. Chem.*, 2005, **17**, 469–473.
38. P. Ballester, M. A. Sarmentero, Hybrid cavitand-resorcin[4]arene receptor for the selective binding of choline and related compounds in protic media, *Org. Lett.*, 2006, **8**, 3477–3480.
39. A. Lledó, R. J. Hooley, J. Rebek Jr., Recognition of guests by water-stabilized cavitand hosts, *Org. Lett.*, 2008, **10**, 3669–3671.
40. S. M. Biros, E. C. Ullrich, F. Hof, L. Trembleau, J. Rebek Jr., Kinetically stable complexes in water: the role of hydration and hydrophobicity, *J. Am. Chem. Soc.*, 2004, **126**, 2870–2876.
41. A. Mehdizadeh, M. C. Letzel, M. Klaes, C. Agena, J. Mattay, Chiral discrimination on the host–guest complexation of resorcin[4]arenes with quaternary amines, *Eur. J. Mass Spectrom.*, 2004, **10**, 649–655.
42. A. Arduini, A. Pochini, A. Secchi, Rigid calix[4]arene as a building block for the synthesis of new quaternary ammonium cation receptors, *Eur. J. Org. Chem.*, 2000, 2325–2334.
43. D. Garozzo, G. Gattuso, F. H. Kohnke, A. Notti, S. Pappalardo, M. F. Parisi, I. Pisagatti, A. J. P. White, D. J. Williams, Inclusion networks of a calix[5]arene-based exoditopic receptor and long-chain alkylammonium ions, *Org. Lett.*, 2003, **5**, 4025–4028.
44. D. Garozzo, G. Gattuso, A. Notti, A. Pappalardo, S. Pappalardo, M. F. Parisi, M. Perez, I. Pisagatti, A calix[5]arene based heterotetratopic host for molecular recognition of long-chain, ion-paired-α,ω-alkanediylammonium salts, *Angew. Chem. Int. Ed.*, 2005, **44**, 4892–4896.
45. A. Shivanyuk, E. F. Paulis, V. Böhmer, Guest-controlled formation of a hydrogen-bonded molecular capsule, *Angew. Chem. Int. Ed.*, 1999, **38**, 2906–2909.
46. H. Mansikkamäki, M. Nissinen, K. Rissanen, Noncovalent $\pi \ldots \pi$-stacked exo-functional nanotubes : subtle control of resorcinarene self-assembly, *Angew. Chem. Int. Ed.*, 2004, **43**, 1243–1246.
47. L. Avram, Y. Cohen, Self-assembly of resorcin[4]arene in the presence of small alkylammonium guests in solution, *Org. Lett.*, 2008, **10**, 1505–1508.
48. L. Frish, M. O. Vysotsky, S. E. Matthews, V. Böhmer, Y. Cohen, Tropylium cation capsule of hydrogen-bonded tetraurea calix[4]arene dimers, *J. Chem. Soc., Perkin Trans. 2*, 2002, 88–93.
49. J. Lagona, P. Mukhopadhyay, S. Chakrabarti, L. Isaacs, The cucurbit[n]uril family, *Angew. Chem. Int. Ed.*, 2005, **44**, 4844–4870.
50. W. L. Mock, N. Y. Shih, Structure and selectivity in host–guest complexes of cucurbituril, *J. Org. Chem.*, 1986, **51**, 4440–4446.
51. C. Márquez, R. R. Hudgins, W. M. Nau, Mechanism of host–guest complexation by cucurbituril, *J. Am. Chem. Soc.*, 2004, **126**, 5806–5816.
52. W. Ong, M. Gómez-Kaifer, A. E. Kaifer, Cucurbit[7]uril: a very effective host for viologens and their cation radicals, *Org. Lett.*, 2002, **4**, 1791–1794.
53. R. Eckel, R. Ros, B. Decker, J. Mattay, D. Anselmetti, Supramolecular chemistry at the single-molecule level, *Angew. Chem. Int. Ed.*, 2005, **44**, 484–488.
54. A. Arduini, D. Demuru, A. Pochini, A. Secchi, Recognition of quaternary ammonium cations by calix[4]arene derivatives supported on gold nanoparticles, *Chem. Commun.*, 2005, 645–647.
55. N. Korbakov, P. Timmerman, N. Lidich, B. Urbach, A. Sa'ar, S. Yitzchaik, Acetylcholine detection at micromolar concentrations with the use of an artificial receptor-based fluorescence switch, *Langmuir*, 2008, **24**, 2580–2587.

56. D. M. Bailey, A. Hennig, V. D. Uzunova, W. M. Nau, Supramolecular tandem enzyme assays for multiparameter sensor arrays and enantiomeric excess determination of amino acids, *Chem. Eur. J.*, 2008, **14**, 6069–6077.
57. V. Balzani, A. Credi, M. Venturi, Adressing energy and signals by processing molecular and supramolecular systems, *Chem. Eur. J.*, 2008, **14**, 26–39.
58. a) C. Schäfer, R. Eckel, R. Ros, J. Mattay, D. Anselmetti, Photochemical single-molecule affinity switch, *J. Am. Chem. Soc.*, 2007, **129**, 1488–1489; b) D. Anselmetti, F. W. Bartels, A. Becker, B. Decker, R. Eckel, M. McIntosh, J. Mattay, P. Plattner, R. Ros, C. Schäfer, N. Sewald, Reverse engineering of an affinity-switchable molecular interaction characterized by atomic force microscopy single-molecule force spectroscopy, *Langmuir*, 2008, **24**, 1365–1370.
59. L. Grubert, W. Abraham, unpublished results.
60. L. Grubert, W. Abraham, Photoswitchable calix[4]arenes bearing dihydroacridine substituents at the upper rim, *Tetrahedron*, 2007, **63**, 10778–10787.

13

Proteins as Host for Enantioselective Catalysis: Artificial Metalloenzymes Based on the Biotin–Streptavidin Technology

Jincheng Mao and Thomas R. Ward
Department of Chemistry, University of Basel, CH-4056 Basel

13.1 Introduction

In order to meet the ever increasing demand for enantiopure compounds, heterogeneous-, homogeneous- and enzymatic catalysis have evolved independently in the past.[1,2] Although all three approaches have yielded industrially viable processes, the latter two are the most widely used and can be regarded as complementary in many respects.[3]

From the viewpoint of molecular encapsulation, enzymes have evolved to accommodate and orient their substrate within a well-defined environment provided by the protein, Figure 13.1a. In many cases, the substrate is embedded within the enzyme. In contrast, homogeneous catalysts typically possess a (chelating) ligand which occupies one hemisphere around the metal, while the substrate occupies the other hemisphere. The main effort in catalyst optimization has thus been to extend the reach of the selectivity-orienting ligand, Figure 13.1b.

With the aim of complementing the well established fields of homogeneous and enzymatic catalysis, organocatalysis and artificial metalloenzymes have witnessed a recent revival.[4,5] Artificial metalloenzymes, result from combining an active but unselective organometallic moiety with a macromolecular host (protein or DNA), Figure 13.2.

Both Kaiser and Whitesides suggested the possibility of creating artificial metalloenzymes in the late 1970s already.[6,7] However, there existed a widespread belief that proteins

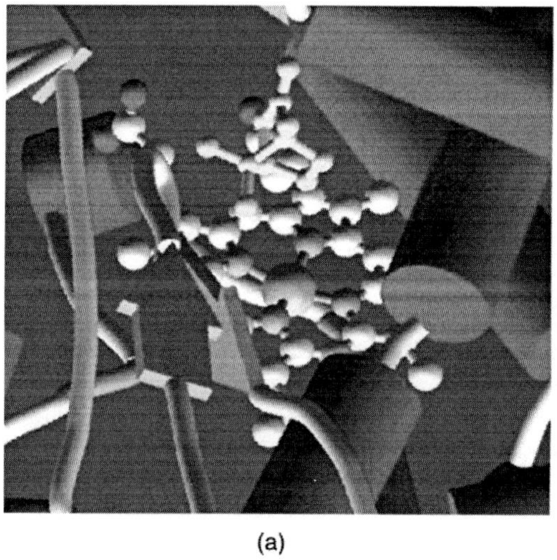

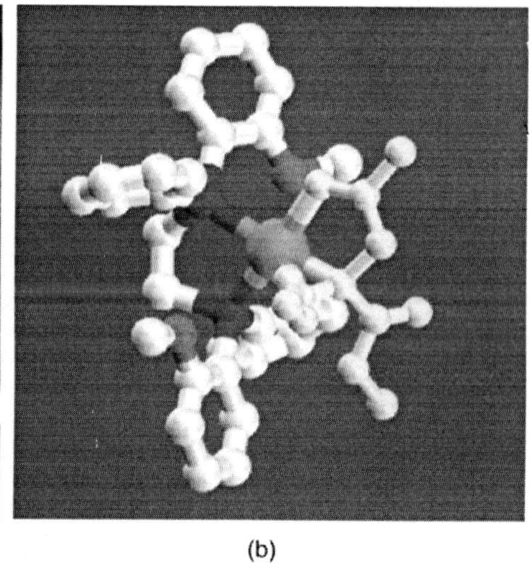

Figure 13.1 *Structure of cytochrome P450 cam with its substrate (camphor, yellow), emphasizing the 'lock and key' complementarity between the enzyme and its substrate (pdb ref. code: 2CPP) (a); Structure of Knowles's {Rh(dipamp)} + bound to its substrate (yellow), emphasizing the limited space (hemisphere) occupied by the enantiopure dipamp ligand (CSD ref. code: VERYAS) (b)*

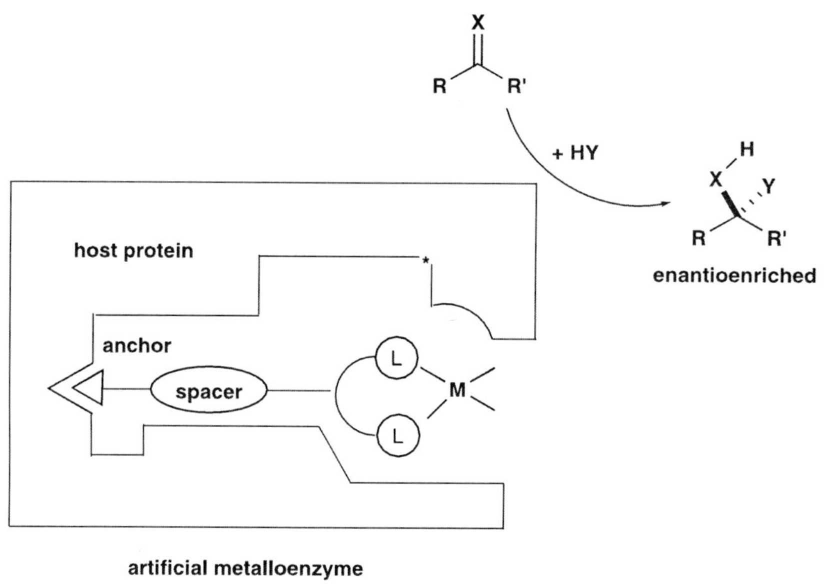

Figure 13.2 *Anchoring a catalytically active but unselective organometallic catalyst within a host protein affords artificial metalloenzymes for enantioselective catalysis. Chemical optimization can be achieved either by varying the spacer (oval) or the chelating ligand. Site-directed or random mutagenesis at a position close to the metal moiety (*) can be used for the genetic optimisation. Reprinted with permission of Swiss Chemical Society*

and organometallic catalysts are incompatible with each other. This severely hampered research in this area at the interface between homogeneous and enzymatic catalysis.[8] Since 2000 however, there has been an increasing interest in the field of artificial metalloenzymes for enantioselective catalysis.[9–13]

For the localization of an organometallic moiety within a macromolecular host, three anchoring strategies can be envisaged: covalent, dative or supramolecular.[5] Using such approaches, enantioselective artificial metalloenzymes have been created for hydrolysis,[12] hydrogenation,[14–17] transfer hydrogenation,[18] allylic alkylation,[19] sulfoxidation,[20–24] epoxidation,[25,26] dihydroxylation,[27] Diels–Alder,[28,29] trans-amination,[30] Michael addition[31] and fluorination.[32] Comparatively, supramolecular anchoring appears as the most appealing, since its allows separate variation of both biological and chemical components, followed by straightforward combination of the organometallic moiety and the macromolecular host. In addition, no chemical modification step is required after the incorporation of the catalyst precursor, thus ensuring the integrity of the organometallic species. In recent years, we have been exploiting the biotin–avidin technology for the creation of artificial metalloenzymes for enantioselective catalysis. Herein we summarize the challenges and our progress in this area.

13.2 The Biotin–Avidin Technology

The biotin–avidin technology, often referred to as molecular Velcro, relies on the very high affinity of biotin for either avidin or streptavidin – (strept)avidin hereafter.[33] The versatility of this technology can be traced back to the following features:

i) The nearly irreversible interaction between biotin and (strept)avidin relies solely on hydrogen bonds and hydrophobic interactions and thus can reliably take place in a variety of environments (cell, blood, surfaces, etc.) under mild conditions.
ii) A straightforward amide bond-forming reaction allows to append a biotin moiety to nearly any probe. This derivatization of the valeric acid side chain of biotin does not significantly affect the high affinity of the biotin–avidin couple.[34]
iii) (Strept)avidin possesses a homotetrameric β-barrel structure with four equivalent biotin-binding sites. This allows to bring into proximity different moieties via the biotin–avidin technology.
iv) (Strept)avidin is a very robust protein scaffold which can be easily handled by unexperienced chemists without the risk of loosing biotin-binding activity.

In the past four decades, this versatile tool has found numerous applications in affinity chromatography, signal amplification, drug delivery, immunoassays, immobilization etc.[35]

Inspired by the pioneering work of Whitesides,[7,17] we reasoned that we could exploit the biotin–avidin technology for the creation and optimization of artificial metalloenzymes. Introduction of a biotin anchor on a chelating ligand ensures the localization of an organometallic moiety within a chiral environment provided by (strept)avidin. We endeavored to use the modern tools of genetic and chemical optimization to fine-tune the second coordination sphere of the resulting artificial metalloenzymes. The main stages include (i) mutation and overexpression of the (strept)avidin isoform in a suitable host, (ii) isolation, purification and quantification of the activity of the mutated (strept)avidin,

(iii) synthesis of the biotinylated catalyst precursor and incorporation within (strept) avidin, (iv) catalysis, and (v) product isolation and analysis.

One of the most attractive feature of this approach is the diversity generated by combining both chemical and genetic optimization strategies.[36] Following this approach, we have been able to develop artificial metalloenzymes for the hydrogenation, the allylic alkylation and the transfer hydrogenation reaction.

13.2.1 The Chemical Optimization Dimension

In the spirit of homogeneous catalyst optimization, variation of the first coordination sphere provided by a chelating ligand often has a significant impact on the performance of the resulting catalyst. Inspired by some of the best homogeneous catalysts, a series of aminodiphosphines and amidosulfonamides were synthesized and coupled with the biotin-anchor. We reasonned that introduction of a spacer between the biotin and the chelating ligand would project a different chiral environment for each biotin–spacer–ligand combination, thus offering a straightforward optimization strategy. The most versatile ligands and spacers are sketched in Figure 13.3.

13.2.2 The Genetic Optimization Dimension

Considering the ease of recombinant protein production in various hosts, we set out to produce streptavidin (Sav hereafter) in *Escherichia coli* and avidin (Avi hereafter) in *Pichia pastoris*.[37,38] Although in our hands both proteins could be overexpressed to high levels (230 mg l^{-1} cell culture and 330 mg l^{-1} cell culture respectively) streptavidin was eventually preferred over avidin for the following reasons:

i) In contrast to streptavidin, avidin is glycosylated. This is a source of inhomogeneity and thus may cause reproducibility concerns. In addition, the inhomogeneity renders crystallization more challenging.
ii) Each avidin monomer possesses a single intramolecular disulfide bridge. As a consequence, the protein cannot be denatured and renatured, which is a limitation for purification purposes.
iii) *E. coli* is a well behaved and a very easy to handle organism for the overexpression of recombinant proteins. However, it should be emphasized that *P. pastoris* secretes the recombinant proteins in the culture medium, which could be a major asset for future directed evolution studies.

In light of the above reasons and considering that our initial success was obtained with streptavidin rather than avidin,[14] we rapidly focused on streptavidin and mutants thereof as host for artificial metalloenzymes.

Docking studies performed with either [Rh(COD)(Biot–**1**)]$^+$ or [Ru(η^6-arene)H(Biot–IIIp–**2**)] within WT-Sav revealed that the potential energy surface describing the localization of the metal within the protein environment was flat and the metal adopted different positions within a very small energy window.[39,40] Analysis of the C_α–M distances revealed that the asymmetric carbon (C_α) of residue S112 Sav lied systematically closest to the metal center. Based on these *in silico* studies, we set out to produce all twenty Sav isoforms resulting from saturation mutagenesis at position S112X Sav.[41] Except for S112P

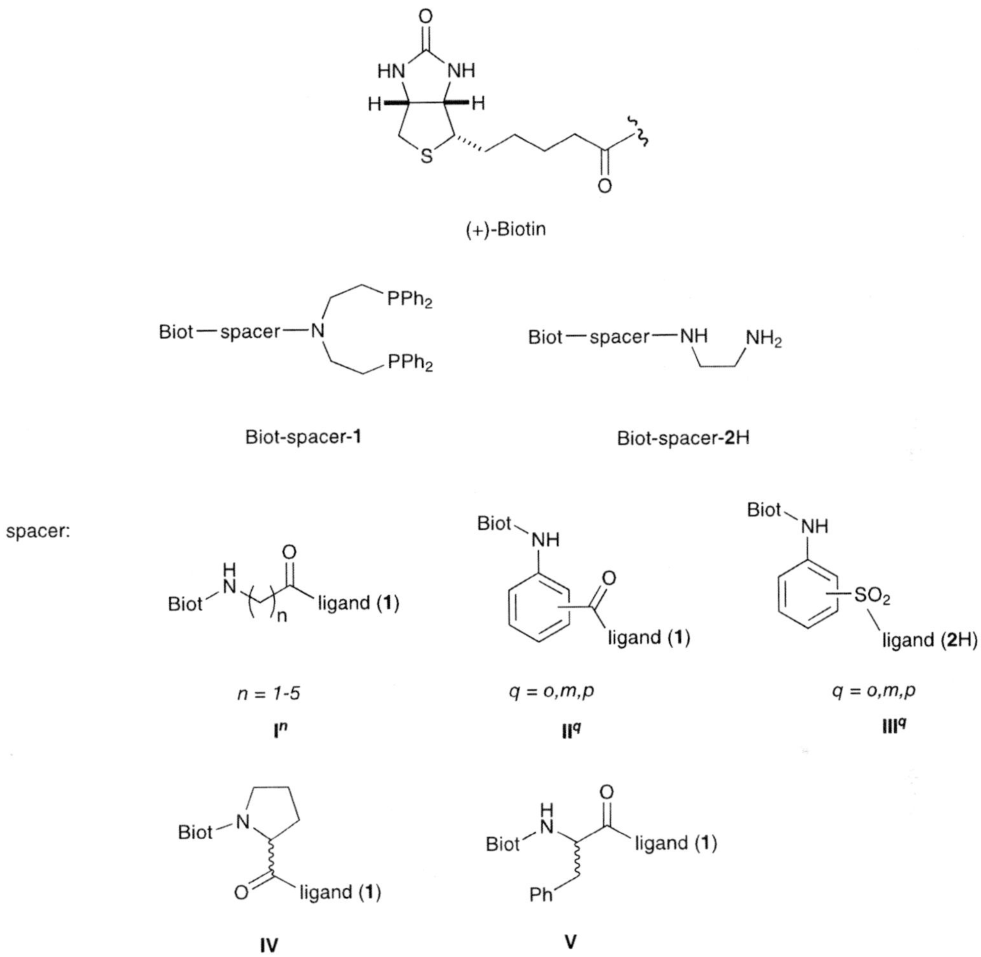

Figure 13.3 Chemical diversity generated by combining different biotinylated spacers and ligands. After addition of a suitable metal source, the resulting biotinylated organometallic catalyst precursors are combined with (strept)avidin to afford artificial metalloenzymes. Reprinted with permission of Swiss Chemical Society

Sav (which afforded low levels of pure protein), all mutants could be produced in high yield and purified from 10 litre *E. coli* fermentations.

13.3 Artificial Hydrogenases

Having synthesized a library of biotinylated diphosphine ligands, these were reacted with $[Rh(COD)]^+$ and combined with a Sav isoform. The resulting artificial metalloenzymes were screened for the reduction of *N*-acetamidodehydroalanine and *N*-acetamidodehydrophenylalanine to afford *N*-acetamidoalanine (*N*-AcAla) and

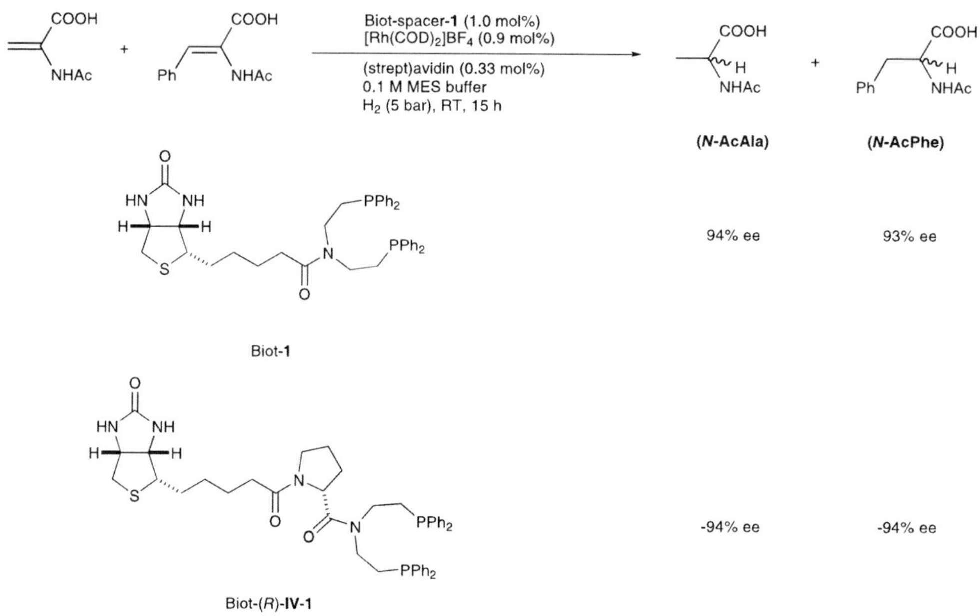

Figure 13.4 *Substrates and operating conditions used for the hydrogenation reaction catalyzed by artificial metalloenzymes based on the biotin–avidin technology*

Table 13.1 *Selected results from the asymmetric hydrogenation reactions*[a]

Entry	Protein	Ligand	ee% (Conv.%) N-AcAla	ee% (Conv.%) N-AcPhe
1	WT Sav	Biot-**1**	94 (quant.)	93 (84)
2	S112A	Biot-**1**	93 (quant.)	94 (94)
3[b]	S112W	Biot-(*R*)-**IV-1**	−94 (98)	−94 (86)
4[c]	WT Sav	Biot-(*R*)-**IV-1**	−87 (quant.)	−91 (quant.)
5[d]	WT Sav	Biot-(*R*)-**IV-1**	−87 (quant.)	−86 (94)
6	S112E	Biot-(*R*)-**IV-1**	−79 (quant.)	−55 (16)

[a] Positive *ee* values in favour of the (*R*)-enantiomer, negative *ee* values in favour of the (*S*)-enantiomer, quant. = Quantitative.
[b] Biphasic;
[c] 27% DMSO;
[d] 45% DMSO.

N-acetamidophenylalanine (*N*-AcPh) respectively, Figure 13.4. Overall, twenty two ligands were screened in the presence of the twenty Sav isoforms and WT Avi. Noteworthy results are collected in Table 13.1.[14,41–45] Based on these data, the following trends can be identified:

i) both (*R*)- and (*S*)-enantiomers of the product can be obtained in good enantioselectivities (>90% ee, Table 13.1 entries 1–6)
ii) chemical diversity (e.g. variation of the spacer) is a more versatile means for optimization that genetic diversity. This latter can be regarded as a fine tuning step.

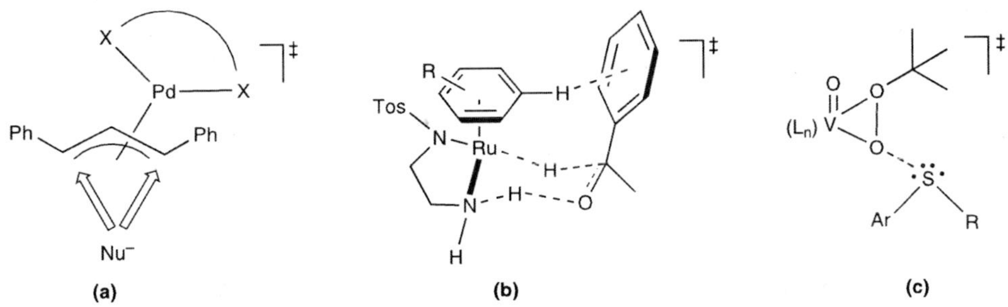

Figure 13.5 *Postulated transition states of homogeneous enantioselective reactions which proceed without coordination of one of the substrates to the metal centre: allylic alkylation (a), transfer hydrogenation (b) and sulfoxidation (c). Reprinted with permission of Swiss Chemical Society*

iii) introduction of conformationally constrained spacers (eg. proline **IV**) confers significant organic solvent tolerance to the systems. Both biphasic and DMSO/water mixtures yield good conversions and selections (Table 13.1 entries 3–5).[45]

iv) the artificial hydrogenases follow Michaelis-Menten kinetics. Introduction of the catalyst within streptavidin contributes to improve both the activity (eg. phenomenon of protein accelerated catalysis, compare [Rh(COD)(Biot–**1**)]$^+$ k_{cat} 3.06 min^{-1}, K_M 7.38 mM, with [Rh(COD)(Biot–(*R*)-**IV–1**)]$^+$ $\subset$ WT Sav k_{cat} 12.30 min^{-1}, K_M 4.36 mM) and the selectivity.[42]

v) although most ligand–protein combinations afford comparable levels of conversion and enantioselectivity for both substrates, certain combinations yield substrate specific artificial hydrogenases (Table 13.1 entry 6).

Having demonstrated the potential of artificial metalloenzymes for the reduction of *N*-protected dehydroaminoacids, we turned our attention towards organometallic-catalyzed reactions where the enantiodiscrimination step occurs without coordination of one of the reactants to the metal centre. We anticipated that incorporation of the metal complex within a protein environment may steer the enantioselection without requiring transient coordination to the metal. In this context, we selected the palladium-catalyzed asymmetric allylic alkylation, the ruthenium-catalyzed transfer hydrogenation as well as the vanadyl-catalyzed sulfoxidation reaction. Indeed, these reactions are believed to proceed without prior coordination of the soft nucleophile, the prochiral ketone or the prochiral sulfide respectively, Figure 13.5.

13.4 Artificial Allylic Alkylases

In the field of C—C bond forming reactions, palladium occupies a privileged position.[46] In this context, the asymmetric allylic alkylation attracted our interest due to its unique outersphere attack of the malonate on the η^3-coordinated 1,3-diphenylallyl moiety.[47]

Following the above-described chemogenetic optimization procedure, we screened twenty one biotinylated ligands in conjunction with twenty two (strept)avidin isoforms for the allylic alkylation of 1,3-diphenylallyl acetate. Addition of didodecyldimethylammonium bromide (DMB) allowed to significantly improve the yield of alkylation product – over the hydrolysis product, 1,3-diphenylallylalcohol, formed predominantly in the absence of DMB (Figure 13.6). This screening revealed the following trends:[19]

i) The nature of the spacer plays a critical role in conferring activity to the artificial alkylase. In most cases and despite the addition of DMB, hydrolysis product was formed predominantly. The best spacers are conformationally constrained: either proline **IV** or *o*-aminobenzoate **IIo**, Table 13.2, entries 1–3.
ii) Despite the presence of chaotropic agents (surfactant and basic pH), the artificial metalloenzyme behave remarkably well. While the high pH is required for the reaction to proceed, the reaction carried out in the absence of DMB leads to a significant erosion of the conversion, with only a modest increase in selectivity, Table 13.2, entries 3, 4.
iii) As previously observed for the hydrogenation reaction, conformationally constrained ligands confer improved organic solvent tolerance. Reactions with [Pd(η^3-allyl)(Biot–**IIo–1**)]$^+$ $\subset$ S112A Sav can be performed in 45% DMSO without addition of DMB, Table 13.2, entry 5.[48]
iv) Again here, the versatility of the chemical optimization allowed to rapidly identify an active biotinylated ligand. The genetic optimization can be regarded as a fine tuning step: whereas most streptavidin isoforms yielded (*R*)-alkylation products in

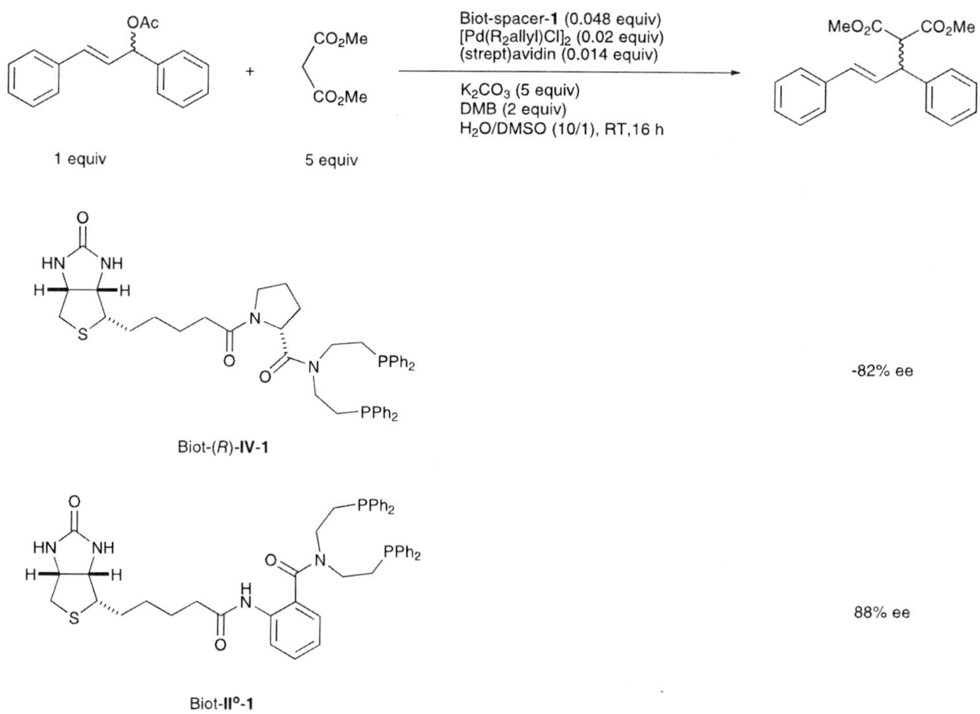

Figure 13.6 *Operating conditions used for the asymmetric allylic alkylation catalyzed by artificial metalloenzymes based on the biotin–avidin technology. Reprinted with permission of Swiss Chemical Society*

Table 13.2 *Selected results from the asymmetric allylic alkylations*

Entry	Protein	Ligand	Conv. (%)	ee (%)[a]
1	S112G V47G	Biot-(*R*)-**IV-1**	92	−82
2	S112G	Biot-**IIo-1**	95	88
3	S112A	Biot-**IIo-1**	95	90
4[b]	S112A	Biot-**IIo-1**	20	93
5[b,c]	S112A	Biot-**IIo-1**	90	95
6	S112Q	Biot-**IIo-1**	96	-31

[a] Positive *ee* values in favour of the (*R*)-enantiomer, negative *ee* values in favour of the (*S*)-enantiomer.
[b] No DMB was added.
[c] The reaction was performed in 45% DMSO.

conjunction with [Pd(η^3-allyl)(Biot–**IIo–1**)]$^+$, S112Q clearly stands out as it affords (*S*)-alkylation product, Table 13.2 entry 6.

13.5 Artificial Transfer Hydrogenase

The asymmetric transfer hydrogenation of ketones is an effective way to prepare enantiopure alcohols.[49] We were attracted to this reaction as we anticipated that one could exploit the reversibility of the reaction to perform either for the enantioselective reduction or for the kinetic resolution of racemic alcohols via oxidation.[50] This behaviour is reminiscent of alcohol dehydrogenases which can operate either as oxidases or reductases.[2]

Incorporation of a biotinylated d6-piano-stool complex bearing a biotinylated aminosulfonamide ligand into streptavidin afforded artificial transfer hydrogenases for the reduction of acetophenone derivatives. A buffered solution (pH 6.25) containing sodium formate as a hydrogen source proved best in terms of conversion and of selectivity. Unexpectedly, the performance of the artificial metalloenzyme was improved by raising the temperature to 55 °C.[18,39,51] A combined chemogenetic procedure combining 21 piano stool complexes with the twenty S112X Sav isoforms allowed to optimize the activity and the selectivity of these hybrid catalysts for the reduction of prochiral ketones, Figure 13.7. The following trends were identified:[52]

i) Only catalysts bearing the *para*-substituted spacer biot–**III^p–2** afforded any significant level of conversion.
ii) Ruthenium-based catalysts [Ru(η^6-arene)H(Biot–**III^p–2**)] outperformed either rhodium or iridium based systems [M(η^5-C_5Me_5)H(Biot–**III^p–2**)].
iii) Under otherwise identical conditions, the benzene [Ru(η^6-*p*-benzene)H(Biot–III^p–**2**)] and the *p*-cymene [Ru(η^6-*p*-cymene)H(Biot–III^p–**2**)] complexes afforded opposite enantiomers, Table 13.3, entries 1, 2.
iv) Various acetophenone derivatives could be reduced with good enantioselectivities up to 92% ee for (*S*)-products and 97% ee for (*R*)-products, Table 13.3, entries 3, 4.
v) In contrast, the selectivities obtained for the reduction of dialkylketones remained modest, Table 13.3, entries 5, 6.

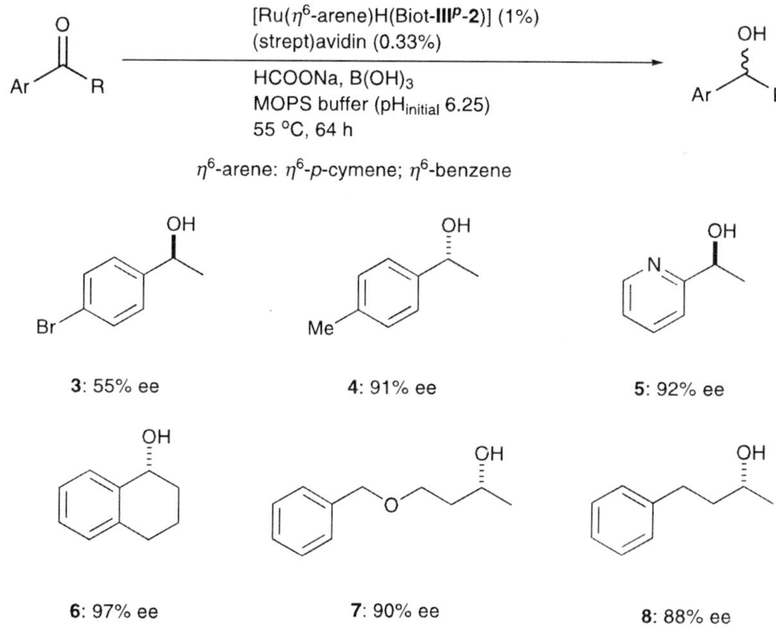

Figure 13.7 *Substrates and operating conditions used for the transfer hydrogenation catalyzed by artificial metalloenzymes based on the biotin–avidin technology. Reprinted with permission of Swiss Chemical Society*

Table 13.3 *Selected results from the asymmetric transfer hydrogenation using {Ru(η^6-arene)Cl(Biot–IIIp–**2**)}*

Entry	η^6-Arene	Protein	Product	Conv. (%)	ee (%)[a]
1	benzene	S112T	**3**	90	−55
2	*p*-cymene	S112A	**4**	98	91
3	*p*-cymene	S112A K121N	**5**	quant.[b]	−92
4	*p*-cymene	S112Y	**6**	79	97
5	*p*-cymene	S112A	**7**	97	69
6	*p*-cymene	S112A	**8**	98	48
7	*p*-cymene	S112A K121T	**7**	quant.[b]	90
8	*p*-cymene	S112A K121T	**8**	99	88

[a] Positive and negative *ee* values correspond to the *R* and *S* enantiomers, respectively.
[b] Quantitative.

In the context of transfer-hydrogenation, dialkylketones are challenging substrates as the major contribution to the enantiodiscrimination event is believed to be C—H······π interactions between the η^6-bound arene and the phenyl moiety on the prochiral substrate, Figure 13.5.[53–55] We anticipated that the introduction of additional sites of mutation in the proximity of the piano-stool moiety may allow to address this challenge. For this purpose, we invested considerable effort into crystallizing the most promising (*R*)- and (*S*)-selective artificial metalloenzymes: [Ru(η^6-benzene)Cl(Biot–**III**^{p}–**2**)] $\subset$ S112K Sav and [Ru(η^6-*p*-cymene)Cl(Biot–**III**^{p}–**2**)] $\subset$ S112A Sav. We were fortunate to succeed in crystallizing the

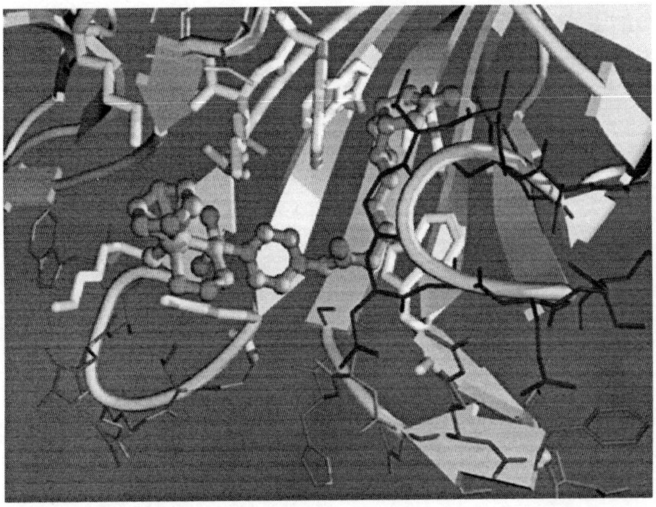

Figure 13.8 *Close-up view of the X-ray crystal structure of [Ru(η^6-p-benzene)Cl(Biot–IIIp–2)] ⊂ S112K Sav. The closest lying residues to the ruthenium are highlighted in yellow (K112A,$_B$ and K121A,$_B$) and orange (L124) respectively. The ruthenium position is partially occupied (20%). The absolute configuration at Ru is (S)*

former and its structure is depicted in Figure 13.8.[51] Based on this structure, we performed an additional round of mutations, concentrating on the close lying K121 and L124 residues and using WT Sav, S112K and S112A as background, thus affording a total 120 Sav isoforms. In order to speed up the optimization process, a straightforward Sav immobilization procedure was implemented.[51] For this purpose, commercially available biotinylated sepharose was added to the supernatant of the *E. coli* lysed cell cultures. After centrifugation, the immobilized Sav mutants, which on average possessed three remaining biotin-binding sites, are treated with an excess of [Ru(η^6-arene)Cl(Biot–**III**^{p}**–2**)], and washed and tested in the presence of Sav. This second optimization round allowed to identify artificial transfer hydrogenases which displayed much improved selectivities towards dialkylketones, Table 13.3, entries 7, 8.

Having developed an efficient artificial transfer hydrogenase, we attempted to apply the same methodology to the reverse reaction: the kinetic resolution of racemic alcohols. To our disappointment, we were forced to use strong oxidizing agents (eg. *t*-BuOOH rather than acetone, in the spirit of an Oppenauer-type mechanism) to drive the reaction to completion.[56] We speculate that, in the presence of water, the ruthenium is unable to abstract the β-hydrogen on the prochiral alcohol.

Despite the excitement generated by the X-ray structure, the localization of the biotinylated metal complex within the protein environment was a source of disappointment. Indeed, we anticipated that the active catalyst would be encapsulated within the biotin-binding pocket of Sav rather than on the surface of the protein. To overcome this problem we tested whether the polar residues present within the biotin-binding pocket itself may be capable of binding a polar coordination complex. With this goal in mind, we tested the catalytic potential of vanadyl as a catalyst.[57]

13.6 Enantioselective Sulfoxidation Based on Vanadyl-loaded Streptavidin

Addition of vanadyl sulfate to Sav affords an artificial oxidase for the sulfoxidation of prochiral sulfides. The system is most efficient at low pH (2.2) and organic peroxides are required for efficient catalysis, Figure 13.9. From the screening, the following trends emerge:[21]

i) A variety of prochiral sulfides are oxidized to the corresponding sulfoxides with no overoxidation.
ii) At room temperature and relying on 0.02 equivalents vanadium, the conversions are complete in most cases and the (*R*)-selectivities often exceed 90%, Table 13.4, entries 1-5.
iii) Since there is no ligand, the system cannot be optimized chemically.
iv) In the presence of biotin-loaded Sav, the sulfoxidation with $VOSO_4$ affords racemic product, suggesting that vanadium cannot compete for binding with biotin, Table 13.4, entry 6.

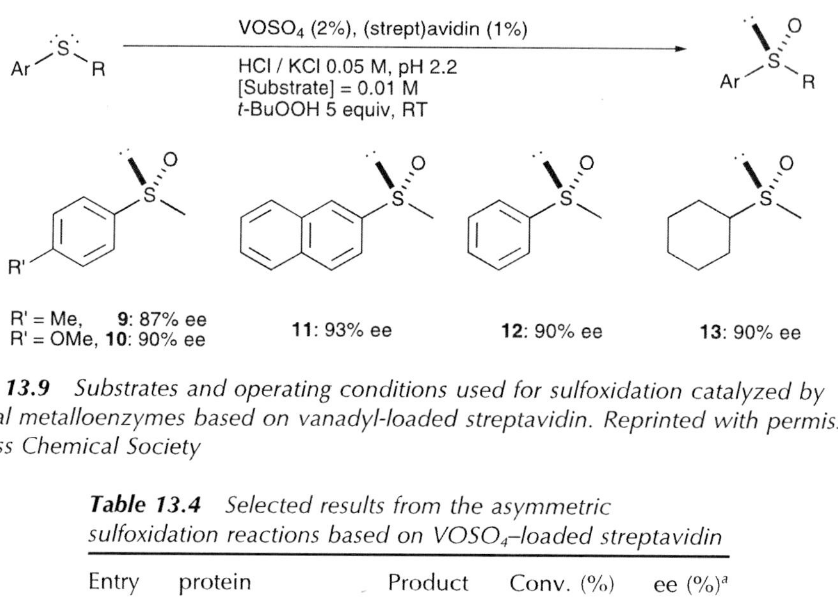

Figure 13.9 *Substrates and operating conditions used for sulfoxidation catalyzed by artificial metalloenzymes based on vanadyl-loaded streptavidin. Reprinted with permission of Swiss Chemical Society*

Table 13.4 *Selected results from the asymmetric sulfoxidation reactions based on $VOSO_4$–loaded streptavidin*

Entry	protein	Product	Conv. (%)	ee (%)[a]
1	WT Sav	**9**	96	87
2	WT Sav	**10**	quant.[b]	90
3	WT Sav	**11**	53	93
4	WT Sav	**12**	96	90
5	WT Sav	**13**	61	86
6	WT Sav + biotin	**10**	96	0
7	D128A	**10**	97	0
8	Aviloop[c]	**13**	54	90

[a] (*R*)-enantiomer produced preferentially.
[b] Quantitative.
[c] see reference [59].

v) It is widely recognized that residue D128 Sav is most important in biotin-binding.[58] The mutant D128A Sav affords racemic sulfoxide, thus suggesting that the vanadyl moiety is located within the biotin-binding site, Table 13.4, entry 7.
vi) Replacement of the L3,4-loop of Sav by the corresponding L3,4-loop of Avi affords a chimaeric protein (Aviloop)[59] with improved properties for the sulfoxidation of dialkylsulfides, Table 13.4 entries 8.

13.7 Conclusions and Outlook

As summarized here, artificial metalloenzymes based on the biotin–avidin technology have developed into a versatile approach to enantioselective catalysis. In many cases, selectivities exceeding 92% ee could be obtained, relying on a chemogenetic optimization strategy. Such hybrid catalysts display features which are reminiscent of both homogeneous and enzymatic catalysis.

Comparing alternative anchoring strategies (i.e. dative, covalent or supramolecular),[5] we speculate that a dative anchoring between an amino acid side chain and the catalytically active metal moiety may provide an attractive means to ensure a better encapsulation. In this context, one should mention the interesting results recently disclosed by the groups of Kazlauskas and Soumillon.[25,26] Substitution of Zn by Mn in carbonic anhydrase yields an artificial epoxidase (up to 55% ee (*R*)). Along similar lines, Sheldon and co-workers investigated the sulfoxidation properties of a vanadium-incorporated phytase. After incorporation of vanadate into phytase from *A. ficuum*, the resulting semisynthetic peroxidase displayed moderate enantioselectivity (up to 66% ee (*S*)) for the sulfoxidation of thioanisole derivatives in the presence of hydrogen peroxide, with very little detectable overoxidation.[20] In contrast to our systems, the above two examples of artificial metalloenzymes display a very narrow substrate scope. This suggests that the dative anchoring strategy provides a more 'enzyme-like' second coordination sphere environment (i.e. Lock-and-Key) for the substrate.

Although the reactions implemented so far are limited to model systems, we believe that artificial metalloenzymes will reveal their full potential for transformations for which there exists to date no good homogeneous catalyst. In this context, our current efforts are directed towards the hydroxylation of alkanes (where overoxidation is difficult to prevent with homogeneous systems) as well as highly sequence-specific DNA hydrolysis.

Acknowledgements

This work was funded by the Swiss National Science Foundation (Grants FN 200021-105192 and 200020-113348), the Roche Foundation as well as the FP6 Marie Curie Research Training Network (MRTN-CT-2003-505020) and the Canton of Neuchâtel. We thank Umicore Precious Metals Chemistry for a loan of ruthenium as well as C. R. Cantor for the streptavidin gene.

References

1. *Comprehensive Asymmetric Catalysis*, E. N. Jacobsen, A. Pfaltz, H. Yamamoto (Eds.), Springer, Berlin, 1999.
2. K. Faber, *Biotransformations in Organic Chemistry*, 5th ed., Springer, Berlin, 2004.
3. M. Breuer, K. Ditrich, T. Habicher, B. Hauer, M. Kesseler, R. Stürmer, T. Zelinski, Industrial methods for the production of optically active intermediates, *Angew. Chem. Int. Ed.*, 2004, **43**, 788–824.
4. R. Berkessel, H. Gröger, *Asymmetric Organocatalysis: From Biomimetic Concepts to Applications in Asymmetric Synthesis*, Wiley-VCH, Weinheim, 2005.
5. C. Letondor, T. R. Ward, Artificial metalloenzymes for enantioselective catalysis: recent advances, *ChemBioChem*, 2006, **7**, 1845–1852.
6. E. T. Kaiser, D. S. Lawrence, Chemical mutation of enzyme active sites, *Science*, 1984, **226**, 505–511.
7. M. E. Wilson, G. M. Whitesides, Conversion of a protein to a homogeneous asymmetric hydrogenation catalyst by site-specific modification with a diphosphinerhodium(I) moiety, *J. Am. Chem. Soc.*, 1978, **100**, 306–307.
8. O. Pàmies, J.-E. Bäckvall, Combination of enzymes and metal catalysts. A powerful approach in asymmetric catalysis, *Chem. Rev.*, 2003, **103**, 3247–3262.
9. T. Ueno, T. Koshiyama, S. Abe, N. Yokoi, M. Ohashi, H. Nakajima, Y. Watanabe, Design of artificial metalloenzymes using non-covalent insertion of a metal complex into a protein scaffold, *J. Organomet. Chem.*, 2007, **692**, 142–147.
10. Y. Lu, Biosynthetic inorganic chemistry, *Angew. Chem. Int. Ed.*, 2006, **45**, 5588–5601.
11. G. Roelfes, DNA and RNA induced enantioselectivity in chemical synthesis, *Mol. Biosyst.*, 2007, **3**, 126–135.
12. D. Qi, C.-M. Tann, D. Haring, M. D. Distefano, Generation of new enzymes via covalent modification of existing proteins, *Chem. Rev.*, 2001, **101**, 3081–3112.
13. C. M. Thomas, T. R. Ward, Artificial metalloenzymes: proteins as hosts for enantioselective catalysis, *Chem. Soc. Rev.*, 2005, **34**, 337–346.
14. J. Collot, J. Gradinaru, N. Humbert, M. Skander, A. Zocchi, T. R. Ward, Artificial metalloenzymes for enantioselective catalysis based on biotin–avidin, *J. Am. Chem. Soc.*, 2003, **125**, 9030–9031.
15. M. T. Reetz, J. J.-P. Peyeralans, A. Maichele, Y. Fu, M. Maywald, Directed evolution of hybrid enzymes: evolving enantioselectivity of an achiral Rh-complex anchored to a protein, *Chem. Commun.*, 2006, 4318–4320.
16. H. Yamaguchi, T. Hirano, H. Kiminami, D. Taura, A. Harada, Asymmetric hydrogenation with antibody-achiral rhodium complex, *Org. Biomol. Chem.*, 2006, **4**, 3571–3573.
17. C.-C. Lin, C.-W. Lin, A. S. C. Chan, Catalytic hydrogenation of itaconic acid in a biotinylated Pyrphos–rhodium(I) system in a protein cavity, *Tetrahedron Asymmetry*, 1999, **10**, 1887–1893.
18. C. Letondor, N. Humbert, T. R. Ward, Artificial metalloenzymes based on biotin–avidin technology for the enantioselective reduction of ketones by transfer hydrogenation, *Proc. Natl. Acad. Sci. USA*, 2005, **102**, 4683–4687.
19. J. Pierron, C. Malan, M. Creus, J. Gradinaru, I. Hafner, A. Ivanova, A. Sardo, T. R. Ward, Artificial metalloenzymes for asymmetric allylic alkylation on the basis of the biotin–avidin technology, *Angew. Chem. Int. Ed.*, 2008, **47**, 701–705.
20. F. van de Velde, L. Könemann, F. v. Rantwijk, R. A. Sheldon, Enantioselective sulfoxidation mediated by vanadium-incorporated phytase: a hydrolase acting as a peroxidase, *Chem. Commun.*, 1998, 1891–1892.
21. A. Pordea, M. Creus, J. Panek, C. Duboc, M. Novic, D. Mathis, T. R. Ward, Artificial metalloenzyme for enantioselective sulfoxidation based on vanadyl-loaded streptavidin, *J. Am. Chem. Soc.*, 2008, **130**, 8085–8088.
22. A. Mahammed, Z. Gross, Albumin-conjugated corrole metal complexes: extremely simple yet very efficient biomimetic oxidation systems, *J. Am. Chem. Soc.*, 2005, **127**, 2883–2887.

23. M. Ohashi, T. Koshiyama, T. Ueno, M. Yanase, H. Fujii, Y. Watanabe, Preparation of artificial metalloenzymes by insertion of chromium(III) schiff base complexes into apomyoglobin mutants, *Angew. Chem. Int. Ed.*, 2003, **42**, 1005–1008.
24. J. R. Carey, S. K. Ma, T. D. Pfister, D. K. Garner, H. K. Kim, J. A. Abramite, Z. Wang, Z. Guo, Y. Lu, A site-selective dual anchoring strategy for artificial metalloprotein design, *J. Am. Chem. Soc.*, 2004, **126**, 10812–10813.
25. K. Okrasa, R. J. Kazlauskas, Manganese-substituted carbonic anhydrase as a new peroxidase, *Chem. Eur. J.*, 2006, **12**, 1587–1596.
26. A. Fernandez-Gacio, A. Codina, J. Fastrez, O. Riant, P. Soumillon, Transforming carbonic anhydrase into epoxide synthase by metal exchange, *ChemBioChem*, 2006, **7**, 1013–1016.
27. T. Kokubo, T.Sugimoto, T. Uchida, S. Tanimoto, M. Okano, The bovine serum albumin–2-phenylpropane-1,2-diolatodioxo-somium(VI) complex as an enantioselective catalyst for cis-hydroxylation of alkenes, *J. Chem. Soc., Chem. Commun.*, 1983, 769–770.
28. M. T. Reetz, N. Jiao, Copper-phthalocyanine conjugates of serum albumins as enantioselective catalysts in Diels–Alder reactions, *Angew. Chem. Int. Ed.*, 2006, **45**, 2416–2419.
29. A. J. Boersma, J. E. Klijn, B. L. Feringa, G. Roelfes, DNA-based asymmetric catalysis: sequence-dependent rate acceleration and enantioselectivity, *J. Am Chem. Soc.*, 2008, **130**, 11783–11790.
30. R. S. Roy, B. Imperiali, Pyridoxamine-amino acid chimeras in semisynthetic aminotransferase mimics, *Prot. Engineering*, 1997, **10**, 691–698.
31. D. Coquière, B. L. Feringa, G. Roelfes, DNA-based catalytic enantioselective Michael reactions in water, *Angew. Chem. Int. Ed.*, 2008, **46**, 9308–9311.
32. N. Shibata, H. Yasui, S. Nakamura, T. Toru, DNA-mediated enantioselective carbon-fluorine bond formation, *Synlett*, 2007, 1153–1157.
33. M. Wilchek, E. A. Bayer, *Methods in Enzymology*, Vol. 184, Academic Press, San Diego, 1990.
34. A. Loosli, U. E. Rusbandi, J. Gradinaru, K. Bernauer, C. W. Schläpfer, M. Meyer, S. Mazurek, M. Novic, T. R. Ward, (Strept)avidin as host for biotinylated coordination complexes: stability, chiral discrimination, and cooperativity, *Inorg. Chem.*, 2006, **45**, 660–668.
35. R. J. McMahon, *Methods in Molecular Biology*, Vol. 418, J. M. Walker (Ed.), Humana Press, Totowa, 2008.
36. M. Creus, T. R. Ward, Designed evolution of artificial metalloenzymes: protein catalysts made to order, *Org. Biomol. Chem.*, 2007, **5**, 1835–1844.
37. N. Humbert, A. Zocchi, T. R. Ward, Electrophoretic behavior of streptavidin complexed to a biotinylated probe: a functional screening assay for biotin-binding proteins, *Electrophoresis*, 2005, **26**, 47–52.
38. A. Zocchi, A. M. Jobé, J.-M. Neuhaus, T. R. Ward, Expression and purification of a recombinant avidin with a lowered isoelectric point in Pichia pastoris, *Prot. Expr. Purif.*, 2003, **32**, 167–174.
39. C. Letondor, A. Pordea, N. Humbert, A. Ivanova, S. Mazurek, M. Novic, T. R. Ward, Artificial transfer hydrogenases based on the biotin-(strept)avidin technology: fine tuning the selectivity by saturation mutagenesis of the host protein, *J. Am. Chem. Soc.*, 2006, **128**, 8320–8328.
40. S. Mazurek, T. R. Ward, M. Novic, Counter propagation artificial neural networks modeling of an enantioselectivity of artificial metalloenzymes, *Mol. Divers.*, 2007, **11**, 141–152.
41. G. Klein, N. Humbert, J. Gradinaru, A. Ivanova, F. Gilardoni, U. E. Rusbandi, T. R. Ward, Tailoring the active site of chemzymes by using a chemogenetic-optimization procedure: towards substrate-specific artificial hydrogenases based on the biotin–avidin technology, *Angew. Chem. Int. Ed.*, 2005, **44**, 7764–7767.
42. U. E. Rusbandi, C. Lo, M. Skander, A. Ivanova, M. Creus, N. Humbert, T. R. Ward, Second generation artificial hydrogenases based on the biotin–avidin technology: improving activity, stability and selectivity by introduction of enantiopure amino acid spacers, *Adv. Synth. Catal.*, 2007, **349**, 1923–1930.
43. J. Collot, N. Humbert, M. Skander, G. Klein, T. R. Ward, Artificial metalloenzymes for enantioselective catalysis: the phenomenon of protein accelerated catalysis, *J. Organomet. Chem.*, 2004, **689**, 4868–4871.

44. T. R. Ward, Artificial metalloenzymes for enantioselective catalysis based on the noncovalent incorporation of organometallic moieties in a host protein, *Chem. Eur. J.*, 2005, **11**, 3798–3804.
45. M. Skander, C. Malan, A. Ivanova, T. R. Ward, Chemical optimization of artificial metalloenzymes based on the biotin–avidin technology: (*S*)-Selective and solvent-tolerant hydrogenation catalysts via the introduction of chiral amino acid spacers, *Chem. Commun.*, 2005, 4815–4817.
46. J. Tsuji, *Topics in Organometallic Synthesis*, Vol. 14, Springer, Berlin, Heidelberg and New York, 2005.
47. B. M. Trost, M. L. Crawley, Asymmetric transition-metal-catalyzed allylic alkylations: applications in total synthesis, *Chem. Rev.*, 2003, **103**, 2921–2944.
48. C. Lo, T. R. Ward, *unpublished results*, 2009.
49. K.-J. Haack, S. Hashiguchi, A. Fujii, T. Ikariya, R. Noyori, The catalyst precursor, catalyst, and intermediate in the Ru(II)-promoted asymmetric hydrogen transfer between alcohols and ketones, *Angew. Chem. Int. Ed. Engl.*, 1997, **36**, 285–288.
50. S. Hashiguchi, A. Fujii, K.-J. Haack, K. Matsumura, T. Ikariya, R. Noyori, Kinetic resolution of racemic secondary alcohols by Ru-catalyzed hydrogen transfer, *Angew. Chem. Int. Ed. Engl.*, 1997, **36**, 288–290.
51. M. Creus, A. Pordea, T. Rossel, A. Sardo, C. Letondor, A. Ivanova, I. Le Trong, R. E. Stenkamp, T. R. Ward, X-ray structure and designed evolution of an artificial transfer hydrogenase, *Angew. Chem. Int. Ed.*, 2008, **47**, 1400–1404.
52. A. Pordea, T. R. Ward, Chemogenetic protein engineering: an efficient tool for the optimization of artificial metalloenzymes, *Chem. Commun.*, 2008, 4239–4349.
53. M. Yamakawa, I. Yamada, R. Noyori, CH/π Attraction: the origin of enantioselectivity in transfer hydrogenation of aromatic carbonyl compounds catalyzed by chiral η^6-arene-ruthenium(II) complexes, *Angew. Chem. Int. Ed.*, 2001, **40**, 2818–2821.
54. A. Schlatter, M. K. Kundu, W.-D. Woggon, Enantioselective reduction of aromatic and aliphatic ketones catalyzed by ruthenium complexes attached to β-cyclodextrin, *Angew. Chem. Int. Ed.*, 2004, **43**, 6731–6734.
55. A. Schlatter, W.-D. Woggon, Enantioselective transfer hydrogenation of aliphatic ketones catalyzed by ruthenium complexes linked to the secondary face of β-cyclodextrin, *Adv. Synth. Catal.*, 2008, **350**, 995–1000.
56. C. M. Thomas, C. Letondor, N. Humbert, T. R. Ward, Aqueous oxidation of alcohols catalyzed by artificial metalloenzymes based on the biotin–avidin technology, *J. Organomet. Chem.*, 2005, **690**, 4488–4491.
57. K. A. Jorgensen, Transition-metal-catalyzed epoxidations, *Chem. Rev.*, 1989, **89**, 431–458.
58. J. DeChancie, K. N. Houk, The origins of femtomolar protein–ligand binding: hydrogen-bond cooperativity and desolvation energetics in the biotin–(strept)avidin binding site, *J. Am Chem. Soc.*, 2007, **129**, 5419–5429.
59. Y. Eisenberg-Domovich, Y. Pazy, O. Nir, B. Raboy, E. A. Bayer, M. Wilchek, O. Livnah, Structural elements responsible for conversion of streptavidin to a pseudoenzyme, *Proc. Natl. Acad. Sci. USA*, 2004, **101**, 5916–5921.

14

Chemical Reactions with RNA and DNA Enzymes

Andres Jäschke
Institut für Pharmazie und Molekulare Biotechnologie (IPMB), Universität Heidelberg, Im Neuenheimer Feld 364, D-69120 Heidelberg, Germany

14.1 Introduction

One of the most fascinating questions at the interface of chemistry and biology is how enzymes achieve the catalysis of chemical reactions. These biocatalysts bring about tremendous rate accelerations under mild conditions with often extreme chemo-, regio- and stereoselectivities. For decades, generations of biochemists, enzymologists and structural biologists have been attracted to this phenomenon, and consequently, a rather thorough understanding has been achieved regarding the catalytic mechanisms of protein enzymes.

As enzymes encapsulate multiple functionalities within their catalytic cavity, they have also served as a major source of inspiration for the fields of biomimetic chemistry and supramolecular catalysis. Early mechanistic theories about how enzymes work have prompted scientists from various fields to explore similar approaches for synthetic systems. One of these approaches is host-guest catalysis, where one or more substrates are bound in a cavity next to the catalytically active site.

Enzymes are, however, much more than just a combination of substrate binding site and the catalytically active site. Important contributions to enzymatic catalysis arise from substrate preorganization, restriction of substrate motion, catalyst dynamics, transition state binding, and desolvation of the substrates, and natural evolution has used these strategies over long periods of time to fine-tune catalytic properties. In synthetic systems, such effects could so far only be utilized to a lesser extent.

Since the early 1980s it has become clear that the exclusive assignment of enzymatic activity to the chemical substance class of proteins in incorrect, as ribonucleic acids were discovered that possess such activities. Most of these naturally occurring 'ribozymes' (shortened for '*ribo*nucleic acid en*zymes*' catalyse (at least to our present-day knowledge) the hydrolysis and formation of phosphodiester bonds, and for some of these systems, high-resolution crystal structures and mechanistic investigations provide a basis for an understanding of how these ribozymes work.[1,2] Combinatorial chemistry on the other hand has allowed to isolate (or generate) RNA and also DNA catalysts with activities not known to exist in biology. It turns out that – although equipped with a much less varied arsenal of chemically different building blocks – nucleic acids can form tertiary structures with binding pockets and catalytic centres very much like proteins.

The history of ribozymes, in the true sense of the word, is connected to the so-called RNA world,[3,4] a hypothetic prebiotic stage in the evolution of life, where the majority of catalytic functions were effected by RNA, and proteins not yet existed. The idea of an RNA world alone implies that RNA should be practically omnipotent in the diversity of its catalytic potential, despite having been outperformed by proteins later on. The investigation of RNA catalysis provides a simplified and contrasting perspective on macromolecular catalytic mechanisms compared with protein enzymes, and an improved understanding of the chemical origins of ribozyme catalysis may add to our knowledge on macromolecular catalysis in general.

While enzymological research over several decades has provided a rather detailed understanding of how certain enzymes accelerate chemical reactions, knowledge about ribozyme mechanisms and catalytic strategies is scarce in comparison. For some natural ribozymes, mechanistic and structural studies were carried out, but artificial ribozymes remain largely uncharacterized. Therefore it is unclear in many cases whether these catalysts are passive containers working solely by substrate pre-organization, or whether there are specific effects on the activation parameters of the reaction.

In this review, I will first give a brief overview on the types of chemical reactions catalysed by naturally occurring ribozymes, including some mechanistic considerations. This is followed by a discussion of the methods used for generating new nucleic acid enzymes, and a compilation of artificial ribozymes and deoxyribozymes. Finally, I will discuss the current knowledge on the arguably best-characterized artificial ribozymes, namely RNA enzymes that catalyse the formation of C—C bonds by Diels–Alder reaction, with special emphasis on crystallographic and complementary mechanistic investigations.

14.2 Catalysis by Naturally Occurring Ribozymes

In the contemporary biosphere, ribozymes carry out a limited range of chemical reactions, mostly involving phosphoryl transfer, and in particular transesterification and hydrolysis reactions at internucleotide phosphodiester bonds (Table 14.1). They are widespread in nature, from bacteria and their phages, archaea, to yeasts and fungi and higher eukaryotes, and are used by these organisms mostly for processing of certain ribonucleic acids. They either cut oligomeric replication products to unity length, a crucial process in viral repli-

Table 14.1 *Classes of natural ribozymes*

Ribozyme	Reaction	Mechanism
Hammerhead	viral RNA cleavage	Transesterification (adjacent 2′-OH)
Hairpin	viral RNA cleavage	Transesterification (adjacent 2′-OH)
Hepatitis Delta Virus	viral RNA cleavage	Transesterification (adjacent 2′-OH)
Varkud Satellite	viral RNA cleavage	Transesterification (adjacent 2′-OH)
glmS	mRNA cleavage	Transesterification (adjacent 2′-OH)
Group I Intron	RNA self-splicing	Nucleotidyl transfer (external guanosine)
Group II Intron	RNA self-splicing	Nucleotidyl transfer (remote 2′-OH)
RNase P	tRNA processing	Hydrolysis
Ribosome	Peptide biosynthesis	Peptidyl transfer

cation, they trim the ends of the precursors of tRNA, they remove an internal sequence fragment (intron) from a precursor transcript, or they destroy an mRNA in response to a signal. All these reactions are highly site- and sequence-specific. In some cases, these ribozymes act on themselves (self-cleavage or *in-cis* cleavage, i.e., they are enzyme and substrate at the same time), in others they act as true enzymes on external RNA substrates (*in-trans* cleavage). The RNA-cleaving enzymes can be grouped by the nature of the nucleophile they use for RNA cleavage (Table 14.1): They either utilize the 2′-hydroxyl group of the ribose adjacent to the phosphodiester bond to be cleaved (small viral and viroidal RNAs), or a distant 2′-OH group (in splicing, group II introns), a water molecule (Ribonuclease P), or a small organic molecule (group I introns). Finally, the peptidyl transferase activity of the ribosome is also ribozyme-based and catalyses what is arguably the most important reaction of the cell, the condensation of amino acids into polypeptides.

Just as protein enzymes must be correctly folded into the conformation required for catalytic activity, so must RNA. In general, the precise nature of Watson–Crick base pairing leads to the relatively easy formation of secondary structure. Most critical for catalysis and substrate binding is, however, the formation of the tertiary structure. The negatively charged phosphate backbone of RNA results in a strong electrostatic contribution to this process, and therefore metal ions play important roles in RNA structure formation. The folded RNA structure can bind metal ions, either site-specifically or diffusely, and these bound ions can play a direct role in catalysis.[1,2]

Compared to proteins, RNA has only very limited resources to build a catalytic site. Only four monomeric building blocks with low structural and chemical diversity can be utilized. The nucleobases have hydrogen bond donors and acceptors that can be used to bind the substrate, and potentially to stabilize a transition state. In principle they could also act as general acids and bases, but their pK_a values are either too high or too low to make them significant contributors at physiological pH values. This situation can, however, be changed by neighbouring phosphate groups or metal ions. The second potential players are metal ions, and their associated water molecules. Metal ions can act as Lewis acids, or provide electrostatic stabilization of negative charge in a transition state. Water molecules contained within the inner coordination sphere may participate in general acid–base catalysis. In addition to chemical participants, RNA can also potentially exploit its structure to contribute to catalysis. Substrate binding can result in acceleration of reaction velocity due to proximity and orientation, together with structural stabilization of the transition state.

Given the relative paucity of potential catalytic groups present in RNA molecules, ribozymes achieve some impressive rate accelerations. Typical values for the RNA-cleaving ribozymes reach about 10^6-fold over the background reaction, with rates of ~1 min^{-1}, and some variants were reported to reach over 10 s^{-1}.

The field of natural ribozymes has been extensively reviewed.[1,2]

14.3 How to Generate Artificial RNA and DNA Catalysts

RNA molecules with new catalytic activities were first reported in the 1990s. These were isolated by a technique termed *in vitro* selection or SELEX (systematic evolution of ligands by exponential enrichment), starting from synthetic combinatorial RNA libraries (pools), which typically contain 10^{14}–10^{15} different sequences, i.e., 10^{14}–10^{15} chemically different compounds. These enormous complexities can only be handled since nucleic acid molecules are genetically encoded, i.e., they carry the information for their own replication. This property allows for an iterative deconvolution of the libraries over several rounds. Such SELEX experiments have led to nucleic acids that catalyse a broad range of chemical transformations, ranging from cleavage of amide- or carboxylic ester bonds to amide-bond-, C—C- and C—S-bond forming reactions to the catalysis of redox reactions (see below).

The vast majority of in vitro selection experiments, however, did not focus on catalysts but on RNA molecules that specifically bind a given target molecule, so-called aptamers.[5] The most crucial step in any variant of this technique is selection, meaning an effective distinction between desired and undesired species. It is in the selection step that the most decisive advances in ribozyme discovery were made, as will become clear in the discussion below. The standard selection event in aptamer SELEX is a binding event, and consequently the desired molecules can be distinguished from the rest by physical separation, namely by affinity chromatography.

The basic (aptamer) SELEX protocol starts with a chemically synthesized DNA library (Figure 14.1). By using monomer (phosphoramidite) mixtures in solid phase oligonucleotide synthesis, huge complexities can be easily achieved. If, for example, a stretch of 20 nucleotides is synthesized completely random by using a mixture of the four standard monomers, a mixture of 4^{20} or 10^{12} different DNA sequences is generated. In the second step, the single-stranded DNA is made double-stranded, and several copies are made by using polymerase chain reaction (PCR). Using a DNA-dependent RNA polymerase, a single-stranded RNA library is finally generated from the double-stranded DNA library that is used as input for the first round of selection. The target that the RNA molecules are expected to bind is immobilized on a solid support, and the RNA library is added and allowed to interact for a certain period of time. Non-binding RNAs are removed by washing, while binders are recovered. They are then converted into cDNA by reverse transcription, amplified by PCR, and this cycle is repeated several times until a significant fraction of the RNA binds to the immobilized target with high affinity. The enzymatic steps require the incorporation of certain constant sequence elements (primer binding sites, promoters) into the design of the starting library.

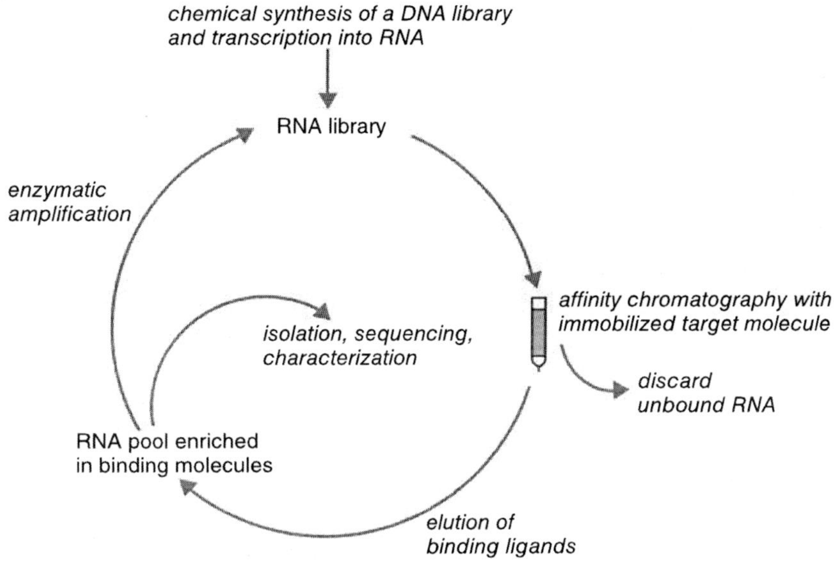

Figure 14.1 *General scheme for the in vitro selection of RNA aptamers*

Conceptually starting from this basic SELEX protocol, several strategies to create RNAs with catalytic properties were developed. The first one relies on transition state analogues (TSAs) and had been emulated from the catalytic antibodies field. Aptamers against transition state analogs of different reactions were raised and tested for catalytic activity. The Schultz group reported catalysis of a biphenyl isomerisation[6] and porphyrin metalation reaction.[7] Cholesterol esterase activity of yet another RNA isolated by the TSA approach was found by Chun *et al.*[8] While all these RNAs act as catalysts, the rate acceleration was rather low (below 500-fold in most cases), as compared to 10^3–10^5 for most TSA antibodies and up to 10^{18} for natural enzymes. Other attempts to isolate catalytic RNAs via TSAs, for example for Diels–Alder reactions, have thus far remained at the stage of aptamer binding.[9]

Better acceleration rates and catalysis of more complex reactions were obtained by direct selection. In this strategy, selection is not based on a binding event; rather it is based directly on the desired catalytic event. For example, if the to-be-catalysed reaction is phosphodiester hydrolysis, the RNA library is immobilized on solid support. The selection event is thus directly tailored to hydrolysis: active molecules cleave their covalent linkage to the solid support and thus dissociate from it. The catalytic event thus enables spatial separation of active and inactive molecules. Direct selection has therefore allowed to isolate catalysts for numerous RNA-modifying reactions, from hydrolysis and ligation to RNA alkylation, acylation, phosphorylation and many others (see below).

The use of reactants other than RNA in direct selection allows targeting of more complex reactions. An essential prerequisite is the attachment of a (non-RNA) reactant X to the RNA molecules, i.e., conversion of the RNA library into a RNA-reactant library. Assuming a desired RNA catalyst of the chemical reaction $X \rightarrow Z$, which may include simple addition as in $X + Y \rightarrow Z$, the selection has to be designed in a way to isolate

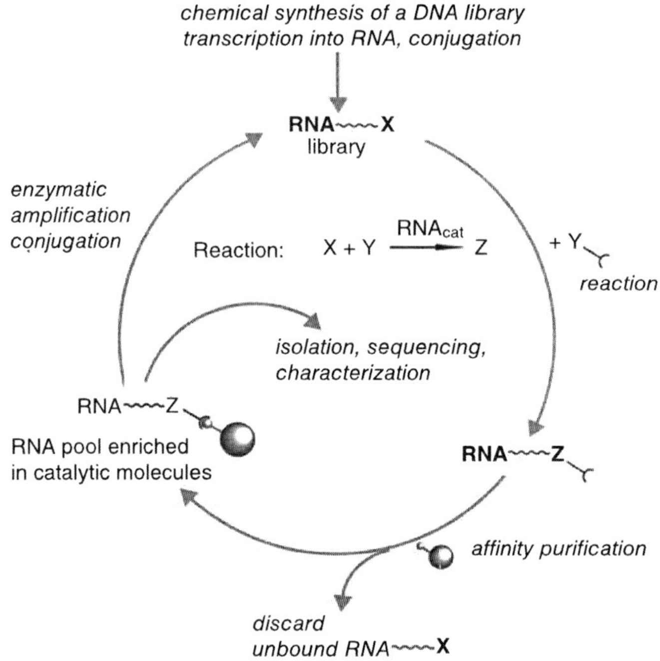

Figure 14.2 *Direct selection with tethered reactants for the isolation of RNA catalysts*

all RNA library members carrying the target reaction product Z. This can be through chemical trapping or affinity purification of Z directly, or by designing Y in a way that the final product contains a functional moiety of Y which allows facile affinity purification.

An exciting example for the former case is the ribocatalytic oxidation of benzyl alcohol to benzaldehyde using NAD^+ as a cofactor: the desired aldehyde product was chemically trapped by a biotin-conjugated hydrazine derivative.[10] This case will be more explicitly discussed below. The latter case has been successfully demonstrated by several labs employing a biotin-conjugated reactant Y thus allowing purification of the reactive species by immobilized streptavidin, a biotin-binding protein. A variant of direct selection uses tethered reactants. This design, tailored to find catalysts for the chemical addition of compound Y to compound X to yield the addition product Z, is depicted in Figure 14.2. Reactant X is now attached to the RNA via a long poly(ethylene glycol) tether, allowing it to interact with potential catalytic pockets in the RNA that are spatially remote from either extremity. In our lab, using tethered reactants, a ribozyme was selected that catalyses a Diels–Alder reaction. Anthracene tethered to RNA via a PEG-linker corresponds to reactant X, in this case a diene, and biotin-maleimide was employed as reactant Y, the free dienophile.[11] This design allowed easy isolation of the reaction product Z by streptavidin agarose.

Depending on the reactivity of reactant Y towards the functional groups contained in RNA, its conjugation to species of the RNA library will occur with more or less site-specificity to the conjugated X-reactant. More reactive Ys may add to RNA at positions other than the desired reactant X, thus also fulfilling the practical requirements for selection and amplification. Results of such selections were occasionally reported in the literature.[12] A strategy to circumvent non-specific addition involves the incorporation of a chemically orthogonal cleavage site between reactant X and the RNA library. After immobilization as a primary selection event, a secondary selection event is created by a treatment targeted specifically to the cleavage site. If immobilization of a given RNA sequence is based on addition of reactant Y to reactant X with correct site specificity, the cleavage will release the desired sequence into solution for amplification. However, addition of Y to sites other than X anywhere in the RNA will result in persistent immobilization during the secondary selection event, thus effectively removing unwanted species from the amplification cycle. The chemistry used for the cleavage event in the secondary selection step should not interact in any way with the rest of the conjugated library. Orthogonal approaches described to-date include a disulfide bridge embedded within the tether, and photocleavable *o*-nitrobenzyl groups.

The strategies to generate catalytic DNA molecules are almost identical and start with the same type of libraries. While one would intuitively expect that these selections are easier to perform since transcription and reverse transcription are eliminated from the selection cycles, in practice they are not, since the double-stranded DNA libraries resulting from PCR-type amplifications are catalytically incompetent, and their conversion into high-quality single-stranded DNA libraries is not trivial.

14.4 The Catalytic Spectrum of Artificial Ribozymes

In the following the scope of *in vitro* selected ribozymes will be described with special regard to organic chemistry. Several highlights will be discussed in detail. In Table 14.2 the variety of reactions catalysed by ribozymes is summarized.

The catalytic spectrum of RNA can not be discussed without the background of a hypothetical prebiotic RNA world. As stated in the introduction, the concept of an RNA world means that RNA, at one point, may have performed almost all catalytic functions necessary for survival. As Woese points out,[25] the term survival does not necessarily apply to a discrete, living entity or even to distinct species, be they single molecules or single cells. Therefore, and because it is unclear which chemical reactions exactly might have been catalysed by RNA or other matter like catalytic surfaces, a complete picture of the RNA's catalytic potential does not come with the RNA world hypothesis. Yet, there is a high number of chemical reactions and properties which have been predicted to occur in an RNA world. For a large number of the most important reactions, proof of the catalytic potential has meanwhile been given, including for example ribonucleotide polymerization, aminoacylation, and peptide bond formation.[26]

Although the RNA world presumably existed almost four billion years ago, X-ray-structures of the ribosome seem to exhibit fragments of this ancestral era. The structures revealed that RNA-mediated catalysis plays an important role in the peptide synthesis of the ribosome. The key step in translation is catalysed only by the ribonucleic acid component of the ribosome, without any direct contribution of proteins from the spatial vicinity. That impressively demonstrates the catalytic potential of RNA in a biochemical reaction that may arguably be called the most important ever.

Table 14.2 Catalytic activities of artificial ribozymes (selected examples)

Reaction type	References
C—C bonds	
Diels-Alder reaction	[11, 13]
Aldol reaction	[14]
Claisen condensation	[15]
biphenyl isomerization	[6]
C—N bonds	
amide bond formation	[16]
N-glycosidic bond formation	[17]
N-alkylation	[12]
C—S bonds	
Michael addition	[18]
S-acylation	[19]
S-alkylation	[20]
C—O bonds	
transesterification	[21]
carbonate ester hydrolysis	[8]
P—O bonds	
RNA polymerization	
phosphorylation	[22]
RNA ligation	[23]
C-metal bonds	
porphyrin metallation	[7]
Redox reactions	
oxidation and reduction	[10, 24]

An artificial ribozyme mimics this translation step of the ribosome.[27] The specifity of this selected ribozyme is based on the recognition of an adenosine moiety of the amino acid ester and allows the utilization of leucine- and phenylalanine- as well as methionine-derivatized substrates. This tolerance for various amino acids indicates the possibility of selecting more general ribozymes for protein synthesis. Furthermore, a related ribozyme efficiently catalyses the synthesis of ~30 different dipeptides from an aminoacyl-adenylate substrate. Ribozyme-mediated synthesis of uncoded peptides might have been an important step in the transition from a RNA to a peptide world before the emergence of the ribosome.[28]

Another important development was the isolation of a ribozyme which performs nucleotide synthesis by forming a glycosidic linkage from activated ribose (pRpp)[17] in a way similar to the modern biosynthesis of nucleotides.

Before the discovery of catalytic RNA, the principal indications of a possible RNA world had been the role of tRNA and ribosomal RNA in translation, the use of RNA as genetic material in retroviruses, and the ubiquitous occurrence of RNA-related enzymatic co-substrates like GTP, ATP, AMP, cAMP, SAM, $FADH_2$ and NAD^+ in all major metabolic pathways. Clearly, the capability of RNA to employ these and other ubiquitous co-substrates in catalysis must be expected.[26]

Definitively a highlight in ribozyme research is the *in vitro* evolution of a ribozyme that oxidizes an alcohol in a NAD^+ dependent manner. The resulting RNA-aldehyde was trapped via a chemoselective modification with biotin hydrazide. The function of this ribozyme is analogous to the natural alcohol dehydrogenase enzyme (ADH) and depends also on the same cofactors.[10] Furthermore this ribozyme was coupled with an electron transfer process between NADH and FAD. Thus a NAD^+ regeneration system is constituted. Interestingly, the reverse reaction, the RNA-catalysed reduction of the aldehyde, is also possible in presence of NADH.[24] This is the first clear-cut demonstration of a typical redox reaction catalysed by RNA.

Further artificial ribozymes are known to react with cosubstrates, for example acylating the thiol group of tethered Co-enzyme A with the AMP-activated biotin. These ribozymes also produce the crucial metabolic intermediates acetyl-CoA and butyryl-CoA at substantial reaction rates. For the selection of this ribozyme, the employed RNA pool had been coupled at its 5′-end to CoA by a previously isolated capping ribozyme.[19] Ribozymes for a number of related reactions were isolated, and the reader is referred to an excellent review discussing in detail the current knowledge on ribozyme catalysis in the context of the RNA world hypothesis.[26]

The current hypothesis predicts evolution of the RNA world into the modern DNA–RNA–protein world, thereby DNA taking over the storage of genomic information. One advantage of DNA over RNA as genetic material is the better chemical and enzymatic stability of deoxynucleic acids. That same advantage also comes to bear as a property of deoxyribozymes, of which a wide range has been selected by now. Despite this, there is little evidence that ribozymes from the RNA world might have been replaced by deoxyribozymes. Rather, proteins have taken over the vast majority of catalytic functions.

The intrinsically restricted functionality of nucleic acids as compared to proteins is a serious shortcoming for the expression of catalytic potential. A possible remedy is the introduction of additional, non-natural functional groups *via* the incorporation of modified nucleotides.[29] The tolerance of the employed RNA polymerase towards the modifications limits the general use of this technique for the generation of modified ribozymes. Many such ribozymes with modified bases catalyse RNA cleavage or ligation. Further catalytic activities concern metalation of *N*-methylmesoporphyrin, formation of a phosphodiester bond with a deoxynucleotide, and cleavage of phosphodiester *in trans* under simulated physiological conditions.

The incorporation by T7 RNA polymerase and successful use in SELEX of a number of uridine derivatives was reported by Eaton's laboratory, resulting in the development of a cupric ion-dependent modified ribozyme with Diels–Alderase activity.[13] To current knowledge, the Diels–Alder reaction does not play a major role in any biochemical pathway, and the use of modified nucleotides to expand the catalytic repertoire of RNA was certainly not undertaken with the RNA world as principal motivation. Rather, these approaches underscore the potential for application of catalytic RNA in modern fields of biotechnology and organic synthesis. The use of proteic enzymes in organic synthesis has become a commonplace strategy where applicable, the limiting factor being the choice of suitable enzymes and their respective substrate specificities. From this perspective, several ribozymes display catalytic activity of high interest to the organic chemist (see Table 14.2). Importantly, most of these ribozymes have been obtained using a direct selection protocol, where the choice of reactants has dictated the substrate specificity of the resulting ribozymes.

Thus, the direct selection method holds the alluring possibility to engineer substrate-specific ribozymes which are tailored to the particular reaction an organic chemist might wish to carry out. Since nucleic acids are chiral, one can even anticipate stereoselectivity in the would-be custom-made catalysts. Although this remains a vision yet to be accomplished, several cases of ribozymes with promising properties have been reported, accelerating certain redox reactions, Michael additions and cycloadditions. Already mentioned was the oxidation of a benzyl alcohol to the corresponding aldehyde, and the reverse reaction from Suga`s laboratory. The redox ribozymes still require their substrate to be covalently bound, meaning they act *in cis*, thus performing neither true catalysis nor multiple turnover.[10]

In the Famulok laboratory, a ribozyme was isolated which promotes a reaction corresponding to the first step of the formation of dTMP from dUMP in proteic thymidylate synthases. This ribozyme mediates Michael-adduct formation at a Michael-acceptor substrate. The reaction is accelerated by a factor of nearly 10^5. The selected ribozyme could be engineered to act in an intermolecular reaction on a substrate tethered to an RNA oligomer. The demonstration of RNA catalysis of this reaction has bearings on the RNA world hypothesis, as well as implications for possible synthetic applications.[18]

14.5 Deoxyribozymes – DNA Molecules with Catalytic Properties

While there is still no demonstration that DNA carries out catalytic functions in present-day biology, numerous DNAzymes have been developed by in vitro selection. While the spectrum is not as wide as for ribozymes, it clearly goes far beyond what was imagined a few years back (Table 14.3).[30]

One class of deoxyribozymes has found routine practical applications in molecular biology. These are deoxyribozymes that cleave target RNAs in a highly predicable, sequence-specific manner, thereby acting as designer nucleases. These DNAzymes generally utilize as nucleophile the 2′-OH group on the RNA substrate adjacent to the phosphodiester bond to be cleaved and generate cleavage products with a 5′-OH and a 2′,3′-cyclophosphate terminus, respectively, which can be easily processed by other enzymes.

Table 14.3 *Representative deoxyribozymes*

Reaction type	Reference
RNA cleavage	[32]
DNA cleavage (oxidative)	[33]
RNA ligation	[30]
DNA ligation	[34]
DNA phosphorylation	[35]
DNA adenylation	[36]
DNA depurination	[37]
Thymine dimer photoreversion	[38]
Porphyrin metalation	[39]
Diels-Alder reaction	[40]
Nucleopeptide formation	[41]

DNAzymes can also cleave DNA. This is, however, achieved by an oxidative mechanism involving freely diffusible hydroxyl radicals. The site and sequence specificity of cleavage is low, limiting the synthetic utility of these enzymes.

Many DNAzymes catalyse RNA ligation reactions to yield linear, branched, and lariat-type reaction products. The ligation of DNA strands as well as the phosphorylation of DNA or RNA oligonucleotides was described. Some notable extensions beyond phosphodiester chemistry include a photoreversion reaction, a deglycosylation, porphyrin metalation, nucleopeptide bond formation, and finally a Diels–Alder reaction. This latter reaction is essentially the same studied by the Jäschke lab with RNA as a catalyst (as discussed below in more detail), and the information published to-date indicates that the catalytic proficiency of DNA and RNA enzymes for Diels–Alder reactions is very similar.

Like in the case of RNA, DNAzymes were created with a variety of 'side chains', ranging from imidazol groups to primary amines and imidines.[31]

14.6 Catalysis of C—C Bond Formation by Diels–Alderase Ribozymes

In this last section, I will discuss structure and mechanism of one artificial ribozyme in more detail. This ribozyme, selected in my laboratory, is the only RNA catalyst for small-molecule chemistry with a known spatial structure, and due to extensive studies, it is arguably the best-characterized artificial ribozyme known to-date. These data provide for the first time an insight into how a small RNA can accelerate reactions different from phosphodiester chemistry, and what structural prerequisites are required.

14.6.1 *In Vitro* Selection

The selection started with an RNA library of ~2 × 10^{14} sequences of 160 nucleotide length.[11] After tethering the RNA to anthracene via long flexible PEG chains, this library was allowed to react for with an excess of biotinylated maleimide. Biotinylated RNA was recovered using immobilized strepavidin, and selection pressure was gradually increased by reducing reaction time and maleimide concentration. After 10 iterations of selection and amplification, the enriched library showed ~6,500-fold rate acceleration, compared to the starting library. Individual members of this enriched library were sequenced and assayed for activity.

32 of the 35 catalytically active RNAs were found to share a common motif (Figure 14.3): Central element is an asymmetric internal bubble comprised of the two consensus sequences UGCCA and AAUACU, framed by two helices (helix II and III, light and dark gray, respectively). From helix II, another helix (helix I) continues in which one strand is formed by nucleotides from the conserved 5′-primer binding site. In all sequences containing the motif, the 5′-terminal GGAG is left (formally) unpaired. The first G is the site where the tethered anthracene is attached, suggesting an important structural role of this tetranucleotide. The respective ends of helix I and III are connected by loops L1 and L2 of varying size and sequence. A small 49-mer RNA containing this motif only (with the ends of helix I and III closed by stabilizing tetranucleotide loops) accelerated the reaction between the tethered anthracene and biotin maleimide about 18,500-fold, confirming the findings of sequence analysis.[11] All characterization studies that follow have

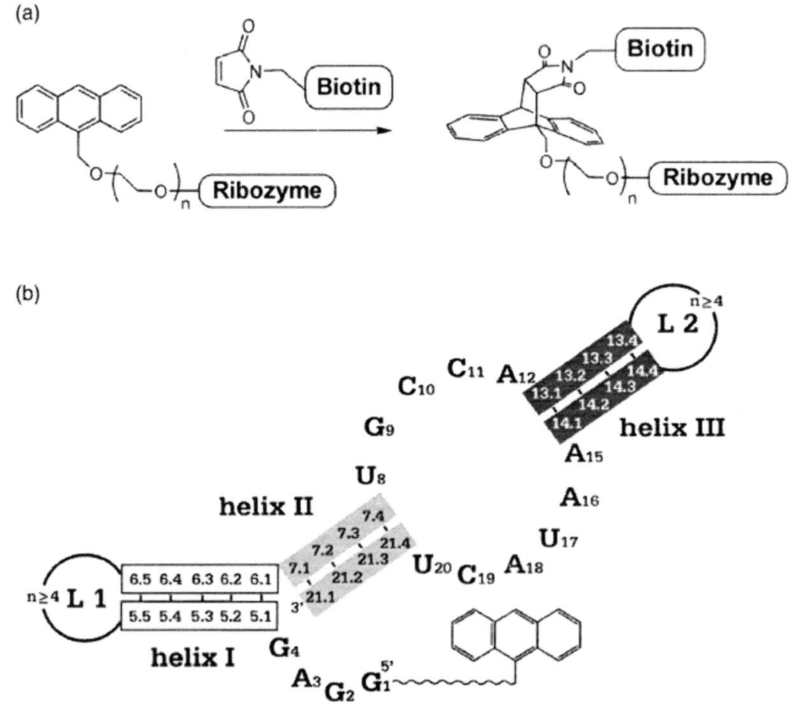

Figure 14.3 *(**a**) RNA-catalyzed Diels–Alder reactions utilizing a tethered diene; (**b**) Secondary structure motif responsible for catalysis and numbering scheme*[11]

been performed on the minimal 49-mer derived from the conserved motif, or on variants of this minimal ribozyme.

14.6.2 Overall Catalytic Properties

The ribozymes were selected to catalyse the Diels–Alder reaction between a covalently tethered anthracene and a biotinylated maleimide, which is a single-turnover format. However, the selected Diels–Alderase ribozymes and the minimal 49-mer could be shown to accelerate C—C bond formation in a true bimolecular fashion (*in trans* reaction, Figure 14.4).[42] Substrate molecules as small as 9-hydroxymethylanthracene and N-ethylmaleimide are specifically recognized by the ribozymes, followed by conversion to the respective Diels–Alder products and product dissociation from the catalyst. The 49-mer ribozyme performs the reaction with fast multiple turnovers, and a k_{cat} of 21 min^{-1} was measured. Saturation type kinetics with respect to both reactants ($K_{M,Diene} = 370\,\mu M$, $K_{M,Dienophile} = 8\,mM$) as well as product inhibition were observed. The 'effective molarity' as a measure for the entropic gain of the ribozyme-catalysed reaction was determined from the ratio k_{cat}/k_{uncat} to be 6.6 M, giving a lower estimate for the rate acceleration of 1,100-fold.[42]

Another characteristic feature of enzymatic catalysis was demonstrated for the Diels–Alderase ribozymes, namely enantioselective bond formation. While the uncatalysed

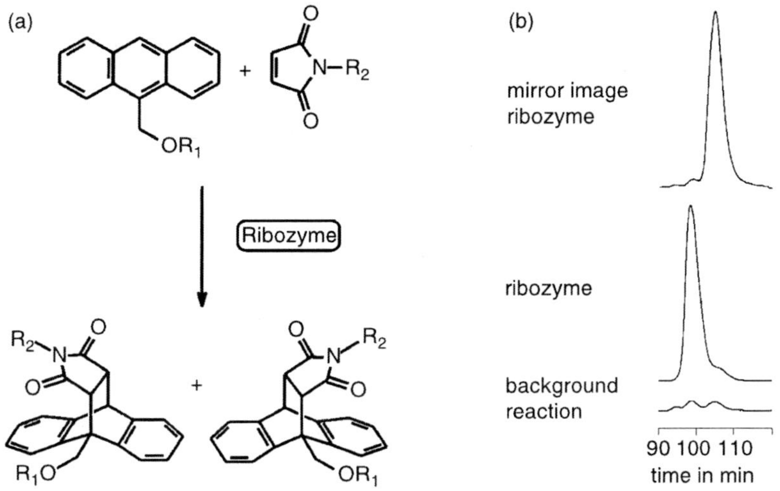

Figure 14.4 *Ribozyme catalysis of a Diels–Alder reaction.* ***a:*** *Chemistry of the reaction.* ***b:*** *Chiral HPLC analysis of products obtained in reactions either without catalyst (background reaction), with the catalytic 49mer minimum ribozyme or its synthetic mirror-image L-RNA version ($R_1 = (C_2H_4O)_6$-H; $R_2 = (CH_2)_5COOCH_3$)*[42]

reaction produces racemic product mixtures, the ribozyme-catalysed conversion shows an enantioselectivity of over 95% *ee*. The enantioselectivity was shown to be primarily dependent on the size of the substituent at the anthracene ring system, and the enantiomer of this ribozyme (chemically synthesized from the unnatural L-ribonucleotides) showed the opposite stereoselectivity (Figure 14.4).[42]

14.6.3 Overall Structure of the Ribozyme

The three-dimensional structure of the Diels–Alderase ribozyme could be solved both in form of the Apo-enzyme and in complex with the Diels–Alder cycloaddition product.[43] The molecule adopts an overall topology that resembles the greek letter lambda λ (Figure 14.5), featuring a special kind of three-way junction. Stems II and III are co-linearly stacked (best seen in panel b), bridged by a zippered-up asymmetric bubble, while stem I is extended by two base pairs and branches off at the bubble at ~60 ° angle. In the centre of the molecule there is extra electron density that is not related to nucleotides and can be traced back to the *S*, *S* stereoisomer of the Diels–Alder product, tightly bound into a pocked as a single, well-defined conformer (Figure 14.5c).

The 5′-GGAG tetranucleotide plays a critical role in shaping both the RNA scaffold and the catalytic pocket. G1 and G2 form Watson-Crick pairs with C11 and C10, while A3 and G4 pair with U20 and C19, thereby generating the nested pseudoknot topology within the RNA scaffold (dotted lines in Figure 14.5A, 14.5B). ‘Pseudoknot’ denotes a secondary structure resulting from nucleotides in the loop of an RNA stem-loop structure with nucleotides outside of the stem-loop. The 5′-terminal GGAG serves to clamp together the opposite sites of the asymmetric internal bubble, and the direct connection of four helices without interjecting spacers can impose severe strain on the system. Such an

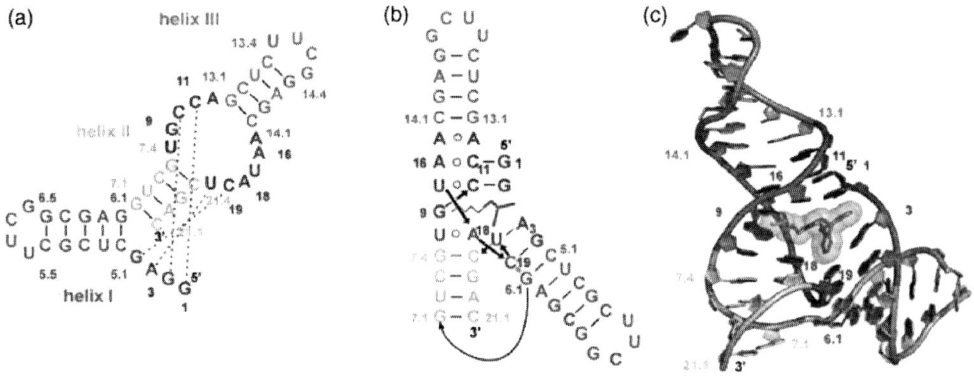

Figure 14.5 *Crystal structure of the Diels–Alderase ribozyme.* ***a****) Ribozyme secondary structure with helices I, II and III, the asymmetric bubble, and the conserved 5′end. Dotted lines represent pseudoknot Watson-Crick type interactions.* ***b****) Tertiary fold and* ***c****) three-dimensional topology in the crystal structure of the ribozyme – product complex. One enantiomer of the Diels–Alder product (sticks with transparent spheres) is bound into the catalytic pocket of the ribozyme*[43]

architecture appears to be an efficient way to achieve a rather dense and stable packing of helical elements.

14.6.4 Architecture of the Catalytic Pocket

The catalytic pocket is wedge-shaped and bracketed by the Watson-Crick A_3•U_{20} pair (bottom), the U_{17}•(G_2•C_{10}) triple (upper right), and the U_8•A_{18} reverse Hoogsteen pair (left, Figure 14.6), and is lined by an intricate network of hydrogen bonds (dotted lines). The (former) maleimide ring is stacked over C_{10}, and apparently oriented through a pair of hydrogen bonds involving one of its two carbonyl oxygens, while its side chain runs inside a hydrophobic canyon. One of the aromatic rings of the Diels–Alder product is sandwiched between purines G_2 and A_3, while the other is wedged between the base and sugar components of A_{18} and U_{20} (Figure 14.5). The catalytic pocket, which is primarily formed by base edges with minimal contribution by the sugar-phosphate backbone, is accessible to the product from the front while a narrow orifice is visible in the back. It should be noted that the catalytic pocket contains not only a cavity to accommodate the cycloaddition product but also a direction-specific surface channel to accommodate the maleimide side chain, which provides a convenient explanation for the observed stereoselectivity. There are no Mg^{2+} cations positioned within the immediate vicinity of the pocket that could participate in the catalytic process.

The structure of the free Diels–Alderase ribozyme was found to be virtually identical to the structure of the product complex, providing strong support for the concept of a preformed catalytic pocket, established by chemical probing.[44] The anthracene-RNA conjugate could also be crystallized, however, the anthracene module and its linker were found to be disordered in the crystal. The RNA mapped well with the other crystal structures.

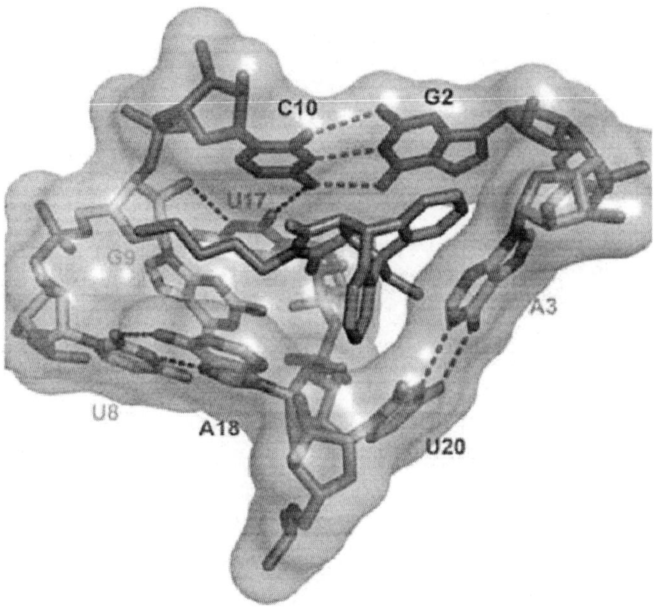

Figure 14.6 Surface representation of the catalytic pocket with stick representations of the nucleotides. Dotted lines indicate confirmed interactions.

14.6.5 Interactions Between the Ribozyme and its Substrates and Products

The crystallographic information about the architecture of the catalytic pocket could be compared with an earlier study that elucidated the interactions of the ribozyme with its substrates and products by chemical substitution analysis using 44 different, systematically varied analogues.[43] In that study, RNA-diene interaction was found to be governed by stacking interactions, while hydrogen bonding and metal ion coordination appeared to be less important. The diene has to be an anthracene derivative, and substituents at defined positions are permitted, thereby shedding light on the geometry of the binding site (Figure 14.7). The crystal structure shows that there in no space to accommodate substituents at positions 2, 3, 6 and 7 (shown in outline letters), as these would clash into the walls of the catalytic pocket. Some space for small substituents is found at positions 4, 5 and 10, pointing towards the back of the pocket. Large substituents are tolerated at positions 1, 8, and 9, pointing out of the pocket. Interestingly, the polyethylene glycol tether used in the selection does not make any contribution to binding and can be removed or shifted to another position (e.g., position 1) without penalty.

The dienophile must be a five-membered maleimidyl ring with an unsubstituted reactive double bond. Substituents would again collide with the walls of the catalytic pocket. A hydrophobic side chain (alkyl or aryl) makes a major contribution to RNA binding. Branching is not allowed at the alpha position (next to the maleimide ring), due to steric clash with the walls. The ribozyme distinguishes between different enantiomers of chiral substrates and accelerates cycloadditions with both enantio- and diastereoselectivity.

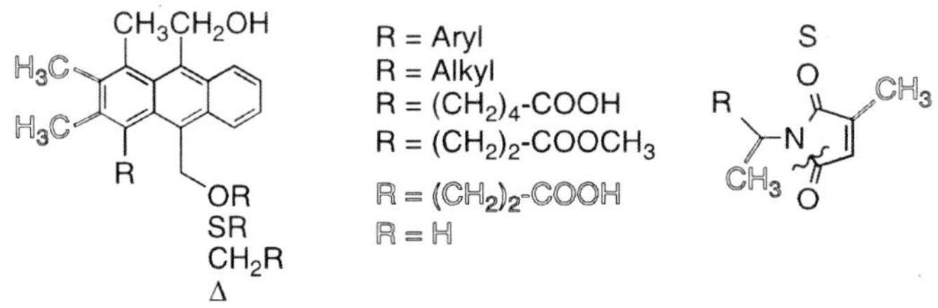

Figure 14.7 *Summarized structural requirements of dienes and dienophiles for acceptance by the Diels–Alderase ribozyme. Substitutions shown in black are tolerated, while those in outline letters are deleterious. Δ: removal of whole substituent.*[45]

The stereochemistry of the reaction is controlled by RNA-diene interactions. The RNA interacts strongly and stereoselectively with the cycloaddition products, requiring several structural features to be present. Strong and stereoselective product inhibition is observed.[45]

The crystal structure shows three H-bonds between the reaction product and RNA. One that involves an ether oxygen of the polyethylene glycol tether is irrelevant for catalysis as the tether can be removed entirely without effect on catalysis. The other two involve the maleimide's carbonyl oxygen and could be mechanistically meaningful. The importance of these bonds is, however, not yet completely established, and atomic mutagenesis is currently used to probe these interactions.

14.6.6 Conformational Dynamics and the Roles of Metal Ions

The investigation of this ribozyme's kinetics, reaction mechanism and folding by conventional chemical and biochemical methods revealed quite early complex relationships and suggested the existence of various conformational forms of this RNA molecule. Single-molecule fluorescence resonance energy transfer (FRET) experiments were carried out using dye-labeled ribozyme molecules.[46] The results show that the DAse ribozyme possesses three structurally different states that are distinguishable by their FRET efficiency distributions, the unfolded (U), intermediate (I) and folded (F) states. Mg^{2+} titration allowed to measure changes in their equilibrium populations and to observe the collapse of the intermediate state. Both effects were described with a thermodynamic model, revealing how the F state is stabilized by successive binding of Mg^{2+} ions. Furthermore, continuous fluctuations were observed between the I and F states on the 100 ms timescale.

To further elucidate the interactions of the Diels–Alderase ribozyme with divalent metal ions in solution, electron paramagnetic resonance (EPR) spectroscopy was applied using paramagnetic manganese instead of magnesium ions.[47] Manganese ion titrations revealed five high-affinity Mn^{2+} binding sites with an upper K_d of 0.6 μM. In order to characterize each binding site individually, EPR-silent Cd^{2+} ions were used to saturate the other binding sites. This cadmium-induced EPR silencing showed that the Mn^{2+} binding sites possess different affinities. In addition, these binding sites could be assigned to three different types, including innersphere, outersphere, and a Mn^{2+} dimer.

14.6.7 Mechanistic Considerations

The X-ray crystal structure of the ribozyme-product complex,[43] coupled with chemical[45] and biochemical[44] experiments, suggests that the ribozyme should bind the diene and dienophile in a precisely defined steric orientation within a wedge-shaped catalytic pocket, thereby facilitating the reaction by reducing translational and rotational degrees of freedom. The anthracene substrate could be bound by stacking interactions between G2 and the A3-U20 pair. The reaction requires that the maleimide substrate be stacked on top of the anthracene, and parallel to it, at a distance of about 3.5 Å. In this position, maleimide can form hydrogen bonds between its carbonyl-oxygen and the exocyclic amino group of G9 and the 2′-OH of U17. The N-alkyl side chain can be placed within the hydrophobic canyon (Figure 14.8), which would allow the approach of the maleimide to the bound anthracene only from one direction, thereby providing the stereoselectivity of the reaction. In the transition state of the reaction, two new single-bonds are partly formed (bond lengths ~2.2 Å), and the anthracene ring system is bent out of planarity to

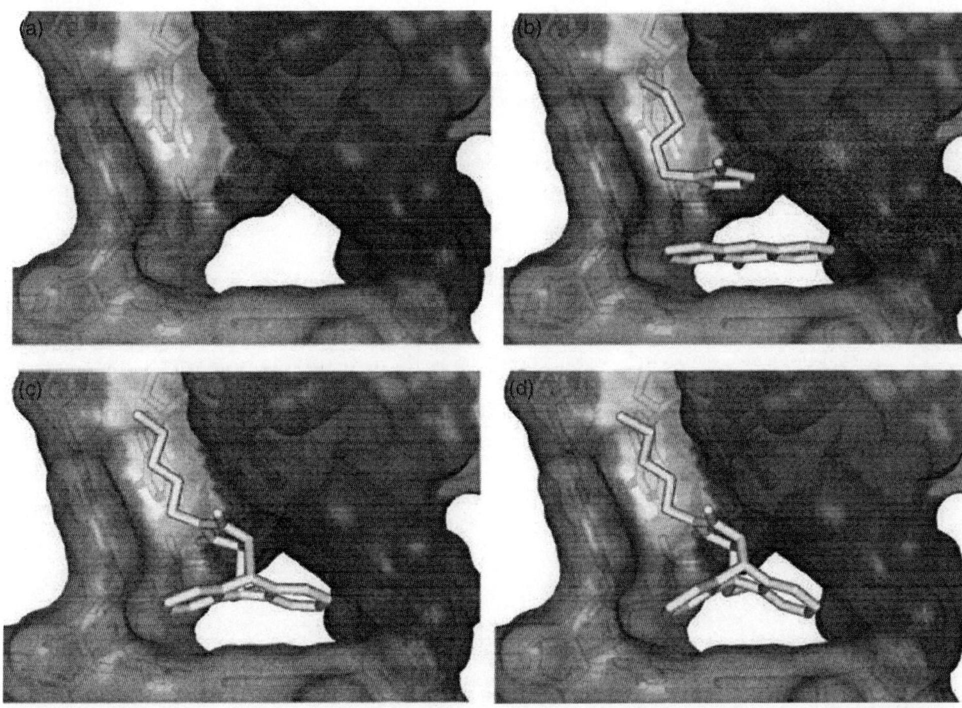

Figure 14.8 Proposed model for the catalytic mechanism of the Diels–Alderase ribozyme. a. Empty catalytic pocket. b. Michaelis complex with both substrates bound. c. Transition state inside the catalytic pocket, d. Ribozyme-bound product. Panels a and d are derived directly from the crystallographic data, panels b and c were obtained by manually docking the two substrates and the transition state, respectively, into the pocket.

yield an angle of about 154° (Figure 14.8), therefore further reducing the minimal unoccupied space underneath the bridgehead carbons seen in the structure of the ribozyme-product complex. Remarkably, the transition state has near-perfect shape complementarity with the catalytic pocket, a feature of prime importance in antibody catalysis of the Diels–Alder reactions.[48,49] In addition to steric factors, both stacking and hydrogen bonding may contribute the energetics of the reaction. Stacking of the anthracene with nucleotides A3 and U20 could accelerate the reaction by increasing the diene's electron density, while hydrogen bonding of the maleimide would make it more electron-deficient, thereby increasing catalytic reactivity. Thus, it appears that ribozyme-based catalysis of Diels–Alder reactions reflects a combination of proximity, shape complementarity, and energetic contributions to the catalytic process.

The catalytic parameters of the Diels–Alderase ribozymes were analyzed by Houk and coworkers and were found to compare favorably with those of catalytic Diels–Alderase antibodies, cyclodextrins, and synthetic capsules.[50] From the measured parameters, these authors concluded that most of the rate acceleration arises from the conversion of a second-order reaction of diene and dienophile into a first-order reaction of the termolecular complex ribozyme, diene, and dienophile, and calculated an extra transition state stabilization (relative to the ground state) of only $-1.1\,\text{kcal mol}^{-1}$, which is an order of magnitude lower than typical values for natural (protein) enzymes. Computational analysis of the ribozyme-catalysed Diels–Alder reaction by Zhang and Bruice using QM/MM methods confirmed the concerted nature of the cycloaddition and found that the proficiency of the RNA-catalysed reaction originates from the active site holding the two reactants in reactive conformations, in which the reacting atoms are brought together in van der Waals distances and reactants approach each other at an appropriate angle.[51]

14.7 Conclusion

Despite their limited set of functional groups, ribozymes can accelerate complex organic transformations like Diels–Alder reactions between small molecules in a way similar to protein enzymes or supramolecular catalysts featuring multiple turnover, substrate specificity and stereoselectivity. The three-dimensional structure shows striking similarities with proteins evolved for similar reactions, and the catalytic strategies used appear to be similar as well.

References

1. J. C. Cochrane, S. A. Strobel, Catalytic strategies of self-cleaving ribozymes, *Acc. Chem. Res.*, 2008, **41**, 1027–1035.
2. D. M. J. Lilley, F. Eckstein (Ed.) *Ribozymes and RNA catalysis*, RSC Publishing, Cambridge, 2008.
3. W. Gilbert, The RNA world, *Nature*, 1986, **319**, 618.
4. C. R. Woese, *The Genetic Code: The Molecular Basis for Genetic Expression*, Harper & Row, New York, 1967.
5. S. Klussmann (Ed.) *The Aptamer Handbook – Functional Oligonucleotides and Their Applications*, Wiley VCH, Weinheim, 2006.

6. J. R. Prudent, T. Uno, P. G. Schultz, Expanding the scope of RNA catalysis, *Science*, 1994, **264**, 1924–1927.
7. M. M. Conn, J. R. Prudent, P. G. Schultz, Porphyrin metalation catalysed by a small RNA molecule, *J. Am. Chem. Soc.*, 1996, **118**, 7012–7013.
8. S.-M. Chun, S. Jeong, J.-M. Kim, B.-O. Chong, Y.-K. Park, H. Park, J. Yu, Cholesterol esterase activity by in vitro selection of RNA against a phospate transition-state analogue, *J. Am. Chem. Soc.*, 1999, **121**, 10844–10845.
9. K. N. Morris, T. M. Tarasow, C. M. Julin, S. L. Simons, D. Hilvert, L. Gold, Enrichment for RNA molecules that bind a Diels–Alder transition state analog, *Proc. Natl. Acad. Sci. USA*, 1994, **91**, 13028–13032.
10. S. Tsukiji, S. B. Pattnaik, H. Suga, An alcohol dehydrogenase ribozyme, *Nat. Struct. Biol.*, 2003, **10**, 713–717.
11. B. Seelig, A. Jäschke, A small catalytic RNA motif with Diels–Alderase activity, *Chem. Biol.*, 1999, **6**, 167–176.
12. C. Wilson, J. W. Szostak, In vitro evolution of a self-alkylating ribozyme, *Nature*, 1995, **374**, 777–782.
13. T. M. Tarasow, S. L. Tarasow, B. E. Eaton, RNA-catalysed carbon-carbon bond formation, *Nature*, 1997, **389**, 54–57.
14. S. Fusz, A. Eisenfuhr, S. G. Srivatsan, A. Heckel, M. Famulok, A ribozyme for the aldol reaction, *Chem. Biol.*, 2005, **12**, 941–950.
15. Y. Ryu, K. J. Kim, C. A. Roessner, A. I. Scott, Decarboxylative Claisen condensation catalysed by in vitro selected ribozymes, *Chem. Commun.*, 2006, 1439–1441.
16. T. W. Wiegand, R. C. Janssen, B. E. Eaton, Selection of RNA amide synthases, *Chem. Biol.*, 1997, **4**, 675–683.
17. P. J. Unrau, D. P. Bartel, RNA-catalysed nucleotide synthesis, *Nature*, 1998, **395**, 260–263.
18. G. Sengle, A. Eisenfuhr, P. S. Arora, J. S. Nowick, M. Famulok, Novel RNA catalysts for the Michael reaction, *Chem. Biol.*, 2001, **8**, 459–473.
19. V. R. Jadhav, M. Yarus, Acyl-CoAs from coenzyme ribozymes, *Biochemistry*, 2002, **41**, 723–729.
20. M. Wecker, D. Smith, L. Gold, In vitro selection of a novel catalytic RNA: characterization of a sulfur alkylation reaction and interaction with a small peptide, *RNA*, 1996, **2**, 982–994.
21. M. Illangasekare, G. Sanchez, T. Nickles, M. Yarus, Aminoacyl-RNA synthesis catalysed by an RNA, *Science*, 1995, **267**, 643–647.
22. J. R. Lorsch, J. W. Szostak, In vitro evolution of new ribozymes with polynucleotide kinase activity, *Nature*, 1994, **371**, 31–36.
23. E. H. Ekland, J. W. Szostak, D. P. Bartel, Structurally complex and highly active RNA ligases derived from random RNA sequences, *Science*, 1995, **269**, 364–370.
24. S. Tsukiji, S. B. Pattnaik, H. Suga, Reduction of an aldehyde by a NADH/Zn^{2+} -dependent redox active ribozyme, *J. Am. Chem. Soc.*, 2004, **126**, 5044–5045.
25. C. R. Woese, On the evolution of cells, *Proc. Natl. Acad. Sci. USA*, 2002, **99**, 8742–8747.
26. X. Chen, N. Li, A. D. Ellington, Ribozyme catalysis of metabolism in the RNA world, *Chem. Biodivers.*, 2007, **4**, 633–655.
27. B. Zhang, T. R. Cech, Peptide bond formation by in vitro selected ribozymes, *Nature*, 1997, **390**, 96–100.
28. L. Sun, Z. Cui, R. L. Gottlieb, B. Zhang, A selected ribozyme catalyzing diverse dipeptide synthesis, *Chem. Biol.*, 2002, **9**, 619–628.
29. S. Verma, S. Jager, O. Thum, M. Famulok, Functional tuning of nucleic acids by chemical modifications: tailored oligonucleotides as drugs, devices, and diagnostics, *Chem. Rec.*, 2003, **3**, 51–60.
30. S. K. Silverman, Catalytic DNA (deoxyribozymes) for synthetic applications-current abilities and future prospects, *Chem. Commun.*, 2008, 3467–3485.
31. M. Hollenstein, C. J. Hipolito, C. H. Lam, D. M. Perrin, A self-cleaving DNA enzyme modified with amines, guanidines and imidazoles operates independently of divalent metal cations (M^{2+}), *Nucleic Acids Res.*, 2009, **37**, 1638–1649.

32. R. R. Breaker, G. F. Joyce, A DNA enzyme that cleaves RNA, *Chem. Biol.*, 1994, **1**, 223–229.
33. N. Carmi, L. A. Shultz, R. R. Breaker, In vitro selection of selfcleaving DNAs, *Chem. Biol.*, 1996, **3**, 1039–1046.
34. B. Cuenoud, J. W. Szostak, A DNA metalloenzyme with DNA ligase activity, *Nature*, 1995, **375**, 611–614.
35. Y. Li, R. R. Breaker, Phosphorylating DNA with DNA, *Proc. Natl. Acad. Sci. USA*, 1999, **96**, 2746–2751.
36. Y. Li, Y. Liu, R. R. Breaker, Capping DNA with DNA, *Biochemistry*, 2000, **39**, 3106–3114.
37. T. L. Sheppard, P. Ordoukhanian, G. F. Joyce, A DNA enzyme with N-glycosylase activity, *Proc. Natl. Acad. Sci. USA*, 2000, **97**, 7802–7807.
38. D. J. Chinnapen, D. Sen, A deoxyribozyme that harnesses light to repair thymine dimers in DNA, *Proc. Natl. Acad. Sci. USA*, 2004, **101**, 65–69.
39. Y. Li, D. Sen, A catalytic DNA for porphyrin metallation, *Nat. Struct. Biol.*, 1996, **3**, 743–747.
40. M. Chandra, S. K. Silverman, DNA and RNA can be equally efficient catalysts for carbon-carbon bond formation, *J. Am. Chem. Soc.*, 2008, **130**, 2936–2937.
41. S. M. Knudsen, A. D. Ellington. Aptazymes: Allosteric Ribozymes and Deoxyribozymes as Biosensors, in *The Aptamer Handbook*, S. Klussmann (Ed.), Wiley-VCH, Weinheim, 290–310, 2006.
42. B. Seelig, S. Keiper, F. Stuhlmann, A. Jäschke, Enantioselective ribozyme catalysis of a bimolecular cycloaddition reaction, *Angew. Chem. Int. Ed. Engl.*, 2000, **39**, 4576–4579.
43. A. Serganov, S. Keiper, L. Malinina, V. Tereshko, E. Skripkin, C. Hobartner, A. Polonskaia, A. T. Phan, R. Wombacher, R. Micura, Z. Dauter, A. Jäschke, D. J. Patel, Structural basis for Diels–Alder ribozyme-catalysed carbon-carbon bond formation, *Nat. Struct. Mol. Biol.*, 2005, **12**, 218–224.
44. S. Keiper, D. Bebenroth, B. Seelig, E. Westhof, A. Jäschke, Architecture of a Diels–Alderase ribozyme with a preformed catalytic pocket, *Chem. Biol.*, 2004, **11**, 1217–1227.
45. F. Stuhlmann, A. Jäschke, Characterization of an RNA active site: interactions between a Diels–Alderase ribozyme and its substrates and products, *J. Am. Chem. Soc.*, 2002, **124**, 3238–3244.
46. A. Y. Kobitski, A. Nierth, M. Helm, A. Jäschke, G. U. Nienhaus, Mg^{2+}-dependent folding of a Diels–Alderase ribozyme probed by single-molecule FRET analysis, *Nucleic Acids Res.*, 2007, **35**, 2047–2059.
47. N. Kisseleva, S. Kraut, A. Jäschke, O. Schiemann, Characterizing multiple metal ion binding sites within a ribozyme by cadmium-induced EPR silencing, *HFSP J.*, 2007, **1**, 127–136.
48. J. Chen, Q. Deng, R. Wang, K. Houk, D. Hilvert, Shape complementarity, binding-site dynamics, and transition state stabilization: a theoretical study of Diels–Alder catalysis by antibody 1E9, *Chembiochem*, 2000, **1**, 255–261.
49. J. Xu, Q. Deng, J. Chen, K. N. Houk, J. Bartek, D. Hilvert, I. A. Wilson, Evolution of shape complementarity and catalytic efficiency from a primordial antibody template, *Science*, 1999, **286**, 2345–2348.
50. S. P. Kim, A. G. Leach, K. N. Houk, The origins of noncovalent catalysis of intermolecular Diels–Alder reactions by cyclodextrins, self-assembling capsules, antibodies, and RNAses, *J. Org. Chem.*, 2002, **67**, 4250–4260.
51. X. Zhang, T. C. Bruice, Diels–Alder ribozyme catalysis: a computational approach, *J. Am. Chem. Soc.*, 2007, **129**, 1001–1007.

15

Reactions in Supramolecular Systems

Lucia Zakharova, Alla Mirgorodskaya, Elena Zhiltsova, Ludmila Kudryavtseva, and Alexander Konovalov

A.E. Arbuzov Institute of Organic and Physical Chemistry of the Russian Academy of Science, ul. Akad. Arbuzov 8, 420088 Kazan (Russia)

15.1 Introduction

The catalysis of reactions in organized media is of current interest.[1–5] The compartmentalization of reagents within nanosized aggregates results in a sharp increase in their local concentrations (the concentration factor) and affects their microenvironment (the microenvironment factor), thereby influencing their reactivity. Since 1980 we have been studying nucleophilic substitution at phosphorus and carbon in organized solutions based on surfactants. The transfer of phosphoryl and carbonyl groups is one of the most fundamental chemical and biochemical reactions.[6] Phosphorus acid esters are compounds with interesting biological and pharmacological properties and are widely used as pesticides, drugs, and nerve gases.[7] Their accumulation and their impact on the environment are of paramount importance, therefore chemical decontamination of toxic compounds is of increasing interest.

At first, the catalytic effect of the single micellar solutions based on surfactants of different structure was studied. The data covering this period are reviewed in refs.[8,9] The regularities when generalized made it possible to proceed with the more complicated systems, in particular, the binary surfactant solutions, the polymer-surfactant and polymer-cyclophane (calixarene, pyrimidinophane, and etc.) systems. For these systems, the terms supramolecular systems (ensembles, compositions, complexes) are further used, since they can cover ensembles of different morphology formed through diverse noncovalent interactions of both intra- and intermolecular nature, as well as through inclusion interactions involving the cyclophane cavity. In these systems structural transitions induced by changes in solution conditions (concentration, temperature, ionic force, pH, and etc.) are observed. Along with the above factors of concentration and microenvironment these structural transitions can affect the reaction rate. Therefore the investigation of the catalytic effect of the polycomponent systems requires knowledge of the structural behaviour of the systems, in particular, packing modes, morphology of aggregates, surface potential, etc.

15.2 The Single Micellar Systems: Factors of Concentration and Micellar Microenvironment

The physico-chemical fundamentals of micellar catalysis and quantitative evaluation of basic factors responsible for rate alteration in organized media were developed by Berezin et al.[1,10] Within the frame of pseudo phase approach, the micellar solution is assumed to consist of micellar and bulk pseudo phases, with reagents partitioned between them. In general, reaction occurs in both pseudo phases simultaneously. For the case of bimolecular reactions the observed rate constant can be expressed as follows:

$$k'_{obs} = \frac{k_{2,0} + \frac{k_{2,m}}{V} K_S K_{Nu} C}{(1+K_S C)(1+K_{Nu} C)} \tag{15.1}$$

where k'_{obs} is the second order rate constant obtained by the division of the observed pseudo first rate constant (k_{obs}) by the total nucleophile concentration; $k_{2,0}$ is the second order rate constant of the reaction in the water; k_m (= $k_{2,m}/V$) represents the reactivity in the micellar phase where the reaction occurs; K_s and K_{Nu} are binding constants of substrate and nucleophile; C is the total surfactant concentration minus the critical micelle concentration (cmc) of the surfactant mixture. The approach developed by Berezin makes it possible to differentiate the factors responsible for the micellar effects, i.e. the micellar microenvironment factor (F_m) and the factor of the concentrating of reagents in micelles (F_c).

Different modifications of the pseudo phase model and their applications for quantitative treatment of kinetics of bimolecular reactions are reviewed by Bunton and Romsted.[2,3] A wide variety of ion-molecular reactions has been covered in these comprehensive reviews, and a generalization was drawn on the prevalent contribution of the concentrating of reagents to the rate effect and negative influence of micellar microenvironment on reactivity. Our systematic investigations of the kinetics of basic hydrolyses of phosphorus acid esters in aqueous and organic solutions of ionic surfactants of different structure showed that the above tendency of a decrease in the reactivity on the transfer of the reaction from the bulk pseudo phase to the micellar interior is only typical for the systems with effective binding of reagents. In fact, this is observed for the reactions involving hydrophobic compounds and occurring in aqueous solutions of surfactants with effective micellization capacity. In this case, the k_{obs}–[surfactant] kinetic dependences demonstrate a marked maximum. Most kinetic data reviewed in ref.[2,3] are related to these systems. Indeed, investigations of the basic hydrolyses of phosphonates **1–11** (Scheme 15.1) in

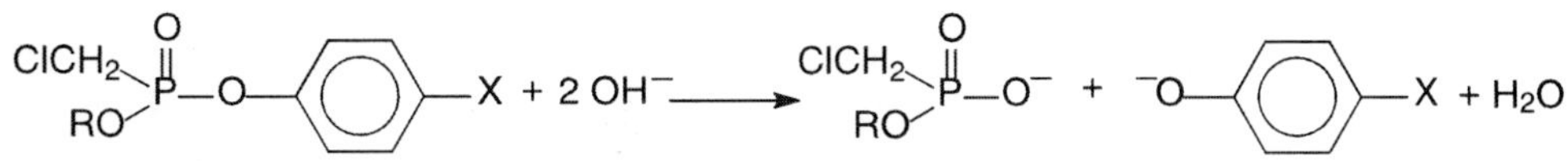

X=NO$_2$: R=C$_2$H$_5$ (**1**); n-C$_4$H$_9$ (**2**); n-C$_6$H$_{13}$ (**3**); n-C$_8$H$_{17}$ (**4**)
R=C$_2$H$_5$: X=Br (**5**); H (**6**); C$_2$H$_5$ (**7**); n-C$_4$H$_9$ (**8**); n-C$_8$H$_{17}$ (**9**); i-C$_{12}$H$_{25}$ (**10**)
X=H, R=n-C$_6$H$_{13}$ (**11**);

Scheme 15.1 *The schematic representation of hydrolyses of phosphonic acid esters*

aqueous solutions of cationic surfactants, in particular cetyltrimethylammonium bromide (CTAB)[11] and cetylpyridinium bromide (CPB)[12] have revealed that the factor of concentration reached two orders of magnitude, while the transfer of the reaction to the micellar microenvironment resulted in a marked decrease in the reactivity of the phosphonates (Table 15.1).

The transfer from CTAB and CPB (cmc 0.00085 M) to dodecylpyridinium bromide (DDPB; cmc 0.018 M) results in a marked increase in the F_m value for substrate **1**, followed by changes from the extremum type of kinetic dependences to S-shaped curves.[11] While in the case of the more hydrophobic substrate **3** in the DDPB micellar solution, the maximum type kinetic curve and the negative effect of the micellar microenvironment ($F_m < 1$) are preserved (Table 15.1).

The investigation of the kinetics of the solvolysis of *p*-nitrophenyl bis(chloromethyl) phosphinate **12** in the direct micelles of sodium dodecylsulfate (SDS) in ethylene glycol have demonstrated that the ratio of the factors F_c and F_m in this system differ from that in aqueous micelles, namely, the positive predominant contribution of the factor of microenvironment to the micellar rate effect is observed (Table 15.1).[13] In the solution of ethylene glycol the micellization is much less effective as compared to the aqueous solution (cmc of SDS in ethylene glycol is equal to 0.18 M, while in water it is 0.0085 M) and the micelles formed have low aggregation numbers, a loose structure and a weak solubilization capacity (the binding constants of substrates are lower by two orders of magnitude as compared to those in aqueous micellar solutions).[13]

The tendency of decreasing the contribution of the concentration factor and increasing the contribution of the factor of the micellar microenvironment is also observed for the reverse nonaqueous micellar solutions. For the case of aminolysis of phosphonates **1–4** in the polyethylene glycol-600-monolaurate (PM) reverse micelles, both factors are shown to contribute positively to the micellar rate effect, with F_m higher than F_c (Table 15.1).[14,15] In nonaqueous systems micellization is mainly contributed by dispersion interactions and results in the formation of small aggregates with a low solubilization capacity towards the substrates.[16]

In work,[11] the mechanism of the alteration of the reactivity in direct micelles is investigated. The kinetics of hydrolyses of phosphonates **1** and **3** in the DDPB micellar solution is measured at different temperatures and the activation parameters of the reaction, the enthalpy of activation and the entropy of activation in water in the absence of a surfactant ($\Delta H^{\neq}_0$, $\Delta S^{\neq}_0$) and in the micellar pseudo phase ($\Delta H^{\neq}_0$, $\Delta S^{\neq}_0$) are estimated. The transfer of the ion-molecular reactions from water to the less polar pseudo phase is shown to result in some decrease in the enthalpy of activation (from 46.2 to 39. 3 kJ mol^{-1} for **1** and from 40.8 to 30.0 kJ mol^{-1} for **3**). This is in line with the Ingold–Hughes principle postulating

Table 15.1 *The results of quantitative treatment of kinetic data for nucleophilic substitution in phosphorus acid esters in terms of equation 15.1*[a]

Substrate	System parameters	$k_{2,m}/M^{-1}s^{-1}$	K_S/M^{-1}	K_{Nu}/M^{-1}	$(k_{obs}/k_0)^g$ max	F_c [f]	F_m [f]	Ref.
CPB/water								
1	0.0005M NaOH	0.78	2020	80	35	187	0.19	12
2	0.0005M NaOH	1.07	3000	115	75	265	0.30	12
3	0.0005M NaOH	0.62	4720	307	105	650	0.19	12
4	0.0005M NaOH	0.38	7600	337	95	766	0.14	12
CTAB/water								
1	0.0005M NaOH	0.10	1190	145	7.0	266	0.026	11
1	0.001M NaOH	0.055	1780	240	7.5	432	0.014	11
5	0.001M NaOH	0.016	945	135	5.5	240	0.028	11
6	0.001M NaOH	0.010	400	90	5.0	144	0.042	11
7	0.001M NaOH	0.0098	800	75	7.0	146	0.053	11
8	0.001M NaOH	0.0082	1680	87	9.0	190	0.051	11
9	0.001M NaOH	0.0043	2350	350	22	604	0.037	11
10	0.001M NaOH	0.0031	1490	470	24	640	0.039	11
11	0.001M NaOH	0.014	2800	365	74.2	657	0.120	11
DDPB/water								
1	0.001M NaOH	4.0	219	4.2	9.5	10.8	1.0	11
3	0.001M NaOH	0.90	527	24	20	55	0.36	11
SDS/ethylene glycol								
12[b]	pH 6.5	3×10^{-5}	6.70	2.3	31.5	3.1	10.4	13
PM/nucleophile/toluene								
1[d]	0.05M hexylamine	0.029	5.2	2.3	11.5	2.7	4.2	14
2[d]	0.05M hexylamine	0.020	6.7	2.4	11.6	3.1	3.7	14
3[d]	0.05M hexylamine	0.033	8.2	1.2	13.3	2.1	6.7	14
4[d]	0.05M hexylamine	0.025	21.0	1.2	17.6	2.5	6.3	14
12	0.00725 M PEI_5	0.16	97	19	1.8	31	0.05	61

CTAB/$C_{14}E_9$/0.001M NaOH								
1	$\alpha_1 = 1.0$	0.62	593	69	20.3	128	0.16	34
1	$\alpha_1 = 0.83$	2.90	398	15.8	32.4	36.7	0.76	34
1	$\alpha_1 = 0.67$	2.10	494	21.9	30.0	49.8	0.54	34
1	$\alpha_1 = 0.60$	0.75	600	50	19.0	100	0.19	34
1	$\alpha_1 = 0.50$	0.20	1280	149	14.4	276	0.052	34
1	$\alpha_1 = 0.33$	0.07	1840	265	8.2	464	0.018	34
SDS/PEI/water								
1	0.1M PEI-30	0.0005	3630	73	25	187	0.130	44
1 [b]	0.1M PEI-30	0.0014	1800	56	17	134	0.130	44
1 [e]	0.1M PEI-30	0.0023	935	83	21	163	0.130	44
1 [f]	0.1M PEI-30	0.0047	254	82	14	111	0.125	44
3 [e]	0.1M PEI-30	0.0006	629	183	13	257	0.051	44
5 [e]	0.1M PEI-30	0.0003	424	117	8	168	0.049	44
CTAB/PEI/water								
1 [c]	0.02 M PEI-10	0.0070	250	220	626	195	2.86	48
3 [c]	0.02 M PEI-10	0.0043	950	320	1500	590	1.77	48
5 [c]	0.02 M PEI-10	0.0005	4600	90	420	230	1.8	48
6 [c]	0.02 M PEI-10	0.0040	1690	30	4	75	0.06	48
1	0.02M PEI-30	0.0011	1920	251	21	451	0.05	47
3	0.02M PEI-30	0.0004	3070	270	11.5	535	0.01	47
16	0.02M PEI-30	0.00006	2700	430	14.3	900	0.013	47
CDAB/PEI/water								
1	0.02M PEI-30	0.0008	1040	460	24.7	554	0.03	47
3	0.02M PEI-30	0.0008	2290	535	36.5	811	0.03	47
16	0.02M PEI-30	0.00014	5420	260	16.6	580	0.03	47

[a]for 25 °C, **b** for 35 °C, **c** for 30 °C, **d** for 40 °C, **e** for 45 °C, **f** for 55 °C, **g**; The second order rate constants of the basic hydrolysis of substrates ($k_{2,0}$, $M^{-1}s^{-1}$) at 25 °C are equal to 4.0 (**1**), 3.6 (**2**), 3.3 (**3**), 2.7 (**4**), 0.55 (**5**), 0.24 (**6**), 0.19 (**7**), 0.16 (**8**), 0.12 (**9**), 0.08 (**10**), 0.12 (**11**).

f – calculated with the use of the equation: $(k_{obs}/k_0)_{max} = \frac{k_{2,m}}{k_{2,0}} \times \frac{K_S K_{Nu}}{V\left(\sqrt{K_S} + \sqrt{K_{Nu}}\right)^2}$; the term on the left expressed as the ratio between the pseudo first rate constants in the micellar system and water describes the maximum acceleration of the reaction. The first term on the right is associated with the influence of the micellar microenvironment (F_m) and the second term reflects the concentrating of reagents in micelles (F_c).

that a decrease in the polarity, in particular on transfer from water to the micellar microenvironment should favour ion-molecular reactions since they are characterized by some delocalization of the charge in the transition state as compared to the initial state. Concurrently the transfer from the dilute aqueous solution of reagents to the concentrated micellar pseudo phase is followed by a decrease in the entropy of activation by 22.0 $J\,mol^{-1}\,K^{-1}$ for **1** and by 45.6 $J\,mol^{-1}\,K^{-1}$ for **3**, which is probably due to losses in the entropy of mixing, the decrease in the degree of the freedom of reagents, and the alteration of their orientation. In the case of substrate **1**, changes in enthalpy and entropy of activation compensate each other, so that the reactivity is practically identical in both pseudo-phases ($F_m = 1$, Table 15.1). An increase in the concentration effect for substrate **3** results in greater losses in the entropy of activation exceeding the gain in the enthalpy of activation, which is expressed in the apparent failure of the Ingold–Hughes principle ($F_m < 1$, Table 15.1). Therefore, one can conclude that the reactivity in the micellar systems is controlled by changes in the entropy of activation on the transfer of the reaction from the bulk solution to the micellar interior.

15.3 The Role of the Structural Factor in Supramolecular Catalytic Systems

15.3.1 The Influence of the 'Sphere–Rod' Micellar Transition on the Reactivity

The addition of electrolytes to the ionic micellar systems decreases the surface potential of micelles thereby decreasing the catalytic effect.[1–3,8,9] Besides such 'trivial' inhibition, electrolytes may exert an indirect influence on the reaction rate in micellar systems. At a certain concentration of counterions there occurs the 'sphere–rod' micellar transition, which is analytically indicated by the break in the 'property' *versus* 'logarithm of the counterion concentration' plot (Figure 15.1)[17–19]. Changes in the structural characteristics as a result of this transition may also affect the reactivity of compounds in micelles. To verify this assumption, the catalytic properties in parallel with the structural behaviour are studied for the DDPB micelles in the presence of sodium salicylate.[19] The treatment of the kinetic data of basic hydrolysis of **1** in semi-logarithmic coordinates provides the critical concentration of sodium salicylate (NaSal) ~0.05 M defined as a break point in the $k_{obs} - \lg C_{Sal}$ plot. The size and shape of the micelles are estimated by the methods of ^{1}H NMR spectroscopy based on the self-diffusion coefficients of DDPB and EPR spectroscopy based on the correlation time of spin-label rotation. Near the concentration of 0.05 M a change is shown to occur from a spherical to a rod-like shape of aggregates.[19] This provides evidence for the correlation between the above break points on the kinetic plot and the micellar 'sphere–rod' transitions, and hence, for the influence of these structural rearrangements on the reactivity. This approach is used for the treatment of the salt effect on the structural behaviour and catalytic effect of the conventional cationic surfactants CTAB and CPB, as well as gemini cationic surfactants and the hydroxyalkylated single-head cationic surfactants.[18,20,21]

15.3.2 The Effect of the Clustering of Reverse Micelles on the Reactivity

For reverse micellar systems, the phenomenon of electrical percolation is observed. This is manifested by the sharp increase in the conductivity of the systems by three-four

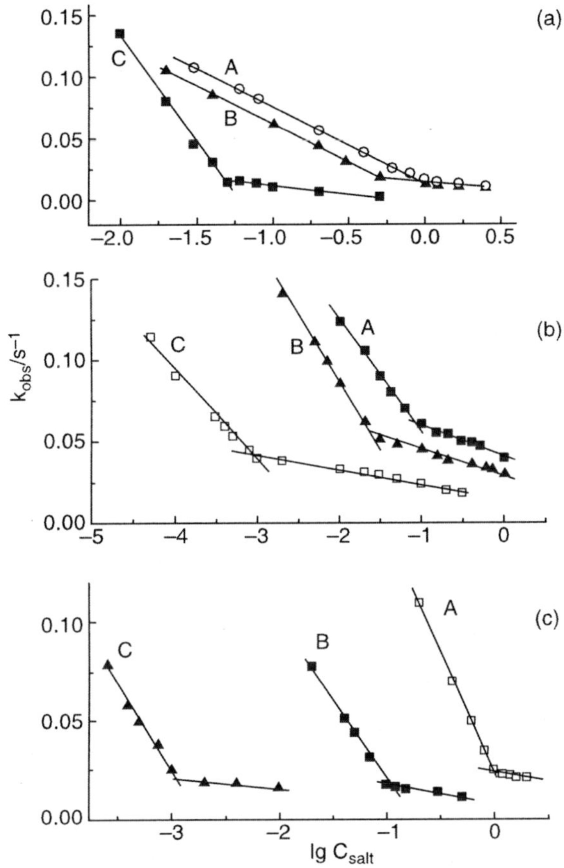

Figure 15.1 *The dependence of the observed rate constant of basic hydrolysis of **1** on the logarithm of the concentration of electrolytes in the micellar solutions of cationic surfactants; a – in the decylpyridinium chloride system and b – in the CPB system in the presence of NaCl (A), NaBr (B), NaSal (C); c – in the decylpyridinium chloride system in the presence of NaBr (A), NaSal (B); and in the CTAB micelles in the presence of NaSal (C)*

orders of magnitude with an increase in the temperature or in volume fraction of the dispersed phase.[22] In works,[23–25] the influence of this phenomenon on the reactivity of **1** in the sodium bis(2-ethylhexyl)sulfosuccinate (AOT)-n-alkane-water reverse micellar system was studied with different parameters Z = [alkane]/[AOT] (mol) and W = [H_2O]/[AOT] (mol) (Figure 15.2). Based on the data of the conductivity of solutions, the values of the temperature of the percolation threshold are estimated and the experimental conditions for the kinetic runs below and above the temperature induced percolation are chosen.

Unlike aqueous solutions, a typical Arrhenius dependence for the reverse systems is maintained only before a certain critical temperature (T_{cr}), while above this temperature a decrease in k_{obs} occurs with temperature. The value of T_{cr} is found to coincide with the percolation threshold temperature.[24] We assume that the same temperature induced structural changes of the reverse micelles are responsible for the clustering of the reverse

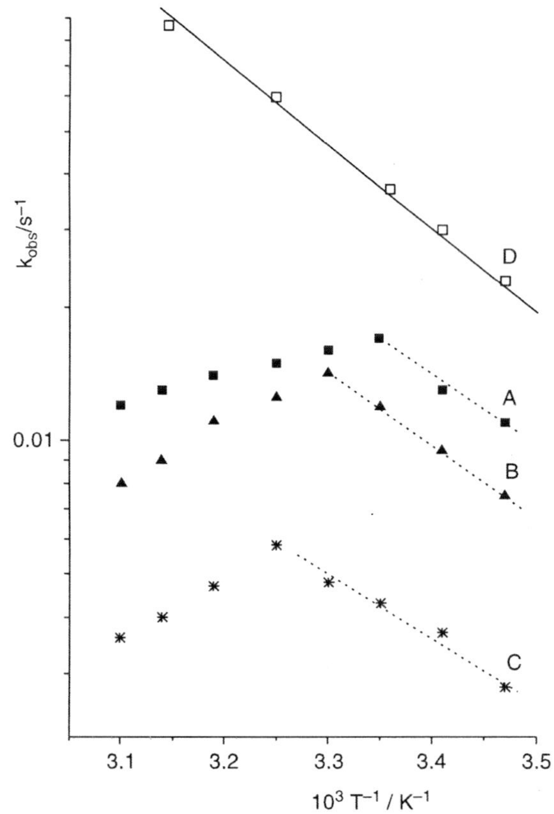

Figure 15.2 *The arrhenius dependence of the observed rate constant of the basic hydrolysis of **1** in the AOT – nonane – water reverse micellar system at different values of W: 9.8 (A), 15.1 (B), 20.0 (C) and in water (D); 0.01 M NaOH; 0.40 M AOT*

micelles as well as for the change in the slope of the Arrhenius plot. Probably, the above structural changes can be accompanied by the alteration in the properties of the micellar microenvironment of reagents, such as micropolarity, the surface potential, and etc., which can affect the reactivity. The investigation of the absorption spectra of substrate **1** and the products of its hydrolysis validated that the relocation of these compounds towards the less polar microenvironment occurred above the percolation threshold. Granting this, the effect of the temperature on the chemical kinetics in the reverse micellar systems can be divided into two contributions. One is responsible for the standard temperature activation of chemical processes, while the other accounts for the change in the localization of reagents and their partition between phases due to modification of the properties of aggregates. The typical trend of the Arrhenius dependence below the percolation threshold T_{cr} (Figure 15.2) provides evidence that the first contribution is prevalent. In this case the slope of the lg k_{obs} *vs.* 1/T plot is determined by the energy of the activation (E_a) of basic hydrolysis of **1**. Above the T_{cr} value the competitive tendency is probably predominant. Therefore the trend of the Arrhenius dependence is determined by the change in the local properties of the microenvironment of reagents with temperature. In this range, a decrease in k_{obs} occurs with temperature.

15.3.3 The Hydrolysis of Carbonic Acid Esters in Microemulsions under Phase Inversion Conditions

Microemulsions with high surfactant concentrations provide a wide-range variation in the component ratio, which can result in essential changes in their structural behaviour and properties, in rearrangements of the interface, and in phase inversions. As a result, transitions become possible between three different phases: (i) reverse microemulsions (water/oil; w/o) composed of water microdroplets dispersed in a continuous oil phase and stabilized by the surfactant monolayer including a co-surfactant if necessary; (ii) bicontinuous systems with both water and oil as continuous phases and surfactant molecules intertwined in a three-dimensional network; (iii) direct microemulsions (oil/water; o/w) composed of the oil microdroplets dispersed in a continuous aqueous phase.[26]

Basic hydrolyses of *p*-nitrophenyl acetate **13** and laurate **14** are studied in the CTAB/butanol/hexane/water system under the phase transitions induced by changes in the water/oil ratio.[27–30] The structural behaviour of the systems is examined by methods of conductometry, NMR self-diffusion, EPR spine probe spectroscopy, UV-Vis solvatochromic probe $E_T(30)$ spectroscopy. This makes it possible to characterize some micro- and macroscopic properties of the systems and to define the points of phase transitions in the series w/o-bicontinuous-o/w microemulsions. The dependence of the observed rate constant on the volume fraction of water in microemulsions also demonstrated three different regions with boundaries coinciding with the above structural transitions in the systems.[28] The character of the k_{obs} *vs*. 'water fraction' dependence is mainly influenced by the local medium polarity and the activity of hydroxide-ions. In the w/o microemulsion at low water content, the bound water (hydrated water) is predominantly present in the micellar core. When the water core increases, there appears the bulk water with a higher polarity, which favors the dissociation processes and increases the activity of hydroxide ions. This results in the increase in the k_{obs} value. In a bicontinuous system an increase in k_{obs} is less pronounced, since the aqueous phase is concentrated in narrow channels thereby preventing the growth of the volume fraction of water. Besides, the considerable increase in microviscosity also contributes to the observed rate constants of the hydrolysis as a diffusion term. Further, the transition 'bicontinuous – o/w' system results once again in a sharp increase in the reaction rate with the dispersion phase volume. Water is assumed to regulate the partition of butanol between phases, retaining constant its concentration in the dispersion phase, which is supported by the data of NMR self-diffusion.[27] This decreases the content of butanol at the interface increasing the positive charge density of aggregates and the surface concentration of hydroxide-ions.

15.4 Binary Surfactant Systems

15.4.1 Aqueous Binary System Ionic–Nonionic Surfactants

In works,[31–35] the hydrolysis of **1** is studied in the binary systems based on CTAB and a series of nonionic surfactants: i.e. polyoxyethylene(10)monooleinic ether (Brij-97), polyoxyethylene(23)monododecyl ether (Brij-35), polyoxyethylene(9)monotetradecyl ether ($C_{14}E_9$), polyoxyethylene(9.5)mono-4-isooctylphenyl ether (Triton-X-100) exhibiting a wide range of hydrophilic-lipophilic balance and micellizating properties (cmc of the surfactants above are equal to 0.000017, 0.00031, 0.000052, 0.00022 M).

The micellization of binary surfactant solutions is studied by the method of tensiometry at the variation of the fraction of ionic surfactant α_1 within the range 0–1.0. The quantitative treatment of the surface tension isotherms within the framework of the phase separation model[36] showed that the values of the interaction parameter β in these systems lay within the range (–0.85)–(–4.6). This indicates that in these systems synergetic behaviour occurs, i.e. the mixed aggregates are formed. The surface potential is calculated for all the systems. A decrease in the potential is shown to occur with a decrease in α_1.[31,32,34] In the single CTAB micelles, up to a 20-fold acceleration of the hydrolysis of **1** is observed. The nonionic Brij-97[31] and $C_{14}E_9$[34] micelles do not influence the reaction rate, while the Brij-35[35] and Triton-X-100[33] micelles retard the reaction by 3–5 times.

Figure 15.3 shows that a decrease in the fraction of ionic surfactant α_1 results in a decrease in the catalytic effect. In the case of the Brij-35 and Triton-X-100 micelles

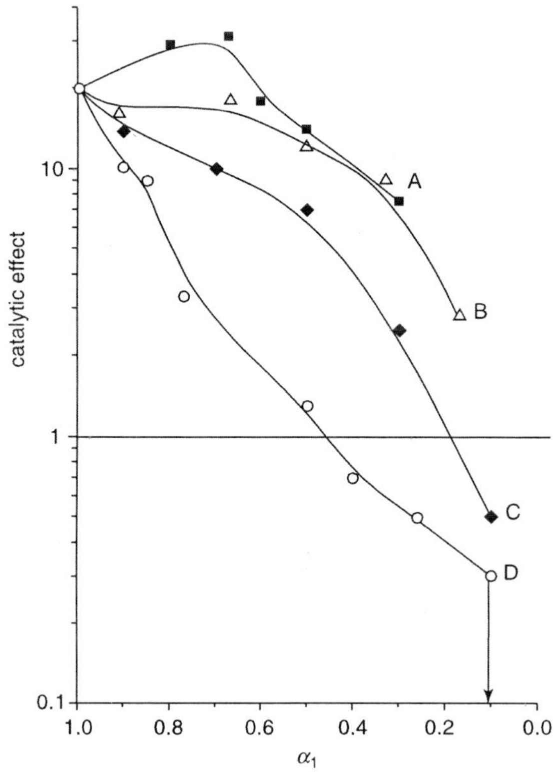

Figure 15.3 *The catalytic effect of the binary surfactant solutions CTAB/$C_{14}E_9$ (A), CTAB/Brij-97 (B), CTAB/Triton-X-100 (C), CTAB/Brij-35 (D) on the basic hydrolysis of phosphonate **1** as function of the molar fraction of CTAB α_1; 0.001 M NaOH, 25 °C*

an inversion from the catalysis to the inhibition of the processes occurs. At a certain concentration of Brij-35 in the mixture actually an absolute termination of the reaction is observed. The elucidation of the main factors responsible for the alteration of the reaction rate in micelles, i.e. the factor of concentration (F_c) and the factor of microenvironment (F_m) (Table 1, the data are exemplified by the CTAB/$C_{14}E_9$ system[34]) makes it possible to derive the driving force of the catalysis in binary systems. The compensatory change of the F_c and F_m values occurs within the whole range of the surfactant ratio. However, with small additives of nonionic surfactants, an increase in the factor of the micellar microenvironment is a prevalent tendency, controlling the increase of the resulting catalytic effect. Further decrease in α_1 results in the inversion of the trend, so that an increase in F_c is accompanied by a decrease in F_m (Table 15.1). In these surfactant mixtures, the contribution of the negative influence of the factor of micellar microenvironment remains dominating, which provides reasons for the reduction of the catalytic efficiency. Thus, the catalytic effect of the binary surfactant systems is determined by the tendency of changing the factor of micellar microenvironment throughout the whole surfactant proportions. This indicates that unlike surfactant-electrolyte systems,[17–19] the surface potential controlling the concentration factor plays only an auxiliary role in mixed micelles.

15.4.2 The Binary Surfactant Mixtures in Non-aqueous Media

In works,[37–39] the catalytic effect of the CTAB and PM reverse micelles on the reaction of pyrimidinophanes (PPh) (Scheme 15.2) and their acyclic analogue *N,N′*-bis(2-methylthio)-6-methylpyrimidin-4-yl)hexamethylenediamine with phosphinate **12** in chloroform is studied. The formation of the CTAB–PM mixed aggregates is proved by the NMR self-diffusion, by dielcometric titration, and by dynamic light scattering.[37,39] The micellar catalytic effect is shown to change with the size of macrocycle in the series PPh-4 < PPh-3 < PPh-2 < PPh-1. The maximum acceleration of the reaction observed at a 5 : 1 CTAB/PM molar ratio exceeds five orders of magnitude.

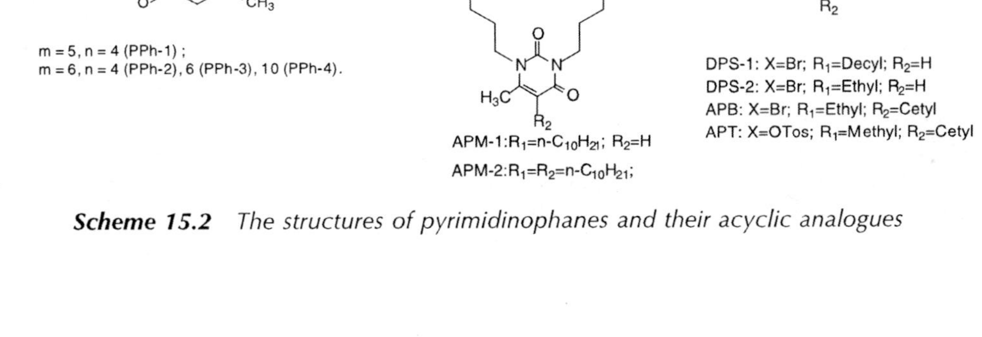

Scheme 15.2 *The structures of pyrimidinophanes and their acyclic analogues*

15.5 Polycomponent Catalytic Systems Based on Amphiphiles and Polymers

15.5.1 The Conventional Surfactant/Polyethyleneimine Systems

Since the self-organized catalytic systems are considered to be the simplest models of enzyme catalysts,[1–3] we made an effort to model one of the basic principles of biocatalysis, namely, the polyfunctional mechanism of the enzyme action. Therefore, the main idea is to design the polycomponent catalytic ensembles by including ingredients with their own catalytic, sorption or solubilizing activities. For this purpose the step-by-step complication of supramolecular systems is carried out beginning with the aqueous polyethyleneimine (PEI) solution. The components of supramolecular systems were chosen in accordance with the following: PEI is known to accelerate the hydrolysis of phosphorus acid esters via the general basic catalysis.[40] Besides, it contributes to the catalysis of the reaction due to the sorption of reagents and to the formation of the catalytic complex substrate-polymer. The addition of a surfactant or other self-assembling amphiphiles to the systems opens up additional opportunities for binding the reagent due to the solubilization mechanism together with the sorption pattern. The modification of the system with metal ions makes possible the contribution of the electrophilic catalysis to the catalytic mechanism, since a considerable acceleration of nucleophilic substitution at phosphorus by metal ions is well known.[6,41]

15.5.1.1 The Single PEI Solution

There is much evidence that the amines including polyamines accelerate hydrolyses of phosphorus acid esters via the general basic catalysis.[40,42] In the unbuffered solution, a small amount of protonated amino groups are present in the PEI molecules (<0.5–5.0% depending on the PEI molecular mass[43]; PEI with the molecular mass from 800 (PEI-800) to 60000 (PEI-60) were studied). This means that under these conditions PEIs behave as neutral polymers rather than polyelectrolytes. The pH kinetic profile demonstrates that up to pH 9.2 the observed rate constant of hydrolyses of **1** and **3** in 0.05 M solution of PEI-50 does not exceed the limits of $5 \times 10^{-5}\ s^{-1}$, while a further increase in the pH results in a *ca.* tenfold increase in the hydrolysis rate of both phosphonates.[43] This pH dependence is rather typical and reflects the growth of the rate constant in the field of accumulation of catalytic species, namely nonprotonated amino groups. In addition, the contribution of basic hydrolysis to k_{obs} values at high pH should also be taken into account.

To elucidate the influence of the molecular mass of the polymer on reactivity, the kinetics of the hydrolyses of phosphonates **1** and **3** was measured in the presence of ethylenediamine (ED), tetraethylenetetramine (TETA), PEI-800, PEI-25, and PEI-50 at spontaneous solution pH equal to 10.0–10.8 depending on the molecular mass of the PEI.[43] For all the systems studied, an increase in k_{obs} with the amine concentration occurs due to general basic catalysis. Starting from TETA the curves are S-shaped, which is in line with the preliminary equilibrium process such as the formation of the catalytic complex substrate-polymer. Therefore, the reagent binding by oligomerous and polymeric amines provides an additional contribution to the rate effect. An increase in molecular mass of amines results in a slight increase in the catalytic effect per amino group. The nature of substrate does not strongly affect the reactivity. Some higher k_{obs} values were found for the less hydrophobic phosphonate **1**.

15.5.1.2 The SDS/PEI Supramolecular Systems

In works,[43,44] the self-organization is studied in the binary SDS/PEI systems by methods of tensiometry, conductometry, viscosimetry, potentiometry and dynamic light scattering. A marked decrease in cmc is observed when transferred from the single SDS solution (cmc 0.0085 M) to the SDS/PEI system (critical aggregation concentration (cac) 0.00065 M, the data for PEI-25[43]), which provides evidence for the formation of the mixed surfactant-polymer ensembles. In the SDS/PEI-30 systems the kinetics of hydrolyses of phosphonates **1**, **3**, **5** is studied under the variation of the molecular mass of polymer.[44] The acceleration of the reactions is observed as compared to the aqueous PEI solution. The reactivity of substrates decreases in the series **1** > **3** > **5**. The same tendency occurs in the alkali water solution and in the single PEI system. The k_{obs} *vs.* [SDS] kinetic curves demonstrate an extreme type dependence unlike those in single PEI solutions. The kinetic curves with a maximum are typical for micellar catalysis and indicate the essential contribution of the solubilization pattern to the mechanism of reagent binding together with the sorption pattern involved in the single PEI solution. In the SDS/PEI system the higher reactivity is observed for the less hydrophobic substrate **1** as compared to **3**. The rate effect is shown to be due to concentrating the reagents in the polymer-colloid complexes for all the systems studied (Table 15.1). To elucidate the structuring role of PEI and contribution of the polymer catalysis to the summary rate effect, the kinetics of the hydrolyses of phosphonates **1** and **3** was examined in the micellar SDS solution in the presence of ED. In this case, a more than 2-fold retardation of the reaction occurs. Such a difference in the catalytic effect of monomeric and polymeric amines supports the above assumption that both the general basic catalysis and binding the substrates by the PEI molecules contribute to the acceleration of the reaction.

For substrate **1** the kinetics of hydrolysis is measured in the SDS/PEI-30 system at different temperatures,[44] and then the activation parameters for the reaction in different pseudo phases are calculated. For micellar pseudo-phase the enthalpy of activation $\Delta H^{\neq}_{m} = 55.2$ kJ mol^{-1} and entropy of activation $\Delta S^{\neq}_{m} = -121.7$ J mol^{-1} K^{-1} are calculated, which are in agreement with the bimolecular mechanism of the reaction. These values differ little from the activation parameters obtained for the reaction in the aqueous pseudo-phase ($\Delta H^{\neq}_{0} = 56.6$ kJ mol^{-1}, $\Delta S^{\neq}_{0} = -100.3$ J mol^{-1} K^{-1}). This probably provides evidence that the mechanism of catalysis remains unchanged when the process is transferred from water to aggregates, i.e. the general basic catalysis occurs in both cases.

The next stage of the study was on the kinetics of hydrolyses of **1**, **3** and di-*p*-nitrophenyl phosphate **15** in the ternary SDS/PEI/La(III) system at spontaneous solution pH (pH $\approx$ 8.7).[43] In this system there occurs a high performance in the reactivity of phosphonates and its growth. The maximum catalytic effect exceeding four orders of magnitude is found for the substrate **15** as compared to the single PEI solution. Comparison with the basic hydrolysis of **15** reveals the rate effect over six orders of magnitude.

15.5.1.3 The NonIonic-Surfactant/PEI System

Unlike the SDS/PEI system the cmc value does not decrease markedly in the solutions of nonionic surfactant $C_{12}E_{10}$ and Triton-X-100 when the polymer is added. This is in line with the assumption on the weak affinity between the nonionic surfactant and polymers.[45] However, as will be shown below a rather high probability occurs that the mixed nonionic surfactant/PEI aggregation does take place. In work,[46] a step-by-step investigation of the aggregation and catalytic effect towards the cleavage of phosphorus acid esters **1** and **3** has been performed in the following series: the single nonionic surfactant micelles → the binary nonionic surfactant/PEI micellar system → the ternary nonionic surfactant/PEI/La(III) micellar system. The single $C_{12}E_{10}$ micelles do not influence the hydrolysis rate, while Triton-X-100 solutions slightly retard the reaction.[33] In the binary nonionic/PEI solutions there occurs the rate enhancement of hydrolyses of phosphonates **1** and **3**. For example, in the $C_{12}E_{10}$/PEI-60 system 3.6-fold and 4.7-fold accelerations of the hydrolyses of **1** and **3** respectively occur as compared to the single PEI solution (Figure 15.4a). When the low molecular mass amines are used instead of PEI, the k_{obs} *versus* surfactant concentration dependence assume quite a different shape (Figure 15.4b). Thus, provided that there occurs only the general basic catalysis of the hydrolysis of the phosphonates due to aminogroups of PEI, the inhibition of the reaction might be observed. These data provide evidence for the formation of a new type of aggregates as compared to the single $C_{12}E_{10}$ micelles, i.e. they are indirect arguments in favor of the fact that mixed aggregation occurs

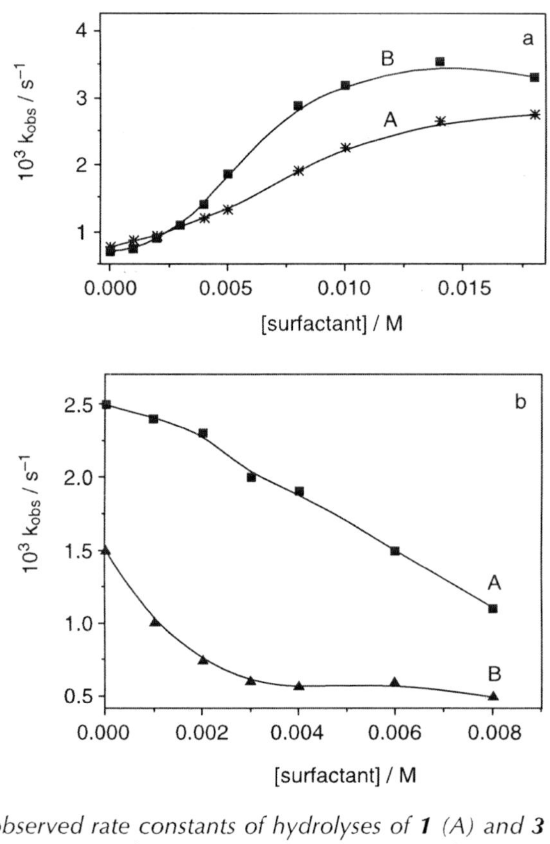

Figure 15.4 *The observed rate constants of hydrolyses of **1** (A) and **3** (B) as functions of the surfactant concentration in the $C_{12}E_{10}$/PEI-60/water (a) and $C_{12}E_{10}$/ethylenediamine/water (b) ternary systems at fixed amine concentration of 0.05 M; 25 °C. Reprinted with permission from [46]. Copyright 2007 American Chemical Society*

in the nonionic surfactant/polymer system. The examination of the $C_{12}E_{10}$/PEI/La(III) ternary system with a polarizing microscope revealed the lyotropic mesomorphism with fan texture typical for the hexagonal packing. The concentration and temperature ranges of the hexagonal phase were determined. The ternary system is found to be an effective nanosized catalyst in both micellar and mesophase regions. It demonstrates the complex mechanism of the catalytic effect under mild conditions (25 °C, pH ~8.0) and the substrate specificity towards the more hydrophobic phosphonate. A marked (*ca.* 10^3-fold) acceleration of the reaction studied as compared to the basic hydrolysis of the substrates has been attained.[46]

15.5.1.4 The Cationic Surfactants/PEI Supramolecular Catalytic Systems

The cationic surfactant-polymer systems are much less studied than those based on anionic or nonionic surfactants. In works,[47,48] the catalytic effect of the cationic surfactant/PEI systems towards hydrolyses of **1**, **3**, **5** and **6** is studied, in cases when the molecular mass of PEI and the structure of the head group of surfactants experience changes. In the CTAB/PEI-10 system the acceleration of hydrolyses of all the substrates studied is observed. The catalytic effect depends to a great extent on the structure of phosphonates and varies within the interval 4–1500 (Table 15.1), with the maximum acceleration occurring for phosphonate **3**.[48] The observed rate constant in the binary CTAB/PEI-10 system decreases in the series **3** > **1** > **6** > **5**. In the water alkali solutions in the absence of the surfactant and PEI the k_{obs} of the basic hydrolysis changes as follows: **1** > **3** > **5** > **6**, while in the CTAB micellar solution with no PEI added the order is **3** > **1** > **5** > **6**.[11] These differences in the reactivity are due to the fact that the basic factors controlling the reactivity in the surfactant free solutions are the electronic and steric effects of substituents at the reaction centre, while in the organized solutions a marked contribution to the catalytic effect is made by binding the reagent in nanoaggregates, which correlates with the hydrophobicity of compounds. The above inversion of reactivity in the pairs of substrates **1**, **3** and **5**, **6** in the mixed polymer-surfactant systems as compared to aqueous basic hydrolysis is due to (i) the change in the mechanism of the hydrolysis in the presence of PEI, i.e. the transfer from the specific to general basic catalysis and (ii) the peculiarity of the catalysis in constrained systems, in which the main contributor is the concentrating of reagents due to their binding by supramolecular ensembles. For the substrates **1** and **3** the influence of substituents in aqueous and micellar pseudo phase is similar, which is evident from the decrease of $k_{2,m}$ with the length of the alkyl radical at reaction centre (Table 15.1). However, there occurs a considerable increase in the binding constant of the more hydrophobic substrate ($k_{2,m}$ = 0.0070 and 0.0043 $M^{-1} s^{-1}$; K_S = 250 and 950 M^{-1} for **1** and **3** respectively), which is just responsible for the higher observed rate constants for phosphonate **3**.

In work,[47] the hydrolyses of phosphonates **1**, **3** and *p*-nitrophenyldiphenyl phosphate (**16**) in the systems based on PEI-30 and two cationic surfactants CTAB and cetyldimethylethylammonium bromide (CDAB) are studied. The nature of substrate and the structure of the head group is found to markedly influence the reaction rate in these systems. The maximum value of the observed rate constant of **1** in the CTAB/PEI-30 system is twofold higher than that in the CDAB based system, while for the phosphonate **3** k_{obs} is threefold higher in the CDAB/PEI system. These differences are controlled by the $k_{2,m}$ values for the substrates in both systems (Table 15.1). The observed rate constants of **16** are much lower than those for **1** and **3**, which is due to the low $k_{2,m}$ value (Table 15.1). Unlike the phosphonates **1**, **3** a little difference in the reactivity of **16** in CTAB and CDAB is observed. In the CDAB aqueous solutions the hexagonal liquid crystalline mesophase (E-phase) is formed above the concentration of 0.4 M. The kinetic study of the reactions when the E-phase is formed demonstrates a decrease in the observed rate constant of the hydrolyses of the substrates with the ordering of solution.

In works,[49,50] the formation of the supramolecular systems based on branched and hydrophobized (alkyl-substituted) PEIs is revealed in both the absence and presence of cationic surfactants including those with hydroxyalkylated head groups. The latter are capable of specific interactions with other components of the systems, in particular, polymer and reagents. Hydrophobization of PEIs is found to result in a sharp increase in their solubilizing and catalytic efficiency. Besides, a marked increase in the catalysis occurs on transfer from conventional cationic surfactants to hydroxyalkylated analogues. By means of the optimization of the catalytic composition and reaction conditions a substantial 700–2000-fold acceleration of the phosphonate **1** is reached at neutral pH.[50]

15.5.2 The Pyrimidinic Surfactant Based Systems

In works,[51–53] the above step-by-step approach was extended to a new type of building blocks, namely pyrimidinic dimeric surfactants. The self-organization of new amphiphilic pyrimidinic macrocycles of dimeric structure APM-1 and APM-2 (Scheme 15.2) has been studied in aqueous solutions.[51] The complex examination of the APM based systems with the help of tensiometry, conductometry, dynamic light scattering, UV-, and NMR-spectroscopy provides evidence for their aggregation. Calculations based on the surface tension isotherms and on the packing parameter considerations (P ~1) make it possible to assume a lamellar packing of macrocycles when aggregating. Marked differences in the aggregation behaviour of APM-1 and APM-2 have been found. In the tensiometry study, the lower cmc and higher cross-section area per molecule are found for APM-2 as compared to APM-1. The additives of PEI do not exert much influence on the cmc of APM-1. A pronounced decrease in cmc and also a ca. twofold decrease in the surface area per molecule occur in the APM-2/PEI systems. This can indicate the transition from the folded to the extended packing mode. The APM ensembles are explored as nanoreactors for the hydrolyses of phosphonic acid esters. The kinetic study reveals a minor rate effect of the APM-1 based systems on hydrolyses of phosphonates **1** and **3**. Opposite to this the acceleration of the hydrolyses of both phosphonates occurs in the APM-2 based systems as compared to the uncatalyzed process. The dramatic differences in the catalytic activity of the two macrocycles are assumed to be underlain by the different morphology of aggregates. Within the APM-2→APM-2/PEI→APM-2/PEI/La(III) series, due to the cooperative contributions of the micellar, polymer and homogeneous catalysis an increase in the catalytic effect is observed from 30 times to three orders of magnitude as compared to the basic hydrolysis.

In work,[52] two dimeric pyrimidinic surfactants (DPS), acyclic analogues of APM-1 were synthesized: a gemini surfactant DPS-1, and a typical bolaform surfactant DPS-2 (Scheme 15.2). The catalytic effect of the single surfactant solutions and their mixtures with PEI towards hydrolyses of **1** and **3** was studied. They exhibit specific aggregation and catalytic properties. The more hydrophobic surfactant DPS-1 demonstrates behaviour typical for cationic surfactants, in particular, it shows catalytic activity towards hydrolyses of phosphonates, which increases with the substrate hydrophobicity. In the DPS-2 based systems, a 25-fold inhibition of the reaction occurs, with no substrate specificity. Such behaviour is rather unusual for a cationic surfactant. The differences in the aggregation and catalytic properties can be due to the differences in the packing mode between the DPS-1 and DPS-2 molecules when self-assembling.

Recently, new amphiphilic pyrimidinic compounds with different kinds of counterions, inorganic bromide anions (APB) and hydrophobic tosylate anions (APT) were synthesized,[53] which showed intermediate hydrophobicity between DPS-1 and DPS-2. The catalytic effect towards basic hydrolyses of **1** and **3** is examined. For comparison, the rate effect of conventional surfactants CTAB and cetyltrimethylammonium tosylate (CTAT) with similar counterions is studied as well. Both CTAB and APB solutions accelerate hydrolyses of phosphonates **1** and **3**. Surfactants with hydrophobic anions demonstrate a different behaviour, namely, hydrolysis of phosphonate **3** is accelerated by CTAT and APT micelles, while the less hydrophobic substrate **1** is retarded by the APT micelles unless alkali excess is used. This deviation from the typical behaviour for cationic surfactants is due to the decrease in solution pH with APT concentration above the cmc. Detailed self-organization studies by methods of tensiometry, conductometry, potentiometry and NMR spectroscopy were conducted. A marked difference in the structural behaviour between the AP-type and conventional CTA surfactants is found: (i) for both pairs APB/CTAB and APT/CTAT studied the counterion binding is stronger for the conventional cationic surfactants as compared to dimeric pyrimidinic surfactants; (ii) although a structural transition is probably observed in the APT micelles at $C \geq 0.01\,M$, no sharp micellar growth occurs with the APT concentration, as distinct from the CTAT micelles; (iii) the cmc's of dimeric surfactants are only a little lower than those of CTA analogues. The geometry of AP compounds may be assumed to be mainly responsible for the above result. A branched molecular architecture prevents the close packing of the monomers in the bulk solution and at the interface and produces a steric hindrance around the head groups. This effect can be responsible for the high surface potential, and hence for the decrease in solution pH due to the polarization of the hydrated water molecules and their further deprotonation.

15.5.3 The Calixarene Based Systems

Recently, new building blocks, in particular, amphiphilic calixarenes were explored for the design of polycomponent catalysts.[54–57] We intended testing the contribution of the inclusive interactions to the self-assembling of components. In works,[54,55] the aggregation behaviour and catalytic activity of PEI and sulfonatomethylated calix[4]resorcinarene (SCA) (Scheme 15.3) based systems are studied by methods of the NMR and UV-vis spectroscopy. The mixed PEI–SCA assembling is shown to occur in the concentration range lower than the cmc indicated by the tensiometry data (cmc = 0.011 M) due to the inclusive type of the PEI–SCA interactions. Different modes of self-organization assuming the PEI inclusion into the calixarene cavity are discussed: (i) the layered ensembles separated by the PEI chains dipped into the cavity, and (ii) the SCA molecules penetrated

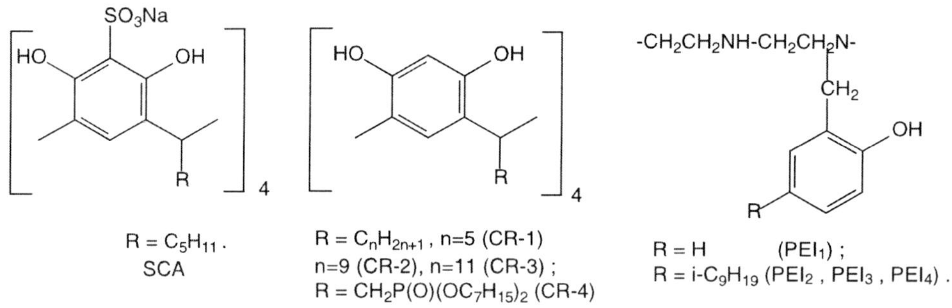

Scheme 15.3 *The chemical structures of calixarenes and substituted PEIs*

with PEI of the pseudorotaxane type. The latter is in good agreement with the extensive NMR and calculating data for this system.

In the kinetic studies the step-by-step investigation of the catalytic activity towards the hydrolytic cleavage of phosphonate **1** and *p*-nitrophenyl dimethyl phosphate (**17**) has been carried out in the single PEI system, the binary PEI/SCA system and the ternary PEI/SCA/La(III) system. Two reaction routes of hydrolysis of **17** (the leaving of aromatic versus alcohol fragment) are revealed in the single PEI solution and the binary PEI/SCA systems by both NMR and UV-vis studies, while in the presence of La(III) only the departure of *p*-nitrophenolate group occurs. The addition of noncatalytic SCA to the PEI solution is shown to result in a tenfold acceleration of the reaction below the cmc due to the catalytic action of the PEI/SCA inclusive type ensembles. The ternary PEI/SCA/La(III) system demonstrates a marked catalytic activity attaining two (for **1**) and four (for **17**) orders of magnitude due to superposition of several factors of catalysis.

In work,[56] the aggregation behaviour and catalytic activity of supramolecular systems composed of three basic components in different combinations are investigated. They are amphiphilic calix[4]resorcinarene CR-1 (Scheme 15.3), anionic surfactant SDS and PEI-25, either of which is capable of self-assembling in solution and interacting through a 'guest–host' mechanism. The evaluations of the size of aggregates, cloud points and catalytic properties are in agreement with the assumption on the mixed aggregation in the systems studied. The step-by-step investigation of the catalytic effect towards hydrolyses of phosphonates **1** and **3** is performed in the series of supramolecular systems CR-1/SDS; CR-1/PEI; SDS/PEI; CR-1/SDS/PEI. The change from ca. 20-fold inhibition of the reaction in ensembles of anionic amphiphiles CR-1 and SDS to the tenfold catalysis occurs due to the immobilization of anionic nanoreactors at the catalytically active polymer matrix, PEI. The possibility for the transfer from the microheterogeneous to the heterogeneous system and for the separation of the catalyst and products by the cloud point strategy is presented.

15.5.4 The Non-Aqueous Supramolecular Systems

In works,[58–60] the influence of cationic and nonionic surfactants, CTAB and PM on the reaction of phosphinate **12** with branched PEI-10 bearing *o*-hydroxybenzyl groups at the tertiary and secondary nitrogen (Scheme 15.3) in chloroform is investigated. The average

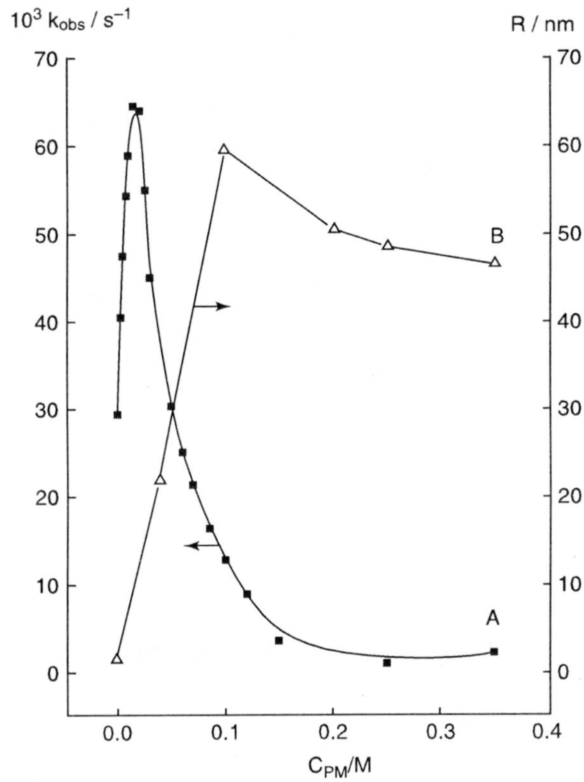

Figure 15.5 *The dependence of the observed rate constant of the reaction of PEI_2 with phosphinate **12** (A) and the effective radius (R) of the PEI_2-PM ensembles in chloroform (B) on the PM concentration; 25 °C; 0.01 M PEI_2*

ratio between unsubstituted and hydroxybenzylated fragments amounts to 0.3; 0.12; 0.16 and 0.3 for PEI_1, PEI_2, PEI_3 and PEI_4 respectively. The formation of the mixed PEI/CTAB and PEI/PM ensembles are supported by dielcometric titration, dynamic light scattering, viscosimetry and the solubilization of *p*-nitrophenol measurements. The character and extent of their influence on the phosphorylation of PEIs is found to depend on the PEI/surfactants ratio, on the component structure, and on the temperature. In the case of CTAB an acceleration of the reaction occurs throughout the whole concentration range studied,[58] while in the PEI/PM system an increase in the surfactant concentration at first results in the increase in the observed rate constant and then in its decrease.[59] The inversion of the rate effect is probably due to the rearrangement of the supramolecular ensembles. Indeed, the concentration range corresponding to the inhibition of the processes coincides with the formation of the lager aggregates (Figure 15.5).[59] The surfactant rate effects on the phosphorylation of PEI equal the ca. 6–7-fold acceleration (for CTAB) or the inhibition of the same magnitude (for PM).[58,59] The use of the CTAB/PM mixture provides a marked increase in the catalytic effect, which depends on the molar ratio of the surfactant, reaching ca. 24-fold acceleration.[60] The rate enhancement increases in the series $PEI_1 < PEI_4 < PEI_3 < PEI_2$. The quantitative treatment of the kinetic data in terms of pseudo phase model demonstrates that small as it is the factor of concentration of reagents mainly contributes to the rate effect. Indeed, the microenvironment factor may vary from the negative influence ($F_m = 0.32$ for PEI_1) to 1.0 for PEI_4, while the concentration factor F_c>1 in all cases, thereby provides the acceleration of the reaction.

In work,[61] the reaction of alkylated branched PEI-10 (APEI-10) bearing tetradecyl substituents (PEI_5) with phosphinate **12** is studied in the presence of PM in toluene. The k_{obs} *versus* C_{PM} dependence has an extremum. A maximum twofold acceleration of the reaction is observed, followed by a *ca.* 5–7-fold inhibition with the further increase in the surfactant concentration. Similar to the above phosphorylation reaction, the unfavorable influence of the microenvironment and the prevalent positive contribution of the concentration factor occur in this system. In particular, the values $F_m = 0.55$ and $F_c = 31.0$ are obtained when fitting the kinetic data in terms of Equation 15.1 (Table 15.1).

In work,[62] the complexation of APEI-10 bearing *n*-tetradecyl (PEI_5, PEI_6) and *n*-nonyl (PEI_7) radicals (the extent of substitution of 0.3; 0.6 and 0.26 respectively) with functionalized calix[4]resorcinarenes (CR) in chloroform is revealed by methods of UV-Vis spectrophotometry, dielcometric titration and dynamic light scattering. The CRs alkylated or phosphorylated at low rim are studied (Scheme 15.3). The complexation of PEI and CR is shown to change the nucleophilicity of the polymer, which is exemplified by the reaction between APEI and phosphinate **12**.[63] Based on the NMR ^{31}P data a conclusion is drawn that PEI rather than CR functions as a nucleophile towards the substrate in the PEI/CR/chloroform system. CR mediates this reaction as a catalytic species involved in the mixed aggregation. The rate effect depends on the ratio of components and their structure. A decrease in the hydrophobicity and increase in the extent of substitution of PEI result in the enhancement of the catalysis. The highest *ca.* 35-fold acceleration of the phosphorylation occurs at the low CR and polymer concentrations. An additional 6–7-fold increase in the catalytic effect can be attained by the addition of surfactants. For example, the PEI_5/CR-2/CTAB-chloroform system demonstrates a 45-fold acceleration of the reaction under study.[64]

15.6 Conclusions

The catalytic effect of supramolecular systems based on amphiphilic compounds of different structure (surfactants, hydrophobized polymers, calixarenes, pyrimidinophanes) on the nucleophilic substitution at phosphorus and carbon has been investigated in both aqueous and nonaqueous media. They demonstrate a wide-range catalytic activity from the deep retardation of the reactions to the multiple accelerations to the point of several orders of magnitude. The factors determining the rate effect have been analyzed. Two main points should be emphasized, i.e. (i) the crucial role of the structural factors in the catalysis has been revealed; (ii) the catalytic effect of polycomponent ensembles is markedly influenced by the nature of components and their ratio. This indicates that reactivity in supramolecular systems can be controlled by the purposeful selection of catalytic species and by inducing structural transition in the systems. Besides, such kind of control requires knowledge of the structural behaviour of the systems, in particular, packing modes, morphology of aggregates, surface potential, etc.

Acknowledgements

We thank the Russian Academy of Sciences (Chemistry and Material Science Division) and the Russian Foundation for Basic Researches (grant 09-03-00572) for their financial support.

Dedication

This work is dedicated to the memory of Lyudmila Andreevna Kudryavtseva who had made an important contribution to the development of catalysis in organized systems.

References

1. I. V. Berezin, K. Martinek, A. K. Yatsimirskii, Physicochemical foundations of micellar catalysis, *Russ. Chem. Rev.*, 1973, **42**, 787–802.
2. C. A. Bunton, G. Savelli, Organic reactivity in aqueous micelles and similar assemblies, *Adv. Phys. Org. Chem.*, 1986, **22**, 213–309.
3. L. S. Romsted, Micellar effects on reaction rates and equilibria, *Surfactants in Solution*, Vol. 2, K. L. Mittal, B. Lindman (Eds.), Plenum Press, New-York-London, 1984.
4. K. Holmberg, Organic reactions in microemulsions, *Curr. Opin. Colloid Interface Sci.*, 2003, **8**, 187–196.
5. T. Dwars, E. Paetzold, G. Oehme, Reactions in micellar systems, *Angew. Chem. Int. Ed.*, 2005, **44**, 7174–7199.
6. W. Jencks, *Catalysis in Chemistry and Enzymology*, Courier Dover Publications, New-York, 1987.
7. Y. C. Yang, J. A. Bakker, J. R. Ward, Decontamination of chemical warfare agents, *Chem. Rev.*, 1992, **92**, 1729–1743.
8. E. P. Tishkova, L. A. Kudryavtseva, Reactions of esters of tetracoordinated phosphorus acids with nucleophilic reagents in highly organized media, *Russ. Chem. Bull.*, 1996, **45**, 284–299.
9. L. Ya. Zakharova, A. B. Mirgorodskaya, E. P. Zhil'tsova, L. A. Kudryavtseva, A. I. Konovalov, Catalysis of nucleophilic substitution reactions in supramolecular systems, *Russ. Chem. Bull.*, 2004, **53**, 1385–1401.
10. Yu. L. Khmel'nitskii, A. V. Levashov, N. L. Klyachko, K. Martinek, Microheterogeneous medium for chemical (enzymic) reactions based on a colloidal solution of water in an organic solvent, *Russ. Chem. Rev.*, 1984, **53**, 319–331.
11. R. A. Shagidullina, L. Ya. Zakharova, F. G. Valeeva, L. A. Kudryavtseva, The reactivity of phosphonic esters in aqueous micellar solutions of cationic surfactants, *Russ. Chem. Bull.*, 2001, **50**, 1181–1186.
12. R. A. Shagidullina, L. Ya. Zakharova, L. A. Kudryavtseva, The effects of cetylpyridinium bromide micelles on the kinetics of alkaline hydrolysis of alkyl *p*-nitrophenyl chloromethylphosphonate, *Russ. Chem. Bull.*, 1999, **48**, 278–283.
13. L. Ya. Zakharova, F. G. Valeeva, L. A. Kudryavtseva, Yu. F. Zuev, Micelle forming properties and catalytic effect of the system sodium dodecyl sulfate–ethylene glycol in solvolysis of *p*-nitrophenyl bis(chloromethyl)phosphinate, *Russ. J. Gen. Chem.*, 2002, **72**, 1367–1372.
14. R. A. Shagidullina, E. P. Zhil'tsova, L. A. Kudryavtseva, Aminolysis of O-alkyl-O–4-nitrophenyl chloromethylphosphonates in reverse micellar solutions of the nonionic surfactant, *Russ. Chem. Bull.*, 1998, **47**, 1686–1690.
15. L. Ya. Zakharova, R. A. Shagidullina, F. G. Valeeva, E. P. Zhil'tsova, L. A. Kudryavtseva, Kinetics of nucleophilic substitution reactions in ordered media, *Russ. J. Phys. Chem.*, 2000, **74**, 1821–1824.

16. E. P. Zhil'tsova, A. P. Timosheva, L. A. Kudryavtseva, Yu. I. Sal'nikov, I. S. Ryzhkina, V. E. Kataev, R. A. Shagidullina, Micellization and catalytic properties of polyethylene glycol–600-monolaurate, *Russ. J. Gen. Chem.*, 2002, **72**, 1017–1022.
17. L. Ya. Zakharova, L. A. Kudryavtseva, A. I. Konovalov, Mechanism of the inhibiting action of electrolytes on the micellar effect in alkaline hydrolysis of *p*-nitrophenyl ethyl chloromethylphosphonate, *Mendeleev Commun.*, 1998, **8**, 163–165.
18. L. Ya. Zakharova, D. B. Kudryavtsev, F. G. Valeeva, L. A. Kudryavtseva, Inhibition of alkaline hydrolysis of ethyl *p*-nitrophenyl (chloromethyl)phosphonate in the system cationic surfactant-water-electrolyte, *Russ. J. Gen. Chem.*, 2002, **72**, 1215–1221.
19. L. Ya. Zakharova, D. B. Kudryavtsev, L. A. Kudryavtseva, Yu. F. Zuev, N. L. Zakhartchenko, N. N. Vylegzhanina, Z. Sh. Idiyatullin, V. D. Fedotov, Effect of electrolytes on the catalytic properties and structural characteristics of dodecylpyridinium bromide micelles, *Russ. J. Gen. Chem.*, 2002, **72**, 426–431.
20. A. B. Mirgorodskaya, L. R. Bogdanova, L. A. Kudryavtseva, S. S. Lukashenko, A. I. Konovalov, Role of surface potential in the catalytic action of micelles of cationic surfactant with a hydroxyalkyl fragment in the head group, *Russ. J. Gen. Chem.*, 2008, **78**, 163–170.
21. A. B. Mirgorodskaya, L. A. Kudryavtseva, V. A. Pankratov, S. S. Lukashenko, L. Z. Rizvanova, A. I. Konovalov, Geminal alkylammonium surfactants: aggregation properties and catalytic activities, *Russ. J. Gen. Chem.*, 2006, **76**, 1625–1631.
22. Yu. Feldman, N. Kozlovich, I. Nir, N. Garti, V. Archipov, Z. Idiyatullin, Yu. Zuev, V. Fedotov, Mechanism of transport of charge carriers in the sodium bis(2-ethylhexyl) sulfosuccinate-water-decane microemulsion near the percolation temperature threshold, *J. Phys. Chem.*, 1996, **100**, 3745–3748.
23. L. Ya. Zakharova, F. G. Valeeva, L. A. Kudryavtseva, A. I. Konovalov, N. L. Zakharchenko, Yu. F. Zuev, V. D. Fedotov, Alkaline hydrolysis of ethyl *p*-nitrophenyl chloromethylphosphonate in the reverse micellar AOT – decane – water system, *Russ. Chem. Bull.*, 1999, **48**, 2240–2244.
24. L. Ya. Zakharova, A. R. Ibragimova, F. G. Valeeva, V. M. Zakharov, L. A. Kudryavtseva, A. I. Konovalov, N. L. Zakharchenko, Yu. F. Zuev, Influence of temperature on the reactivity of phosphorus acid esters in reverse micellar systems based on sodium bis(2-ethylhexyl)sulfosuccinate, *Russ. Chem. Bull.*, 2005, **54**, 1449–1557.
25. L. Ya. Zakharova, F. G. Valeeva, L. A. Kudryavtseva, Yu. F. Zuev, Factors determining the micellar effect in nucleophilic substitution reactions, *Russ. J. Phys. Chem.*, 2000, **74**, 1825–1829.
26. *Microemulsions: Structure and Dynamics*. S. E. Friberg, P. Bothorel (Eds.). CRC Press, Inc., Boca Raton, Florida, 1987.
27. Yu. F. Zuev, A. B. Mirgorodskaya, B. Z. Idiatullin, Structural properties of microheterogeneous surfactant-based catalytic system. Multicomponent self-diffusion NMR approach, *Appl. magnetic resonance*, 2004, **27**, 489–500.
28. Yu. F. Zuev, A. B. Mirgorodskaya, L. A. Kudryavtseva, B. Z. Idiyatullin, R. N. Khamidullin, Influence of the structure of cetyltrimethylammonium bromide based microemulsions on base hydrolysis of carboxylic acid esters, *Russ. J. Gen. Chem.*, 2004, **74**, 1051–1056.
29. A. B. Mirgorodskaya, L. A. Kudryavtseva, L. S. Shtykova, I. V. Bogomolova, S. N. Shtykov, Aminolysis of carboxylic acid esters in direct, bicontinual, and inverse microemulsions based on cetyltrimethylammonium bromide, *Russ. J. Gen. Chem.*, 2005, **75**, 1108–1112.
30. A. B. Mirgorodskaya, L. A. Kudryavtseva, Yu. F. Zuev, B. Z. Idiyatullin, V. D. Fedotov, Structure and properties of water-oil microemulsions, *Russ. J. Gen. Chem.*, 2002, **72**, 1007–1011.
31. L. Zakharova, F. Valeeva, A. Zakharov, A. Ibragimova, L. Kudryavtseva, H. Harlampidi, Micellization and catalytic activity of the cetyltrimethylammonium bromide-Brij–97-water mixed micellar system, *J. Colloid Interface Sci.*, 2003, **263**, 597–605.
32. L. Ya. Zakharova, F. G. Valeeva, A. V. Zakharov, A. B. Mirgorodskaya, L. A. Kudryavtseva, A. I. Konovalov, Micellization properties and the catalytic effect of the aqueous cetyltrimethylammonium bromide-Triton-X–100 binary system in nucleophilic substitution in esters of phosphorous acids, *Kin. Catal.*, 2007, **48**, 221–227.
33. A. B. Mirgorodskaya, L. Ya. Zakharova, F. G. Valeeva, A. V. Zakharov, L. Z. Rizvanova, L. A. Kudryavtseva, H. E. Harlampidi, A. I. Konovalov, The regulation of the catalytic effect

of the cetyltrimethylammonium bromide micelles with the addition of the background electrolyte and nonionic surfactant, *Russ. Chem. Bull.*, 2007, **56**, 2000–2007.
34. L. Ya. Zakharova, F. G. Valeeva, A. R. Ibragimova, L. A. Kudryavtseva, N. N. Valeev, T. L. Didenko, V. I. Kovalenko, A. I. Konovalov, Micelle-forming, liquid-crystalline properties and catalytic effect of the mixed cetyltrimethylammonium bromide – polyethylene glycol(~9) monoalkanoate(~14) – water micellar system, *Russ. Chem. Bull.*, 2002, **51**, 2176–2182.
35. L. Ya. Zakharova, F. G. Valeeva, A. R. Ibragimova, V. M. Zakharov, L. A. Kudryavtseva, Yu. G. Elistratova, A. R. Mustafina, A. I. Konovalov, S. N. Shtykov, I. V. Bogomolova, Properties of a sodium dodecyl sulfate-Brij–35 binary micellar system and their effect on the alkaline hydrolysis of O-ethyl-O-p-nitrophenylchloromethylphosphonate, *Colloid J.*, 2007, **69**, 718–725.
36. D. N. Rubingh, Mixed micelle solutions, in *Solution Chemistry of Surfactants*, Vol. **1**, K. L. Mittal (Ed.), Plenum Press, New York, 1979.
37. E. P. Zhil'tsova, L. A. Kudryavtseva, G. A. Gainanova, A. P. Timosheva, A. S. Mikhailov, R. H. Giniyatullin, A. A. Naficova, V. S. Reznik, A. I. Konovalov, Phosphorylation of pyrimidinophane in the reversed micelles of cationic and nonionogenic surfactants, *Russ. J. Gen. Chem.*, 2005, **75**, 1729–1734.
38. E. P. Zhil'tsova, L. A. Kudryavtseva, A. S. Mikhailov, V. E. Semenov, V. S. Reznik, A. I. Konovalov, Reverse micellar catalysis of phosphorylation of pyrimidinophane, *Russ. J. Gen. Chem.*, 2008, **78**, 50–56.
39. S. V. Kharlamov, E. P. Zhil'tsova, G. A. Gainanova, L. A. Kudryavtseva, A. P. Timosheva, A. V. Aganov, Sh. K. Latypov, Aggregation in a mixture of cetyltrimethylammonium bromide and polyoxyethylene–600-monolaurate solutions, *Colloid Journal*, 2006, **68**, 504–510.
40. R. F. Bakeeva, L. A. Kudryavtseva, V. E. Belskii, B. E. Ivanov, Hydrolysis of di(*p*-nitrophenyl) methylphosphonate in presence of polyethyleneimine, *Zhurn. Obshch. Khim.*, 1983, **43**, 1058–1061.
41. J. Chin, Developing artificial hydrolytic metalloenzymes by a unified mechanistic approach, *Acc. Chem. Res.*, 1991, **24**, 145–152.
42. M. L. Bender, *Mechanisms of homogeneous catalysis from protons to proteins*. Wiley-Interscience, New York, 1971.
43. L. Ya. Zakharova, A. R. Ibragimova, F. G. Valeeva, A. V. Zakharov, A. R. Mustafina, L. A. Kudryavtseva, H. E. Harlampidi, A. I. Konovalov, Nanosized reactors based on polyethyleneimines: from microheterogeneous systems to immobilized catalysts, *Langmuir*, 2007, **23**, 3214–3224.
44. L. Ya. Zakharova, F. G. Valeeva, D. B. Kudryavtsev, A. V. Bilalov, A. Ya. Tret'yakova, L. A. Kudryavtseva, A. I. Konovalov, V. P. Barabanov, Sodium dodecyl sulfate-polyethyleneimine-water system. Self-organization and catalytic activity, *Russ. Chem. Bull.*, 2005, **54**, 641–649.
45. K. Holmberg, B. Jönsson, B. Kronberg, B. Lindman, *Surfactants and Polymers in Aqueous Solution*, John Wiley and Sons Ltd, London, 2002.
46. L. Ya. Zakharova, A. R. Ibragimova, F. G. Valeeva, L. A. Kudryavtseva, A. I. Konovalov, A. V. Zakharov, N. M. Selivanova, V. V. Osipova, M. V. Strelkov, Y. G. Galyametdinov, The self-organization and the catalytic activity of the polyethyleneglycol(10)monododecyl ether/ polyethyleneimine/lanthanum nitrate system, *J. Phys. Chem. C*, 2007, **111**, 13839–13845.
47. R. F. Bakeeva, D. B. Kudryavtsev, L. Ya. Zakharova, A. Raevska, V. F. Sophin, Micellar, liquid crystalline and polymer systems based on surfactant and polyethyleneimine as nanoreactors for the transfer of phosphoryl group, *Mol. Cryst. Liq. Cryst.*, 2001, **367**, 585–597.
48. L. Ya. Zakharova, F. G. Valeeva, D. B. Kudryavtsev, A. R. Ibragimova, L. A. Kudryavtseva, A. P. Timosheva, V. E. Kataev, Catalytic effect of multicomponent supramolecular systems in phosphoryl-group transfer reactions, *Kin. Catal.*, 2003, **44**, 547–551.
49. E. M. Kosacheva, D. B. Kudryavtsev, R. F. Bakeeva, A. I. Kuklin, A. Kh. Islamov, L. A. Kudryavtseva, V. F. Sopin, A. I. Konovalov, The aggregation of branched polyethyleneimine and cationic surfactants in aqueous systems, *Colloid J.*, 2006, **68**, 713–720.
50. S. S. Lukashenko, A. V. Yurina, T. N. Pashirova, D. B. Kudryavtsev, E. M. Kosacheva, L. A. Kudryavtseva, A. I. Konovalov, Systems based on the hydrophobically modified polyethyleneimines and surfactants: aggregation and catalysis, *Colloid J.*, 2008, **70**, 317–326.

51. L. Ya. Zakharova, V. E. Semenov, M. A. Voronin, F. G. Valeeva, A. R. Ibragimova, R. Kh. Giniatullin, A. V. Chernova, S. V. Kharlamov, L. A. Kudryavtseva, Sh. K. Latypov, V. S. Reznik, A. I. Konovalov, Nanoreactors based on amphiphilic uracilophanes: the self-organization and the reactivity study, *J. Phys. Chem. B.*, 2007, **111**, 14152–14162.
52. L. Ya. Zakharova, V. E. Semenov, M. A. Voronin, F. G. Valeeva, R. Kh. Giniatullin, L. A. Kudryavtseva, V. S. Reznik, A. I. Konovalov, Supramolecular catalytic systems based on dimeric pyrimidinic surfactants and polyethyleneimine, *Mendeleev Commun.*, 2008, **18**, 158–160.
53. L. Ya. Zakharova, V. V. Syakaev, M. A. Voronin, V. E. Semenov, F. G. Valeeva, A. R. Ibragimova, A. V. Bilalov, R. Kh. Giniyatullin, Sh. K. Latypov, V. S. Reznik, A. I. Konovalov, New self-assembling systems based on bola-type pyrimidinic surfactants, *J. Colloid Interface Sci.*, 2010, **342**, 119–127.
54. L. Ya. Zakharova, F. G. Valeeva, A. R. Ibragimova, M. A. Voronin, L. A. Kudryavtseva, V. V. Syakaev, E. Kh. Kazakova, N. A. Makarova, Yu. E. Morozova, N. B. Mel'nicova, O. E. Zemniakova, A. I. Konovalov, Supramolecular catalytic systems based on anionic amphiphiles and polyethyleneimine for hydrolytic cleavage of phosphorus ester bonds, *Izvestia Academii Nauk, Seria Khimicheskaya*, 2008, 366–373.
55. L. Ya. Zakharova, V. V. Syakaev, M. A. Voronin, F. G. Valeeva, A. R. Ibragimova, Yu. R. Ablakova, E. Kh. Kazakova, Sh. K. Latypov, A. I. Konovalov, The NMR and spectrophotometry study of the supramolecular catalytic system based on polyethyleneimine and amphiphilic sulfonatomethylated calix[4]resorcinarene, *J. Phys. Chem. C.*, 2009, **113**, 6182–6190.
56. L. Ya. Zakharova, A. R. Mustafina, F. G. Valeeva, A. R. Ibragimova, Yu. R. Ablakova, Yu. G. Elistratova, V. V. Syakaev, L. A. Kudryavtseva, A. I. Konovalov, Supramolecular catalytic systems based on calix[4]resorcinarene for nucleophilic substitution in phosphorous acid esters, *Colloid J.*, 2008, **70**, 444–454.
57. T. N. Pashirova, S. S. Lukashenko, E. M. Kosacheva, M. V. Leonova, L. I. Vagapova, A. R. Burilov, M. A. Pudovik, L. A. Kudryavtseva, A. I. Konovalov, Aggregation behavior and catalytic properties of systems based on aminomethylated calix[4]resorcinarenes and polyethyleneimines, *Russ. J. Gen. Chem.*, 2008, **78**, 202–209.
58. E. P. Zhil'tsova, L. A. Kudryavtseva, G. A. Gainanova, A. P. Timosheva, S. S. Lukashenko, A. I. Konovalov, Inverse micellar catalysis of phosphorylation of hydroxybenzilated polyethyleneimines, *Russ. J. Gen. Chem.*, 2005, **75**, 1603–1609.
59. G. A. Gainanova, E. P. Zhil'tsova, L. A. Kudryavtseva, S. S. Lukashenko, A. P. Timosheva, V. E. Kataev, A. I. Konovalov, Aggregation and catalysis in a nonionic surfactant–polyethyleneimine–chloroform system, *Colloid J.*, 2006, **68**, 533–540.
60. G. A. Gainanova, E. P. Zhil'tsova, L. A. Kudryavtseva, S. V. Kharlamov, Sh. K. Latypov, A. P. Timosheva, A. I. Konovalov, Mixed micelles of cetyltrimethylammonium bromide and poly(ethylene glycol)–600-monolaurate as catalysts of polyethyleneimine phosphorylation in chloroform, *Russ. Chem. Bull.*, 2006, **57**, 1411–1418.
61. E. P. Zhil'tsova, D. B. Kudryavtsev, L. Ya. Zakharova, S. S. Lukashenko, I. S. Ryzhkina, L. A. Kudryavtseva, Reactivity of polyethyleneimines in reversed and normal micellar solutions of surfactants, *Russ. J. Phys. Chem.*, 2002, **76**, 1853 –1857.
62. G. A. Gainanova, E. P. Zhil'tsova, L. A. Kudryavtseva, S. S. Lukashenko, A. P. Timosheva, I. R. Knyazeva, A. R. Burilov, A. I. Konovalov, Complex formation and self-association in alkylated polyethyleneimine-calix[4]resorcinarene–chloroform system, *Russ. J. Gen. Chem.*, 2006, **76**, 1788–1794.
63. G. A. Gainanova, E. P. Zhil'tsova, L. A. Kudryavtseva, S. S. Lukashenko, I. R. Knyazeva, A. R. Burilov, V. I. Kovalenko, L. V. Avvakumova, A. I. Konovalov, Phosphorylation of polyethyleneimines in chloroform in the presence of calix[4]resorcinarenes, *Russ. J. Gen. Chem.*, 2007, **77**, 40–46.
64. G. A. Gainanova, E. P. Zhil'tsova, L. A. Kudryavtseva, S. S. Lukashenko, A. P. Timosheva, A. I. Konovalov, Phosphorylation of alkylated polyethyleneimine in chloroform in the presence of cetyltrimethylammonium bromide and calix[4]resorcinarene, *Russ. Chem. Bull.*, 2007, **56**, 953–958.

16

Encapsulation Processes by Bilayer Vesicles

Marc C. A. Stuart and Jan B. F. N. Engberts
Groningen Biomolecular Sciences and Biotechnology Institute and Stratingh Institute of Chemistry, University of Groningen, Nijenborgh 4, 9747 AG Groningen, The Netherlands

16.1 Introduction

16.1.1 Vesicular Aggregates

Surfactants, or amphiphiles, comprise a large class of compounds, characterized by one (or more) ionic or highly polar headgroups and one, two, or rarely more, hydrophobic alkyl tails (usually n-C_8 to n-C_{22}, the most common being C_{12}, C_{14}, C_{16}, C_{18}, and C_{20}). In aqueous media, above a critical concentration, these systems cope with their dual structural properties by forming aggregates in which the headgroups are in contact with water while the alkyl chains reside in the core of the aggregate, avoiding interactions with the aqueous medium.

A variety of different morphologies are possible for these aggregates, depending on the molecular structure of the single amphiphilic molecule, including their size and shape. But other factors like ionic strength, temperature, pressure and pH also play a role.[1]

Vesicles (from the Latin vesicula, small bubble) belong to the most extensively studied amphiphilic aggregates.[2] Not only because the bilayer membrane is the building block of cell membranes, but also because vesicles or liposomes offer the unique possibility for solutes to bind to the outer- and inner leaflets of the bilayer whereas the hydrophobic interior of the bilayer can be used to entrap hydrophobic solutes and also membrane proteins.

The Gibbs energy of a vesicle depends on a compromise of attractive and repulsive interactions. The main contributions come from hydrophobic interactions between the tails, the major driving force for aggregation, and from repulsive interactions between the headgroups at the aqueous interface, which, in case of ionic surfactants, are in part compensated by counterion binding. This counterion binding is usually stronger for vesicles than for micelles.[3] Other factors also play a role, one of them being the loss of conformational entropy of the hydrocarbon tails in the vesicular core.

Water is the essential medium for surfactant aggregation, although association can occasionally also occur in polar solvents with a high cohesive energy density.[4]

Which factors determine the packing efficiency of the amphiphiles in the aggregate? The dimensionless shape factor or packing parameter (P) provides a relation between the molecular shape of the amphiphile and the preferred morphology of the aggregate in dilute aqueous solution at low ionic strength and at ambient temperatures:[5]

$$P = V/a_0 \times l_c$$

Herein, V is the volume of the hydrophobic chain, a_0 the optimal cross-sectional surface area of the headgroup, and l_c the length of the all-*trans* alkyl tails. It will be clear that the curvature of the hydrophilic/hydrophobic interface is related to the value of P. The relation between the aggregate morphology and P is given in Figure 16.1. The packing parameter concept provides a reasonable and useful rationalization of the rich polymorphism of surfactant aggregates, but it has also been criticized. A significant issue is, of course, that the values of a_0 and l_c are not independent of each other.[6] Nevertheless, extensive studies of a large series of amphiphiles in which the structure was gradually modified showed that the approach worked satisfactorily.[7,8]

It was Bangham who, in the sixties of the past century discovered that liposomes could be made from egg yolk.[9] But phospholipids are by no means unique in this behaviour. Kunitake[10] showed that bilayer vesicles can be obtained from synthetic surfactants in case $0.5 < P < 1.0$, while for $P = 1.0$ a flat bilayer is formed.[11] Bilayer morphologies, usually preferred by double-tailed amphiphiles, are shown in Figure 16.2 and are called lamellar phases. The bilayer consists of two opposing layers of amphiphiles, with the aligned tails facing each other and the headgroups situated at the aqueous interface. In case the two ends of the bilayer close to form a global structure with an aqueous inner compartment, the aggregate is called a vesicle, or a liposome if the amphiphilic constituent is a phospholipid molecule. Mammalian cell membranes may contain microdomains ('rafts') which can also be formed in multicomponent liposomes.[12,13] For the structures of the most common phospholipids in cell membranes, the reader is referred to appropriate biochemistry textbooks.

Vesicles and particularly liposomes are highly useful, although severely simplified, models for biological membranes.

Conventional molecular dynamic (MD) simulations of the formation of small dipalmitoylphosphatidylcholine (DPPC) liposomes in atomistic detail (1017 DPPC molecules in 106.563 waters) have provided mechanistic insights into some general features of liposome formation.[14] In the very rapid initial stages of the aggregation process micellar-like structures are formed. Lipid bridging between these micelles leads to larger, lamellar morphologies, which are, in fact, curved bilayers. Spontaneous development of curvature, resulting from minimization of edge energy, is then claimed to induce formation of

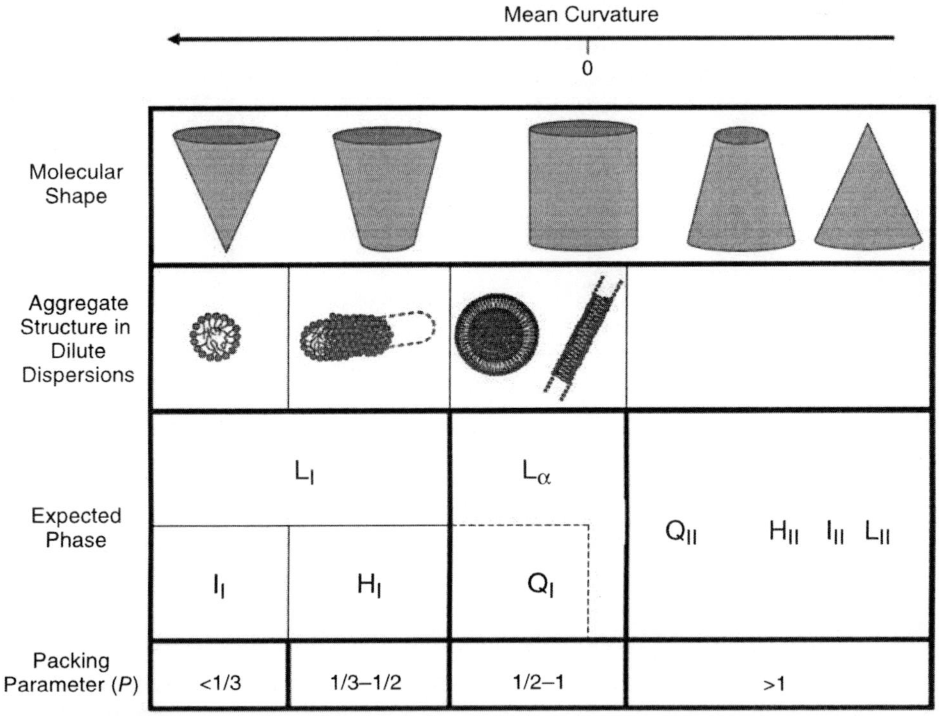

Figure 16.1 *Relationship between molecular shape, aggregate structure in dilute dispersions, phase behavior and packing parameter. Micellar phase (L_I), cubic micellar phase (I), hexagonal phase (H), bicontinuous cubic phase (Q), L_α lamellar phase. Subscripts I and II indicate normal and inverted phases, respectively. From: M. Scarzello, Aggregation Properties of Amphiphilic DNA-Carriers for Gene Delivery, Ph. D. Thesis University of Groningen, p 6, 2006*

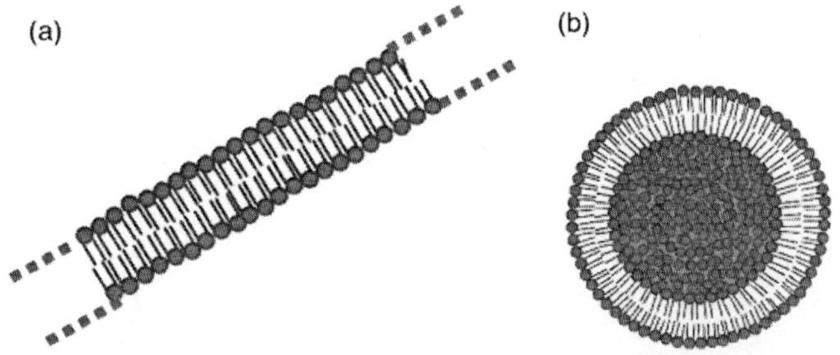

Figure 16.2 *Schematic representation of two possible morphologies of L_α: flat bilayer (a) and a vesicle (b). From: M. Scarzello, Aggregation Properties of Amphiphilic DNA-Carriers for Gene Delivery, Ph. D. Thesis, University of Groningen, p 8, 2006*

liposome-like shapes. These contain water pores with diameters of 1–4 nm, allowing relatively slow equilibration of lipid molecules between the inner and outer leaflets by diffusion. This so-called 'flip-flop' has also been studied in mature vesicles already many years ago.[15] Finally these processes result in collapse of the water pores and formation of the most stable liposomal aggregate.[16,17] Spontaneous formation of vesicles from mixed amphiphiles as well as the interaction of vesicles with other molecules (including cholesterol[18]) has also been investigated using MD techniques.[19]

With a few exceptions (catanionic vesicles, for example[20,21]), vesicles are metastable aggregates and posses a tendency to undergo aggregation followed by merging of the bilayers and eventually precipitation. Therefore vesicles can be best coined as 'colloidal structures'.[22] Apart from catanionic vesicles and contrary to the suggestion by some MD simulations,[14] closed vesicle formation from single amphiphilic molecules is not spontaneous but requires input of mechanical energy such as shaking, stirring, extrusion or sonication, inducing formation of closed vesicles from bilayer sheets and increasing the bilayer curvature and reducing the vesicular size. In fact, the method of vesicle preparation has an effect on the exact lamellar structure and on the vesicular size.[23,24] A bending energy can be defined for formation of a closed vesicle from a bilayer sheet. The curvature stress energy for a bilayer can be calculated with the (simplified) Helfrich Gibbs energy equation.[25] In contrast to micelles, vesicles tend to separate from their dispersions, but this process may take a considerable time span (ranging from seconds to months).

To prove the formation of vesicles a number of indirect techniques can be used such as dynamic light scattering, the use of fluorescent probes and pulsed field gradient NMR self-diffusion measurements. Some more direct techniques such as freeze-fracture and negative staining electron microscopy are less biased by the interpretation of the scientist, but also these methods have their limitations. Cryo-electron microscopy, as introduced by Dubochet in the 80s, [26] is the method of choice when it comes to visualization of small colloidal structures.[27] Recent developments in the vitrification of specimens make it now possible to observe vesicles and other aggregated structures artifact free.

Vesicles with sizes varying from nanometers to micrometers usually contain large numbers of surfactant molecules. Large unilamellar vesicles (LUVs) possess diameters which may vary considerably (50–500 nm) and have a large encapsulation efficiency (5–20 $l\,mol^{-1}$). Small unilamellar vesicles (SUVs) with diameters of 20–50 nm encapsulate smaller volumes (0.5–1.0 $l\,mol^{-1}$) Under special conditions, giant vesicles may be formed (diameter in the order of 1–500 μm) with a low bilayer curvature, akin to that of a flat bilayer.[28,29] These giant vesicles are able to eject smaller vesicles by a budding process.[30]

In vesicles, the fluidity of the bilayer[31] and the associated stiffness and order of the alkyl tails can vary significantly as a function of temperature, but, of course, also depending on the detailed structure (straight or branched), size, and the presence of one or more double or triple carbon–carbon bonds in the tails. At lower temperatures, the bilayers reside in the gel phase (L_β) with the strongest packing of the all-*trans* alkyl chains and with the largest order and rigidity. Upon increasing the temperature, a cooperative phase transition occurs to the L_α or liquid crystalline phase with a more liquid-like core, allowing fast diffusion of the amphiphilic components, and in which gauche conformations of the alkyl tails are allowed. As expected this main phase transition temperature (T_c) becomes lower if the alkyl chains are shortened or if unsaturation is introduced into the tails.

Sometimes an intermediate 'rippled' phase ($P_{\beta'}$)[32] is first formed from the L_β phase, which transforms to the L_α phase at higher temperatures. At still more elevated temperatures, the lamellar phase may change into inverted cubic (I_β) or related isotropic phases.[33,34] These highly complex aggregates have been proposed as intermediates during membrane fusion. A further temperature increase may lead to formation of inverted hexagonal phases (H_{II}), in which the weakly hydrated amphiphiles are packed in inverted micellar-like cylinders, with the headgroups facing inside and in contact with water in the enclosed compartment.

Vesicles possess many interesting potentials for applications, one of the most important being encapsulation of solutes, a phenomenon that has been investigated in detail and has been utilized in different ways. To get insight into the possibilities for encapsulation, unavoidably involving disturbances of the bilayer structuring and to predict which binding sites are available for solubilizates of different chemical composition, we can have a look at the lateral pressure profile, extensively investigated by Cantor.[35,36] The lateral pressure (LP) is defined as $d\pi(z)$ acting within a thin slice of the bilayer with thickness dz. Positive values of the LP indicate repulsion, negative values stand for attraction. The lateral pressure density is then given by $p(z) = d\pi/dz$, with dimensions of bulk pressures. At the aqueous interface of the membrane, the headgroups repel each other (positive LP), whereas at the interface somewhat deeper into the bilayer, the tails attract each other through hydrophobic interactions leading to a large, negative LP. Still deeper in the membrane, there is little or no water and the enforced partial ordening of the tails is accompanied by a considerable loss of entropy due to their reduced conformational freedom. This means repulsion and a positive LP in the central part of the core of the bilayer. The variation of the LP with the depth in the membrane is an important issue and has been applied in detail in the field of bilayer-protein interactions.[37]

Figure 16.3 shows a LP profile for one mono-layer, the profile for the opposite monolayer is the same. We see how the LP varies with depth in the membrane. Since the formation of a vesicular membrane is associated with a favorable Gibbs energy, the total lateral pressure in the membrane is zero (or nearly so).

The lateral pressure concept provides useful insight into the spatial distribution of forces and is a sum over terms corresponding to layers of finite thickness (dz).[38] Quantitative measurements of the LP remain difficult and even the concept has been criticized.[39]

Chain order parameters, measured by ^{2}H-NMR studies on amphiphiles with specifically deuterium-labelled methylene units in the tails,[40–42] gave a similar view of the distribution of the chain attractions/repulsions as a function of position in the tails. Increasing quadrupole splittings, indicated by an increasing distance between the two peaks in the ^{2}H-NMR spectrum, are associated with increasing chain order.

In sum, a larger repulsion between specific parts of the alkyl chains in a bilayer is reflected in a higher lateral pressure, a decrease in chain order and more area available for that part of the tails. Such changes can be induced by, for example, a higher temperature or introduction of a *cis* C=C bond into the tail or via *trans–cis* isomerization of double-tailed, azobenzene-substituted amphihiles.[43]

In recent years, several computational methods have been developed to calculate LP profiles in membranes. Among them, molecular dynamics (MD) simulations have been particularly successful. Timescales extending to hundreds of nanoseconds are now computationally feasible, but the simulations are still restricted to relatively small system

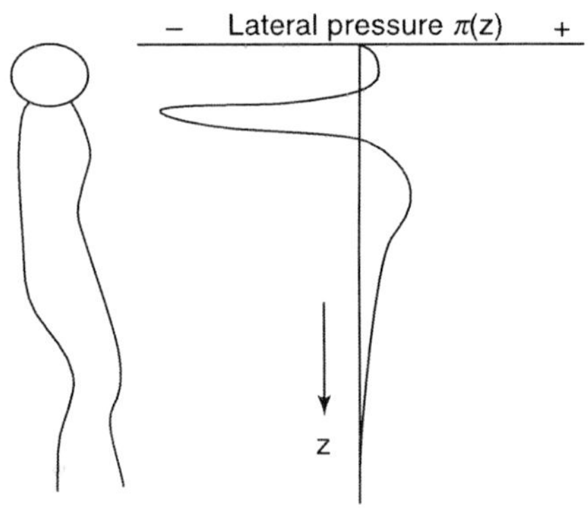

Figure 16.3 *Schematic representation of a lateral pressure profile in a vesicular bilayer. Only one monolayer is shown here, the same profile applies for the other monolayer. The lateral pressure π depends on the depth (Z) in the monolayer. From: M. Kuiper, Azobenzene-substituted Phosphate Amphiphiles : Effect of Light-induced* trans–cis *Isomerisation on Vesicular Properties and the Channel Protein MscL, Ph. D. Thesis, University of Groningen, p 12, 2005*

sizes. MD is also useful in studies of undulatory and thickness fluctuation modes of bilayers.[44]

Finally we note that, apart from double-tailed surfactant molecules, vesicular aggregates can also be formed from a number of other building blocks, including nonionic amphiphiles to form niosomes,[45] single-tailed surfactants[46,47] and complex (co)polymers, polypeptides, and dendrimers.[48] In this chapter we restrict our discussion to encapsulation processes by vesicles formed from synthetic surfactants and phospholipids.

16.1.2 Solute Encapsulation by Vesicles

It will be obvious that the encapsulation of a solute molecule into a vesicular membrane will affect the interchain interactions in the bilayer.[49] Overall, the Gibbs energy for binding should be negative, otherwise the solute would remain in bulk solution. Therefore, partitioning of a relatively apolar solute into the membrane will be most favorable at a location with the best balance between solubilizate-tail and solubilizate-headgroup interactions on one hand and changes in tail-tail interactions on the other hand. These counteracting contributions will also significantly depend on hydration changes. On the other hand, the (solvation) properties of the solubilizate are also different from those in dilute aqueous solution as a result of binding in the vesicular bilayer.

In this brief review we will describe some important encapsulation processes by vesicular aggregates. Encapsulation will be broadly interpreted. Apart from solubilization in the aqueous pool inside the vesicle, the term will also encompass binding of solubilizates to all binding sites available in the vesicular system. We will not discuss bilayer vesicles formed from amphiphiles further functionalized by receptor molecules, such as amphiphilic cyclodextrins.[50]

Instead of a comprehensive review of the vast literature on solute encapsulation, a selection will be made of various illustrative binding possibilities in or at the bilayer.

Although binding to vesicles has been less extensively studied than binding to micelles, attempts will be made to rationalize why specific solute molecules bind to particular binding locations.

16.1.3 Binding Locations

Let us now look at the different locations for spontaneous encapsulation of solutes of different chemical composition. First of all, it goes without saying that encapsulation experiments should be carried out under well-defined conditions. Often vesicles, after their preparation, and subsequently after binding of the solubilizate, need time to arrive at their most stable configurations (size, tail ordering) and adequate attention should be given to this issue. This process may sometimes take days at room temperature. Also the method used for their preparation may exert a definite effect on the vesicular properties. Thus, for a honest comparison of encapsulation abilities, it is recommended that the vesicles are prepared under exactly the same conditions and that their properties are checked by a proper physical technique, such as cryo-electron microscopy. Also temperature control needs attention since the gel and liquid-crystalline phase exhibit different binding efficiencies of solutes.

The binding locations can be charaterized as follows:

(a) In the outer Stern region, where the solute interacts almost exclusively with the polar or ionic headgroups. In case of ionic headgroups, relatively hydrophilic solutes, carrying opposite charges (preferably double-charged) are the best candidates for strong binding. Interaction is mainly governed by electrostatic forces. Close approach between oppositely charged ionic moieties is necessarily accompanied by partial dehydration of the ions, an effect that may in part counteract the attractive interactions.

Hydrophilic solutes like carbohydrates, i.e. glucose, sucrose, and trehalose also bind close to the charged headgroups, replacing hydration water, as laid down in the 'water-replacement hypothesis'.[51] They act as cryoprotectants and help to stabilize the bilayer in case of insufficient water.

There is now compelling evidence that selective hydrogen-bonding interactions between functional groups of hydrophobically-buried molecules residing near the vesicle-water interface are able to overcome the otherwise dominant hydrogen bonding to water molecules, present in high excess. A recent example is the self-adhesion among phospholipid vesicles carrying adhesive agents.[52] Such associations in the aqueous interface are largely governed by entropic factors, as discussed in a beautiful review.[53] The entropic costs for association of ligands tethered to the surface of vesicles are smaller than those for interaction of similar species diffusing independently in three dimensions in an aqueous medium.

A highly unexpected result, first reported in 2003, was the efficient adsorption of hydroxide ions at pH values slightly above 7.5 to (almost) uncharged hydrophobic surfaces[54] and vesicular surfaces formed by reduced-sugar based gemini surfactants.[55,56] The origin of this phenomenon has been investigated in some detail.[57]

(b) In the 'inner' Stern layer, that is at binding locations with negative LP values, at a few methylene units from the headgroups. These are the most frequently encountered binding sites where relatively hydrophobic solutes preferably interact with the partly hydrated alkyl tails through hydrophobic interactions. Often these solutes also possess a hydrophilic substituent, necessary for sufficient water solubility, and this substituent will be preferentially directed towards the more aqueous region of a bilayer for beneficial contact with water. Incorporation of solutes may disturb the tail–tail interactions to some extent. An example is provided by the anchoring of hydrophobically modified poly(sodium acrylate)s into di-*n*-dodecylphosphate (DDP) vesicles. In case the polymer contains *n*-dodecyl chains, the bilayer is hardly affected as indicated by detailed DSC measurements. By contrast, for *n*-nonyl and *n*-octadecyl chains the hydrophobic mismatch leads to lower main phase transition temperatures and reduced enthalpies of transition per mole of DDP monomer. Another notable consequence of binding of the hydrophobically modified polymer into the vesicle bilayer is the increase of the cooperativity of the melting process, probably resulting from the presence of larger 'patches' with a lower curvature.[58] Another interesting effect induced by hydrophobic mismatching is the clustering of membrane proteins in case their hydrophobic transmembrane domains do not match with the surrounding lipid bilayer.[59]

Interestingly, transvesicular reactions between substrates bound in the exovesicular and endovesicular leaflets have been studied, particularly at relatively low vesicular fluidities.[60]

(c) Deep in the interior of the bilayer. These are the preferred binding locations for strongly hydrophobic molecules. The chain order is relatively low in this region and the LP is positive. There is also more free volume available for incoming hydrophobic solutes. These binding processes will affect bilayer dynamics.

It has recently been shown that introduction of a small, hydrophobic molecule like isoprene may lead to a more ordered and better packed lipid membrane as revealed by MD simulations. The stabilized membrane is then protected against temperature-induced disordering of the tails.[61] Similar thermoprotective effects are induced by cholesterol which also increases the molecular packing of the tails and, in contrast to isoprene, affects the dynamics of the lipids in the bilayer.[62]

(d) In the aqueous inner compartment of the vesicle. Sufficiently hydrophilic solutes can be entrapped in the aqueous medium inside the vesicular system and may favourably interact with the headgroups stacking into this medium.[63] In contrast to the protocols used for binding at the sides labelled (a)–(c), binding in the water pool is realized by preparation of the vesicle dispersion in a medium already containing the solute. During the formation of the vesicular aggregate, the solute becomes entrapped in the inner aqueous compartment. Of course, a prerequisite for this type of encapsulation is that leakage over the bilayer is slow. Carboxyfluorescein, a highly water-soluble fluorescent dye, has been extensively used to probe leakage over vesicular bilayers.[63] Strong bilayer packing, as induced by encapsulated cholesterol, slows down leakage processes.

Finally, we stress that the bound solutes are often certainly not localized at a single, precisely defined, position but may exchange rapidly between various positions of similar binding characteristics and partition coefficients.

As we will discuss below, microencapsulation offers interesting possibilities for reactivity control.[64]

16.1.4 Experimental Techniques for Measuring Encapsulation Processes

It is obvious that binding of a solute to a vesicle will induce changes in the physical properties of both solute and the amphiphilic molecules that make up the bilayer aggregate. Two sorts of information can be obtained from these changes. First, the strength of the interaction, which can be quantified by the solvent/bilayer partition coefficient of the solute. Second, in many cases information can be obtained regarding the location of the binding site although the exact interpretation of the experimental data can be quite challenging. Both types of data are of immediate importance for applications of the particular vesicular aggregates.

A brief summary will now follow of some of the most frequently employed experimental techniques. Several of these have been particularly applied for micellar aggregates, but their usage can often be easily extended to vesicles.

16.1.4.1 Microcalorimetry

The greatly improved sensitivity of the modern microcalorimeters for both differential scanning microcalorimetry (DSC) and isothermal titration microcalorimetry (ITC) has made their application highly useful. Titration microcalorimetry yields binding constants and, when applied as a function of temperature, full thermodynamic details for the binding process.[65] DSC provides details about the disturbance of the bilayer system by the solubilized guest from which conclusions may be drawn about the preferred binding locations.

16.1.4.2 NMR Spectroscopy

Ring current effects have already been examined more than twenty years ago and indicated which groups are in close proximity to an aromatic ring in the solute or in the amphiphile. This is useful information but the exact positioning of these interaction sites within the bilayer may remain a problem.[66,67] Nuclear Overhauser effects have also been employed frequently.

Paramagnetic relaxation enhancement experiments measure the distribution coefficients of solubilizates.[68,69] A comparison is made between the proton spin–lattice relaxation rates of the solubilizate in the absence and presence of noncomplexing paramagnetic ions of the same sign of charge as the headgoups. Such paramagnetic ions reside necessarily in the bulk solvent and their effect on the nuclei of the solubilizate depend on the efficiency of penetration of the solubilizate into the core of the bilayer. The method is applicable for a variety of NMR-active nuclei in the guest molecule. These results can be compared with those obtained from Fourier transform pulsed gradient spin-echo NMR self-diffusion measurements.

16.1.4.3 Phosphorescence, Fluorescence and Steady-state Absorption Techniques

These techniques can be employed for quantitative studies of the distribution constants of solubilizates as well as for measuring the kinetics of solubilization. Binding at the surface or in the core of the aggregate can be distinguished under favorable conditions by fluorescence probing.[70–72]

Solvatochromic probes like Nile Red allow characterization of temperature-dependent phase transitions. This is possible due to the large difference in polarity between the ground and excited state of the probe.[73] At binding sites of significantly different polarity the probe exhibits selective excitation properties and excitation-dependent emission maxima.

16.1.4.4 Combined Fluorescence Quenching Experiments and Paramagnetic Resonance

Data obtained by this technique allow rather detailed conclusions about the location of the solubilizate in the aggregate. Fluorescence quenching rate constants vary with temperature, local microviscosity and quencher concentration.[74–76] Encapsulation of spin probes has been examined in detail by EPR spectroscopy.[77]

16.1.4.5 Cryo-Electron Microscopy

Apart from (bio-)chemical data, direct structural information of vesicles and micelles can be obtained by cryo-electron microscopy. Cryo-electron microscopy as introduced by Jaques Dubochet,[26] has now become a standard technique to gain artifact-free structural information within the natural solvent. In most cases water [78] is used but also other solvents like polar organic solvents[79] and apolar solvents[80] can be employed. Thin layers of solvent with colloidal suspended material can be vitrified by rapid cooling in liquid ethane. By remaining at low temperature inside an electron microscope the solvent will not evaporate in the high vacuum. This allows to gain high resolution structural information.

16.1.4.6 Miscellaneous Techniques

Other techniques that can be used to measure distribution coefficients of a guest molecule between the bulk aqueous phase and a micellar or vesicular aggregate include gel filtration, electromotive force measurements, solubility and vapor pressure measurements, muon spin rotation experiments and headspace gas chromatography.[81] Theoretical models are, of course, also helpful, as, for example, in studies of the enveloping of charged proteins by lipid bilayers.[25]

16.2 Catalysis by Vesicles. Encapsulation of Reactants

Micellar catalysis and inhibition of organic reactions has been studied in great detail.[82] Much less work has been performed on vesicular catalysis and inhibition. The fact that most vesicles are meta-stable aggregates is a reason for this lack of popularity. In order to obtain satisfactory reproducibility of the rate constants, adequate methods for vesicle formation are required and kinetic measurements have to be performed with special care. On the other hand, if the reaction is strongly medium dependent and rate constants can be measured accurately, rather subtle changes in substrate binding location can be nicely revealed. We note here that vesicular catalysis in water can be performed at lower surfactant concentrations than in case of micellar catalysis, which is a significant advantage from the view point of green chemistry.

It will be clear that encapsulation of the reactant(s) is a prerequisite for vesicle-induced rate effects. In case of unimolecular organic reactions, the change in rate constant relative to the rate constant in bulk aqueous solution, is determined by the change in reaction environment going from water to the reactant binding sites in the vesicle. Several studies have suggested that the reaction environment in the vesicular phase is often less polar than that in micelles.[83] Of course, the kinetic effect is a function of the distribution of the reactant over the aqueous and vesicular pseudophases. If the medium effects on the reaction are understood in some detail, the vesicle-induced rate effects provide information about the nature of the reactant encapsulation process.

For bimolecular organic reactions the situation is more complicated. We have now two reactants that bind to the vesicle (sometimes one may stay in the aqueous phase), not necessarily at similar sites in the aggregate. Vesicular catalysis/inhibition now depends on the same environmental factors as described for unimolecular reactions, but also on the binding efficiency, determining the reactant concentrations in the vesicular reaction volume. If reactant binding is sufficiently strong, the local reactant concentrations in the vesicular reaction volume may become greater than those in water, with the concominant higher rate resulting from this 'concentration effect'. Even in the case of a vesicular rate constant smaller than the one in water, catalysis may nevertheless be observed due to the overruling effect of the higher reactant concentrations.

Encapsulation of the reactants in the vesicle can affect rate constants in a number of ways. These include the following: (a) a reduced medium polarity in the Stern region and particularly deeper in the bilayer; (b) a lower water concentration; (c) for charged vesicles: a local electric charge at the surface of the vesicle due to an incomplete counterion binding and a significant ionic strength in the Stern region; (d) rates may respond to the bilayer fluidity which is different below and above the main phase transiton temperature, and finally, the vesicle size[84] may also play a role. We also note that giant vesicles provide interesting microchemical vessels for various types of reactions, including decontamination of mustard-like compounds and polymerization processes.[82]

16.2.1 Unimolecular Decarboxylation of 6-NBIC

The strongly solvent-dependent, unimolecular decarboxylation of sodium 6-nitrobenzisoxazole-3-carboxylate (6-NBIC, Figure 16.4) has been a popular kinetic probe for aqueous media containing surfactant aggregates.

After pioneering work by Kemp and Paul,[85] the main noncovalent interactions which govern the medium effects have been elucidated in some detail by Hilvert *et al.*[86]

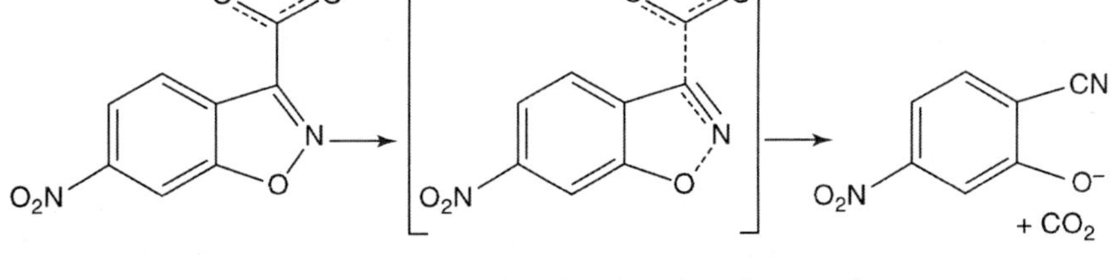

Figure 16.4 The unimolecular decarboxylation of 6-NBIC

Hydrogen-bonding stabilization of the carboxylate group as well as ion-pair formation (particularly in apolar, aprotic solvents) retard the reaction by reactant stabilization. Breaking-up these interactions and stabilization of the highly polarizable activated complex (AC) by London dispersion interactions dramatically accelerate the decarboxylation. As anticipated, 6-NBIC binds to positively charged surfaces of surfactant aggregates with the aromatic part of the molecule engaging in hydrophobic interactions with methylene groups of the alkyl tails residing near or at the aggregate's surface.[87,88]

Reaction rates in the presence of cationic surfactant aggregates can be successfully analyzed using the pseudophase model, providing the rate constants for reaction within the aggregate (for vesicles, k_{ves}, taking into account ion exchange processes[89]). Generally, reaction rates increase with decreasing water penetration into the aggregate and with decreasing fluid-like character of the surface of the aggregate:

$$\text{Micelles} < \text{bilayers}\,(T > T_c) < \text{bilayers}\,(T < T_c)$$ [88]

The kinetics of the decarboxylation of 6-NBIC has also been examined in liposome dispersions[83] and in spontaneously formed catanionic vesicles.[90] For equimolar mixtures of *n*-cetyltrimethylammonium bromide (CTAB) and sodium *n*-heptyl sulfate, k_{ves} ($8.29 \times 10^{-4}\,s^{-1}$) is higher than that in pure CTAB ($k_{ves} = 6.76 \times 10^{-4}\,s^{-1}$). Part of this rate increase may be due to the large concentration of NaBr in the solution.

Cationic vesicles, for example those formed from di-*n*-hexadecyldimethylammonium bromide (DHAB) accelerate the decarboxylation by a factor of about 1000 relative to pure water.[91] Dehydration of the carboxylate group at the binding sites is most likely the main factor behind the catalysis. Different isokinetic temperatures (obtained from linear plots of enthalpies *vs.* entropies of activation) have been observed above and below the main phase transition temperature. These excellent isokinetic relationships indicate that the catalytic effects are caused by a single important interaction mechanism.[92]

From the view point of encapsulation, the effects of additives on the nature of vesicular binding sites are of special interest. An important additive is, of course, cholesterol, that plays an important role in life processes. It exerts important effects on the biophysical properties of biomembranes including their fluidity and the formation of microdomains but also on membrane proteins, thereby regulating their functions. Interestingly, there is compelling evidence that phospholipid bilayer systems can form distinct cholesterol-rich and cholesterol-poor domains.[93] Extensive differential scanning microcalorimetric studies on vesicles formed from both synthetic amphiphiles and phospholipids have shown significant effects of cholesterol on the main phase transition temperature, governed by changes in bilayer packing.[94] Compensating enthalpy and entropy effects on melting were found and cholesterol also influences the cooperativity of bilayer melting.

Coming back to the decarboxylation of 6-NBIC, cholesterol was found to reduce k_{ves} for DHAB vesicles by a factor of 3 when at 50 mol % in the bilayer. As a result of its appreciable hydrophobic surface area, it penetrates significantly into the bilayer, thereby decreasing the inter-amphiphile interactions. Under these conditions, the hydration of the interface is increased, and the reactant is stabilized. On the other hand, for trehalose as the additive, the value of k_{ves} is slightly increased. This is in accord with the notion that binding to the bilayer surface leads to replacement of water molecules from the vesicular interface with a concomitant destabilization of bound 6-NBIC.

16.2.2 The Kemp Elimination: Rate-limiting Proton Transfer

A bimolecular reaction, with a substrate that is structurally related, albeit uncharged, and rather similar to the one examined in the previous paragraph, is the Kemp elimination. It involves the rate determining hydroxide-ion induced deprotonation of 5-nitrobenzisoxazole (5-NBI, **1**, Figure 16.5). The high solvent sensitivity of this reaction primarily depends on the reactivity of the OH^- ion.[95] The more weakly the hydroxide ion is solvated, particularly in apolar, aprotic solvents, the stronger the base catalysis. For example, the bimolecular rate constant (k_2) is even 457 times higher in ethanol than in water. In cationic micelles (such as formed from *n*-C_{12}PyrI, DTAB, CTAB, CTACl and OTACl) the organic substrate will bind close to the aqueous interface, participating in both hydrogen bonding and hydrophobic interactions.[96] But what are exactly the properties of this interfacial region as a reaction environment? Indeed, k_2 is also about 400 times higher than in water, in accord with the about similar dielectric constants of ethanol and of the Stern region of the micelles. Plots of k_{obs} *vs*. the surfactant concentration show the typical biphasic pattern for a bimolecular reaction with both substrates partitioning between the bulk aqueous and micellar pseudophase. Hydroxide-ion binding to the micelles depends on the ability of the hydroxide ion to replace the counterion of the cationic micelles.

A kinetic study was also performed in a variety of vesicular solutions (DDAB, DODAB, DODAC; [NaOH] = 2.25 mM, 25 °C).[96] Interestingly, the vesicles possess stronger catalytic reaction environments than the micelles.[97] The rate-determining proton transfer from carbon to the hydroxide ion was accelerated up to 850 fold in di-*n*-dodecyldimethylammonium bromide (DDAB) vesicles. This is evidence that the reaction sides are less aqueous than those in micelles, as anticipated. Application of the pseudophase model afforded the bimolecular rate constants in the vesicles (k_{ves}). For the different vesicles, k_{ves} is significantly higher (*ca*. 12 times for DODAB) than the second-order rate constant in water. This shows that the catalysis is due to both a medium effect and a concentration effect. It was assumed that there was a fast equilibrium for substrate binding to the inner and outer leaflets of the bilayer, in accord with the fact that no two-phase kinetics were found.

Vesicle catalysis followed the order DDAB > DODAC > DODAB, with k_{max}/k_w values of about 850, 550, and 160, respectively. This looked an unexpected result because the rate constants did not respond in the usual way to an increasing chain length of the tails. But it was recognized that the high catalytic efficiency of DDAB with the shortest alkyl chains was the result of vesicles being in the liquid-crystalline state at the reaction

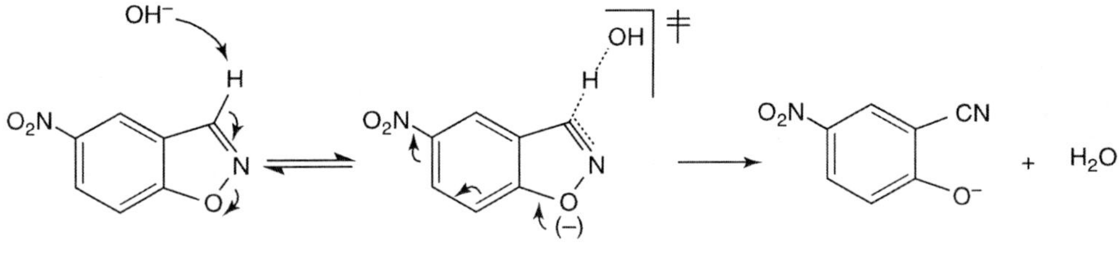

Figure 16.5 *The hydroxide-ion catalyzed Kemp elimination of 5-NBI*

temperature while the other bilayers were in the gel state. A similar effect has been observed before[98].

Addition of cholesterol leads to two counteracting effects on the rate constants. The first is a smaller counterion binding, reducing the rate constants. The other is a rate enhancing effect resulting from the less polar vesicular binding sides. The overall effect depends on the exact reaction conditions.[96]

In a further detailed study the effects of several added long-tailed alcohols and *n*-alkyl pyranosides were examined.[99] The overall results showed that minor structural changes in the additive and concentration of the additive exerted significant changes in the properties of the interfacial region and it was suggested that similar effects can be anticipated for the much more complex biological cell membranes.

Perhaps most interesting was the kinetic response upon addition of *n*-dodecyl-β-glucoside (C_{12}Glu) and *n*-dodecyl-β-maltoside (C_{12}Mal). For C_{12}Glu a rate increase was found, largely due to an increase of k_{ves} since substrate binding (K_s) shows only a minor increase. Using a number of spectroscopic probes, the normalized polarity of the interfacial region was estimated and the data indicated only a minor difference with bulk water.[99] All evidence pointed to a partial dehydration of the interfacial reaction sites induced by C_{12}Glu and a concomitant dehydration of the OH^- ion. Therefore this effect was taken as the origin of the kinetic effects.

Addition of sodium di-*n*-decyl phosphate (DDP) to DODAB vesicles leads to the formation of catanionic vesicles and rather drastic changes in the propertes of the interfacial region. Again the Kemp elimination was employed to probe these changes, in combination with cryo-electron microscopy, DSC, and measurements of the surface charge density and zeta-potentials.[97] Binding site polarties were assessed using Reichardt's E_T-30 probe and pyrene.

Whereas for di-*n*-octadecyldimethylammonium chloride (DODAB) vesicles k_{ves} is about 65 times larger than the water rate constant, addition of DDP decreases the vesicular catalysis. Experiments with the polarity probes indicated that the polarity of the bilayer surface hardly responded to addition of DDP and also the binding constant of 5-NBI was not affected. It was argued that the reduction of the catalytic effect was due to a decrease in counterion binding resulting from addition of DDP.[97] Catalysis could be turned into rate inhibition as was observed for negatively charged vesicles containing 70 mol% DDP in the bilayer. Interestingly, DSC experiments revealed the presence of neutral microdomains ('rafts') in case of 5 and 30 mol % DDP in the bilayer.

16.2.3 Bimolecular Nucleophilic Substitution

Another bimolecular single-step organic reaction which can be used to probe binding sites in vesicular aggregates is the S_N2 reaction of a series of aromatic alkylsulfonates (MNBS; AlkONs, Alk = Me, Et, *n*-Pr, *n*-Hex) with water (Figure 16.6).[100] Now the water is a nucleophile instead of a hydration agent as in the previous probe reactions.

The rates of these hydrolysis reactions (Nu = H_2O) can be compared with nucleophilic substitution by bromide ions (Nu = Br^-). A kinetic study has been made of these reactions in the presence of vesicles formed from synthetic amphiphiles, phospholipids, and mixtures of both types of amphiphiles.[100] Particular attention was paid to the effect of addition of *n*-dodecyl-β-glucoside (C_{12}Glu) as a mimic for glycolipids. Kinetic data were

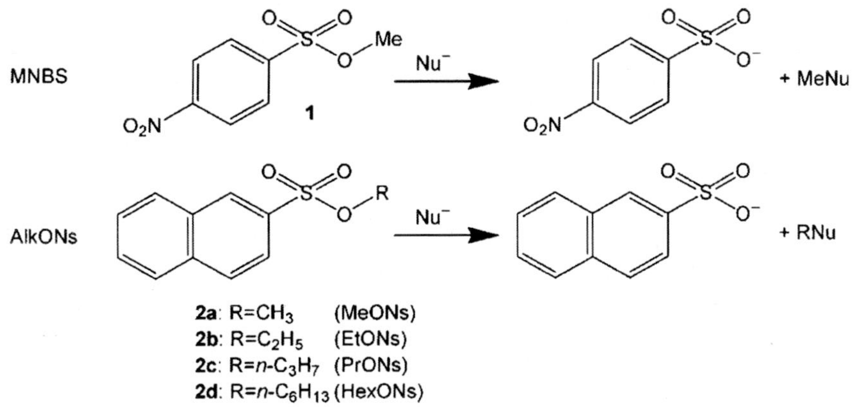

Figure 16.6 *The S_N2 reaction of MNBS and AlkONs with nucleophiles*

analyzed using the pseudo-phase model to afford rate constants for the S_N-2 process in the aqueous phase and in the vesicular pseudo-phase. The experimental data can be reconciled with the reactions occurring at the vesicle–water interphase, with the nature of the binding sites being modified by the presence of C_{12}Glu. It was calculated that for vesicles containing 50 mol% of C_{12}Glu, the vesicular surface is covered for about 34% by C_{12}Glu.

Rate constants for the reactions in the presence of overall positively charged vesicles are about ten times larger than those in the absence of vesicles. The effect was ascribed to an increase in the reactivity of water. In case the water molecules at the vesicular interphase are in part replaced by the glucose groups of C_{12}Glu, the catalytic efficiency of the vesicles decreases significantly.

The bimolecular rate constants for the reaction with bromide ions are smaller at the vesicular interphase, and independent of the presence of C_{12}Glu. These results indicate that this reaction, not involving water as a reactant, is not sensitive towards partial dehydration of the binding sites of the organic substrate.

16.3 Liposomal Encapsulation in Drug Delivery

Since the discovery of liposomal vesicles[9] it has been envisioned that the closed inner compartment of a vesicle can be used to trap potentially harmful substances or to protect unstable compounds from decomposition by shielding them from the outside.[101,102] Despite the high potential of liposomal formulations in drug delivery only a few applications have made it into approved therapeutic drugs. The key to success is the stability of the formulation and the ability to retain the encapsulated compound for a long time once it is loaded into a vesicle. The low solubility and high toxicity and/or unwanted site effects of many anti-cancer drugs make this class of compounds a logic choice for the use of encapsulation. Furthermore the discontinuous vascular blood vessels of tumours are leaky for small 200–1200 nm particles, and liposomes of 100–200 nm readily extravasate at the tumour site.[103]

Efficient liposomal therapeutics also require long circulation times. Normal liposomes are cleared from the blood rapidly by the reticuloendothelial system, but circulation times can be easily increased by giving the liposomes a so-called stealth character.[104] Stealth liposomes are sterically stabilized by lipids with a long polyethylene glycol unit attached to the headgroup, usually phosphatidylethanolamine (PE).[105]

Good encapsulations begin with vesicles in which any compound can be retained for a long time. One way to achieve this is by making the bilayer less permeable for the encapsulated compound. Although a phospholipid membrane is semipermeable, small molecules and monovalent ions can leak in and out, typical at a time span of hours. In the presence of serum, vesicles are usually more leaky than in buffer without serum albumin. Very often cholesterol is added to phospholipids vesicles to make them less permeable as was demonstrated by leakage experiments with carboxyfluorescein (CF) and calcein.[106] Instead of membranes based on (phospho)lipids, the use of block copolymers, with the same basic architecture as lipids but with larger masses, that can form vesicles[107,108] (polymersomes) gains popularity.[109] Another way of making less permeable membranes, involves the use of newly discovered lipids like the laderanes from the anammoxosomes,[110,111] unique organelles in anammox bacteria, and multimethyl branched lipids from the membranes of archeabacteria.[112] Although results on their capabilities are limited at the moment,[113] the synthesis of these lipids became recently available [114–116] allowing large scale experiments on their possible usage as drug delivery capsules.

16.3.1 Encapsulated Drugs

One of the first encapsulated drugs that was reported was liposomal insulin for oral administration instead of the now used method of subcutaneous injection. Insulin, when administrated orally, is broken down in the digestive track, making it useless for sugar uptake from the blood. By encapsulation in bilayer vesicles the idea was that insulin could successfully be administrated orally. The shielding effect of the bilayer to prevent decomposition could facilitate the delivery of intact insulin. The results, however, were inconsistent and an application was never launched. Although the basic idea was good, this method faced the major problems still involved in efficient encapsulation. To retain any encapsulated compound effectively is by making use of the low solubility limit of the compound either by itself or by gelating it with another compound.[117] This was nicely demonstrated for vesicles loaded with the anti-cancer drug doxorubicin or the antibiotic ciprofloxacin.[118] Doxorubicin was actively loaded by a sulphate gradient.[119] Due to the low solubility of the precipitated doxorubicin (Figure 16.7) no leakage was observed even in serum containing media. On the other hand, ciprofloxacin, with a much higher solubility limit leaked out of the vesicles on an hour's time scale.[118] Doxorubicin-loaded vesicles are by now one of the best characterized liposomal drug formulations[120] approved for the treatment of several solid tumours. By encapsulation of the doxorubicin the effective dose could be increased whereas the large site effects, such as the breakdown of cardiac tissue, could be avoided.[121]

Cisplatin is an effective and frequently used anti-cancer drug against a variety of solid tumours. The first trials of encapsulation faced the problem of low drug to lipid ratio due to the limited solubility of cisplatin resulting in low bioavailability.[122]

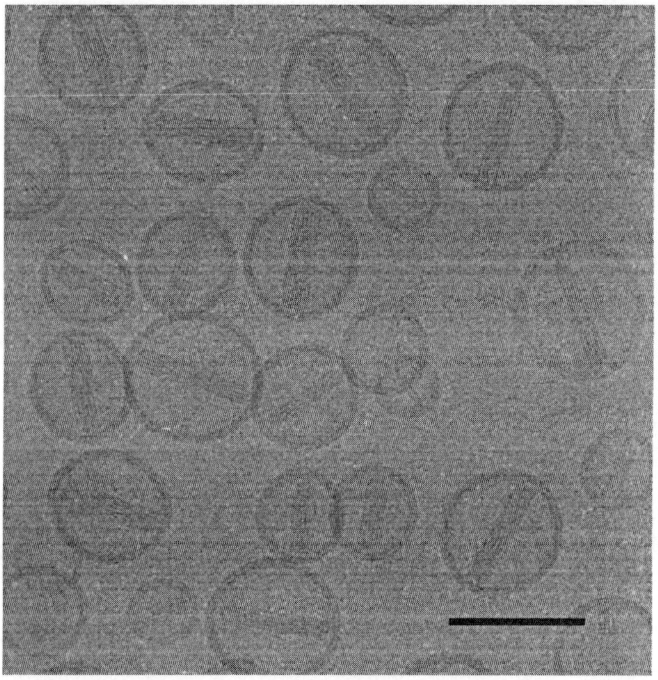

Figure 16.7 *Cryo-electron micrograph of doxorubicin-loaded vesicles. The drug is precipitated into needle-like crystals. Courtesy by Dr. P. M. Frederik, University of Maastricht, The Netherlands. Bar 100 nm*

Effective encapsulation of cisplatin was achieved by multiple freeze thawing steps of neutral and cationic soluble cisplatin in the presence of anionic phospholipids like phosphatidylserine (PS).[123] By freezing the cisplatin is concentrated because the solute is expelled from the ice crystals. When the concentration limit is reached the neutral cisplatin, with the lowest solubility limit, aggregates followed by aggregation of positively-charged cisplatin when the solution further concentrates during the freezing. The negatively-charged PS collapses on the small particles resulting in small aggregates covered by lipids.[124] In this way very high cisplatin to lipid ratios were achieved.

Today more than ten liposomal drug formulations have been approved for clinical use[125,126] and the number is growing. Ideally a liposomal encapsulated drug exhibits hardly any drug release in the blood plasma, but once at the designated site all its content should be released.

To trust only on mechanical instability by accumulation on a desired spot for the release of pay-load from vesicles is an imperfection of the system. By the incorporation of channel proteins in the bilayer of vesicles that can open or close upon a variety of signals[127] smart release of encapsulated substances is brought closer. Especially the modified mechano-sensitive channel protein MscL[128,129] from Lactococcus lactis can be remotely controlled. The MscL can be modified to be responsive to light or pH[130–132] for the release of encapsulated compounds.

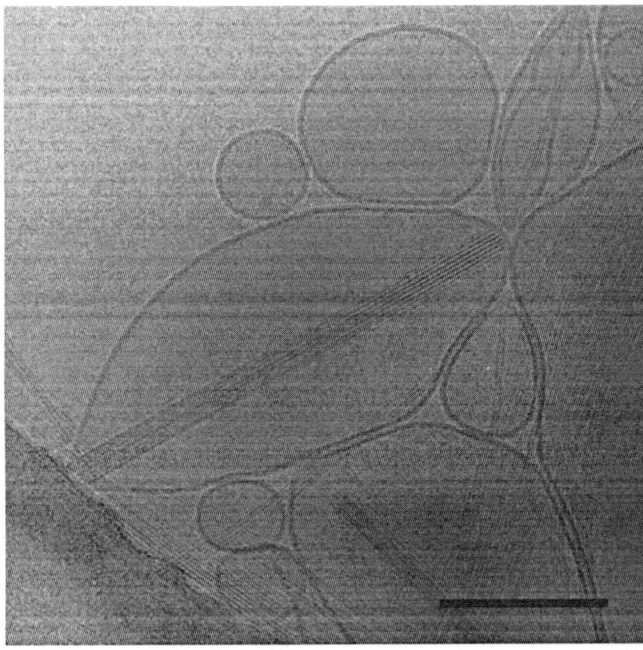

Figure 16.8 Gellosomes, growth of gel-fibers directly in the aqueous compartement of a phospholipid vesicle. Reproduced with permission from Wiley-VCH Verlag GmbH & Co. KGaA

An alternative for encapsulation in vesicles is the use of low molecular weight hydrogelators. Gels can retain solvents and solutes in the space between a dense network of intertwined fibres. Similar to encapsulated material by vesicles, gels can retain compounds. Aqueous hydrogelators based on cyclohexyl-tris-amino acid can self-assemble in one direction through the establishment of H-bonds, leading to the formation of a fibre network and consequently macroscopic gels.[133] The formation of these gel networks was found to be independent from the self-assembly of phospholipids into vesicles which is driven by hydrophobic interaction. By encapsulating a gel network into liposomes (Figure 16.8) the advantages of both liposomes and gel networks can be combined in a single system.[134]

Despite the high expectations only a handful of applications of liposome- encapsulated drugs is available. Nevertheless the high potency of liposomes is still considered as very useful in this field and research is progressing.

16.4 Vesicle–Nucleic Acid Interactions: Gene Transfer Using Lipoplexes

Genetic modification of living cells can be accomplished by delivering exogenous genetic material into the cell thereby replacing a missing or deficient gene with the accompanying therapeutic effects and improved cell biological functions. Medical applications in the clinique have already been successful using viral vectors for DNA trafficking[135] During the past decades, extensive research has been performed on the synthesis and testing of *non-viral* vectors for delivering nucleic acids into living cells, both *in vitro* and *in vivo*. Examples include DOTMA (*N*-(1-(2,3-dioleoyloxy)propyl)-*N*,*N*,*N*-trimethylammonium chloride), DOTAB (*n*-dodecyl-trimethylammonium bromide), SAINT-2 (*N*-methyl-4-(dioleoyl)methylpyridinium chloride and sugar-based gemini surfactants. This work makes use of some of the most sophisticated applications of bilayer encapsulation. The vectors include cationic, bilayer-forming amphiphiles,[136–138] which are viewed as useful alternatives for virus-based delivery agents which may be associated with mutational and immunological hazards. These carriers are often used in combination with specific phospholipids. The transfection efficiencies of different commercially available cationic lipid-based transfection kits have been compared.[139] A number of biodegradable pyridinium amphiphiles have also been synthesized and exhibited remarkably high transfection efficiencies.[140]

The literature on gene transfection is enormous and the results have been reviewed.[137,141,142] We will focus here exclusively on nucleic acid encapsulation by vesicles and discuss some of the factors governing the interactions. The resulting lipoplexes may contain entire genes, (antisense) oligonucleotides[143–145] or RNA and small interference RNA (siRNA) and are in most cases obtained by adding the nucleic acid to preformed vesicles.

16.4.1 Lipoplex Formation

Lipoplex formation is very efficient (occurring on an ms time scale) with *cationic* amphiphiles, indicating that the binding is primarily electrostatically driven. The ratio amphiphile/DNA is such that all negative charges of the DNA are involved in binding to positively charged headgroups of the amphiphiles, thereby leading to a concomitant condensation of DNA to a compact toroidal structure. There should be sufficient additional amphiphile to provide the lipoplex with an overall positive charge. These lipoplexes can then favorably interact with the negatively charged cell surface, for example with anionic proteoglycans.

Liposomes formed from natural phospholipids are either zwitterionic or carry a net negative charge. Using them as gene carriers, nucleic-acid binding must now rely mainly on entrapment into the aqueous pool of the liposomes.

There is compelling evidence that the efficiency of the delivery vehicle depends strongly on the ability of the lipoplex to interact with the cell surface, necessary for induction of transfer of the gene into the cellular interior. As will be discussed later, the cationic amphiphiles should possess the necessary properties for affecting transport *across* the membrane and they should allow destabilization of the endosomal membrane and subsequent release of DNA into the cytosol for trafficking to the nuclear membrane. The latter step is a key factor in gene transfection. A variety of issues play a role here, including a favorable chemical constitution of the amphiphilic carrier, environmental factors, and intermingling of the vesicular phase of the lipoplex with selected phospholipids that can provide endosomal membrane destabilization and a concomitant release of the bound DNA. *Endocytosis* is often considered to be the major entry pathway for this process. It has been shown that the complex internalization process can occur via the cholesterol-dependent clathrin-mediated pathway of endocytosis.[146,147] Evidence includes the observation that transfection is greatly inhibited when plasma membrane cholesterol is depleted with methyl-β-cyclodextrin whereas cell-association remains unchanged.

Then, when delivered into the cytosol, the gene has to find its way to the cell nucleus, the site where transcription and replication will occur. A detailed recent study showed the crucial importance of nuclear transcription efficiencies.[148] In fact there are many individual steps that make up the gene transfection and mechanistic studies are performed to identify these steps and to find ways to facilitate these steps.

16.4.2 Lipoplex Structure

Lipoplexes can have several morphologies each possessing different transfection efficiencies.[142] The most frequently encountered structural phase is lamellar (L^c_α) in which the DNA is sandwiched between the lamellae (Figure 16.9), as indicated by high resolution cryo-electron microscopy and SAXS measurements[149] as well as by X-ray diffraction.[150] Their average size is about 200 nm, allowing a few of these lipoplexes to fit within an endosomal compartment. Mixing the cationic carrier with the phospholipid DOPE (mole fraction >0.4) leads to formation of a highly curved mesomorphic morphology, in this case an inverted hexagonal structure (H^c_{II}) containing DNA rods coated with a monolayer of the amphiphile arranged on a hexagonal lattice. This phase is the favourable one for inducing efficient release of the DNA cargo from an (early) endosomal compartment in vitro. The hydrophobic outer-surface of inverted hexagonal lipoplexes make them not suitable for *in vivo* use. It has also been established that the H^c_{II} phase plays an important role in destabilizing the endosomal membrane, thereby inducing efficient translocation of DNA across the endosomal membrane into the cytosol.[151]

The effect of the helper lipid on the lipoplex morphology depends on control of the spontaneous radius of curvature of the bilayers and is dependent on its mole fraction (mf) in the bilayer. For example, for DOTAB/DOPE cocktails, a pure lamellar phase is formed when mf(DOPE) is <0.41. A pure H^c_{II} phase is formed when mf(DOPE) is >0.75,

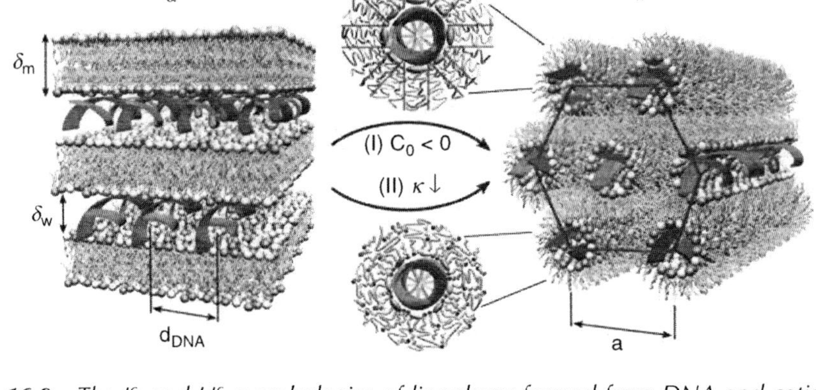

Figure 16.9 *The L^c_α and H^c_{II} morphologies of lipoplexes formed from DNA and cationic bilayer-forming amphiphilles. Reprinted with permission from AAAS*

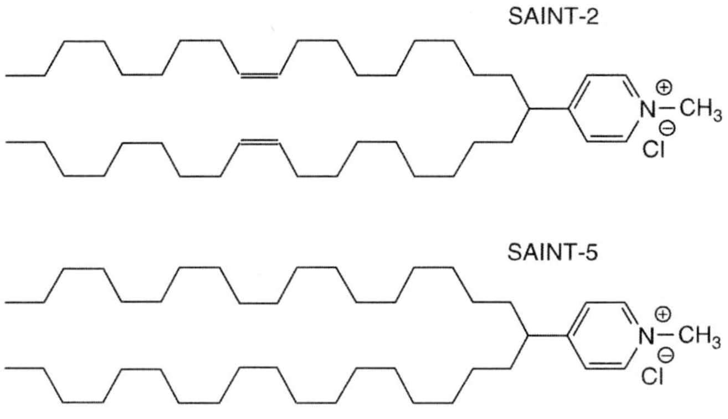

Figure 16.10 *Molecular structures of SAINT-2 and SAINT-5*

whereas both morphologies coexist for $0.41 < mf(DOPE) < 0.75$. Increasing mole fractions of DOPE lead to gradual dehydation of the DNA bases. Transformation of DOTAP/DOPE systems into the H_{II}^{c} phase was evidenced by small angle X-ray scattering (SAXS) and optical microscopy.[152] For SAINT-2/DOPE cocktails a similar phase change, but now at high salt concentrations, was shown by NMR spectroscopy and SAXS measurements.[153,145,154]

The transition from L_{α}^{c} to H_{II}^{c} can be rationalized by considering the packing parameter P (section 16.1.1). For bilayer forming carrier molecules, P is between 0.5 and 1.0, often close to 1.0; by contrast the phospholipid DOPE is cone-shaped and P is > 1, leading to a preferred formation of an inverted hexagonal phase.

There is evidence that a transient spaghetti-like morphology can be formed between the L_{α}^{c} and H_{II}^{c} phases, acting as a precursor to the H_{II}^{c} morphology.[155]

We like to emphasize here, that the efficiency of a helper lipid like DOPE is, however, influenced by the packing of the alkyl tails in the membrane formed by the cationic carrier amphiphile. This was demonstrated[156] by a comparative study of two structurally related bilayer-forming surfactants, SAINT-2 (with two C18:1 tails) and SAINT-5 (with two C18:0 tails) (Figure 16.10). Both carriers display transfection activity, but DOPE exerts a positive effect on SAINT-2-mediated transfection, but has a negligible effect on transfection mediated by SAINT-5. Interestingly, DOPE effectively enhances DNA dissociation from the lipoplexes formed from both carriers. Most likely, membrane stiffness plays an important role here. Since the bilayer composed of SAINT-5 is more rigid than that formed from SAINT-2, because of the absence of unsaturation in the tails, the plasmid DNA becomes less effectively condensed, and the lipoplex is structurally deformed. This has no effect on cellular uptake but reduces the efficiency of translocation of the plasmid across the membranes of the endosome and/or of the cell nucleus.

In contrast with supercoiled DNA, steric factors most likely prevent translocation of the uncondensed DNA from the endosome into the cytosol. Consistent with this interpretation, the much smaller oligonucleotides are effectively translocated into cells by lipoplexes formed from both SAINTs. In case the plasmid is stabilized by condensation with poly-L-lysine, the transfection by SAINT-5/DOPE is greatly improved. The observed phenomena illustrate that the structural shape of the plasmid is a substantial factor in transfection processes. This factor has been considered previously but without reaching consensus.[157]

A third morphology that has recently been identified is an intercalated hexagonal structure in which three honeycombs of amphiphile micelles cover the DNA rods, thereby forming a normal, hexagonal lattice (H_{I}^{c}). This was found for the two sugar-based gemini surfactants GS1 and GS2 (Figure 16.11), for which the lipoplex forms an L_{α}^{c} phase at physiological pH values as indicated by cryo-EM and SAXS[158] measurements. The phase transition after internalization of the lipoplex into the endosomes with a mildly acidic pH was examined using the solvatochromic fluorescent probe Nile Red.[73] In the resulting H_{I}^{c} phase the polar headgroups of the geminis are exposed on the outside in contact with water.[159] This in contrast to the earlier proposed H_{II}^{c} phase for the lipoplexes.[158] As a consequence, these particles exhibit an unusually high colloidal stability (as shown by turbidity measurements) facilitating their application in *in vivo* gene delivery experiments.[160] In the H_{II}^{c} phase the polar headgroups of the amphiphile participate in

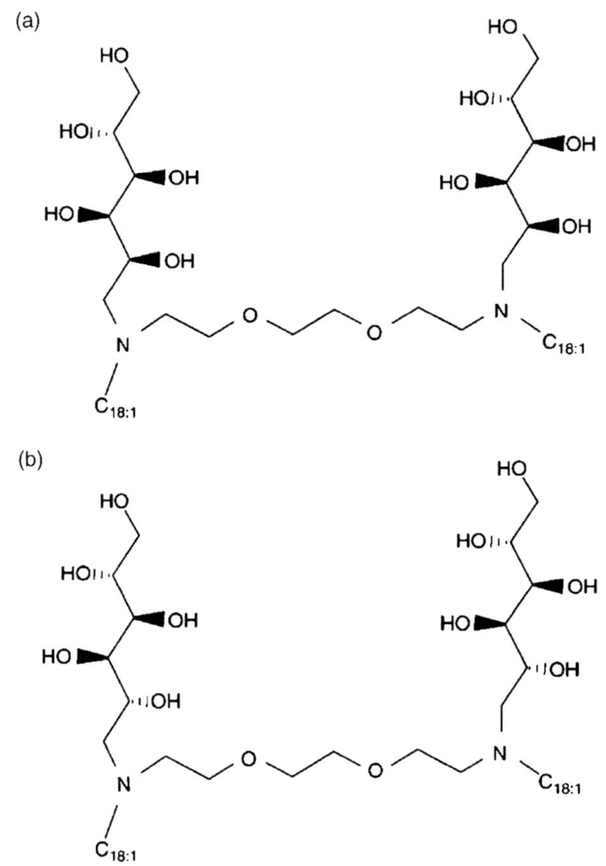

Figure 16.11 *Molecular structures of the reduced sugar-based gemini surfactants GS1 (a) and GS2 (b). Note the different stereochemistry of the sugar moieties*

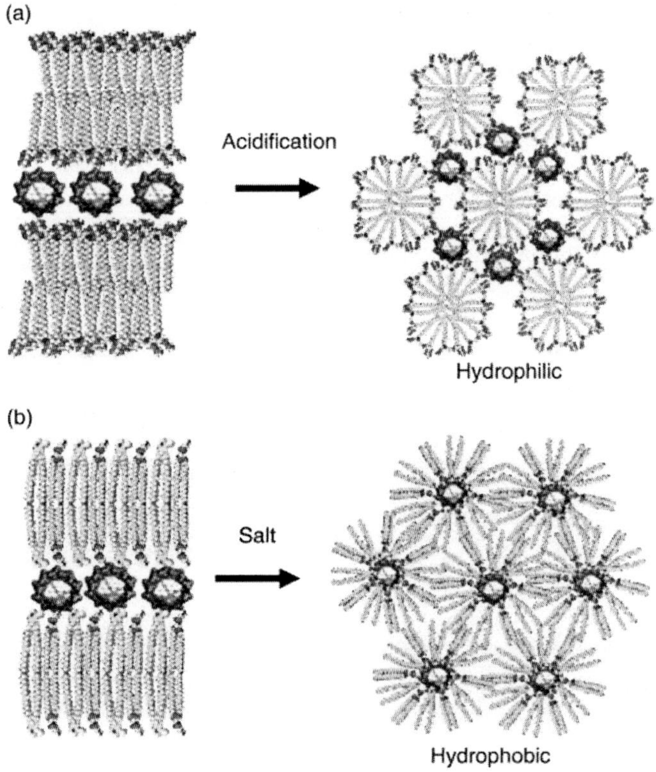

Figure 16.12 *The H_I (GS1,2) (a) and H_{II} (SAINT-2/DOPE with added salt) phases (b) lipoplex morphologies. Reprinted with permission of Elsevier. Copyright 2006*

electrostatic interactions with the DNA and the hydrophobic tails are exposed on the outside where they can easily induce aggregation behaviour and capture in the lung endothelium capillaries. This problem has been tackled by coating the lipoplexes with PEG-lipids that enhance the blood circulation times.

In Figure 16.12 the morphology of the H_I^c phase is compared with that of the H_{II}^c phase formed from SAINT-2 and DOPE. A similar H_I^c phase has previously been reported for lipoplexes composed of the single-tailed, micelle-forming surfactant CTAB.[161] In this phase the amphiphile monomers are dynamic and able to move in and out of the micelles. These properties allow translocation of endosomal lipids into the lipoplex thereby stimulating release of DNA into the cytosol.

Oligonucleotides (ODNs) can function as effective gene-specific regulators and have considerable therapeutic potential. Although small quantities of ODNs can be endocytosized via adsorption, it is necessary to prevent their sequestering in endocytic compartments. It has been shown[144] that SAINT-2 can act as an effective carrier for specific antisense ODNs to target mRNA. No problems with cytotoxicity were found. Using Chinese hamster ovary cells, the protein levels of the receptor for the neuropeptide corticotropin- releasing factor were assessed. ODN release did not depend on the size of the lipoplex and the presence of serum. However, in case serum proteins are incorporated into the lipoplex, the lipoplex membrane is stabilized, hampering ODN release.

For *in vivo* applications, the circulation time of the ODN-carrier complex has to be increased. This was accomplished for the same cell type by incorporating complexes of PEG with either phosphatidylethanolamine (PE) or ceramide.[145] The data suggested that cytosolic release of the ODNs from the endosomal compartment was inhibited by the PEG-lipid through stabilization of the lamellar phase of the lipoplexes. For a more detailed discussion of this important field, the interested reader is referred to recent reviews.[162,143]

Rather surprisingly, it was found possible to form a lipoplex from DNA and a zwitterionic multilamellar phase at high lipid/DNA weight ratios.[163] The impetus for these studies was the lower cytotoxicity of neutral (and also negatively charged) liposomes. Encapsulation of DNA by soya bean diacylphosphatidylcholine (PC) was carried out by mixing the lipid (containing a small amount of a helper surfactant) and short DNA fragments (*ca.* 150 base pairs) in excess water and subsequent freeze-drying. The resulting dry powder was then hydrated with deionized water. SAXS measurements on these systems were in accord with a multilamellar structure with intercalated monolayer DNA between the neutral lipid layers, despite the absence of electrostatic binding interactions. The absence of significant DNA-lipid interactions was experimentally verified and as a consequence the DNA molecules possess more motional freedom than in the conventional L_α^c complexes formed from cationic carrier systems.

Another remarkable lipoplex morphology was proposed in experiments in which short DNA fragments (either pure or marked with a fluorescent dye) were locally injected, using a micropipette, into a part of the membrane of a giant unilamellar vesicle (GUV, diameter *ca.* 100 μm). These vesicles were formed from phosphatidylcholines and upto 33 mol% of a cationic sphingosine by electroformation.[164] Membrane topology was observed in phase contrast, DNA distributions by fluorescence spectroscopy. Local DNA-lipid interactions in the membrane induced endocytosis, which needed a minimum concentration of D-erythrosphingosine. At lower concentrations only lateral adhesions between neighboring vesicles were found upon local addition of DNA. The size and shapes of the endosomes were dependent on the kind of DNA and the initial GUV membrane tension.

Although the evidence was not fully compelling, it was suggested that DNA-lipid interactions involved DNA encapsulation within a cylindrical inverted micelle, included in the lipid membrane (Figure 16.13).

Gene expression was verified in cell-sized, giant vesicles formed from DOPC/DOPG (10 : 1). In individual vesicles expression was found of red-shifted green fluorescent protein (rsGFP) using fluorescence spectroscopy.[165] Particularly in the early stage of the reaction, expression inside the vesicle was significantly higher than that in bulk aqueous solution. Interestingly, the rsGFP synthesized in the vesicles is protected from attack by proteinase K that was added to the external aqueous medium.

A lot of work has been done to find out how lipoplexes and lipoplex-cell interactions respond to the presence of serum. Many relevant references are cited in a paper published in 2002[166] that reports how lipoplex stability and processing are affected by serum. The cationic surfactant carrier was SAINT-2 using DOPE as the helper lipid. Previous studies had already shown that transfection efficiencies are reduced in the presence of serum, in

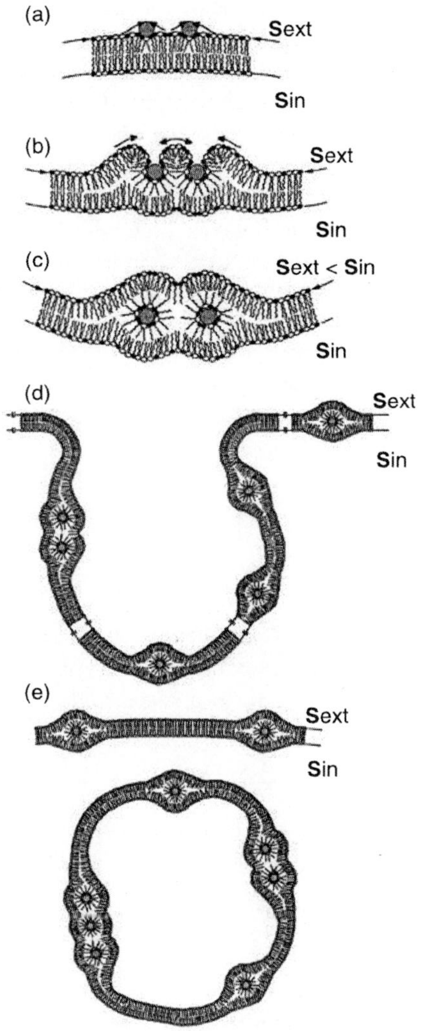

Figure 16.13 *Suggested mechanism for endosome formation of short DNA and GUV membranes. (a) DNA adsorption to the planar GUV membrane (dashed circles represents transverse sections of the DNA molecules). (b) Lateral diffusion and increase of the Sph+ concentration, decoupling of both monolayers, and external monolayer 'rolling' up on the DNA molecules. (c) Topological transformation of the external lipid monolayers and encapsulation of DNA within a cylindrical inverted micellar structure. Membrane asymmetry is created ($S_{ext} < S_m$). (d) Membrane invaginates at a scale of a few micrometers. (e) Formation of the endosome. Reprinted with kind permission of Springer Science+Business Media*

accord with the results for SAINT-2 lipoplexes. In the absence of serum, clustered lipoplexes ($^{-}$FBS lipoplexes, FBS is fetal bovine serum) were observed. Clustering did *not* occur in the presence of serum ($^{+}$FBS lipoplexes) or when serum was present during lipoplex formation (FBS lipoplexes). Interestingly, the topology of DNA in FBS lipoplexes changes from a supercoiled conformation (as also in $^{-}$FBS lipoplexes) to a predominantly open-circular conformation. This leads to faster digestion by DNase. Most significant was the observation that internalization of $^{-}$FBS and $^{+}$FBS lipoplexes is about three times slower than that of FBS lipoplexes although their transfection efficiencies are about five times higher. The data indicate that smaller serum protein-penetrated particles cannot fast enough release their DNA from pre-lysosomal endocytic compartments and are delivered to lysosomes where they are prone to degradation pathways. Size and morphology of the lipoplex govern their ability to interact with and perturb cell membranes, processes that are needed for release of the gene. Serum regulates these processes in an amphiphile-dependent manner through complex 'penetration' and modulation of the DNA conformation.

16.4.3 Future Prospects

It will be clear that DNA encapsulation by vesicles is the starting point for gene transfection. In a complex follow-up, DNA is finally transported to the cell nucleus. The variety of factors determining transfection efficiencies, including lipoplex size and zeta potential, incubation time, cytotoxicity and lipoplex morphology, both *in vitro* and *in vivo*, are still under active investigation. The final goal is a successful and safe application in the clinique. Our attention here was only focused on lipoplex formation, their structure and properties. The results obtained so far overwhelmingly show the complex features of DNA encapsulation by the cationic amphiphilic carrier systems. Insight into these issues as embedded in a realistic mechanistic picture of the whole transfection process might ultimately give the desired success.

References

1. D. F. Evans, H. Wennerstrom, *The Collodial Domain: Where Physics, Chemistry and Biology Meet*, Wiley, New York, 1999.
2. A. M. Carmonaribeiro, Synthetic amphiphile vesicles, *Chem. Soc. Rev.*, 1992, **21**, 209–214.
3. M. V. Scarpa, F. A. Maximiano, H. Chaimovich, I. A. Cuccovia, Interfacial concentrations of chloride and bromide and selectivity for ion exchange in vesicles prepared with dioctadecyldimethylammonium halides, lipids, and their mixtures, *Langmuir*, 2002, **18**, 8817–8823.
4. D. F. Evans, D. D. Miller, Organized solutions and their manifestations in polar solvents, in *Organized Solutions. Surfactants in Science and Technology*, S. E. Friberg, B. Lindman (Eds.) Surfactant Science Series, Dekker, New York, 1992.
5. J. N. Israelachvili, D. J. Mitchell, B. W. Ninham, Theory of self-assembly of hydrocarbon amphiphiles into micelles and bilayers, *J. Chem. Soc. Farad. Trans. II*, 1976, **72**, 1525–1568.
6. R. Nagarajan, Molecular packing parameter and surfactant self-assembly: the neglected role of the surfactant tail, *Langmuir*, 2002, **18**, 31–38.
7. J. J. H. Nusselder, J. B. F. N. Engberts, A search for a relation between aggregate morphology and the structure of 1,4-dialkylpyridinium halide surfactants, *J. Org. Chem.*, 1991, **56**, 5522–5527.

8. J. J. H. Nusselder, J. B. F. N. Engberts, Surfactant structure and aggregate morphology – the urge for aggregate stability, *J. Am. Chem. Soc*, 1989, **111**, 5000–5002.
9. A. D. Bangham, R. W. Horne, Negative staining of phospholipids and their structural modification by-surface active agents as observed in the electron microscope, *J. Mol. Biol.*, 1964, **8**, 660–668.
10. T. Kunitake, Y. Okahata, A totally synthetic bilayer membrane, *J. Am. Chem. Soc*, 1977, **99**, 3860–3861.
11. T. Kunitake, Synthetic bilayer-membranes – molecular design, self-organization, and application, *Angew. Chem. Int. Ed.*, 1992, **31**, 709–726.
12. M. Hirai, H. Iwase, T. Hayakawa, M. Koizumi, H. Takahashi, Determination of asymmetric structure of ganglioside-DPPC mixed vesicle using SANS, SAXS, and DLS, *Biophys. J.*, 2003, **85**, 1600–1610.
13. C. Dietrich, L. A. Bagatolli, Z. N. Volovyk, N. L. Thompson, M. Levi, K. Jacobson, E. Gratton, Lipid rafts reconstituted in model membranes, *Biophys. J.*, 2001, **80**, 1417–1428.
14. A. H. de Vries, A. E. Mark, S. J. Marrink, Molecular dynamics simulation of the spontaneous formation of a small DPPC vesicle in water in atomistic detail, *J. Am. Chem. Soc*, 2004, **126**, 4488–4489.
15. R. D. Kornberg, H. M. Mcconnel, Inside-outside transitions of phospholipids in vesicle membranes, *Biochemistry*, 1971, **10**, 1111–1120.
16. A. H. de Vries, A. E. Mark, S. J. Marrink, The binary mixing behaviour of phospholipids in a bilayer: a molecular dynamics study, *J. Phys. Chem. B*, 2004, **108**, 2454–2463.
17. S. J. Marrink, A. E. Mark, Molecular dynamics simulation of the formation, structure, and dynamics of small phospholipid vesicles, *J. Am. Chem. Soc*, 2003, **125**, 15233–15242.
18. A. M. Smondyrev, M. L. Berkowitz, Structure of dipalmitoylphosphatidylcholine/cholesterol bilayer at low and high cholesterol concentrations: molecular dynamics simulation, *Biophys. J.*, 1999, **77**, 2075–2089.
19. S. C. Ji, J. D. Ding, Spontaneous formation of vesicles from mixed amphiphiles with dispersed molecular weight: Monte Carlo simulation, *Langmuir*, 2006, **22**, 553–559.
20. E. W. Kaler, A. K. Murthy, B. E. Rodriguez, J. A. N. Zasadzinski, Spontaneous vesicle formation in aqueous mixtures of single-tailed surfactants, *Science*, 1989, **245**, 1371–1374.
21. R. T. Buwalda, M. C. A. Stuart, J. B. F. N. Engberts, Interactions of an azobenzene-functionalized anionic amphiphile with cationic amphiphiles in aqueous solution, *Langmuir*, 2002, **18**, 6507–6512.
22. R. G. Laughlin, Equilibrium vesicles: fact or fiction? *Colloid Surf. A-Physicochem. Eng. Asp.*, 1997, **128**, 27–38.
23. E. Abuin, E. Lissi, D. Aravena, A. Zanocco, M. Macuer, A fluorescence probe study of the effect of size on the properties of dioctadecyldimethylammonium chloride vesicles, *J. Colloid Interface Sci.*, 1988, **122**, 201–208.
24. F. Szoka, D. Papahadjopoulos, Comparative properties and methods of preparation of lipid vesicles (liposomes), *Annu. Rev. Biophys. bioeng.*, 1980, **9**, 467–508.
25. D. Harries, A. Ben-Shaul, I. Szleifeo, Enveloping of charged proteins by lipid bilayers, *J. Phys. Chem. B*, 2004, **108**, 1491–1496.
26. J. Dubochet, M. Adrian, J. J. Chang, J. C. Homo, J. Lepault, A. W. McDowall, P. Schultz, Cryo-electron microscopy of vitrified specimens, *Q. Rev. Biophys.*, 1988, **21**, 129–228.
27. M. Almgren, K. Edwards, G. Karlsson, Cryo transmission electron microscopy of liposomes and related structures, *Colloid Surf. A-Physicochem. Eng. Asp.*, 2000, **174**, 3–21.
28. L. A. M. Rupert, D. Hoekstra, J. B. F. N. Engberts, Fusogenic behaviour of didodecyldimethylammonium bromide bilayer vesicles, *J. Am. Chem. Soc*, 1985, **107**, 2628–2631.
29. F. M. Menger, K. D. Gabrielson, Cytomimetic organic-chemistry – early developments, *Angew. Chem. Int. Ed.*, 1995, **34**, 2091–2106.
30. F. M. Menger, N. Balachander, Chemically-induced aggregation, budding, and fusion in giant vesicles – direct observation by light-microscopy, *J. Am. Chem. Soc*, 1992, **114**, 5862–5863.
31. W. J. Vanblitterswijk, R. P. Vanhoeven, B. W. Vandermeer, Lipid structural order parameters (reciprocal of fluidity) in biomembranes derived from steady-state fluorescence polarization measurements, *Biochim. Biophys. Acta*, 1981, **644**, 323–332.

32. M. J. Janiak, D. M. Small, G. G. Shipley, Temperature and compositional dependence of the structure of hydrated dimyristoyl lecithin, *J. Biol. Chem.*, 1979, **254**, 6068–6078.
33. D. P. Siegel, Inverted micellar intermediates and the transitions between lamellar, cubic, and inverted hexagonal lipid phases. 1. Mechanism of the L-alpha, HII phase-transitions, *Biophys. J.*, 1986, **49**, 1155–1170.
34. D. P. Siegel, Inverted micellar intermediates and the transitions between lamellar, cubic, and inverted hexagonal lipid phases. 2. Implications for membrane-membrane interactions and membrane-fusion, *Biophys. J.*, 1986, **49**, 1171–1183.
35. R. S. Cantor, The lateral pressure profile in membranes: a physical mechanism of general anesthesia, *Biochemistry*, 1997, **36**, 2339–2344.
36. R. S. Cantor, Lateral pressures in cell membranes: a mechanism for modulation of protein function, *J. Phys. Chem. B*, 1997, **101**, 1723–1725.
37. H. D. Hong, L. K. Tamm, Elastic coupling of integral membrane protein stability to lipid bilayer forces, *Proc. Natl. Acad. Sci. USA*, 2004, **101**, 4065–4070.
38. R. H. Templer, S. J. Castle, A. R. Curran, G. Rumbles, D. R. Klug, Sensing isothermal changes in the lateral pressure in model membranes using di-pyrenyl phosphatidylcholine, *Faraday Discuss.*, 1998, 41–53.
39. S. M. Oversteegen, P. A. Barneveld, F. A. M. Leermakers, J. Lyklema, On the pressure in mean-field lattice models, *Langmuir*, 1999, **15**, 8609–8617.
40. E. van den Brink-van der Laan, V. Chupin, J. A. Killian, B. de Kruijff, Small alcohols destabilize the KcsA tetramer via their effect on the membrane lateral pressure, *Biochemistry*, 2004, **43**, 5937–5942.
41. E. van den Brink-van der Laan, V. Chupin, J. A. Killian, B. de Kruijff, Stability of KcsA tetramer depends on membrane lateral pressure, *Biochemistry*, 2004, **43**, 4240–4250.
42. H. Heerklotz, T. Wieprecht, J. Seelig, Membrane perturbation by the lipopeptide surfactin and detergents as studied by deuterium, *J. Phys. Chem. B*, 2004, **108**, 4909–4915.
43. J. M. Kuiper, M. C. A. Stuart, J. B. F. N. Engberts, Photochemically induced disturbance of the alkyl chain packing in vesicular membranes, *Langmuir*, 2008, **24**, 426–432.
44. E. Lindahl, O. Edholm, Mesoscopic undulations and thickness fluctuations in lipid bilayers from molecular dynamics simulations, *Biophys. J.*, 2000, **79**, 426–433.
45. D. A. van Hal, J. A. Bouwstra, A. van Rensen, E. Jeremiasse, T. de Vringer, H. E. Junginger, Preparation and characterization of nonionic surfactant vesicles, *J. Colloid Interface Sci.*, 1996, **178**, 263–273.
46. Y. Q. Liang, L. X. Wu, Y. C. Tian, Z. Q. Zhang, H. D. Chen, Structure control of synthetic bilayer membranes from single-chain amphiphiles containing the Schiff base segment. 1. Conformation control and spectral characterization, *J. Colloid Interface Sci.*, 1996, **178**, 703–713.
47. H. Hoffmann, D. Grabner, U. Hornfeck, G. Platz, Novel vesicles from single-chain surfactants, *J. Phys. Chem. B*, 1999, **103**, 611–614.
48. M. Antonietti, S. Forster, Vesicles and liposomes: a self-assembly principle beyond lipids, *Adv. Mater.*, 2003, **15**, 1323–1333.
49. S. Marcelja, Toward a realistic theory of the interaction of membrane inclusions, *Biophys. J.*, 1999, **76**, 593–594.
50. C. W. Lim, B. J. Ravoo, D. N. Reinhoudt, Dynamic multivalent recognition of cyclodextrin vesicles, *Chem. Commun.*, 2005, 5627–5629.
51. J. H. Crowe, J. S. Clegg, L. M. Crowe, Anhydrobiosis: the water replacement hypothesis, in *The Properties of Water in Foods*, D. S. Reid (Ed.), Chapman and Hall, New York, 1998.
52. F. M. Menger, H. L. Zhang, Self-adhesion among phospholipid vesicles, *J. Am. Chem. Soc*, 2006, **128**, 1414–1415.
53. M. Mammen, S. K. Choi, G. M. Whitesides, Polyvalent interactions in biological systems: implications for design and use of multivalent ligands and inhibitors, *Angew. Chem. Int. Ed.*, 1998, **37**, 2754–2794.
54. R. M. Pashley, Effect of degassing on the formation and stability of surfactant-free emulsions and fine teflon dispersions, *J. Phys. Chem. B*, 2003, **107**, 1714–1720.

55. M. Johnsson, A. Wagenaar, M. C. A. Stuart, J. B. F. N. Engberts, Sugar-based gemini surfactants with pH-dependent aggregation behaviour: vesicle-to-micelle transition, critical micelle concentration, and vesicle surface charge reversal, *Langmuir*, 2003, **19**, 4609–4618.
56. M. Johnsson, A. Wagenaar, J. B. F. N. Engberts, Sugar-based gemini surfactant with a vesicle-to-micelle transition at acidic pH and a reversible vesicle flocculation near neutral pH, *J. Am. Chem. Soc*, 2003, **125**, 757–760.
57. J. E. Klijn, M. C. A. Stuart, M. Scarzello, A. Wagenaar, J. B. F. N. Engberts, pH-dependent phase behaviour of carbohydrate-based gemini surfactants. The effects of carbohydrate stereochemistry, head group hydrophilicity, and nature of the spacer, *J. Phys. Chem. B*, 2007, **111**, 5204–5211.
58. J. Kevelam, J. B. F. N. Engberts, M. J. Blandamer, B. Briggs, P. M. Cullis, Anchoring of hydrophobically modified poly(sodium acrylate)s into DDP vesicle bilayers: hydrophobic match and mismatch, *Colloid Polym. Sci.*, 1998, **276**, 190–194.
59. U. Schmidt, G. Guigas, M. Weiss, Cluster formation of transmembrane proteins due to hydrophobic mismatching, *Phys. Rev. Lett.*, 2008, **101**.
60. R. A. Moss, S. Swarup, Transvesicular reactions of thiols with Ellman reagent, *J. Org. Chem.*, 1988, **53**, 5860–5866.
61. M. E. Siwko, Disturb or stabilize? Effects of different molecules on biological membranes, Thesis/Dissertation, University of Groningen, 2008.
62. A. Filippov, G. Oradd, G. Lindblom, The effect of cholesterol on the lateral diffusion of phospholipids in oriented bilayers, *Biophys. J.*, 2003, **84**, 3079–3086.
63. F. Szoka, D. Papahadjopoulos, Procedure for preparation of liposomes with large internal aqueous space and high capture by reverse-phase evaporation, *Proc. Natl. Acad. Sci. USA*, 1978, **75**, 4194–4198.
64. R. A. Moss, S. Swarup, H. M. Zhang, Reactivity control by microencapsulation in simple ammonium ion vesicles, *J. Am. Chem. Soc*, 1988, **110**, 2914–2919.
65. M. J. Blandamer, B. Briggs, P. M. Cullis, K. D. Irlam, J. B. F. N. Engberts, J. Kevelam, Titration microcalorimetry of adsorption processes in aqueous systems – Interaction of sodium dodecylsulfate and sodium decylsulfate with poly(*N*-vinylpyrrolidone), *J. Chem. Soc. Faraday Trans.*, 1998, **94**, 259–266.
66. J. Ulmius, B. Lindman, G. Lindblom, T. Drakenberg, H–1,C–13,Cl–35, and Br–81 NMR of aqueous hexadecyltrimethylammonium salt-solutions – solubilization, viscoelasticity, and counterion specificity, *J. Colloid Interface Sci.*, 1978, **65**, 88–97.
67. C. A. Bunton, C. P. Cowell, The binding of phenols and phenoxide ions to cationic micelles, *J. Colloid Interface Sci.*, 1988, **122**, 154–162.
68. N. R. Jagannathan, K. Venkateswaran, F. G. Herring, G. N. Patey, D. C. Walker, Localization of methanol, ethanol, and 2-propanol at micelles in water – an NMR T1-relaxation study, *J. Phys. Chem.*, 1987, **91**, 4553–4555.
69. Z. S. Gao, R. E. Wasylishen, J. C. T. Kwak, An NMR paramagnetic relaxation method to determine distribution coefficients of solubilizates in micellar systems, *J. Phys. Chem.*, 1989, **93**, 2190–2192.
70. K. Kachel, E. suncion-Punzalan, E. London, The location of fluorescence probes with charged groups in model membranes, *Biochim. Biophys. Acta-Biomembr.*, 1998, **1374**, 63–76.
71. M. Almgren, F. Grieser, J. K. Thomas, Dynamic and static aspects of solubilization of neutral arenes in ionic micellar solutions, *J. Am. Chem. Soc*, 1979, **101**, 279–291.
72. H. Cang, D. D. Brace, M. D. Fayer, Dynamic partitioning of an aromatic probe between the headgroup and core regions of cationic micelles, *J. Phys. Chem. B*, 2001, **105**, 10007–10015.
73. M. C. A. Stuart, J. C. Van de Pas, J. B. F. N. Engberts, The use of Nile Red to monitor the aggregation behaviour in ternary surfactant-water-organic solvent systems, *J. Phys. Org. Chem.*, 2005, **18**, 929–934.
74. N. Lebedeva, R. Ranganathan, B. L. Bales, Location of spectroscopic probes in self-aggregating assemblies. II. The location of pyrene and other probes in sodium dodecyl sulfate micelles, *J. Phys. Chem. B*, 2007, **111**, 5781–5793.

75. N. Lebedeva, B. L. Bales, Location of spectroscopic probes in self-aggregating assemblies. I. The case for 5-doxylstearic acid methyl ester serving as a benchmark spectroscopic probe to study micelles, *J. Phys. Chem. B*, 2006, **110**, 9791–9799.
76. R. Ranganathan, C. Vautier-Giongo, B. L. Bales, Toward a hydrodynamic description of bimolecular collisions in micelles. An experimental test of the effect of the nature of the quencher on the fluorescence quenching of pyrene in SDS micelles and in bulk liquids, *J. Phys. Chem. B*, 2003, **107**, 10312–10318.
77. M. Peric, M. Alves, B. L. Bales, Precision parameters from spin-probe studies of membranes using a partitioning technique. Application to two model membrane vesicles, *Biochim. Biophys. Acta-Biomembr.*, 2005, **1669**, 116–124.
78. P. M. Frederik, D. H. W. Hubert, Cryoelectron microscopy of liposomes, in *Liposomes*, N. Duzgunes (Ed.), Methods in enzymology, 391, Academic Press, New York, 2005.
79. M. C. A. Stuart, J. van Esch, J. C. V. De Pas, J. B. F. N. Engberts, Chain-length and solvent dependent morphological changes in sodium soap fibers, *Langmuir*, 2007, **23**, 6494–6497.
80. H. Cui, T. K. Hodgdon, E. W. Kaler, L. Abezgauz, D. Danino, M. Lubovsky, Y. Talmon, D. J. Pochan, Elucidating the assembled structure of amphiphiles in solution via cryogenic transmission electron microscopy, *Soft Matter*, 2007, **3**, 945–955.
81. M. F. Vitha, A. J. Dallas, P. W. Carr, Study of water-sodium dodecyl sulfate micellar solubilization thermodynamics for several solute homolog series by headspace gas chromatography, *J. Phys. Chem.*, 1996, **100**, 5050–5062.
82. S. J. Lee, J. S. Keiper, Giant vesicles as microchemical vessels, in *Reactions and Synthesis in Surfactant Systems*, J. Texter (Ed.), Surfactant Science Series, 100, Marcel Dekker, New York, 2001.
83. M. V. Scarpa, P. S. Araujo, S. Schreier, A. Sesso, A. G. Oliveira, H. Chaimovich, I. M. Cuccovia, Effect of vesicles of dimethyldioctadecylammonium chloride and phospholipids on the rate of decarboxylation of 6-nitrobenzisoxazole–3-carboxylate, *Langmuir*, 2000, **16**, 993–999.
84. M. K. Kawamuro, H. Chaimovich, E. B. Abuin, E. A. Lissi, I. M. Cuccovia, Evidence that the effects of synthetic amphiphile vesicles on reaction-rates depend on vesicle size, *J. Phys. Chem.*, 1991, **95**, 1458–1463.
85. D. S. Kemp, D. D. Cox, K. G. Paul, Physical organic-chemistry of benzisoxazoles. 4. Origins and catalytic nature of solvent rate acceleration for decarboxylation of 3-carboxybenzisoxazoles, *J. Am. Chem. Soc*, 1975, **97**, 7312–7318.
86. J. W. Grate, R. A. Mcgill, D. Hilvert, Analysis of solvent effects on the decarboxylation of benzisoxazole–3-carboxylate ions using linear solvation energy relationships – relevance to catalysis in an antibody-binding site, *J. Am. Chem. Soc*, 1993, **115**, 8577–8584.
87. C. A. Bunton, M. J. Minch, Micellar catalyzed decarboxylation of 6-nitrobenzisoxazole–3-carboxylate ion, *Tetrahedron Lett.*, 1970, **11**, 3881-3884.
88. T. Kunitake, Y. Okahata, R. Ando, S. Shinkai, S. Hirakawa, Decarboxylation of 6-nitrobenzisoxazole–3-carboxylate catalyzed by ammonium bilayer-membranes – a comparison of the catalytic behaviour of micelles, bilayer-membranes, and other aqueous aggregates, *J. Am. Chem. Soc*, 1980, **102**, 7877–7881.
89. J. H. Fendler, W. L. Hinze, Reactivity control in micelles and surfactant vesicles – kinetics and mechanism of base-catalyzed-hydrolysis of 5,5′-dithiobis(2-nitrobenzoic acid) in water, hexadecyltrimethylammonium bromide micelles, and dioctadecyldimethylammonium chloride surfactant vesicles, *J. Am. Chem. Soc*, 1981, **103**, 5439–5447.
90. R. Talhout, J. B. F. N. Engberts, Self-assembly in mixtures of sodium alkyl sulfates and alkyl trimethylammonium bromides: aggregation behaviour and catalytic properties, *Langmuir*, 1997, **13**, 5001–5006.
91. M. G. M. Jongejan, J. E. Klijn, J. B. F. N. Engberts, Vesicular catalysis of the decarboxylation of 6-nitrobenzisoxazole–3-carboxylate. The effects of sugars, long-tailed sugars, cholesterol and alcohol additives, *J. Phys. Org. Chem.*, 2006, **19**, 249–256.
92. M. S. Patel, K. Bijma, J. B. F. N. Engberts, Decarboxylation of 6-nitrobenzisoxazole–3-carboxylate – enthalpy-entropy compensation in micellar and vesicular catalysis – a novel analysis of contrasting rate variations, *Langmuir*, 1994, **10**, 2491–2492.

93. T. P. W. McMullen, R. N. McElhaney, Physical studies of cholesterol-phospholipid interactions, *Curr. Opin. Colloid Interface Sci.*, 1996, **1**, 83–90.
94. M. J. Blandamer, B. Briggs, P. M. Cullis, B. J. Rawlings, J. B. F. N. Engberts, Vesicle-cholesterol interactions: effects of added cholesterol on gel-to-liquid crystal transitions in a phospholipid membrane and five dialkyl-based vesicles as monitored using DSC, *Phys. Chem. Chem. Phys.*, 2003, **5**, 5309–5312.
95. D. S. Kemp, M. L. Casey, Physical organic-chemistry of benzisoxazoles. 2. Linearity of Brönsted free-energy relationship for base-catalyzed decomposition of benzisoxazoles, *J. Am. Chem. Soc*, 1973, **95**, 6670–6680.
96. J. Perez-Juste, F. Hollfelder, A. J. Kirby, J. B. F. N. Engberts, Vesicles accelerate proton transfer from carbon up to 850-fold, *Org. Lett.*, 2000, **2**, 127–130.
97. J. E. Klijn, J. B. F. N. Engberts, Kemp elimination in membrane mimetic reaction media: probing catalytic properties of catanionic vesicles formed from double-tailed amphiphiles, *J. Am. Chem. Soc*, 2003, **125**, 1825–1833.
98. T. Kunitake, H. Ihara, Y. Okahata, Phase-separation and reactivity changes of phenyl ester substrate and imidazole catalyst in the dialkylammonium bilayer-membrane, *J. Am. Chem. Soc*, 1983, **105**, 6070–6078.
99. J. E. Klijn, J. B. F. N. Engberts, The Kemp elimination in membrane mimetic reaction media. Probing catalytic properties of cationic vesicles formed from a double-tailed amphiphile and linear long-tailed alcohols or alkyl pyranosides, *Org. Biomol. Chem.*, 2004, **2**, 1789–1799.
100. J. E. Klijn, J. B. F. N. Engberts, Vesicular catalysis of an S_N2 reaction: toward understanding the influence of glycolipids on reactions proceeding at the interface of biological membranes, *Langmuir*, 2005, **21**, 9809–9817.
101. G. Gregoria, E. J. Wills, C. P. Swain, A. S. Tavill, Drug-carrier potential of liposomes in cancer chemotherapy, *Lancet*, 1974, **1**, 1313–1316.
102. G. Sessa, G. Weissman, Phospholipid spherules (liposomes) as a model for biological membranes, *J. Lipid Res.*, 1968, **9**, 310-318.
103. S. K. Hobbs, W. L. Monsky, F. Yuan, W. G. Roberts, L. Griffith, V. P. Torchilin, R. K. Jain, Regulation of transport pathways in tumor vessels: role of tumor type and microenvironment, *Proc. Natl. Acad. Sci. USA*, 1998, **95**, 4607–4612.
104. T. M. Allen, The use of glycolipids and hydrophilic polymers in avoiding rapid uptake of liposomes by the mononuclear phagocyte system, *Adv. Drug Deliv. Rev.*, 1994, **13**, 285–309.
105. D. D. Lasic, D. Needham, The 'Stealth' liposome: a prototypical biomaterial, *Chem. Rev.*, 1995, **95**, 2601 2628.
106. T. M. Allen, L. G. Cleland, Serum-induced leakage of liposome contents, *Biochim. Biophys. Acta.*, 1980, **597**, 418–426.
107. D. M. Vriezema, P. M. L. Garcia, N. S. Oltra, N. S. Hatzakis, S. M. Kuiper, R. J. M. Nolte, A. E. Rowan, J. C. M. van Hest, Positional assembly of enzymes in polymersome nanoreactors for cascade reactions, *Angew. Chem. Int. Ed.*, 2007, **46**, 7378–7382.
108. D. E. Discher, A. Eisenberg, Polymer vesicles, *Science*, 2002, **297**, 967–973.
109. T. Smart, H. Lomas, M. Massignani, M. V. Flores-Merino, L. R. Perez, G. Battaglia, Block copolymer nanostructures, *Nano Today*, 2008, **3**, 38–46.
110. J. S. S. Damste, M. Strous, W. I. C. Rijpstra, E. C. Hopmans, J. A. J. Geenevasen, A. C. T. van Duin, L. A. van Niftrik, M. S. M. Jetten, Linearly concatenated cyclobutane lipids form a dense bacterial membrane, *Nature*, 2002, **419**, 708–712.
111. E. F. Delong, Microbiology – all in the packaging, *Nature*, 2002, **419**, 676–677.
112. D. E. Minnikin, L. Kremer, L. G. Dover, G. S. Besra, The methyl-branched fortifications of Mycobacterium tuberculosis, *Chem. Biol.*, 2002, **9**, 545–553.
113. H. Boumann, M. Longo, P. Stroeve, M. Jetten, B. Poolman, J. S. Damste, S. Schouten, Biophysical properties of ladderane lipids derived from anammox bacteria, *Chem. Phys. Lipids*, 2007, **149**, S11-S11.
114. B. ter Horst, B. L. Feringa, A. J. Minnaard, Catalytic asymmetric synthesis of phthioceranic acid, a heptamethyl-branched acid from Mycobacterium tuberculosis, *Org. Lett.*, 2007, **9**, 3013–3015.

115. B. ter Horst, B. L. Feringa, A. J. Minnaard, Catalytic asymmetric synthesis of mycocerosic acid, *Chem. Commun.*, 2007, 489–491.
116. E. Casas-Arce, B. ter Horst, B. L. Feringa, A. J. Minnaard, Asymmetric total synthesis of PDIM A: a virulence factor of Mycobacterium tuberculosis, *Chem.-Eur. J.*, 2008, **14**, 4157–4159.
117. D. D. Lasic, P. M. Frederik, M. C. A. Stuart, Y. Barenholz, T. J. Mcintosh, Gelation of liposome interior – a novel method for drug encapsulation, *FEBS Letters*, 1992, **312**, 255–258.
118. D. D. Lasic, B. Ceh, M. C. A. Stuart, L. Guo, P. M. Frederik, Y. Barenholz, Transmembrane gradient driven phase-transitions within vesicles – lessons for drug-delivery, *Biochim. Biophys. Acta-Biomembr.*, 1995, **1239**, 145–156.
119. D. D. Lasic, Doxorubicin in sterically stabilized liposomes, *Nature*, 1996, **380**, 561–562.
120. S. A. Abraham, D. N. Waterhouse, L. D. Mayer, P. R. Cullis, T. D. Madden, M. B. Bally, The liposomal formulation of doxorubicin, in *Liposomes*, N. Duzgunes (Ed.), Methods in enzymology, Academic Press, New York, 2005.
121. G. Berry, M. Billingham, E. Alderman, P. Richardson, F. Torti, B. Lum, A. Patek, F. J. Martin, The use of cardiac biopsy to demonstrate reduced cardiotoxicity in AIDS Kaposi's sarcoma patients treated with pegylated liposomal doxorubicin, *Ann. Oncol.*, 1998, **9**, 711–716.
122. K. J. Harrington, C. R. Lewanski, A. D. Northcote, J. Whittaker, H. Wellbank, R. G. Vile, A. M. Peters, J. S. W. Stewart, Phase I-II study of pegylated liposomal cisplatin (SPI–077 (TM)) in patients with inoperable head and neck cancer, *Ann. Oncol.*, 2001, **12**, 493–496.
123. K. N. J. Burger, R. W. H. M. Staffhorst, H. C. de Vijlder, M. J. Velinova, P. H. Bomans, P. M. Frederik, B. de Kruijff, Nanocapsules: lipid-coated aggregates of cisplatin with high cytotoxicity, *Nat. Med.*, 2002, **8**, 81–84.
124. A. I. P. M. De Kroon, R. W. H. M. Staffhorst, B. de Kruijff, K. N. J. Burger, Cisplatin nanocapsules, in *Liposomes*, N. Duzgunes (Ed.), Methods in enzymology, 391, Academic Press, New York, 2005.
125. Y. Barenholz, Relevancy of drug loading to liposomal formulation therapeutic efficacy, *J. Liposome Res.*, 2003, **13**, 1–8.
126. W. C. Zamboni, Liposomal, nanoparticle, and conjugated formulations of anticancer agents, *Clin. Cancer Res.*, 2005, **11**, 8230–8234.
127. B. Poolman, J. J. Spitzer, J. A. Wood, Bacterial osmosensing: roles of membrane structure and electrostatics in lipid-protein and protein-protein interactions, *Biochim. Biophys. Acta-Biomembr.*, 2004, **1666**, 88–104.
128. J. H. A. Folgering, P. C. Moe, G. K. S. Wolters, P. Blount, B. Poolman, Lactococcus lactis uses MscL as its principal mechanosensitive channel, *J. Biol. Chem.*, 2005, **280**, 8784–8792.
129. J. H. A. Folgering, J. M. Kuiper, A. H. de Vries, J. B. F. N. Engberts, B. Poolman, Lipid-mediated light activation of a mechanosensitive channel of large conductance, *Langmuir*, 2004, **20**, 6985–6987.
130. A. Kocer, A remote controlled valve in liposomes for triggered liposomal release, *J. Liposome Res.*, 2007, **17**, 219–225.
131. A. Kocer, M. Walko, E. Bulten, E. Halza, B. L. Feringa, W. Meijberg, Rationally designed chemical modulators convert a bacterial channel protein into a pH-sensory valve, *Angew. Chem.-Int. Edit.*, 2006, **45**, 3126–3130.
132. A. Kocer, M. Walko, W. Meijberg, B. L. Feringa, A light-activated nanovalve derived from a channel protein, *Science*, 2005, **309**, 755–758.
133. A. Heeres, C. van der Pol, M. C. A. Stuart, A. Friggeri, B. L. Feringa, J. van Esch, Orthogonal self-assembly of low molecular weight hydrogelators and surfactants, *J. Am. Chem. Soc*, 2003, **125**, 14252–14253.
134. A. Brizard, M. Stuart, K. van Bommel, A. Friggeri, M. de Jong, J. van Esch, Preparation of nanostructures by orthogonal self-assembly of hydrogelators and surfactants, *Angew. Chem. Int. Ed.*, 2008, **47**, 2063–2066.
135. M. Cavazzana-Calvo, S. Hacein-Bey, C. D. Basile, F. Gross, E. Yvon, P. Nusbaum, F. Selz, C. Hue, S. Certain, J. L. Casanova, P. Bousso, F. Le Deist, A. Fischer, Gene therapy of human severe combined immunodeficiency (SCID)-X1 disease, *Science*, 2000, **288**, 669–672.

136. J. B. F. N. Engberts, D. Hoekstra, Vesicle-forming synthetic amphiphiles, *Biochim. Biophys. Acta-Rev. Biomembr.*, 1995, **1241**, 323–340.
137. I. S. Zuhorn, J. B. F. N. Engberts, D. Hoekstra, Gene delivery by cationic lipid vectors: overcoming cellular barriers, *Eur. Biophys. J. Biophys. Lett.*, 2007, **36**, 349–362.
138. I. van der Woude, A. Wagenaar, A. A. P. Meekel, M. B. A. TerBeest, M. H. J. Ruiters, J. B. F. N. Engberts, D. Hoekstra, Novel pyridinium surfactants for efficient, nontoxic in vitro gene delivery, *Proc. Natl. Acad. Sci. USA*, 1997, **94**, 1160–1165.
139. D. Simberg, D. Hirsch-Lerner, R. Nissim, Y. Barenholz, Comparison of different commercially available cationic lipid-based transfection kits, *J. Liposome Res.*, 2000, **10**, 1–13.
140. A. Roosjen, J. Smisterova, C. Driessen, J. T. Anders, A. Wagenaar, D. Hoekstra, R. Hulst, J. B. F. N. Engberts, Synthesis and characteristics of biodegradable pyridinium amphiphiles used for in vitro DNA delivery, *Eur. J. Org. Chem.*, 2002, 1271–1277.
141. C. Madeira, L. M. S. Loura, M. Prieto, A. Fedorov, M. R. Aires-Barros, Liposome complexation efficiency monitored by FRET: effect of charge ratio, helper lipid and plasmid size, *Eur. Biophys. J. Biophys. Lett.*, 2007, **36**, 609–620.
142. B. C. Ma, S. B. Zhang, H. M. Jiang, B. D. Zhao, H. T. Lv, Lipoplex morphologies and their influences on transfection efficiency in gene delivery, *J. Control. Release*, 2007, **123**, 184–194.
143. F. X. Shi, D. Hoekstra, Effective intracellular delivery of oligonucleotides in order to make sense of antisense, *J. Control. Release*, 2004, **97**, 189–209.
144. F. X. Shi, A. Nomden, V. Oberle, J. B. F. N. Engberts, D. Hoekstra, Efficient cationic lipid-mediated delivery of antisense oligonucleotides into eukaryotic cells: down-regulation of the corticotropin-releasing factor receptor, *Nucleic Acids Res.*, 2001, **29**, 2079–2087.
145. F. X. Shi, L. Wasungu, A. Nomden, M. C. A. Stuart, E. Polushkin, J. B. F. N. Engberts, D. Hoekstra, Interference of poly(ethylene glycol)-lipid analogues with cationic-lipid-mediated delivery of oligonucleotides; role of lipid exchangeability and non-lamellar transitions, *Biochem. J.*, 2002, **366**, 333–341.
146. I. S. Zuhorn, R. Kalicharan, D. Hoekstra, Lipoplex-mediated transfection of mammalian cells occurs through the cholesterol-dependent clathrin-mediated pathway of endocytosis, *J. Biol. Chem.*, 2002, **277**, 18021–18028.
147. J. S. Shin, S. N. Abraham, Cell biology – Caveolae – Not just craters in the cellular landscape, *Science*, 2001, **293**, 1447–1448.
148. S. Hama, H. Akita, R. Ito, H. Mizuguchi, T. Hayakawa, H. Harashima, Quantitative comparison of intracellular trafficking and nuclear transcription between adenoviral and lipoplex systems, *Mol. Ther.*, 2006, **13**, 786–794.
149. D. D. Lasic, H. Strey, M. C. A. Stuart, R. Podgornik, P. M. Frederik, The structure of DNA-liposome complexes, *J. Am. Chem. Soc*, 1997, **119**, 832–833.
150. J. O. Radler, I. Koltover, T. Salditt, C. R. Safinya, Structure of DNA-cationic liposome complexes: DNA intercalation in multilamellar membranes in distinct interhelical packing regimes, *Science*, 1997, **275**, 810–814.
151. I. S. Zuhorn, U. Bakowsky, E. Polushkin, W. H. Visser, M. C. A. Stuart, J. B. F. N. Engberts, D. Hoekstra, Nonbilayer phase of lipoplex-membrane mixture determines endosomal escape of genetic cargo and transfection efficiency, *Mol. Ther.*, 2005, **11**, 801–810.
152. I. Koltover, T. Salditt, J. O. Radler, C. R. Safinya, An inverted hexagonal phase of cationic liposome-DNA complexes related to DNA release and delivery, *Science*, 1998, **281**, 78–81.
153. M. Scarzello, V. Chupin, A. Wagenaar, M. C. A. Stuart, J. B. F. N. Engberts, R. Hulst, Polymorphism of pyridinium amphiphiles for gene delivery: influence of ionic strength, helper lipid content, and plasmid DNA complexation, *Biophys. J.*, 2005, **88**, 2104–2113.
154. J. Smisterova, A. Wagenaar, M. C. A. Stuart, E. Polushkin, G. ten Brinke, R. Hulst, J. B. F. N. Engberts, D. Hoekstra, Molecular shape of the cationic lipid controls the structure of cationic lipid/dioleoylphosphatidylethanolamine-DNA complexes and the efficiency of gene delivery, *J. Biol. Chem.*, 2001, **276**, 47615–47622.
155. S. May, A. Ben-Shaul, Modeling of cationic lipid-DNA complexes, *Curr. Med. Chem.*, 2004, **11**, 151–167.

156. I. S. Zuhorn, V. Oberle, W. H. Visser, J. B. F. N. Engberts, U. Bakowsky, E. Polushkin, D. Hoekstra, Phase behaviour of cationic amphiphiles and their mixtures with helper lipid influences lipoplex shape, DNA translocation, and transfection efficiency, *Biophys. J.*, 2002, **83**, 2096–2108.
157. R. P. Balasubramaniam, M. J. Bennett, A. M. Aberle, J. G. Malone, M. H. Nantz, R. W. Malone, Structural and functional analysis of cationic transfection lipids: the hydrophobic domain, *Gene Ther.*, 1996, **3**, 163–172.
158. P. C. Bell, M. Bergsma, I. P. Dolbnya, W. Bras, M. C. A. Stuart, A. E. Rowan, M. C. Feiters, J. B. F. N. Engberts, Transfection mediated by gemini surfactants: engineered escape from the endosomal compartment, *J. Am. Chem. Soc*, 2003, **125**, 1551–1558.
159. L. Wasungu, M. C. A. Stuart, M. Scarzello, J. B. F. N. Engberts, D. Hoekstra, Lipoplexes formed from sugar-based gemini surfactants undergo a lamellar-to-micellar phase transition at acidic pH. Evidence for a non-inverted membrane-destabilizing hexagonal phase of lipoplexes, *Biochim. Biophys. Acta-Biomembr.*, 2006, **1758**, 1677–1684.
160. L. Wasungu, M. Scarzello, G. van Dam, G. Molema, A. Wagenaar, J. B. F. N. Engberts, D. Hoekstra, Transfection mediated by pH-sensitive sugar-based gemini surfactants; potential for in vivo gene therapy applications, *J. Mol. Med.*, 2006, **84**, 774–784.
161. R. Ghirlando, E. J. Wachtel, T. Arad, A. Minsky, DNA packaging induced by micellar aggregates – A novel invitro DNA condensation system, *Biochemistry*, 1992, **31**, 7110–7119.
162. M. A. Lysik, S. Wu-Pong, Innovations in oligonucleotide drug delivery, *J. Pharm. Sci.*, 2003, **92**, 1559–1573.
163. T. Pott, D. Roux, DNA intercalation in neutral multilamellar membranes, *FEBS Lett.*, 2002, **511**, 150–154.
164. M. I. Angelova, N. Hristova, I. Tsoneva, DNA-induced endocytosis upon local microinjection to giant unilamellar cationic vesicles, *Eur. Biophys. J.*, 1999, **28**, 142–150.
165. S. Nomura, K. Tsumoto, T. Hamada, K. Akiyoshi, Y. Nakatani, K. Yoshikawa, Gene expression within cell-sized lipid vesicles, *ChemBioChem*, 2003, **4**, 1172–1175.
166. I. S. Zuhorn, W. H. Visser, U. Bakowsky, J. B. F. N. Engberts, D. Hoekstra, Interference of serum with lipoplex-cell interaction: modulation of intracellular processing, *Biochim. Biophys. Acta-Biomembr.*, 2002, **1560**, 25–36.

17

Reactions in Liposomes

Pasquale Stano and Pier Luigi Luisi
Biology Department; University of Roma Tre; 00146 Rome, Italy

17.1 Introduction

Compartmentalized reactions belong to a class of molecular reacting systems, which play a fundamental role in our understanding of cellular life. In fact, the most striking property of living systems is their cellular nature, and the fact that life is based on reactions that are organized by a functional as well as spatial point of view. In the last decades, cellular biology has provided a great amount of evidences about the structure and dynamics of living cells, by *in vivo* or *ex vivo* investigations. Such approach – that can be defined as 'analytical' – has been germane to our comprehension of several mechanisms, but is, in some aspects, not sufficient. There are two reasons that lie at the basis of modern studies on compartmentalized reactions: the first is related to the need of a complementary approach, that we may call 'synthetic' or 'constructive',[1–3] since it tries to understand cells by constructing more or less elaborated models of them. This point of view can be summarized by the famous Feynman's sentence 'What I cannot create, I do not understand'. The second reason is related to the tradition of artificial life and origins of life studies, where great efforts have been carried out to produce synthetic analogues of living cells, or to reproduce – at least as proof of principle – the steps that led to the transition from inanimate matter to the first living protocell. It is within this conceptual framework that the study of reactions in compartments (and inside lipid vesicles, in particular) have currently witnessed an enormous rise of interest, that has also catalyzed an ever-growing number of scientists involved in such research. Synthetic biology,[4–6] a new-born subject that combines molecular biology with an engineering approach, can be perhaps considered as the proper field where microcompartmentalized reactions belong to, even if some conceptual differences have been pointed out.[1,3] More in general, however, we can look at these studies as investigations on molecular self-organization and on the physical-molecular basis of living process.

In this chapter, we will introduce the basic concepts of reactions inside vesicles, describing first the self-assembly nature of lipid and fatty acid vesicles, which are very important cell models (it has been suggested that fatty acid vesicles represent the most plausible candidates for protocells). Then, the experimental strategies and some theoretical aspects will be presented and discussed, in order to give a general view on the subject. In the last part, we will discuss some specific examples of reactions in liposomes, giving much emphasis to the technical advancements and open questions, but commenting also on the relevance of such studies for synthetic biology and origins of life. More in particular, we will focus on the issue of 'semi-synthetic minimal cell',[7] that we – and others – are currently pursuing and developing.

17.2 Lipid Vesicles (Liposomes)

17.2.1 What are Liposomes (and Fatty Acid Vesicles)

Lipid vesicles (namely, liposomes), as well as fatty acid vesicles, form spontaneously in aqueous solution by a self-assembly process. The building blocks of liposomes are natural or synthetic phospholipids (Figure 17.1A), which are water-insoluble amphiphilic molecules formed generally by two hydrophobic chains and a polar (charged) headgroup. The formation of bilayers by phospholipids is qualitatively predicted by geometrical considerations based on the so-called surfactant parameter.[8] In the case of fatty acids (Figure 17.1B), bilayers form only within a determined pH range (for oleic acid vesicles, around pH 8.5 ± 1 pH unit), because the protonation state of the carboxylate group strongly affects the self-assembly properties of such molecules. At high pH, fatty acids, being deprotonated, assemble as micelles; at low pH, fatty acids separate from the solution as oil droplets. At intermediate pH values, fatty acids and their salts coexist in the membrane, which results stabilized by a network of hydrogen bonds.[9] The fact that fatty acids assemble as bilayers at intermediate pH and as micelles at high pH can be viewed and explained as a change of surfactant parameter (from cylidrical to conical shape). Phospholipid vesicles and fatty acid vesicles have several common properties, but differ in some key aspects: for example, both form only above critical aggregate concentrations (typically 10^{-10} M for phospolipids, and around 10^{-4}–10^{-2} M for fatty acids, depending on their chain length). Fatty acids, in fact, are more soluble than phospholipids, due to their molecular structure, which presents only one hydrophobic chain instead of two, and a polar and exposed ionic head group. This difference affects also the molecular dynamics of the systems: phospholipid membranes are more 'frozen', whereas fatty acid membranes are more 'dynamic'.

It is important to remark that the formation of membrane bilayers is guided by the entropically driven hydrophobic effect.[10] Spherical or spheroidal liposomes, which form

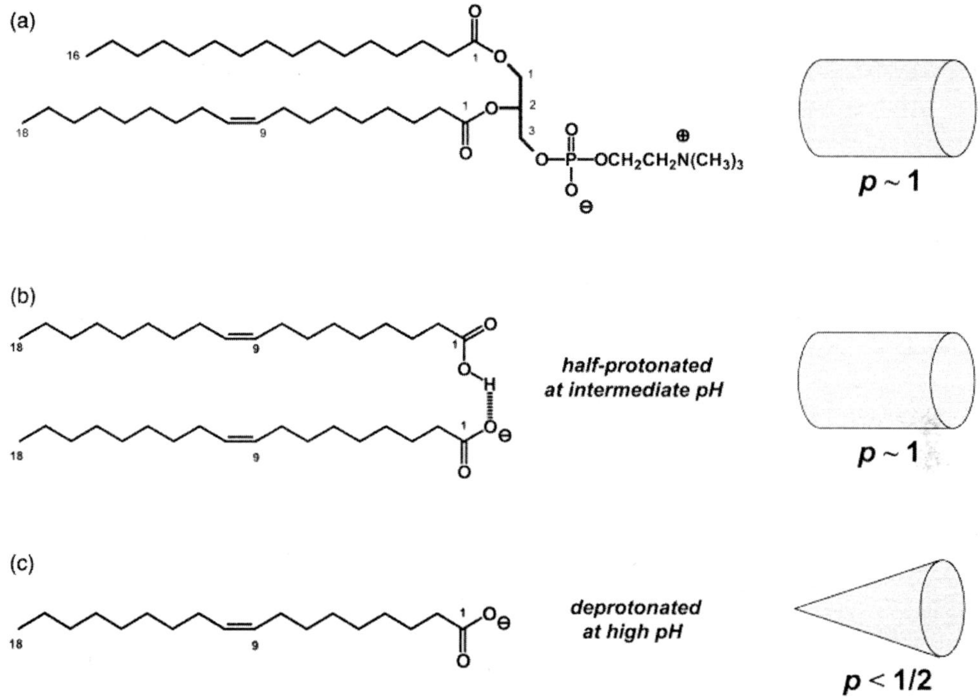

Figure 17.1 Molecular structures of 1-palmitoyl-2-oleoyl-sn-glycero-3-phosphatidylcholine (A); oleic acid/oleate dimer (through hydrogen bond) (B); oleate (C). The values of surfactant parameters are shown on the right, as well as the approximate surfactant shape. Surfactants with cylindrical geometry ($p \sim 1$) generally form bilayer, and therefore vesicles; surfactants with conical geometry ($p < 0.5$) generally form micelles. Notice that oleate may have different p values according to the pH. When the ionization degree is about 50%, oleate bilayers are stabilized by a network of intermolecular hydrogen bonds. Other fatty acids behave similarly

by different mechanisms,[11] differ in their size and morphology, mainly depending on the chemical nature of lipids, physico-chemical parameters (temperature, ionic strenght, pH, etc), and also according to the method of liposome preparation, as will be discussed later. In Figure 17.2, some typical size/shape of liposomes are shown.

Two important aspects of liposome formation must be emphasized here. First, in some cases we observe the formation of ordered supramolecular structures starting from a chaotic disordered mixture of surfactants (as in the ethanol injection method[12]). As noticed before, this increase of order is attended by a simultaneous increase of water entropy and a decrease of overall free energy (lipids *and* solvent). Secondly, every time a liposome forms, there is the 'emergence' of a division, with an inside world that is different from the external environment, even if the two worlds actually interact with each other. The discrimination between inside and outside, applicable to lipid vesicles, is the first structural pre-requisite for the living cell. It is therefore clear that lipid or fatty acid vesicles may be considered relevant experimental model of simplified cells, and their role on

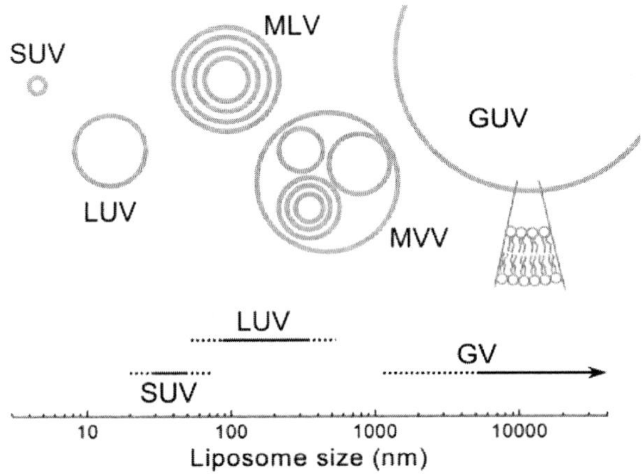

Figure 17.2 *Size and morphology of vesicles. Small unilamellar vesicles (SUVs) are generally small (<50 nm); large unilamellar vesicles (LUVs) have size between 50 nm and few hundreds nanometers; multilamellar vesicles (MLVs) have an indefinite number of concentric lamellae and are generally larger than LUVs; multivesicular vesicles (MVVs) are formed by several smaller vesicles (uni-, oligo-, or multilamellar) within the cavity of a larger (uni-, oligo-, or multilamellar) vesicle. Giant vesicles are very large vesicles (in the micrometer range) of various lamellarity. If unilamellar, they are called giant unilamellar vesicles (GUVs). It is often impossible to fix a size limit to each vesicle class; different size and morphologies are often possible*

origins of life is today greatly acknowledged. Further discussion on the emergence of compartments and their relevance to the origins of life, can be find in Luisi (2006).[1]

17.2.2 Morphology and Methods of Preparation

Liposomes can be prepared according to different methods. Classical discussions[11,13] focus on the preparation of phospholipid vesicles, whereas the preparation of fatty acid vesicles has been reviewed only recently.[14]

The key point is that different preparation methods produce different vesicle samples, in terms of size, morphology and size distribution (this observation provides a further confirmation on the non-equilibrium nature of vesicle samples). A collection of vesicle types is shown in Figure 17.2. Unilamellar vesicles (UVs) are generally classified as small, large or giant ones (SUVs, LUVs, or GUVs, respectively); SUVs have very small diameters (<50 nm), whereas LUVs' diameters span from 50 to about 400 nm. Notice that it is difficult to produce large vesicles which are truly unilamellar; very often they are bi- or oligo-lamellar. Giant vesicles are very large vesicles (>1 μm) that can be visualized by a light microscope; they are generally oligo- or multilamellar, but thanks to special methods,[15] it is possible to produce GUVs. Unilamellar vesicles are the most useful vesicles since the dependence of their properties from the number of lamella has been eliminated. As a consequence, it is expected that their physico-chemical properties depend mainly from their size. Moreover, diffusion of solutes inside unilamellar vesicles is simpler than in other cases, and can be therefore clearly explained. Unfortunately, most of the common preparation methods – as we will see later – do not produce unilamellar vesicles. The typical outcome of lipid hydration procedures is a sample of oligo- or multilamellar vesicles (OLVs, MLVs), or more complex structures as multivesicular vesicles (MVVs). Such particles, which have a variable morphology and an unpredictable number of lamellae can be used for some applications, but in most refined studies are first converted into unilamellar ones. It is remarkable that unilamellar vesicles have of course higher total entrapped volume when compared to OLVs, MLVs, or MVVs of same size, since the architecture of each compartment involves the smallest possible lipid quantity, and therefore the number of compartments is higher than the opposite case (in MLVs, several lipid bilayer are 'wasted' to build the compartment boundary). It is therefore clear that unilamellar vesicles are the preferred containers to carry out microcompartmentalized reactions.

Let us discuss briefly the main preparation methods of phospholipid vesicles (Figure 17.3). The formation of vesicles takes place when lipids are dispersed in water; the different methods differ for the way this dispersion occurs.

In the thin film hydration method, lipids are first stratified on a very large surface, in order to make a very thin film (generally in a round-bottom glass flask). Within the thin film, which form by evaporation from a lipid/solvent solution, lipids are already organized as bilayers. After hydration of the film, large vesicle (MLVs) arise, entrapping solvent and solutes in their aqueous core and between lamellae. Vesicles are then processed in order to reduce lamellarity and size, to produce SUVs or LUVs. This process is generally done by freezing and thawing the vesicles in liquid nitrogen, followed by extrusion; alternatively, and at the same aim, vesicles can also be treated with ultrasounds. If the lipid film is done on rod electrodes, and hydration is accomplished by applying alternate voltage, GUVs emerge from the electrodes surface.[15,16]

The reverse-phase evaporation method[17] starts from water-in-oil (w/o) compartments, obtained by dispersing a small amount of an aqueous solution (the same that needs to be entrapped in the vesicles) in a lipid-containing solution, such an ether/chloroform solution. Lipids stabilize the formation of such droplets. After solvent evaporation the compartments collapse to give first a gel and then a dispersion of uni- and oligolamellar vesicles, with size ranging from 100 to 500 nm. The advantage of this method is the high entrapment yield (notice that the starting point is already a '100% inside' compartmentalized state); the disadvantage is the use of organic solvents, which may be dangerous for some solutes (e.g. folded proteins).

The detergent depletion method also starts from a special organization form: the mixed micelles. When lipids are mixed with excess detergent in aqueous solution, lipids are 'solubilized' in the detergent micelles. When detergent is slowly removed by gel filtration chromatography or dialysis, the mixed micelles transform into phospholipid vesicles. The method works because the detergent molecules are also present as monomer, in equilibrium with the micelle, and the exchange of aqueous phase brings to a slow detergent depletion via monomer removal. The method produces SUVs or LUVs, entrapping the solutes present in the aqueous phase used to remove the detergent. The high concentration of detergents, often needed during the preparation step, may be a disadvantage when sensitive solutes must be entrapped.

With the ethanol (or methanol) injection method[12] a small aliquot of an alcoholic lipid solution (where lipids are likely present as monomer species) is simply injected into a

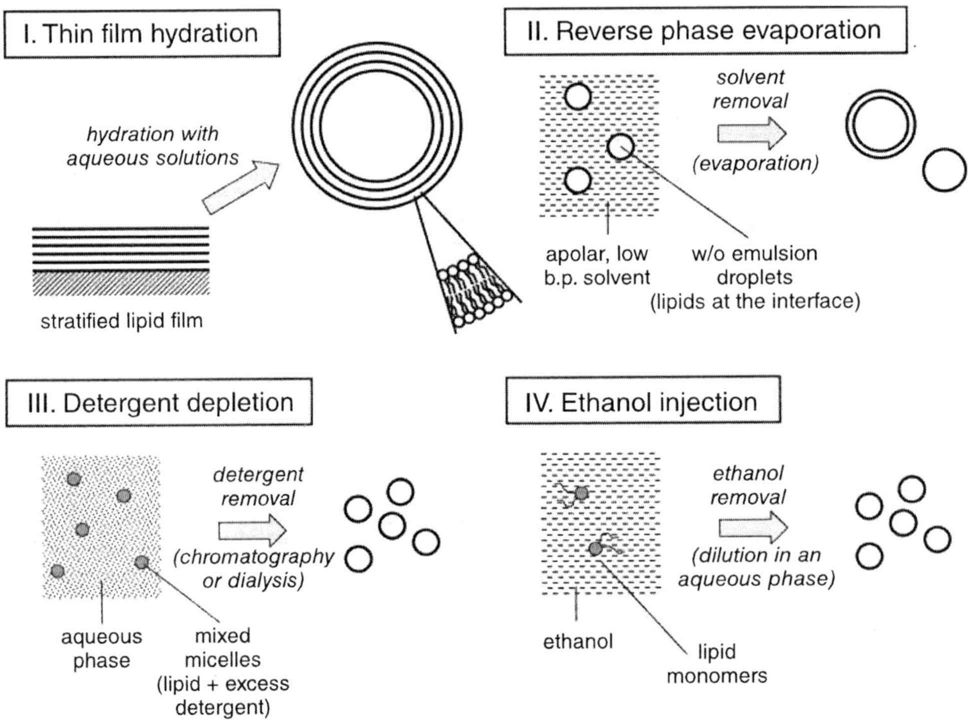

Figure 17.3 *Four classical methods of liposome preparation. (I) Thin film hydration method. Lipids are solubilized in apolar solvents (chloroform, ether) and stratified as dry thin film on a solid support (generally glass). After hydration and shaking, liposomes form (very often MLVs MVVs). Formation of GVs by natural swelling method also require the formation of stratified lipids on the wall of container; GVs electroformation starts from stratified lipids on electrodes; (II) Reverse phase evaporation. To a solution of lipids in chloroform/ether, a small aliquot of aqueous phase is added and sonicated, in order to form a water-in-oil emulsion. After controlled solvent evaporation, a gel form, that collapses upon vortexing to a suspension of vesicles. The overall process leads to high entrapment yields. Intermediate-size vesicles are often formed. (III) Detergent depletion method. A solution of mixed micelles (lipid plus excess detergent) is processed so that the detergent amount is reduced, by dialysis, gel filtration chromatography or by dilution. Liposomes form slowly due to the low final detergent/lipid ratio. Typically, SUVs or LUVs are generated. This method is often used to reconstitute membrane proteins in liposomes, starting from mixed micelles containing a membrane protein. (IV) Ethanol injection method. A small aliquot (1–5% v/v) of a lipid solution in ethanol or methanol in injected in an aqueous solution. Liposomes form immediately due to ethanol dilution in water phase. The liposome size depends on the lipid concentration in alcohol. Working with diluted solution, it is possible to form SUVs. In all methods, liposomes entrap the water-soluble compounds that present in the aqueous solution. High entrapment yields are obtained by rehydrating freeze-dried liposomes. This method is reminescent of method I, but starts from lipid preorganized in different way (freeze-dried liposomes probably have partially conserved lipid bilayers)*

large amount of aqueous solution, containing the solute(s) to be entrapped. Ethanol mixes with water and liposome forms within the mixing time. Interestingly, such method provides a one-step way to form unilamellar vesicles (SUVs or LUVs), and the main factor governing the size and morphology of resultant vesicles is the lipid concentration in ethanol.[18,19] The presence of ethanol (or methanol) in the final dispersion may be disadvantageous, but in many cases it is limited to few units percent.

In addition to these classical methods, two special methods must be mentioned, due to their advantageous entrapment yield. The first is based on the rehydration of freeze-dried liposomes, which are in turn prepared from 'empty' liposomes or from liposomes that already contain the solute(s) of interest.[20–22] Upon rehydration, freeze-dried vesicles reform and entrap the solute in high concentration. The second is a new method for producing GVs starting from large w/o emulsion droplets, which are centrifuged through a lipid-rich macroscopic oil–water interface. The w/o droplets become covered with an external lamella during the passage through the oil–water interface, and are transformed into GVs, with the advantage of keeping inside their content.[23] Many of the most recent works on biocomparmentalized reactions are carried out in lipid vesicles formed by the latter two methods.

Fatty acid vesicles, which are stable only in a rather restricted pH range, can be prepared: (1) by direct dispersion of fatty acids or their salts in an aqueous phase; (2) by the film method (when feasible); or (3) by acidification of an alkaline fatty acid salt micelles solution. In general, all preparation methods lead to a heterogeneous vesicle population; in special conditions, however, a narrowly sized vesicle population can be obtained by an autocatalytic process.[24,25]

As a final remark, it is clear that the choice of liposome preparation method depends on the experiment's aim. Since we are dealing with compartmentalized reactions, it is often convenient to reach the maximum entrapment efficiency, in order to have 'filled' vesicles. We will see in next section what are the general strategies to entrap chemicals inside vesicles, and how to feed them with an externally added reactant. Another important issue, discussed below, is the chemical compatibility between the lipids, the preparation method, and the (bio)chemicals used in the experiments.

We conclude this survey on vesicle preparation methods by noticing that there are not standard method to prepare unilamellar vesicles in the range 0.5–1 μm, although this size would be rather convenient for carrying out compartmentalized reactions (large trapped volume, possibility to employ modern visualization techniques, or high-throughput techniques).

17.2.3 Chemical Compatibility

There are two aspects pertinent to chemical compatibility and the design and the realization of reactions inside vesicles. The first concerns the chemical structure of lipids (or fatty acids, or other co-surfactants, as ammonium salts, derivatized lipids, or sterols), the second is related to the preparation method.

Among phospholipids, phosphatidylcholine (lecithin) is certainly the less reactive, whereas anionic phospholipids such as phosphatidylglycerol, phosphatidylserine, and phosphatidic acid are of course sensitive to cations (e.g., Ca^{2+}, Mg^{2+}; phosphatidic acid is sensitive to H^+ too), to polyamines (spermine, spermidine, etc.), or basic proteins. Fatty acids vesicles, stable at weakly alkaline pH, are also negatively charged, and therefore sensitive to H^+ (a pH increase or decrease transform fatty acid vesicles into micelles and oil droplets, respectively), to divalent cations, and to basic proteins.[26] The latter point deserves a particular mention since it may prevent the use of fatty acid vesicles as compartment for Mg^{2+}-dependent biochemical reactions. To date, the best way to make fatty acid vesicles compatible with such kind of chemicals is the addition of fatty acids–glycerol monoesters.[27,28] Cationic surfactant, generally based on ammonium salts, may be not compatible with anionic solutes (calcein, a classical water-soluble negatively charged probe used in vesicle research, interact with some cationic surfactant), and with polyanions like RNAs and DNAs. It has been recently reported that cell-free protein expression is inhibited by cationic liposomes[29] in a dose-dependent manner.

The preparation method may also present some constrains. For example, the thin film hydration is the only method which does not require the presence of additional compounds (reverse phase evaporation is carried out in the presence of organic solvents; the detergent depletion in the presence of detergent; and the ethanol injection, in the presence of ethanol) and it is therefore the only method of general use. However, it requires additional steps for vesicle processing because it produces a heterogeneous population of large MLVs. The reverse phase evaporation method, which is very convenient in terms of entrapment efficiency (10–40%), suffers of the presence of organic solvents, so that proteins or other bio-supramolecular complex may lose their activity. Such preparation method, however, is applicable to inorganic or simple organic reactions. Similar considerations apply to detergent-based methods and ethanol injection procedures. It has been reported that ethanol (3.5% v/v) and methanol (5% v/v) halves the yield of cell-free protein expression.[30]

In conclusion, chemical compatibilities must be considered at the very beginning in the process of designing a microcompartmentalized bioreactor, and this will affect the choice of lipids as well as the preparation procedure.

17.3 Experimental Strategies and Theoretical Aspects

17.3.1 Basic Strategies

The result of the preparative methods sketched above is a vesicle suspension, where individual vesicles may or may not contain the solute(s) of interest. We should now perform a reaction inside the vesicle, and at this point, two different approaches must be distinguished. One is the case where the resulting vesicles still miss one (or more) reactant(s) necessary for the reaction under study (Figure 17.4A), whereby the missing reactant(s) is deliberately not included in the hydrating solution. The other is the case where all reactants are present in the hydrating solution, so that fully 'functional' vesicles may be present in the sample (Figure 17.4B). In the first case the compartmentalized reaction cannot start until the missing reactant is added; in the second case, when all reactants are provided from the beginning, one must find some way so that the reaction does not start right away, namely it should start only when the compartmentation of all components has taken place. Notice, in fact, that in order to follow the reaction inside vesicles, in both strategies, the external – the non-entrapped – material should be removed, destroyed or inhibited.

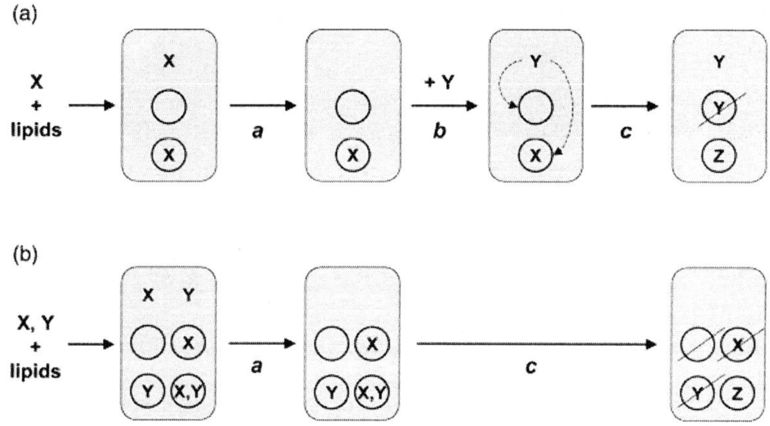

Figure 17.4 *Two general strategies to carry out a compartmentalized reaction 'X + Y = Z'. (A) Liposomes are formed in the presence of X, which is entrapped; then the free (unentrapped) X is removed, and later the second reactant Y is added. (B) Liposomes are formed in the presence of reactants X and Y, unentrapped material is removed (or a not-permeant inhibitor is added). The reaction is then started by physical means (e.g., temperature, light). X and Y may represent a single chemical specie, or a collection of reactants. When two or more reactants are simultaneously entrapped, the issue of composition of individual liposome may become relevant for interpretation of data. In panel B, for example, not all liposomes contain both X and Y (this condition may become critical for a large number of reactants)*

Let us discuss a general case where the reaction under study is modeled by the equation $X + Y \rightarrow Z$.

The strategy indicated in Figure 17.4A involve the vesicle formation in the presence of the reactant X (let us suppose, for the sake of simplicity; that X is a single chemical species). Depending on the conditions, not all vesicles may contain X, and certainly the numerical distribution of X inside the vesicles is centered around an average value, so that vesicles should be considered heterogeneous in terms of internal composition. We will discuss this theoretical aspect later. The first step is the removal of free (the non encapsulated) X from the system, and this is generally accomplished by dialysis, gel filtration chromatography, ultracentrifugation or ultrafiltration. For successful procedures, the external X and the vesicles are separated from each other due to different physical properties, such as molecular weight, size, or density. An alternative strategy involves the addition of a component that is able to inhibit the external (and only the external) reaction.

Now, in order to make the reaction possible, the second reactant Y must be added. In Figure 17.4A it is shown the addition of Y in the external phase, followed by the diffusion of Y inside the vesicles. The final state is the product Z, synthesized inside the vesicle. This way is the simplest one, but requires that the Y permeability coefficient differs from zero. Moreover, if the entrance of Y in the vesicles becomes rate-limiting, the observed reaction rate will be limited by this slow step.

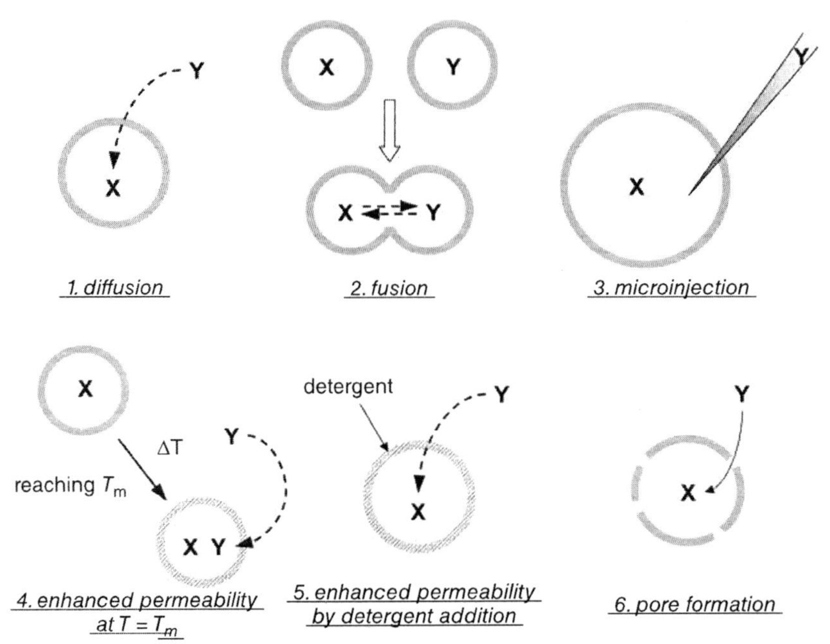

Figure 17.5 *Experimental methods to delivery a second reactant Y inside a X-containing vesicle. (1) Free (passive) diffusion of Y from outside to vesicle inside. (2) Fusion between two or more vesicles. (3) Microinjection of Y inside a giant vesicle. (4) Keeping the vesicles at the phase transition temperature T_m (or by thermal cycles around T_m). The permeability of lipid membranes is generally maximal at T_m. (5) Adding detergents at sublytic concentration, so that the membrane permeability is increased (especially for small solutes) without dramatic changes of membrane integrity. (6) Incorporation of pore-forming compounds in liposomes (α-hemolysin, OmpF porin, ...)*

Figure 17.5 shows some possibilities, alternative to the passive diffusion (Figure 17.5.1). The diffusion of Y into a vesicle which contains already X is useful when the permeability of Y is much higher than X's one. A typical example is the entrapment of an enzyme inside vesicles, and the addition of its substrate(s) – generally small molecules – from outside. For a review on enzymes inside vesicles, strategies, and applications, see.[31]

Fusion among two or more vesicles, each containing different solutes, can be a way to reach the desired reactant mixtures (Figure 17.5.2). Vesicles can fuse under the effect of electrostatic forces (as in the case of cationic and anionic vesicles[32,33]), or by some bridging agent such as Ca^{2+} for phosphatidylserine vesicles,[34] or dehydrating agents like poly(ethylenglycol).[35] Although fusion rarely occurs at high yield, the idea of using vesicle fusion to build artificial cells has been already suggested[1,36] and recently implemented by means of fusogenic peptides.[37]

The technique of microinjection (Figure 17.5.3) bypasses the problem of low membrane permeability, or even the entire entrapment step, by direct injecting the reactants inside GVs. This technique has been used in combination with GVs electroformation, as an approach to minimal cells.[16]

Another method relies on thermal properties of lipid membranes. Every lipid bilayer is characterized by a phase transition, from solid-like to liquid-like states. It is known that at the transition temperature (called 'melting' temperature, T_m), the bilayer permeability is maximal.[38] It is then possible to exploit this higher permeability to allow solutes pass through the membrane (Figure 17.5.4). It has been reported, for example, that ATP permeability increase by a factor ~100 in dimyristoyl phosphatidylcholine vesicles,[39] and that such vesicles can therefore be fed with ATP by keeping them at the T_m (23.3 °C). It has been suggested that membrane defects are involved in the mechanism of enhanced permeability. Recent applications of this method have been reported.[40,41] This physical principle can be generalized – for instance – as thermal cycles around T_m.[42] Notice that the application of freeze-thawing cycles (in liquid nitrogen) is a common laboratory practice for equalizing internal and external liposome content. Although it likely proceeds by rough membrane integrity destruction, the concept of permeation of solute through 'damaged' membrane is somehow similar to the case of permeation at the melting point.

Another possible way to reduce the barrier to free diffusion provided by lipid bilayer involves the use of detergents at sub-lytic concentrations, as membrane dopants (Figure 17.5.5). An example is given by the cholate-induced ADP[43] or glucose-1-phosphate entrance[44] inside 1-palmitoyl-2-oleoyl-phosphatidylcholine (POPC) vesicles. For example, the permeability of ADP can be increased by a factor 10, when cholate/POPC = 4/5 (mol/mol).

Finally, a more elaborate molecular approach has been recently developed, as shown in Figure 17.5.6. It makes use of natural pore-forming molecules, such as porins,[45–48] or α-hemolysin.[49]

The advantage of such approach is the possible selective effects deriving from the molecular weight cut-off, set by the pore size. It is then possible that small molecules may penetrate and react into the vesicles, whereas large macromolecules (such as enzymes, nucleic acids) are retained inside.

After this discussion on the methods to feed vesicles by externally added reactants, let us focus again of Figure 17.4B where an alternative way to the first route (Figure 17.4A) is shown: the encapsulation – from the beginning – of all required reactants inside vesicles. As in the previous case (Figure 17.4A), the first step is the removal of external reactants, or the inhibition of external reaction. Notice that this step becomes now critical, since all reactants are present in the vesicles *as well as* in the external phase. It is clear that such approach can be followed only if there is a way to trig the reaction by some physical mean. A classical example is the vesicle entrapment of the whole biochemical machinery for transcription and translation reactions (from DNA to messenger RNA, and from messenger RNA to protein, respectively). The reactions do not proceed at any significant extent at low temperatures (0–5 °C) and can be triggered by incubating the sample at 37 °C. Then the sample, kept at low temperature, is quickly manipulated by adding an inhibitor (that is not capable to enter the vesicle), or by exchanging the external phase by spin column gel chromatography, and later incubated at high temperature. In addition to this example of thermal trig, when possible, photochemical triggering can also be an opportunity.

The final step of every experiment with compartmentalized reactions is the product detection. This depends on the nature of the products, and in some cases by those of the reactants. It is evident that any analytical technique can be used in this respect, and direct chemical analysis should be preferred. When, however, real-time analysis is required, only non-destructive methods can be applied, and in this respect the use of optical methods is very intriguing, but it also requires some caution. Lipid and fatty acid vesicles, in fact, strongly interfere with the spectrophotometric or spectrofluorimetric detection of analytes. The discussion of this issue lies outside the purpose of this chapter, and the reader is referred to specialized literature.

17.3.2 Theoretical Aspects

In this section we will discuss briefly some theoretical aspects related to the design and realization of compartmentalized reactions, mainly derived from our experience with biochemical systems. As a general premise, we remark that although an ever increasing number of published work on this subject (mono- or multi-enzymatic reactions inside liposomes, fatty acid vesicles, polymersomes,[50,51] as well as w/o[52] or w/o/w compartments[53]) a full understanding of the physico-chemical details of compartmentalized reactions has not been reached. A related field, which has a longer tradition, i.e. that one of enzymes in reverse micelles has revealed interesting compartmentation-dependent mechanisms.[54]

The main question is if – and to what extent – a reaction occurring inside a micrometric or submicrometric compartments differs from the same reaction in the bulk, and whether the formation of the compartment itself brings about novel effects. Clearly, these theoretical questions have great relevance in basic science, since the basic unit of life is itself a compartment (the biological cell), as well as in applications of compartmentalized systems for biotechnology (e.g. artificial cells, biosensors, etc.).[55]

We have recently discussed some of these theoretical issues in a publication.[36] Here we present some arguments on: (*i*) the local concentration of compartmentalized substances; (*ii*) the entrapment of solutes; (*iii*) the reactivity of solutes inside compartments. The reader should understand these discussions not as conclusive, but as a contribution to the ongoing research, and maybe feel stimulated to make research efforts in these directions.

(*i*) *Local concentration.* When a real vesicle sample is prepared in order to carry out internal molecular reactions, it is often useful to know what is the concentration of solutes inside the aqueous phase present in the interior of vesicles. The experimentally accessible quantity, however, is generally the total amount of solute (e.g., an enzyme) entrapped inside the total volume of all vesicles. In fact, with the exception of very large compartments as GVs and w/o or w/o/w compartments, it is generally difficult to determine the individual composition of each vesicle. Electron microscopy may help in some special cases,[56,57] and modern flow-cytometric techniques have been applied only very recently, and for large vesicles.[58] Complex analytical methods have been also used;[59] whereas the large size of GVs favours the direct analysis, so that concentration hetereogeneity has been promptly observed.[60] The interest toward the determination of individual vesicle content is justified by the fact that the theoretical treatment based on the concept of 'water-pool' – or pseudo-phase – introduced for reverse micelle systems, cannot be applied in the case of vesicles. In fact, whereas in reverse micelles the highly dynamic nature of such compartment (that allows solute exchange) average out the individual micelle diversity, liposomes and fatty acid vesicles are characterized by a proper individuality. Far for being a disadvantage, such individuality is an important prerequisite for their use as cell models and for their relevance as plausible protocells. Liposomes, in fact, do not fuse in absence of specific triggers,[61] and do not exchange their content. This fact ultimately derives from the well know kinetically-trapped nature of lipid vesicles,[62,63] in contrast to reverse micelles, which are classical equilibrium systems. As a main consequence of this behaviour, it is expected (and observed) that a population of vesicles is heterogenous in terms of internalized solutes. The experimentally observed quantity, however, is often the total amount of entrapped solute, for example: 50 pmoles of a certain enzyme have been entrapped in 1 mL of 10 mM phosphatidylcholine liposomes (radius 50 nm). The *average* enzyme concentration inside liposomes can be calculated, estimating the total entrapped volume of such liposomes (31 μL/mL), to give *ca.* 1.6 μM. However, this average concentration corresponds to an *average* number of 0.5 enzyme molecules/vesicle. Clearly, no half-enzyme molecule will be present in any compartment, and a rough estimate may say that 50% of the compartments will contain no enzyme and the other 50% will contain just one enzyme. The expected number of enzymes inside vesicles can be predicted by means of the Poisson distribution,[30,58,64] which gives, in this case, the probability values reported in Figure 17.6A. We can translate this values in the following physical meaning: ca. 60% of liposomes will be empty, *ca.* 30% will contain 1 enzyme, *ca.* 8% will contain 2 enzymes and so on. The analysis of the reaction, e.g. in this case the enzyme-catalyzed conversion of a substrate into a product, if carried out by averaging techniques ('batch' photometry or fluorimetry, or chemical analysis) cannot reveal this vesicle heterogeneity.

A second theoretical issue on local concentration is trickier and can be discussed as it follows. Chemists normally express molecular concentrations in terms of mole per litre (molarity), a 1 M solution being composed by 6.023×10^{23} molecules L^{-1}. Molar concentration can be *calculated* from the number of molecules present is a certain volume. So, for instance, we can refer again to the above mentioned case, e.g., a system of 2 enzyme molecules in a vesicle volume of 0.5 aL (10^{-18} L); where the *calculated* enzyme concentration is 6.3 μM. We may ask, however, if this tiny 6.3 μM enzyme solution behaves really as a 'normal' bulk solution of the same concentration, but containing of course a much larger number of molecules (10^{18} molecules/L). Clearly, when the number of molecules is so small, and moreover the system is 'confined', classical views on molecular reactivity may require a change in our attitude. For example, in the case of enzymatic reactions, one issue is the frequency of collision between a single enzyme and its substrate (which may also be another macromolecule present itself in few copies), which might result in different reaction rates. It is clear that the chemistry of large numbers breaks down when small compartments are considered. Stochastic simulations clearly show the relevance of fluctuations when we move from the continuum regime of Avogadro-like numbers to the realm of a few molecules[65,66] Discreteness of matter becomes important in compartmentalized reactions, and some unexpected behaviours have already been described (see below).

(*ii*) *Entrapment of solutes*. The discussion on Poisson distribution of previous paragraph can be used as starting point for the issue of solutes entrapment. More in particular, we may ask whether the Poissonian model, which is based on the hypothesis of independent events, apply to real mechanism of vesicle formation and entrapment. When a strong interaction between solutes and the membrane is present (e.g., electrostatic or hydrophobic one), the answer is of course negative, due to the preferential accumulation (or deplenishment) of solutes near the bilayers during the liposome formation. However, the

(a)

μ = 0.5

k	0	1	2	3	4	5	6	7	8
P(k, 0.5)	0.607	0.303	0.076	0.013	0.002	2.E-04	1.E-05	9.E-07	6.E-08

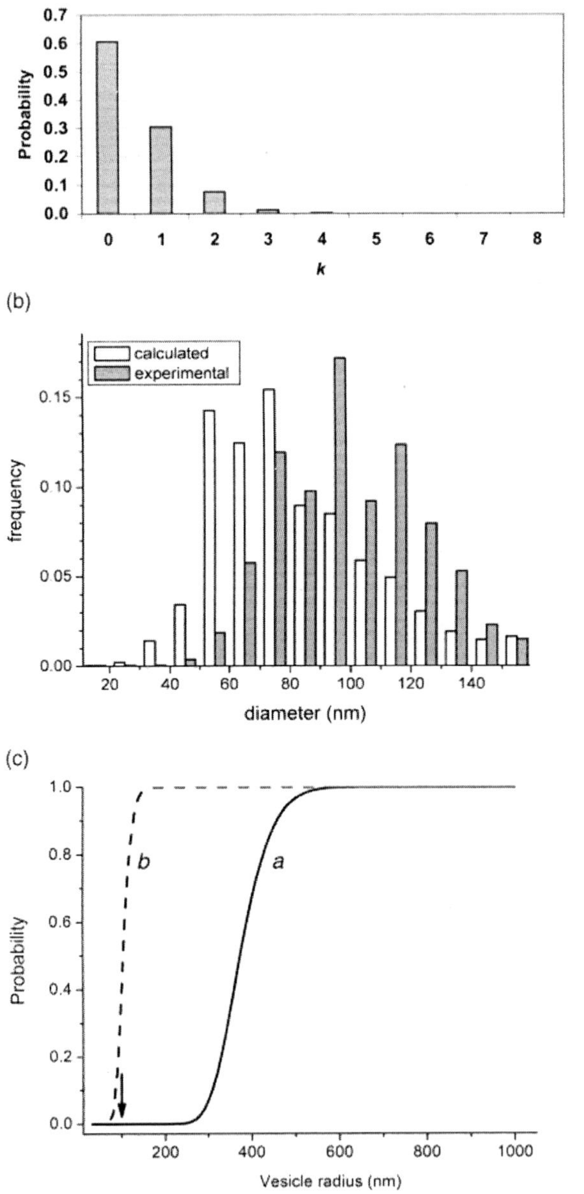

Figure 17.6 *Entrapment statistics. (A) A Poisson distribution may be used to describe the number of solutes entrapped into liposomes. Here it is reported the distribution of a solute which is expected to be find on average as 0.5 molecule/vesicle. Notice that there is about 10% probability of finding* more than two *solute molecule inside the vesicle. (B) Comparison between calculated and experimental distribution of a water-soluble marker (ferritin) inside POPC vesicles. Detailed data analysis shows that in some cases ferritin can be entrapped with efficiency higher than what expected on theoretical basis (Poisson distribution). Data taken from Berclaz et al.*[67] *(C) Probability of co-entrapment of all macromolecular components of transcription-translation kit inside lipid vesicles of a given radius. The entrapment of each molecule is modelled as a poissonian process, and the cumulative probability is calculated as product of probabilities of independent events. The curve (a) indicates the probability of entrapping at least one copy of each molecular specie inside the same vesicle. The curve (b) indicates the probability of entrapping at least one copy of each molecular species under the hypothesis that their concentrations are all 50 times higher than the nominal (bulk) concentrations. Adapted from Souza et al*[30]

question may be relevant also in the absence of strong solute/bilayer interaction. An analyical discussion is very difficult in this case, since – for example – details of liposome formation mechanisms are still missing. Lasic, in his monography,[11] classifies liposome formation mechanisms into two broad classes: (1) budding-off, and (2) collapse/closure of small bilayer fragments. Entrapment of water-soluble molecules occurs during these processes, and therefore further investigations on such dynamics – at the molecular level – would be highly useful for a deeper understanding of entrapment mechanism. From the empirical viewpoint, however, there are a number of experimental evidences suggesting that more intriguing mechanism may bring to *enhanced* entrapment of solutes, caused by possibly cooperative effects.

We were first interested in this idea and we have carried out a careful evaluation of ferritin entrapment in phosphatidylcholine vesicles.[56,57,67] Ferritin is an iron rich protein that can be easily visualized by electronmicroscopy, so that the composition of each vesicles can be simply obtained by visual inspection of microscopic images. The analysis of thousand vesicles leads to the unexpected observation that the numerical distribution of ferritin molecules inside vesicles significantly diverges from theoretical one (Poisson model). Figure 17.6B shows a typical result. More recently, Yomo and co-workers[68] reported that in some condition of liposome formation, genetic material (in form of plasmid) may be found inside liposomes in an amount higher than expected; and a recent report by Keating[69] suggests that molecular crowding may favour super-entrapment inside GVs.

A very interesting case, not yet fully clarified, concerns the simultaneous entrapment of several (>80) macromolecular compounds (the whole transcription-translation machinery) inside submicrometric lipid vesicles (radius ~100 nm).[30] In fact, under the conditions of the experiment, the Poisson probability to find a small vesicles containing more than 80 different compounds is 10^{-26}, i.e., critically close to zero. However, the experimental results indicate a low but well measurable yield of protein produced by the entrapped molecular machinery. Now, the calculated cumulative probability for the entrapment of *ca.* 80 molecules in a vesicle should be the product of 80 independent

Poissonian distribution, and should be size-dependent, as shown in Figure 17.6C, curve a. Large vesicles, in fact, may contain all required ingredients (and be therefore functional), whereas small ones are expected to be not functional. Since the experimental results clash with theoretical predictions, a possible physical explanation suggests that local (internal) enzymes and ribosomes concentrations are enhanced by a factor >10 (Figure 17.6C, curve b). Further studies are in progress in order to clarify the nature of such effect.

(*iii*) *Reactivity inside compartments*. What are the characteristics of the reactivity inside compartments? Can this reactivity be simply the reflection of the enormous body of *in vitro* biochemical data? To date, it is not possible to give a clear answer to this question, since, as evident from the above-mentioned discussion, we still miss a theoretical understanding of the compartmentalized systems, and in particular of the effect deriving from their small size. In this respect, recently some interesting effects have been described, such as the enhanced protein synthesis rate (2× in the first 3 hours) in GVs,[70] or the enhanced filament assembly based on DNA-protein association rate.[40] In the case of protein synthesis in small vesicles (radius ~100 nm), it has been reported that the inner yield is about 6 times the yield of the same reaction in the bulk.[30] These deviations from expected results may perhaps derive from molecular crowding and/or in vitro confinement effects, as also suggested by reports on gene expression in moderately crowded model systems (gene brushes),[71] or enhanced protein association in PEG-based crowded environments, based on depletion forces.[72] In addition to these effects, that may be ascribed to the nature of entrapped solution, it has been also reported that lipid itself (surface effects?) may enhance (1.6×) the *in vitro* gene expression.[73] Although not discussed from the theoretical viewpoint, all these reports indicate a possible existence of a specific effect deriving from confinement and/or crowding. Interestingly, as discussed at the point (*ii*) of this paragraph, enhanced entrapment may also be an outcome of liposome formation and entrapment, at least in some cases. A very complex (yet fascinating) scenario may emerge from these new studies on compartmentalized system. Such intriguing scenario may change our view of reactions in small compartments, and, more in general, of cellular biochemistry. We therefore believe that the next generation of studies in microcompartmentalized systems must take into account the combined effects of single-molecule reactivity, confinement, molecular crowding, excluded volume, and molecular correlation, as well as not-well investigated proteins (or nucleic acids)/membrane interactions.

17.4 A Theoretical Framework for Complex Reactions in Liposomes

After the technical excursus presented in the last paragraphs, we now turn back to important microcompartmentalized biochemical reactions, in particular those relevant for constructing cell models. The reader interested in a more detailed discussion may refer to recently published reviews[7,74]

The research on biochemical reactions inside vesicles has at least three genealogical lines. The first is related to origins of life studies, where vesicles (in particular fatty acid vesicles) are taken as model of primitive cells. In this studies, vesicles are *per se* interesting, since their behaviour may reveal interesting pattern that might be at the basis of the transition from non-living entities to living ones. Moreover, by creating cellular models based on biochemical reactions inside vesicles, it is possible to understand the basic principles of compartmentalized systems, and again get insights into the roots of cellular life. The second genealogical line, which is also somehow related to the first, is more general and deals with the need of realizing (bio)chemical autopoietic systems. Autopoiesis (from Greek 'auto': self; 'poiesis': creation, production) describe the nature of life as 'self-producing' systems, also endowed of basic cognition processes. Autopoietic theory was developed by the two Chilean neurobiologists Humberto Maturana and Francisco Varela in the Seventies, as a theory that aims to describe life emphasizing the very organization of its processes, i.e., the organizational invariance despite the continuous building and destruction of its structural parts; and the concept of autonomy.[75–78] With the realization of surfactant-based autopoietic chemical systems, mainly by us in the early Nineties,[79–81] the concept of creating an autopoietic minimal cell[7,74,82,83] started to attract the interest of several research group, although in a diversified way. The third line is more modern, and deals with the new and broad research avenue that is called synthetic biology. By this constructive approach, scientists tries to understand biological *systems* by building them instead of analyzing them. Today, we are witnessing a great interest in synthetic biology, perhaps due to its biotechnological promises, but we also think that the new constructive view can be useful for basic science and for pure understanding of living principles.

Let us start the discussion from a theoretical perspective, i.e., the autopoietic system, displayed in Figure 17.7a. Here, a general autopoietic system is represented by a circle, the components A being the constituents of autopoietic system. The autopoietic system is characterized by an 'inside' and an 'outside', and these two worlds are separated by a boundary, which is itself generated from the autopoietic system (generation from within). Thanks to its inner organization (not shown), the autopoietic system can process X, present outside initially, but that can enter the autopoietic unit, so that new A is produced, becoming a new part of the autopoietic organization. This first step correspond to generation of autopoietic component from within. In addition to this route, A can be removed from the autopoietic system (A being transformed into W); this second transformation correspond to decay of autopoietic components into non-autopoietic ones. It is remarkable that notwithstanding with these two opposite routes, the autopoietic unit – and, more importantly, its organization, is conserved. When the two routes are also balanced, we can speak of homeostasis. Clearly, when one of the two routes prevails, growth or decay of the autopoietic unit is foreseeable. Moreover, it is clear that the autopoietic system interacts with its environment. Undoubtedly, the autopoietic unit describes how living cells work. But, thanks to this description, it allows to define what are the minimal functions a living cell must have in order to be recognized as alive (for a recent discussion on autopoiesis as necessary and sufficient – or not sufficient – condition for life, see Bitbol and Luisi[78]). It follows that it is a system that '(1) builds its own boundaries, in a way that this construction is due (2) to reaction(s) (activity) taking place within the system, and that it is (3) performed through reactions determined by the system itself'.[84] Thus, such system is autopoietic, and Figure 17.7a, despite its simplicity, gives a very general description of an autopoietic unit. However, in order to translate this description in (bio) chemical terms, we have to think in molecular terms, and we need a more detailed description.

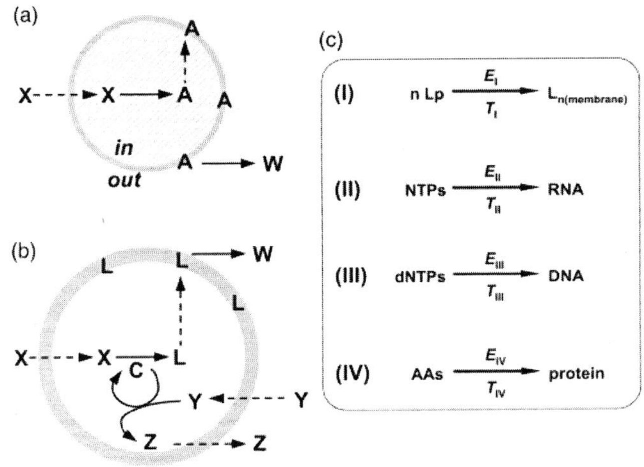

Figure 17.7 *Autopoietic systems and complex biochemical reactions in liposomes. (a) The general scheme for an autopoietic system. The autopoietic unit is composed by the components A, which constitute the autopoietic unit, boundary included. The external component X is incorporated within the autopoietic unit, transformed into an autopoietic element A, which takes now part within the autopoietic organization. According to a decay process, some autopoietic elements A are transformed into W, which does not take part to the autopoietic organization. The autopoietic system has a boundary that discriminates between the unit and the environment. However, thanks to the above mentioned interactions, the autopoietic unit interacts with the environment. (b) A more specific autopoietic design based on surfactant self-assembled structures. The boundary-forming molecular component L is generated by internal transformations starting from a precursor X by means of an internal catalyst or catalytic network C. The subsystem C is also self-reproducing, by transforming building blocks Y into C, whereas Z ones are discarded. A decay reaction also consumes L to give W. This general scheme also describes the cell metabolism in an homeostatic state. (c) Some important biopolymerization reactions which are implemented in minimal cell models, also based on autopoiesis: I. formation of membrane from lipid precursors (a sort of non-covalent polymerization); II. formation of RNA from nucleotides (generally indicated as NTPs); III. formation of DNA from deoxynucleotides (indicated as dNTPs); IV: formation of proteins from amino acids (AAs). These four reactions have been already implemented inside lipid vesicles, see also Table 17.1*

For example, the X to A transformation of Figure 17.7a is described without details, whereas in chemical terms, several molecular transformations require an additional reactant, and in biology, almost all transformation are catalyzed by enzymes. It follows that a slightly more complex description of autopoietic unit is required. In particular, if the X to A transformation (Figure 17.7a) refers to molecular systems, a catalyst C must be introduced in the autopoietic organization. Clearly, the catalyst C, being a single molecule, *or a network of processes carried out by several molecules*, must also be the product of its own activity, as required by autopoiesis.

In Figure 17.7b we represent an autopoietic unit composed by boundary-forming molecules L and by a catalytic system C. The precursor(s) X are now transformed into L molecules thanks to the catalytic activity of C, which is also reproducing itself by uptaking the precursor(s) Y. Notice that all component of the systems (L and C) are produced from within, the systems is self-bounded, and its behaviour is determined by internal laws. Therefore the system in Figure 17.7b is autopoietic. Also in this case, this system interacts with the environment by taking up building blocks (X and Y) and releasing waste products (W and Z).

Thanks to autopoiesis, we can define minimal functions, implemented them in a synthetic construct, and study the behaviour of such system. Clearly, the system depicted in Figure 17.7b can be realized at different complexity levels, also depending on the chemical nature of the components. Living cells are autopoietic systems where L are of course the lipids (and proteins) constituting the boundary, whereas C is the *whole* metabolism (including genetic material), that produce itself and the boundary.

In recent years, after the achievement of simple chemical autopoietic systems based on surfactant self-assemblies[79–81] there has been an increasing interest in achieving the first minimal cell by implementing biochemical reactions inside vesicles. Such reactions have as final goal the synthesis of an autopoietic molecular system as that one represented in Figure 17.7b, although by stepwise advancements. To this aim, several groups started to carry out molecular biology reactions inside vesicles, in particular simple enzymatic reactions – which will be not discussed here – and polymerization reactions (Figure 17.7c). The latter ones have great relevance since (1) all functional biological molecules (proteins, RNA, DNA) are biopolymers (the membrane itself can be considered a non-covalent lipid polymer); (2) polymerization reactions have also a key role in origins of life studies, since they make possible the emergence of first catalysts (such as ribozymes, or the first catalytic peptides) out from small molecules. Notice here that the term polymers is restrictive, as enzymes and nucleic acids are actually co-polymers, being formed by several different co-monomers. The synthesis of copolymers, and particularly of multiple identical copies of ordered sequences, is a very complex problem, and one which is not yet solved in the origin of life field-nor experimentally or theoretically.

Going back to the question of autopoiesis, Figure 17.7c also indicates possible *experimental* schemes (indicated as type I to IV systems), and although their implementation may be very difficult, there is no doubt that the interest of the scientific community in this kind of approaches is continuously growing. We are, however, still far from the achievement of a synthetic autopoietic cell, and the current research is addressing the basic properties of compartmentalized reactions. In the following, we would like to comment in details some relevant cases of the four systems listed in Figure 17.7c. A full list, with short comments, is given in Table 17.1.

17.5 Four Cases of Compartmentalized Reactions

Type I system. Self-reproduction of vesicles. The starting point is the concept of vesicle self-reproduction. This important vesicle pattern was observed for the first time in 1994[81,85] and thoroughly studied till recently, being one of the most important reactive behaviour of fatty acid vesicles. Figure 17.8a shows the experimental model for autopoietic

Table 17.1 *Biopolymerizations inside vesicles (production of RNAs, DNAs, proteins)*

#	Authors (year)	Description	Type	E	T	Ref.
1	Schmidli *et al.* (1991)	Four enzymes, required for the lecithin biosynthesis are reconstituted in a soybeanPC proteoliposome. Short- and long-chain lecithin can be synthesized within such liposomes, that were prepared by the detergent depletion method.	I	4 enzymes	not needed	[88]
2	Walde *et al.* (1994)	PNPase has been entrapped inside extruded fatty acid vesicles, which are fed by externally added ADP and oleic anhydride. ADP is polymerized inside the vesicles, producing poly(A), whereas – simultaneously – oleate vesicles self-reproduce at the expenses of oleic anhydride.	II	PNPase	none	[98]
3	Chakrabarti *et al.* (1994)	PNPase has been entrapped inside extruded DMPC liposomes, which are fed by externally added ADP. In order to increase the entrance of ADP, liposomes were kept at 23 °C (i.e., at the T_m).	II	PNPase	none	[97]
4	Oberholzer *et al.* (1995)	Qβ replicase has been entrapped within extruded oleate vesicles, together with a template RNA and NTPs. Vesicles are fed by externally added oleic anhydride, so that self-reproduction occurs. Simultaneously, RNA is replicated in the vesicle core. This is an example of core-and-shell reproduction.	II	Qβ replicase	RNA	[82]
5	Oberholzer *et al.* (1995)	DNA polymerase has been entrapped within POPC extruded vesicles, together with a template DNA, primers, and dNTPs. Thermal cycles allow the polymerase chain reaction occurs within vesicles.	III	DNA polymerase	DNA	[100]
6	Oberholzer *et al.* (1999)	Ribosomes were entrapped inside POPC vesicles together with poly(U), acting as template mRNA, t-RNA^{Phe}, Phe-tRNA-synthetase, Phe, and all low mw compounds required for the translation. Poly(Phe) is produced accordingly inside extruded vesicles.	IV	Ribosomes	RNA	[110]
7	Tsumoto *et al.* (2001)	RNA polymerase was entrapped within DOPC/DOPG (10/1) giant vesicles, together with DNA template and NTPs. RNA was synthesized within such vesicles, which were prepared by the natural swelling method.	II	RNA polymerase	DNA	[111]

8	Yu *et al.* (2001)	DNA, coding for the GFP, was introduced in liposomes composed by a phospholipid mixture (*a*), together with the whole T&T machinery (T7 RNA polymerase and *E. coli* cell extracts). GFP was synthesized inside large MLVs prepared by the dehydration/rehydration method.	IV	T&T cell extracts	DNA	[112]
9	Fischer *et al.* (2002)	T7 RNA polymerase was encapsulated, by microinjection, inside GVs, together with DNA template and NTPSs (microinjected as well). The corresponding RNA was synthesized inside GVs prepared by the electroformation method.	II	RNA polymerase	DNA	[113]
10	Treyer *et al.* (2002)	PNPase has been entrapped inside extruded POPC liposomes, and fed by externally added ADP. In order to increase the entrance of ADP, liposomes were treated with cholate. Optimal permeability was achieved at cholate/POPC = 4/5 mol/mol.	II	PNPase	none	[43]
11	Oberholzer and Luisi (2002)	DNA, coding for the GFP was introduced in liposomes composed by POPC, together with the whole T&T machinery (T7 RNA polymerase and *E. coli* cell extracts). GFP was synthesized inside vesicles prepared by the ethanol injection method.	IV	T&T cell extracts	DNA	[114]
12	Monnard and Deamer (2002)	T7 RNA polymerase has been entrapped inside DMPC liposomes, together with DNA template, and fed by externally added NTPs. In order to favour the entrance of NTPs, liposomes were incubated at cycling temperature (from 23 to 37 °C). RNA was therefore then observed within DMPC vesicles.	II	RNA polymerase	DNA	[42]
13	Nomura *et al.* (2003)	DNA, coding for the GFP, was introduced in liposomes composed by DOPC:DOPG (10:1), together with the whole T&T machinery (T7 RNA polymerase and *E. coli* cell extracts). GFP was synthesized inside GVs prepared by the natural swelling method.	IV	T&T cell extracts	DNA	[70]
14	Ishikawa *et al.* (2004)	DNA, coding for GFP and T7 RNA polymerase, was introduced in liposomes composed by a phospholipid mixture (*a*), together with the T&T machinery (SP6 RNA polymerase and *E. coli* cell extracts). T7 RNA polymerase was synthesized firstly (the *rnapol* gene was under SP6 promoter), and in turn it allowed the synthesis of GFP (the *gfp* gene was under T7 promoter). This is an example of cascade genetic reactions inside large MLVs prepared by the dehydration/rehydration method.	II + IV	T&T cell extracts	2 DNA	[115]

(*continued overleaf*)

Table 17.1 *(continued)*

#	Authors (year)	Description	Type	E	T	Ref.
15	Noireaux and Libchaber (2004)	DNA, coding for GFP and α-hemolysin, was introduced in GVs composed by eggPC, together with the whole T&T machinery (T7 RNA polymerase and *E. coli* cell extracts). α-hemolysin, once synthesized, self-assembles at the membrane to form a pore (cut off ~3kDa), allowing small solutes enter from outside. This bioreactor was able to sustain GFP expression for four days. GVs were prepared by centrifugation of a pre-formed w/o emulsion.	IV + IV	T&T cell extracts	2 DNA	[49]
16	Sunami *et al.* (2006)	DNA, coding for GFP, was introduced in liposomes composed by a phospholipid mixture (*b*), together with the whole T&T machinery (PURESYSTEM). GFP was synthesized within large liposomes prepared by the dehydration/rehydration method.	IV	T&T purified kit	DNA	[58]
17	Shohda and Sugawara (2006)	*E. coli* Klenow fragment was entrapped in liposomes composed by the lipid mixture (*c*), together with template DNA and dNTPS. DNA was then produced within vesicles prepared by the dehydration-rehydration method.	III	Klenow fragment (DNA polymerase)	DNA	[116]
18	Zepik *et al.* (2007)	Thioglutamic acid was entrapped within DDAB vesicles, together with Glu_{10}, acting as a primer. Spontaneous elongation occurs.	IV	None	None	[117]
19	Murtas *et al.* (2007)	DNA, coding for the GFP was introduced in liposomes composed by POPC, together with the whole T&T machinery (PURESYSTEM). GFP was synthesized inside large vesicles prepared by the film hydration method.	IV	T&T purified kit	DNA	[107]
20	Saito *et al.* (2007)	DNA, coding for the GFP, was introduced in liposomes composed by eggPC, together with the whole T&T machinery (PURESYSTEM). GFP was synthesized inside giant vesicles prepared by centrifugation of a pre-formed w/o emulsion.	IV	T&T purified kit	DNA	[118]
21	Monnard *et al.* (2007)	T7 RNA polymerase has been entrapped inside DMPC liposomes, together with DNA template, and fed by externally added NTPs. In order to increase the entrance of NTPs, liposomes were kept at 23 °C (at the T_m). RNA was therefore synthesized within DMPC vesicles prepared by the dehydration-rehydration and extrusion method.	II	RNA polymerase	DNA	[41]
22	Rajamani *et al.* (2008)	AMP/UMP were entrapped within lipid vesicles composed by POPC or POPA or LPC, and polymerized by dehydration/rehydration cycles (carried out at 90 °C). RNA-like oligomers are produced in the process, and found partially entrapped within vesicles.	II	None	None	[119]

23	Kuruma *et al.* (2009)	DNA, coding for G3PAT and LPAAT, was introduced in liposomes composed by lipid mixture (*d*), together with the whole T&T machinery (PURESYSTEM). Both membrane enzymes were synthesized in active forms. The two-steps reaction that leads to phosphatidic acid, starting from G3P and two acylCoA was carried out in two discrete steps, due to different optimal activity conditions of the two enzymes.	I + IV	T&T purified kit	2 DNA	[89]
24	Kita *et al.* (2008)	(+)RNA, coding for Qβ replicase, was introduced in liposomes composed by lipid mixture (*c*), together with the whole T&T machinery (PURESYSTEM). Qβ replicase was first synthesized, allowing the replication of (+)RNA to (-)RNA, which in turn codifiess for β-galactosidase. Produced β-galactosidase was detected by a co-entrapped fluorogenic substrate. It is then shown that in large vesicles, prepared by dehydration-rehydration method, RNA may function as genetic polymer, with the two RNA strands codifying for two functions (replication and an enzymatic activity).	II	T&T purified kit	RNA	[99]
25	Hosoda *et al.* (2008)	DNA, coding for the β-glucuronidase, was introduced in liposomes composed by lipid mixture (*c*), together with the whole T&T machinery (PURESYSTEM). Enzymatic activity was monitored by co-entrapped fluorogenic substrates. β-glucuronidase was synthesized inside large MLVs and MVVs prepared by dehydration-rehydration method.	IV	T&T purified kit	DNA	[68]
26	Mansy *et al.* (2008)	DNA template and primers were entrapped within fatty acid vesicles (*e*). The substrates, in form of phosphorimidazoyl nucleotides, were added externally and permeate through the vesicle membrane. DNA was therefore duplicated within unextruded fatty acid vesicles in the absence of polymerase.	III	None	DNA	[101]
27	Souza *et al.* (2009)	DNA, coding for the GFP, was introduced in POPC liposomes, together with the whole T&T machinery (T7 RNA polymerase and *E. coli* cell extract, or PURESYSTEM). GFP was synthesized inside LUVs (radius 100 nm).	IV	T&T purified kit	DNA	[30]

(*a*) EggPC : cholesterol : DSPE-PEG5000 (1.5 : 1 : 0.08 mol : mol)
(*b*) POPC/PLPC/SOPC/SLPC/cholesterol/DSPE–PEG5000 129 : 67 : 48 : 24 : 180 : 14)
(*c*) POPC/cholesterol / DSPE-PEG5000 (58 : 39 : 3 mol : mol)
(*d*) POPC : POPE : POPG : cardiolipin (50.8 : 35.6 : 11.5 : 2.1 mol : mol)
(*e*) myristoleic acid : GMM (2 : 1), decanoic acid : DOH : GMD (4 : 1 : 1), myristoleic acid : farnesol (2 : 1)

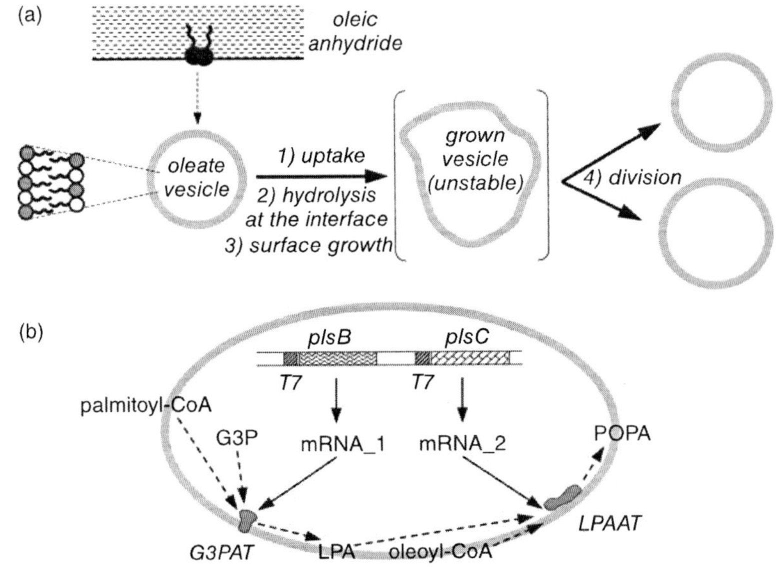

Figure 17.8 *A lipid-synthesizing liposome. (a) Classical studies on self-reproduction of oleate vesicles at the expenses of oleic anhydride. Pre-formed oleate vesicles (the grey and the white headgroup represent the ionized and non-ionized forms of fatty acids, respectively) are put in contact with oleic anhydride, which form a second phase. Oleic anhydride is taken up by oleate vesicles, being hydrolyzed at the membrane interface. The resulting surface increase brings about a destabilization of vesicle, which grows in non spherical manner and divide in two or more daughter vesicles.*[81] *Similar results have been achieved by using oleate micelles instead of oleic anhydride. (b) The experimentally realized lipid-producing liposomes. Two membrane proteins, glycerol-3-phosphate acyl transferase (G3PAT) and lysophosphtatic acid acyl transferase (LPAAT) are expressed inside a vesicle, starting from the corresponding genes. The two enzymes, working in series, can transform the water soluble compound glycerol-3-phosphate into the membrane forming compound phosphatidic acid*[89]

self-reproduction (compare Figure 17.7a). Oleic anhydride is stratified over a alkaline solution containing pre-formed oleate vesicles. Due to basic pH values, the anhydride is hydrolyzed giving two oleate molecules. In the absence of pre-formed oleate vesicles, this reaction would occur at the (macroscopic) anhydride/buffer interface, proceeding in a sluggish way. By contrast, when oleate vesicles are present, the anhydride is rapidly hydrolyzed thanks to vesicle catalysis. In particular, anhydride is first solubilised in the membrane of oleate vesicles, and then hydrolyzed at the membrane/buffer interface. The product of such hydrolysis (oleate molecules) are actually the membrane forming molecules, so that an autopoietic process takes place. Following the surface growth, the vesicle become unstable, and in turn it may divide in two or more *daughter* vesicles. These vesicles can start again the cycle and convert more anhydride into new oleate vesicles; notice that the process is overall *autocatalytic* and that oleate vesicles follow a sigmoid growth profile.

Recently, by using oleate micelles instead of oleic anhydride, it has been possible to apply optical methods to investigate the mechanism of vesicle self-reproduction, leading to the discovery of the 'matrix effect', namely the self-reproduction of oleate vesicles in a way that the average vesicles size is more or less conserved among vesicle generations.[24,25,57,86,87] The self-reproduction of vesicles, as shown in Figure 17.8a, closely resemble the general autopoietic mechanism (Figure 17.7a), however, the vesicle itself act as a sort of catalyst, favoring the transformation of the precursor X (the oleic anhydride) into the component of autopoietic unit A (the membrane-forming oleate molecule). In more elaborated models, this transformation should be achieved thank to an internal catalysts (compare Figure 17.7b), which should in turn be reproduced in a true and complete autopoietic system.

Such complex design has not been realized yet, however there have been two relevant attempts finalized at this goal. The first, not discussed here, was based on the lipid vesicle entrapment of four enzymes, namely the enzymes required to produce phosphatidylcholine starting from water-soluble compounds.[88] The aim was to create a lipid-producing liposome. The conversion of water-soluble precursors into phosphatidylcholine was around 10%, but no morphological changes were observed. A more recent report, based on the above mentioned work, is based on protein expression inside liposomes (Figure 17.8b). In this study, the two genes *plsB* and *plsC*, codifyng, respectively, for the enzymes glycerol-3-phosphate acyltransferase (G3PAT, an integral membrane protein), and lysophosphatidic acid acyltransferase (LPAAT, a periferal membrane protein), are entrapped in a lipid vesicles together with a transcription-translation machinery.[89] The latter – known as PURESYSTEM[90] – is assembled from purified components, and consists in a set of 36 enzymes, ribosomes, tRNAs, and all the low molecular-weight components (amino acids, GTP, etc.). The composition of lipid membrane (see Table 17.1 entry 23) is optimized in terms of entrapment efficiency, protein expression yield and – more importantly – activity of the two *in situ* synthesized membrane enzymes. This work demonstrates that two membrane enzymes can be synthesized in active form inside liposomes. In fact, by supplying the substrates *sn*-glycerol-3-phosphate and palmitoyl-CoA, the first enzyme (G3PAT) could produce 1-palmitoyl-*sn*-glycerol-3-phosphate (a lysophosphatidic acid) in 15% yield (% reacted/entrapped). The low yield is due to the low activity of the freshly synthesized G3PAT, which is highly hydrophobic, and localizes and folds (in this synthetic system) without assistance of dedicated macromolecules. The second enzyme (LPAAT) leads to the final product by adding an oleyl group to the free 2-hydroxy group of lysophosphatidic acid, to give 1-palmitoyl-2-oleoyl-sn-glycerol-3-phosphate, i.e. a phospholipid that insert in the liposome membrane. Although the yield of the reaction catalyzed by LPAAT is rather high (~90%, reacted/entrapped), the overal conversion (13.5%) is not very high, also because the two enzymes have different optimal redox conditions.

If we compare the experimental setup (Figure 17.8b) with the autopoietic model (Figure 17.7b), it is easy to recognize how the different autopoietic functions are implemented experimentally. The key transformation ($X \rightarrow L$) consists in the *stepwise* reaction between *sn*-glycerol-3-phosphate and palmitoyl-CoA, firstly, and oleoyl-CoA, later. The product readily inserts in the membrane, as required in Figure 17.7b. The catalysts C required to these transformations are two membrane enzymes (G3PAT and LPAAT), which are produced by a metabolic systems, starting from the corresponding DNA genes. What is missing, in the experimental setup, it is the reproduction of the catalysts, of the machinery, and of the genes that are involved in the generation of C (namely, the self-reproduction of the PURESYSTEM itself and the replication of genes). This has not been achieved yet, in any systems (see below for additional discussion on the consequences of this fact). Of course, the low yield of lipid production does not favour the observation of autopoietic behaviour, ideally the self-reproduction as in the case of anhydride hydrolysis. Moreover, with respect with the open system character of autopoietic cells, the substrates X are delivered from outside – as required by the model, whereas the building blocks Y (amino acids, NTPs, etc.) are co-entrapped with the PURESYSTEM.

Type II system. RNA replication inside self-reproducing vesicles. The biosynthesis of RNA molecules inside vesicles is also a very important target. It is commonly accepted, in fact, that RNA may play a key role in the early phase of life scenario, as epitomized in the RNA world hypothesis[91–94] an hypothesis that was considerably strengthened by the discovery of ribozymes.[95,96] Two early studies (1994) were focused on this topic. Poly(adenilic acid) was synthesized inside lipid vesicles by Joyce and Deamer, by exploiting the polynucleotide phosphorilase (PNPase) catalyzed-polymerization of ADP.[97] Independently, Walde *et al.* reported on the same system, but compartmentalized in self-reproducing oleate vesicles.[98] In 1995, Oberholzer *et al.*[82] reported the replication of RNA strands as catalyzed by Qβ replicase, with the simultaneous oleate vesicle reproduction (Figure 17.9). This system will be discussed in detail since it represent an interesting case of core-and-shell reproduction, although partially successful.

Pre-formed oleate vesicles are filled with a template RNA, Qβ replicase and nucleotides triphosphate (NTPs). Then, they are put in contact with a second oil phase consisting in oleic anhydride. Two reactions occurs simultaneously. Located at the membrane, the solubilization of oleic anhydride and its alkaline hydrolysis takes place, leading to vesicle growth and division, as discussed in the previous section. In the aqueous vesicle core, on the other hand, Qβ replicase, being a RNA-dependent RNA polymerase, makes additional copies of the RNA template at the expenses of NTPs, which are also present in the vesicle. Therefore, the number of RNA molecules increase due to the replication reaction. Following one or more vesicle division cycles, the content of the vesicle is redistributed among the *daughter* vesicles, presumably randomly. This is an example of simultaneous (although not *coupled*) core-and-shell reproduction. The core reproduction is not complete, however, since the catalyst (Qβ replicase) does not reproduce itself, the number of such molecules remaining constant. It is therefore foreseeable that at a certain point some vesicles 'die' due to the dilution of the catalyst, i.e., no catalytic molecule is present in the aqueous core so that the vesicle can undergo to shell reproduction only, without replicating its inner RNA. When compared with theoretical model (Figure 17.7b), the system depicted in Figure 17.9 shows that shell self-reproduction occurs at the membrane interface, and it is not catalyzed by an internal autopoietic element. On the other hand, internalized components are replicated (RNA molecules), but the catalyst is not (Qβ replicase). Moreover, the precursor X (oleic anhydride) is delivered from the environment, whereas the building blocks Y (NTPs in this case), are co-entrapped with the template and the enzyme inside the vesicles, from the beginning (the system is therefore semi-open, as in the previous case). Very recently, Yomo and co-workers[99] reported on the Qβ replicase-catalyzed replication of RNA coding for Qβ replicase itself – inside liposomes. In this autocatalytic system, RNA is replicated by Qβ replicase, and Qβ

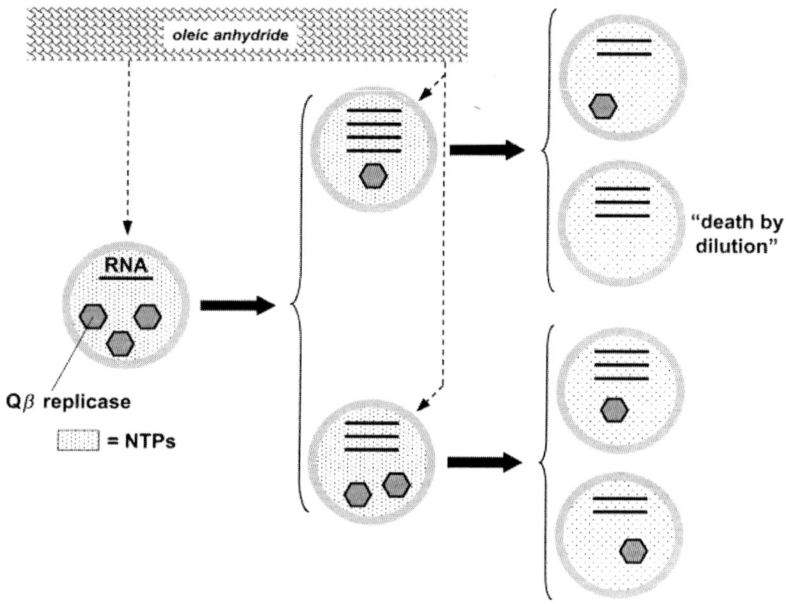

Figure 17.9 RNA replication inside self-reproducing oleate vesicles. Qβ replicase, an RNA-dependent RNA polymerase is entrapped inside oleate vesicles, in the presence of RNA template and NTPs. Copies of the RNA template are produced, and, simultaneously, oleate vesicles grow and divide according to the mechanism shown in Figure 17.6. In this way, internal RNA is replicated within self-reproducing vesicles. The mechanism stops when a daughter vesicles, produced by division of a previously grown vesicle, does not contain any Qβ replicase molecule. This vesicle 'dies' due to the dilution of the catalyst.[82] Notice that if solutes are always retained in every division step (i.e., without any loss in the environment) when only one Qβ replicase is present in a certain grown vesicle, the division will produce one functional daughter vesicle

replicase is produced from RNA. The number of two components increases, giving rise to a dual numerical and functional replication. Also in this case, however, the system does not fully reproduce itself since lipids do not reproduced and – more importantly – the catalysts for the transformation RNA → Qβ replicase (i.e. ribosomes, tRNAs, etc.) do not as well.

In Table 17.1, there are additional examples of RNA synthesis inside vesicles, generally done by means of a DNA-dependent RNA polymerase.

Type III system. DNA synthesis inside fatty acid vesicles. DNA is today the cellular macromolecule where the information about the proteins sequence is stored as ordered nucleobases sequence. DNA is thought to appear later than RNA in the molecular evolution, mainly with the function of storage. Its minor reactivity (due to loss of 2′-OH group) favour its chemical stability. The use of DNA polymerization in vesicle-based cell models aims to demonstrate how nucleic acids can be synthesized inside compartments, again in an origin-of-life perspective. However, due to the recent developments of DNA-based biotechnologies, it is possible that DNA-reactions in micro- or submicro-compartments may achieve a technological relevance. Our first report (1995) on DNA synthesis in liposomes[100] by means of polymerase chain reaction (PCR) demonstrated how all common processes involved in this very important process are compatible with phospholipid vescles (DNA melting, primer annealing, and DNA extension). More recently, Szostak and co-workers showed that polymerase-free DNA extension takes place inside fatty acid vesicles.[101] In this system, the composition of vesicle membrane was optimized in order to display resistance to Mg^{2+} ions (needed for DNA oligomerization), as well as maximal permeability to reactants, which were added externally. Three different vesicles were studied: (*i*) myristoleate : glycerol monomyristoleate (2 : 1) vesicles; (*ii*) decanoate:1-decanol : glycerol monodecanoate (4 : 1 : 1); and (*iii*) myristoleate : farnesol (2 : 1). In all cases, the properties of mixed membranes were superior to those of membrane composed by pure components. A DNA template was entrapped inside vesicles, together with a primer, and the nucleotides (in form of activated nucleotides, i.e., 2′-amino-2′,3′-dideoxyguanosine-5′-phosphorimidazolide) were added externally. It has been demonstrated that activated nucleotides could permeate the vesicle membrane and react inside vesicles, giving rise to DNA extension in the absence of catalyst. The product, which has high molecular weight (a DNA oligomer 15 bp long) does not escape from the vesicles. When compared with the general autopoietic scheme (Figure 17.7b), the current work shows how building blocks can enter the compartment and fuel the internal oligomerization reaction, driven by thermodynamics thanks to the activation of nucleotides. Boundary molecules are not reproduced, however, and no reaction 'rules' are provided by the internal organization.

Type IV system. Protein synthesis inside lipid vesicles. The field of protein expression inside compartments is certainly the most advanced one in context of compartmentalized reactions. To date, in fact, it represents the state of the art in the construction of synthetic cells (more properly, semi-synthetic cells). This interest is due to the above mentioned convergence of some aspects of synthetic biology and experimental studies on the origins of life. A comprehensive review of studies on semi-synthetic cells has been recently published by us,[7] and other reviews have discussed this subject;[55,102] here we will comment on a couple of relevant examples and highlighting some trends and future perspectives. The complete list of published works can be found in Table 17.1.

In their 2004 paper, Noireaux and Libchaber describe a very intriguing vesicle-based system based on the entrapment of cell extracts within giant vesicles.[49] This work is of particular interest because it introduces a conceptual advancement when compared with the previous studies. In particular, a plasmid encoding for two proteins is first entrapped in giant vesicles, and the two genes are expressed by the transcription/translation machinery, which was also entrapped together with the plasmid (Figure 17.10a). The two proteins were, respectively, the green fluorescent protein (GFP), which acts as reporter, and α-hemolysin, a water soluble protein that can self-assemble at the membrane into heptamers, which constitute a pore (cut-off ~3 kDa). Hemolysin has the function to create a pore, allowing the nutrients (such as additional amino acids, NTPs, and other low molecular weight compounds, present in the environment) entering in the liposome, so that additional GFP (and hemolysin as well) are produced. In this way, it was shown that protein synthesis could be extended up to four days (whereas in 'closed' vesicles, it generally stops after few hours). Even if not stated in the original paper, the formation of the pore also allows the escape from vesicle interior of reaction byproducts. This paper is important

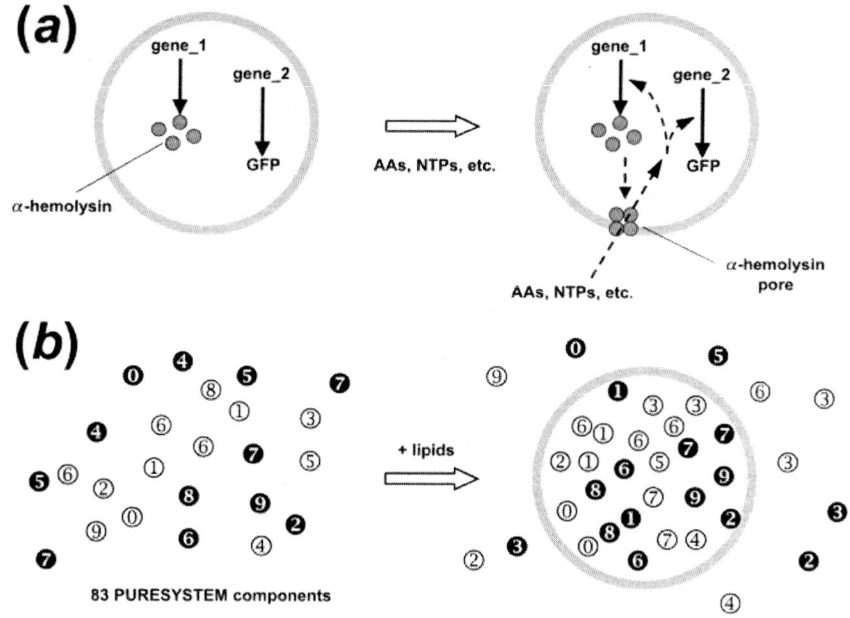

Figure 17.10 Protein expression inside vesicles and the issue of multiple entrapment. (a) An advanced cell model based on protein expression. Two genes, codifying for green fluorescent protein (GFP) and α-hemolysin, are entrapped inside GVs together with cell extracts. The in situ produced α-hemolysin molecules self assemble as heptamer in the membrane, forming a pore that allows nutrients to enter the GV, so that a very prolonged (4 days) protein expression is observed fluorimetrically.[49] (b) The expression of GFP inside 100 nm (radius) vesicles requires the simultaneous presence of more than 80 different macromolecular compounds in the same confined space. However, the Poisson statistics shows that this is a very improbable event, so that the accumulation of solutes inside small vesicles has been recently postulated.[30]

because it is the first example of a synthetic cell that interacts with its environment due to the presence of an internally synthesized compound. It is easy to imagine that such process may be also triggered by an external chemical signal, that may enter in the vesicle and stimulate the production of the pore-forming compound, starting in this way a sort of communication between the synthetic cell and the environment.[103] Alternatively, an internalized byproduct may activate a similar process. If we compare this work with the general mechanism (Figure 17.7b), it is remarkable that the Noireaux and Libchaber work[49] fulfils the conditions of cell/environment communication, and has a complex metabolism inside. It does not reproduce the membrane, however, and the internalized components are also not reproduced.

The second example deals with our contribution to the development of semi-synthetic minimal cells, that in the recent years has been very technical, but with important conceptual and physical spin-off, as we will shortly illustrate here. Cell-free protein expression is a well established procedure, born more than forty years ago[104] and the first studies on protein expression inside liposomes made use of home-made or commercially available cellular extracts. Very recently, however, Ueda and co-workers developed a new protein expression kit consisting of purified components.[90,105] Such kit, now commercially available with the trademark PURESYSTEM® (Post Genome Institute Co., Ltd., Tokyo), i.e. PUrified Recombinant Enzymes SYSTEM. As already shortly discussed before, PURESYSTEM is composed by 36 individually purified (and His_6-tagged) enzymes, tRNAs mixture, and purified ribosomes. It has been shown that the PURESYSTEM is the smallest set of macromolecules required to synthesize proteins *in vitro*. It follows, then, that the genes codifying for the 36 PURESYSTEM proteins, plus the genes codifying for the 55 macromolecular components of the ribosomes, and those codifying for a minimal tRNA set (for a total of about 150 genes), represent a good starting point to define operationally the *minimal genome*. In a recent study, the group of Moya, based on comparative genomics, indicated that the minimal genome should correspond to a definite set of around 200 genes.[106] It is therefore important to develop minimal cells models by using the PURESYSTEM because: (1) from a theoretical point of view, the components of PURESYSTEMS and the corresponding genes are *the* parts of a minimal semi-synthetic cell, representing, respectively, the minimal number of catalytic macromolecules and the minimal genome (notice that a small number of additional functions are possibly required); (2) from the 'constructive' point of view, it is very valuable carrying out experiments with a reconstructed mixture of macromolecules, with known concentrations, instead of using 'black boxes' as cellular extracts (notice that one of the features of synthetic biology is the 'standardization' of molecular parts, see for example http://parts.mit.edu). We therefore started to systematically use PURESYSTEM in compartmentalized protein synthesis studies;[107] other research groups are also shifting to this molecular tool.[58] The use of PURESYSTEM has another advantage. Since the concentration of components are known, it is possible to evaluate some physico-chemical parameters that were not available in cell extracts. For example, in a recent study, we have established a method for protein synthesis inside small (100 nm radius) vesicles,[30] consisting in forming small vesicles in one step, by the already mentioned ethanol injection method. By this method, liposomes form spontaneously by lipid self-assembly, and in the process of their formation, solutes are entrapped inside liposomes. After entrapment, it has been possible to detect the synthesis of green fluorescent protein within this very tiny compartments, albeit in low yield. When this yield was compared with the theoretical expected value, it resulted greatly *enhanced*, since the calculated yield is about zero. In fact, thanks to the knowledge of concentration of PURESYSTEM components, it is possible to show that the Poisson probability of finding a 100 nm (radius) vesicle containing all required macromolecular components (and therefore capable to host protein synthesis) is about 10^{-26}. But, since protein synthesis did occur in small vesicles, it has been suggested that the entrapment inside lipid vesicles may diverge significantly from the expected behaviour, so that inner solute concentration may overcome the external one by a factor >10 (see Figure 17.10b). To these super-concentrated vesicles, which may represent a very small fraction of the whole population, can be ascribed the observed protein synthesis. When compared with EGFP synthesis in bulk water, it also follows that the internalized components display an activity which is at least one order of magnitude higher. This result paves the way to new and more detailed investigation on the physics of vesicle entrapment, so that a new theoretical understanding of this basic phenomenon may appear in the future.

Compartmentalized protein synthesis, as noticed above, combines cell-free expression with liposome technology. Several important systems have been designed and experimentally done, as shown by the several reports collected in Table 17.1. There are, however, goals not yet achieved, such as the simultaneous protein synthesis (inside the vesicle) and vesicle self-reproduction, as done in the case of $Q\beta$ replicase experiment.[82] Moreover, these two processes should occur not only simultaneously, but must be functionally *coupled*, i.e. the first process should affect the second, as tentatively done in the case of lipid-synthesizing liposomes.[88,89] However, this would not suffice. In order to make an autopoietic minimal cell, the self-reproduction (or self-replication) of all internalized metabolic components is required, so that the issue of 'death by dilution' is avoided.

17.6 Conclusion

The importance of compartmentalized reactions is constantly growing and evolving. After about 20 years from the first reports on enzymes inside liposomes, current research deals also with the assembly of semi-synthetic cells.

In this chapter, we have firstly discussed the general issue of reactions in liposomes from a technical viewpoint. Starting from this basic knowledge, that is general and can be applied to whatever chemical system, we have then discussed some relevant examples of complex biochemical reactions in liposomes and fatty acid vesicles. This choice reflects the authors' interest, more than representing the large spectrum of possible applications,[108] or the more general field of water-soluble enzymes entrapped inside liposomes (recently reviewed by Walde and Ichikawa[31]). The use of liposomes as a tool to study membrane-proteins has also been skipped.[109]

Despite the advanced of recent years, several open questions are still under debate, and future work will maybe clarify some of the important aspects we have discussed in this chapter. There are also need of technical advancements, for example to improve the reproducibility of vesicle preparation for carrying out compartmentalized reactions.

In conclusion, reactions inside vesicles are a challenging and attractive field of inquiry, requiring often a multidisciplinary approach, technical skills, efforts and curiosity. The most advanced studies on construction of semi-synthetic minimal cells, for example, that have also important philosophical aspects, witnessed recently an abrupt rise of interest. It appears that one additional reason for this may lie in a sense of confidence that the minimal cell is indeed an experimentally accessible target.

Acknowledgements

This work is performed within the SYNTHCELLS project (Approaches to the Bioengineering of Synthetic Minimal Cells, EU Grant #FP6–043359), and further supported by the Human Frontiers Science Program (RGP0033/2007-C).

Abbreviations

(+)RNA/(−)RNA plus/minus (sense and antisense) RNA strands
AcylCoA acyl coenzymeA
ADP adenosine diphosphate
DDAB didodecyl dimethyl ammonium bromide
DMPC 1,2-dimyristoyl-*sn*-glycero-3-phosphatidylcholine
DNA deoxynucleic acid
dNTPs deoxynucleotide triphosphates
DOH 1-decanol
DOPC 1,2-dioleoyl-*sn*-glycero-3-phosphatidylcholine
DOPG 1,2-dioleoyl-*sn*-glycero-3-phosphatidylglycerol
DSPE-PEG5000 1,2-distearoyl-*sn*-glycero-3-phosphatidylethanolamine-PEG5000
EggPC phosphatidylcholine extracted from egg yolk
G3P glycerol-3-phosphate
G3PAT glycerol-3-phosphate acyltransferase
GFP green fluorescent protein
Glu_{10} decamer of glutamic acid
GMD glycerol monodecanoate
GMM glycerol monomyristoylate
GVs giant vesicles
LPAAT lysophosphatidic acid acyltransferase
LPC lysophosphatidylcholine
LUVs large unilamellar vesicles
MLVs multilamellar vesicles
mRNA messenger RNA
MVVs multivesicular vesicles
mw molecular weight
NTPs nucleotide triphosphates
PC phosphatidylcholine
Phe phenylalanine
PLPC 1-palmitoyl-2-lauryl-*sn*-glycero-3-phosphatidylcholine
PNPase polynucleotide phosphorilase
Poly(A) poly(adenilic acid)
Poly(Phe) poly(phenylalanine)
poly(U) poly(uridylic acid)
POPA 1-palmitoyl-2-oleoyl-*sn*-glycero-3-phosphatidic acid
POPC 1-palmitoyl-2-oleoyl-*sn*-glycero-3-phosphatidylcholine
POPE 1-palmitoyl-2-oleoyl-*sn*-glycero-3-phosphatidylethanolamine
POPG 1-palmitoyl-2-oleoyl-*sn*-glycero-3-phosphatidylglycerol
PURESYSTEM protein synthesis using recombinant elements (PURE) system
RNA ribonucleic acid
SLPC 1-stearoyl-2-lauril-*sn*-glycero-3-phosphatidylcholine
SOPC 1-stearoyl-2-oleoyl-*sn*-glycero-3-phosphatidylcholine
soybeanPC phosphatidylcholine extracted from soybean
T&T transcription and translation

T7 promoter T7
T_m lipid phase transition temperature
t-RNA^{Phe} phenylalanine-transferRNA

References

1. P. L. Luisi, *The Emergence of Life: From Chemical Origin to Synthetic Biology*, Cambridge University Press, Cambridge, 2006.
2. K. Kaneko, *Life: Introduction to Complex Systems Biology*, Springer-Verlag, Berlin Heidelberg, 2006.
3. P. L. Luisi, Chemical aspects of synthetic biology, *Chem. Biodiv.*, 2007, **4**, 603–621.
4. S. A. Benner, Act natural, *Nature*, 2003, **421**, 118–118.
5. S. A. Benner, A. M. Sismour, Synthetic biology, *Nat. Rev.*, 2005, **6**, 533–543.
6. D. Endy, Foundations for engineering biology, *Nature*, 2005, **438**, 449–453.
7. P. L. Luisi, F. Ferri, P. Stano, Approaches to semi-synthetic minimal cells: a review, *Naturwissenschaften*, 2006, **93**, 1–13.
8. J. N. Israelachvili, D. J. Mitchell, B. W. Ninham, Theory of self-assembly of lipid bilayers and vesicles, *Biochim. Biophys. Acta*, 1977, **470**, 185–201.
9. T. H. Haines, Anionic lipid headgroups as a proton-conducting pathway along the surface of membranes: a hypothesis, *Proc. Natl. Acad. Sci. USA*, 1983, **80**, 160–164.
10. C. Tanford, The hydrophobic effect and the organization of living matter, *Science*, 1978, **200**, 1012–1018.
11. D. D. Lasic, *Liposomes: From Physics to Applications*, Elsevier, Amsterdam, 1993.
12. S. Batzri, E. D. Korn, Single bilayer liposomes prepared without sonication, *Biochim. Biophys. Acta*, 1973, **298**, 1015–1019.
13. R. R. C. New, Preparation of liposomes, in *Liposomes: A practical approach*, R. R. C. New (Ed.), Oxford University Press, Oxford, 33–104, 1990.
14. P. Walde, T. Namani, K. Morigaki, H. Hauser, Formation and properties of fatty acid vesicles (liposomes), in *Liposome Technology*, 3rd ed., Vol. **I**, G. Gregoriadis (Ed.), Informa Healthcare, New York, 1–19, 2006.
15. M. I. Angelova, D. S. Dimitrov, Liposome electroformation, *Faraday Discuss. Chem. Soc.*, 1986, **81**, 303–311.
16. P. Bucher, A. Fischer, P. L. Luisi, T. Oberholzer, P. Walde, Giant vesicles as biochemical compartments: the use of microinjection techniques. *Langmuir*, 1998, **14**, 2712–2721.
17. F. Szoka Jr., D. Papahadjopoulos, Procedure for preparation of liposomes with large internal aqueous space and high capture by reverse-phase evaporation. *Proc. Natl. Acad. Sci. USA*, 1978, **75**, 4194–4198.
18. A. S. Domazou, P. L. Luisi, Size distribution of spontaneously formed liposomes by the alcohol injection method. *J. Liposome Res.*, 2002, **12**, 205–220.
19. P. Stano, S. Bufali, C. Pisano, F. Bucci, M. Barbarino, M. Santaniello, P. Carminati, P. L. Luisi, Novel camptothecin analogue (gimatecan)-containing liposomes prepared by the ethanol injection method, *J. Liposome Res.*, 2004, **14**, 87–109.
20. C. Kirby, G. Gregoriadis, Dehydration–rehydration vesicles: a simple method for high yield drug entrapment in liposomes, *Nature BioTechnology*, 1984, **2**, 979–984.
21. R. Shew, D. Deamer, A novel method for encapsulation of macromolecules in liposomes, *Biochim. Biophys. Acta*, 1983, **816**, 1–8.
22. H. Kikuchi, N. Suzuki, K. Ebihara, H. Morita, Y. Ishii, A. Kikuchi, S. Sugaya, T. Serikawa, K. Tanaka, Gene delivery using liposome technology, *J. Control. Release*, 1999, **62**, 269–277.
23. S. Pautot, B. J. Frisken, D. A. Weitz, Production of unilamellar vesicles using an inverted emulsion, *Langmuir*, 2003, **19**, 2870–2879.
24. E. Blochliger, M. Blocher, P. Walde, P. L. Luisi, Matrix effect in the size distribution of fatty acid vesicles, *J. Phys. Chem.*, 1998, **102**, 10383–10390.

25. S. Lonchin, P. L. Luisi, P. Walde, B. H. Robinson, A matrix effect in mixed phospholipid/fatty acid vesicle formation, *J. Phys. Chem. B*, 1999, **103**, 10910–10916.
26. K. Ruiz-Mirazo, P. Stano, P. L. Luisi, Lysozyme effect on oleic acid/oleate vesicles, *J. Lipos. Res.*, 2006, **16**, 143–154.
27. P.-A. Monnard, C. L. Apel, A. Kanavarioti, D. W. Deamer, Influence of ionic inorganic solutes on self-assembly and polymerization processes related to early forms of life: implications for a prebiotic aqueous medium, *Astrobiology*, 2002, **2**, 139–152.
28. I. A. Chen, K. Salehi-Ashtiani, J. W. Szostak, RNA catalysis in model protocell vesicles, *J. Am. Chem. Soc.*, 2005, **127**, 13213–13219.
29. R. Tachibana, H. Harashima, T. Ishida, Y. Shinohara, M. Hino, H. Terada, Y. Baba, H. Kiwada, Effect of cationic liposomes in an in vitro transcription and translation system, *Biol. Pharm. Bull.*, 2002, **25**, 529–531.
30. T. P. Souza, P. Stano, P. L. Luisi, The minimal size of liposome-based model cells brings about a remarkably enhanced biological activity. *ChemBioChem*, 2009, **10**, 1056–1063.
31. P. Walde, S. Ichikawa, Enzymes inside lipid vesicles: preparation, reactivity and applications. *Biomol. Eng.*, 2001, **18**, 143–177.
32. D. P. Pantazatos, R. C. MacDonald, Directly observed membrane fusion between oppositely charged phospholipid bilayers, *J. Membr. Biol.*, 1999, **170**, 27–38.
33. C. F. Thomas, P. L. Luisi, Novel properties of DDAB: matrix effect and interaction with oleate, *J. Phys. Chem. B*, 2004, **108**, 11285–11290.
34. J. Wilshut, N. Duzgunes, R. Fraley, D. Papahadjopoulos, Studies on the mechanism of membrane fusion: kinetics of calcium ion induced fusion of phosphatidylserine vesicles followed by a new assay for mixing of aqueous vesicle contents, *Biochemistry*, 1980, **19**, 6011–6021.
35. J. Lee, B. R. Lentz, Secretory and viral fusion may share mechanistic events with fusion between curved lipid bilayers, *Proc. Natl. Acad. Sci. USA*, 1998, **95**, 9274–9279.
36. P. L. Luisi, T. Souza, P. Stano, Vesicle behaviour: in search of explanations, *J. Phys. Chem. B*, 2008, **112**, 14655–14664.
37. K. Tsumoto, K. Kamiya, T. Yoshimura, Membrane fusion between a giant vesicle and small enveloped particles: possibilities for the application to construct model cells. Micro-NanoMechatronics and Human Science, 2006, International Symposium on 2006, pp. 1–6.
38. G. Cevc, Solute transport across bilayers, in *Phospholipid Handbook*, G. Cevc (Ed.), Marcel Dekker, Inc., New York, 639–661, 1993.
39. P.-A. Monnard, D.W. Deamer, Nutrient uptake by protocells: a liposome model system, *Orig. Life Evol. Biosph.*, 2001, **31**, 147–155.
40. I. Cisse, B. Okumus, C. Joo, T. Ha, Fueling protein–DNA interactions inside porous nanocontainers, *PNAS*, 2007, **104**, 12646–12650.
41. P.-A. Monnard, A. Luptak, D. W. Deamer, Models of primitive cellular life: polymerases and templates in liposomes, *Phil. Trans. R. Soc. B*, 2007, **362**, 1741–1750.
42. P.-A. Monnard, D. W. Deamer, Membrane self-assembly processes: steps toward the first cellular life, *Anat. Rec.*, 2002, **268**, 196–207.
43. M. Treyer, P. Walde, T. Oberholzer, Permeability enhancement of lipid vesicles to nucleotides by use of sodium cholate: Basic studies and application to an enzyme-catalyzed reaction occurring inside the vesicles. *Langmuir*, 2002, **18**, 1043–1050.
44. T. Oberholzer, E. Meyer, I. Amato, A. Lustig, P.-A. Monnard, Enzymatic reactions in liposomes using the detergent-induced liposome loading method, *Biochim. Biophys. Acta*, 1999, **1416**, 57–68.
45. M. Winterhalter, C. Hilty, S. M. Bezrukov, C. Nardin, W. Meier, D. Fournier, Controlling membrane permeability with bacterial porins: application to encapsulated enzymes, *Talanta*, 2001, **55**, 965–971.
46. V. Vamvakaki, D. Fournier, N. A. Chaniotakis, Fluorescence detection of enzymatic activity within a liposome based nano-biosensor, *Biosens. Bioelectr.*, 2005, **21**, 384–388.
47. A. Graff, M. Winterhalter, W. Meier, Nanoreactors from polymer-stabilized liposomes, *Langmuir*, 2001, **17**, 919–923.

48. M. Yoshimoto, S. Wang, K. Fukunaga, D. Fournier, P. Walde, R. Kuboi, K. Nakao, Novel immobilized liposomal glucose oxidase system using the channel protein OmpF and catalase, *Biotechnol. Bioeng.*, 2005, **90**, 231–238.
49. V. Noireaux, A. Libchaber, A vesicle bioreactor as a step toward an artificial cell assembly, *PNAS*, 2004, **101**, 17669–17674.
50. H.-J. Choi, C. D. Montemagno, Light-driven hybrid bioreactor based on protein-incorporated polymer vesicles, *IEEE Trans. Nanotechnol.*, 2007, **6**, 171–176.
51. D. M. Vriezema, P. M. L. Garcia, N. S. Oltra, N. S. Hatzakis, S. M. Kuiper, R. J. M. Nolte, A. E. Rowan, J. C. M. van Hest. Positional assembly of enzymes in polymersome nanoreactors for cascade reactions, *Angew. Chem. Int. Ed.*, 2007, **46**, 7378–7382.
52. D. S. Tawfik, A. D. Griffiths, Man-made cell-like compartments fro molecular evolution. *Nat. Biotechnol.*, 1998, **16**, 652–656.
53. K. Bernath, M. Hai, E. Mastrobattista, A. D. Griffiths, S. Magdassi, D. S. Tawfik, In vitro compartmentalization by double emulsions: sorting and gene enrichment by fluorescence activated cell sorting, *Anal. Biochem.*, 2004, **325**, 151–157.
54. P. L. Luisi, B. Straub (Eds.), *Reverse Micelles. Biological and Technological Relevance of Amphiphilic Structures in Apolar Media*, Plenum Press, New York, 1984.
55. A. Pohorille, D. Deamer, Artificial cells: prospects for biotechnology, *Trends Biotechnol.*, 2002, **20**, 123–128.
56. N. Berclaz, M. Müller, P. Walde, P. L. Luisi, Growth and transformation of vesicles studied by ferritin labeling and cryotransmission electron microscopy, *J. Phys. Chem. B*, 2001, **105**, 1056–1064.
57. N. Berclaz, E. Blöchliger, M. Müller, P. L. Luisi, Matrix effect of vesicle formation as investigated by cryotransmission electron microscopy, *J. Phys. Chem. B*, 2001, **105**, 1065–1071.
58. T. Sunami, K. Sato, T. Matsuura, K. Tsukada, I. Urabe, T. Yomo, Femtoliter compartment in liposomes for in vitro selection of proteins, *Anal. Biochem.*, 2006, **357**, 128–136.
59. B. Sun, D. Chiu, Determination of the encapsulation efficiency of individual vesicles using single-vesicle photolysis and confocal singlemolecule detection. *Anal. Chem.*, 2005, **77**, 2770–2776.
60. L. M. Dominak, C. D. Keating, Polymer encapsulation within giant lipid vesicles, *Langmuir*, 2007, **23**, 7148–7154.
61. Z. Cheng, P. L. Luisi, Coexistence and mutual competition of vesicles with different size distribution, *J. Phys. Chem. B*, 2003, **107**, 10940–10945.
62. D. D. Lasic, On the thermodynamic stability of liposomes, *J. Colloid Interface Sci.*, 1990, **140**, 302–304.
63. P. L. Luisi, Are micelles and vesicles chemical equilibrium systems, *J. Chem. Educ.*, 2000, **78**, 380–384.
64. E. Boukobza, A. Sonnenfeld, G. Haran, Immobilization in surface-tethered lipid vesicles as a new tool for single biomolecule spectroscopy, *J. Phys. Chem. B*, 2001, **105**, 12165–12170.
65. N. G. van Kampen, *Stochastic Processes in Physics and Chemistry*, Elsevier, Amsterdam, 2007.
66. F. Mavelli, S. Piotto, Stochastic simulations of homogeneous chemically reacting systems, *J. Mol. Struct. (THEOCHEM)*, 2006, **771**, 55–64.
67. N. Berclaz, Studies on the formation and transformation of protein-containing vesicles using cryo-transmission electron microscopy. Ph.D. dissertation Nr. 13493, ETH Zurich, 1999.
68. K. Hosoda, T. Sunami, Y. Kazuta, T. Matsuura, H. Suzuki, T. Yomo, Quantitative study of the structure of multilamellar giant liposomes as a container of protein synthesis reaction, *Langmuir*, 2008, **24**, 13540–13548.
69. L. M. Dominak, C. D. Keating, Macromolecular crowding improves polymer encapsulation within giant lipid vesicles, *Langmuir*, 2008, **24**, 13565–13571.
70. S. M. Nomura, K. Tsumoto, T. Hamada, K. Akiyoshi, Y. Nakatani, K. Yoshikawa, Gene expression within cell-sized lipid vesicles, *ChemBioChem*, 2003, **4**, 1172–1175.
71. A. Buxboim, S. S. Daube, R. Bar-Ziv, Synthetic gene brushes: a structure–function relationship, *Mol. Syst. Biol.*, 2008, **4**, 181.

72. N. Kozer, Y. Y. Kuttner, G. Haran, G. Schreiber, Protein-protein association in polymer solutions: from dilute to semidilute to concentrated, *Biophys. J.*, 2007, **92**, 2139–2149.
73. H. T. Bui, H. Umakoshi, K. X. Ngo, M. Nishida, T. Shimanouchi, R. Kuboi, Liposome membrane itself can affect gene expression in the escherichia coli cell-free translation system. *Langmuir*, 2008, **24**, 10537–10542.
74. P. L. Luisi, Toward the engineering of minimal living cells, *Anat. Rec.*, 2002, **268**, 208–214.
75. F. Varela, H. R. Maturana, R. B. Uribe, Autopoiesis: the organization of living system, its characterization and a model, *Biosystems*, 1974, **5**, 187–196.
76. H. Maturana, F. Varela, *Autopoiesis and Cognition: The Realization of the Living*, Reidel, Dordrecht, 1980.
77. P. L. Luisi, Autopoiesis: a review and a reappraisal, *Naturwissenschaften*, 2003, **90**, 49–59.
78. M. Bitbol, P. L. Luisi, Autopoiesis with or without cognition: defining life at its edge, *J. R. Soc. Interface*, 2004, **1**, 99–107.
79. P. A. Bachmann, P. Walde, P. L. Luisi, J. Lang, Self-replicating reverse micelles and chemical autopoiesis, *J. Am. Chem. Soc.*, 1990, **112**, 8200–8201.
80. P. A. Bachmann, P. L. Luisi, J. Lang, Autocatalytic self-replicating micelles as models for prebiotic structures, *Nature*, 1992, **357**, 57–59.
81. P. Walde, R. Wick, M. Fresta, A. Mangone, P. L. Luisi, Autopoietic self-reproduction of fatty acid vesicles, *J. Am. Chem. Soc.*, 1994, **116**, 11649–11654.
82. T. Oberholzer, R. Wick, P. L. Luisi, C. K. Biebricher, Enzymatic RNA replication in self-reproducing vesicles: an approach to a minimal cell, *Biochem. Biophys. Res. Comm.*, 1995, **207**, 250–257.
83. P. L. Luisi, Th. Oberholzer, A. Lazcano, The notion of a DNA minimal cell: a general discourse and some guidelines for an experimental approach. *Helv. Chim. Acta*, 2002, **85** (6), 1759–1777.
84. F. Varela, *El fenómeno vida*. Dolmen Esayo, Santiago, Chile, 2000.
85. K. Morigaki, S. Dallavalle, P. Walde, S. Colonna, P. L. Luisi, Autopoietic self-reproduction of chiral fatty acid vesicles, *J. Am. Chem. Soc.*, 1997, **119**, 292–301.
86. S. Rasi, F. Mavelli, P. L. Luisi, Cooperative micelle binding and matrix effect in oleate vesicle formation, *J. Phys. Chem. B*, 2003, **107** (50), 14068–14076.
87. P. Stano, E. Wehrli, P. L. Luisi, Insights on the oleate vesicles self-reproduction, *J. Phys. Condens. Matter*, 2006, **18**, S2231–S2238.
88. P. K. Schmidli, P. Schurtenberger, P. L. Luisi, Liposome-mediated enzymatic synthesis of phosphatidylcholine as an approach to self-replicating liposomes, *J. Am. Chem. Soc.*, 1991, **113**, 8127–8130.
89. Y. Kuruma, P. Stano, T. Ueda, P. L. Luisi, A synthetic biology approach to the construction of membrane proteins in semi-synthetic minimal cells, *Biochim. Biophys. Acta*, 2009, **1788**, 567–574.
90. Y. Shimizu, A. Inoue, Y. Tomari, T. Suzuki, T. Yokogawa, K. Nishikawa, T. Ueda, Cell free translation reconstituted with purified components, *Nat. Biotechnol.*, 2001, **19**, 751–755.
91. F. H. C. Crick, The origin of the genetic code, *J. Mol. Biol.*, 1968, **38**, 367–379.
92. L. E. Orgel, Evolution of the genetic apparatus, *J. Mol. Biol.*, 1968, **38**, 381–393.
93. C. R. Woese, *The Genetic Code: The Molecular Basis for Genetic Expression*, Harper & Row, New York, 1967.
94. W. Gilbert, The RNA world, *Nature*, 1986, **319**, 618.
95. K. Kruger, P. J. Grabowski, A. J. Zaug, J. Sands, D. E., Gottschling, T. R. Cech, Self-splicing RNA. Autoexcision and autocyclization of the ribosomal RNA intervening sequence of Tetrahymena, *Cell*, 1982, **31**, 147–157.
96. C. Guerrier-Takada, K. Gardiner, T. March, N. Pace, S. Altman, The RNA moiety of ribonuclease P is the catalytic subunit of the enzyme, *Cell*, 1983, **35**, 849–857.
97. A. C. Chakrabarti, R. R. Breaker, G. F. Joyce, D. W. Deamer, Production of RNA by a polymerase protein encapsulated within phospholipid vesicles, *J. Mol. Evol.*, 1994, **39**, 555–559.

98. P. Walde, A. Goto, P. A. Monnard, M. Wessicken, P. L. Luisi, Oparin's reactions revisited: enzymatic synthesis of poly (adenylic acid) in micelles and self-reproducing vesicles, *J Am. Chem. Soc*, 1994, **116**, 7541–7544.
99. H. Kita, T. Matsuura, T. Sunami, K. Hosoda, N. Ichihashi, K. Tsukada, I. Urabe, T. Yomo, Replication of genetic information with self-encoded replicase in liposomes, *ChemBioChem*, 2008, **9**, 2403–2410.
100. T. Oberholzer, M. Albrizio, P. L. Luisi, Polymerase chain reaction in liposomes. *Chem. Biol.*, 1995, **2**, 677–682.
101. S. Mansy, J. P. Schrum, M. Krishnamurthy, S. Tobé, D. A. Treco, J. W. Szostak, Template-directed synthesis of a genetic polymer in a model protocell, *Nature*, 2008, **454**, 122–125.
102. A. C. Forster, G. M. Church, Towards synthesis of a minimal cell, *Mol. Syst. Biol.*, 2006, **2**, 45.
103. P. Stano, F. Ferri, P. L. Luisi, From the minimal genome to the minimal cell: theoretical and experimental investigations, in '*Life As We Know It*', J. Seckbach (Ed.), Springer, Berlin, 181–198, 2006.
104. M. W. Nirenberg, J. H. Matthaei, The dependence of cellfree protein synthesis in E. coli upon naturally occurring or synthetic polyribonucleotides, *Proc. Natl. Acad. Sci. USA*, 1961, **47**, 1588–1602.
105. Y. Shimizu, T. Kanamori, T. Ueda, Protein synthesis by pure translation systems, *Methods*, 2005, **36**, 299–304.
106. R. Gil, F. J. Silva, J. Peretó, A. Moya, Determination of the core of a minimal bacteria gene set, *Microbiol. Mol. Biol. Rev.*, 2004, **68**, 518–537.
107. G. Murtas, Y. Kuruma, P. Bianchini, A. Diaspro, P. L. Luisi, Protein synthesis in liposomes with a minimal set of enzymes, *Biochem. Biophys. Res. Commun.*, 2007, **363**, 12–17.
108. D.D. Lasic, Y. Barenholz (Eds.), *Handbook of Nonmedical Applications of Liposomes*, CRC, Press, Boca Raton, 1996.
109. J. L. Rigaud, B. Pitard, D. Levy, Reconstitution of membrane-proteins into liposomes – application to energy-transducing membrane-proteins, *Biochem. Biophys. Acta*, 1995, **1231**, 223–246.
110. T. Oberholzer, K. H. Nierhaus, P. L. Luisi, Protein expression in liposomes, *Biochem. Biophys. Res. Commun.*, 1999, **261**, 238–241.
111. K. Tsumoto, S. M. Nomura, Y. Nakatani, K. Yoshikawa, Giant liposome as a biochemical reactor: transcription of DNA and transportation by laser tweezers, *Langmuir*, 2001, **17**, 7225–7228.
112. W. Yu, K. Sato, M. Wakabayashi, T. Nakatshi, E. P. Ko-Mitamura, Y. Shima, I. Urabe, T. Yomo, Synthesis of functional protein in liposome, *J. Biosci. Bioeng.*, 2001, **92**, 590–593.
113. A. Fischer, A. Franco, T. Oberholzer, Giant vesicles as microreactors for enzymatic mRNA synthesis, *ChemBioChem*, 2002, **3**, 409–417.
114. T. Oberholzer, P. L. Luisi, The use of liposomes for constructing cell models, *J. Biol. Phys.*, 2002, **28**, 733–744.
115. K. Ishikawa, K. Sato, Y. Shima, I. Urabe, T. Yomo, Expression of a cascading genetic network within liposomes, *FEBS Lett.*, 2004, **576**, 387–390.
116. K. Shohda, T. Sugawara, DNA polymerization on the inner surface of a giant liposome for synthesizing an artificial cell model. *Soft Matter*, 2006, **2**, 402–408.
117. H. H. Zepik, S. Rajamani, M.-C. Maurel, D.W. Deamer, Oligomerization of thioglutamic acid: encapsulated reactions and lipid catalysis. *Orig. Life Evol. Biosph.*, 2007, **37**, 495–505.
118. H. Saito, A. Yamada, R. Ohmori, Y. Kato, T. Yamanaka, K. Yoshikawa, T. Inoue, Towards constructing synthetic cells: RNA/RNP evolution and cell-free translational systems in giant liposomes, in *Micro-NanoMechatronics and Human Science, 2007. MHS '07 (International Symposium on) 2007*, pp. 286–29.
119. S. Rajamani, A. Vlassov, S. Benner, A. Coombs, F. Olasagasti, D. Deamer, Lipid-assisted Synthesis of RNA-like Polymers from Mononucleotides, *Orig. Life Evol. Biosph.*, 2008, **38**, 57–74.

Index